Second Edition

Restoration of Boreal and Temperate Forests

Integrative Studies in
Water Management and Land Development

Series Editor
Robert L. France

Published Titles

Second Edition

Restoration of Boreal and Temperate Forests

Edited by
John A. Stanturf

CRC Press
Taylor & Francis Group
Boca Raton London New York

CRC Press is an imprint of the
Taylor & Francis Group, an **informa** business

CRC Press
Taylor & Francis Group
6000 Broken Sound Parkway NW, Suite 300
Boca Raton, FL 33487-2742

First issued in paperback 2019

ISBN-13: 978-1-4822-1196-2 (hbk)
ISBN-13: 978-0-367-86882-6 (pbk)

Library of Congress Cataloging-in-Publication Data

Restoration of boreal and temperate forests. -- Second edition / John A. Stanturf, editor.
    pages cm. -- (Integrative studies in water management & land development ; 13)
  "A CRC title."
  Includes bibliographical references and index.
  ISBN 978-1-4822-1196-2 (alk. paper)
  1. Reforestation. 2. Afforestation. 3. Forest ecology. 4. Forest restoration. I. Stanturf, John A., editor.

SD409.R485 2016
634.9'56--dc23   2015004799

Visit the Taylor & Francis Web site at
http://www.taylorandfrancis.com

and the CRC Press Web site at
http://www.crcpress.com

Contents

Preface

This second edition of *Restoration of Boreal and Temperate Forests* builds on the first edition by bringing many chapters up-to-date and adding some additional ecosystems. All authors in the first edition were given the option to revise their chapters or write something new. Some authors declined to do either; not surprisingly, some authors had retired or taken new positions and could not devote the time to revising. New chapters add new perspectives and set this edition apart from the first; these new chapters include the effects of ungulates (Rooney et al.), site preparation techniques (Löf et al.), natural regeneration for restoration (Fischer et al.), watershed restoration in China (Liu et al.), oak woodlands and savannah in the Midwestern United States (Dey and Kabrick), Andean oak forests in Colombia (Avella et al.), and recovery from severe insect outbreaks (Burton et al.).

The idea for the first edition arose on a field trip to view restoration research in Iceland that led to a conference that was held in Vejle, Denmark in 2002 under the auspices of the International Union of Forest Research Organizations (IUFRO). The editors of that edition (Stanturf and Madsen) and many of the authors continue to be active in IUFRO and maintain the network of forest restoration researchers. The editors used this IUFRO network to send out the call for authors to contribute to this revised second edition. Unfortunately, Palle Madsen—one of the editors of the first version—regrettably had to decline editorial participation in this second edition because of other commitments.

Many people and organizations richly deserve my thanks for their support of both editions of this book. The US Forest Service Southern Research Station has supported me throughout. In particular, I wish to thank Dr. Peter Roussopoulos, retired Director of the Southern Research Station, and Prof. Niels Elers Koch, retired Dean and Director, University of Copenhagen, for their enthusiastic support from the very beginning. This book is a collaboration of many scientists and practitioners from around the world. I wish to thank those who contributed directly to this volume as authors, coauthors, and reviewers; and the many others whose field excursions, conference presentations, and thoughtful discussions broadened my knowledge and sharpened the concepts presented here. The authors are to be commended for their diligence, perseverance, and commitment to producing a quality addition to the forest restoration literature.

I relied on many external reviewers who read, commented, and offered valuable suggestions for improving individual chapters. The reviewers include Christian Ammer, Andreas Bolte, Richard Bradshaw, Philip Burton, Wayne Clatterbuck, Catherine Colet, Robert Deal, Daniel Dey, Emile Gardiner, Eric Gustafson, Björn Hånell, Allison Hester, Todd Hutchinson, Mikko Hypponen, Theresa Jain, Kier Klepzig, Brian Roy Lockhart, Magnus Löf, William (Bill) Mason, Kevin O'Hara, Chadwick D. Oliver, Yeong-dae Park, David L. Peterson, Nelson Thiffault, Jean-Pierre Tremblay, Carl Trettin, Alberto Vilagrosa, Hiromi Yamagawa, and Toshiya Yoshida.

Eileen, my long-suffering wife, deserves the most gratitude for putting up with yet another book.

John A. Stanturf
Athens, Georgia

Editor

John A. Stanturf is a senior scientist (research ecologist) at the Center for Forest Disturbance Science, Southern Research Station, US Forest Service in Athens, Georgia. Dr. Stanturf earned his BS in plant and soil science from Montana State University and his MS and PhD in forest soils from Cornell University. He was a Lady Davis Fellow at the Technion in Haifa, Israel and was awarded with an honorary doctorate from the Estonian University of Life Sciences. Before joining the Forest Service as a project leader for the Center for Bottomland Hardwoods Research in Stoneville, Mississippi, he worked in industrial forestry research and on the faculty of Cornell, University of Pittsburgh, and Pennsylvania State University. He currently is an adjunct professor of forestry at Auburn and Mississippi State Universities.

His research focuses on forest restoration, climate change mitigation and adaptation, and disturbance ecology. Dr. Stanturf actively participates in international activities, serving as the Research Group Leader for Restoration of Degraded Sites (RG 1.06.00) for the International Union of Forest Research Organizations and is Deputy of RG 1.01.00 Temperate and Boreal Silviculture and WP 1.01.12 Silviculture and Ungulates. Among other international activities, he provides technical assistance on climate change adaptation, REDD+, and biodiversity conservation to the US Agency for International Development missions in several African countries (Ghana, Malawi, Zambia, Liberia, Sierra Leone, Guinea, and Côte d'Ivoire).

Contributors

Vanda Acácio
Universidade Téchnica de Lisboa
Lisbon, Portugal

Andres Avella
Universidad Nacional de Columbia
Bogotá, Colombia

Patrick J. Baker
University of Melbourne
Melbourne, Australia

Philippe Balandier
Irstea
Nogent-sur-Vernisson, France

Yves Bergeron
Université du Québec en
    Abitibi-Témiscamingue
Rouyn-Noranda, Québec, Canada

Richard H.W. Bradshaw
University of Liverpool
Liverpool, United Kingdom

Dale G. Brockway
US Forest Service
Auburn, Alabama

Philip J. Burton
University of Northern British Columbia
Terrace, British Columbia, Canada

Rita Buttenschøn
University of Copenhagen
Aarhus, Denmark

Kenneth A. Byrne
University of Limerick
Limerick, Ireland

Isabel Cañellas
INIA-CIFOR
Madrid, Spain

Luis Mario Cárdenas
Pontificia Universidad Javeriana
Bogotá, Colombia

K. David Coates
Ministry of Forests
Smithers, British Columbia, Canada

Daniel C. Dey
US Forest Service
Northern Research Station
Columbia, Missouri

Back Tomas Ersson
Swedish University of Agricultural Sciences
Umeå, Sweden

Mike Fenger
Mike Fenger and Associates Ltd.
Victoria, British Columbia, Canada

Holger Fischer
Technische Universität Dresden
Tharandt, Germany

Søren Fodgaard
Danish Forest Association
Fredriksberg, Denmark

Guillermo Gea-Izquierdo
CEREGE
Aix-en-Provence, France

Kurt W. Gottschalk
US Forest Service
Morgantown, West Virginia

Tanis L. Gower
Fernhill Consulting
Courtney, British Columbia, Canada

Russell T. Graham
US Forest Service
Rocky Mountain Research Station
Moscow, Idaho

Ulrike Hagemann
Technische Universität Dresden
Tharandt, Germany

Jörg Hansen
Albert-Ludwigs-University
Freiburg im Briesgau, Germany

Ralph Harmer
Forest Research
Surrey, United Kingdom

Joakim Hjältén
Swedish University of Agricultural Sciences
Umeå, Sweden

Franka Huth
Leibniz Centre for Agricultural Landscape
    Research
Müncheberg, Germany

Theresa B. Jain
US Forest Service
Rocky Mountain Research Station
Moscow, Idaho

Finn A. Jensen
Hedeselskabet
Viborg, Denmark

Kalev Jõgiste
Estonian University of Life Sciences
Tartu, Estonia

Everett E. Johnson
Auburn University
Andalusia, Alabama

John M. Kabrick
US Forest Service
Columbia, Missouri

Daniel Kneeshaw
Université du Québec à Montréal
Montreal, Québec, Canada

Timo Kuuluvainen
University of Helsinki
Helsinki, Finland

Don Koo Lee
Yeungnam University
Gyeongsan, Republic of Korea

Sandra Liebal
Technische Universität Dresden
Tharandt, Germany

Shirong Liu
Chinese Academy of Forestry
Beijing, China

Magnus Löf
Swedish University of Agricultural Sciences
Alnarp, Sweden

Jiangming Ma
Guangxi Normal University
Guilin, China

Palle Madsen
University of Copenhagen
Vejle, Denmark

Doug McCreary
University of California-Berkeley
Browns Valley, California

Marek Metslaid
Estonian University of Life Sciences
Tartu, Estonia

Ning Miao
Sichuan University
Chengdu, China

Takuo Nagaike
Yamanashi Forest Research Institute
Yamanashi, Japan

Tomas Nordfjell
Swedish University of Agricultural
    Sciences
Umeå, Sweden

Kevin L. O'Hara
University of California-Berkeley
Berkeley, California

Carsten Riis Olesen
Danish Hunter's Association
Rønde, Denmark

Juan A. Oliet
Technical University of Madrid
Madrid, Spain

Chadwick D. Oliver
Yale University
New Haven, Connecticut

Kenneth W. Outcalt
US Forest Service
Southern Research Station
Athens, Georgia

Pil Sun Park
Seoul National University
Seoul, Republic of Korea

Yeong Dae Park
Daegu University
Gyeongsan, Republic of Korea

Bernard Prévosto
Irstea
Aix-en-Provence, France

Fernando Pulido
Universidad de Extremadura
Plasencia, Spain

Duncan Ray
Forest Research
Midlothian, United Kingdom

Florence Renou-Wilson
University College Dublin
Dublin, Ireland

Thomas P. Rooney
Wright State University
Dayton, Ohio

Alejandro A. Royo
US Forest Service
Northern Research Station
Irvine, Pennsylvania

Mariola Sánchez-González
INIA-CIFOR
Madrid, Spain

Heinrich Spiecker
Albert-Ludwigs-University
Freiburg im Briesgau, Germany

John A. Stanturf
US Forest Service
Athens, Georgia

Susan L. Stout
US Forest Service
Northern Research Station
Irvine, Pennsylvania

Miroslav Svoboda
Czech University of Life Sciences
Praha, Czech Republic

Donald J. Tomczak
US Forest Service
Atlanta, Georgia

Selene Torres
Universidad de los Andes
Bogotá, Colombia

Veiko Uri
Estonian University of Life Sciences
Tartu, Estonia

Sven Wagner
Technische Universität Dresden
Tharandt, Germany

Kevin Watts
Forest Research
Surrey, United Kingdom

Norbert Weber
Technische Universität Dresden
Tharandt, Germany

Ian Willoughby
Forest Research
Alice Holt Lodge
Surrey, United Kingdom

1

What Is Forest Restoration?

John A. Stanturf

CONTENTS

1.1 Introduction

The need to repair habitat and restore forest structure and function is recognized throughout the temperate and boreal zones as a component of sustainable forest management (Dobson et al. 1997; Krishnaswamy and Hanson 1999; Minnemayer et al. 2011). Forest restoration is a complex task, complicated by diverse ecological and social conditions, which challenges our understanding of forest ecosystems. The term "restoration" is used indiscriminately and it is difficult to define in such a way that encompasses all situations found in the literature and in practice (Lamb et al. 2012; Stanturf et al. 2014b). Generally, restoration is seen as symmetric with degradation (Putz and Redford 2010; Simula and Mansur 2011): An undisturbed forest in a natural or historical condition can be degraded, and a degraded forest can be restored to that natural or historical condition. As it will become apparent, reality is more complicated and the fully restored state is probably unattainable (Cairns 1986; Stanturf and Madsen 2002; Hobbs et al. 2011). Terminology, however, is not merely an academic issue; definitions related to forestry and restoration are used under several international conventions such as climate change and biodiversity where distinctions and nuance have important policy implications (FAO 2002). The objective of this chapter is to provide a conceptual framework for the terms used throughout this book, to facilitate understanding of the diverse cultural and ecological contexts for restoration of temperate and boreal forests. This chapter has three parts: a historical context for restoration, which differs geographically; a conceptual framework for understanding the relationship between degradation and restoration; and an attempt to define restoration terms within that framework.

1.2 Historical Context

Throughout history, forests have been a residual land use; external pressures such as expanding human populations have caused forests to be cleared, usually for agriculture (Noble and Dirzo 1997; Ellis et al. 2013). The conversion of forest habitat to other uses has occurred at different rates and different times in history (Goudie 1986; Dobson et al. 1997; Flinn and Vellend 2005; Kareiva et al. 2007). For example, most of Europe and Asia were settled millennia ago, while human occupation of the Americas is much shorter. Nevertheless, the transformation of land use is not unidirectional; wars, plagues, population movement and fluctuations, and climate changes cause agricultural abandonment and reversion to forests. Significant changes have occurred within the last 200 years as developed nations shifted from a biomass energy economy to fossil fuels (Clawson 1979; Ericsson et al. 2000; Johan et al. 2004). Further changes are likely in the industrialized nations of the temperate zone, as changing policies for agriculture and nature conservation provide incentives for land-use shifts from agriculture to forest.

Forest restoration in the broad sense is not a new endeavor. Agricultural abandonment and natural recolonization from remnant forests is a passive form of restoration that continues to occur (McIver and Starr 2001), notably in some former communist countries within the Commonwealth of Independent States (FAO 2001). Active restoration also has a long history; if the indirect effects of efforts to restore productivity to degraded land can be considered unintentional restoration (Stanturf et al. 2014b). The development of secondary spruce forests in central Europe is an example of the complex pathway of degradation and restoration (Johan et al. 2004). Similarly, the extensive loblolly pine forests of the southeastern United States were established to protect water and soil resources (USDA Forest Service 1988; Stanturf et al. 2003). In many countries, coastal dunes and heathland were planted to reclaim wasteland (e.g., Denmark; see Madsen et al. 2015). Active but unintentional restoration was motivated by the threat of timber scarcity and movements to improve nature, often with sociopolitical overtones such as providing employment and patriotic duty (Heske 1938; Orni 1969).

Forest restoration in industrialized countries at the beginning of the twenty-first century emerges from these earlier, more utilitarian concerns but with greater emphasis on restoring more natural forests (Farrell et al. 2000; Burton and Macdonald 2011; Stanturf et al. 2014b). Natural has multiple meanings (Cole and Yung 2010) and idealistically may mean a pristine ecosystem lacking any effect of humans, akin to the North American view of wilderness. A related view allows for minimal human influence but free from intentional human control (Cole 2000). The idealized natural state has been called variously the climax, urwald, pristine, or old-growth; this is the forest that develops without human influence and may persist within a landscape mosaic of actively and passively managed (erroneously termed "unmanaged") forests. Whether or not this idealized state exists now or within the last 25,000 years is debatable (e.g., Denevan 1992; Ellis et al. 2013). Certainly old forests with minimal human interference exist that provide examples of more natural conditions than intensively managed or secondary forests (e.g., Matuszkiewicz et al. 2013). Therefore, the defining characteristics of "naturalness" are the lack of major human interference for all or most of the lifespan of the oldest trees; complex vegetative structures; native species composition; and historical fidelity in terms of disturbance regimes and proportion in the landscape (Hunter 1996).

A concept which is often associated with the idealized view of natural ecosystems is that systems with minimal human influence are stable or in equilibrium (Perry 2002) and self-regulating (Middleton 1999). Thus, restoration to a more natural state under this steady-state view of forest ecosystems means returning to a condition of historical fidelity (Bradshaw 2015), with similar species composition and structure as before significant human intervention (Higgs 2003) or at least within the range of historic variability (Keane et al. 2009). Since the 1980s, recognition of the role of disturbances in forest ecosystem development and maintenance, including the importance of dead wood and other legacies of the predisturbance ecosystem, has gained appreciable momentum (Pickett and White 1985; Sprugel 1991; Beatty and Owen 2005; Oliver and O'Hara 2005; Turner 2010). Nevertheless, these dynamic views of forest ecosystems are less well accepted among the public and some practitioners (Oliver and O'Hara 2005).

We view a gradient of naturalness from the degraded forest to the idealized state of one without any human influence, indigenous or otherwise. In short, we regard naturalness as a continuum and not a binary state (Stanturf and Madsen 2002). Understanding and defining the various states along the continuum relies on ecological understanding, which has changed over time and now regards forested landscapes as open rather than closed systems, as dynamic rather than steady-state systems (Oliver and O'Hara 2005). The characteristics that define naturalness, however, must be defined in local terms that reflect local values (Frelich and Reich 2003; Laarmann et al. 2009). For example, how much human influence is allowed? Even forests that ecologists and the public regard as the most natural have legacies of the past human influence (Krech III 1999; Ellis et al. 2013). Are the conditions under which the old forest was initiated and developed still operating today (Bradshaw 2015; Millar and Woolfenden 1999; Millar 2014)? In many places, conditions have changed (e.g., Sprugel 1991), sometimes as a result of human alterations. For example, many lowland landscapes have been drastically altered and regional and local drainage patterns disrupted or inundation regimes changed by drainage, levees, or dams (e.g., Mississippi River floodplain, see Stanturf et al. 2000; European rivers, see Hughes et al. 2012). Furthermore, the notion of native species is somewhat mutable; for example, the postglacial dispersal of major trees species in northern Europe is still underway (Bradshaw 2015).

Practitioners within the restoration ecology community (Davis and Slobodkin 2004; Hobbs 2004) and other resource professionals (Wagner et al. 2000; Stanturf et al. 2014b) have challenged the notion of naturalness as a restoration objective. The crux of the debate is whether naturalness represents a scientifically defensible concept (Anderson 1991) or is simply a statement of a preference for one kind of ecosystem over another (Hobbs 2004). Some restoration ecologists are moving away from the purist position, especially the more ideological views that set the goal of restoration to be an idealized pristine state, which implies a static view of ecosystems (Davis and Slobodkin 2004; Hobbs 2004). Nevertheless, a lively debate has ensued over whether the endpoint of restoration can be set in a way that is free of values (Winterhalder et al. 2004).

Forests today are human-dominated systems (Noble and Dirzo 1997; Vitousek et al. 1997; Ericsson et al. 2000), although there is a tendency in the Americas to underestimate the extent to which indigenous peoples influenced the forests as described by the first European naturalists (Stanturf et al. 2002). If the starting point is a degraded forest only slightly removed in time from a natural or seminatural forest, suitable reference stands may be available for setting restoration goals or endpoints. Suitable reference conditions are unavailable and are likely to be unknowable if a forest minimally influenced by humans is hundreds or thousands of years distant from the starting point (Wagner et al. 2000; Hobbs 2004). Recognizing the difficulty of setting restoration goals on the basis of recreating past conditions (Sprugel 1991; Hobbs and Norton 1996; Parker and Pickett 1997; Bradshaw 2015) leads to the conclusion that the endpoint is a sociopolitical decision (Hobbs 2004) that can be informed by science (Keddy and Drummond 1996), but cannot be determined by science alone. What constitutes successful restoration will be defined within a cultural and ecological context, including financial costs and unexpected consequences (Anderson and Dugger 1998; Holl and Howarth 2000; Palik et al. 2000; Anand and Desrochers 2004). The appropriate intervention will be determined largely by the degree of degradation (Hobbs 2004) and the likelihood of success (Stanturf et al. 2014a).

1.3 Degradation and Restoration Processes

Forest condition is dynamic, subject to natural developmental processes (Oliver et al. 2015) as well as natural and anthropogenic disturbances (Covington et al. 1997; Angelstam 1998; Turner et al. 1998; Beatty and Owen 2005; Stanturf et al. 2014b). Degradation results from changes to forest structure or function that lowers its productive capacity (FAO 2002), including limited biodiversity. Degradation is not synonymous with disturbance; disturbance becomes degradation, however, when it crosses a threshold beyond the natural resilience of a forest type. The simplest conceptualization of the relationship between degradation and recovery processes is to place a forest on a continuum from natural to degraded (Bradshaw 1997; Harrington 1999). Levels of state factors such as biomass or biodiversity in a forest subjected to degradation follow a presumably linear trajectory. At any point along the trajectory, recovery toward a natural forest can be initiated once the stress or disturbance abates.

1.3.1 Degradation Processes

The dynamic relationship between processes degrading and restoring forests is more easily understood if considered in light of two dimensions: changes in land cover, land use, or both (Stanturf and Madsen 2002). Taking as the starting point the undisturbed, idealized natural mature forest (Westhoff 1983; Goudie 1986), conversions to other land use such as agriculture (cultural landscape) or pasture (seminatural landscape) are through deforestation (Figure 1.1). Relatively frequent but moderate disturbance (plowing, herbicides, and grazing) maintains the nonforest cover. Similarly, a change in both land cover and land use occurs when forests are removed and the land is converted to urban areas, flooded by dams, or removed along with topsoil and overburden by mining and extractive activities. Such drastic degradation involves deforestation, usually accompanied by ongoing disturbance. The nonforest cover is maintained more or less permanently by structures, more so than by cultural activities (Figure 1.1). Agricultural land can also be converted to urban use.

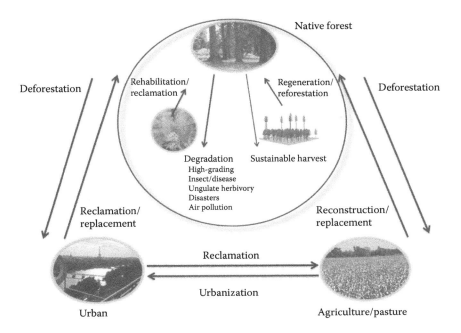

FIGURE 1.1
Forest restoration begins with forests that have been degraded (rehabilitation) or after deforestation and conversion to other land uses (reconstruction or reclamation). Self-renewal processes operate within forests that are disturbed (e.g., sustainable harvesting) but not degraded (regeneration/reforestation).

Harvesting a mature forest in a sustainable manner is a change of land cover but not land use (FAO 2002). A new, young forest will result from natural regeneration or by reforestation (within the envelope of forest cover in Figure 1.1). Unsustainable harvesting without securing adequate regeneration, however, may degrade stand structure or diversity. Outbreaks of insects or diseases (especially exotic species), fire suppression and disruption of natural fire regimes, invasion by aggressive exotic plants or disasters such as hurricanes or wildfires can degrade forest stands and change attributes of land cover (Stanturf et al. 2014b), but these stressors do not change land use (Figure 1.1). Chronic low-level pollutant loading may degrade a forest without changing land use, although heavy loading may deforest an area and change use into wasteland.

1.3.2 Restoration Processes

Forests are resilient: given sufficient time and the cessation of disturbances, agricultural, and urbanized land will revert to forest. Abandonment and reversion to forests, although secondary, seminatural, or degraded forest types, will be on a timescale of a few decades to centuries as existing forests expand into nonforest areas, or natural recolonization occurs. Human intervention, however, can accelerate the reversion process (Ferris-Kaan 1995). Restoring forests through afforestation of agricultural land may consist of simply planting trees, although more intensive techniques are available (Stanturf et al. 2014a). Restoration of urbanized land usually requires extensive modification, including stabilization of spoil banks or removal of water control structures, followed by tree planting. Severe site degradation may limit the possibility of restoring to native forest conditions, thus nonnative species may be used (e.g., Parrotta et al. 1997).

1.3.3 Naturalness and Ecological Theory

Intervention can facilitate recovery from disturbance or degradation. For convenience, intervention can be divided into three levels of increasing effort: self-renewal, rehabilitation, and reconstruction/reclamation (Stanturf et al. 2001). In the self-renewal phase, resistance and resilience mechanisms maintain or return the forest more or less to its original state, without human intervention, in a relatively short time. Sustainably managed forests rely on self-renewal processes, for example, naturally regenerated forests managed for timber. Intervention at this stage will be to ensure that composition and structure meet management objectives. Plantations of native species can be within the scope of self-renewal, where intervention (reforestation) is undertaken to control species and stocking. At intermediate to intense levels of disturbance, beyond the self-renewal phase, degradation occurs. If a forest is degraded but remains in forest land use, which means that it is not deforested, it can be rehabilitated to a forest condition that is within the range of self-renewal mechanisms. Recovery to a more natural forest will take longer, but the time required can be shortened by human intervention. Rehabilitation by reforestation of forests consumed by severe wildfire is an example (e.g., Kaufmann et al. 2005). In the most degraded state, forest cover is removed and the land is converted to another use—this is deforestation (FAO 2002; Stanturf et al. 2014b). In this chapter, restoration encompasses all interventions into degraded forests, those stands disturbed beyond the range of self-renewal processes. Rehabilitation refers to restoration of degraded forests; reconstruction and reclamation encompass restoration of forests from nonforest land uses. A consistent terminology will be introduced later, after a brief exploration of the dominant restoration paradigms.

1.4 Restoration Paradigms

1.4.1 Revegetation: A Utilitarian Response

Historically, active restoration began as early attempts in restoring ecosystems by focusing primarily on revegetation without much regard for nativeness of species or structural diversity. Early examples of revegetation were primarily aimed at restoring productive functions or avoiding further soil erosion. More modern examples can be found in the efforts to restore mined land without regard for structure, composition, or disturbance processes. The primary ecological goal of these early restoration programs was revegetation but occasionally a nationalistic or social motivation was included, such as providing employment. Examples are restoration of heathland in western Denmark (Madsen et al. 2015), afforestation in Israel (Orni 1969), watershed restoration in the southern United States (USDA Forest Service 1988), or tree planting during the Dustbowl of the 1930s to mitigate soil erosion and respond to declining timber resources (Dumroese et al. 2005).

Revegetation has limited value as a contemporary forest restoration paradigm, with the exception of drastically disturbed sites. Simply providing a forest cover, while beneficial in terms of soil protection, may provide few other ecosystem services. Nevertheless, on sites degraded physically, chemically, or both by surface mining, severe soil erosion, or radioactive fallout, for example, the site has been so altered that native vegetation likely will fail. Thus, the revegetation approach fits within the restoration framework as the special case of reclamation and revegetation with exotic species may be a transitory phase (Parrotta et al. 1997; Lamb et al. 2005).

1.4.2 Ecological Restoration: Ecological Determinism

Ecological restoration "is an intentional activity that initiates or accelerates recovery of an ecosystem with respect to its health, integrity and sustainability" (SER 2002, p. 1). Explicit in this definition is that ecological restoration is relative to reference conditions (Wagner et al. 2000; Perrow and Davy 2002a,b; SER 2002), which are related to a notion of natural conditions for a site (Hobbs and Norton 1996; Egan and Howell 2001). Thus, ecological restoration has the most ambitious goal of returning to a prior ecosystem condition (Van Diggelen et al. 2001; SER 2002), some natural state that often presumes an absence of human disturbance (Hobbs and Norton 1996). Ecological restoration, in many ways, is the antithesis of revegetation because its goal is more than simply to revegetate, but rather includes specific goals for composition and structure, an approach dominated by restoration to past conditions, as exemplified by reference sites.

Reference conditions in North America were often the presumed historic conditions before European settlement, which were believed to represent minimal human influence. Insight into how and why ecological restoration projects currently are undertaken is gained from the literature (Burton and Macdonald 2011) and reports from active restoration projects (Clewell and Aronson 2006; Hallett et al. 2013). The relationships among restoration motivations, attributes of restored ecosystems (goals, or definitions of success) and broad societal goals for restoration illustrate the preponderance of repairing ecosystem function (Stanturf et al. 2014b). Any single project is likely to have multiple goals arising from the motivations of those involved.

Challenges to the notion of one historic past resulted in the concept of multiple possible reference conditions that existed within the range of historical variability (Landres et al. 1999; Keane et al. 2009). Even though the requisite condition of minimal human influence was relaxed to accommodate landscapes heavily influenced by millennia of human intervention (such as Europe, Asia, and Africa), an ecological imperative focused on stable ecosystems (Clewell and Aronson 2013) still dominates the ecological restoration paradigm (Perrow and Davy 2002a,b; SER 2002; Burton and Macdonald 2011; Hobbs 2013). In spite of the increasing recognition that restoration to past conditions is generally infeasible, the ecological restoration paradigm remains focused on historical conditions, on stand-level activity, and on "natural" processes and structures (Clewell and Aronson 2013; Hobbs 2013).

1.4.3 Forest Landscape Restoration: Sustainable Development

Forest landscape restoration (FLR) differs from site-level restoration because it seeks to restore ecological processes that operate at larger landscape-level scales (Mansourian and Vallauri 2005), to regain ecological integrity, and to enhance human well-being (Maginnis and Jackson 2007). Thus, FLR not only broadens the scope of restoration to consideration of the entire landscape but also explicitly incorporates human activities and needs; FLR is a decision-making process and not simply a series of ad hoc treatments that eventually cover large areas (Lamb et al. 2012). FLR involves choices about how much and where restoration is undertaken, as well as the technical question of how to restore (Palik et al. 2000; Lindenmayer et al. 2008). Not only must restoration be feasible in terms of conforming to the ecological conditions of particular sites but the restoration techniques used and outcomes desired must meet the socioeconomic constraints of the (often multiple) landowners, land users, or stakeholders (Clement and Junqueira 2010; Shinneman et al. 2012).

Finding ways to implement restoration at large or landscape scale is a challenge of FLR (Frelich and Reich 2003; Brudvig 2011). This is done by taking a strategic approach: key locations are targeted that return the greatest social benefit (e.g., Mercer 2005; Maron and Cockfield 2008; Wilson et al. 2012) rather than relying on the individual decisions of separate landholders. The decision-making framework can use relatively informal techniques (Boedhihartono and Sayer 2012; Sayer et al. 2013) or more complex computer decision models (Pullar and Lamb 2012). Nevertheless, inequities must be avoided and care must be taken to see that restoration is not carried out at the expense of some landowners or land users, particularly the poorest or least vocal elements of society.

Advantages of the FLR paradigm over the ecological restoration paradigm include the expanded focus on landscape-level restoration and the explicit inclusion of meeting human livelihood needs. By recognizing livelihood and food security needs, FLR is more appropriate in the developing world than ecological restoration, which often has a restore-then-preserve underpinning (e.g., Stanturf et al. 2001). The potential disadvantage of FLR is that it may narrowly focus on current local needs, ignoring broader social needs unless they are included within the mandate of the funding authority (Stanturf et al. 2014b).

1.4.4 Functional Restoration: A Pragmatic Response

Functional restoration emphasizes the restoration of abiotic and biotic processes in degraded ecosystems. While ecosystem structure and function are closely connected, functional restoration focuses on the underlying processes that may be degraded, regardless of the structural condition of the ecosystem. As such, a functionally restored ecosystem may have different structure and composition than the historical reference condition if some threshold of degradation has been crossed (Whisenant 2002) or the environmental drivers, such as climate, that influenced structural and (especially) compositional development have changed. Structural restoration is more focused on a static view of an ecosystem, for example, historical reference condition, while functional restoration focuses on the dynamic processes that drive structural and compositional patterns (King and Hobbs 2006). One example is the popular attempts to reintroduce fire into longleaf pine (*Pinus palustris*) forests where long-term fire suppression has allowed a broadleaf mid-story to develop (Brockway et al. 2015; Schwilk et al. 2009; Phillips et al. 2012).

1.4.5 Intervention Ecology: A Transformative Approach

The challenges of continuing global change and impending climate variability render the goal of restoring to some past conditions ever more unachievable (Harris et al. 2006). Recognition that restoration must take place within the context of rapid environmental change has begun to redefine the goals of restoration toward future adaptation rather than a return to historic conditions (Choi 2007; Choi et al. 2008). This redefinition of restoration removes the underpinning of an ecological imperative (Angermeier 2000; Burton and Macdonald 2011) and underscores the importance of clearly defined goals focused on functional ecosystems. Recognizing the increasing difficulty of returning ecosystems to historic states, and the dangers of creating false expectations and failing to deliver, some have proposed a transformational approach to restoration, intervention ecology (Sarr and Puettmann 2008; Hobbs et al. 2011). Intervention ecology incorporates ecological and socioeconomic aspects as well as anticipating the need to intervene in governance systems (Hobbs et al. 2011).

1.5 Restoration Terminology

Commonly used restoration terms can be understood within a conceptual framework (Figure 1.2) that takes into account the relationships between changes in forest cover and land use (Figure 1.1). This is not an attempt to standardize these terms, but to harmonize (FAO 2002; Hasenauer 2004) them for consistent use in the chapters that follow. The degradation trajectory begins with the idealized forest at Ω as the starting point (Figure 1.2). This beginning point is culturally and situationally determined. In some contexts, it may represent an actual historical reality, or it may be a conceptual model of the potential natural vegetation for an area. The degradation trajectory moves toward a degraded endpoint (A in Figure 1.2). The possible endpoints are shown in Figure 1.1; the most degraded states will include deforestation and conversion to nonforest land use. The intermediate points B_1 to B_3 represent forests degraded by air pollution, exploitive harvesting, natural disasters, and so on. These degraded forests, as well as nonforest conditions (A), represent starting points for restoration trajectories. For ease of representation, the A to Ω trajectory is presented as linear; in reality it is probably more complex (Anand and Desrochers 2004; Stanturf et al. 2014b).

The path extending from A to Ω, labeled as Recreate, represents the strictly defined ecological or historical restoration (SER 2002). Recreating the ideal natural or historical forest ecosystem is unlikely to be successful over large areas (Van Diggelen et al. 2001) and will certainly be expensive. The reconstruction pathway refers to restoration of agricultural land to forest conditions (Stanturf et al. 2014a), through afforestation or natural recolonization. In our terminology, afforestation is distinct from reforestation; the latter being the normal forestry practice of establishing a new stand following the removal of the previous stand without an interval of another land use. The endpoints for reconstruction (B_1 to B_3) may be a less diverse natural forest (B_2) or a mixed species plantation of native species (B_3).

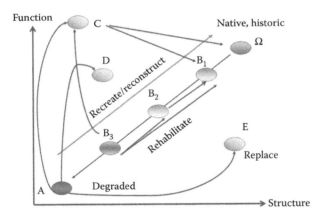

FIGURE 1.2
A conceptual framework for forest restoration has a starting point of a degraded forest (A) and an idealized endpoint of a restored native, historical forest (Ω). The symmetric degradation/recreation trajectories have intermediate points that represent starting or ending points (B_1 to B_3) for reconstruction or reclamation of severely degraded forests (deforested and converted to other land uses) or less severely degraded forests that are restored using rehabilitation strategies. Replacement trajectories denote functionally restored forests that lack the structure or species composition of native forests. Note that a replacement forest (C or E) can be the subsequent starting point for restoration to a more natural forest.

Active reconstruction by planting restricted areas, such as is done for stabilizing sand dunes or stream banks or protecting riparian areas with buffer strips, may not be called afforestation but they amount to the same thing—reconstructing a forest where it has been absent for some time and the land has been in other uses. Some definitions of afforestation set a time criteria (e.g., 25 years) for the nonforest interlude to accommodate mixed uses such as swidden agriculture (Lund 1999). Agroforestation is a related strategy to introduce trees onto farms, primarily for carbon sequestration (van Noordwijk et al. 2008).

A site may be so degraded that native species are replaced by exotics; this pathway (to E) would be termed as "replacement," a term that also may be used to describe assisted migration of species as an adaptation response to climate change (Williams and Dumroese 2013; Stanturf et al. 2014a,b). Reclamation begins with urban or built land use or highly degraded sites such as eroded pastures and may require land stabilization as well as afforestation (Arnalds et al. 1987; Parrotta et al. 1997; Lamb et al. 2005; Stanturf et al. 2014a; Renou-Wilson et al. 2015). These reclamation sites are shown as C, D, and E in Figure 1.2. On such sites, it may be necessary to ameliorate soil conditions with amendments to raise pH, add nutrients, and modify bulk density. For both reconstruction and reclamation, continuing intervention over time may move the forest condition closer to the natural endpoint (shown as lines from C to B_1 and Ω in Figure 1.2).

Rehabilitation of degraded forests has one of the intermediate conditions (B_1–B_3) as a starting point: forest cover has been removed or degraded but no change to nonforest land use has occurred (Stanturf 2014a). Rehabilitation encompasses many techniques to restore stand structure, species composition, natural disturbance regimes, or to remove exotic plants. Specific forms of rehabilitation are termed conversion (Nyland 2003; Spiecker et al. 2004) or transformation (Kenk and Guehne 2001). While these approaches share some characteristics, conversion seems to apply to wholesale removal of an existing overstory and replacement with other species. Transformation applies to a more extended process of partial removals and species replacement but obviously the demarcation between these approaches is fuzzy. For example, selection harvests over time could transform a forest lacking desired structure with the advantage of maintaining a continuous forest cover (O'Hara 2001; O'Hara and Ramage 2013). Alternatively, a conversion approach could call for clear felling followed by natural or artificial regeneration (Hansen and Spiecker 2015).

1.6 Transforming Temperate and Boreal Forest Landscapes

Forests are human-dominated ecosystems (Noble and Dirzo 1997). Reconstructing forests where they are now absent, as well as altering existing forests to more natural conditions, are important aspects of sustainable forest management. Important tasks for forest restorationists are to understand how ecosystems were degraded, how to reverse degradation processes, and how to efficiently initiate recovery processes (Hobbs and Norton 1996). Silviculturists and forest ecologists have important tools—diagnostic and predictive skills and effective intervention techniques—that are critical to successful restoration of complex ecosystems (Stanturf et al. 2014a). These tools must be used appropriately, however, within diverse sociopolitical, ecological, and historic contexts. A broad conception of restoration allows more diverse goals (endpoints) than the narrow construct of recreating particular, preexisting ecosystem states (reference conditions). Greater flexibility in setting restoration objectives is not a retreat from basing restoration on ecological science (Wagner et al. 2000). Rather, it is recognition that incomplete knowledge of past ecosystem states (Hobbs 2004), changes in the global environment (Vitousek et al. 1997), costs, and the scale of degradation argue for a pragmatic approach.

The dualistic notion of degradation and restoration as opposing trajectories of forest development leads to an understanding of restoration in a broader context than ecological restoration (SER 2002). In this view, the restored forest that results from reconstruction, reclamation, or rehabilitation may never recreate the original state for all functions (Cairns 1986; Bradshaw 1997; Harrington 1999). Any endpoint within the natural range of managed forests where self-renewal processes operate is acceptable as restoration. Thus, restoration to an early seral stage would be acceptable for a forest that is likely to attain a more complex structure through typical stand dynamics. How quickly the forest moves to the self-renewal phase is a function of forest type, site resources, and the amount invested to overcome the degraded conditions. This model offers a broader context for restoration on private land; landowners with management objectives other than preservation are able to contribute to ecosystem restoration (Farrell et al. 2000; Stanturf et al. 2001).

References

Anand, M. and Desrochers, R.E., Quantification of restoration success using complex systems concepts and models, *Restor. Ecol.*, 12, 117, 2004.

Anderson, J.E., A conceptual framework for evaluating and quantifying naturalness, *Conserv. Biol.*, 5, 347, 1991.

Anderson, D.H. and Dugger, B.D., A conceptual basis for evaluating restoration success, in *Transactions of 63rd North American Wildlife and Natural Resources Conference*, pp. 376–383, Orlando, FL, Wildlife Management Institute, Washington, DC, vol. 111, 1998.

Angelstam, P., Maintaining and restoring biodiversity by developing natural disturbance regimes in European boreal forests, *J. Veg. Sci.*, 9, 593, 1998.

Angermeier, P.L., The natural imperative for biological conservation, *Conserv. Biol.*, 14, 373, 2000.

Arnalds, O., Aradóttir, A.L., and Thorsteinsson, I., The nature and restoration of denuded areas in Iceland, *Arctic Alpine Res.*, 19, 518, 1987.

Beatty, S.W. and Owen, B.S., Incorporating disturbance into restoration, in *Restoration of Boreal and Temperate Forests*, Stanturf, J.A. and Madsen, P., Eds., CRC Press, Boca Raton, FL, p. 61, 2005.

Boedhihartono, A.K. and Sayer, J., Forest landscape restoration: Restoring what and for whom? in *Forest Landscape Restoration*, Stanturf, J.A., Lamb, D., and Madsen, P., Eds., Springer, New York, p. 309, 2012.

Bradshaw, A.D., What do we mean by restoration? in *Restoration Ecology and Sustainable Development*, Urbanska, K.M., Webb, N.R. and Edwards, P.J., Eds., Cambridge University Press, Cambridge, UK, p. 8, 1997.

Bradshaw, R.H.W., What is a natural forest? in *Restoration of Boreal and Temperate Forests*, 2nd ed., Stanturf, J.A., Ed., CRC Press, Boca Raton, FL, p. 17, 2015.

Brockway, D.G., Outcalt, K.W., Tomczak, D.J., and Johnson, E.E., Restoring longleaf pine forest ecosystems in the Southern United States, in *Restoration of Boreal and Temperate Forests*, 2nd ed., Stanturf, J.A., Ed., CRC Press, Boca Raton, FL, p. 445, 2015.

Brudvig, L.A., The restoration of biodiversity: Where has research been and where does it need to go? *Am. J. Bot.*, 98, 549, 2011.

Burton, P.J. and Macdonald, S.E., The restorative imperative: Challenges, objectives and approaches to restoring naturalness in forests, *Silva Fenn.*, 45, 843, 2011.

Cairns, J. Jr., Restoration, reclamation, and regeneration of degraded or destroyed ecosystems, in *Conservation Biology*, Soule, M.E., Ed., Sinauer Publishers, Ann Arbor, MI, p. 465, 1986.

Choi, Y.D., Restoration ecology to the future: A call for new paradigm, *Restor. Ecol.*, 15, 351, 2007.

Choi, Y.D., Temperton, V.M., Allen, E.B., Grootjans, A.P., Halassy, M., Hobbs, R.J. et al. Ecological restoration for future sustainability in a changing environment, *Ecoscience*, 15, 53, 2008.

Clawson, M., Forests in the long sweep of American history, *Science*, 204, 1168, 1979.

Clement, C.R. and Junqueira, A.B., Between a pristine myth and an impoverished future, *Biotropica*, 42, 534, 2010.

Clewell, A.F. and Aronson, J., Motivations for the restoration of ecosystems, *Conserv. Biol.*, 20, 420, 2006.

Clewell, A. and Aronson, J., The SER primer and climate change, *Ecological Manage. Restor.*, 14, 182, 2013.

Cole, D.N., Paradox of the primeval: Ecological restoration in wilderness, *Ecol. Restor.*, 18, 77, 2000.

Cole, D.N. and Yung, L., Eds., *Beyond Naturalness: Rethinking Park and Wilderness Stewardship in an Era of Rapid Change*, Island Press, Washington, DC, 2010.

Covington, W.W., Fulé, P.Z., Moore, M.M., Hart, S.C., Kolb, T.E., Mast, J.N. et al. Restoring ecosystem health in ponderosa pine forests of the southwest, *J. For.*, 95, 23, 1997.

Davis, M.A. and Slobodkin, L.B., The science and values of restoration ecology, *Restor. Ecol.*, 12, 1, 2004.

Denevan, W.M., The pristine myth: The landscape of the Americas in 1492. *Ann. Assoc. Am. Geograph.*, 82, 369, 1992.

Dobson, A.P., Bradshaw, A.D., and Baker, A.J.M., Hopes for the future: Restoration ecology and conservation biology, *Science*, 277, 515, 1997.

Dumroese, R.K., Landis, T.D., Barnett, J.P., and Burch, F., Forest Service Nurseries: 100 years of ecosystem restoration, *J. For.*, 103, 241, 2005.

Egan, D. and Howell, E.A., Eds., *The Historical Ecology Handbook: A Restorationist's Guide to Reference Ecosystems*, Island Press, Washington, DC, 2001.

Ellis, E.C., Kaplan, J.O., Fuller, D.Q., Vavrus, S., Goldewijk, K., and Verburg, P.H., Used planet: A global history, *Proc. Natl. Acad. Sci.*, 110, 7978, 2013.

Ericsson, S., Östlund, L., and Axelsson, A.-L., A forest of grazing and logging: Deforestation and reforestation history of a boreal landscape in central Sweden, *New Forest*, 19, 227, 2000.

Farrell, E.P., Führer, E., Ryan, D., Andersson, F., Hüttl, R., and Piussi, P., European forest ecosystems: Building the future on the legacy of the past, *For. Ecol. Manage.*, 132, 5, 2000.

Ferris-Kaan, R., Ed., *The Ecology of Woodland Creation*, John Wiley & Sons, Chichester, UK, 1995.

Flinn, K.M. and Vellend, M., Recovery of forest plant communities in post-agricultural landscapes. *Front. Ecol. Environ.*, 3, 243, 2005.

Food and Agriculture Organization (FAO), Global Forest Resources Assessment 2000, FAO Forestry Paper 140, Rome, 2001.

Food and Agriculture Organization (FAO), *Proceedings Expert Meeting on Harmonizing Forest-related Definitions for Use by Various Stakeholders*, Rome, 22–25 January 2002, FAO, Rome, 2002.

Frelich, L.E. and Reich, P.B., Perspectives on development of definitions and values related to old-growth forests. *Environ. Rev.*, 11(S1), S9, 2003.

Goudie, A., *The Human Impact on the Natural Environment*, MIT Press, Cambridge, MA, 1986.

Hallett, L.M., Diver, S., Eitzel, M.V., Olson, J.J., Ramage, B.S., Sardinas, H. et al. Do we practice what we preach? Goal setting for ecological restoration, *Restor. Ecol.*, 21, 312, 2013.

Hansen, J. and Spiecker, H., Conversion of Norway spruce (*Picea abies* [L.] Karst.) forests in Europe, in *Restoration of Boreal and Temperate Forests*, 2nd ed., Stanturf, J.A., Ed., CRC Press, Boca Raton, FL, p. 355, 2015.

Harrington, C.A., Forests planted for ecosystem restoration or conservation, *New Forest*, 17, 175, 1999.

Harris, J.A., Hobbs, R.J., Higgs, E., and Aronson, J., Ecological restoration and global climate change, *Restor. Ecol.*, 14, 170, 2006.

Hasenauer, H., Glossary of terms and definitions relevant for conversion, in *Norway Spruce Conversion—Options and Consequences*, Spiecker, H., Hansen, J., Klimo, E., Skovsgaard, J.P., Sterba, H. and von Teuffel, K., Eds., Brill, Boston, p. 5, 2004.

Heske, F., *German Forestry*, Yale University Press, New Haven, CT, 1938.

Higgs, E., *Nature by Design: People, Natural Process, and Ecological Design*, MIT Press, Cambridge, MA, 2003.

Hobbs, R.J., Forum: Restoration ecology: The challenge of social values and expectations, *Front. Ecol.*, 2, 43, 2004.

Hobbs, R.J., Grieving for the past and hoping for the future: Balancing polarizing perspectives in conservation and restoration, *Restor. Ecol.*, 21, 145, 2013.

Hobbs, R.J., Hallett, L.M., Ehrlich, P.R., and Mooney, H.A., Intervention ecology: Applying ecological science in the twenty-first century, *BioSci.*, 61, 442, 2011.

Hobbs, R.J. and Norton, D.A., Towards a conceptual framework for restoration ecology, *Restor. Ecol.*, 4, 93, 1996.

Holl, K.D. and Howarth, R.B., Paying for restoration, *Restor. Ecol.*, 8, 260, 2000.

Hughes, F.M.R., González del Tánago, M., and Mountford, J.O., Restoring floodplain forests in Europe, in *A Goal-Oriented Approach to Forest Landscape Restoration*, Stanturf, J.A., Madsen, P., Lamb, D., Eds., Springer, New York, p. 393, 2012.

Hunter, M., Editorial: Benchmarks for managing ecosystems: Are human activities natural? *Conserv. Biol.*, 10, 695, 1996.

Johan, E., Agnoletti, M., Axelsson, A.-L., Bürgi, M., Östlund, L., Rochel, X. et al. History of secondary spruce forests in Europe, in *Norway Spruce Conversion—Options and Consequences*, Spiecker, H., Hansen, J., Klimo, E., Skovsgaard, J.P., Sterba, H. and von Teuffel, K., Eds., Brill, Boston, p. 25, 2004.

Kareiva, P., Watts, S., McDonald, R., and Boucher, T., Domesticated nature: Shaping landscapes and ecosystems for human welfare, *Science*, 316, 1866, 2007.

Kaufmann, M.R., Fulé, P.Z., Romme, W.H., and Ryan, K.C., Restoration of ponderosa pine forests in the interior western U.S. after logging, grazing, and fire suppression, in *Restoration of Boreal and Temperate Forests*, Stanturf, J.A. and Madsen, P., Eds., CRC Press, Boca Raton, FL, p. 481, 2005.

Keane, R.E., Hessburg, P.F., Landres, P.B., and Swanson, F.J., The use of historical range and variability (HRV) in landscape management, *For. Ecol. Manage.*, 258, 1025, 2009.

Keddy, P.A. and C.G. Drummond, Ecological properties for the evaluation, management, and restoration of temperate deciduous forest ecosystems, *Ecol. Appl.*, 6, 748, 1996.

Kenk, G. and Guehne, S., Management of transformation in central Europe, *For. Ecol. Manage.*, 151, 107, 2001.

King, E.G. and Hobbs, R.J., Identifying linkages among conceptual models of ecosystem degradation and restoration: Towards an integrative framework, *Restor. Ecol.*, 14, 369, 2006.

Krech III, S., *The Ecological Indian: Myth and History*, WW Norton & Company, New York, 1999.

Krishnaswamy, A. and Hanson, A., Eds., *Our Forests, Our Future: Summary Report, World Commission on Forests and Sustainable Development*, Cambridge University Press, Cambridge, UK, 1999.

Laarmann, D., Korjus, H., Sims, A., Stanturf, J.A., Kiviste, A., and Köster, K., Analysis of forest naturalness and tree mortality patterns in Estonia, *For. Ecol. Manage.*, 258, S187, 2009.

Lamb, D., Erskine, P.D., and Parrotta, J.A., Restoration of degraded tropical forest landscapes. *Science*, 310, 1628, 2005.

Lamb, D., Stanturf, J.A., and Madsen, P., What is forest landscape restoration? in *A Goal-Oriented Approach to Forest Landscape Restoration*, Stanturf, J.A., Madsen, P., Lamb, D., Eds., Springer, New York, p. 3, 2012.

Landres, P.B., Morgan, P., and Swanson, F.J., Overview of the use of natural variability concepts in managing ecological systems, *Ecol. Appl.*, 9, 1179, 1999.

Lindenmayer, D., Hobbs, R.J., Montague-Drake, R., Alexandra, J., Bennett, A., Burgman, M. et al. A checklist for ecological management of landscapes for conservation, *Ecol. Lett.*, 11, 78, 2008.

Lund, H.G., A "forest" by any other name, *Environ. Sci. Pol.*, 2, 125, 1999.

Madsen, P., Jensen, F., and Fodgaard, S., Afforestation in Denmark, in *Restoration of Boreal and Temperate Forests*, 2nd ed., Stanturf, J.A., Ed., CRC Press, Boca Raton, FL, p. 201, 2015.

Maginnis, S. and Jackson, W., What is FLR and how does it differ from current approaches? in *The Forest Landscape Restoration Handbook*, Rietbergen-McCracken, J., Maginnis, S., and Sarre, A., Eds., Earthscan, London, UK, 2007.

Mansourian, S. and Vallauri, D., *Forest Restoration in Landscapes: Beyond Planting Trees*, Springer, New York, 2005.

Maron, M. and Cockfield, G., Managing trade-offs in landscape restoration and revegetation projects, *Ecol. Appl*, 18, 2041, 2008.

Matuszkiewicz, J.M., Kowalska, A., Kozłowska, A., Roo-Zielińska, E., and Solon, J., Differences in plant-species composition, richness and community structure in ancient and post-agricultural pine forests in central Poland, *For. Ecol. Manage.*, 310, 567, 2013.

McIver, J. and Starr, L., Restoration of degraded lands in the interior Columbia River basin: Passive versus active approaches, *For. Ecol. Manage.*, 153, 15, 2001.

Mercer, D.E., Policies for encouraging forest restoration, in *Restoration of Boreal and Temperate Forests*, Stanturf, J.A. and Madsen, P., Eds., CRC Press, Boca Raton, FL, p. 97, 2005.

Middleton, B.A., *Wetland Restoration, Flood Pulsing, and Disturbance Dynamics*, Wiley, New York, 1999.

Millar, C.I., Historic variability: Informing restoration strategies, not prescribing targets, *J. Sustain. For.*, 33(suppl 1), S28, 2014.

Millar, C.I. and Woolfenden, W.B., The role of climate change in interpreting historical variability, *Ecol. Appl.*, 9, 1207, 1999.

Minnemayer, S., Laestadius, L., and Sizer, N., *A World of Opportunity*, World Resource Institute, Washington, DC, 2011.

Noble, I.R. and Dirzo, R., Forests as human-dominated ecosystems, *Science*, 277, 522, 1997.

Nyland, R.D., Even- to uneven-aged: The challenges of conversion, *For. Ecol. Manage.*, 172, 291, 2003.

O'Hara, K.L., The silviculture of transformation—a commentary, *For. Ecol. Manage.*, 151, 81, 2001.

O'Hara, K.L. and Ramage, B.S., Silviculture in an uncertain world: Utilizing multi-aged management systems to integrate disturbance, *Forestry*, 86, 401, 2013.

Oliver, C.D. and O'Hara, K.L., Effects of restoration at the stand level, in *Restoration of Boreal and Temperate Forests*, Stanturf, J.A. and Madsen, P., Eds., CRC Press, Boca Raton, FL, p. 31, 2005.

Oliver, C.D., O'Hara, K.L., and Baker, P.J., Effects of restoration at the stand level, in *Restoration of Boreal and Temperate Forests*, 2nd ed., Stanturf, J.A., Ed., CRC Press, Boca Raton, FL, p. 37, 2015.

Orni, E., *Afforestation in Israel*, Keren Kayemeth Leisrael, Jerusalem, 1969.

Palik, B.J., Goebel, P.C., Kirkman, L.K., and West, L., Using landscape hierarchies to guide restoration of disturbed ecosystems, *Ecol. Appl.*, 10, 189, 2000.

Parker, V.T. and Pickett, S.T.A., Restoration as an ecosystem process: Implications of the modern ecological paradigm, in *Restoration Ecology and Sustainable Development*, Urbanska, K.M., Webb, N.R. and Edwards, P.J., Eds., Cambridge University Press, Cambridge, UK, p. 17, 1997.

Parrotta, J.A., Turnbull, J.W., and Jones, N., Catalyzing native forest regeneration on degraded tropical lands, *For. Ecol. Manage.*, 99, 1, 1997.

Perrow, M.R. and Davy, A.J., Eds., *Handbook of Ecological Restoration 1: Principles of Restoration*, Cambridge University Press, Cambridge, UK, 2002a.

Perrow, M.R. and Davy, A.J., Eds., *Handbook of Ecological Restoration 2: Restoration in Practice*, Cambridge University Press, Cambridge, UK, 2002b.

Perry, G.L., Landscapes, space and equilibrium: Shifting viewpoints, *Prog. Phys. Geogr.*, 26, 339, 2002.

Phillips, R.J., Waldrop, T.A., Brose, P.H., and Wang, G.G., Restoring fire-adapted forests in eastern North America for biodiversity conservation and hazardous fuels reduction, in *A Goal-Oriented Approach to Forest Landscape Restoration*, Stanturf, J.A., Madsen, P., Lamb, D., Eds., Springer, New York, p. 187, 2012.

Pickett, S.T.A. and White, P.S., Eds., *The Ecology of Natural Disturbance and Patch Dynamics*, Academic Press, New York, 1985.

Pullar, D. and Lamb, D., A tool for comparing alternative forest landscape restoration scenarios, in *A Goal-Oriented Approach to Forest Landscape Restoration*, Stanturf, J.A., Madsen, P., and Lamb, D., Eds., Springer, New York, p. 3, 2012.

Putz, F.E. and Redford, K.H., The importance of defining "forest": Tropical forest degradation, deforestation, long-term phase shifts, and further transitions, *Biotropica*, 42, 10, 2010.

Renou-Wilson, F. and Byrne, K.A., Irish peatland forests: Lessons from the past and pathways to a sustainable future, in *Restoration of Boreal and Temperate Forests*, 2nd ed., Stanturf, J.A., Ed., CRC Press, Boca Raton, FL, p. 321, 2015.

Sarr, D.A. and Puettmann, K.J., Forest management, restoration, and designer ecosystems: Integrating strategies for a crowded planet, *Ecoscience*, 15, 17, 2008.

Sayer, J., Sunderland, T., Ghazoul, J., Pfund, J.-L., Sheil, D., Meijaard, E. et al. Ten principles for a landscape approach to reconciling agriculture, conservation, and other competing land uses, *Proc. Natl. Acad. Sci.*, 110, 8349, 2013.

Schwilk, D.W., Keeley, J.E., Knapp, E.E., McIver, J., Bailey, J.D., Fettig, C.J. et al. The national fire and fire surrogate study: Effects of fuel reduction methods on forest vegetation structure and fuels. *Ecol. Appl.*, 19, 285, 2009.

Shinneman, D.J., Palik, B.J., and Cornett, M.W., Can landscape-level ecological restoration influence fire risk? A spatially-explicit assessment of a northern temperate-southern boreal forest landscape, *For. Ecol. Manage.*, 274, 126, 2012.

Simula, M. and Mansur, E., A global challenge needing local response, *Unasylva*, 62, 238, 2011.

Society for Ecological Restoration Science and Policy Working Group (SER), The SER Primer on ecological restoration, 2002, (http://www.ser.org/Primer.pdf Viewed February 24, 2004).

Spiecker, H., Hansen, J., Klimo, E., Skovsgaard, J.P., Sterba, H., and von Teuffel, K., Eds., *Norway Spruce Conversion—Options and Consequences*, Brill, Boston, 2004.

Sprugel, D.G., Disturbance, equilibrium, and environmental variability: What is natural vegetation in a changing environment? *Biol. Conserv.*, 58, 1, 1991.

Stanturf, J.A., Gardiner, E.S., Hamel, P.B., Devall, M.S., Leininger, T.D., and Warren, M.E., Restoring bottomland hardwood ecosystems in the Lower Mississippi Alluvial Valley, *J. For.*, 98, 10, 2000.

Stanturf, J.A., Kellison, R., Broerman, F.S., and Jones, S.B., Pine productivity: Where are we and how did we get here? *J. For.*, 101, 26, 2003.

Stanturf, J.A. and Madsen, P., Restoration concepts for temperate and boreal forests of North America and Western Europe, *Plant Biosyst.*, 136, 143, 2002.

Stanturf, J.A., Palik, B.J., and Dumroese, R.K., Contemporary forest restoration: A review emphasizing function, *For. Ecol. Manage.*, 331, 292, 2014a.

Stanturf, J.A., Palik, B.J., Williams, M.I., Dumroese, R.K., and Madsen, P., Forest restoration paradigms, *J. Sust. For.*, 33 (suppl 1), S161, 2014b.

Stanturf, J.A., Schoenholtz, S.H., Schweitzer, C.J., and Shepard, J.P., Achieving restoration success: Myths in bottomland hardwood forests, *Restor. Ecol.*, 9, 189, 2001.

Stanturf, J.A., Wade, D.W., Waldrop, T.A., Kennard, D.K., and Achtemeier, G.L., Fire in southern forest landscapes, in *Southern Forest Resource Assessment*, Wear, D.M. and Greis, J.G., Eds., US Department of Agriculture, Forest Service, Southern Research Station, Asheville, NC, General Technical Report. SRS-53, p. 607, 2002.

Turner, M.G., Disturbance and landscape dynamics in a changing world, *Ecology*, 91, 2833, 2010.

Turner, M.G., Baker, W.L., Peterson, C.J., and Peet, R.K., Factors influencing succession: Lessons from large, infrequent natural disturbances, *Ecosystems*, 1, 511, 1998.

US Department of Agriculture, Forest Service, The Yazoo-Little Tallahatchie flood prevention project, US Department of Agriculture, Forest Service, Southern Region, Atlanta, GA, Forestry Report R8-FR 8, 1988.

Van Diggelen, R., Grootjans, Ab P., and Harris, J.A., Ecological restoration: State of the art or state of the science? *Restor. Ecol.*, 9, 115, 2001.

van Noordwijk, M., Suyamto, D.A., Lusiana, B., Ekadinata, A., and Hairiah, K., Facilitating agroforestation of landscapes for sustainable benefits: Tradeoffs between carbon stocks and local development benefits in Indonesia according to the FALLOW model, *Agric. Ecosys. Environ.*, 126, 98, 2008.

Vitousek, P.M., Mooney, H.A., Lubchenco, J., and Melillo, J.M., Human domination of Earth's ecosystems, *Science*, 277, 494, 1997.

Wagner, M.R., Block, W.M., Geils, B.W., and Wenger, K.F., Restoration ecology: A new forest management paradigm, or another merit badge for foresters? *J. For.*, 98, 23, 2000.

Westhoff, V., Man's attitude toward vegetation, in *Man's Impact On Vegetation*, Holzner, W., Werger, M.J.A. and Ikusima, I., Eds., Junk, The Hague, p. 7, 1983.

Whisenant, S.G., Terrestrial systems, in *Handbook of Ecological Restoration, Volume 1: Principles of Restoration*, Perrow, M.R. and Davy, A.J., Eds., Cambridge University Press, Cambridge, 2002.

Williams, M.I. and Dumroese, R.K., Preparing for climate change: Forestry and assisted migration, *J. For.*, 114, 287, 2013.

Wilson, K.A., Lulow, M., Burger, J., and McBride, M.F., The economics of restoration, in *Forest Landscape Restoration*, Stanturf, J.A., Lamb, D., and Madsen, P., Eds., Springer, New York, p. 215, 2012.

Winterhalder, K., Clewell, A.F., and Aronson, J., Values and science in ecological restoration—A response to Davis and Slobodkin, *Restor. Ecol.*, 12, 4, 2004.

2

What Is a Natural Forest?

Richard H.W. Bradshaw

CONTENTS

2.1 Introduction

A true "natural" forest can be defined as an idealized virgin forest condition that is uninfluenced by large-scale, systematic human activity; yet human activities have been so widespread and taken place over such a long period of time that there is probably little, if any, strictly natural forest remaining on planet Earth. Nevertheless it is a valuable exercise to reconstruct as many properties of natural forest as possible in the boreal and temperate zones to provide a reference for conservation, restoration, and silviculture and to help identify the best examples of near natural forests throughout these biomes. Near-natural forests are important research sites for both conservation biologists and silviculturalists. All silvicultural systems are modifications of natural systems to a greater or lesser extent, and studies of the dynamic processes within natural forest can serve as guidelines and as sources of inspiration for the development of "close-to-nature" silvicultural systems (Bradshaw et al. 1994; Larsen and Nielsen 2007). As society and forest managers seek systems that safeguard ecosystem services such as biodiversity and minimize long-term environmental impact, it seems helpful to reexamine the concept of natural and its realization in reference sites.

In this chapter, I review the probable broad-scale distribution of natural forest within the boreal and temperate zones and discuss some of the defining characteristics of natural forest and the fresh insight into the history of natural forest stands provided by the recent application of molecular biological techniques to forest genetics, species composition and

the genetic structure of tree populations (Hu et al. 2009). I consider in some detail disturbance regimes, climatic change, and other factors that drive dynamic processes, generate
stand structure, and control continuity, and briefly describe some other important natural
forest characteristics including carbon storage, other aspects of biogeochemistry, hydrology, and deadwood. The chapter finishes with a few comments on the restoration of natural forest properties or "rewilding."

2.2 Distribution of Natural Forest within the Boreal and Temperate Zones

Peterken (1996) described the continuum from natural to managed forest on which individual forests can be placed (Figure 2.1). While it is not usually difficult to discover or
estimate the intensity of forest management and place managed sites on this continuum,
the recognition of natural values often requires specialist knowledge of species composition of diverse organisms and the natural drivers of long-term forest dynamics. Natural
forests are rarely "climax" ecosystems of great antiquity, but more often dynamic systems
incorporating a variety of successional stages responding to natural disturbance processes
and changing climate (Figure 2.2). This diversity of states complicates their identification
in the field (Jones 1945; Kuuluvainen and Aakala 2011).

The degree of naturalness varies greatly by region among the boreal and temperate forests of the Northern Hemisphere. Old World forests have a long history of human interference. Evidence for hunting can be traced back several hundreds of thousands of years
in central Europe (Thieme 1997), but major disruption of natural forests in Europe dates
from the origins of organized agriculture. The Neolithic culture spread to the north and
west in Europe following the first cultivation of crops and domestication of livestock in the
near East, beginning around 11,500 years ago (Zeder 2011). In central Europe, this process
began about 7500 years ago, but in the boreal regions of Scandinavia, only as recently as
1000 years ago. Agriculture was introduced to Denmark about 5500 years ago and the
major structure of the cultural landscape was formed by 1000 BC (Odgaard and Rasmussen

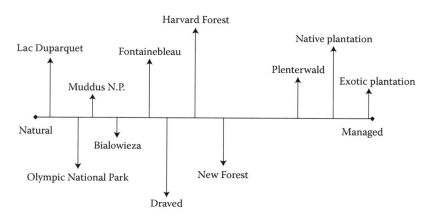

FIGURE 2.1
A hypothetical gradient of forests from truly natural to totally managed. A selection of forest areas and man-
agement systems are positioned on the gradient. Forests listed are as follows: Lac Duparquet, Quebec, Canada;
Olympic National Park, USA; Muddus National Park, Sweden; Białowieża National Park, Poland, Fontainebleau
Reserve, France; Draved Forest, Denmark; Harvard Forest, USA; New Forest, England.

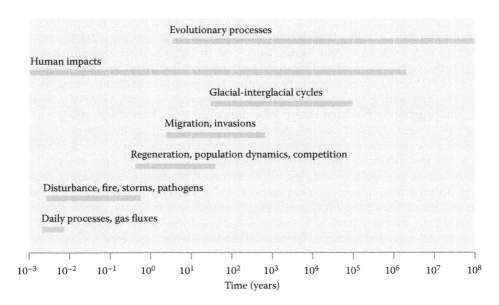

FIGURE 2.2
Temporal scale of some important ecosystem processes that affect forests.

2000). The advent of farming initiated forest fragmentation, altered fire regimes, affected ungulate populations, and caused increasingly rapid species turnover. The changeover from hunter-gatherer societies to settled agriculture is a convenient point to define the onset of large-scale, systematic human activity in forests that become part of organized agricultural systems, such as in Europe. Many forests were cleared for arable land or used as hay-meadows and subsequently abandoned. Larger areas were used for extensive grazing of domestic animals and collection of firewood and construction material. Thus, the least modified forest areas, and therefore best candidates for natural forest, were located in remote regions that were unsuitable for agriculture, where population densities have always been low. These included large areas of northern and eastern Europe and Asia, and the mountains of central Europe. During the 1800s, industrial forestry reached many of these areas, and natural forests declined rapidly in area and quality. Natural forest survived only in regions and stands of little economic value or where timber extraction costs were high. Two consequences of this exploitation history are that the surviving Eurasian natural forests are often inappropriate references for forests on fertile, lowland areas with a high production potential and secondly, more boreal natural forests survive than temperate natural forests. The most natural forest areas in the Old World lie, therefore, in remote, inhospitable areas bypassed by settled agriculture or commercial forestry. In Europe, even the few large, relatively intact areas that survived have experienced disrupted ungulate–predator systems that are an integral part of the forest ecosystem (Vera 2000). The *Betula*-dominated forests along the Scandes Mountains in Scandinavia comprise a large, near-natural forest region that has escaped the attention of farmers and foresters alike, although reindeer (*Rangifer tarandus*) husbandry and suppression of predators has exerted an influence that is unnatural. There is a large, relatively intact mixed coniferous–deciduous forest ecosystem in the southeast Carpathian Mountains of Romania where wolves (*Canis lupus*) are still active, but this is now being exploited for timber. The largest, unexploited forest resources in the Old World, both mixed-deciduous and boreal, lie within the northern and eastern regions of the former Soviet Union, but their future as natural forests is uncertain. Notable smaller natural forests survive in border areas—regions where political unrest discouraged permanent settlement and agriculture, and in large private hunting reserves and estates. Of the latter, the forests that survived the best belonged to stable organizations such as monarchies or churches. Every country has examples of these areas, such as The New Forest, England, Fontainebleau, France, Neuburger Urwald, Germany, Białowieża, Poland, and Siggaboda, Sweden (Koop 1989; Peterken 1996; Parviainen 2000).

In the New World, "presettlement" forests (preEuropean settlers but not before aboriginal settlement) are often taken as a working reference for natural forest. There is widespread debate about the influence and timing of aboriginal peoples and their impact on North American forests, particularly with regard to alteration of the fire and grazing regimes but historical documentation, among other sources, gives a useful impression of the extent and location of these forests. Large fragments of presettlement forest have survived in relatively unmodified form in many parts of Canada, Alaska, and the Pacific Northwest, and thus the widely accepted standard reference for natural forest is better described for boreal and montane North America than for Europe. As in Europe, the temperate forests have a longer history of human modification than do coniferous and boreal forests. Most of the Appalachian range in the Eastern United States has become widely reestablished with hardwood forest following abandonment of the agricultural exploitation from the early European settlers and only fragments of old-growth forest remain in remote upland areas (Foster and O'Keefe 2000). At first glance, the rather inaccessible west coast of South Island, New Zealand, contains large tracts of natural temperate *Nothofagus* forest and the island has only experienced human impact for about 750 years. Yet a combination of ruthless hunting, burning, introduced alien species, forestry, and sheep farming have made significant impact on species composition and forest structure, even though remote Fiordland forest landscapes offer some of the closest to natural forest conditions remaining on planet Earth. So extant references for virgin natural forests are elusive, because human impact has been long and varied and reached into all terrestrial ecosystems. Any natural reference would be dynamic in any case and it is now too late in Earth history to separate fully human impact from natural forest dynamics. We must piece together a model of natural forest from a combination of observation, experiment, and modeling.

2.3 Species Composition and Genetic Structure of Natural Forest

A natural forest concept provides the regional list of native forest species and thus also defines exotic and alien species. The list of native species varies over long timescales because natural extinctions occur and invading species become naturalized (Bradshaw 1995). There has been a loss of tree species diversity during the last 2 million years within northwest Europe (Figure 2.3) that is less apparent in North America and temperate China. The best explanation for this loss is the small and fragmented nature of the glacial refugia in Europe and the speed and severity of the climatic fluctuations (Huntley 1993; Bradshaw 1995; Hewitt 1996). The native tree species list for a country can assume an exaggerated importance in conservation debates where exotic species are regarded as potential threats to natural forest communities. This discussion may seem irrelevant to the shorter timescales of forest management, but the native status of *Acer pseudoplatanus* or *Larix* are important conservation issues in some European countries, where forest conservation and management of indigenous species are prioritized. Exotics are less of an issue in the

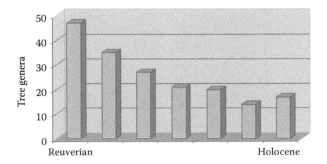

FIGURE 2.3
The number of tree genera recorded as fossils during some Quaternary interglacials. The timescale is nonlinear but begins in the Reuverian ca. 2.4 million years ago.

forestry of more species-rich forest regions. The take-home message is that the species content of natural forest is dynamic, particularly during periods of rapid climatic change such as the present. Most data are available for the dominant tree species, but as these structure the forest environment and act as host to many associated insect, moss, and fungal species, it is likely that a limited turnover of all forest species is a characteristic of natural forest.

Advances in our understanding of the genetic structure of forest tree populations at the continental scale in Europe have emphasized, through molecular genetics, the importance of glacial refugia as centers of genetic diversity and reservoirs for the long-term survival of forest species (Petit et al. 2002). The dynamic aspects of natural forest suggested by paleoecological analyses are also reflected in genetic structures of populations at a continental scale. For example, the pattern of chloroplast DNA variation across Europe displays the same pattern of postglacial species migration as recorded through pollen analysis for many species, and this relationship is particularly apparent within European white oaks (*Quercus robur* and *Quercus. petraea*) (Brewer et al. 2002; Petit et al. 2002). Many northern tree populations are recently (in geological and evolutionary terms) derived from southern areas. Classical pollen studies have mapped how *Carya* and *Castanea* populations spread north and west in North America during recent millennia (Figure 2.4) (Davis 1976), but subsequent genetic studies have modified this generalization with evidence for northern glacial tree refugia and multiple centers for spread too small to detect using conventional pollen data (Anderson et al. 2006). Tree distributions have been dynamic in the past and are forecast to alter rapidly in the future in response to climate change (Hickler et al. 2012), so the notion of the importance of local provenance in forest management and conservation needs careful evaluation (Haskell 2001; Wilkinson 2001).

2.4 Disturbance Regimes

Natural forest dynamics are driven by infrequent but significant disturbance events such as wind, fire, drought, flooding, land-slippage, and disease (Pickett and White 1985; Peterken 1996; Gardiner and Quine 2000). These driving forces interact with the continuous disturbance of browsing and grazing animals and the localized processes of single tree replacement by seedling establishment, growth, competitive interactions, and senescence

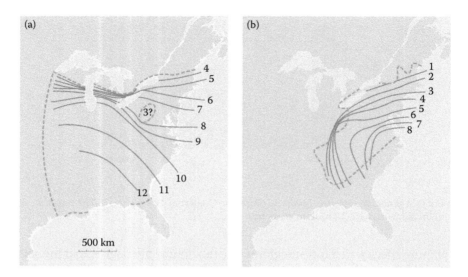

FIGURE 2.4
Holocene spreading patterns of (a) hickory (*Carya*) and (b) chestnut (*Castanea*) in North America. The isochrone lines show the location of the spreading front (thousands of years before present), reconstructed from pollen data. (After Davis, M.B., *Geosci. Man.* 13, 13, 1976.)

(Falinski 1986; Woods 2000). The integration of these factors in the disturbance regime, which creates the conditions necessary for regeneration, generates deadwood, influences the size and age structure of a stand, and controls the continuity of local niches and habitats. The disturbance regime is at the center of many of the dynamic processes and characteristics associated with natural forest.

2.4.1 Fire

The pivotal role of fire in many natural boreal ecosystems is widely acknowledged and has been incorporated into forest management (Johnson 1992; Goldammer and Furyaev 1996; Bradshaw et al. 1997; Angelstam 1998; Halme et al. 2013). Fires have always been important in boreal regions, although fire regimes have varied with prevailing climatic conditions. Individual forest fires tend to be very large in Canada and Siberia, but have been smaller in Scandinavia, at least during the last few centuries, as reconstructed by dendrochronological methods (Niklasson and Granström, 2000). It is still an open question whether this difference in size of fires is due to landscape factors and the distribution of natural firebreaks, fuel characteristics influenced by forest structure and composition, or whether cultural factors play a role. A combination of changes in land use, property development, road-building, forestry, and fire suppression activities have largely removed fire from Fennoscandian forests during recent centuries.

Fire is one of the main disturbance factors in natural boreal forest and there is considerable interest in natural fire regimes for conservation and restoration management that can only be derived from studies of the past. Fire-scar analyses reach back a few centuries, but do not extend beyond time periods with recognized human impact. Sedimentary records from small forest hollows throughout southern Scandinavia contain abundant charcoal from the past, but charcoal is less abundant in recent centuries as is also seen in fire-scar records. Charcoal from pure temperate deciduous forests in southern Scandinavia appears

FIGURE 2.5
Sediment core from a small forest hollow in southern Sweden. Note the charcoal bands indicating local fire.

to be linked with human activities during recent millennia, while the charcoal records from the more coniferous boreal forests indicate longer fire histories, where burning was less controlled by human activities in earlier millennia (Figure 2.5) (Bradshaw et al. 2010). A 5000-year record from Vesijako in the southern boreal zone of Finland contained regular peaks of charcoal in the sediment recurring on average every 430 years between 5000 and 2000 years ago (Figure 2.6). The charcoal record was quite different between 2000 and 700 years ago with larger peak values and with the average fire return interval reduced to just 180 years (Clear et al. 2013). While it is always difficult to assign forest fire evidence directly to human activities, even today, several other studies from the region have concluded that humans began to take control of the fire regime about 2000 years ago, exploiting fire for slash-and-burn agriculture and for the improvement of forest grazing (Bradshaw et al. 2010). Charcoal is completely absent from the recent part of the Vesijako record, which covered the period of recent fire suppression (Figure 2.6).

Comparable fire histories have been reconstructed from seasonally dry coniferous forests in North America. Heinselman (1973) worked in northeastern Minnesota in the wild Boundary Waters Canoe Area bordering Canada and established that wildfire was a natural ecosystem process that had operated for at least hundreds of years before fire suppression during the twentieth century led to a new, unnatural forest composition. Heinselman can be credited with bringing the ecological importance of natural fire in a boreal system to the attention of forest managers and conservation biologists. He concluded long ago that "fire should soon be reintroduced through a program of prescribed fires and monitored lightning fires. Failing this, major unnatural, perhaps unpredictable, changes in the ecosystem will occur" (Heinselman 1973, p. 329).

Fire suppression had become the widely accepted management plan for public land in the United States for several compelling practical reasons and from its origin in 1944, the very effective Smokey Bear publicity campaign to engage the public in fire prevention became the longest running public-service advertising campaign in North American history. However the paleoecological evidence for fire in natural forest has led to its restoration in several areas including *Pinus palustris* forests of northern Florida and Yellowstone National Park. In this Park, natural fires have been allowed to burn under controlled conditions since 1972, and this management appeared to be a satisfactory way of reintroducing the ecosystem benefits of fire while controlling fuel supply to avoid catastrophic megafires. However, a major drought in 1988 resulted in about 55% (500,000 ha) of Yellowstone

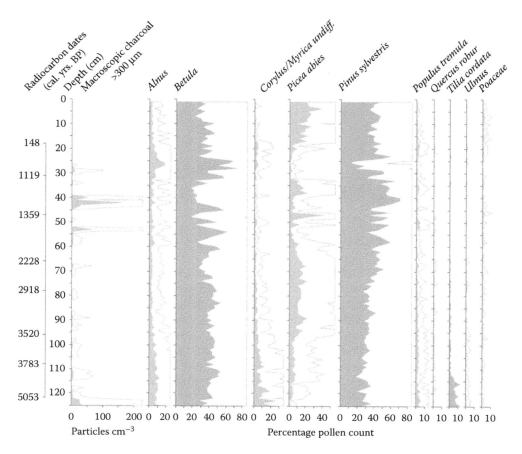

FIGURE 2.6
Charcoal and major pollen types during the last 6000 years in Vesijako, southern Finland. (After Clear, J.C. et al., *Biol. Conserv.*, 166, 90, 2013.)

being affected by a series of large, hot fires. Considerable critical media comment followed this event, but despite major efforts at control there was little that could be done until autumnal rain and snow eventually extinguished the flames. The damage might well have been worse if there had not been the earlier limited tolerance of wildfire to reduce fuel loads. The Park has recovered quickly from the 1988 fires and paleoecological research has shown that fires of this magnitude have been recorded from the past, with three periods of widespread burning during the last 750 years, suggesting that the 1988 fire was not a unique occurrence (Higuera et al. 2011).

What are the benefits of restoring earlier, more natural fire regimes to modern ecosystems? The Yellowstone example showed that a single fire season could affect much of a National Park, and such sudden change is always hard to justify to the general public. However, in southeastern United States fire suppression had caused large-scale ecosystem conversion with considerable reduction in the area of *Pinus palustris* forest. The bringing back of fire to ecosystems exemplifies a new approach to conservation, in which natural processes are of equal or greater importance than the more traditional focus on species preservation.

2.4.2 Browsing Animals

In North America, as in Europe and Asia, the megafaunal extinctions at the end of the last glaciation must have impacted forests. Gill et al. (2009) reported evidence for increased burning and development of new forest types in eastern North America at this time and ancient DNA studies throw some doubt about the scale of human involvement in these significant ecosystem dynamics. Lorenzen et al. (2011) found positive correlations for horse, reindeer, bison, and musk ox between genetically estimated population size and the area of climatically suitable habitat in Eurasia and North America throughout the last 50,000 years. They concluded that climate change had been the major driving force for the population dynamics they observed, arguing against traditional hypotheses of a hunting "overkill" or the spread of infectious disease following first human contact, but they did acknowledge likely human influence in the decline of Eurasian horse and bison. These types of observations illustrate how dynamic forest ecosystems can be over long timescales and also, as with fire, how difficult it is to ascertain the extent of human impact in significant changes in forest properties.

The potential role of browsing and grazing mammals in relation to stand dynamics and ground flora in natural forest has been emphasized by Vera (2000). The Holocene paleoecological record documents large changes in the fauna of boreal and temperate forests and a history of anthropogenic modification of fauna through hunting, herding, and later domestication of large mammals. Both natural and anthropogenic factors have affected the Holocene fauna of northwest Europe. By the early Holocene (9000 ybp), the forests of northwest Europe had been recolonized not only by species remnant in the region, such as red (*Cervus elaphus*) and roe deer (*Capreolus capreolus*), but also by moose (*Alces alces*), aurochs (*Bos primigenius*), bear (*Ursus arctos*), wolf (*Canis lupus*), lynx (*Lynx lynx*), beaver (*Castor fiber*), and others (Aaris-Sørensen 1998; Yalden 1999). However, this fauna was impoverished compared to previous interglacials, specifically in large herbivores and carnivores. Many species disappeared after the Neolithic agricultural revolution or even in very recent centuries, as a result of habitat clearance and hunting. In theory, these species still belong to natural forests of the region. Models for their persistence exist in Poland and Russia, but crucial to the survival is that they are embedded in very large areas of habitat. The mammals form a seminatural metapopulation, which can survive as a whole even if local areas become unsuitable or unavailable for various reasons (Bradshaw et al. 2003). Ancient DNA analyses can now provide paleopopulation estimates, which indicate a European breeding population for wild horse (*Equus ferus*) of between 35,000 and 50,000 individuals 6000 years ago (Lorenzen et al. 2011). This population was dispersed over a very large area (ca. 9 million km²), so population densities were low with little impact on forest cover. Past European bison (*Bison bonasus*) populations have perhaps been underestimated (Kuemmerle et al. 2012), but unfortunately there are not yet any estimates of past population size based on genetic data for bison or aurochs.

The Holocene records for large mammals in Sweden permit the development of models describing changing grazing regimes during the Holocene (Liljegren and Lagerås 1993) (Figure 2.7). There have been continual fluctuations in the balance between browsers and grazers, chiefly due to the early Holocene local extinction of bison and aurochs (Ekström 1993), and the subsequent introduction of domestic cattle nearly 3000 years later. The removal of domestic cattle from southern Swedish forests during the last 100 years was a significant event in long-term grazing–vegetation interactions and has had a major influence on forest composition and structure (Andersson et al. 1993). The changing balance between grazers and browsers in southern Swedish forests suggests that the present large populations of roe deer and moose are a recent development.

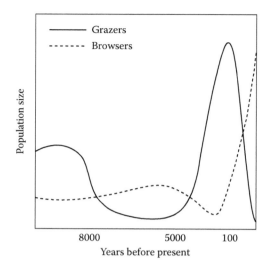

FIGURE 2.7
Conceptual model showing the probable relative changes in population size of large, grazing ungulates (e.g., aurochs, bison, domestic cow, etc.) and browsing ungulates (e.g., moose, roe-deer, etc.) during the Holocene in southern and central Sweden. The time-scale is nonlinear. (Data taken from many sources.)

The much higher ungulate diversity and presumably total biomass of past interglacials strongly suggests a diverse vegetational environment that varied between interglacials, with elephants, rhinos, and many species of deer that must have been niche-separated. Climatic and edaphic factors are the major determinants of the primary habitat structure (Bradshaw et al. 2003). It is unlikely that ungulate communities caused the apparent differences in habitat diversity between past interglacials themselves (Vera 2000; Svenning 2002). Grazers, by definition, are adapted to more open conditions and can hinder woody colonization, but are unsuited for major forest clearance (Lister, personal communication). The large-scale, anthropogenic alteration of landscape structure in northwest Europe makes the present-day situation significantly different from previous interglacials or even from the early- to mid-Holocene. As with fire, we may gain a new insight into processes of ungulate–vegetation interactions from paleoecological data; but the current situation is unique when viewed from the long-term perspective. Studies of the past reconstruct the trajectory of events that created the present condition and provide an insight into processes, but they also show that no single set of equilibrium "base-line" conditions can be recognized in the recent geological past; hence, there is no secure reference from the past to use as a model for restoration or future management.

2.4.3 Storm

Hurricanes and wind damage are one of the most widespread types of disturbance to impact terrestrial ecosystems, affecting areas up to 100,000 km² (Table 2.1; Foster et al. 1998a). Several studies have examined the short-term impacts of hurricanes in the boreal and temperate zones showing that tree size and species are important factors influencing the extent of storm damage. Natural forest mortality rates range between 0.5% and 3% per year (Runkle 1985; Peterken 1996), and natural forests tend to receive less storm damage

TABLE 2.1

Characteristics of Large, Infrequent Disturbances

Characteristic	Volcanic Eruptions	Forest Pathogens	Tornadoes	Forest Fires	Hurricanes	Riverine Floods
Duration of event	Hours	Years	Minutes	Weeks	Hours	Weeks
Return interval (yrs)	10^2–10^3	10^1–10^3	100–300	75–500	60–200	50–200
Size of event (km²)	5–100	1–100,000	5–100	50–20,000	50–100,100	50–50,000
Location	Volcanic mountains	Forest	Inland	Inland	Warm coasts	Riparian areas
Variables Affecting Severity	**Volcanic Eruptions**	**Forest Pathogens**	**Tornadoes**	**Forest Fires**	**Hurricanes**	**Riverine Floods**
Climatic factors	No	Yes	Yes	Yes	Yes	Yes
Season	Yes	Yes	Yes	Yes	Yes	Yes
Topography	Yes	No	No	Yes	Yes	Yes
Vegetation structure	Yes	Occasional	No	Yes	Yes	Yes

Source: After Foster, D.R., Knight, D.H. and Franklin, J.F., *Ecosystems*, 1, 497, 1998a.

than nearby uniform plantations. Wind damage in natural forest tends to cause less tree mortality than hot fires or pathogen outbreaks, so the subsequent forest regeneration can originate more from the recovery of damaged stems or release of suppressed individuals than from seedling recruitment, and the eventual impact on vegetation composition will be less (Martin and Ogden 2006).

Severe storms in Western Europe over the last 30 years have caused extensive damage, especially to conifer plantations, but not exclusively. In December 1999, the most powerful hurricane ever recorded in Denmark (Danmarks Meteorologiske Institut 1999) caused the greatest destruction of forest volume during the 1900s (Fodgaard 2000). Draved Forest is a seminatural deciduous forest in southern Jutland with long-term observations of tree growth and mortality. Through time, storms were the major cause of mortality of large trees. Mortality varied by species (Figure 2.8); *Betula, Fagus,* and *Tilia* were mainly windthrown, whereas *Alnus* and *Fraxinus* were as likely to be standing dead trees (Wolf et al. 2004). In Draved Forest, 4% of all trees larger than 10 cm dbh were damaged by the storm and many of these had resprouted a year after the event, so eventual mortality was considerably lower. Damage was almost total in neighboring commercial coniferous plantations.

Harvard Forest, Massachusetts, United States, is another research forest in a coastal region that has periodically experienced hurricane-force winds. The New England hurricane of 21 September 1938 was one of the most catastrophic in the history of the United States, with sustained wind speeds over 50 m s^{-1} and gusts of over 80 m s^{-1}. The hurricane caused variable damage in a strip of New England 100 km wide and 300 km long (Foster et al. 1998b). While the weaker Danish hurricane of 1999 raced over a flat landscape, slope and aspect influenced where the New England hurricane wreaked most havoc (Foster and Boose 1992). Uprooting was the main type of damage and was correlated with both tree height and species composition. As in Denmark, evergreen conifers such as white pine (*Pinus strobus*) suffered greater damage than deciduous trees (Foster and Boose 1992). Hurricanes in temperate regions tend to occur after deciduous trees have lost their leaves, which is a factor in their greater storm resistance. The biological legacies left in natural forest by hurricanes include a patchwork of age and height structures, standing broken

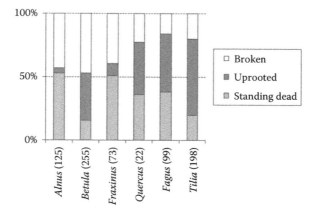

FIGURE 2.8
Cause of mortality for different species in Draved Forest, Denmark during the last 50 years. Sample size is shown in parentheses. *Alnus glutinosa, Betula pubescens, Fraxinus excelsior, Quercus robur, Fagus sylvatica, Tilia cordata.*

TABLE 2.2

Selected Direct and Indirect Anthropogenic Factors That Have Modified Natural Forests

Direct Factors	Indirect Factors
Temporary clearance for agriculture	Hunting
Modified fire regime	Gathering of berries and fungi
Grazing animals	Regional drainage
Drainage ditches	Intentional or unintentional species introductions
Gathering firewood	Invasive aliens
Selective felling	Habitat fragmentation
Planting	Greenhouse gas emissions

trees, coarse and fine woody debris, and soil disturbance such as wind-throw mounds and mass movement of soil following storm-related intensive precipitation events (Foster et al. 1998b).

2.4.4 Human Influences

Pests and disease, flooding, and land-slippage are other natural disturbances that influence the structure and composition of natural forests; but humans are the cause of the greatest disturbances affecting forests in the boreal and temperate zones. Forests have been cleared for agriculture for several millennia in Europe; even if agriculture is abandoned, the break in forest continuity can affect species composition of subsequent forests for a long time. One area of northeastern France was farmed under Roman rule for just 200 years between AD 20 and 250 and then abandoned (Dambrine et al. 2007), yet the legacy of this brief period of land use can still be observed in the ground flora, with less woodland species and more plants characteristic of disturbed ground, despite almost 2000 years of subsequent continuous forest cover. Dambrine et al. (2007) suspected that the earlier period of cultivation irreversibly altered the local soil structure and increased the soil phosphorus content. It seems likely that similar alterations of forest soil properties are widespread around the world, but this has barely been investigated. The adequate description of natural disturbance regimes and full recognition of how they have been modified, both directly and indirectly by anthropogenic activities, is one of the major challenges in the identification and study of natural forest ecosystems (Table 2.2).

2.5 Carbon Storage in Natural Forests

Commercial forestry has altered the size structure of seminatural forests throughout the boreal and temperate zones, removing old, large trees, and shortening rotation lengths (Linder and Östlund 1992) (Figure 2.9), yet a considerable area (6×10^8 ha) of unmanaged primary forest has been estimated to survive in northern hemisphere boreal and temperate forests (Luyssaert et al. 2008). Recent research has questioned the traditional view that these forests are carbon neutral and suggested that they make a significant contribution to global net ecosystem productivity, as well comprising enormous carbon stores (Luyssaert et al. 2008; Keith et al. 2009). *Eucalyptus regnans* forest in southeastern Australia has been

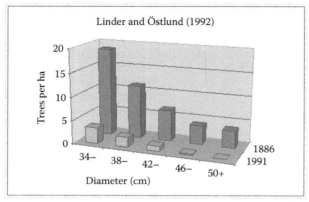

Swedish boreal forest

FIGURE 2.9
Number of trees per hectare and their diameter distribution from Orsa "besparingsskog," Sweden in 1886 and 1991.

proposed to have the highest biomass carbon density in the world with a recorded value of 1867 Mg ha^{-1} of carbon (Keith et al. 2009). Parts of these carbon stocks are occasionally released in fire.

2.6 Other Important Characteristics of Natural Forest Systems

I have highlighted tree species composition and disturbance regime as defining features of natural and near-natural forests. This is, of course, a broad generalization, and many other factors and nuances will be identified in other chapters (Peterken 1996). Three other important characteristics that help characterize natural forests deserve brief mention. The hydrological regime of natural forests in Europe has not been studied in detail, but regional and local drainage of forest and agricultural areas has been relentlessly pursued in the temperate and boreal zones for many decades (Møller 2000). Regional pollen diagrams indicate a general reduction in abundance of *Alnus* and *Salix* on the landscape, and field experience supports the conclusion that systematic alteration of microhabitats has decreased biodiversity in ways that we are just beginning to recognize. Landscape drainage has occurred in North America on a smaller scale relative to the natural forest resources remaining (Figure 2.10) but can be of great regional importance (Gardiner and Oliver 2005). One underappreciated aspect of hydrology is the impact of reduced populations of beaver (*Castor canadensis*) since European settlement (Rudemann and Schoonmaker 1938; Naiman et al. 1986).

Forest soils and the biogeochemical cycling of their chemical constituents have received much attention, particularly with regard to the effects of atmospheric deposition of first sulfates and subsequently nitrates. Nevertheless, soil change is slower than vegetation change and soil profiles may retain relict morphology, and processes such as decomposition may not be in equilibrium with current climate and litter inputs (Willis et al. 1997; Wardle et al. 2004). Careful interpretation of soil features may identify past disturbances

FIGURE 2.10
Acer macrophyllum dripping with bryophytes in undrained, humid temperate rain forest within the Olympic National Park, USA.

such as breaks in forest continuity that are not apparent from present vegetation or hydrology (Dambrine et al. 2007). Windthrow mounds are a feature of many natural forests in regions that are affected by windthrow and these, together with rooting animals, such as wild boar, are important in soil turnover in natural forests. Lack of pit and mound topography may be an indicator of past agricultural use.

Standing deadwood and woody debris are important for forest biodiversity and associated aquatic systems and their occurrence and ecological role has been widely discussed (e.g., Müller and Bütler 2010). The natural level and dynamics of deadwood vary greatly in space and time and deadwood is generated as a consequence of the disturbance regimes discussed above.

2.7 Restoration of Natural Forest

Restoration or "rewilding" activities can help identify which lost features of natural forest ecosystems are regarded as the most valuable. Such restoration has focused both on processes such as fire, browsing, and raised water tables and on components such as deadwood and species (Halme et al. 2013). Ecosystem engineers such as beaver have been a priority in parts of Europe (Nummi and Kuuluvainen 2013) and the value of wolves in the generation of trophic cascades is widely discussed (Bouchard et al. 2013). In any restoration program there is a need to consider the long-term perspective and estimate how far the system has deviated from a natural condition. Jackson and Hobbs (2009) presented different ecosystem trajectories through time and argued that where ecosystems have deviated far from reference states it is impractical to carry out major restoration programs (Figure 2.11). While types 1 and 2 can be found in parts of North America and Australia, examples of severely altered type 3 ecosystems are only too frequent, particularly where previously important species such as large herbivores are now extinct. In other situations, native species of forest animals and plants have been so comprehensively altered by the introduction of invasive aliens from other continents that restoration is impractical, as is the case in parts of New Zealand. Compared with

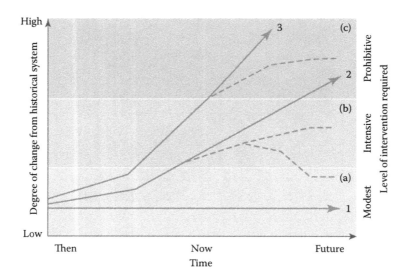

FIGURE 2.11
Contrasting ecosystem trajectories from historic through present to future configurations, indicating degree of change from the historic ecosystem (e.g., physical environment and species pool). Dashed lines (a–c) represent realistic management interventions of varying ambition and cost. (After Jackson, S.T. and Hobbs, R.J., *Science*, 325, 567, 2009.)

more densely populated European countries, Sweden has many ecosystems that could be restored to near natural conditions with just moderate intervention. In parts of the world with long records of human modification of forest ecosystems, including Europe, any potential baseline or presettlement reference state can be so far in the past that this type of ecosystem restoration is even less feasible: less elements of the "undisturbed" reference state remain and the biota will have been influenced by climate history in a great many ways since a natural state existed.

2.8 Conclusions

This chapter opened with the conclusion that truly natural, virgin forests probably do not exist. The many examples from Western Europe of attempts to identify natural forest condition in remote forests support this conclusion. The shift from hunting and gathering to sedentary agriculture is often associated with a significant development from natural to managed forest, but even in areas where sedentary agriculture has not dominated, we cannot discount the effect of fire as used by nonagricultural cultures. Changing fire management through time has influenced forests in many regions, particularly fire suppression in boreal forests during recent years. Why is it important to create a natural forest concept and speculate over the virgin state? Firstly, it satisfies human curiosity and makes us aware of the ecological consequences of socioeconomic development. Secondly, the natural forest concept provides tangible goals for management of wilderness areas, and restoration of overexploited forest. Many people would like to know what their region would look like in a truly natural state and protect a small sample for posterity. However, probably the strongest motivation is that all silvicultural systems are modifications of the natural state adapted to local needs. In a changing society and a changing environment, it is appropriate to develop these systems by adapting natural species and processes that have survived the climatic extremes of geological time. Nevertheless, we should recognize that what is natural forest is a dynamic range of vegetation types, successional stages and disturbance regimes, which at first site may not fit to widely preconceived ideas of virgin forest (Jones 1945).

Acknowledgment

I thank many colleagues for discussions over the years on natural forests, particularly George Peterken and Peter Friis Møller.

References

Aaris-Sørensen, K., *Danmarks Forhistoriske Dyreverden*, Gyldendal, Copenhagen, 232 pp., 1998 (in Danish).

Andersson, L., Appelqvist, T., Bengtsson, O., Nitare, J. and Wadstein, M., *Betespräglad äldre bondesog*, Skogsstyrelsen, Jönköping, 110 pp., 1993.

Anderson, L.L., Hu, F.S., Nelson, D.M., Petit, R.J. and Paige, K.N., Ice-age endurance: DNA evidence of a white spruce refugium in Alaska, *Proc. Natl. Acad. Sci.*, 103, 12447, 2006.

Angelstam, P., Maintaining and restoring biodiversity in European boreal forests by developing natural disturbance regimes, *J. Veg. Sci.*, 9, 593, 1998.

Bouchard, K., Wiedenhoeft, J.E., Wydeven, A.P. and Rooney, T.P., Wolves facilitate the recovery of browse-sensitive understory herbs in Wisconsin forests, *Boreal Env. Res.*, 18 (suppl. A), 43, 2013.

Bradshaw, R.H.W., The origins and dynamics of native forest ecosystems: Background to the use of exotic species in forestry, *Búvísindi*, 9, 7, 1995.

Bradshaw, R.H.W., Gemmel, P. and Björkman, L., Development of nature-based silvicultural models in southern Sweden: The scientific background, *For. Landscape. Res.*, 1, 95, 1994.

Bradshaw, R.H.W., Hannon, G.E. and Lister, A.M., A long-term perspective on ungulate-vegetation interactions, *For. Ecol. Manage.*, 181, 267, 2003.

Bradshaw, R.H.W., Lindbladh, M. and Hannon, G.E., The role of fire in southern Scandinavian forests during the late Holocene, *Int. J. Wildland Fire.*, 19, 1040, 2010.

Bradshaw, R.H.W., Tolonen, K. and Tolonen, M., Holocene records of fire from the boreal and temperate zones of Europe, in *Sediment Records of Biomass Burning and Global Change*, Clark, J.S., Cachier, H., Goldammer, J.G., and Stocks, B.J., Eds. Springer-Verlag, Berlin, 347, 1997.

Brewer, S., Cheddadi, R., Beaulieu, J.L., Reille, M. and data contributors, The spread of deciduous *Quercus* throughout Europe since the last glacial period, *For. Ecol. Manage.*, 156, 27, 2002.

Clear, J.C., Seppä, H., Kuosmanen, N. and Bradshaw, R.H.W., Holocene fire frequency variability in Vesijako, Strict Nature Reserve, Finland, and its application to conservation and management, *Biol. Conserv.*, 166, 90, 2013.

Dambrine, E., Dupouey, J.-L. and Laut, L., Present forest biodiversity patterns in France related to former Roman agriculture, *Ecology*, 88, 1430, 2007.

Danmarks Meteorologiske Institut, *Rapport Orkanen over Danmark den 3–4 December 1999*, Danmarks Meteorologiske Institut, 1999 (in Danish).

Davis, M.B., Pleistocene biogeography of temperate deciduous forests, *Geosci. Man*, 13, 13, 1976.

Ekström, J., *The late Quaternary history of the Urus (Bos primigenius Bojanus 1827) in Sweden*, Ph.D. thesis, Lund University, Department of Quaternary Geology, 129 pp., 1993.

Falinski, J.B., *Vegetation Dynamics in Temperate Lowland Primeval Forests*, Kluwer, Dordrecht, 537 pp., 1986.

Fodgaard, S., Stormfald gennem tiderne, *Skoven*, 3, 144, 2000.

Foster, D.R. and Boose, E.R., Patterns of forest damage resulting from catastrophic wind in central New England, USA, *J. Ecol.*, 80, 79, 1992.

Foster, D.R., Knight, D.H. and Franklin, J.F., Landscape patterns and legacies resulting from large, infrequent forest disturbances, *Ecosystems*, 1, 497, 1998a.

Foster, D.R., Motzkin, G. and Slater, B., Land-use history as long-term broad-scale disturbance, regional forest dynamics in central New England. *Ecosystems*, 1, 96, 1998b.

Foster, D.R. and O'Keefe, J.F., *New England Forests Through Time*, Harvard University Press, Cambridge, 67 pp., 2000.

Gardiner, E.S. and Oliver, J.M., Restoration of bottomland hardwood forests in the Lower Mississippi Alluvial Valley, USA., in *Restoration of Boreal and Temperate Forests*, Stanturf, J.A. and Madsen, P., Eds., CRC Press, Boca Raton, 235–251, 2005.

Gardiner, B.A. and Quine, C.P., Management of forests to reduce the risk of abiotic damage: A review with particular reference to the effects of strong winds, *For. Ecol. Manage.*, 135, 261, 2000.

Gill, J.L., Williams, J.W., Jackson, S.T., Lininger, K.B. and Robinson, G.S., Pleistocene megafaunal collapse, novel plant communities, and enhanced fire regimes in North America, *Science*, 326, 1100, 2009.

Goldammer, J.G. and Furyaev, V.V., *Fire in Ecosystems of Boreal Eurasia*, Kluwer, Dordrecht, 528 pp., 1996.

Halme, P., Allen, K.A., Aunin, A., Bradshaw, R.H.W., Brūmelis, G., Čada, V., Clear, J.L., et al., Challenges of ecological restoration: Lessons from forests in northern Europe, *Biol. Conserv.*, 167, 248, 2013.

Haskell, J., The latitudinal gradient of diversity through the Holocene as recorded by fossil pollen in Europe, *Evol. Ecol. Res.*, 3, 345, 2001.

Heinselman, M.L., Fire in the virgin forests of the Boundary Waters Canoe Area, Minnesota, *Quat. Res.*, 3, 329, 1973.

Hewitt, G.M., Some genetic consequences of ice ages, and their role in divergence and speciation, *Biol. J. Linn. Soc.*, 58, 247, 1996.

Hickler, T., Vohland, K., Feehan, J., Miller, P.A., Smith, B., Costa, L., Giesecke, T. et al., Projecting the future distribution of European potential natural vegetation zones with a generalized, tree species-based dynamic vegetation model, *Global Ecol. Biogeogr.*, 21, 50, 2012.

Higuera, P.E., Whitlock, C. and Gage, J.A., Linking tree-ring and sediment-charcoal records to reconstruct fire occurrence and area burned in subalpine forests of Yellowstone National Park, USA, *Holocene*, 21, 327, 2011.

Hu, F.S., Hampe, A. and Petit, R.J., Paleoecology meets genetics: Deciphering past vegetational dynamics, *Front. Ecol. Environ.*, 7, 371, 2009.

Huntley, B., Species-richness in north-temperate zone forests, *J. Biogeogr.* 20, 163, 1993.

Johnson, E.A., *Fire and Vegetation Dynamics. Studies from the North American Boreal Forest*, Cambridge University Press, New York, 1992.

Jackson, S.T. and Hobbs, R.J., Ecological restoration in the light of ecological history, *Science*, 325, 567, 2009.

Jones, E.W., The structure and reproduction of the virgin forest of the north temperate zone, *New Phytol.*, 44, 130, 1945.

Keith, H., Mackey, B.G. and Lindenmayer, D.B., Re-evaluation of forest biomass carbon stocks and lessons from the world's most carbon-dense forests, *Proc. Natl. Acad. Sci.*, 106, 11635, 2009.

Koop, H., *Forest Dynamics*, Springer-Verlag, Berlin, 229 pp., 1989.

Kuemmerle, T., Hickler, T., Olofsson, J., Schurgers, G. and Radeloff, V.C., Reconstructing range dynamics and range fragmentation of European bison for the last 8000 years, *Diversity Distrib.*, 18, 47, 2012.

Kuuluvainen, T. and Aakala, T., Natural forest dynamics in boreal Fennoscandia: A review and classification. *Silva Fenn.*, 45, 823, 2011.

Larsen, J.B. and Nielsen, A.B., Nature-based forest management—Where are we going? Elaborating forest development types in and with practice. *For. Ecol. Manage.*, 238, 107, 2007.

Liljegren, R. and Lagerås, P., *Från mammutstäpp till kohage. Djurens historia i Sverige*, Wallin and Dalholm, Lund, 48 pp., 1993.

Linder, P. and Östlund, L., *Förändringar i Sveriges Boreala Skogar 1870–1991, Report 1*, Swedish University of Agricultural Sciences, Department of Forest Ecology, 32 pp., 1992 (in Swedish).

Lorenzen, E.D., Nogués-Bravo, D., Orlando L. et al., Species-specific responses of Late Quaternary megafauna to climate and humans, *Nature*, 479, 359, 2011.

Luyssaert, S., Schulze, E.-D., Börner, A., Knohl, A., Hessenmöller, D., Law, B.E., Ciais, P. and Grace, J., Old-growth forests as global carbon sinks, *Nature*, 455, 213, 2008.

Martin, T.J. and Ogden, J., Wind damage and response in New Zealand forests: A review, *NZ. J. Ecol.*, 30, 295, 2006.

Møller, P.F., *Vandet i skoven—hvordan får vi vandet tilbage til skoven?* GEUS, Copenhagen, 60 pp., 2000 (in Danish).

Müller, J. and Bütler, R., A review of habitat thresholds for dead wood: A baseline for management recommendations in European forests, *Eur. J. Forest Res.*, 129, 981, 2010.

Naiman, R.J., Melillo, J.M. and Hobbie, J.E., Ecosystem alteration of boreal forest streams by beaver (*Castor Canadensis*), *Ecology*, 67, 1254, 1986.

Niklasson, M. and Granström, A., Numbers and sizes of fires: Long-term spatially explicit fire history in a Swedish boreal landscape, *Ecology*, 81, 1484, 2000.

Nummi, P. and Kuuluvainen, T., Forest disturbance by an ecosystem engineer: Beaver in boreal forest landscapes, *Boreal Env. Res.*, 18 (suppl. A) 13, 2013.

Odgaard, B.V. and Rasmussen, P., Origin and development of macro-scale vegetation patterns in the cultural landscape of Denmark, *J. Ecol.*, 88, 733, 2000.

Parviainen, J., *Forest Reserves Research Network*, European Commission, Brussels, 377 pp., 2000.

Peterken, G., *Natural Woodland*, Cambridge University Press, Cambridge, 522 pp., 1996.

Petit, R.J., Csaikl, U.M., Bordacs, S., Burg, K., Coart, E., Cottrell, J., van Dam, B. et al., Chloroplast DNA variation in European white oaks—Phylogeography and patterns of diversity based on data from over 2600 populations, *For. Ecol. Manage.*, 156, 5, 2002.

Pickett, S.T.A. and White, P.S., *The Ecology of Natural Disturbance and Patch Dynamics*, Academic Press, Orlando, 1985.

Rudemann, R. and Schoonmaker, W.J., Beaver dams as geological agents, *Science*, 88, 523, 1938.

Runkle, J.R., Disturbance regimes in temperate forests, in *The Ecology of Natural Disturbance and Patch Dynamics*, Pickett, S.T.A. and White, P.S., Eds., Academic Press, Orlando, 17, 1985.

Svenning, J.-C., A review of natural vegetation openness in northwestern Europe, *Biol. Conserv.*, 104, 133, 2002.

Thieme, H., Lower Palaeolithic hunting spears from Germany, *Nature*, 385, 807, 1997.

Vera, F.W.M., *Grazing Ecology and Forest History*, CABI Publishing, Wallingford, 506 pp., 2000.

Wardle, D.A., Walker, L.R. and Bardgett, R.D., Ecosystem properties and forest decline in contrasting long-term chronosequence, *Science*, 305, 509, 2004.

Wilkinson, D.M., Is local provenance important in habitat creation? *J. Appl. Ecol.*, 38, 1371, 2001.

Willis, K.J., Braun, M., Sümegi, P. and Tóth, A., Does soil cause vegetation change or vice versa? A temporal perspective from Hungary, *Ecology*, 78, 740, 1997.

Wolf, A., Møller, P.F., Bradshaw, R.H.W. and Bigler, J., Storm damage and long-term mortality in a semi-natural, temperate deciduous forest, *For. Ecol. Manage.*, 188, 197, 2004.

Woods, K.D., Dynamics in late-successional hemlock-hardwood forests over three decades, *Ecology*, 81, 110, 2000.

Yalden, D., *The History of British Mammals*, T. and A.D. Poyser, London, 305 pp., 1999.

Zeder, M.A., The origins of agriculture in the Near East, *Curr. Anthropol.*, 52, S221, 2011.

3

Effects of Restoration at the Stand Level

Chadwick D. Oliver, Kevin L. O'Hara, and Patrick J. Baker

CONTENTS

3.1 Introduction

The conservation of forests for timber, wildlife, and water became a large concern in many countries over 100 years ago, and a culture of conservation started being accepted, institutionalized, and organized. Recently, concern for forests and other ecological systems has become part of the global concern for sustainability and is building on the earlier concern for conservation (Brundtland 1987). Both conservation and sustainability contain the idea that each place and people should obtain no more or less than its fair share of values from its environment and should leave to future generations an environment that is in at least as good a condition, if not better, than at present (Oliver et al. 2002; Oliver 2003).

    This chapter describes the role of forest restoration at the stand level. The stand level is fundamental to any forest activity. A stand is an area of relatively uniform conditions—soils, species composition, age distribution, stand structure, and history. It usually varies from 2 to 200 ha in size and by definition is treated in the same ways at the same times. This chapter first describes the current understanding of the dynamic nature of forest stands and the opportunities and limitations of approaching restoration at the stand level.

Afterwards, it describes specific activities and pathways that can be applied to stands for restoring forest ecosystems. The chapter concludes with a brief discussion of how this restoration can be incorporated into socioeconomic frameworks.

3.2 People and Forests

Forest ecosystems potentially occupy between approximately 30% and 50% of the Earth's land surface area, 4.1–6.7 billion ha (Miller 1996; Tallis 1991), depending on how forests are distinguished from marginal tree-shrub land. The total forest area was much smaller during the past glacial maximum, 18,000 years ago and much larger approximately 5000–7000 years ago (Tallis 1991).

Forests, like other ecosystems, have evolved over hundreds of millions of years so that they now provide the diversity of life, quality of water and air, products of wood and other substances, and other "ecosystem services" that enable human life and prosperity. Nearly three quarters of the area that could presently be in forests are in intact forests. The other potentially forested area is in farms, residences, urban areas, or similar creations (Tallis 1991). The remaining forests are expected to sustain the ecosystem services once provided by the larger area. Part of the remaining intact forests have been degraded through inappropriate harvesting as well as anthropogenic (Sample and Bixler 2014) and other changes in climates, pests, and species compositions (Stanturf et al. 2014a,b). Appropriate, careful measures will be needed first to restore and then to sustain the ecosystem functions of the remaining forests in their fragmented, altered, and constricted state so they provide as many of these ecosystem services as possible.

Stanturf and Madsen (2002) and others (Stanturf et al. 2014a,b) have proposed that this restoration will entail rehabilitation, reconstruction, reclamation, and replacement strategies. Such restoration will involve several considerations as follows:

- Most people live in potentially forested areas and are highly unlikely to move to areas that are not potentially forested, even if that were desirable. Instead, restoration will need to find a way to accommodate people as part of forest ecosystems, ecosystems that are forested or potentially forested.

- The world's forest area can be divided into hundreds of ecosystems (Bailey 1983), with some species unique to each ecosystem. Consequently, each forest ecosystem will need to be restored and sustained if the goal of biodiversity is to be realized; sacrificing one ecosystem to save another will not achieve the goal of sustainability and is probably not necessary.

- Not all forest ecosystems contain species that are endangered, and species become endangered for many reasons. Some are endangered because their habitats cover a small area and others because they are severely impacted by people, and still others are endangered for a variety of other reasons.

- Different forest ecosystems have been impacted by different natural and human influences and thus will require different restoration activities. Some forest ecosystem areas contain nearly all of their intact forests, while others have been largely converted to farms, housing developments, and cities with much less intact forest remaining. The area of intact forests and human-built structures in each ecosystem area is constantly changing.

- Forest restoration in most places will require an initial input of money to develop plans, create access, obtain equipment, and begin operations.

To achieve sustainability, all forest ecosystems will need to be addressed—with a focus on processes that occur both within and between ecosystems. Most intact forests are smaller than their potential areas, with people living within the natural boundaries of forest ecosystems, and thus preventing the intact forests from expanding to their potential. Fortunately, certain features common to forest ecosystems make it possible to restore forests to sustainable conditions under these constraints. Otherwise, people would need to make the very difficult choices of deciding what forest ecosystems to restore at the expense of others and/or how to reduce current human population levels.

3.3 Changing Perspectives of Forest Stands

3.3.1 Early Ecological Paradigm

Certain uniformities of processes occur within forest ecosystems throughout the world that make understanding their complexity more tractable. These uniformities exist despite the hundreds of forest ecosystems (which can be subdivided into thousands of subecosystems), the wide evolutionary origins of forest plants and animals, the large variety of influences on different ecosystems, and the many plants and animals that live in forests. This understanding has emerged as the result of over 100 years of scientific inquiry and a major paradigm shift, as will be described.

Intensive scientific examination of forest processes during the late nineteenth and early twentieth centuries led to ecological theories that governed further inquiry as well as forest management for many decades and continue to influence the way forests are regarded today (Raup 1964; Boyce and Oliver 1999). Ecological communities were regarded as ecological systems (ecosystems) and were defined, bounded, and studied for both their behavior and their emergent properties (Tansley 1935; Odum 1971; Whittaker 1975). Forest ecosystems were assumed to be closed systems where each plant and animal had a necessary function which, if eliminated, would lead to the collapse of the ecosystem. At its extreme, the forest was considered analogous to a super organism, in which each species had coevolved with all the others (for further discussion, see Boyce and Oliver 1999). Elimination of one part would be similar to removing a part from a machine, and the result would be complete collapse of the forest ecosystem. The natural forest was considered to be a forest undisturbed by people, or at least nonindigenous people, and was assumed to be in a steady-state climax condition in which the large, old trees were gradually but continuously dying and being replaced by younger trees growing from beneath (Figure 3.1).

According to this steady-state theory, if a disturbance did occur to these forests, a predictable succession of species would invade the disturbed site, with progressively more shade-tolerant species gradually replacing less tolerant ones (Braun 1950; Odum 1971; Whittaker 1975; Oliver and Larson 1996).

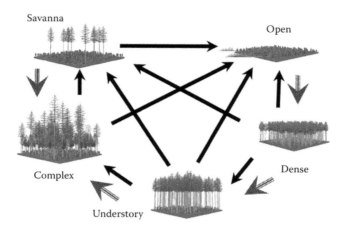

FIGURE 3.1
Stand structures common to many forest ecosystems in the world. The early dominant paradigm assumed that "natural" forests existed only in the "complex" (also known as "climax" or "old growth") structure. It is now accepted that forests change structures through time and by disturbances. Different species depend on each structure within a forest ecosystem. (Copyright, Oliver, C.D., Reproduced in Stanturf J.A. and Marsden P., Eds., *Restoration of Boreal and Temperate Forests*, CRC Press, Boca Raton, 2005.)

Early conservation efforts made use of this steady-state theory to attempt to protect native species. Representative tracts of climax forests were set aside as preserves while forests elsewhere were exploited (Henry and Swan 1974; Botkin 1990). These forests were assumed to provide habitats for all native species, since this climax forest was allegedly the natural forest and therefore the only kind of forest that native species needed to survive. Since the natural or climax forest was assumed to be innately stable, these forests were expected to remain in the steady-state structure forever.

Conservation-oriented forest management also used this idea of a steady-state climax forest to develop silvicultural prescriptions under the assumption that each stand was all-aged and could be sustained in its natural state through selection harvest/regeneration methods. The selection methods were approved of by ecologists and silviculturists because they felt it was natural, scientifically based, and sustainable. And, they were supported by loggers because it meant that they only had to harvest the large, valuable trees without expending time and resources cutting the others (Boyce and Oliver 1999; O'Hara 2002).

3.3.2 Changing Paradigm

In the late 1940s and early 1950s, forest managers throughout the United States and Europe began to realize that the selectively harvested forests were not sustaining the species and wood quality that had been in the original forest. They also noticed that nearby even-aged stands that had grown on old fields or following catastrophic wildfires were rapidly growing trees of high wood quality and appropriate species (Boyce and Oliver 1999). To ensure they could provide timber sustainably, foresters quickly changed to even-aged management, describing the selection methods as mistakes because the forest was not sustaining itself in tree species, growth rates, or vigor (O'Hara 2002).

As long as the steady-state forest was considered the natural forest, and forest managers were only successful in providing timber from forests that were considered artificial, managing for multiple benefits could not be considered synergistic. That is, managing for timber would presumably only be at the expense of providing habitat and other values naturally provided by the forest, a zero-sum outcome. This assumed zero-sum outcome led to polarization between those wanting to sustain commodities and those wanting to sustain habitats and other natural forest values.

Accumulating evidence gradually led scientists and professionals to abandon the steady-state view of forests. The 1938 hurricane in New England blew over all trees in preserves that had been expected to remain in the climax condition perpetually, forcing a critical re-evaluation of a the idea of a stable climax (Henry and Swan 1974). Other presumably climax forests that were set aside were also not remaining stable in structure or species composition, but were gradually changing (Botkin 1990). During this time, studies began demonstrating that the forest was far more resilient than had been assumed; it was not a closed system, but rather was impacted by external forces such as natural disturbances, species immigrations and extinctions, and climate changes (Oliver and Larson 1996). Additionally, these studies indicated that the successional sequence of younger trees and species replacing older ones within a stand was not the common pattern of forest development (Oliver 1978, 1980b). Scientists began to realize that some native animal and plant species could not exist in climax forests; and some species became endangered, and possibly extinct, because there were not enough other forest conditions (e.g., Kirtland's warbler, *Dendroica kirtlandii* (Walkinshaw 1983), lotus blue butterfly, *Lycaeides argyrognomon* (Fry and Money 1994) and others (Young 1992; Oliver et al. 1997)). By 1990, the Ecological Society of America had shifted away from the steady-state paradigm to a more dynamic paradigm of forest development (Stevens 1990).

3.3.3 Stand Development Stages and Stand Structures

The realization that a forest is an open system, not a closed one, means that its behavior is an aggregation of processes that work opportunistically together to provide various, emergent properties (Johnson 2002). Some of these emergent properties can be classified as physical stand structures (Oliver and Larson 1996) that are the result of stand development processes. (For more discussion of the relation of stand structures and development stages, see Oliver and Larson 1996; Camp and Oliver 2004.) The stand structures can be associated with other emergent properties (Peterken 1996) such as fire behavior, evaporation rates, and habitats. Similar stand structure patterns are repeated in many forest ecosystems throughout the world and reflect an underlying uniformity of process (Zebrowski 1997).

A common method of classifying the stand structures as they follow stand development stages is described below and will be used in this chapter (Figure 3.1; Oliver and Larson 1996). A natural forest is sometimes impacted by a stand-replacing disturbance that creates an area devoid of most trees, the *stand initiation* development stage, or the *open* or *savanna* structures (Figure 3.1). As new trees are reoccupying the area, many shrub and herb species also invade. The invading species first expand until they occupy all above- and below-ground growing space, at which time there may be many species of plants and many animal species feeding on the diverse and short (easily accessible) plants.

As the growing space fills, there is intense competition from the existing plants, resulting in the elimination of many individuals and some species, curtailment of new species invasions, and crowding of the remaining plants. Woody plants generally retain the below- and above-ground growing space that they garnered in previous growing seasons, so they gradually take over the stand. Except on very droughty sites where soil-growing space is occupied long before canopy closure, there is generally a separation of the photosynthesizing canopy layer from the ground, and very little sunlight reaches the forest floor. Any plants surviving near the forest floor grow extremely slowly. This stage of tight control of growing space by the canopy and/or roots is known as the *stem exclusion* stage, and creates the *dense* structure. It is characterized by a relative paucity of plant and animal species. During this stage, some tree species can out-compete others and occupy the upper canopy layer, relegating others to understory or midstory where they grow little (Figure 3.2). These species in the understory often appear younger than their overstory contemporaries because they grow so little once overtopped and shaded.

Eventually, small disturbances or natural processes of differentiation, mortality, and wind abrasion can allow more sunlight onto the forest floor and make below-ground growing space available. Then, new herbaceous and woody plants, including trees, begin growing. These plants increase the stand's plant species diversity as well as the habitat value of the forest for animals. This time of regrowing of the understory is referred to as the *understory reinitiation* stage, and the structure created is commonly termed the *understory* structure.

When a stand is in the dense, understory, or complex structure, disturbances can kill some trees in the forest, allowing lower strata, forest floor, or newly regenerating plants to grow upward, where they either reach the overstory or become suppressed in the shade of other, taller trees. This structure can contain very large, magnificent trees of old ages, as well as smaller trees, a diversity of plant species, and habitats for relatively specialized

FIGURE 3.2

Mixed-species, single-cohort stand with oaks in the upper stratum and maples, birches, and other species in the lower strata. Despite the large differences in sizes, virtually all trees in this stand are between 65 and 75 years old and grew following a clear-cut 75 years ago (Yale-Meyers Forest, Connecticut, United States). (Copyright, Oliver, C.D., Reproduced in Oliver, C.D., *Forest Stand Dynamics*, McGraw-Hill, New York, 1990; Update Edition, Wiley, 1996). (Reproduced in Stanturf, J.A. and Marsden, P., Eds., *Restoration of Boreal and Temperate Forests*, CRC Press, Boca Raton, 2005.)

animal species. This structure is referred to as a *complex* structure and had previously been assumed to be the climax forest; however, the forest usually achieves this stage and structure as an open system, impacted by external forces such as disturbances. This condition has been referred to as transition and *old growth* stages of development, but it is still in a very dynamic condition (Franklin et al. 2002).

Other, similar classifications have been used to describe the changing forests (e.g., O'Hara et al. 1996; Carey et al. 1999; Franklin et al. 2002) and are helpful for specific objectives.

As each stand is disturbed, different species can invade, become dominant, and maintain their dominance for the life of the tree species, often hundreds of years. Consequently, there is no species that is naturally predetermined to dominate a site; rather, any of several species can dominate it given a particular combination of initial conditions (i.e., stand structure and composition) and disturbances (Henry and Swan 1974; Oliver and Stephens 1977; Oliver 1980b; Oliver and Larson 1996).

3.4 Restoration and Silviculture

For restoration to be effective at the stand level, it must incorporate a dynamic understanding of forests. A stand cannot be restored to a specific structure and be expected to remain in that condition. Instead, the stand must be restored to a trajectory of changing structures that, in concert with other stands in the landscape, will maintain the appropriate mix of structures across that landscape overtime.

Forest restoration can build on the knowledge gained about silviculture and stand development over the past 100 years. At the stand level, restoration and silviculture will be most effective if they are considered interchangeable, with the following caveats:

- Although most definitions of silviculture consider its objectives to be broad (Ford-Robertson 1971; Daniel et al. 1979; Nyland 1996; Smith et al. 1997; Helms 1998), some definitions assume that the objective of silviculture is simply to provide timber (Toumey 1947). When considered in the context of restoration, the broader definition should be applied.

- To withstand the test of time, restoration will eventually need to adopt many of the management features learned by silviculturists. These features include the need to frame the actions in an economic perspective; the need to establish distinct, measurable objectives; and the need to plan for each stand changing its structure over time—with or without human intervention.

Much of the accumulated knowledge from silviculture can immediately be applied to the practice of restoration.

A reasonable working objective of restoration can be sustainable forestry, which itself can be defined as ensuring that each forest ecosystem in the world provides its fair share of values over time and space. A reasonable working set of values has been provided by various processes and initiatives (Burley 2001) as criteria. As a beginning, the Montreal Process criteria can be used and include: biodiversity, productive capacity, forest health, soil and water, carbon sequestration, socioeconomic considerations, and the infrastructure to provide other values (Fujimori 2001; Oliver et al. 2002; Oliver 2003).

3.5 Landscape Patterns

A landscape, a subcomponent of an ecoregion, is a contiguous land area containing many stands and usually bounded by natural or artificial features such as ridgelines, rivers, landform changes, or property boundaries (Oliver et al. 2012).

Each forested landscape could naturally maintain a variety of structures and species over time, or it could fluctuate widely from a predominance of one structure, and group of species, to another if all stands are impacted by the same disturbance. In the past the large diversity of species was maintained because there were always some stands of each structure for the dependent species to utilize over the large, natural intact forest of each ecosystem. The proportion of different structures may have fluctuated with time and disturbances, allowing species needing some structures to be confined to isolated refugia when their structure was minimal—and then expanding across the landscape if a disturbance or forest regrowth increased their needed structure (Oliver et al. 1998).

With people occupying part of the potentially forested area of many forested ecosystems, the area of remaining intact forests can be too small or fragmented to allow the wide fluctuations in stand structures that occurred naturally and still maintain all structures and species. Consequently, restoration will probably be necessary to ensure that all habitats are maintained in sufficient abundance over many of these constricted intact forested landscapes (Stanturf et al. 2012).

Active management can provide the sustainability values by restoring and maintaining a diversity of stand structures across the landscape (Hunter 1990; Oliver 1992). Active management can mimic, avoid, and recover from natural disturbances to sustain the variety of structures and thus provide habitats for biodiversity, employment, and products during the harvesting to create the different structures that maintain the forest health and soil and water quality. Interspersed with this active management, forests reserved from management can be identified within each ecoregion to ensure that processes currently unknown are not lost (Seymour and Hunter 1999).

3.6 Stand Operations and Pathways

To maintain the diversity of stand structures across the landscape that is needed for the many forest values, restoration efforts will first need to determine the possible ways that a stand can change over time, both with and without human intervention. Then, the appropriate actions, or inaction, can be taken to ensure that the desired changes occur. When ecologists and silviculturists believed that a stand would develop through a predictable successional sequence of species replacing others following a disturbance, a quite straightforward set of silvicultural systems was developed for regenerating and therefore sustaining each species on a site. For example, to regenerate and thus sustain pioneer species, those assumed to regenerate immediately after a stand-replacing disturbance, the clearcutting system was suggested. To regenerate and sustain mid-seral species, a shelterwood system was prescribed to leave some standing trees for protection for the first few years. To regenerate and sustain climax species, selection systems were prescribed. The difficulty in obtaining expected results when applying these silvicultural systems helped lead to the paradigm shift described earlier.

Recognition of the more dynamic nature of forests has led scientists to realize that each species has regrown in a variety of partial and stand-replacing disturbances, and thus there are a variety of ways in which each species can be sustained in a stand. This variety has led silviculturists to begin describing stand management through *operations* and *pathways* instead of through silvicultural systems.

3.6.1 Silvicultural Operations

An operation is a specific activity done to change a stand within a relatively short time. Operations include sowing seeds or planting seedlings, weeding unwanted competing plants, removing some trees through thinning or various partial or complete harvesting methods, fertilizing, pruning, and others (Fujimori 2001). Successful operations generally include the following ones:

- A biological component—understanding the physiological sensitivities and needs of the plants being manipulated.
- Mechanical and labor components—organizing the appropriate machines, equipment, and people to accomplish the operation.
- A logistical component—ensuring that needed activities are done in the proper sequence and time for the components to converge and accomplish the targeted operation.

For example, a successful planting operation would require that the seedlings of the desired species and genetic background be grown and transported to the target site at the same time that the planting labor was available, the weather was appropriate, and the site was properly prepared to receive the planted seedlings.

Silvicultural operations attempt to mimic natural processes to various extents, at times mimicking disturbances and at other times mimicking regeneration patterns. Some operations such as burning beneath a stand mimic natural ground fires quite closely, whereas chemical applications of herbicides are quite unlike natural weeding processes. Natural disturbances varied in frequency and magnitude in most of the world's forest types, and most silvicultural operations lie within the natural range of variability of these processes. Historically, however, silvicultural operations in each forest ecosystem applied a narrower range of activities than the natural variability of disturbances and regeneration patterns and created a narrow range of stand structures and values associated with these structures. For example, forests in ecosystems such as the tall Eucalyptus (*Eucalyptus* species) forests of southwestern Australia or the Douglas-fir (*Pseudotsuga menziesii* (Mirb.) Franco) forests of western Oregon and Washington have primarily been managed by clear-cutting, and the diversity of structures and species maintained by more variable natural disturbances— or that could be maintained by more variable harvest operations—was lost. Other forest ecosystems, such as conifers in the inland western United States and hardwoods in the southeastern United States, have primarily been managed by selective harvesting or "high grading" and there is a shift to more trees of shade-tolerant species and a decline in the open structure (Oliver et al. 2005). Most of the world's ecosystems have been impacted by both stand-replacing disturbances and partial disturbances. Consequently, a greater diversity of operations will be needed to restore the landscape-scale variability in stand structures.

Silviculturists plan operations around windows of opportunity—times of the year when it is biologically and physically possible to accomplish an operation successfully.

For example, seedlings are most easily planted when they are dormant; thus, the planting window in many temperate climates is in the spring after the soil has thawed and before the seedlings have broken dormancy. A harvesting window often depends on the soil conditions; the window for harvesting stands on easily compacted soils is commonly during dry seasons or when the soil is frozen. An additional window to be avoided when thinning stands is during the spring when the tree cambium is active; any scraping of a nonharvested tree during this period can readily scar the stem and allow stem-decay fungi to enter. In forests in which postharvest, prescribed fires are used to prepare the seedbed, there will be a window of opportunity for burning during which the conditions are neither too hot and dry to contain the fire safely, nor too wet and cool to get sufficient burning of the slash necessary for successful establishment of new trees.

A special skill of silviculturists and others engaged in stand-level restoration is to combine the understanding of biology, windows of opportunity, and operational techniques to perform the most efficient actions to achieve the desired results. Consequently, a variety of operations and strategies is needed for stand management, from minimizing regeneration costs by managing forests on a long-rotation basis (Larsen 1995), to heavily investing in regeneration but managing on a relatively short rotation (Phillips 2000).

Silviculturists have generally become quite skilled in performing operations, although some activities, such as successfully regenerating particular species, still elude them. New techniques, such as creating living and dead trees with suitable shapes and cavities for use by targeted wildlife species, are being developed as new values are desired for the forest. Later chapters in this book will describe some of these successful operations, as well as new operational techniques.

3.6.2 Silvicultural Pathways

Without disturbances, each stand will grow in a pattern or trajectory that is determined by the stand's structure (number, spatial arrangement, physical size, species, and genetic makeup of its component species), its soil, and climate media (Oliver and Larson 1996). The stands change in structure and provide different values as individual trees grow and die, as described earlier. Each large or small disturbance alters the configuration of plants, soil, and/or microclimate and thus causes the stand to grow along a new trajectory. Silvicultural operations planned and implemented at specific times can cause a stand to grow along different trajectories, barring natural disturbances. The change in a stand over time caused by a combination of growth and specific operations at specific times is referred to as a pathway (Figure 3.3).

A stand can potentially follow many pathways (Figure 3.4). By understanding the potential trajectories of a stand and the ways in which various silvicultural operations can change these trajectories, a stand's pathway can be designed to provide various structures and values at different future times. Many of the values provided by a landscape will depend on the pathways followed by the component stands, with the spatial arrangement of the stands' pathways also influencing some values. If all stands initiate at or around the same time and follow similar pathways, they will pass through similar structures at the same time and provide each of the values in a pulse (Oliver and Larson 1996). If stands across a landscape follow different pathways or follow the same pathway but start at different times, the resulting variety of structures and values can be distributed more broadly over time, creating a landscape that is dynamic because each stand's structure is changing, but stable because the changes in stands are asynchronous, so that all structures are present at all times—but in different places within the landscape. Careful coordination of

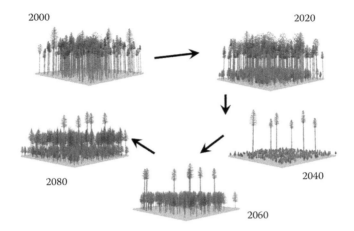

FIGURE 3.3
Schematic silvicultural pathway for a Douglas-fir stand. Operations were done as follows: Thinning and regeneration (2005); Retention harvest and regeneration (2035); Thinning of lower stratum and regeneration (2065). (Copyright, Oliver, C.D., Reproduced in Stanturf, J.A. and Marsden, P., Eds., *Restoration of Boreal and Temperate Forests*, CRC Press, Boca Raton, 2005.)

stand pathways in a landscape (Oliver et al. 2009) can allow the structures to be provided uniformly or in pulses over time (Oliver et al. 1998), as desired.

Each stand cannot be changed to any other structure instantaneously. Instead, the potential pathways that a stand can follow and the time when it can provide future values, if at all, are constrained by its present structure and the soil and climate in which it is growing. Stands also have windows of opportunity during which a pathway can be effectively changed by an appropriate operation. For example, Wilson and Oliver (2000)

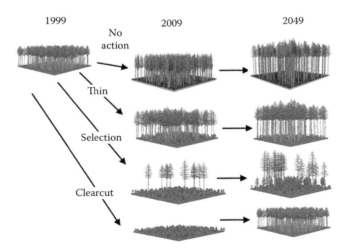

FIGURE 3.4
A stand can potentially follow many different pathways, depending on its present condition and the species and site conditions, as well as the type and timing of silvicultural operations. The manager decides which pathway each stand follows depending on the stand's potential, the objectives of management, and the conditions of other stands in the landscape. (Copyright, Oliver, C.D., Reproduced in Stanturf, J.A. and Marsden, P., Eds., *Restoration of Boreal and Temperate Forests*, CRC Press, Boca Raton, 2005.)

found that Douglas-fir stands can be effectively thinned without high risk of windthrow if the trees are already stable (height–diameter ratio <100) or less than 10 m tall at the time of thinning. Above this tree height, unstable trees will not regain stability for many years and the probable result of thinning will be windthrow of some or all of the remaining trees; and the window for successful thinning will have been missed. Efforts to restore stand structural diversity at the broader landscape scale may be impeded by silvicultural prescriptions that sacrifice operational flexibility (i.e., larger windows of opportunity) for operational efficiency (Wilson and Baker 2001).

Silvicultural pathways can be classified in many different ways, such as according to species mix, operations done, or age distribution. The following discussion will describe various silvicultural pathways according to several, overlapping classifications: "no action," "single-species and single-cohort," "mixed-species and single-cohort," and "multiple-cohort" pathways. A "cohort" is defined as all trees that begin following a single disturbance (Oliver and Larson 1996).

3.6.2.1 No-Action Pathway

The no-action pathway, in which no active silvicultural operations are performed, should always be considered when prescribing pathways, to determine if any actions are really necessary to achieve the stand's objectives. The effects of this pathway depend on the condition of the stand and the objectives. A young stand with many species crowded in the same canopy stratum (i.e., brushy stage, *sensu* Gingrich 1971) may require no actions if the desired tree species will stratify naturally to the upper canopy (Oliver 1980a). Studies have documented predictable patterns of stratification among oaks (*Quercus* species) in the eastern United States; Figure 3.2 (Oliver 1978; Bowling and Kellison 1983; Clatterbuck et al. 1985; O'Hara 1986; Clatterbuck et al. 1987; Clatterbuck and Hodges 1988; Kittredge 1988), Douglas-firs and associated species in the Pacific Northwestern United States (Stubblefield and Oliver 1978; Wierman and Oliver 1979; Cobb et al. 1993), Sitka spruces (*Picea sitchensis* Carriere) and western hemlocks (*Tsuga heterophylla* (Raf.) Sarg.) in Southeast Alaska (Deal et al. 1991), and mountain ash (*Eucalyptus regnans*) and *Acacia* spp. (Ashton 2000; Simkin and Baker 2008) in Australia; however, local professional experience is often needed to determine or ensure applicability of these patterns (O'Hara and Oliver 1999).

A young single-species, single-cohort stand may also differentiate readily with no intervention, allowing some trees to dominate and grow large in diameter and become wind and insect resistant while their neighbors become suppressed and commonly die (Figure 3.5a). In this case, thinning the stand would eliminate the mortality, reduce the fire risk from dead trees, and provide some timber, but would have little effect on the stand's stability. Another stand of the same age, species, and site may not readily differentiate but instead may have all trees grow nearly equally, resulting in trees with large height–diameter ratios that are susceptible to wind, snow, and insect infestations (Figure 3.5b). In this case, thinning the stand early could allow the remaining trees to become stable, while a late thinning would simply accelerate the rate of windthrow in the stand (Wilson and Oliver 2000). Subtleties of the stand's structure such as the relative uniformity of spacing, ages, and sizes of trees can indicate the ability of a stand to differentiate naturally (O'Hara and Oliver 1999). Various tools such as density management diagrams can help indicate when a stand will become susceptible to either differentiation and mortality or stagnation (Reineke 1933; Gingrich 1967; Drew and Flewelling 1979; McCarter and Long 1986). On the other hand, growth models can be poor at predicting a stand's propensity to differentiate if the mortality rate or maximum height/diameter ratio is fixed in the model.

(a)

(b)

FIGURE 3.5
(a) A well-differentiating stand of Douglas-fir, Pacific Northwestern United States. (b) A poorly differentiating stand of loblolly pine (*Pinus taeda* L.), southeastern United States. (Copyright, Oliver, C.D., Reproduced in Oliver, C.D., *Forest Stand Dynamics*, McGraw-Hill, New York, 1990; Update Edition (Wiley, 1996). Reproduced in Stanturf, J.A. and Marsden, P., Eds., *Restoration of Boreal and Temperate Forests*, CRC Press, Boca Raton, 2005.)

3.6.2.2 Single-Species, Single-Cohort Pathways

Single-species, single-cohort stands can develop from natural or artificial stand-replacing disturbances. Many trees and other plant species living in the region can potentially invade the area, and the type of disturbance and other conditions at the time of the disturbance help determine which species actually comprise the new stand.

The age range in single-species, single-cohort stands may be only one or two years where planting or advance regeneration, accompanied by suitable site conditions allowed trees to reoccupy the growing space rapidly; or, it can be several decades where the stand established from windblown seeds on a poor site in which the growing space was not fully occupied for many years (Oliver and Larson 1996).

Stands with wide age ranges or irregular spacings commonly differentiate well, allowing a relatively stable stand of trees to develop with a large range of sizes (Figure 3.5a). The smaller, suppressed trees have higher height–diameter ratios than the dominant trees (Wonn and O'Hara 2001). If these stands begin at a narrow spacing, considerable mortality of small trees will occur as the stand is differentiating. The remaining trees commonly have small lower limbs and knots in their lower boles because of the initial crowding. At wide spacings, large-diameter, stable trees with large limbs and knots will result.

Stands with narrower age distributions and more regular spacing can occur either naturally or with planting. These stands grow more readily to trees of uniform sizes with less tendency to differentiate (Figure 3.5b). Instead, they somewhat uniformly slow in diameter growth at a predictable size for a given spacing and species (Oliver and Larson 1996); without a timely thinning, these trees will become susceptible to wind and snow damage, or insect infestations. On the other hand, stands where trees are spaced less uniformly or at lower densities have less need of thinning and other operations done at precise times, but do not grow the uniform products that can be harvested very efficiently compared to more uniformly grown stands (Wilson and Oliver 2000).

If the diameter at which the trees slow is less than a merchantable size in a uniformly spaced stand, the stand will need to be thinned at a cost (precommercial thinning) to achieve large tree sizes and to avoid stand health problems. Delaying this thinning will generally make the thinning less effective because the suppressed trees, the target trees for removal, will not become larger and more valuable, and the dominant trees are generally becoming less stable because they are not differentiating well. A general issue when establishing single-species plantations is tree spacing. Trying to ensure that the trees are merchantable when the first thinning occurs means forecasting the minimal merchantable sizes for one or several decades into the future. Some people prefer to plant narrowly and plan on a costly precommercial thinning to ensure small branches and a choice of trees to leave for future stand composition. Others prefer to plant widely to avoid the higher cost of planting more trees, the higher cost and added commitment of a precommercial thinning, and to obtain the benefit of slightly larger trees. Sometimes, the problem of large limbs in plantations of wide spacings can be overcome by pruning (Hanley et al. 1995).

Thinning single-cohort, single-species stands is usually most effective if the more dominant trees are left, although some trees in the dominant crown class can be removed if necessary. The dominant trees are the most stable (against wind and snow breakage) and rapidly growing. Thinning can accomplish many objectives simultaneously (Figure 3.6):

- Depending on the intensity of the thinning, it can change the stand from the dense to the understory or savanna structures much sooner than would occur without thinning.
- It sometimes provides income as well as timber for utilization.
- It prevents potentially susceptible stands from succumbing to wind or snow breakage or insect attacks.
- It allows the remaining trees to grow more vigorously, be more stable, and reach larger sizes.
- It allows the growth to be concentrated on trees of desirable characteristics, such as straight stems are usually desired.

If desirable, a stable, single-cohort stand can be converted to a multiple-cohort stand and eventually to a complex stand. It can also be clear-cut to create a more open structure, or a few large trees can be retained to create a savanna structure (Figure 3.4).

The volume achieved in single-species, single cohort stands varies dramatically with species (Figure 3.7). Within a species, the volume may vary at different times with stand densities and silvicultural pathways that include various thinning operations. The general pattern of stand growth in practically all cases is to increase tree volume and size asymptotically during the first few years or decades. In regions with subsistence rural conditions, restoring forests is difficult because local people require fuel wood constantly and often cut young trees of little volume. If such forests are protected, other forests are put under even greater pressure. An alternative may be to provide these people with alternative forms of energy until the forests are old enough to grow high wood volumes.

FIGURE 3.6
(a) Twenty-year-old Douglas-fir stand in Pacific Northwestern United States just before pruning and thinning. (b) Same stand 3 years later. (Copyright, Oliver, C.D., Reproduced in Stanturf, J.A. and Marsden, P., Eds., *Restoration of Boreal and Temperate Forests*, CRC Press, Boca Raton, 2005.)

3.6.2.3 Mixed-Species, Single-Cohort Pathways

Mixed-species, single-cohort stands can develop large ranges in tree sizes and stratify into complex vertical structures (Figure 3.2) (Oliver and Larson 1996). As a result, these stands are often assumed to be multiple-cohort stands, because the species of smaller sizes in the understory appear to be younger, later stages of the classical succession process. Many mixed-species stands have been shown to be single-cohort stands, in which all stems initiated shortly after a stand-replacing disturbance and, after a period of intense competition, stratified into canopy layers by species. As the mixed-species, single-cohort stand development pattern is becoming better understood; it is proving to be quite useful for a variety of commodity and noncommodity objectives.

If the desired species is known to stratify naturally into the dominant canopy stratum, it is necessary only to ensure that appropriate numbers of trees of this species are in the initiating stand. If another, less desired species will out-compete the target one, an operation will be needed to prevent the less desired species from initiating or an intermediate

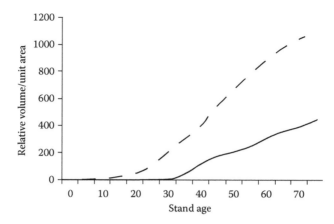

FIGURE 3.7
Different species can grow dramatically different volumes of wood, as shown in this simulation comparing eastern white pine (*Pinus strobus* L., dashed line) and mixed hardwood stands (solid line) in western North Carolina, United States. (Reprinted from Stanturf, J.A. and Marsden, P., Eds., *Restoration of Boreal and Temperate Forests*, CRC Press, Boca Raton, 2005, with permission.)

cleaning will be needed to eliminate it. For example, in the Pacific Northwestern United States, red alders (*Alnus rubra* Bong.) will commonly dominate Douglas-firs if they grow close together on productive sites (Newton et al. 1968; Stubblefield and Oliver 1978). To allow the Douglas-firs to dominate, either the red alders need to be removed by early cleaning operations or prevented from initially regenerating by minimizing mineral soil seedbeds for their light seeds.

Growing mixed-species, single-cohort stands avoid many of the concerns and commitments of single-species stands while providing additional benefits as well. The natural stratification pattern allows the target crop trees to begin in a crowded condition that encourages clear stems; however, instead of needing to remove the competing trees with a costly thinning to allow the crop trees to grow large and remain stable, the competing trees of other species naturally become relegated to lower canopy strata and allow the crop trees to spread their crowns above the others, grow large, and remain stable. For species that develop epicormic sprouts after being released by thinning in single-species stands, maintaining a crowded understory of another species will help produce clear stems (McKinnon et al. 1935). Mixed-species stands may also help prevent some of the insect and pathogen concerns associated with single-species stands (Chandler 1990).

Afforestation or reforestation with appropriate species mixtures can rapidly create a forest with a layered canopy and a variety of tree sizes that benefit some wildlife species, while giving the timber production efficiencies described above. For example, planting mixtures of Douglas-firs and western redcedars (*Thuja plicata* Donn) in western Washington, United States, allow the Douglas-firs to outgrow the redcedars and produce high-quality Douglas-firs and western redcedars with little thinning, at the same time creating a diverse canopy (Figure 3.8). Similarly, when a few cherrybark oaks (*Quercus pagoda* Raf.) grow with sweetgums (*Liquidambar styraciflua* L.) on old fields in the southeastern United States, a similar stratification occurs and yields high-quality oaks and a diverse canopy (Clatterbuck et al. 1985; Clatterbuck and Hodges 1988) (Figure 3.9).

If lower stratum species are spaced correctly at an early age, they can also become merchantable. Volume growth of mixed-species stands can vary from intermediate between

FIGURE 3.8
When Douglas-fir and western redcedar are planted together during afforestation or reforestation in Washington (United States), a predictable stratification occurs that gives benefits for both wildlife and timber production. (a) Douglas-fir (foreground) and redcedar (beside forester) 3 years after planting. (b) Redcedar (left row) and Douglas-fir (right row), 15 years after planting. Pruning was done for visibility. (c) Douglas-fir (center, behind forester and redcedar, surrounding) 70 years after natural regeneration following a clear-cut. (Copyright, Oliver, C.D., Reproduced in Stanturf, J.A. and Marsden, P., Eds., *Restoration of Boreal and Temperate Forests*, CRC Press, Boca Raton, 2005.)

the volume growths of each species grown in a pure stand to greater than the volume growth of either species in a pure stand, depending on the species in the mixture (Figure 3.10) (Kelty 1986, 1989). Harvesting of mixed-species stands can occasionally be done by removing the dominant canopy and allowing lower strata trees to grow. In many cases, however, the overstory trees are so large and/or the understory trees are of such poor vigor and timber quality that removing just the overstory results in a stand of deformed trees of low vigor and a shift in species composition to more shade-tolerant species.

In regions with subsistence rural conditions, restoring forests to mixed-species stands may be advantageous if one species is fast growing and can be harvested for fuelwood when young while allowing the other species to grow longer to more valuable crops.

The structure of even quite young mixed-species stands has many elements of a complex forest: the large mixture of species, vertical depth and layers of foliage, large tree size range, and frequently large diameters of the dominant trees.

Many mixed-species stands were harvested through selective cutting when the allaged succession, steady-state, and climax theories were assumed to be true and the stands were

FIGURE 3.9
Cherrybark oak (being touched by forester) with sweetgum in an old-field abandoned several decades before, Mississippi, United States. (Copyright, Oliver, C.D., Reproduced in Stanturf, J.A. and Marsden, P., Eds., *Restoration of Boreal and Temperate Forests*, CRC Press, Boca Raton, 2005.)

assumed to be all-aged. When silviculturists recognized their lack of success in managing forests through selective harvesting, they tried to grow stands through clear cutting. Loggers, however, frequently resisted clear-cutting because of the extra cost of removing the understory trees of low timber value. Consequently, these loggers defined "clear-cutting" as removing the most valuable trees and leaving the remaining ones because they were "unmerchantable" or considered "advance regeneration" (Figure 3.11a) (Boyce and Oliver 1999). The practice was done in mixed deciduous and coniferous forests in the eastern and western United States and in mixed deciduous forests in the Russian Far East. The remaining trees were not vigorous, often crooked and scarred, and of shade-tolerant species. They grew relatively slowly and their shade commonly killed any shade-intolerant species trying to regenerate. The result has been a shift in species composition in these forests to more shade-tolerant species (Oliver et al. 2005), a reduction in the vigor and timber value of the stands, a reduction in wildlife value of the stands in the eastern United States, and an increase in fire, insect, and disease susceptibility in these stands in the western United States coniferous forests (Sampson and Adams 1994; Oliver et al. 1997). Considerable effort, expertise, time, and money will be needed to restore these stands to their preharvest species compositions and structures.

To distinguish between these loggers' practices and clear-cutting that could sustain the species composition, timber, and structure of the forest in the long term, silviculturists

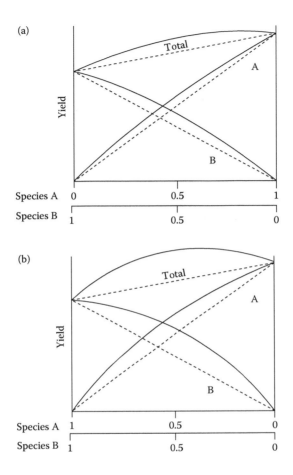

FIGURE 3.10
Potential relations of volume growth to proportions of species in mixed species stands. (a) Mixed volume is less than the volume of one species in a pure stand. (b) The mixed volume is greater than the volume of either species in a pure stand. (Reprinted from The Development and Productivity of Hemlock-Hardwood Forests in Southern New England, Kelty, M.J., PhD dissertation, Yale University, New Haven, CT, 1984, With permission. Reproduced in Stanturf, J.A. and Marsden, P., Eds., *Restoration of Boreal and Temperate Forests*, CRC Press, Boca Raton, 2005.)

termed the removal of only the best trees in a stand as "high grading," removal of only the merchantable trees while leaving the unmerchantable, usually unvigorous ones as "commercial clear-cutting" or "conditional clear-cutting" (in the Russian Far East), and cutting of all stems including the unmerchantable stems to replicate a stand-replacing disturbance as "silvicultural clear-cutting" or "clean-cutting" (Figure 3.11b) (Helms 1998). Selection harvest methods, leading to multiple-cohort stands, can be done under certain circumstances in a sustainable manner as will be discussed further; however, they are subtly different in practice and dramatically different in outcomes from the common practice of high grading, commercial clear-cutting, or conditional clear-cutting.

3.6.2.4 *Multiple-Cohort Pathways*

When a disturbance does not kill all trees but kills enough trees to make growing space (i.e., soil nutrients, water availability, sunlight) newly available, the remaining plants may

FIGURE 3.11
(a) "Commercial clear-cutting" in which the unvigorous, crooked, unmerchantable trees of shade-tolerant species are left. They shade and exclude regeneration of shade-intolerant species. (b) "Silvicultural clear-cut" in which unmerchantable trees are cut, so that new, vigorous stems of a variety of species can regenerate. (Copyright, Oliver, C.D., Reproduced in Stanturf, J.A. and Marsden, P., Eds., *Restoration of Boreal and Temperate Forests*, CRC Press, Boca Raton, 2005.)

expand or new plants may invade or both (Oliver and Larson 1996). If the disturbance is sufficiently large, the remaining plants are sufficiently unvigorous, and/or the new plants are sufficiently vigorous, a new cohort of plants can become established. Commonly, trees are part of this new cohort; and stands with two or more cohorts of trees are referred to as "multiple-cohort" or "multiaged" stands.

Where there is a shortage of stands of complex structure, conversion of single-cohort stands to multiple-cohort stands can help provide many complex structural features. For example, the multiple-cohort Douglas-fir stand in Figure 3.12b was similar to the single-cohort stand in Figure 3.12a until a partial cutting 17 years before the photographs were taken allowed a new cohort to develop. As discussed earlier, reforesting or afforesting with mixed species can also provide many complex forest structural features.

A multiple-cohort stand commonly appears to have a complex structure that can be highly variable (Figures 3.12b through 3.14). A few older trees can exist in the older cohorts, with the majority of the stand being in a younger cohort (Figure 3.13). Alternatively, the opposite can occur with only a few trees in a younger cohort. Shade from much taller overstory trees (i.e., high shade) has different effects on tree growth than the shade from trees of the same size or only slightly taller.

The relation of overstory to understory tree growth in multiple-cohort stands is of theoretical and practical importance. For example, a larger allocation of growing space to the overstory will reduce the growth of the understory and vice versa. O'Hara (1998) presented this as a simple trade-off between the two canopy strata in two-cohort stands. Assuming that tree density by various measures is directly related to the amount of growing space occupied, various possible relationships of overstory and understory density can occur. The possible scenarios can be roughly classified into three overstory behaviors and three understory behaviors, making a total of nine standgrowth behaviors (Figure 3.15). Patterns G, H, and I are most likely to occur because of the greater growth rates of overstory trees (O'Hara 1996; Kollenberg and O'Hara 1999; O'Hara et al. 1999; Seymour and Kenefic 2002). The Multiaged Stocking Assessment Model (MASAM) (O'Hara 1996; O'Hara et al. 2003)

(a)

(b)

FIGURE 3.12
(a) Douglas-fir and western hemlock stand in the Pacific Northwest with one cohort that began 62 years before photograph. (b) Nearby stand that began at same time as (a), but overstory was reduced beginning at age 45 to ensure that the understory grows vigorously (overstory currently is 35 trees/ha.). (Copyright, Oliver, C.D., Reproduced in Stanturf, J.A. and Marsden, P., Eds., *Restoration of Boreal and Temperate Forests*, CRC Press, Boca Raton, 2005.)

was developed to permit land managers to evaluate the results of different growing space allocations in multiple-cohort stands.

The total stand productivity is probably similar for single-cohort and multiple-cohort stands, given equal total site occupancy and species composition. In ponderosa pine (*Pinus ponderosa* Lawson & C. Lawson) in the western United States, O'Hara and Nagel (2006) examined a number of physiological differences between single-cohort and multiple-cohort stands and concluded that the differences related to productivity were less than the operational differences between implementing these two types of systems.

The relationships in Figure 3.15 have practical significance in determining how much forest to maintain in the complex, or multiple-cohort structure in any given forest ecosystem type. If pattern A were occurring, stands in the complex structure would be growing the most volume per ha; therefore, it would take less forest area of stands in this structure to grow and harvest a targeted volume of timber sustainably. It may be appropriate to manage the landscapes within this ecosystem with proportionately more stands following

FIGURE 3.13
A multiple cohort stand in central Turkey, where trees invaded a former "savanna" structure when grazing was stopped (Soguksu National Park, Turkey). (Copyright, Oliver, C.D., Reproduced in Stanturf, J.A. and Marsden, P., Eds., *Restoration of Boreal and Temperate Forests*, CRC Press, Boca Raton, 2005.)

FIGURE 3.14
Multiple cohort stand in Mississippi River floodplains. Dominance of shade-tolerant species (Sugarberry; *Celtis laevigata* Willd.) and crooked stems result from the death of shade-intolerant species and frequent suppression and release of shade-tolerant trees. Suppression and release cause sugarberry trees to leave crooks in their stems when they are released by a partial cutting. (By contrast the dominant trees in Figures 3.2 and 3.9 grew without overstory shade and thus retained their straight stems.) (Copyright, Oliver, C.D., Reproduced in Stanturf, J.A. and Marsden, P., Eds., *Restoration of Boreal and Temperate Forests*, CRC Press, Boca Raton, 2005.)

multiple-cohort pathways to create this complex structure and greater volume. Conversely, if pattern I were occurring within a forest ecosystem type, it may be appropriate to manage more stands in single-cohort trajectories with more dense and understory structures to provide the forest ecosystem type's fair share of timber commodities, leaving fewer multiple-cohort stands, but enough to provide habitat and other values. It is possible that all the relationships depicted in Figure 3.15 exist in different forest types or that several can occur in a single forest type. Figure 3.15 demonstrates the various silvicultural pathways that can be taken to achieve similar objectives; there are a variety of forms of multiple-cohort structures that are sustainable, not just one.

All species decline in growth with increasing overstory shade (Wampler 1993; G, H, and I, Figure 3.15). Shade-tolerant species do not decline as much, however (Oliver et al. 2005; Figure 3.16). A very light overstory will allow all species to grow; however, even a slightly denser overstory will generally favor more shade-tolerant species (Oliver and Larson 1996). As the older cohorts regrow following a disturbance, they increasingly shade

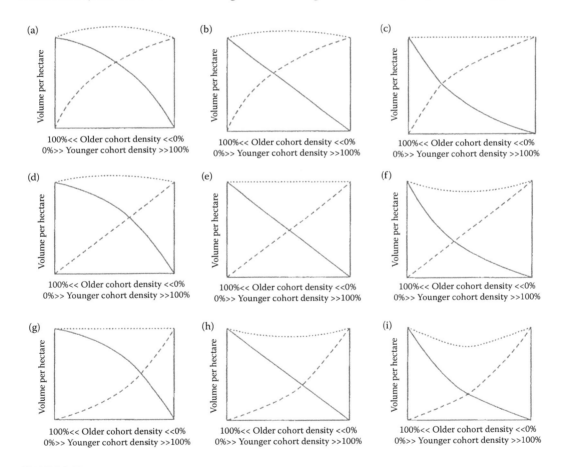

FIGURE 3.15
Possible theoretical relationships of older and younger cohorts in multiple-cohort stands. The relations of each cohort could be positive, neutral, or negative with respect to growth of the other cohorts, resulting in different growth rates for each cohort and the stand as a whole. (Solid line, older cohort; dashed line, younger cohort; dotted line, total stand.) (Reprinted from Stanturf, J.A. and Marsden, P., Eds., *Restoration of Boreal and Temperate Forests*, CRC Press, Boca Raton, 2005, with permission.)

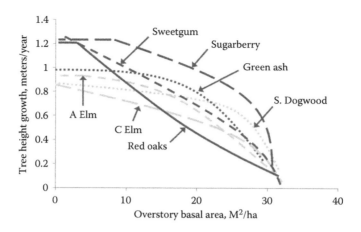

FIGURE 3.16
Maximum growth of many tree species relative to overstory density in the alluvial floodplains of the southeastern United States. (Red oaks, *Quercus pagoda* Raf., *Q. phellos* L., *Quercus nigra* L., *Q. texana* Buckley; sugarberry, *Celtis laevigata* Willd.; Green ash, *Fraxinus pennsylvanica* Marshall; S. (swamp) dogwood, *Cornus foemina* Mill.; A (American) elm, *Ulmus americana* L.; C (cedar) elm, *Ulmus crassifolia* Nutt.) (Reprinted from Oliver, C.D., Burkhardt, E.C. and Skojac, D.A., The increasing scarcity of red oaks in Mississippi River floodplain forests: Influence of the residual overstory, *Forest Ecology and Management*, 210, 393–414, 2005, with permission from Elsevier.)

the younger cohorts, causing all trees in the younger cohorts to slow in growth. More shade-tolerant species, however, will not slow as rapidly nor die as readily as more shade-intolerant species; consequently, there is generally a shift to the more shade-tolerant species (O'Hara 2014). Stands having characteristics described as old growth generally contain many shade-tolerant trees because these stands have endured one or more partial disturbances and their younger cohorts are prone to contain more shade-tolerant trees (Park 2001). The most successful mixed-species, multicohort stands will be those that include species with similar shade tolerances. The success with the Plenterwald in central Europe is aided by the shade tolerance of the three main species: Norway spruce (*Picea abies* [L.] Karst), European silver fir (*Abies alba* A.), and European beech (*Fagus sylvatica* L.). When restoring a landscape to a dynamic distribution of species, maintaining a large amount of complex forests will tend to promote trees of shade-tolerant species.

At increasing levels of overstory shade, nearly all tree species will slow in height growth and assume a flat-topped physiognomy characteristic of their species (Busgen and Munch 1929; Oliver and Larson 1996). Many conifer species will maintain a central stem but slow in height growth even more than they slow in lateral branch growth. The central stem of other conifer species and most deciduous species will also grow more slowly when shaded, but will also bend to a horizontal position and create a bush-like upper stem. Some species, such as beech (*Fagus* sp.), readily lose their upward terminal upon shading while others such as sugar maple can maintain the terminal under relatively intense shade. When released from overhead shade, all trees will resume height growth unless the initial shock of full sunlight kills them. Trees that had maintained their central vertical stem when in the understory will continue growing with a straight stem, but a dense group of branches will mark where they had been suppressed. Trees that have lost their central stem will usually have a crook, fork, or branch where the stem turned vertical again or one or two lateral branches had assumed the dominant position. For example, the stands in Figures and 3.9 and 3.14 grew on similar alluvial soils in Mississippi; however, the stand in Figure 3.9 began as a clear-cut and is dominated by straight oak trees, while the oaks were removed from the stand in Figure 3.14 and the remaining sugarberry trees are crooked because they had been suppressed.

Effective management of multiple-cohort stands for timber requires several considerations as follows:

- Multiple-cohort pathways are much easier to maintain if the species selected for management maintain a strong central stem when beneath high shade; the trees should also not readily develop woodrot if injured during partial logging operations.
- The overstory should be kept at a low enough density through thinnings to keep the younger cohorts alive and growing vigorously.
- The less vigorous trees in each cohort should be removed with each harvesting operation to prevent the result from being a high-graded stand (similar to Figure 3.11a), described earlier.
- The younger cohorts may need to be thinned to prevent overcrowding, shortening of crowns, and loss of vigor.
- Multiple-cohort mixtures of shade-tolerant and shade-intolerant species will tend to encourage tolerant species without purposeful attempts to favor the intolerant species.

Without extreme care, those stands comprised of species that readily become flattopped do not provide very high-quality timber in multiple-cohort stands because of their many stem crooks. Without this care, there will need to be a balance within a forest ecosystem between stands managed in a multiple-cohort manner and those managed otherwise to provide straight trees for timber and to maintain shade-intolerant species within the ecosystem.

In regions with subsistence rural conditions, selective forms of harvesting may be useful as a means of maintaining continuous forest cover on an area for some wildlife habitats, but at the same time obtaining some fuel wood. On the other hand, it is often quite difficult to prevent selectively harvested forests from being high-graded under such circumstances; and the income and employment values from timber would not be sustained if the forest were high-graded.

3.7 Coordinating Restoration of Many Stands

Many different stand structures, each of which has value for sustaining forests, can exist within a landscape. Therefore, instead of a single pathway or structure being the target of restoration, a combination of pathways and structures, the edges between the structures, and the timber harvested during the transition from one structure to another will need to be the target of restoration (Oliver 1992; Boyce 1995).

A landscape may contain continuous forests, forests fragmented by humans, or natural nonforest areas such as water bodies, agriculture fields, prairies, shrub lands, deserts, or urban areas. Stands dominated by the same tree species are usually found in similar topographic positions in many landscapes within a landform because of the similar edaphic, climatic, and land use patterns (Oliver et al. 2012). Discontinuous forests are generally found in similar topographic positions within a landform for similar reasons.

On a landscape of many stands, the species composition also depends on the pattern of disturbances and regrowth at the stand-level (Oliver 1980b; Harris 1984; Camp et al. 1997; Oliver et al. 1998). If trees in many stands on the landscape are knocked over in a single disturbance, the resulting landscape can be dominated by the open structure and by species such as butterflies and deer that benefit from this structure. As these stands regrow to the dense structure, the open species can be greatly reduced, survive in other landscapes, or become locally eliminated or extinct. If species depending on the understory or complex structures had survived the disturbance, they can expand as other forests regrow to this structure.

Some animal species depend on edges between structures, while others depend on interiors of open forests, complex forests, or other structures, away from edges (Hunter 1990). As a landscape changes with time, a boundary between two stands can appear less as a sharp edge between open (or savanna) and closed (dense, understory, or complex) structures (Figure 3.17) and more as a variable canopy within an interior closed forest.

Similar topographic positions in a landscape generally grow vegetation and are disturbed in similar ways, leading to similar stand structures and stand development pathways. Other emergent patterns of landscapes such as water availability, corridors and access, local weather cycles, and land uses are also dependent on the topography and geomorphologic origin (i.e., its landform; Oliver et al. 2012).

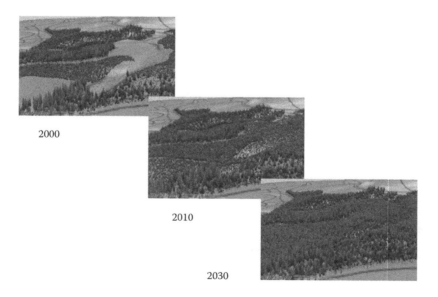

2000

2010

2030

FIGURE 3.17
A forest with many sharp edges between stands at one time (shown in 2000) can lose these sharp edges and become more hospitable to "interior closed canopy" species as the forest regrows. Variability in structure within an "interior closed" forest can be achieved by edges between component stands. Simulation and visualization is of University of Washington Pack Forest inventory and management. (Copyright, Oliver, C.D., Reproduced in Stanturf, J.A. and Marsden, P., Eds., *Restoration of Boreal and Temperate Forests*, CRC Press, Boca Raton, 2005.)

Sustaining plant species through the changing climate can also be addressed by ensuring that some stands of young plants are always present. These young plants are most adapted to the current climate, and consequently most likely to produce progeny of genotypes that are incrementally adapted to survive in future openings no matter which direction the climate changes (Oliver 2014b; Sample and Bixler 2014).

The challenge for restoration is not simply the placement of different stand structures across a landscape. These structures change with time through growth and disturbances; therefore, the changes will need to be coordinated so that some stands are developing into each structure as others are moving out of it (Figure 3.17).

Computer analytical tools have been developed that can help store and analyze the large amount of inventory, map, growth projection, and other data needed for coordination of stands across the landscape (e.g., McGaughey 1997; McCarter et al. 1998; Wilson and McGaughey 2000; Oliver et al. 2009; Khadka et al. 2014; see also: http://landscape-managementsystem.org). These management tools enable managers to follow modern management procedures of first designing an intended outcome and then identifying the necessary silvicultural pathways required to achieving it.

3.8 Paying for Forest Restoration

Restoration will need financial stimuli to get started in most forests (Björklund et al. 2009; von Weizsacker et al. 2009). Various trading mechanisms and incentives have been proposed to provide these stimuli (Peters et al. 1989; Gottfried 1992; Oliver and Lippke 1995; Gottfried et al. 1996; Lippke and Oliver 1993; Houghton and Mendelsohn 1996; Daily and Ellison 2002). In addition, a government policy could provide the stimuli, just as its policy of promoting railroads stimulated the economy in the mid-nineteenth century (Meinig, 1993) and the "GI Bill" (Doan 1996) promoted the housing industry following World War II.

Once restoration has begun, the activities of sustaining a diversity of structures will provide a flow of wood that can stimulate utilizing, manufacturing, and trading of wood—with each step of the value chain adding more jobs and more turnover (Dimand 1988). Greater use of wood for construction will provide the environmental service of avoiding greenhouse gas emissions and fossil fuel used in steel and concrete construction (Oliver et al. 2014a). Consequently, the costs of most stimuli could probably be offset by the increased economic activity to both forest-related communities and other people (Oliver 2014a,b).

References

Ashton, D.H., The Big Ash forest, Wallaby Creek, Victoria—changes during one lifetime, *Aust. J. Bot.*, 48, 1, 2000.

Bailey, R.G., Delineation of ecosystem regions, *Environ. Manage.*, 7, 365, 1983.

Björklund, G., Connor, R., Goujon, A., Hellmuth, M., Moriarty, P., Rast, W., Warner, K. and Winpenny, J., Chapter 2: Demographic, economic, and social drivers, in *World Water Assessment Programme*. The United Nations World Water Development Report 3: Water in a Changing World. UNESCO, Paris, Earthscan, London, pp. 29–40, 2009.

Botkin, D.B., *Discordant Harmonies: A New Ecology for the Twenty-First Century*, Oxford University Press, New York, 1990.

Bowling, D.R. and Kellison, R.C., Bottomland hardwood stand development after clearcutting, *South. J. Appl. For.*, 7, 110, 1983.

Boyce, S.G., *Landscape Forestry*, John Wiley & Sons, New York, 1995.

Boyce, S.G. and Oliver, C.D., The history of research in forest ecology and silviculture, in *Forest and Wildlife Science in America: A History*, Steen, H.K., Ed., Forest History Society, North Carolina, 414, 1999.

Braun, E.L., *Deciduous Forests of Eastern North America*, Macmillan, New York, 1950.

Brundtland, G.H., *Our Common Future*, World Commission on Environment and Development, United Nations, Oxford University Press, New York, 1987.

Burley, J., International initiatives for the sustainable management of forests, in *Forests and Landscapes: Linking Ecology, Sustainability, and Aesthetics*, Sheppard, S.R.J. and Harshaw, H.W., Eds., CABI Publishing in Association with The International Union of Forest Research Organizations, New York, p. 95, 2001.

Busgen, M. and Munch, E., *The Structure and Life of Forest Trees* (translated by T. Thompson), 3rd Ed., St. Giles' Works, Norwich, Great Britain, 1929.

Camp, A.E., Oliver, C.D., Hessburg, P. and Everett, R., Predicting late-successional fire refugia predating European settlement in the Wenatchee Mountains, *For. Ecol. Manage.*, 95, 63, 1997.

Camp, A.E. and C.D. Oliver, *Silviculture Forest Dynamics*, Chapter in Encyclopedia of Forest Science, Elsevier Science Ltd., Oxford, 2004.

Carey, A.B., Lippke, B.R. and Sessions, J., Intentional ecosystem management: Managing forests for biodiversity, *J. Sustain. For.*, 9, 83, 1999.

Chandler, P.M., Ecological knowledge in a traditional agroforest management system among peasants in China, Ph.D. Dissertation, University of Washington, Seattle, 161, 1990.

Clatterbuck, W.K. and Hodges, J.D., Development of cherrybark oak and sweetgum in mixed, even-aged bottomland stands in central Mississippi, U.S.A., *Can. J. For. Res.*, 18, 12, 1988.

Clatterbuck, W.K., Hodges, J.D. and Burkhardt, E.C., Cherrybark oak development in natural mixed oak-sweetgum stands-preliminary results, in *Proceedings of the Third Biennial Silvicultural Research Conference*, Shoulders, E., Ed., U.S. Department of Agriculture, Forest Service, Southern Research Station, General Technical Report SO–54, 438, 1985.

Clatterbuck, W.K., Oliver, C.D. and Burkhardt, E.C., The silvicultural potential of mixed stands of cherrybark oak and American sycamore: Spacing is the key, *South. J. Appl. For.*, 11, 158, 1987.

Cobb, D.F., O'Hara, K.L. and Oliver, C.D., Effects of variation in stand structure on development of mixed-species stands in eastern Washington, *Can. J. For. Res.*, 23, 545, 1993.

Daily, G.C. and Ellison, K., *The New Economy of Nature*, Island Press, Shearwater Books, New York, 2002.

Daniel, T.W., Helms, J.A. and Baker, F.S., *Principles of Silviculture*, 2nd Ed., McGraw-Hill, New York, 1979.

Deal, R.L., Oliver, C.D. and Bormann, B.T., Reconstruction of mixed hemlock-spruce stands in coastal southeast Alaska, *Can. J. For. Res.*, 21, 643, 1991.

Dimand, R.W., *The Origins of the Keynesian Revolution*, Stanford University Press, Stanford, CA, 1988.

Doan, M.C., *American Housing Production, 1880–2000: A Concise History*, University Press of America, Lanham, MD, 1996.

Drew, T.J. and Flewelling, J.W., Stand density management: an alternative approach and its application to Douglas-fir plantations, *Forest Sci.*, 25, 518, 1979.

Ford-Robertson, F.C., *Terminology of Forest Science, Technology, Practice, and Products*, Society of American Foresters, Washington, DC, 1971.

Franklin, J.F., Spies, T.A., van Pelt, R., Carey, A.B., Thornburgh, D.A., Berg, D.R., Lindenmayer, D.B. et al. Disturbances and structural development of natural forest ecosystems with silvicultural implications, using Douglas-fir forests as an example, *For. Ecol. Manage.* 155, 399–423, 2002.

Fry, M.E. and Money, N.R., Biodiversity conservation in the management of utility rights of way, in *Proceedings of the 15th Annual Forest Vegetation Management Conference*, Redding, CA, 25–27 January, 94, 1994.

Fujimori, T., *Ecological and Silvicultural Strategies for Sustainable Forest Management*, Elsevier, Amsterdam, 2001.

Gingrich, S.F., Measuring and evaluating stocking and stand density in upland hardwood forests in the Central States, *Forest Sci.*, 13, 38, 1967.

Gingrich, S.F., Management of upland hardwoods, U.S. Department of Agriculture, Forest Service, Research Paper NE-195, 26 pp., 1971.

Gottfried, R., The value of a watershed as a series of linked multi-product assets, *Ecol. Econ.*, 5, 145, 1992.

Gottfried, R., Wear, D. and Lee, R., Institutional solutions to market failure on the landscape scale, *Ecol. Econ.*, 18, 133, 1996.

Hanley, D.P., Oliver, C.D., Maguire, D.A., Briggs, D.B. and Fight, R.D., *Pruning Conifers in Northwestern North America: Opportunities*, Techniques and Impacts, University of Washington, Institute of Forest Resources, Contribution No. 77, 1995.

Harris, L.D., *The Fragmented Forest: Island Biogeography Theory and the Preservation of Biotic Diversity*, University Chicago Press, Illinois, 1984.

Helms, J.A., Ed., *The Dictionary of Forestry*, Society of American Foresters, Bethesda, MD, 1998.

Henry, J.D. and Swan, J.M.A., Reconstructing forest history from live and dead plant material-an approach to the study of forest succession in southwest New Hampshire, *Ecology*, 55, 772, 1974.

Houghton, K. and Mendelsohn, R., An economic analysis of multiple-use forestry in Nepal, *Ambio.*, 25, 156, 1996.

Hunter, M.L. Jr., *Wildlife, Forests and Forestry*, Regent/Prentice Hall, Englewood Cliffs, NJ, 1990.

Johnson, S., *Emergence: The Connected Lives of Ants, Brains, Cities, and Software*, Scribner, New York, 2002.

Kelty, M.J., The Development and Productivity of Hemlock-Hardwood Forests in Southern New England, Ph.D. Dissertation, Yale University, New Haven, CT, 1984.

Kelty, M.J., Development patterns in two hemlock-hardwood stands in southern New England, *Can. J. For. Res.*, 16, 885, 1986.

Kelty, M.J., Productivity of New England hemlock hardwood stands as affected by species composition and canopy structure, *For. Ecol. Manage.*, 28, 237, 1989.

Khadka, A., Chun Fu, Myint M., Oliver, C. and Saiers, J. Effects of land-cover changes and other remediations on hydrology of Xinjiang River sub-watershed using remote sensing and hydrologic modeling. *J. Environ. Sci. Eng.*, B 2, 416–425, 2014.

Kittredge, D.B., The influence of species composition on the growth of individual red oaks in mixed stands in southern New England, *Can. J. For. Res.*, 18, 1150, 1988.

Kollenberg, C.L. and O'Hara, K.L., Leaf area and tree increment dynamics of even-aged and multi-aged lodgepole pine stands in Montana, *Can. J. For. Res.*, 29, 687, 1999.

Larsen, J.B., Ecological stability of forests und sustainable silviculture, *For. Ecol. Manage.*, 75, 85, 1995.

Lippke, B. and Oliver, C.D., Managing for multiple values, *J. For.*, 91, 14, 1993.

McCarter, J.B. and Long, J.N., A lodgepole pine density management diagram, *West. J. Appl. For.*, 1, 6, 1986.

McCarter, J.M., Wilson, J.S., Baker, P.J., Moffett, J.L. and Oliver, C.D., Landscape management through integration of existing tools and emerging technologies, *J. For.*, 96, 17, 1998.

McGaughey, R.J., Visualizing forest stand dynamics using the Stand Visualization System, in *Proceedings ACSM/AS PRS* 4, 248, 1997.

McKinnon, F.S., Hyde, G.S. and Clinie, A.C., Cut-over Old Field White Pine Lands in Central New England, The Harvard Forest, Harvard University, Harvard Forest Bulletin No. 18, 1935.

Meinig, D.W., *The Shaping of America: A Geographical Perspective on 500 Years of History*, Volume 2: Continental America, 1800–1867, Yale University Press, New Haven, 1993.

Miller, E.W., Chapter 1: Forest regions of the world, in *Forests: A Global Perspective*, Majumdar, S.K., Miller, E.W. and Brenner, F.J., Eds., The Pennsylvania Academy of Science, PA, 1996.

Newton, M.B., El Hassan, B.A. and Zavitkovski, J., Role of red alder in western Oregon forest succession, in *Biology of an Alder Trappe*, Franklin, J.M., Tarrant, R.F. and Hansen, G.M., Eds., U.S. Department of Agriculture, Forest Service, Pacific Northwest Forest and Range Experiment Station, Portland, OR, 73, 1968.

Nyland, R.D., *Silviculture: Concepts and Applications*, McGraw-Hill, New York, 1996.

O'Hara, K.L., Development patterns of residual oaks and oak and yellow-poplar regeneration after release in upland hardwood stands, *South. J. Appl. For.*, 10, 244, 1986.

O'Hara, K.L., Dynamics and stocking-level relationships of multi-aged ponderosa pine stands, *Forest Sci. Monogr.*, 33, 1–34, 1996.

O'Hara, K.L., Silviculture for structural diversity: A new look at multi-aged systems, *J. For.*, 96, 4, 1998.

O'Hara, K.L., The historical development of uneven-aged silviculture in North America, *Forestry*, 75, 339, 2002.

O'Hara, K.L., *Multiaged Silviculture: Managing for Complex Stand Structures*, Oxford Univ. Press, Oxford, UK, 2014.

O'Hara, K.L., Lahde, E., Laiho, O., Norokorpin, Y. and Saksa, T., Leaf area and tree increment dynamics on a fertile mixed-conifer site in southern Finland, *Ann. For. Sci.*, 56, 237, 1999.

O'Hara, K.L., Latham, P.L., Hessburg, P. and Smith, B.G., A structural classification for inland northwest forest vegetation, *West. J. Appl. For.*, 11, 97, 1996.

O'Hara, K.L. and L.M. Nagel., A functional comparison of productivity in even-aged and multi-aged stands: A synthesis for *Pinus ponderosa*, *For. Sci.*, 52, 290, 2006.

O'Hara, K.L. and Oliver, C.D., A decision system for assessing stand differentiation potential and prioritizing precommercial thinning treatments, *West. J. Appl. For.*, 14, 7, 1999.

O'Hara, K.L., Valappil, N.I. and Nagel, L.M., Stocking control procedures for multi-aged ponderosa pine stands in the Inland Northwest, *West. J. Appl. For.*, 18, 5, 2003.

Odum, E.P., *Fundamentals of Ecology*, 3rd ed., Saunders, Philadelphia, 1971.

Oliver, C.D., Development of Northern Red Oak in Mixed-species Stands in Central New England, Yale University School of Forestry and Environmental Studies Bulletin No. 91, 1978.

Oliver, C.D., Even-aged development of mixed-species stands, *J. For.*, 78, 201, 1980a.

Oliver, C.D., Forest development in North America following major disturbances, *For. Ecol. Manage.*, 3, 153, 1980b.

Oliver, C.D., A landscape approach: Achieving and maintaining biodiversity and economic productivity, *J. For.*, 90, 20, 1992.

Oliver, C.D., Sustainable forestry. What is it? How do we achieve if? *J. For.*, 101, 8, 2003.

Oliver, C.D., Forest management and restoration at temporal and spatial scales, *J. Sustainable. Forest.*, 33, S123–S148, 2014a.

Oliver, C.D., Mitigating Anthropocene influences in forests in the United States, in *Forest Conservation and Management in the Anthropocene*, Sample, V.A. and Bixler, R.P., Eds., General Technical Report. US Department of Agriculture, Forest Service, Rocky Mountain Research Station, Fort Collins, Colorado, 2014b.

Oliver, C., Adams, D., Bonnicksen, T., Bowyer, J., Cubbage, E, Sampson, N., Schlarbaum, S., Whaley, R., Wiant, H. and Sebelius, J., Report on Forest Health of the United States by the Forest Health Science Panel, Panel chartered by Charles Taylor, Member, United States Congress, 11th District, North Carolina. (Available through U.S. House of Representatives Resources Committee at www. house.gov/resources/105cong/fullconun/apr09.97/taylor.rpt/taylor.htm) Also available as reprint through University of Washington College of Forest Resources CINTRAFOR RE43, 1997.

Oliver, C.D, Burkhardt, E.C. and Skojac, D.A., The increasing scarcity of red oaks in Mississippi River floodplain forests: Influence of the residual overstory, *Forest Ecol. Manage.*, 210, 393–414, 2005.

Oliver, C.D., Burley, J. and Maathai, W., Sustainable Forestry: What is it? How do we achieve it? In Synopsis of May 3, 2002 Seminar, Global Institute or Sustainable Forestry; School of Forestry and Environmental Studies, Yale University, New Haven, CT, 2002.

Oliver, C.D., Camp, A. and Osawa, A., Forest dynamics and resulting animal and plant population changes at the stand and landscape level, *J. Sustain. For.*, 6(3/4), 281, 1998.

Oliver, C.D., Covey, K., Larsen, D., Wilson, J., Hohl, A., Niccolai, A. and McCarter, J.B., Landscape management. Chapter 3, in *Forest Landscape Restoration: Integrating Natural and Social Sciences*, Stanturf, J., Lamb, D. and Madsen, P., Eds., Springer Publishing, New York, pp. 39–65, 2012.

Oliver, C.D. and Stephens, E.P., Reconstruction of a mixed species forest in central New England, *Ecology*, 58, 562, 1977.

Oliver, C.D. and Larson, B.C., *Forest Stand Dynamics*, Updated Edition, John Wiley & Sons, New York, 1996.

Oliver, C.D. and Lippke, B.R., Wood supply and other values and ecosystem management, in western interior forests, in *Ecosystem Management in Western Interior Forests*, Symposium held May 13–15, 1994, Spokane, Washington, Department of Natural Resource Sciences, Washington State University, Pullman, WA, 195, 1995.

Oliver, C.D., McCarter, J.B., Ceder, K., Nelson, C.S. and Comnick, J.M., Simulating landscape change using the Landscape Management System, in *Models for Planning Wildlife Conservation in Large Landscapes*, Millspaugh, J.J. and Thompson, F.R., Elsevier & Academic Press, New York, 339–366, 2009.

Oliver, C.D., Nassar, N.L., Lippke, B.R. and McCarter, J.B., Carbon, fossil fuel, and biodiversity mitigation with wood and forests, *J. Sustainable Forest.*, 33, 248–275, 2014a.

Oliver, C.D., Oliver, F.A.,Yonavjak, L., Saxena, A., Zeng, L. and Zeydanli, U., The transition from timber to multiple value management: A paradigm shift in forestry, Book chapter in *Sustainable Forest management for Multiple Values: A paradigm Shift*, Bhojvaid, P.P. and Khandekar, N., Eds., The Forest Research Institute, Dehradun, India, 2014b.

Park, P.S., Forest Stand Structure Characteristics for the Cispus Adaptive Management Area, Cascade Range, U.S.A.: Implications for Old Growth, Fire Hazard, Silviculture, and Landscape Management, Ph,D. dissertation, University of Washington, Seattle, WA, 2001.

Peterken, G.F., *Natural Woodland: Ecology and Conservation in Northern Temperate Regions*, Cambridge University Press, New York, 1996.

Peters, C., Gentry, A. and Mendelsohn, R., Valuation of an Amazonian rain forest, *Nature*, 339, 655, 1989.

Phillips, C., Identifying problems and challenges associated with private forest lands-overview in Summit 2000: Washington Private Forests Forum, University of Washington, College of Forest Resources, Institute of Forest Resources, Summer 2000.

Raup, H.M., Some problems with ecological theory and their relation to conservation, *J. Ecol.*, 52, 19, 1964.

Reineke, L.H., Perfecting a stand-density index for even-aged forests, *J. Ag. Res.*, 46, 627, 1933.

Sample, V.A. and Bixler, R.P. Eds., *Forest Conservation and Management in the Anthropocene*, Gen. Tech. Report. Fort Collins, CO, USDA Forest Service, Rocky Mountain Research Station, 2014.

Sampson, R.N. and Adams, D.L. Eds., Assessing forest ecosystem health in the Inland West, in *Proceedings of the American Forest Workshop*, Sun Valley, ID, The Haworth Press, Inc, New York, 1994.

Seymour, R.S. and Hunter, M.L., Principles of ecological forestry, in *Maintaining Biodiversity in Forest Ecosystems*, Hunter, M.L., Ed., Cambridge University Press, Cambridge, 1999.

Seymour, R.S. and Kenefic, L.S., Influence of age on growth efficiency of *Thuga canadensis* and *Picea rubens* trees in mixed-species, multi-aged northern conifer stands, *Can. J. For. Res.*, 32, 2032, 2002.

Simkin, R. and Baker, P.J., Disturbance history and stand dynamics in a tall open forest and riparian rainforest in the Central Highlands of Victoria, *Aust. Ecol.*, 33, 747, 2008.

Smith, D.M., Larson, B.C., Kelty, M.J. and Ashton, P.M.S., *The Practice of Silviculture*, John Wiley & Sons, New York, 1997.

Stanturf, J. A. and Madsen, P., Restoration concepts for temperate and boreal forests of North America and Western Europe, *Plant Biosystems*, 136(2), 143–158, 2002.

Stanturf, J.A., Lamb, D. and Madsen, P., Eds., *Forest Landscape Restoration: Integrating Natural and Social Sciences*, Springer Publishing, New York, 2012.

Stanturf, J.A., Palik, B.J. and Dumroese, R.K., Contemporary forest restoration: A review emphasizing function, *For. Ecol. Manage.*, 331, 292–321, 2014a.

Stanturf, J.A., Palik, B.J., Williams, M.I., Dumroese, R.K. and Madsen, P., Forest restoration paradigms, *J. Sustainable Forest.*, 33(sup1), S161–S194, 2014b.

Stevens, W.K., New eye on nature: The real constant is eternal turmoil, New York Times, Science article, Tuesday, July 31, B5–B6, 1990.

Stubblefield, G.W. and Oliver, C.D., Silvicultural implications of the reconstruction of mixed alder/ conifer stands, in *Utilization and Management of Red Alder*, Atkinson, W.A., Briggs, D. and DeBell, D.S., Eds., U.S. Department of Agriculture, Forest Service, General Technical Report PNW-70, 307, 1978.

Tallis, J.H., *Plant Community History: Long-term Changes in Plant Distribution and Diversity*, Chapman & Hall, New York, 1991.

Tansley, A.G., The use and abuse of vegetational concepts and terms, *Ecology*, 16, 284, 1935.

Toumey, J.W., *Foundations of Silviculture, Upon an Ecological Basis*, 2nd ed., John Wiley & Sons, New York, 1947.

von Weizsacker, E., Hargroves, K.C., Smith, M.H., Desha, C. and Stasinopoulos, P., Factor five: Transforming the global economy through 80% improvements in resource productivity. A Club of Rome Report. Earthscan, London, 2009.

Walkinshaw, L.H., *Kirtland's Warbler, Cranbrook Institute of Science*, Bloomfield Hills, MI, 1983.

Wampler, M., Growth of Douglas-fir under Partial 0verstory Retention, Master's thesis, University of Washington College of Forest Resources, 1993.

Whittaker, R.H., *Communities and Ecosystems*, 2nd ed., Macmillan Publishing Company, New York, 1975.

Wierman, C.A and Oliver, C.D., Crown stratification by species in even-aged mixed stands of Douglas-fir/western hemlock, *Can. J. For. Res.*, 9, 1, 1979.

Wilson, J.S. and Baker, P.J. Flexibility in forest management: Managing uncertainty in Douglas-fir forests of the Pacific Northwest, *For. Ecol. Manage.*, 145, 219, 2001.

Wilson, J.S. and McGaughey, R., Presenting landscape-scale forest information: What is sufficient and what is appropriate? *J. For.*, 98, 21, 2000.

Wilson, J.S. and Oliver, C.D., Stability and density management in Douglas-fir plantations, *Can. J. For.* Res., 30, 910, 2000.

Wonn, H.T. and O'Hara, K.L., Height:diameter ratios and stability relationships for four northern Rocky Mountain tree species, *West. J. Appl. For.*, 16, 87, 2001.

Young, M.R., Conserving insect communities in mixed woodlands, in *The Ecology of Mixed-species Stands of Trees*, Cannell, M.C.R., Malcolm, D.C. and Robertson, P.A. , Eds., Blackwood Scientific Publications, London, 277, 1992.

Zebrowski, Jr. E., *Death and Life, in Perils of a Restless Planet: Scientific Perspectives on Natural Disasters*, Cambridge University Press, New York, Ch. 4, 93, 1997.

4

Integrating Ungulate Herbivory into Forest Landscape Restoration

Thomas P. Rooney, Rita Buttenschøn, Palle Madsen, Carsten
Riis Olesen, Alejandro A. Royo, and Susan L. Stout

CONTENTS

4.1 Introduction

It is widely appreciated that forest composition, structure, and function are determined by biogeographic history, climate and soils, and are further modified by past land use and management history, natural disturbance regimes, invasive species, and pathogens. Ungulate herbivores also exert profound changes on forest systems. In recent decades, reports have highlighted the role of hyper abundant ungulate herbivores in altering forested landscapes (Gill 1992; Ammer 1996; Healy 1997; McShea et al. 1997; Horsley et al. 2003; Rooney and Waller 2003; Côté et al. 2004). There are cases where ungulate herbivory changed nutrient cycling (Pastor and Cohen 1997) or gave rise to indirect effects on wildlife communities (deCalesta 1994; Augustine and deCalesta 2003; Martin et al. 2011; Nuttle et al. 2011). In contrast, controlled grazing by large—primarily domestic—grazers is one of the most important management tools in European conservation management (Hester et al. 2000). Controlled grazing is sometimes aimed at maintaining or reestablishing the structure and composition of traditional agro-forest systems, such as the garrigue of the Mediterranean region, the intensive wood pasture systems of the temperate forest zone, and the graminoid-dominated tundra shaped by grazing of semidomesticated reindeer of North-Scandinavia (Bergmeier et al. 2010). In these communities, both the insufficient and too much grazing pressure (e.g., Bråthen et al. 2007) can threaten conservation or management objectives. The use of grazers in conservation management is based on both the direct experiences of practitioners and careful scientific research that demonstrates an increase in biological diversity or forest regeneration in response to increased grazing, up to some optimal level (Mitchell and Kirby 1990; Wallis deVries 1998; Ristau and Horsley 1999; Buttenschøn and Buttenschøn 2013). The role of ungulate herbivores in forest ecosystems is highly complex, as they can have both detrimental and beneficial effects.

When planning forest restoration, it is useful to know whether changes in the types and numbers of ungulate herbivores will be beneficial or detrimental and in what ways. Important factors include the ungulate species present and their diet breadth. Are they highly selective "browsers" or less selective "grazers" (Duncan and Poppi 2008)? Ungulate herbivores can be classified along a continuum between browsers and grazers, according to whether they eat primarily forbs and woody browse or grasses (Hofmann 1989; Searle and Shipley 2008). The feeding behavior among present grazers and browsers differs in part due to the differences in morphology of the digestive system (Sidebar 4.1). They also differ in behavior in terms of the relationship between resource abundance and intake rate (Janis 2008). Despite both eating plants, grazers and browsers have qualitatively different effects on forest ecosystems, and this has important implications for restoration. For example, a restoration plan which adds browsers to a system that actually depends on grazers could fail; browsers and grazers are not functionally equivalent. Additionally, it is important to recognize that grazers and browsers interact with site conditions, land use and management history, and *management goals*. The interaction between ungulate herbivores and management goals is particularly challenging, as it speaks to Bradshaw's (2005) question: "What is a natural forest?" When we adopt a broader time perspective which includes the expanding human influence and the changing impacts of browsers and grazers since the beginning of the Holocene, a satisfactory answer proves elusive. Additionally, it is relevant to consider ungulate effects over longer periods of time when identifying goals of forest and landscape management, as done by Bradshaw and Mitchell (1999). They assess and view the present interactions between vegetation and ungulates at three temporal scales: hundreds, thousands and millions of years. This approach seems highly relevant as a frame and is the key to a better understanding not only of today's management challenges and possible solutions, but also to inform the discussions about what are desirable management goals in a longer-term sustainability context. As pointed out by Stanturf et al. (2014) the "Ending Point" or the management goal for forest landscape restoration efforts—of which conservation management by grazers is one of the commonly used methods—is a "social construct." It is defined by social values and preferences referring to what is viewed as "natural," a "natural forest" or a "natural habitat," or indeed a "cultural landscape" which can also be a strong driver for a "restoration" effort. In North America, the restoration goals are often identified as "natural forest ecosystems" lacking or with very little human influence (Stanturf et al. 2014). In Europe, many of the protected habitats are open habitats with no or low forest canopy density, with some originating from historic land use, the so-called "cultural landscapes" (e.g., Convery and Dutson 2012; Newton 2013).

Integrated forest landscape restoration—a process that aims to regain ecological integrity and enhance human well being in deforested or degraded forest landscapes (Maginnis and Jackson 2007)—offers a promising approach for effectively integrating ungulate herbivores in forest ecosystems. Armed with an understanding of the site history and disturbance regimes, the coevolution of forest plants and herbivores, and their complex interactions and feedbacks, landscape restoration practitioners can explore the interrelations of the system holistically. This perspective can broaden the view of landowners and managers beyond a narrow focus on immediate and seemingly unconnected problems. Instead, the informed landowners, practitioners, and managers ask deeper questions. How land-use change, climatic change, invasive species, and evolution of social attitudes have created the context for the current situation? Which components, or elements of the system should receive primary attention in a restoration effort, and what secondary and tertiary results may arise as these elements are rearranged and rebalanced? Those with a stake in a single element of the system can begin to understand their interests in a broader, holistic context, and better understand the trade-offs which restoration may require.

Our objective in this chapter is to provide historical context, principles, and practices for the use in forest landscape restoration efforts where ungulate herbivory is important. We describe the history and ecological context of forest–ungulate systems, because an understanding of this history and context can inform targets for contemporary restoration efforts. We then highlight principles that can inform forest landscape restoration, goal-setting, approaches, management systems, and methods in systems where herbivores are important to landowner and societal goals.

4.2 The Historical and Ecological Context of Forest Ungulates

When addressing management goals and challenges and trying to solve problems in our existing forest landscapes and urban interfaces with forests, we tend to forget that the species-animals, trees, shrubs, or herbs carry a biological history of adaptation to meet the challenges of an ever-changing environment. Ungulates are often seen as having damaging effects on forest ecosystems: they browse desirable artificial and natural regeneration, strip bark, alter floristic diversity, change soil properties, and reduce populations of other animals in the community (Gill 1992; Côté 2004; Ammer 1996; Rooney 2001; Horsley et al. 2003; Olesen and Madsen 2008). Even with a less value-laden view, ungulates are generally viewed as a significant disturbance factor, as they browse, graze, rub, trample, or otherwise damage some of the trees or herbs that we try to promote through our forest landscape restoration practices. However, it is useful to remember that many of the species we manage in our present ecosystems are adapted to some degree of browsing and grazing (Bradshaw and Mitchell 1999; Vera 2000; Bakker et al. 2004). Indeed, seedling establishment can be significantly increased by the presence of ungulates creating regeneration niches or reducing, for example, broadleaved regeneration which competes with desired species in forest regenerations. Thus, ungulate disturbance can pose opportunities, not just obstacles, when developing management goals, both for conservation management and for productive and profitable forestry.

Through most of their history, the majority of our tree and shrub species may have depended on ungulates to create and maintain open habitats, create regeneration niches, and provide seed dispersal (Vera 2000). Whether this happened at large or small spatial scales, or perhaps in a shifting vegetation mosaic, remains intensively debated (Vera 2000; Bradshaw et al. 2003; Mitchell 2005). Likewise, most contemporary tree, shrub-, and plant-species have lived and survived through the dramatically shifting landscapes of the Quaternary (Huntley 1990; Williams et al. 2004) with a much more diverse ungulate fauna. Furthermore, from 2.5 million years ago until very recently, spectacular megafaunal species like rhino, auroch, bison, camel, hippopotamus, and giant deer occupied Holarctic forest and shrub-steppe. The shifting role and impact of ungulates and other megaherbivores up through the Pleistocene and finally the Holocene is worth considering when evaluating present interactions between ungulates and vegetation. This is not because Pleistocene ecosystems provide a reference condition for restoration. Rather, nearly all of our extant trees and shrubs coevolved with large herbivores, and in landscapes with browsing and grazing-associated disturbances that were more complex than they are today.

The anthropogenic impact on Holarctic landscapes started tens of thousands of years ago, and has increased ever since. Initially, the primary human impact was imposed by hunting and the use of fire. Later, land clearing for agriculture led to the transformative impact we see in our modern world. Throughout history, landscapes have been modified and degraded, often severely. Novel ecosystems have been created. These cultural landscapes include North American preEuropean settlement landscapes maintained by fire, and the cultural heathland landscapes in Europe. Heathland landscapes were primarily managed in part to support wild and/or domestic ungulates. On these lands today, land restoration is often directed to maintain or recreate these cultural landscapes and the unique biodiversity they contained. A key management challenge involves reconstructing the historic role of ungulates in forming and maintaining the cultural landscapes, and determining how this role might be mimicked with management options (including ungulate species) available today.

There is an extensive literature that ignores long-term forest-grazing interactions and instead focuses on the short-term negative effects of ungulates on forests (reviewed in Rooney and Waller 2003; Côté et al. 2004). The short-term fate of vegetation in response to ungulates often dictates the benchmarking restoration goal, and this can lead to an inappropriate short-term evaluation of management success. We caution against this. Ungulate grazing modulates the occurrence, growth, and development of vegetation over broader time scales than we are accustomed to think about. The challenge lies in developing restoration goals which reflect long-term forest–ungulate interactions as far as possible. For example, we can easily collect the field data needed to help us in answering a question such as "do the desired trees and shrubs grow and establish well within a short time period?" The inference is that if they are not establishing within a few years, something is wrong, and intervention is needed. However, what if the absence of a broader perspective leads us to desire and restore the wrong trees and shrubs? Or what if establishment takes more than a few years—maybe even decades? Species dependent on regular browsing, grazing, wallowing, or other ungulate activity might even decline in the short term but persist over longer time scales. Oaks (*Quercus*), rowan (*Sorbus*), apple (*Malus*) and others may appear heavily browsed, and still depend on ungulate disturbance to keep the canopy open enough for persistence. This would necessitate a shifting landscape mosaic of grassland, scrub, open woodland, and forest, based on Vera's (2000) hypothesis of open forest. Kirby (2004) provides a landscape model showing how this can be achieved.

4.3 Contemporary Role of Ungulates in the Landscape

Both grazing and browsing ungulate strongly influence forests through forage uptake. This can vary 100-fold across terrestrial ecosystems, with uptake of less than 1% to over 60% of the accessible forage (McNaughton et al. 1989). Forage selection is determined by both individual and species preferences from the available vegetation, as well as their location and apparency. Some plant species are highly sought, other species are eaten in bulk, and still others are avoided entirely, unless no other forage is available. The variable selectivity of ungulate foraging, coupled with differential plant responses to herbivory, leads to a long-term vegetation change (Weisberg and Bugmann 2003). Plants have developed various strategies which can reduce the negative effect of herbivory on plant fitness (Rosenthal and Kotanen 1994). Herbivore resistance strategies include structural and chemical defenses. Herbivore tolerance strategies include high growth rates and/or compensatory responses of growth to offset harmful effect tissue loss (Skarpe and Hester 2008). Furthermore, an individual plant's tolerance to being eaten depends on its phenological stage, and thus varies with season as well as age. Over time, grazing will change successional pathways (Hidding et al. 2013). For example, populations of some grazing intolerant species may succumb to disturbance, while grazing tolerant species or plant species the animals avoid become more predominant. Ultimately, the effect of ungulate feeding on forest ecosystems depends on many factors, including but not limited to: the differential vegetation removal rate, the timing/season of feeding, the species of ungulates present, and the interaction among ungulate species (van de Koppel and Prins 1998).

Grazing and browsing ungulate feeding also have indirect effects on forests. Foraging can reduce woodland understory and ground-layer biomass (Riggs et al. 2000). This reduction in foliar biomass increases light availability, and changes the physical vegetation structure in the horizon within the reach of the animals. Trampling or heavy feeding can create patches of bare soil, which may serve as seedbeds for light demanding plant species (Collins and Good 1987; Bakker et al. 2004). Most of the foliar nitrogen (frequently, i.e. >90%) consumed by ungulates is returned to the habitat through urination or defecation (Murray et al. 2014). Nitrogen in urine and feces is highly labile, thus readily available for plant growth. This can act to accelerate nutrient cycling (Stark et al. 2000), provided that ungulates are not favoring plant species that produce recalcitrant litter in those environments (Pastor and Naiman 1992). Most ungulates have nonrandom excreta deposition patterns (Murray et al. 2014), and over time, hotspots of soil nitrogen availability may occur in the landscape following that pattern. Nonfeeding activities of ungulates can also be important. For example, trampling is often detrimental to disturbance-sensitive plant species but do also create microhabitat heterogeneity and altered spatial distribution and abundance of other plant populations. Additionally, ungulates may also serve as seed dispersal vectors (Bartuszevige and Endress 2008; Dennis and Westcott 2007).

Throughout temperate forest regions, there is a mix of browsers and mix-feeders, but comparatively few species of grazers. However, because domesticated ungulates (cattle, horses, and sheep) are grazers, they are of disproportionate importance in many areas of the world. Through time, the relative influence of grazers and browsers on the landscape has changed. The domestication of animals, enclosing of pastureland, and intensification of animal husbandry have ensured for centuries that grazing was a primary ungulate influence in large parts of the world. Today, the numbers and influence of grazers are declining in North America and Europe (Luoto et al. 2003), while the numbers and importance of browsers are increasing (Côté et al. 2004; Ward 2005). The recent high densities of ungulate browsers in much of Europe and North America are accordingly considered to be the major cause of large-scale shifts in tree species composition, population structure and dynamics observed in many areas (Côté et al. 2004; Skarpe and Hester 2008). This type of shift in the relative abundance of grazers and browsers has important implications for restoration goals.

SIDEBAR 4.1   Ecological Adaptation of Ungulate Digestive Physiology

Some ungulates, such as horses, have a single stomach. Ruminant ungulates, including deer, bison, and domestic cattle, sheep and goats, have a specialized four-chambered stomach. Mammals with this specialized digestive system regurgitate and chew their food more than once, and are called ruminants. Ruminants are able to digest a major part of the structural carbohydrates (fibers) of plants, through fermentation by bacteria and other single-celled organisms held in the rumen–reticulum. These endosymbionts synthesize the enzymes needed to degrade cellulose. The digestive system of ruminants provides two key advantages, one being the ability of efficient utilization of plant fibers. Another is the ability to fill the rumen quickly when feeding on open land, and to move and ruminate in areas with more cover. This ability to continue feeding in an area less exposed may be an adaptation to reduce predation risk.

Compared to the one stomach digestive system, there are both advantages and disadvantages. In both digestive systems, the quality of the food determines the rate of passage. When food quality is low (the percentage of foliar fiber and lignin is high), passage time will decrease in the one stomach system but increase in the other (ruminant system). Consequently, ruminants are only able to improve their energy gain by searching for and acquiring high-quality food, while the other stomach digesters compensate for low food quality by increasing food intake.

Ruminants are adapted to an environment containing variety of plant foods, primarily those of intermediate quality. They specialize neither on highly digestible foods (fruit, seeds, or berries) nor on the least digestible food (coarse grasses and tree fiber). However, not all species ruminants share an identical digestive anatomy and functionality. The cost of adapting the digestive system to a broad range of plant material is ultimately determined by metabolic energetic requirements. The energetic requirement of a mammal is set by the specific surface to body weight ratio or metabolic weight. Small mammals have large relative surface areas and higher energetic demands per unit of body mass, compared with animals of increasing body size. Accordingly, smaller herbivores with high energetic demands select a diet of high digestibility and protein content, such as fresh shoots, butts, seeds, and fruits. Some larger herbivores, such as red deer, are capable of surviving on a diet of lower digestibility and are often classified as intermediate or typical grazers. Hofmann (1989) provides a detailed treatment of the digestive ecophysiology of ruminants.

4.4 Six Principles of Restoration of Landscapes with Ungulates

We hope that this brief overview of the history and dynamics of ungulates and forest landscape systems gives an indication of how interdependent these systems are. Landscape restoration often focuses on reintroduction, renewal, or persistence of key elements, from plants through ecosystem services. The influence of ungulate herbivores on key elements of forest landscape systems must be considered in restoration goal setting, planning, implementation, and monitoring. What principles will help managers and landowners or restoration teams in establishing goals and identify practices based on this integrated landscape approach?

4.4.1 Consider all Elements of the System

The first principle is to consider all elements of this system when developing restoration targets. We must consider not just the landscape and ungulates, but how these interact with each other and with other system elements through time. This is more challenging than it appears. For example, a number of questions are relevant if we focus our attention on ungulates and the landscape. For example, which resources does the landscape offer to ungulates? Are there multiple ungulate species in the system, and/or are the numbers changing through time? If so, what are their dietary niches (are they browsers or grazers)? How do they overlap or compete? Does the mix of ungulates include domestic and wild animals? Are there predators in the system, be they human hunters or wild predators? What elements of the system do these predators depend upon? How do ungulates affect the ecosystem, especially regeneration, species diversity, and ecosystem services? How do the various landscape elements affect the ungulates? Are there both farm-crop land and more "natural" or seminatural components such as forests, pastures, heathland? The former may serve as highly attractive feeding areas, for example, deer or wild boar in the growing season and influence the habitat and feeding preferences of these highly mobile ungulate species. What are the results for restoration goals when ungulate density is "too low" or "too high?" Indeed, how are such issues incorporated and defined in a meaningful way in relation to the management of ungulates—both native and nonnative ungulate species in landscapes including highly artificial components like modern farm crops as well as the more "natural" elements? What other factors influence the sustainability of those ecosystem elements of interest to society, landowners, or the restoration team?

The answers to these questions are not fixed. The number of ungulate species in the landscape might change through time. Predators can appear and disappear. A landscape element that is not important to ungulates in most years can become a critical resource, for example, during a drought year, or following a forest fire. The same density of ungulates might be too low to achieve restoration goals in a wet year, and too high in a dry year. As we broaden our questions to include more system elements, we gain a deeper appreciation of the challenges of landscape restoration.

4.4.2 Understand How Restoration Might Change Landscape Carrying Capacity

The second principle is to understand and think about carrying capacity of the landscape and how restoration might alter carrying capacity (Sidebar 4.2). The effect that ungulates have on a forested landscape varies not so much by the absolute density of ungulates, but by the density relative to the carrying capacity of the landscape (deCalesta and Stout 1997). Because ungulates can persist for long periods of time on a landscape at levels that challenge sustainability of other resources (Horsley et al. 2003); stakeholders may sometimes underestimate both the impact of ungulates on other ecosystem elements, and on the ungulates themselves. Carrying capacity for ungulates changes through time, and restoration activities may further increase or decrease the carrying capacity. This in turn changes the impact that ungulates have on the landscape. The landscape changes feed back into the ungulate populations, affecting vital rates (McCullough 1979). This can be seen in a landscape-level restoration project in the forests of eastern North America (Stout et al. 2013), for example. This project did not reduce carrying capacity per se, but instead sought to reduce deer numbers relative to the carrying capacity. Over 10 years, habitat quality improved, as did forest regeneration. Improved habitat quality fed back to the ungulate population, resulting in improved growth and trophy scores of those ungulates.

SIDEBAR 4.2 Integrating Ungulate Density and Landscape Heterogeneity

Ecological theory argues that browsing impacts are mediated via scale-dependent foraging decisions influenced by habitat structure, resource availability, and predation risk (Senft et al. 1987; Hobbs 2003). Under this framework, linking ungulate densities to heterogeneity in forest habitats and structure (e.g., higher forage quantity/quality; greater edge density, higher cover; see: Cadenasso and Pickett 2000; Kie et al. 2002; Plante et al. 2004; Forester et al. 2008) at the scale of the resident ungulate population's home range will predict browse impact on vegetation better than ungulate densities alone. Recognition of the importance of landscape structure affecting foraging has led to at least two seemingly differing views regarding the relative influence of ungulate densities and habitat heterogeneity in determining ungulate foraging impacts. One view posits that ungulate populations typically respond positively to increased heterogeneity thereby resulting in increased foraging impact on vegetation, particularly on browse-sensitive species (Alverson et al. 1988; Augustine and Jordan 1998; Kie et al. 2002; Reimoser et al. 2009). From a restoration perspective, this viewpoint suggests that land-use decisions that minimize forest habitat heterogeneity, such as maintaining large, contiguous forested blocks or low-intensity silvicultural techniques (e.g., single-tree selection), may diminish browsing impact on vegetation (Alverson et al. 1988; Nelson and Halpern 2005; Kramer et al. 2006). Alternatively, others conjecture that despite the positive effects of habitat heterogeneity on ungulate populations, variability in habitats at larger scales is critical in modulating ungulate foraging impacts at a local scale (Porter 1992; Johnson et al. 1995; deCalesta and Stout 1997; Takada et al. 2002; Augustine and DeCalesta 2003). This view suggests that in landscapes containing a diverse configuration of habitat patch types and sizes, the per capita impact of ungulates on vegetation may be less than within large blocks of relatively undisturbed forest. Consequently, land management activities that promote habitat heterogeneity may limit localized browse impact on vegetation, even under relatively high deer populations. The practical extension of this view for forests posits that restoration actions that create a diversity of habitats, particularly those that increase forage availability (e.g., timber harvests and grassy openings), will diffuse foraging impact (deCalesta and Stout 1997; Schmitz 2005; Miller et al. 2009).

We present a unified conceptual model illustrating how ungulate density and habitat heterogeneity might jointly influence browse impact at a local level (Figure 4.1). The model predicts that at any given ungulate density, their localized browsing impact on plant communities will decline as habitat heterogeneity increases throughout their range. The shifts in browse impact mediated by habitat heterogeneity predicted by this model assume deer populations within the sample landscape remain relatively stable in the short term. However, any natural or anthropogenic disturbance, which increases productivity and forage availability, may result in greater ungulate carrying capacity and thereby increase deer numbers through greater fecundity and immigration (Gill et al. 1996; Kramer et al. 2006). Consequently, increased habitat heterogeneity may ultimately intensify browse impacts (see dotted line in Figure 4.1). Nevertheless, ungulate populations often build over years as a result of distinct lags in juvenile recruitment and low emigration rates from nearby social groups (McCullough 1979; Fryxell et al. 1991; Gill et al. 1996; Miller et al. 2010). Additionally, marked population increases may be restrained through adaptive hunting regulations and courting hunter participation (Brown et al. 2000; Reitz et al. 2004). Thus, while restoration activities that increase heterogeneity may ultimately intensify browse impact, in the highly dynamic conditions found in disturbed stands, the lagged response in deer population growth creates a temporal reprieve from intense browsing, and may allow regenerating seedlings to grow out of browsing height (Partl et al. 2002).

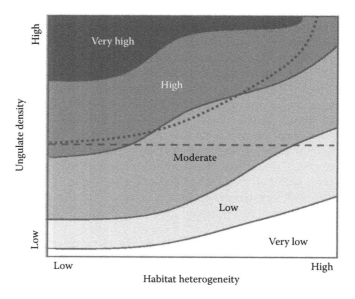

FIGURE 4.1
Conceptual model illustrating local browse impact (shaded isoclines) as a function of ungulate density and habitat heterogeneity. Dashed line illustrates a constant ungulate density exerting high-to-low impact depending on habitat heterogeneity. Dotted line represents an ungulate population that swells as heterogeneity increases, thus nullifying any heterogeneity-mediated reductions in browse impact. (Adapted from Marquis, D.A., Ernst, R.L., Stout, S.L. 1992; Prescribing silvicultural treatments in hardwood stands of the Alleghenies (revised). USDA Forest Service General Technical Report NE-96. USDA Forest Service, Northeastern Forest Experiment Station, Newtown Square, Pennsylvania; Augustine, D.J. and deCalesta, D.S. 2003. *Ecoscience* 10: 472–486; and Reimoser, S. et al. 2009. *Ecological Modelling* 220: 2231–2243.)

4.4.3 Incorporate Landscape Variability into Measures of Restoration Success

The third principle is to incorporate landscape variability into measures of restoration success. Landscapes by definition contain heterogeneity. Some parts of landscapes are better able to sustain grazing, while others are not. Indeed, the composition and configuration of a landscape can modulate grazing effects. For example, the impact of ungulates on the regeneration of a wildflower population in a woodlot surrounded by alfalfa fields is quite different from the impact of the same density of ungulates on the same species of wildflower in a deep forest context (Augustine and Jordan 1998). Vera's (2000) model of herbivore-driven landscape regeneration consists of some patches that thrive under heavy grazing pressure, and others that fare poorly under browsing and grazing. The integrated landscape approach to forest restoration must allow for different goals in different locations or for a purposeful shifting of landscape mosaic patterns to facilitate long-term restoration across an entire landscape (Kirby 2004). In regions where fencing regeneration is part of a restoration strategy, a necessary result will increase ungulate impact on unfenced areas. The state-and-transition model framework as developed by Westoby et al. (1989) could be useful for describing and managing this landscape variability.

4.4.4 Recognize Nonforest Food Sources

The fourth principle is to recognize nonforest food sources for ungulates. The planting of surrogate meadows or grasslands interspersed within the forest landscape can reduce grazing pressure on areas with a closed forest canopy, for example. The creation of open woodland pasture can concentrate grazing in a portion of the landscape, and may help in triggering the return of an ungulate-maintained forest landscape regeneration cycle (Vera 2000). Short-term supplemental feeding can help to redistribute herbivores through the landscape, or alleviate grazing pressure on other resources (Sahlsten et al. 2010). An implicit corollary of this principle is consideration of the effects of restoration activities on neighbors. While it might be tempting to solve an ungulate herbivore problem by incorporating nearby farm fields into the restoration solution for a forested landscape, displacing woodland browsing into crop damage is not necessarily a realistic or sustainable solution.

4.4.5 Include Social Dimensions

The fifth principle is inclusion of social dimensions in integrated forest landscape restoration. While this principle applies to all restoration efforts, there are special considerations in coupled forest–ungulate–social systems. Landowners, neighbors, hunters, and recreationists often hold conflicting views with respect to what constitutes a desirable density and visibility of ungulates, what successful forest regeneration and renewal looks like, and which ecosystem services should be found in a restored landscape. If the restoration team successfully communicates a systems' perspective on the ways elements of this system interact, this may facilitate communication and compromise among different stakeholders. In some regions, there may be a need to restore models of governance, replacing centralized decision-making with broader, participatory comanagement agreements. There are numerous such governance experiments underway in marine fisheries, water management, and sustainable development projects. Berkes (2009) provides an overview of how such comanagement agreements work.

4.4.6 Integrate Natural Disturbance into Restoration Planning

The sixth principle is integration of natural disturbance into forest landscape restoration planning. A variety of natural and anthropogenic disturbance regimes in forest landscapes support and are impacted by ungulate herbivores. These include wind, fire, drought, frost damage, extreme climate events, and land-use change. Recognition of likely future disturbances and their effects on ungulate carrying capacity and ecosystem dynamics will help forest restoration efforts in incorporating the management flexibility needed to succeed as disturbances occur. While little research has been conducted on multiple interacting disturbances, Royo et al. 2010b show this critically important point: such interacting disturbances may produce results that are quite different from the simple, additive impacts of each disturbance. In some landscapes, for example, wind or fire can both stimulate production of forage on the landscape. Restoration in a region where such disturbance is likely may need to include plans that have one or more of these

- Adjustment of anthropogenic disturbances such as timber harvest to reflect the changes in carrying capacity caused by natural disturbance events.
- Use of the full suite of regeneration tools, from natural to direct seeding and/or planting, to restore desired vegetation in disturbed areas.
- Time regeneration establishment (natural or artificial) before or as the ungulate population increases in response to the increased food supply.
- Use of browse tolerant regeneration strategies (including density and species composition) where necessary and desirable (see Sidebar 4.3).
- Manipulation of hunting pressure to maintain relative ungulate density and impact at an acceptable level after natural disturbance.

Where drought may periodically reduce carrying capacity, a restoration plan must acknowledge and include contingencies for this risk. Restoration planners also need to consider likely changes in land use surrounding the target landscape, and how they will influence ungulates, forests, and restoration goals.

SIDEBAR 4.3   Deer Browse Tolerant Regeneration: What Would That Be?

It is crucial to have reliable and cost-effective methods to regenerate desired tree and shrub species given present and future site conditions. No tree and shrub species should be excluded from future forest landscapes—particularly under restoration—just because of challenges presented during the regeneration phase.

Both the literature and practical experience indicate that densely stocked regeneration is more robust against deer browse than sparsely stocked regeneration (Reimoser and Gossow 1996; deCalesta and Stout 1997; Baumhauer et al. 2005; Schmitz 2005; Miller et al. 2009). In particular, Reimoser and Gossow (1996) highlight that dense clusters of regeneration of even highly preferred species like silver fir (*Abies alba*), oaks, and rowan (*Sorbus aucuparia*) have a lower risk of being browsed and better chance of growing out of the reach of browsers than when they grow in sparsely stocked regenerations.

Natural regeneration is often densely stocked, with densities of tens and even hundreds of thousands of young trees per hectare. However, natural regeneration may not be sufficient to support proper restoration of a forest. Important species may not be present and as such the seed sources of such desired tree species may be absent. This becomes even more important when restoration efforts include due attention to expected future climatic conditions as well as changes in ecological services desired from the forests.

Among artificial regeneration methods, planting is the most common method. Planting with stock densities compared to natural regeneration is impossible because of high costs. Plantings are usually established at stock densities lower than 5000 seedlings ha^{-1} and in many parts of the world even below 2000 seedlings ha^{-1}.

Densely stocked regeneration arising from direct seeding appears to be a promising approach for integrated ungulate and forest management. If seed of desired species are available in quantity and are affordable, then stock densities, much greater than those achieved by planting, can be achieved at considerably lower costs (Baumhauer et al. 2005; Madsen and Löf 2005). When successful, such densely stocked young stands provide both food and shelter for deer, which is identified by Reimoser and Gossow (1996) as an important feature of forest structures robust to deer impact. The primary risk of this method (compared to planting) is that the most vulnerable stages of seed and seedling development occur in the forest rather than in the protected environment of a nursery.

A deer browse tolerant regeneration method based on direct seeding is presently being developed, documented and demonstrated in operational scale (Madsen et al. 2009). The method rests on sowing desired tree and shrub species mixtures for the site. Typically pioneer species are included to provide nurse effects for the more shade tolerant species that may be desired. The shade tolerant species are typically sensitive to frost damage, competition, deer browse and rodent damage in the regeneration phase and the nurse species may offer some protection, especially where protection of former stand is completely removed due to lack of stability or wildfire.

4.5 Conclusion: Sustainable Browsing Pressure

All of the above principles come into play as restoration ecologists seek to understand the interplay between their target restoration elements, their target landscape, and the ungulates that live there. Thus, any approach to restoration must include thoughtful recognition of all ungulates, their predators, and their preferred habitats within the landscape. Equally important, however, is understanding the current condition of the landscape with respect to its ecological carrying capacity for ungulates (Sidebar 4.2). This should include an understanding of survival, reproduction, and mortality patterns of ungulates on the landscape as it currently exists, and an empirical assessment of their impact on vegetation dynamics. Such assessments might include erecting small fences designed to exclude ungulates, or direct measurements of browsing and grazing damage, as well as drawing upon all available relevant literature from previous studies. These assessments should be conducted in areas that reflect the various conditions on the landscape—from landscape plantings in yards to farm crops to meadows. Ideally, planning should incorporate knowledge of which habitat conditions the various ungulate species prefer for different phases of their life cycle.

While generalized objective and quantitative standards for sustainable browsing pressure have thus far proven elusive, there is a way forward. The goal for landscape scale ecosystem restoration should be sustainability for target species, habitats, and the ungulates. The sustainability of browsing pressure needs to be assessed in terms of plant demography, affected wildlife, and populations of ungulates themselves. In landscapes with multiple ownerships, the objectives of the various landowners must also be factored into restoration planning. Situations in which individual landowners have objectives related to maximum densities of domesticated or wild ungulates will substantially complicate restoration efforts. So too will situations in which landowner opposition to hunting wild ungulates, as this can limit the success of restoration efforts elsewhere in the adjacent landscape. The successful integration of ungulates into forest landscape restoration is achievable, but it will require broad, creative, and imaginative thinking and planning on the part of all involved.

References

Alverson, W.S., Waller, D.M., and Solheim, S.L. 1988. Forests too deer: Edge effects in northern Wisconsin. *Conservation Biology* 2: 348–358.

Ammer, C. 1996. Impact of ungulates on structure and dynamics of natural regeneration of mixed mountain forests in the Bavarian Alps. *Forest Ecology and Management* 88(1–2): 43–53.

Augustine, D.J. and deCalesta, D.S. 2003. Defining deer overabundance and threats to forest communities: From individual plants to landscape structure. *Ecoscience* 10: 472–486.

Augustine, D.J. and Jordan, P.A. 1998. Predictors of white-tailed deer grazing intensity in fragmented deciduous forests. *Journal of Wildlife Management* 63(3): 1076–1085.

Bakker, E.S., Olff, H., Vandenberghe, C., de Maeyer, K., Smit, R., and Gleichman, J.M. et al. 2004. Ecological anachronisms in the recruitment of temperate light-demanding tree species in wooded pastures. *Journal of Applied Ecology* 41: 571–582.

Bartuszevige, A.M. and Endress, B.A. 2008. Do ungulates facilitate native and exotic plant spread?: Seed dispersal by cattle, elk and deer in northeastern Oregon. *Journal of Arid Environments* 72(6): 904–913.

Baumhauer, H., Madsen, P., and Stanturf, J. 2005. Regeneration by direct seeding—a way to reduce costs of conversion. In: Stanturf, J., Madsen, P. (Eds.), *Restoration of Boreal and Temperate Forests*. CRC Press, Boca Raton, FL, USA.

Bergmeier, E., Petermann, J., and Schröder, E. 2010. Geobotanical surveys of wood-pastures habitats in Europe: Diversity, threats and conservation. *Biodiversity Conservation* 19: 2995–3014.

Berkes, F. 2009. Evolution of co-management: Role of knowledge generation, bridging organizations, and social learning. *Journal of Environmental Management* 90: 1692–1702.

Bradshaw, R.H.W. 2005. What is a natural forest? In: Stanturf, J.A., Madsen P. (Eds.), *Restoration of Boreal and Temperate Forests*. CRC Press, Boca Raton, FL, USA, pp. 15–30.

Bradshaw, R.H.W., Hannon, G.E., and Lister, A.M. 2003. A long-term perspective on ungulate-vegetation interactions. *Forest Ecology and Management* 181(1–2): 267–280.

Bradshaw, R. and Mitchell, F.J.G. 1999. The palaeoecological approach to reconstructiong former grazing-vegetation interactions. *Forest Ecology and Management* 120: 3–12.

Bråthen, K.A., Ims, R., Yoccoz, N., Fauchald, P., Tveraa, T., and Hausner, V. 2007. Induced shift in ecosystem productivity? Extensive scale effects of abundant large herbivores. *Ecosystems* 10: 773–789.

Brown, T.L., Decker, D.J., Riley, S.J., Enck, J.W., Lauber, T.B., Curtis, P.D. et al. 2000. The future of hunting as a mechanism to control white-tailed deer populations. *Wildlife Society Bulletin* 28: 797–807.

Buttenschøn, R.M. and Buttenschøn, J. 2013. Woodland grazing with cattle—results from 25 years of grazing in acidophilus pedunculate oak (Quercus robur) woodland. In: Rotherham, I.D. (Ed.), *Trees, Forested Landscapes and Grazing Animals—A European Perspective on Woodlands and Grazed Treescapes*. Routledge, Oxford, pp. 317–329.

Cadenasso, M.L. and Pickett, S.T.A. 2000. Linking forest edge structure to edge function: Mediation of herbivore damage. *Journal of Ecology* 88: 31–44.

Collins, S.L. and Good, R.E. 1987. The seedling regeneration niche: Habitat structure of tree seedlings in an oak-pine forest. *Oikos* 48: 89–98.

Convery, I. and Dutson, T. 2012. Wild Ennerdale: A cultural landscape. In Stanturf, J.A., Lamb, D., Madsen, P. (Eds.), *Forest Landscape Restoration*. Springer, New York.

Côté, S.D., Rooney, T.P., Tremblay, J.-P., Dussault, C., and Waller, D.M. 2004. Ecological impacts of deer overabundance. *Annual Review of Ecology, Evolution, and Systematics* 35: 113–147.

deCalesta, D.S. 1994. Effect of white-tailed deer on songbirds within managed forests in Pennsylvania. *Journal of Wildlife Management* 58(4): 711–718.

deCalesta, D.S. and Stout, S.L. 1997. Relative deer density and sustainability: A conceptual framework for integrating deer management with ecosystem management. *Wildlife Society Bulletin* 25(2): 252–258.

Dennis, A.J. and Westcott, D.A. 2007. Estimating dispersal kernels produced by a diverse community of vertebrates. In: Dennis, A.J. (Ed.), *Seed Dispersal: Theory and its Application in a Changing World*. CABI, Wallingford, pp. 201–228.

Duncan, A.J. and Poppi, D.P. 2008. Nutritional ecology of grazing and browsing ruminants. In: Gordon, I.J., Prins, H.H.T. (Eds.), *The Ecology of Browsing and Gazing*. Springer, Berlin, pp. 89–1116.

Forester, J.D., Anderson, D.P., and Turner, M.G. 2008. Landscape and local factors affecting northern white cedar (*Thuja occidentalis*) recruitment in the Chequamegon-Nicolet National Forest, Wisconsin (USA). *The American Midland Naturalist* 160: 438–453.

Fryxell, J.M., Hussell, D.J.T., Lambert, A.B., and Smith, P.C. 1991. Time lags and population fluctuations in white-tailed deer. *The Journal of Wildlife Management* 55: 377–385.

Gill, R.M.A. 1992. A review of damage by mammals in north temperate forests: 1. Deer. *Forestry* 65(2): 145–169.

Gill, R.M.A., Johnson, A.L., Francis, A., Hiscocks, K., and Peace, A.J. 1996. Changes in roe deer (*Capreolus capreolus* L.) population density in response to forest habitat succession. *Forest Ecology and Management* 88: 31–41.

Healy, W.M. 1997. Influence of deer on the structure and composition of oak forests in central Massachusetts. In: McShea, W.J., Underwood, H.B., Rappole, J.H. (Eds.), *The Science of Overabundance: Deer Ecology and Population Management*. Smithsonian Institution Press, Washington, DC, pp. 246–266.

Hester, A.J., Edenius, L., Butterschon, R.M., and Kuiters, A.T. 2000. Interactions between forests and herbivores: The role of controlled grazing experiments. *Forestry* 73: 381.

Hidding, B., Tremblay, J.-P., and Côté, S.D. 2013. A large herbivore triggers alternative successional trajectories in the boreal forest. *Ecology* 94: 2852–2860.

Hobbs, N.T. 2003. Challenges and opportunities in integrating ecological knowledge across scales. *Forest Ecology and Management* 181: 223–238.

Hofmann, R.R. 1989. Evolutionary steps of ecophysiological adaptation and diversification of ruminants: A comparative view of their digestive system. *Oecologia* 78: 443–457.

Horsley, S.B., Stout, S.L., and deCalesta, D.S. 2003. White-tailed deer impact on the vegetation dynamics of a northern hardwood forest. *Ecological Applications* 13(1): 98–118.

Huntley, B. 1990. European vegetation history: Paleovegetation maps from pollen data—13 000 yr BP to Present. *Journal of Quaternary Science* 5: 103–122.

Janis, C. 2008. An evolutionary history of browsing and grazing ungulates. In: Gordon, I.D. and Prins, H.H.T. (Eds.), *The Ecology of Browsing and Grazing*. Springer-Verlag, Berlin Heidelberg, pp. 21–45.

Johnson, A.S., Hale, P.E., Ford, W.M., Wentworth, J.M., French, J.R., Anderson, O.F. et al. 1995. White-tailed deer foraging in relation to successional stage, overstory type and management of southern Appalachian forests. *American Midland Naturalist* 133: 18–35.

Kie, J.G., Bowyer, R.T., Nicholson, M.C., Boroski, B.B., and Loft, E.R. 2002. Landscape heterogeneity at differing scales: Effects on spatial distribution of mule deer. *Ecology* 83: 530–544.

Kirby, K.J. 2004. A model of natural wooded landscape in Britain as influenced by large herbivore activity. *Forestry* 77: 405–420.

Kramer, K., Bruinderink, G.G., and Prins, H.H.T. 2006. Spatial interactions between ungulate herbivory and forest management. *Forest Ecology and Management* 226: 238–247.

Luoto, M., Pykälä, J., and Kuussaari, M. 2003. Decline of landscape-scale habitat and species diversity after the end of cattle grazing. *Journal for Nature Conservation* 11: 171–178.

Madsen, P. and Löf, M. 2005. Reforestation in southern Scandinavia using direct seeding of oak (*Quercus robur* L.). *Forestry* 78: 55–64.

Madsen, P., Madsen, T.L., Olesen, C.R., and Buttenschøn, R.M. 2009. Skovforyngelse under højt vildttryk. In: Kanstrup, N., Asferg, T., Flinterup, M., Thorsen, B.J., Jensen, T.S. (Eds.) *Vildt & Landskab, Resultater af 6 års integreret forskning i Danmark 2003-2008*. Miljøministeriet, Skov- og Naturstyrelsen, København, pp. 82–87.

Maginnis, S. and Jackson, W. 2007. What is FLR and how does it differ from current approaches? In: Rietbergen-McCracken, J., Maginnis, S., Sarre, A. (Eds.), *The Forest Landscape Restoration Handbook*. Earthscan, London, UK, pp. 5–20.

Marquis, D.A., Ernst, R.L., Stout, S.L. 1992. Prescribing silvicultural treatments in hardwood stands of the Alleghenies (revised). USDA Forest Service General Technical Report NE-96. USDA Forest Service, Northeastern Forest Experiment Station, Newtown Square, Pennsylvania.

Martin, T.G., Arcese, P., Scheerder, N. 2011. Browsing down our natural heritage: Deer impacts on vegetation structure and songbird populations across an island archipelago. *Biological Conservation* 144: 459–469.

McCullough, D. 1979. *George Reserve Deer Herd: Population Ecology of a K-Selected Species*. Blackburn Press, Caldwell, NJ, 271p.

McNaughton, S.J., Oesterheld, M., Frank, D.A., and Williams, J. 1989. Ecosystem-level patterns of primary productivity and herbivory in terrestrial habitats. *Nature* 341: 142–144.

McShea, W.J., Underwood, H.B., and Rappole, J.H. (Eds.), 1997. *The Science of Overabundance: Deer Ecology and Population Management*. Smithsonian Institution Press, Washington, DC.

Miller, B.F., Campbell, T.A., Laseter, B.R., Ford, W.M., and Miller, K.V. 2009. White-tailed deer herbivory and timber harvesting rates: Implications for regeneration success. *Forest Ecology and Management* 258: 1067–1072.

Miller, B.F., Campbell, T.A., Laseter, B.R., Ford, W.M., and Miller, K.V. 2010. Test of localized management for reducing deer browsing in forest regeneration areas. *The Journal of Wildlife Management* 74: 370–378.

Mitchell, F.J.G. 2005. How open were European primeval forests? Hypothesis testing using palaeoecological data. *Journal of Ecology* 93: 168–177.

Mitchell, F.J.G. and Kirby K.J. 1990. The impact of large herbivores on the conservation of seminatural woods in the British Uplands. *Forestry* 63(4): 333–353.

Murray, B.D., Webster, C.R., and Bump, J.K. 2014. A migratory ungulate facilitates cross-boundary nitrogen transport in forested landscapes. *Ecosystems* 17: 1002–1013.

Nelson, C.R. and Halpern, C.B. 2005. Edge related responses of understory plants to aggregated retention harvest in the Pacific northwest. *Ecological Applications* 15: 196–209.

Newton, A. 2013. Biodiversity conservation and the traditional management of common land: The case of the new forest. In: Rotherham, I.D. (Ed.), *Cultural Severance and the Environment*. Springer, Berlin, pp. 353–370.

Nuttle, T., Yerger, E.H., Stoleson, S.H., and Ristau, T.E. 2011. Legacy of top-down herbivore pressure ricochets back up multiple trophic levels in forest canopies over 30 years. Ecosphere 2:art4. http://dx.doi.org/10.1890/ES10-00108.1

Olesen, C.R. and Madsen, P. 2008. The impact of roe deer (*Capreolus capreolus*), seedbed, light and seed fall on natural beech (*Fagus sylvatica*) regeneration. *Forest Ecology and Management* 255: 3962–3972.

Partl, E., Szinovatz, V., Reimoser, F., and Schweiger-Adler, J. 2002. Forest restoration and browsing impact by roe deer. *Forest Ecology and Management* 159: 87–100.

Pastor, J. and Cohen, Y. 1997. Herbivores, the functional diversity of plants, species, and the cycling of nutrients in ecosystems. *Theoretical Population Biology* 51(3): 165–179.

Pastor, J. and Naiman, R.J. 1992. Selective foraging and ecosystem processes in boreal forests. *American Naturalist* 139: 690–705.

Plante, M., Lowell, K., Potvin, F., Boots, B., and Fortin, M.J. 2004. Studying deer habitat on Anticosti Island, Québec: Relating animal occurrences and forest map information. *Ecological Modelling* 174: 387–399.

Porter, W.F. 1992. Burgeoning ungulate populations in national parks: Is intervention warranted? In: McCollough, D.R., Barrett, R.H. (Eds.), *Wildlife 2001: Populations*. Elsevier Science Publishers, New York, NY, pp. 304–312.

Reimoser, F. and Gossow, H. 1996. Impact of ungulates on forest vegetation and its dependence on the silvicultural system. *Forest Ecology and Management* 88: 107–119.

Reimoser, S., Partl, E., Reimoser, F., and Vospernik, S. 2009. Roe-deer habitat suitability and predisposition of forest to browsing damage in its dependence on forest growth—Model sensitivity in an alpine forest region. *Ecological Modelling* 220: 2231–2243.

Reitz, S., Hille, A., and Stout, S.L. 2004. Silviculture in cooperation with hunters: The Kinzua Quality Deer Cooperative. In: Shepperd, W.D., Eskew, L.G. (Eds.), Silviculture in Special Places: Proceedings of the National Silviculture Workshop RMRS-P-34. USDA Forest Service RMRS, Fort Collins, CO, pp. 110–126.

Riggs, R.A., Tiedemann, A.R., Cook, J.G., Ballard, T.M., Edgerton, P.J., Vavra, M. et al. 2000. Modification of mixed-conifer forests by ruminant herbivores in the Blue Mountains ecological province. Res. Pap. PNW-RP-527. Portland, OR: US Department of Agriculture, Forest Service, Pacific Northwest Research Station. 77p.

Ristau, T.E. and Horsley, S.B. 1999. Pin cherry effects on Allegheny hardwood stand development. *Canadian Journal of Forest Research* 29: 73–84.

Rooney, T.P. 2001. Deer impacts on forest ecosystems: A North American perspective. *Forestry* 74: 201–208.

Rooney, T.P. and Waller, D.M. 2003. Direct and indirect effects of white-tailed deer in forest ecosystems. *Forest Ecology and Management* 181: 165–176.

Rosenthal, J.P. and Kotanen, P.M. 1994. Terrestrial plant tolerance to herbivory. *Trends in Ecology Evolution*, 9(4): 145–148.

Royo, A.A., Collins, R., Adams, M.B., Kirschbaum, C., and Carson, W.P. 2010b. Pervasive interactions between ungulate browsers and disturbance regimes promote temperate forest herbaceous diversity. *Ecology* 91(1): 93–105.

Sahlsten, J., Bunnefeld, N., Månsson, J., Ericsson, G., Bergström, R., and Dettki, H. 2010. Can supplementary feeding be used to redistribute moose *Alces alces*? *Wildlife Biology* 16: 85–92.

Schmitz, O.J. 2005. Scaling from plot experiments to landscapes: Studying grasshoppers to inform forest ecosystem management. *Oecologia* 145: 224–233.

Searle, K.R. and Shipley, L.A. 2008. The comparative feeding behaviour of large browsing and grazing herbivores. In: *The Ecology of Browsing and Grazing*. Springer, Berlin, Heidelberg, pp. 117–148.

Senft, R.L., Coughenour, M.B., Bailey, D.W., Rittenhouse, L.R., Sala, O.E., and Swift, D.M. 1987. Large herbivore foraging and ecological hierarchies. *BioScience* 37: 789–799.

Skarpe, C. and Hester, A.J. 2008. Plant traits, browsing and gazing herbivores, and vegetation dynamics. In: Gordon, I.D. and Prins, H.H.T. (Eds.), *The Ecology of Browsing and Grazing*. Springer, Berlin Heidelberg, pp. 217–261.

Stanturf, J., Palik, B., Williams, M., Dumrose, K.R., and Madsen, P. 2014. Forest restoration paradigms. *Journal of Sustainable Forestry*. Special Issue: Science Considerations in Functional Restoration. Accepted.

Stark, S., Wardle, D.A., Ohtonen, R., Helle, T., and Yates, G.W. 2000. The effect of reindeer grazing on decomposition, mineralization and soil biota in a dry oligotrophic Scots pine forest. *Oikos* 90: 301–310.

Stout, S.L., Royo, A.A., deCalesta, D.S., McAleese, K., and Finley, J.C. 2013. The Kinzua Quality Deer Cooperative: Can adaptive management and local stakeholder engagement sustain reduced impact of ungulate browsers in forest systems? *Boreal Environment Research* 18 (suppl. A): 50–64.

Takada, M., Asada, M., and Miyashita, T. 2002. Cross-habitat foraging by sika deer influences plant community structure in a forest-grassland landscape. *Oecologia* 133: 389–394.

Van de Koppel, J. and Prins, H.H. 1998. The importance of herbivore interactions for the dynamics of African savanna woodlands: An hypothesis. *Journal of Tropical Ecology* 14(5): 565–576.

Vera, F.W.M. 2000. *Grazing Ecology and Forest History*. CABI, Wallingford.

Wallis deVries, W.F. 1998. Large herbivores as key factors for nature conservation. In: Wallis, W.F. deVries, Bakker, J.P., van Wieren, S.E. (Eds.), *Grazing and Conservation Management*. Kluwer Academic Publishers, Dordrecht, Netherlands, pp. 1–20.

Ward, A.I. 2005. Expanding ranges of wild and feral deer in Great Britain. *Mammal Review* 35: 165–173.

Weisberg, P.J. and Bugmann, H. 2003. Forest dynamics and ungulate herbivory: From leaf to landscape. *Forest Ecology and Management* 181(1): 1–12.

Westoby, M., Walker, B., and Noy-Meir, I. 1989. Opportunistic management for rangelands not at equilibrium. *Journal of Range Management* 42: 266–274.

Williams, J.W., Shuman, B.N., Webb, T. III, Bartlein, P.J., Leduc, P.L. 2004. Late-Quaternary vegetation dynamics in North America: Scaling from taxa to biomes. *Ecological Monographs* 74: 309–334.

5

Site Preparation Techniques for Forest Restoration

Magnus Löf, Back Tomas Ersson, Joakim Hjältén,
Tomas Nordfjell, Juan A. Oliet, and Ian Willoughby

CONTENTS

5.1 Introduction

There is a pressing global requirement to combat the conversion of natural habitats into agricultural land or degradation of forests, which threatens biodiversity and erodes environmental services, and contributes to climate change. It has been estimated that forest landscape restoration may be a potential option for more than two billion ha worldwide (Minnemayer et al. 2011). The tree regeneration phase offers the best opportunity to change tree species and forest ecosystem structure, and this period is therefore an important initial step in restoration of forests. Regeneration can be achieved through natural or artificial (planting and direct seeding) means, but planting trees is almost always a key component of restoration activities (Oliet and Jacobs 2012; Stanturf et al. 2014). However, a particular challenge of the regeneration phase is that it is often expensive, and costs for restoration usually increase with the degree of ecosystem degradation (Stanturf et al. 2001; Birch et al. 2010). Thus, it is important to achieve regeneration success, but unless steps are taken to control competing vegetation and improve soil conditions, artificial or natural forest regeneration often results in unacceptably poor seedling survival and growth, (e.g., Balandier et al. 2006; Löf et al. 2012), which in turn may result in substantial economic losses. Site preparation techniques are important tools to counteract this.

Site preparation is usually achieved by using prescribed burning, herbicides, mechanical treatments, or alternative eco-techniques (Prévost 1992; Sutton 1993; Willoughby et al. 2009a; Brose et al. 2013; Piñeiro et al. 2013). These techniques can be used by themselves or in combination. While the effects of various site preparation techniques on regeneration in traditional plantation forestry is well documented, little synthesis has been conducted specific to use of site preparation in forest restoration. Restoration practice often utilizes techniques common to silviculture. No clear line distinguishes ordinary forestry practices from restoration, but the latter is likely to have multiple goals and, require additional activities to repair degraded ecosystems (Stanturf et al. 2014). As site preparation can dramatically influence forest restoration outcomes, the aim of this chapter is to summarize current knowledge on the effects of site preparation on seedling performance, and to discuss economic efficiency and environmental impact of the most important techniques.

This chapter begins with site preparation treatments aimed at reducing influence from competing vegetation (prescribed burning and herbicides), and is followed by treatments aiming at also improving soil conditions (mechanical site preparation (MSP) and eco-techniques in dry areas). Many methods are used to prepare a site for regeneration, and our aim is to cover the most important of these. Treatments where material is added to a site (e.g. mulching, cover crops, hydrogel, and fertilization) are not included. In addition, intensive MSP methods such as deep plowing and terracing, which due to environmental concerns are seldom used nowadays, are touched upon only briefly.

5.2 Prescribed Burning

Fire is important as a natural disturbance in many biomes and it has profound influence on the structure and composition of many forest ecosystems (Brown and Smith 2000; Liu et al. 2010). However, humans have changed the frequency, and size of fire events through various types of land use, for example, forestry and agriculture as well as fire protection programs. Altered fire regimes influence the structure and function of forest ecosystems, as fire is important for ecological processes such as nutrient cycling and vegetation dynamics (Brown and Smith 2000; Brassard and Chen 2006). Thus, altered fire regimes can lead to changes in vegetation composition, elimination of native species, and also invasions of exotic species. Recreating fire regimes can therefore be an important tool for restoration of forest habitats (Figure 5.1).

The use of fire for forest restoration depends on the fire history of specific habitats. These habitats can be divided into fire dependent, fire sensitive, or fire independent habitats. In the first type, species have evolved in the presence of fire, and fire is an essential process for conserving biodiversity (e.g., savannas, temperate coniferous forests). In the second type, species have not evolved in the presence of fire, and fire plays a secondary role in maintaining natural ecosystem structure and function (e.g., tropical moist broadleaf forests). In the third type, fire is not an evolutionary force (e.g., deserts and tundra). However, until recently, fire has primarily been a tool for clearing fallow agricultural land, expanding agriculture into forested areas, and rejuvenating the palatability of pastures. This is still a common practice in parts of Africa, South America, and Asia.

The use of prescribed burning for forest restoration is often very costly. In plantation forestry, soil cultivation methods such as ploughing have been found to be more cost efficient than burning (Ahtikoski et al. 2010). However, for forest restoration the monetary value of resulting benefits (e.g., increased biodiversity, etc.) are difficult to estimate. During restoration of shortleaf pine–bluestem grass ecosystem in Oklahoma using fire to control competing vegetation, increased rotation time was estimated to cost 4.2 million USD per year due to management cost and loss of timber harvest revenue. Alternative methods of controlling competing vegetation were estimated to further increase cost or reduce efficiency (Zhang et al. 2010). Thus, the economy and efficiency (relative to other site preparation techniques) of fire for forest restoration depends on the circumstances and purpose of the treatment and general conclusions about efficiency are therefore difficult to make.

FIGURE 5.1
Prescribed burning of a Scots pine stand in southern Sweden carried out by Sveaskog AB. Here, reducing the risk for wildfire is not an important issue and prescribed fire is often used as a tool to maintain forest biodiversity. (Photo courtesy Magnus Löf.)

Historically, fire has been used in forestry to improve seedling establishment. This method was important before MSP techniques or herbicides were developed (von Hofsten and Weslien 2005). During natural regeneration, successful recruitment of new seedling cohorts often depends on the availability of suitable microsites (Simard et al. 1998; de Chantal et al. 2009). In addition, understory vegetation interferes with seedling survival and growth (Nilsson and Wardle 2005). The use of prescribed fire provides a natural way to remove competing vegetation and provides seeds with suitable microsites for successful germination, especially for pioneer tree species with small seeds (Lampainen et al. 2004; de Chantal et al. 2009). Prescribed fire also releases nutrients from the soil that benefit seedling growth (Brown and Smith 2000). However, fire intensity (i.e., ground or crown) is important and the effect depends on type of seed, for example, whether the seeds belong to the seed bank or are stored in the canopy of the trees (Ryan 2002). Thus, the outcome of fire on seedling establishment is complicated to predict and microsite variation plays a crucial role. Prescribed fire can also be used as a tool to control competing vegetation after seedling establishment. For example, fire is considered a very important factor for regeneration and dynamics of oak forest in North America (e.g., McEwan et al. 2011; Brose et al. 2013). However, the use of prescribed fire to restore oak has provided variable results. Single burns have not always been efficient, whereas repeated fires may be a better tool. Multiple ecosystem drivers, including decreased fire frequency, have contributed to the reduction of oak forests in North America (McEwan et al. 2011). Furthermore, many species of pines are linked to fire disturbance (Rodriguez-Trejo and Fule 2003). Postestablishment burning is also used to control the populations of competing loblolly pine during restoration of longleaf pine stands. It has been suggested that prescribed fire after every two to three years is critical for control of loblolly pine regeneration during restoration of longleaf pine (Knapp et al. 2011).

One of the drawbacks of using fire for forest regeneration is that it can increase the density of pest species at present, and thus reduce seedling survival and growth (Von

Hofsten and Weslien 2005). For example, pine weevil damage on seedlings is often high in the first years after fire and up to 80% of newly established seedlings may be killed (Von Hofsten and Weslien 2005). Another problem is that the effect of fire on tree regeneration is dependent on fire behavior, for example, intensity, duration and if the fire is a ground or a canopy fire (Ryan 2002). In addition, prescribed burning is limited seasonally by weather and fuel conditions. This is not always easy to control in the field and the outcome of using prescribed fire for forest restoration is sometimes difficult to predict. Smoke near urban areas and unintentional effects on nontarget species are further potential disadvantages. A meta-analysis of studies in North America revealed that many birds (34% of all species studied) were disfavored by intensive fires, at least in the short term (Fontaine and Kennedy 2012). Kalies et al. (2010) also reported that high severity burns had a negative impact on 12 bird species even 20 years after the fire events in south western Ponderosa pine and in other conifer forests. Therefore, although high severity burns might be necessary to provide good microsites for tree seedlings, they may have negative effects on some wildlife species. However, many forest species are adapted to fire disturbance and are strongly favored and dependent on it for their long-term survival (Granström and Schimmel 1993; Wikars 2002).

5.3 Herbicides

One of the key problems facing young regenerating tree seedlings is competition from other vegetation for scarce resources such as light, moisture and nutrients. In the worst cases, young tree seedlings do not establish or survive this competition; but even where they are not killed outright, growth can be so severely suppressed that regeneration, and hence restoration, fails (Balandier et al. 2006). Moreover, other vegetation may inhibit growth through allelopathy, and on restoration sites provide an environment suitable for voles and fungi, which may damage seedlings (Löf et al. 2006). Control of competing vegetation is therefore widely accepted as being critical to successful regeneration (e.g., Willoughby et al. 2004).

Although there are instances of natural products being used as fungicides or insecticides in antiquity, agricultural pesticides only became available in significant numbers from the mid-1940s (Aldhous 2000). Since the 1970s, the use of herbicides has been one of the commonest methods of weed control used in forests in many countries, due to the practical advantages offered by chemical weeding compared to other potential approaches (Little et al. 2006). In particular, energy, carbon, and economic costs are usually considerably lower than nonchemical alternatives. Herbicides can be highly effective and efficient weed killers and, if used correctly, impacts on biodiversity and soil resources are usually similar or lower than many other more intensive nonchemical approaches to vegetation management (e.g., Michael et al. 1999; Zimdahl 1999). Once trees have been planted, or tree seed has successfully germinated, herbicides can offer a cheap, effective and safe means of eliminating competing vegetation.

The ability of herbicides to efficiently and cheaply control a wide range of competitive plant species has meant that they have been one of the primary means by which the long term growth potential of forests has been improved in recent years (Wagner et al. 2006). In general, regions with higher labor costs (e.g., North America, Australia, Europe, and South Africa) tend to be more reliant on herbicides to control competing vegetation than countries where labor is currently cheaper (e.g., India and China) (Little et al. 2006). Compared to agricultural situations, a relatively limited range of herbicides are commonly used in forest regeneration. Selection of particular herbicides for any given regeneration situation depends on the nature of the site, competing vegetation, and management objectives for the site. In addition, different legislative regimes and commercial decisions by manufacturers mean that herbicide availability varies in different regions of the world. As the particular substances used vary from country to country, and forest to forest, it is difficult to generalize. However, the broad spectrum herbicide glyphosate appears to be the most commonly used herbicide for forest regeneration worldwide. In North America and Australia 2,4-D (for the control of herbaceous competitors) and triazines (for broad spectrum residual control without harming tree species) are also commonly used (Little et al. 2006; Willoughby et al. 2009a; McCarthy et al. 2011). On all but the most fertile sites, one to two applications more than 2–5 years are normally sufficient to allow trees to establish successfully. In general, for most species of competing vegetation, there exists a herbicide which in theory, if approved, could be used to give sufficient control for trees to establish.

Methods of application of herbicides can also vary considerably. In large scale efforts to restore forests, for example in North or South America, aerial spraying is sometimes used as an economic means of releasing favored trees species from competition. Mechanized spraying is also widely practiced where site access permits (Figure 5.2). On a smaller scale, and on more sensitive or fragile sites, application through hand-held sprayers is the norm, as this offers the opportunity to better target sprays and hence limits the negative impacts of herbicide use.

For initial site preparation when restoring agricultural land to forests, overall sprays of broad spectrum products can be used to effectively kill existing agricultural weeds. Although overall or indiscriminate use of broad spectrum herbicides for site preparation may benefit survival and growth of subsequently planted trees, more targeted approaches, using either more carefully directed sprays of broad spectrum products, or more selective herbicides, usually need to be adopted within existing woodlands where there are functioning woodland ecosystems to protect.

A major disadvantage of herbicides compared to other methods of site preparation for restoration is their potentially very serious negative impacts on environmental and human health if misused. Along with all natural and artificial chemical substances, herbicides

FIGURE 5.2
Mechanized spraying with herbicides for afforestation in Indiana, USA. (Photo courtesy Ron Rathfon.)

have a measurable toxicity to mammals, and hence humans. Although some are considerably less toxic than commonly used household chemicals or foodstuffs, many are potentially toxic to operators, forest users, wildlife and the wider terrestrial and aquatic environment. For this reason, in most developed countries, strict regulation of pesticides exists. Moreover, herbicide use can also be counterproductive as it can damage and kill other nontarget vegetation, and thus contribute to further degradation of a forest if not applied according to instructions and regulations.

Pesticide use in forests, both for site preparation and for subsequent release of planted trees from competition, normally only represents a tiny fraction of that used in food production, industrial, home and garden applications, for example, less than 1% in most European countries (Willoughby et al. 2009a). However, institutional and public attitudes are widely thought to be negative (Wagner et al. 1998; Willoughby et al. 2009a; Wyatt et al. 2011); hence in recent years the policy of many Governments around the world has been to reduce and restrict the use of herbicides to essential situations only (Ammer et al. 2011; Thiffault and Roy 2011). In addition, independent global certification initiatives such as those operated by the Forest Stewardship Council (FSC) and the Programme for Endorsement for Forest Certification schemes (PEFC), require managers to adopt a strategy for the progressive reduction and elimination of all synthetic chemical use, unless the only available alternatives are impractical or excessively costly. Reviews of product safety, commercial decisions by manufacturers based on the increasing costs of product development and registration, and economic pressures to minimize the cost of woodland regeneration operations have also led to a reduction in the amount of pesticides used in forestry. Therefore, despite the apparent advantages of relying on herbicides for site preparation for forest restoration, in recent years there has been renewed interest in adopting alternative, nonchemical methods. In some countries in Europe, for example, Denmark, Sweden and Germany, herbicide use is effectively restricted to afforestation, or privately owned forests (McCarthy et al. 2011). Integrated vegetation management (IPM/IVM) approaches are increasingly being favored (Willoughby et al. 2004, 2009b), which involve understanding the nature and impacts of the competing vegetation, and considering the full range of potential treatment solutions and their likely efficacy and impacts, rather than simply relying on repeated herbicide use as a first resort.

5.4 Mechanical Site Preparation and Planting

Several of the MSP methods commonly used during the last 50 years were already described 150 years ago in detail (e.g., Sutton 1993 and references therein), although at that time, manual or horse power were used instead of heavy machines. With the ready availability of petrol and diesel machinery in the 1940s, MSP became integrated into forestry practices, especially in even-aged management systems. During the same time period, MSP methods were introduced for restoration of former mine sites and of Mediterranean dry sites (e.g., Moffat and Bending 2000; Barberá et al. 2005).

There are several reasons for using MSP methods. Depending on the equipment used, MSP can include cultivating the soil layers, clearing unwanted vegetation, and removing logging slash to facilitate planting or direct seeding (Löf et al. 2012). Because large machinery is often needed for MSP, it is most efficient when treating large areas. For example, restoration of modern surface mining sites can be on an enormous scale, over many square kilometers (Bradshaw and Hüttl 2001). MSP methods are often used in combination to achieve multiple objectives. Sites prepared for conversion of Norway spruce (*Picea abies*) to broadleaves in southern Sweden, for example, are normally treated by piling of slash to ready the site for planting, which is then followed by disk trenching to prepare the soil (Löf et al. 2004). In Spain, the terracing of hill slopes may be followed by subsoiling to improve water infiltration into the soil and plant root development (Querejeta et al. 2001). In afforestation of bottomland pastures or old fields in the United States, the mowing of vegetation may be followed by soil ripping, disking or mounding (Kabrick et al. 2005) (Figure 5.3).

There are three main types of MSP for treatment of soils and vegetation: scarification, mounding, and subsoiling/ripping. Scarification of the soil is done to remove vegetation and the upper organic layers and uncover bare soil. In contrast to scarification techniques, mounding creates elevated planting spots and modifies soil structure, and is often used to improve soil aeration (Sutton 1993).

Subsoiling or ripping is an MSP method used for dry soils, reclaimed mined sites, or for other soils that have a compacted surface layer below the soil surface that restricts root growth and plant development (e.g., Moffat and Bending 2000; Barberá et al. 2005; Palacios et al. 2009). Depending on machinery and attached equipment, they can all be carried out using different intensity: (1) continuous, as connected tracks or areas, (2) intermittent with patches, mounds, or subsoiling done at regular distances, or (3) directed where patches or mounds are placed at certain suitable areas. The choice of method may depend on soil type, topography, and the degree of disturbance needed for regeneration or other restoration or conservation concern.

The primary benefit of MSP is that it improves seedling survival and growth (Löf et al. 2006). However, it often influences both soil conditions and the intensity of competing vegetation, and hence the specific cause of any positive effects on seedling development is difficult to interpret (Löf et al. 2012). In addition, confounding effects such as altered herbivory by insects or browsing and predation from small mammals are also common (Löf 2000; Birkedal et al. 2010). Enhanced seed survival and germination during natural

FIGURE 5.3
Oak seedlings planted on a relatively big mound in wet soil in southern Sweden where a pine stand was converted to oak. (Photo: Magnus Löf.)

regeneration or direct seeding is another example of positive effects of scarification and mounding (Karlsson 2001; Löf and Birkedal 2009), although unwanted competing vegetation may prosper as well.

The economic efficiency of MSP during forest restoration may be difficult to evaluate since it is not only plant growth and survival, but also the degree of soil disturbance and effects on biodiversity, that are important. However, some researchers have found that an intermittent method (spot mounding) resulted in lower total management costs compared to disk trenching due to reduced need for precommercial thinning and increased growth of crop trees (Uotila et al. 2010). In another study conducted in shelterwoods of different densities by Granhus and Fjeld (2008), the time taken for manual planting between two intermittent methods, patch scarification and inverting MSP (mounding, but soil was put back in the dugout hole), and an untreated control were compared. The authors found that MSP greatly reduced the time required for planting; planting was easier and the largest reduction occurred with the inverting MSP. A preferable method to establish trees during forest restoration may be used by MSP in a few spots in combination with the use of large seedlings. Even if larger seedlings cost more to produce and plant, they normally have a significantly higher survival following planting and compete better with vegetation than smaller seedlings.

Depending on the intensity, MSP can lead to impacts over relatively large areas and deep soil disturbance, and there is in general a loss of carbon from soil following MSP (Jandl et al. 2007). Another disadvantage is that most MSP methods can cause soil erosion if not carefully implemented and adapted to the specific site characteristics and climate (Alcázar et al. 2002). For example, in steep forest lands in northern Spain, mechanized forest operations including down-slope ploughing significantly increased soil losses, with effects on site productivity (Edeso et al. 1999). In addition, MSP has the potential to damage ancient remains that may be hidden under the organic soil layer in the forests or under agricultural topsoil (Löf et al. 2012).

When using machinery for MSP, the work has to be done productively enough to compensate for the total cost of machine and operator, thereby reaching an acceptable cost per treated area. The machines must be reliable enough to achieve a high degree of utilization and work long enough to justify their initial capital cost. This reality usually means that the machinery has to be utilized many hours per day to be cost effective. The most common technical principles for conducting MSP are described in the following text. This section also includes mechanized tree planting, which is a natural step in the continuing mechanization of forest restoration management. Such machines often combine MSP and tree planting.

If there are large amounts of logging residues (slash) or bushes present, the first step is usually to remove them when preparing a site for regeneration using machines. Following this operation, the piles of slash are often burned or collected and used for energy purposes (Richardson et al. 2002). If there is much brush vegetation that needs to be treated, the first step can also be to use a forestry mulcher (Jeglum et al. 2003). Commonly, mulchers are horizontally mounted drums, fast-rotating circa 2 m long with teeth that crush bushes and slash, and are attached to bulldozers or other tracked base machines.

Forwarders or skidders are the most commonly used continuously advancing base machines. The most common equipment added to those base machines are disk trenching (scarification) and mounding devices (Figure 5.4). Normal productivity averages 0.8–1.0 ha h^{-1}. A disk trenching device has two to three hydraulically powered disks. Each disk is 1.0–1.4 m in diameter and has 8–10 teeth. They are normally used for continuous soil scarification, but some disk trenching devices can also work intermittently. The

FIGURE 5.4
Two-row disk trencher (a), three-row mounder (b), and rotary cultivator with seeding equipment (c) all three mounted on large-sized forwarders. Agricultural tractor-mounted drill (d). (Photo courtesy Bracke Forest and Tomas Nordfjell.)

rotational speed as well as the disk angle and vertical force can be adjusted in order to choose how deep and wide the prepared area will be. The distance between the discs can also be adjusted. Small disk trenching devices can also be mounted on medium-sized farm tractors.

A mounding device has two to four hydraulically powered mattock wheels. Each mattock wheel is circa 1.4 m in diameter and has three or four tooth-equipped edges. Mounding devices work intermittently, producing a row of soil mounds on the humus layer behind each mattock wheel. The rotational speed is proportional to the driving speed and the vertical force can be adjusted in order to vary the size of the mounds.

Intermittently advancing machines generally use a crane (boom) to prepare the soil and are often ordinary tracked excavators (von der Gönna 1992). However, backhoe loaders, agricultural tractors, and harvesters can also be used. Intermittently advancing machines are flexible and can prepare the soil as, for example, patches, mounds, inverted mounds (i.e., for inverting site preparation/hole planting) or drilled spots. Ordinary or slightly modified excavator buckets can also be used for ditch mounding, a method suitable for waterlogged sites or peat soils where the excavator creates 30–50 cm deep ditches and places the deposits on both sides of the ditch in long mounds (bedding). Drills, usually circa 20 cm wide and either crane-mounted on excavators or rear-mounted on farm tractors, are used to break up the top 40–50 cm of soil in a small spot (Figure 5.4).

Mechanized seeding can be carried out using equipment added to agricultural tractors or the continuously advancing base machines mentioned above (Figure 5.4). Direct seeding (sowing) is commonly used to establish coniferous stands in northern Sweden. Many similar types of equipment have also been developed for the direct seeding of, for example, beech (*Fagus sylvatica*) and oak (*Quercus* spp.) during afforestation and conversion of coniferous to broadleaved forests in central Europe (e.g., Hahn et al. 2005). Seed dispersion density can often be adjusted, and sometimes it is also possible to simultaneously use seeds from at least two different tree species. Some seeding equipment also performs microsite preparation (Figure 5.4).

Tree planting machines perform both the site preparation and the planting and can be categorized as either continuously or intermittently advancing. During afforestation, the machines are often continuously advancing with continuously working planting devices (Figure 5.5). Besides afforestation, these planting devices can also be used during reforestation of sedimentary soils, as long as stumps have been removed since they need obstacle-free terrain to function (Stjernberg 1985). The base machine is usually an agricultural tractor with a planting unit attached to the rear hitch where seedlings can be stored and one to four persons can be seated. Bare-rooted or containerized seedlings are either planted manually in the furrow or manually fed to mechanical planting fingers. Productivity averages around 400–1000 seedlings per hour (Rottensteiner and Stampfer 2009). In the past, there were several continuously advancing machines with intermittently working devices developed to also work on moraine soils, with the Silva Nova being the foremost example (Ersson 2010). This machine could plant >2000 seedlings per hour under ideal conditions (Hallonborg et al. 1997).

Intermittently advancing planting machines are stationary during site preparation and planting. They generally use excavators as base machines and have crane-mounted

FIGURE 5.5
Agricultural tractor-mounted planting device, continuously working during afforestation (a), the Silva Nova carried by a medium-sized forwarder (b), the one-headed Bracke Planter planting device mounted on a tracked excavator (c), and the two-headed EcoPlanter planting device mounted on a medium-sized harvester (d). (Photo courtesy Wolf-Dieter Emmrich, Jan Åhlund, Back Tomas Ersson and Stefan Mattsson/Skogforsk.)

planting devices that mound the soil and plant containerized seedlings. Today, mechanized tree planting typically entails tracked excavators with one-headed devices, but in the past, harvesters with planting devices that scarified using rotovators were also used (Figure 5.5). However, two-headed devices are also available and can be more productive than one-headed devices (averaging 150–350 versus 100–250 seedlings per hour, respectively) as long as the terrain is relatively clear of obstacles (Rantala et al. 2009). Compared to continuously advancing mechanical scarifiers and subsequent manual planting, today's tree planting machines are less cost-efficient because of lower productivity. However, they generally plant seedlings in a better way with increased survival and disturb a smaller proportion of soil (Ersson et al. 2011; Luoranen et al. 2011; Sjögren 2013).

5.5 Eco-Techniques in Dry Areas

Dry lands occupy almost half of the earth's land surface, and many areas show severe levels of degradation (Reynolds 2013). Aridity and degradation interact in dry lands, and the effects of water limitation are often increased by soil infertility and erosion on steep slopes where runoff reduces the water available for plants. Techniques to improve productivity of arid lands are far from new. Documented references to runoff collecting in the desert for agricultural purposes are registered as far back as 2000 years ago, and eco-techniques based on runoff collection have been applied for many centuries in agriculture and forestry (e.g., Ayuso et al. 1982; Tenbergen et al. 1995).

An eco-technique is defined as any structure or device aimed at ameliorating field site conditions, particularly soil and microclimatic limitations (Piñeiro et al. 2013). Some are designed to improve resource availability, such as soil structures to improve runoff harvesting, deeper planting holes and fog capturing equipment. Others are based on improved biotic interactions to facilitate seedling establishment like mycorrhization and facilitation by shrub-cover vegetation (Chirino et al. 2009; Cortina et al. 2011; Prieto et al. 2011; Vallejo et al. 2012). Because there is often considerable variation within dry landscapes, the type of eco-technique needs to be matched to the specific planting microsite (Cortina et al. 2011; Kribeche et al. 2012). Here we will address runoff harvesting and fog catching, which are highly specific techniques for dry lands.

Microcatchments are manually built channels placed up-slope and converging to direct runoff toward the seedling (Kaplan et al. 1970) (Figure 5.6). Sometimes, the soil is manually refined and flattened creating a reverse-slope with a small basin at the uphill edge of the planting spot that increases the storage and infiltration capacity of the microcatchment. In recent years, microcatchments have been improved by introducing additional elements or structures like plastic sheets upstream to generate runoff from light rainfall events, as well as drywells filled with stones (20–25 cm soil depth) in the spot to promote infiltration (Vallejo et al. 2012). These improved microcatchments are designed to harvest minimum rainfall events (1–10 mm), which are an important part of total yearly precipitation in dry lands (Cortina et al. 2011).

Another system to improve water availability is by collecting fog from the atmosphere (Figure 5.7). Two types of systems have been tested in Spain. The first system consisted of an 18 m^2 mesh panel providing water to a drip irrigation system. Approximately 3000 L were harvested during one spring, supplying two 4.5 L pulses per plant in summer (Estrela

FIGURE 5.6
Improved microcatchment in Spain with channels (a) on both sides of the planting spot that direct runoff water toward the seedling. The seedling is also surrounded by a browse protection net and two dry wells (b) that improve infiltration. (Photo courtesy Joan Llovet.)

et al. 2009). The other system was constructed to be placed near each plant with pipes to carry captured water directly to the root zone (Figure 5.7).

The results from studies testing microcatchments in dry lands vary. Even where trials show an improvement in soil water availability, operational effectiveness is often determined by factors such as plant cover of the site, rain intensity, slope or soil properties, as well as microcatchment dimensions (Vallejo et al. 2012). Moreover, microcatchments will only improve water availability if rainfall events generate sufficient surface runoff. To date, microcatchments have improved planting results only at the most arid sites. The first year

FIGURE 5.7
Fog harvesting system for individual seedlings in Spain where water is captured by the mesh and directed to the channel at the bottom. The micro pipe then carries the water to the root zone of the browse-protected seedling. (Photo courtesy David Fuentes.)

after planting is the critical period for seedling establishment in dry lands (Oliet et al. 2002). Microcatchments, however, have a longer life span, and it may therefore be that effects of microcatchment can be accumulative on following year's growth. Individual fog systems have been tested in two areas of East Spain (Vallejo et al. 2012). Significant improvement in growth after planting of seedlings was only found at the harshest site. However, the large panel collector resulted in significantly higher survival (Estrela et al. 2009).

Eco-techniques are still under development, and have so far often been applied without questioning their cost-efficiency. There is a lack of evaluation in the subject. When choosing eco-techniques to improve water availability in dry areas, we should also assess whether the alternative approach of simply replanting failed areas would also guarantee, more cheaply, a minimum survival that achieves restoration goals. For water collecting techniques such as microcatchments or fog trapping structures, cost/benefit analysis should be carried out comparing them with other techniques such as microirrigation. Microcatchments are often considered to be more efficient and less expensive than the other watering systems (Vallejo et al. 2012), but the variability in results is still high and careful observation of characteristics of the site prior to restoration are needed. Fog-harvesting systems are very expensive but designing collectors to feed individual plants would reduce the cost of the irrigation systems (Estrela et al. 2009; Cortina et al. 2011). More information is also needed regarding optimum planting densities, as well as life span of the water collecting structures and the potential effect on survival and growth of seedlings. Moreover, microcatchments or fog capturing structures may have a negative visual impact on the landscape and plastic residues can remain for many years after the structure has ceased effective operation. Proper management of the discarded structures is therefore necessary, which also increases total costs.

5.6 Conclusions

Site preparation is usually achieved by using prescribed burning, alternative eco-techniques, herbicides or MSP, each of which can dramatically influence forest restoration outcomes. Therefore all of these techniques are likely to remain important restoration tools in the future.

However, all of the techniques discussed have potential disadvantages which need to be managed during the restoration process. Prescribed burning is limited seasonally by weather and fuel conditions; it can increase the amount of pest species, may be dangerous and cause smoke problems near urban areas, and creates unintentional effects on non-target species. Eco-techniques in dry areas may affect landscape esthetical values and produce plastic residues that need to be disposed of. A major disadvantage of herbicides compared to other methods of site preparation for restoration is their potentially serious negative impacts on environmental and human health if misused. Finally, MSP can lead to deep soil disturbance and erosion.

Furthermore, some of these techniques can be expensive to implement, although the management goals for forest restoration often differ from those for reforestation. In the latter case, stand replacement is most often scheduled to minimize the rotation age. Site preparation is then important because it is designed to improve early tree survival and growth, which often results in higher land expectation values. During forest restoration, early survival and growth are still critical, but other objectives such as improved soil protection and biodiversity are equally important, thereby making economic evaluations more complex. In addition, subsidies often play an important role during restoration projects, which also increases the complexity in economic evaluations and influences the choice of methods during implementation. However, ecological and economic goals are not mutually exclusive during forest restoration. Public landowners and small nonindustrial landowners are not only interested in multipurpose management including biodiversity and soil protection, but are also interested in timber production. The latter may be capable of producing a financial return sufficient to ensure the long-term maintenance of the restored sites. Therefore, efficient establishment of tree stands and woody vegetation is important to avoid unnecessary costs, and excessively long delays in achieving successful forest restoration.

The various site preparation methods should be implemented carefully to avoid negative side effects, and adapted to site conditions. Seedling performance appears to increase with increased disturbance (water infiltration, soil disturbance and reduced competition). Simultaneously, the negative side effects can also increase. During any development of new site preparation techniques for the future, spot methods (intermittent or directed methods) to achieve a high degree of disturbance but only on a limited, carefully targeted area, seems to be a promising starting point. This may lead to a combination of low overall environmental impact with improved seedling performance. To achieve this goal, new types of machinery and attached equipment may need to be developed. By contrast, prescribed burning cannot be implemented in small spots. Although not discussed in detail here, sheet mulches may provide another alternative. Mulch is a layer of material (often plastic sheeting or bark chips) applied to the soil surface to conserve moisture, improve fertility, and reduces competing vegetation. However, mulching also has negative aspects such as having high costs and creating residues. It has so far predominantly been used in gardens or landscaping projects.

A promising alternative site preparation tool may be the use of nurse vegetation (shrubs or trees). Recently, there has been renewed interest in the use of surrounding vegetation to facilitate survival and growth of target trees, especially in dry areas. The use of two-storied mixed species plantations (fast growing nurse trees and under-planted target trees) may reduce natural competing vegetation that would otherwise limit survival and growth of target trees. Although increased costs are associated with planting and managing mixed-species plantations, they can be used in all vegetation zones. Also, they may offer short-term advantages for rapidly building new forest structure while simultaneously increasing productivity, which might be a cost-effective strategy for forest restoration (Nichols et al. 2006).

References

Ahtikoski, A., Alenius, V., and Makitalo. K., Scots pine stand establishment with special emphasis on uncertainty and cost-effectiveness, the case of northern Finland. *New Forests* 40, 69, 2010.

Alcázar, J., Rothwell, R.L., and Woodward, P.M., Soil disturbance and the potential of erosion after mechanical site preparation. *North. J. Appl. For.* 19, 5, 2002.

Aldhous, J.R., *Pesticides, Pollutants, Fertilizers and Trees: Their Role in Forests and Amenity Woodlands.* Research Studies Press Ltd, Baldock, Hertfordshire, UK, 2000.

Ammer, C., Balandier, P., Scott-Bentsen, N., Coll, L., and Löf, M., Forest vegetation management under debate—an introduction. *Eur. J. For. Res.* 130, 1, 2011.

Ayuso, J., Giráldez, J.V., and Ciria, F., *Perspectivas hidrológicas de las zonas áridas. Seminario sobre zonas áridas.* Instituto de Estudios Almerienses, Almería, Spain, 1982.

Balandier, P., Collet, C., Miller, J.H., Reynolds, P.E., and Zedaker, S.M., Designing forest vegetation management strategies based on the mechanisms and dynamics of crop tree competition by neighbouring vegetation. *Forestry* 79, 3, 2006.

Barberá, G., Martínez-Fernández, F., Alvarez-Rogel, J., Albaladejo, J., and Castillo, V., Short- and intermediate-term effects of site and plant preparation techniques on reforestation of a semi-arid ecosystem with *Pinus halepensis* Mill. *New Forests* 29, 177, 2005.

Birch, J.C., Newton, A.C., Aquino, C.A., Cantarello, E., Echeverría, C., Kitzberger, T., Schiappacasse, I. and Garavito, N.T., Cost-effectiveness of dryland forest restoration evaluated by spatial analysis of ecosystem services. *Proc. Natl. Acad. Sci.* 107, 21925, 2010.

Birkedal, M., Löf, M., Olsson, G.E., and Bergsten, U., Effects of granivorous rodents on direct seeding of oak and beech in relation to site preparation and sowing date. *For. Ecol. Manage.* 259, 2382, 2010.

Bradshaw, A.D. and Hüttl, R.F., Future mine site restoration involves a broader approach. *Ecol. Eng.* 17, 87, 2001.

Brassard, B.W. and Chen, H.Y.H., Stand structural dynamics of North American boreal forests. *Critical Reviews in Plant Sciences* 25, 115, 2006.

Brose, P.H., Dey, D.C., Phillips, R.J. and Waldrop, T.A., A meta-analysis of the fire-oak hypothesis: Does prescribed burning promote oak reproduction in eastern North America? *Forest Science* 59, 322, 2013.

Brown, J.K. and Smith, J., (Eds.), 2000. Wildland fire in ecosystems: Effects of fire on flora. *General Technical Report*, RMRS-GTR-24., Ogden, UT, Forest Service, Rocky Mountains Research Station, USA, 2000.

Chirino, E., Vilagrosa, A., Cortina, J., Valdecantos, A., Fuentes, D., Trubat, R., Luis, V.C., et al. Ecological restoration in degraded drylands: The need to improve the seedling quality and site conditions in the field, in: *Forest Management*, Grossberg, S.P. (Ed.), Nova Publisher, New York, 85p., 2009.

Cortina, J., Amat, B., Castillo, V., Fuentes, D., Maestre, F.T., Padilla, F.M., and Rojo, L., The restoration of vegetation cover in the semi-arid Iberian Southeast. *J. Arid Environ.* 75, 1377, 2011.

de Chantal, M., Lilja-Rothsten, S., Peterson, C., Kuuluvainen, T., Vanha-Majamaa, I., and Puttonen, P., Tree regeneration before and after restoration treatments in managed boreal *Picea abies* stands. *Appl. Veg. Sci.* 12, 131, 2009.

Edeso, J.M., Merino, A., Gonzalez, M.J., and Marauri, P., Soil erosion under different harvesting management in steep forestlands from northern Spain. *Land Degrad. Dev.* 10, 79, 1999.

Ersson, B.T., Bergsten, U., and Lindroos, O., The cost-efficiency of seedling packaging specifically designed for tree planting machines, *Silva Fenn.* 45, 379, 2011.

Ersson, B.T., *Possible Concepts for Mechanized Tree Planting in Southern Sweden—An Introductory Essay on Forest Technology*, Department of Forest Resource Management, SLU, Arbetsrapport 269, 51p, 2010.

Estrela, M.J., Valiente, J.A., Corell, D., Fuentes, D., and Valdecantos, A., Prospective use of collected fog water in the restoration of degraded burned areas under dry Mediterranean conditions. *Agr. Forest Meteorol.* 149, 1896, 2009.

Fontaine, J.B. and Kennedy, P.L., Meta-analysis of avian and small-mammal response to fire severity and fire surrogate treatments in US fire-prone forests. *Ecol. Appl.* 22, 1547, 2012.

Granhus, A. and Fjeld, D., Time consumption of planting after partial harvest. *Silva Fenn.* 42, 49, 2008.

Granström, A. and Schimmel, J. Heat-effects on seeds and rhizomes of a selection of boreal forest plants and potential reaction to fire. *Oecologia* 94, 307, 1993.

Hahn, K., Emborg, J., Larsen, J.B. and Madsen, P., Forest rehabilitation in Denmark using nature-based forestry, in: *Restoration of Boreal and Temperate Forests*, Stanturf, J., and Madsen, P. (Eds), CRC Press, Boca Raton, pp. 299–317, 2005.

Hallonborg, U., von Hofsten, H., Mattsson, S., and Thorsén, Å., Forestry planting machines—A description of the methods and the machines, *Skogforsk, Redogörelse* 7, 24, 1997.

Jandl, R., Lindner, M., Vesterdal, L., Bauwens, B., Baritz, R., Hagedorn, F., Johnson, D.W., Minkkinen, K. and Byrne, K.A., How strongly can forest management influence soil carbon sequestration? *Geoderma* 137, 253, 2007.

Jeglum, J.K., Kershaw, H.M., Morris, D.M. and Cameron, D.A., *Best Forestry Practices: A Guide for the Boreal Forest in Ontario*, Ontario Ministry of Natural Resources, Sault Ste. Marie, Canada, 2003.

Kabrick, J.M., Dey, D.C., Van Sambeek, J.W., Wallendorf, M., and Gold, M.A., Soil properties and growth of swamp white oak and pin oak on bedded soils in the lower Missouri River flood-plain. *For. Ecol. Manage.* 204, 315, 2005.

Kalies, E.L., Chambers, C.L., and Covington, W.W., Wildlife responses to thinning and burning treatments in southwestern conifer forests: A meta-analysis. *For. Ecol. Manage.* 259, 333, 2010.

Kaplan, J., Karschon, R., and Kolar, M., Israel, in: *Afforestation in Arid Zones*, Kaul, R.N. (Ed.), Dr. W. Junk N.V. Publishers, The Hague, 137p., 1970.

Karlsson, M., Natural regeneration of broadleaved tree species in southern Sweden—effects of silvicultural treatments and seed dispersal from surrounding stands. *Silvestria* 196, PhD Thesis, SLU, Alnarp, 2001.

Knapp, B.O., Wang, G.G., Hu, H., Walker, J.L., and Tennant, C., Restoring longleaf pine (*Pinus palustris* Mill.) in loblolly pine (*Pinus taeda* L.) stands: Effects of restoration treatments on natural loblolly pine regeneration. *For. Ecol. Manage.* 262, 1157, 2011.

Kribeche, H., Bautista S., Chirino, E., Vilagrosa, A., and Vallejo, V.R., Effects of landscape spatial heterogeneity on dryland restoration success. The combined role of site conditions and reforestation techniques in Southeastern Spain. *Int. J. Med. Ecol.* 38, 5, 2012.

Lampainen, J., Kuuluvainen, T., Wallenius, T.H., Karjalainen, L., and Vanha-Majamaa, I., Long term forest structure and regeneration after wildfire in Russian Karelia. *J. Veg. Sci.* 15, 245, 2004.

Little, K., Willoughby, I., Wagner, R.G., Adams, P., Frochot, H., Gava, J., Gous, S., Lautenschlager, R.A., Örlander, G., Sankaran, K.V., and Wei, R.P., Towards reduced herbicide use in forest vegetation management. *South Afr. For. J.* 207, 63, 2006.

Liu, Y., Stanturf, J., and Goodrick, S., Trends in global wildfire potential in a changing climate. *For. Ecol. Manage.* 259, 685, 2010.

Löf, M. Influence of patch scarification and insect herbivory on growth and survival in *Fagus sylvatica* L., *Picea abies* L. Karst. and *Quercus robur* L. seedlings following a Norway spruce forest. *For. Ecol. Manage.* 134, 111, 2000.

Löf, M. and Birkedal, M., Direct seeding of *Quercus robur* L. for reforestation: The influence of mechanical site preparation and sowing date on early growth of seedlings. *For. Ecol. Manage.* 258, 704, 2009.

Löf, M., Dey, D.C., Navarro, R.M., and Jacobs, D.F., Mechanical site preparation for forest restoration. *New Forests* 43, 825, 2012.

Löf, M., Isacsson, G., Rydberg, D., and Welander, N.T., Herbivory by the pine weevil (*Hylobius abietis* L.) and short-snouted weevils (*Strophosoma melanogrammum* Forst. and *Otiorhynchus scaber* L.) during the conversion of a wind-thrown Norway spruce forest into a mixed-species plantation. *For. Ecol. Manage.* 190, 281, 2004.

Löf, M., Rydberg, D., and Bolte, A., Mounding site preparation for forest restoration: Survival and growth responses in *Quercus robur* L. seedlings. *For. Ecol. Manage.* 232, 19, 2006.

Luoranen, J., Rikala, R., and Smolander, H., Machine planting of Norway spruce by Bracke and Ecoplanter: An evaluation of soil preparation, planting method and seedling performance. *Silva Fenn.* 45, 341, 2011.

McCarthy, N., Bentsen, N.S., Willoughby, I., and Balandier, P., The state of forest vegetation management in Europe in the 21st century. *Eur. J. For. Res.* 130, 7, 2011.

McEwan, R.W., Dyer, J.M., and Pederson, N., Multiple interacting ecosystem drivers: Toward an encompassing hypothesis of oak forest dynamics across eastern North America. *Ecography* 34, 244, 2011.

Michael, J.L., Webber, E.C. Jr., Bayne, D.R., Fischer, J.B., Gibs, H.L., and Seesock W.C., Hexazinone dissipation in forest ecosystems and impacts on aquatic communities. *Can. J. For. Res.* 29, 1170, 1999.

Minnemayer, S., Laestadius, L., Sizer, N., Saint-Laurent, C., and Popapov, P., *A World of Opportunity*. World Resource Institute, Washington, DC. Available at www.wri.org/restoringforests (2014-04-09), 2011.

Moffat, A.J. and Bending, N.A.D., Replacement of soil and soil-forming materials by loose tipping in reclamation to woodland. *Soil Use Manage.* 16, 75, 2000.

Nichols, J.D., Bristow, M., and Vanclay, J.K., Mixed species plantations: Prospects and challenges. *For. Ecol. Manage.* 233, 383, 2006.

Nilsson, M.C. and Wardle, D.A., Understory vegetation as a forest ecosystem driver: evidence from the northern Swedish boreal forest. *Front. Ecol. Environ.* 3, 421, 2005.

Oliet, J.A. and Jacobs, D.F., Restoring forests: Advances in techniques and theory. *New Forests* 408, 535, 2012.

Oliet, J., Planelles, R., López Arias, M., and Artero, F., Soil water content and water relations in planted and naturally regenerated *Pinus halepensis* Mill. seedlings during the first year in semi-arid conditions. *New Forests* 23, 31, 2002.

Palacios, G., Navarro-Cerrillo, R.M., del Campo, A., and Toral, M., Site preparation, stock quality and planting date effect on early establishment of Holm oak (*Quercus ilex* L.) seedlings. *Ecol. Engineer.* 35, 38, 2009.

Piñeiro, J., Maestre, F.T., Bartolomé, L., and Valdecantos, A., Ecotechnology as a tool for restoring degraded drylands: A meta-analysis of field experiments. *Ecol. Eng.* 61, 133, 2013.

Prévost, M., Effets du scarifiage sur les propriétés du sol, la croissance des semis et la competition: Revue des connaissances actuelles et perspectives de recherches au Québec. *Ann For. Sci.* 49, 277, 1992.

Prieto, I., Padilla, F.M., Armas, C., and Pugnaire, F.I., The role of hydraulic lift on seedling establishment under a nurse plant species in a semi-arid environment. *Perspect. Plant Ecol. Evol. Syst.* 13, 181, 2011.

Querejeta, J.I., Roldán, A., Albaladejo, J., and Castillo, V., Soil water availability improved by site preparation in a *Pinus halepensis* afforestation under semiarid climate. *For. Ecol. Manage.* 149, 115, 2001.

Rantala, J., Harstela, P., Saarinen, V.-M., and Tervo, L., A techno-economic evaluation of Bracke and M-Planter tree planting devices. *Silva Fenn.* 43, 4, 2009.

Reynolds, J.F., Desertification, in: *Encyclopaedia of Biodiversity*, vol. 2, Levin S.A., (Ed.), Academic Press, Waltham, 479p, 2013.

Richardson, J., Björheden, R., Hakkila, P., Lowe, A.T., and Smith, C.T., (Eds.), *Bioenergy from Sustainable Forestry—Guiding Principles and Practice*. Kluwer Academic Publishers, Dordrecht, The Netherlands, Forestry Sciences, vol. 71, 2002.

Rodriguez-Trejo, D.A. and Fule, P.Z., Fire ecology of Mexican pines and a fire management proposal. *Int. J. Wildland Fire* 12, 23, 2003.

Rottensteiner, C. and Stampfer, K., *Mechanisierte Pflanzung von Forstballenpflanzen*, Institut für Forsttechnik, Department für Wald- und Bodenwissenschaften, Universität für Bodenkultur Wien, Report, 2009.

Ryan, K.C., Dynamic interactions between forest structure and fire behaviour in boreal ecosystems. *Silva Fenn.* 36, 13, 2002.

Simard, M.J., Bergeron, Y. and Sirois, L. Conifer seedling recruitment in a southeastern Canadian boreal forest: The importance of substrate. *J. Veg. Sci.* 9, 575–582, 1998.

Sjögren, V., *Natural Regeneration After Site Preparation with Disc Trenchers or Bracke Planter in Småland*. Department of Forest Ecology and Management, SLU, Examensarbeten 2013:2, 23p., 2013.

Stanturf, J.A., Palik, B.J. and Dumroese, R.K., Contemporary forest restoration: A review emphasizing function. *For. Ecol. Manage.* 331, 392, 2014.

Stanturf, J.A., Schoenholtz, S.H., Schweitzer, C.J., and Shepard, J.P., Achieving restoration success: Myths in bottomland hardwood forests. *Restor. Ecol.* 9, 189, 2001.

Stjernberg, E.I., *Tree Planting Machines: A Review of the Intermittent—Furrow and Spot Planting Types*. Forest Engineering Research Institute of Canada, Special Report No. SR-31, 1985.

Sutton, R.F., Mounding site preparation: A review of European and North American experience. *New Forests* 7, 151, 1993.

Tenbergen, B., Günster, A., and Schreiber, K.F., Harvesting runoff: The minicatchment technique: An alternative to irrigated tree plantations in Semiarid regions. *Ambio* 24, 72, 1995.

Thiffault, N. and Roy, V. Living without herbicides in Quebec (Canada): historical context, current strategy, research and challenges in forest vegetation management. *Eur. J. For. Res.* 130, 117, 2011.

Uotila, K., Rantala, J., Saksa, T., and Harstela, P., Effect of soil preparation method on economic result of Norway spruce regeneration chain. *Silva Fenn.* 44, 511, 2010.

Vallejo, V.R., Smanis, A., Chirino, E., Fuentes, D., Valdecantos, A., and Vilagrosa, A., Perspectives in dryland restoration: Approaches for climate change adaptation. *New Forests* 43, 561, 2012.

von der Gönna, M.A., *Excavator Attachments for Site Preparation in British Columbia*. Forest Engineering Research Institute of Canada, Technical Note TN-180, 1992.

von Hofsten, H. and Weslien, J., Temporal patterns of seedling mortality by pine weevils (Hylobius abietis) after prescribed burning in northern Sweden. *Scand. J. For. Res.* 20(2), 130–135, 2005.

Wagner, R.G., Flynn, J., Gregory, R., and Slovic, P., Public perceptions of risk and acceptability of forest vegetation management alternatives in Ontario. *For. Chron.* 74, 720, 1998.

Wagner, R.G., Little, K.M., Richardson, B., and McNabb, K., The role of vegetation management for enhancing the productivity of the world's forests. *Forestry* 79, 57, 2006.

Wikars, L.-O., Dependence on fire in wood-living insects: An experiment with burned and unburned spruce and birch logs. *J. Insect. Conserv.* 6, 1, 2002.

Willoughby, I., Evans, H., Gibbs, J., Pepper, H., Gregory, S., Dewar, J., Nisbet, T., et al. Reducing pesticide use in forestry. *Forestry Commission Practice Guide 15*. Forestry Commission, Edinburgh, UK, 2004.

Willoughby, I., Scott Bentsen, N., McCarthy, N., and Claridge, J., (Eds.), *Forest Vegetation Management in Europe*. COST Office, Brussels, Belgium, 2009a.

Willoughby, I., Wilcken, C., Ivey, P., O'Grady, K., and Katto, F., *FSC Guide to Integrated Pest, Disease and Weed Management in FSC Certified Forests and Plantations*. FSC Technical Series 2009–001. Forest Stewardship Council, Bonn, Germany, 2009b.

Wyatt, S., Rousseau, M.H., Nadeau, S., Thiffault, N., and Guay, L. Social concerns, risk and the acceptability of forest vegetation management alternatives: Insights for managers. *For. Chron.* 87, 274, 2011.

Zhang, D., Huebschmann, M.M. Lynch, T.B., and Guldin. J.M., Forest policy impact assessment in the Ouachita National Forest and the valuation of conserving red-cockaded woodpeckers. *Am. J. Appl. Sci.* 7, 1345, 2010.

Zimdahl, R.L., *Fundamentals of Weed Science*. 2nd Edition. Academic Press, San Diego, USA, 1999.

6

Developing Restoration Strategies for Temperate Forests Using Natural Regeneration Processes

Holger Fischer, Franka Huth, Ulrike Hagemann, and Sven Wagner

CONTENTS

6.1 Introduction

6.1.1 Forest Restoration as a Holistic Challenge

Forest restoration projects have become increasingly common around the world and many studies have accumulated in this field during the last decades. Forest restoration means changing the forest landscape component toward a "more natural" situation (*sensu* Fischer and Fischer 2012). But what is "natural"? Bradshaw (2002) called it the "original ecosystem" and focused on two major attributes: ecosystem structure and ecosystem function, with typical values for both attributes that are reduced by ecosystem "degradation." Ecological restoration is an activity that also ideally results in the return of an ecosystem to an undisturbed status (Palmer and Filoso 2009). On the path back to the original ecosystem state, natural recovery as well as ecological restoration offer many developmental options. As these definitions necessitate a holistic approach, the attention in recent restoration projects has turned to the integration of ecosystem services. Similarly, restoration actions focused on enhancing biodiversity should also support increased provision of ecosystem services

(Rey Benayas et al. 2009). This approach presents an opportunity for enhancing benefits to human livelihood and funding sources as well as generating public support for such initiatives (Trabucchi et al. 2012).

Restoration ecology is "repairing" degraded, damaged, or destroyed ecosystems usually adversely affected by human activity. Restoration activities are often aimed at increasing the land-base of a particular ecosystem; its biodiversity, resilience, and resistance, the provision of ecosystem services and ecosystem sustainability (see Aronson and Alexander 2013). The formulation of sustainability as an integral part of restoration practice is essential at the very least with respect to the status quo of many European forest areas, which have a long history of human intervention. Historic human impacts in these forests range from the elimination of predators, clearing for agricultural use and settlements, the introduction of domestic grazing stock, management for timber production, and management-induced change of tree species with the introduction of nonnative, partly invasive species, which disturb the natural cycle of forest regeneration (Willoughby and Jinks 2009).

Restoration success in general, and regeneration success in particular, are controlled by these historical occurrences and past silvicultural practices within a forest stand, including past and present species composition, stand structure, wildlife, and potential management-related soil disturbance. Further, recent anthropogenic modifications such as atmospheric deposition (e.g., of nitrogen) or climate change (Fischer and Wagner 2009) also modify forest regeneration in time and space.

6.1.2 Tree Regeneration as a Driver in Forest Restoration

Tree regeneration is the most relevant and effective initial step in the context of forest restoration, as every silvicultural action at this stage affects the development at least of the next stand generation, and potentially beyond. From an ecological perspective, discussions about the forest life cycle focus on the processes involved in replacing mature trees with the next tree generation as well as the colonization of new habitats. Out of all forest developmental phases, the regeneration phase offers the best opportunity to manipulate tree species and forest structure, making it a key for achieving restoration objectives. In this regard, Löf et al. (2012) defined the practice of planting trees and shrubs as a key component of forest restoration.

The majority of existing studies on forest regeneration emphasize reforestation following timber harvest for industrial purposes (Oliet and Jacobs 2012). Even today, there are land use practices, such as surface mining (Hüttl and Bradshaw 2001) or remediation of forest soils (Kauppi et al. 2012), that create extremely harsh site conditions for restoration. These sites require amelioration using direct sowing and planting (Josa et al. 2012), often in combination with intensive mechanical site preparation. These activities are all active management interventions (Holl and Aide 2011). However, in almost every area with artificial regeneration, the natural invasion of trees, shrubs, and other autotrophs (Mueller-Dombois and Ellenberg 1974) may also play a role. In forest restoration, some groups advocate relying exclusively on natural regeneration and succession dynamics rather than active regeneration and restoration activities (Hüning et al. 2008; Baasch et al. 2009).

The different regeneration methods, each feature advantages and disadvantages. Although forest restoration and natural succession both lead to ecosystem change, they are quite different with respect to the degree of intentionality and should therefore be clearly differentiated. While forest restoration is the assisted, intentional, guided reconstruction of forests, natural succession is regarded as unintended, neither prescribed nor directed by humans (Ciccarese et al. 2012). Despite public perception, natural regeneration techniques are not innately superior or always more appropriate than artificial techniques for restoring forest ecosystems. Where management objectives require precise timing and a particular tree species composition of the restored forest, active intervention at the regeneration stage is essential. Interventions should be based on management objectives, the evaluation and interpretation of site conditions, and profound silvicultural knowledge and skills. Appropriate regeneration choices must contribute to current management objectives and be flexible enough for retaining future options. As forests are typically long-lived ecosystems, forest managers continuously face the risk that changing conditions in the future will negate the assumptions on which they based their current decisions.

In traditional forest management, regeneration success is defined by a minimum stocking density of genetically adapted, vigorous young trees of a defined species composition with adequate leader shoot growth and high competitive power (McWilliams et al. 1995, Ponder 1997, Wagner et al. 2010) and appropriate root development (Brunner et al. 2009; Bayer et al. 2013). These criteria also apply to regeneration for restoration purposes. Although stocking levels may differ depending on management objectives, the financial return to the landowner is in many cases an important objective of forest restoration, together with other objectives such as biodiversity conservation. Restoration often aims to alter species composition, sometimes formulating additional requirements for a minimum number of individuals of certain species, and stocking levels can thus deviate from values required for maximum timber production or quality. Restoration objectives may also call for changes in stand structure. When the target is an uneven-aged stand structure, defining an appropriate stocking density is somewhat more complex, because the relationship between seedling density at a given time and the recruitment of trees for the upper tree layers is not necessarily straightforward (Lundqvist 1995).

The complex task of choosing regeneration measures can be approached in two steps, analysis and decision. At several scales—from stand to microhabitat—environmental conditions must be evaluated for different regeneration measures to properly predict options and obstacles over the entire regeneration phase (Nyland 1996). This has to be done for each stage of the life cycle in order to identify potential environmental hazards as well as environmental prerequisites. Recommended regeneration measures can then be derived from these options (Perala and Alm 1990; Jobidon 1994; Van Der Meer et al. 1999).

Analysis can be a daunting task, and practical considerations often necessitate compromises and decision-making based on incomplete information. For example, site conditions may be surveyed and analyzed at different scales and levels of intensity (Barnes et al. 1998; Kimmins 1987; Gholz and Boring 1991; Smith et al. 2007). In practice, the quality and intensity of site surveys depends on the available resources and may be constrained by factors such as staff availability (both numbers of individuals and their competence) as well as the willingness of the forest owner to invest in site mapping and analyses. However, the forest manager's knowledge and experience regarding local site conditions and the performance of indicator species may offset the lack of complete information. Such experience may thus be critical in accounting for changes that may occur after the regeneration intervention, such as potential competing vegetation (Wagner et al. 2010). Once the objectives regarding the future provenance, species composition, stand and age structure have been set, the forest manager must decide on the appropriate regeneration method or combination of methods, adapted management intensity, and adequate timing of all interventions.

Forest restoration is in some ways a new application of silvicultural expertise and a challenge for any silviculturist. Therefore, we must remain mindful that widely used regeneration techniques may be underlain by different objectives and assumptions. We may carefully use existing knowledge, but should also be prepared to seek new information or analyses. After all, it is a substantial challenge for forest research to create new and relevant knowledge for forests with a high dynamic potential both in time and space.

The following sections give an overview of the inherent and external limitations of the regeneration process, structured by stage, and the measures available to counteract them. Our aim in this chapter is to point out the relevance of single life cycle stages within the regeneration process in the context of restoration ecology, in particular by illustrating the interactions between these stages, site factors, and the techniques available to intervene in the regeneration process. We therefore restrict the scope of the chapter to ecological aspects.

6.2 The Natural Regeneration Process

The events associated with the regeneration process of different tree species are some of the most striking phenomena within forests. The natural regeneration cycle includes numerous relevant stages, starting with the flowering of mature trees and ending with the

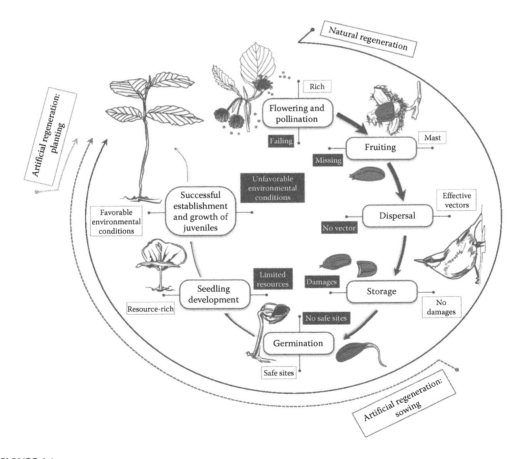

FIGURE 6.1
The regeneration cycle as a cascade of ecological processes and their success and failure in interaction with environmental conditions exemplarily for European beech.

recruitment and establishment of juvenile trees. Figure 6.1 shows a cascade of these eco-logical processes interacting with environmental conditions and their success and failure, illustrated with European beech (*Fagus sylvatica*). Both in theory and practice, forest ecolo-gists have debated for decades over the predominant limiting factors responsible for insuf-ficient natural regeneration within individual forest stands. A vast literature illustrates the complex interactions between mature trees, environmental conditions, and seedling response; but the majority of publications focus only on a few particular phases with high relevance for regeneration success. Ammer et al. (2011), however, underline the need to look at the regeneration cycle in its entirety and to disentangle the multiple factors that are involved throughout the entire process.

Understanding the regeneration cycle is crucial for achieving sustainable forest manage-ment and ensuring "acceptable" forest restoration. From a silvicultural point of view, an acceptable result for each stage can be defined by a minimum density of seeds, seedlings, or saplings required for regenerating a given site; and the limitations are by definition factors causing lower densities than required. In this context, a biological phenomenon will be called "effective" if adequate densities result. This is particularly important in seed dispersal. A detailed analysis of every individual stage will make it easier to evaluate the absence of acceptable natural regeneration density in a specific area, particularly because the failure of any single stage within the complex generative cycle will compromise the overall regeneration success.

6.2.1 Flowering, Pollination and Fruiting as Prerequisites

Reproductive phenology, starting with floral initiation, via full florescence through pol-lination and seed or fruit maturity, is well known for temperate conifers, but less studied for temperate hardwoods (Owens 1995). In order to manipulate the regeneration cycle for specific restoration aims, the drawbacks and opportunities for natural regeneration should be fully understood, because constraints to seed or fruit reproduction are manifold and influence all subsequent stages of the regeneration cycle.

6.2.1.1 Flowering

The majority of seasonally flowering trees flower infrequently. The generative character is caused by a range of developmental, physiological, and environmental conditions such as species-specific pubescence, light intensity, minimum and maximum temperature, water regime, and nutrients. Before the generative character of buds and subsequent flowering becomes visible, particular biochemical changes are measurable (e.g., gibberellins mediate and promote growth and development within the flowering processes of conifers; Bonnet-Masimbert 1987). Many temperate forest trees exhibit so-called "indirect flowering," char-acterized by a period of dormancy between floral initiation and pollination (Owens 1995). In most cases, floral initiation occurs before the onset of winter dormancy. In some spe-cies, the percentage of reproductive buds can be predicted from their external morphology (Owens and Blake 1985).

In silvicultural management, it is a common practice to modify light conditions in order to increase the production of reproductive buds and improve flowering. Such methods have been successfully applied in North American temperate conifers (Ross and Pharis 1985) and European beech stands (Holmsgaard and Olsen 1966), where special thinning regimes, adequate tree spacing, and crown management are promising strategies when nat-ural regeneration is missing. Although higher temperatures at the time of floral initiation may enhance flowering in temperate trees (Ross 1989; Owens 1995), in contrast low winter temperatures enhance flowering in some species such as *Eucalyptus nitens* (Moncur and Hasan 1994). Water stress and associated aspects such as shallow soils, poorly drained soils such as pseudogleys, soil compaction, and anoxia have also been said to enhance flowering in temperate hardwoods and conifers (see Owens 1995 and references therein).

Often successful flowering and later stages are not differentiated because the frequency of (sometimes inconspicuous) flowering may be less obvious than the frequency of seed or fruit production at the end of the reproductive cycle. This is not permissible, however, and may lead to false conclusions and misinterpretations when evaluating natural regeneration success.

6.2.1.2 Pollination

Successful pollination is the next step within the regeneration cycle (Figure 6.1). It tends to be relatively constant from year to year. Although the majority of tree species in temperate forests are wind-pollinated, a few species have other pollination strategies, for example, relying on insects (*Prunus avium*, *Robinia pseudoacacia*, and *Tilia* spec.) or even water (*Abies* spec., see Chandler and Owens 2004). Ashley (2010) demonstrated that pollination and pollen dispersal are complex phenomena which are influenced by many ecological processes. She cited numerous parentage studies showing long-distance dispersal of pollen to be common in both wind- and animal-pollinated tree species, with average pollination distances being hundreds of meters.

Generally, forest trees are highly diverse organisms genetically, and pollen dispersal is usually considered the main driver of genetic relatedness patterns (Sagnard et al. 2011). Genetic variability depends on population size and the spatial distribution of tree individuals. For tree species characterized by a large and continuous range such as European beech, the genetic diversity within populations is typically high while the genetic diversity between populations is comparatively low (Thomsen and Kjær 2002; Buiteveld et al. 2007; Jump and Penuelas 2007; Nyari 2010). Species with small population size in glacial refugia and a low mixing rate between postglacial recolonization lineages typically feature a strong genetic differentiation among but not within populations (e.g., *Fraxinus excelsior*; see Rüdinger et al. 2008; Dobrowolska et al. 2011). However, even within populations or forest stands, genetic diversity is rarely homogeneously distributed, with genetic similarity between individuals often decreasing with increasing distance between them, such as described for the genera *Fagus* (Jump and Penuelas 2007) and *Quercus* (Bacilieri et al. 1994).

The knowledge of spatial genetic patterns resulting from population history and limited gene flow can be important for regeneration management, in order not to misrepresent genetic diversity of species or populations in the progeny population and to avoid autogamy (self-fertilization, measured by the heterozygosity of the resulting juvenile generation) when distances between parental trees are high (Streiff et al. 1998). Due to the impact of lethal regressive genes, germination percent, height growth, or resilience can be lower in the case of self-pollination compared to cross-pollination (Hattemer 2005). Hence, pollen transfer is especially important for small populations whose habitat has been considerably reduced (e.g., Silver fir (*Abies alba*), see Wolf 2003) or fragmented (e.g., genus *Ulmus*, see Goodall-Copestake et al. 2005).

The degree of gene transmission from the parent to the juvenile generation in forest stands depends on the applied silvicultural harvesting and regeneration methods (Finkeldey and Ziehe 2004; Hosius et al. 2006). For example, the removal of a few mature trees from a stand prior to seed production in uneven-aged Plenterwald forests may have no measurable impact on the genetic structure and variability of the naturally regenerated understory (Lefèvre 2004). Similarly, other regeneration strategies such as "Femelschlag" have no negative effects on the progeny generation—the longer the regeneration period, the more uneven-aged the progeny population. In this method, tree cutting intensity is low and explicitly heterogeneous with respect to horizontal structure; the next generation consists of different age classes. This method of harvesting and regeneration is believed to be well-suited to accommodate the variable mast years of different forest tree species (Geburek and Turok 2005). In contrast to these silvicultural regeneration systems, shelterwood and seed tree retention systems are expected to affect genetic diversity to a higher degree, because only a limited number of trees are left for reproduction or are given a limited number of years to pollinate (Buchert 1992). This is especially important if the trees left in the stand are genetically inferior.

6.2.1.3 Fruiting

Although effective pollination tends to be relatively constant from year to year, many tree species fluctuate between years of high and low reproduction, with few "average" years. Even when flowering occurs, fruiting can be erratic. Studies have shown that in harsh climates, single extreme weather events can invalidate supposedly promising background conditions for mast years. Mast years have been variously defined as years where seed production exceeds the long-term mean by some predetermined level (LaMontagne and Boutin 2009 and references therein). Mast seeding is a phenomenon referring to individual trees producing seeds synchronously at "superannual intervals" (Rapp et al. 2013), which leads to large fluctuations in seed production at the population level. Evolutionary ecologists hypothesize that mast seeding occurs because synchronous reproduction among conspecifics is associated with several fitness benefits, including enhanced rates of pollination (Kelly and Sork 2002), increased attraction of seed dispersers (Li and Zhang 2007), and reduced seed predation (Curran and Leigthon 2000; Crone et al. 2011).

Apart from single weather events, the intensity and frequency of fruiting depends on the age and vitality of the individual tree, as well as on the presence or absence of pollinators, or seed and flower predators. Light-demanding pioneer species such as birch (*Betula* spp.), aspen (*Populus tremula*), rowan (*Sorbus aucuparia*), and willow (*Salix* spp.) naturally flower and fruit with high abundance after reaching (early) pubescence. While abundant birch seed crops are observed at 2–3 year intervals in Northern Europe (Hynynen et al. 2010), Central European birch (*Betula pendula*) produces seeds almost annually (Cameron 1996), although with varying quantities (Huth 2009).

Heavy-seeded species such as beech, oak, or chestnut (*Juglans regia*), feature pronounced mast years that usually occur at intervals longer than 5 years (Watt 1923; Sork 1993; Röhrig et al. 2006). Recent observations, particularly from beech forests in Europe (Hilton and Packham 2003), indicate a positive trend in mast frequency: according to the general opinion among foresters in Northern and Central Europe, it has been easier to regenerate common tree species during the last 20 years than previously. An analysis by Övergaard et al. (2007) illustrates the observed trend of generally increasing mast year frequency, associated with increases in mast crops. They found in Sweden that the average interval between mast years of beech has decreased from 4 to 6 years for the period ~1700–1960, to 2.5 years during the most recent 30 years, and there were two consecutive mast years twice during the latter period.

Climatic changes, especially increasing temperatures, may be responsible for the higher frequency of mast years, but increased atmospheric nitrogen deposition may also be a contributing factor (Schmidt 2006). More frequent mast years will likely simplify planning of forest regeneration measures (Övergaard et al. 2007), but it is a legitimate question whether and how this phenomenon is interlinked with tree and stand vitality and thus with sustainable stand development.

Fruiting of species often varies from one year to the next and is not synchronized between species, even though seed rain may be synchronized among cooccurring species of the same genus (Shibata and Nakashizuka 1995). Alternating abundance and composition of seed rain (Lässig et al. 1995) may favor certain species on disturbed sites. In forest stands with advance regeneration or dense ground vegetation, for example, a heavy-seeded species will be favored over a light-seeded species by sudden disturbances such as windthrow (natural), harvesting, and site preparation (anthropogenic), because heavy seeds are not impeded by vegetation and their seedlings can thus occupy local regeneration niches first. As seeds are dispersed over limited distances, seed availability highly depends on the distance and on the strength of the nearest seed source (Clark et al. 1998). Selective removal of seed trees of undesired species prior to site preparation or other regeneration measures can thus be used to promote regeneration of desirable species. For dioecious tree species, removal cutting can be restricted to female individuals.

Once a tree reaches maturity, its contribution to seed production depends on its canopy position, vigor, and genetics. Preparatory cuttings, which are commonly used to precondition overstory trees in shelterwood systems (e.g., for beech), and fertilization aimed to induce flowering and fruiting can improve the generative potential of individual trees. The seed crop per area can be maximized by an optimal combination of individual tree vitality and tree density (Nyland 1996). Due to poorly developed individual tree crowns, this optimum is seldom present in stands of maximum density. Selective thinning aimed at the promotion of trees with good phenotypes has therefore been suggested as a possibility for improving the probability of transferring desirable characteristics to the new stand generation (Nyland 1996). However, losses in genetic variability in stands that were thinned accordingly are barely detected over the course of a single stand generation (Müller-Stark et al. 2005).

6.2.2 Mechanisms of Seed Dispersal and Consequences for Regeneration

The number of seed dispersal studies has grown exponentially during the last two decades (Bullock and Nathan 2008; Schupp et al. 2010). Indeed, dispersal models are indispensable for ecological research, because seed dispersal links the adult reproductive cycle to the seedling stage (Wang and Smith 2002; Rother et al. 2013) and strongly influences the demographic process of plants (Harper 1977).

Forester managers need to know how many seeds trees can produce and how far these seeds are typically dispersed, for example, when initiating natural regeneration by harvesting single trees. For many forest stands, these seemingly simple questions are relatively difficult to answer, because it is challenging to quantify the number of seeds prior to dispersal, and postdispersal dynamics may have an important effect on recruitment and later fertility (Wenny 2000; Birkedal et al. 2009). Although fertility can be inferred from postdispersal seed densities (Schurr et al. 2008), this requires knowledge about the parental tree from which the seeds originate, which is difficult to obtain because the seed rain of neighboring trees typically overlap. The study of such environmental effects seems particularly important for understanding and predicting plant performance in heterogeneous environments and their response to environmental change.

With the help of "inverse modeling" (Ribbens et al. 1994; Clark et al. 1998), tree fertility and seed or fruit dispersal originating from a single individual are estimated simultaneously. Inverse modeling takes advantage of the specific probability that—out of a larger number of trees—a particular tree can be considered the source of seeds caught in single traps placed in defined positions within a forest stand (Stoyan and Wagner 2001). However, this method cannot be simply used to estimate effects of spatially varying environments on fertility and dispersal. The process of seed dispersal can be structured into three components (modified from Millerón et al. (2013):

1. Primary dispersal, which is the movement of seeds and fruits from the tree to the ground by gravity (barochory). Apart from the potential area of plant recruitment, this term also subsumes interlinked processes relevant for spatial patterns of adult plants such as germination, predation, and competition (Nathan and Muller-Landau 2000).
2. Secondary dispersal, which is defined as the removal of seeds and fruits by more or less effective vectors such as animals (zoochory), wind (anemochorie), water (hydrochory), and topographical gradients, once the seed is on the ground (Forget et al. 2005).
3. Effective dispersal, which is the combination of primary and secondary dispersal plus establishment.

In plant populations, recruitment necessitates component 1 (or 2) and 3. For trees in the temperate zone, gravity, wind, and animals (especially via birds and mammals) are the most important dispersal vectors. The term "dispersal range" refers to the distance a seed can move from an existing population or a single adult tree. Seed density on the ground generally decreases monotonically with distance from a seed source (Clark et al. 1999), resulting in very low densities at long distances (Bullock and Clarke 2000). Effective dispersal distances are seldom greater than a few times the height of the seed bearer (Nyland 1996), but differences among species are considerable (Ribbens et al. 1994).

6.2.2.1 *Zoochory*

For some species, invasion into stands of other species can be regularly observed, for example, oak or beech invading pine stands via zoochory of specialized birds (Pons and Pausas 2008; Sheffer et al. 2013). In this context, far-flying jays (*Garrulus glandarius*) are particularly important (Bossema 1979). They are abundant in oak and pine stands of all densities and clearly prefer acorns to beechnuts, whereas nuthatches (*Sitta europaea*) are more abundant in beech stands and prefer beechnuts to acorns (Perea et al. 2011). There are other birds such as great tits (*Parus major*) that remove acorns and beechnuts, especially in stands dominated by oaks, but they do not store the seeds.

It has been shown that zoochorous seed removal is determined by the structure of the dominant vegetation because some habitats are more suitable for the disperser animals. Pure pine stands can be appropriate regeneration sites as early as in the pole stage, when light availability on the ground steadily increases (Sonohat et al. 2006). Densities of more than 2000 oak stems ha^{-1} of acceptable quality can thus develop if deer browsing does not stunt or kill oaks smaller than 1.3 m (Mosandl and Kleinert 1998; Schirmer et al. 1999; Stimm and Knoke 2004). Fencing can further accelerate this process.

The European jay has regenerated several thousand ha of pine stands with oak in Lower Saxony, Germany (Otto 1996). Some foresters deliberately offer acorns to jays in special boxes to facilitate the establishment of oak. Similar observations of succession have been made for rowan, well known in Central and Eastern Europe for its ability to invade pure Norway spruce stands via dispersal by birds (Zywiec et al. 2013) or mammals (Guitián and

Munilla 2010). Carnivorous mammals such as the red fox (*Vulpes vulpes*) and the European pine marten (*Martes martes*) are also the main mammalian frugivore-seed dispersers in temperate Scots pine forests (González-Varo et al. 2013) for rowan. In contrast to Scots pine, however, gaps of a minimum size are essential for rowan to succeed in spruce stands. These findings provide a key starting point for understanding and modeling tree succession in restoration processes that include mammal-mediated seed dispersal, such as connectivity, home range expansion, and recolonization.

In addition to mammals, another effective zoochorous vector in secondary dispersal and predation can be seed removal by rodents. While some rodents feed on acorns thus leading to seed predation, others (mainly scatter-hoarders) can also act as effective dispersers (e.g., for sessile and common oak, see den Ouden et al. 2005; for ash (*Fraxinus excelsior*) and wych elm (*Ulmus glabra*): Hulme and Hunt 1999). The role of rodents in tree and shrub seed removal varies depending on the species, but is affected mainly by seed size and morphology (Jansen et al. 2004) as well as by seed encounter and exploitation (Hulme and Hunt 1999). However, successive dispersal movement and distance to rodent shelter (shrub cover) are more important factors than acorn weight to determine dispersal distance and acorn survival (Perea et al. 2011). The ecological balance between seed predation and effective dispersal is still largely unknown.

Effective dispersal not only depends on the quantity of dispersed seeds but also on the quality of the seed dispersal process. According to Schupp et al. (2010), "quantity" is the number of visits of a dispersal agent multiplied by the number of seeds dispersed per visit, while "quality" is the probability that a dispersed seed survives handling by the dispersal agent in a viable condition (quality of treatment in the mouth and gut) multiplied by the probability that a viable dispersed seed will survive, germinate, and produce a new adult (quality of deposition). These interactions are well analyzed by Gómez et al. (2003) for holm-oak (*Quercus ilex*) and one of its main dispersers, the European jay in a heterogeneous Mediterranean landscape. Moreover, there are animals not only effective for dispersal but also germination (Paulsen and Högstedt 2002).

6.2.2.2 Anemochory

Although seed removal by birds has been demonstrated to play an important role in long-distance dispersal in the context of forest restoration, wind is certainly of special relevance as a widely available seed dispersal vector that can transport many seeds over long distances. Unlike water, wind can transport seeds in all directions and is therefore important for dispersal to upstream wetlands (Soons 2006) and to areas not directly connected to a forest. Compared to animals, wind transports seeds to a wider range of sites, therefore reaching more sites but with lower seed densities. In wind dispersal, the effective dispersal distance is determined by wind speed and direction, which can be influenced by the density and height of the remaining trees (Greene and Johnson 1996; Karlsson 2001). Many pioneer tree species with seeds of very low sinking velocity rely on wind as a seed dispersal vector. Apart from open landscapes, wind-dispersed birch species are thus also able to regenerate in pure conifer stands, particularly favored by heavy disturbances (Perala and Alm 1990) that create large gaps favorable for the growth of pioneer tree species (Huth and Wagner 2006).

6.2.2.3 Dispersal Distances

Forest ecologists commonly differentiate between mean dispersal distance (MDD), long dispersal distance (LDD), and sometimes maximal dispersal density (MAXDD), because these distance measures have different functional relevance in ecosystem development and the spatial management of forests. While the MDD is relevant in conventional silviculture to produce quality timber by means of high stand density, the MAXDD is important for long-term dispersal and invasion scenarios, particularly in fragmented landscapes (Malanson and Armstrong 1996) and for natural species recolonization (Pacala et al. 1996). For example, habitat clustering is a frequent phenomenon within forest restoration, resulting from habitat loss and/or fragmentation processes. Both processes operate at different resolution and with different intensity, for example, wind-throw, forest clearcutting, site degradation, or the segregation of tree species in pure stands due to management. The recurrence of trees at different spatial scales varies according to landscape structure and species dispersal strategies. Disentangling the relative impact of habitat loss and fragmentation on the long-term survival of certain species requires understanding of the interactions of habitat cluster availability and dispersal distance, and how they affect dispersal success. Cattarino et al. (2013) addressed this problem by quantifying the magnitude of these interactions, and emphasized the relevance of long distance dispersal.

The relevance of LDD and MAXDD with respect to restoration is stressed in another example: An understanding of dispersal processes is relevant in the context of preventing and controlling invasive alien tree species or managing the distribution of such species. To predict the seed dispersal potential of *Fraxinus pennsylvanica*, an invasive alien in Germany, Schmiedel et al. (2013) used a stochastic model predicting the number of seeds for a single tree individual. The results were used to calculate species-specific dispersal distances and the effect of wind direction for different assumptions regarding dispersal directionality (isotropic and anisotropic). The topic of invasive species clearly illustrates the need to differentiate between MDD and LDD, because the latter is of paramount importance for invasion dynamics and the rate of colonization; further exemplified by Pairon et al. (2006) for black cherry, a highly invasive forest tree species in Europe.

Knowledge of dispersal distances is required to determine "effective" seed density, which should not focus on the maximum observed distances. In mixed stands, the effective distance to seed trees of all species is of particular interest if aiming to maintain the current species mixture. If effective dispersal is lacking or the rate of recolonization is too slow, for example, at long distances from seed sources, silvicultural techniques such as direct sowing or planting of later successional species with limited dispersal ability can be used to facilitate the establishment of desired tree species.

6.2.3 Costs, Benefits, and Implications of Seed Storage

Spatial discordance between primary and effective seed dispersal in forest stands indicates that postdispersal processes are responsible for differences between seed rain densities and seedling recruitment patterns. Although seed rain is observed mostly below and near canopy trees, saplings are often established far from parental trees (e.g., for *Fagus sylvatica*: Milleron et al. 2013, for maple (*Acer* spp.): Wada and Ribbens 1997). This discordance pattern may be the result of secondary dispersal by animals or density-dependent effects such as the Janzen-Connell effect (Janzen 1970 and Connell 1971). The period between dispersal and recruitment, the so-called storage phase, is characterized by high potential mortality rates and is thus one of the most relevant phases of the regeneration cycle.

For some tree species, seeds are not stored but germinate within the current growing season (e.g., *Acer rubrum*, and *Ulmus laevis*). But for most temperate and boreal tree species, seeds are stored on or within the forest floor because of unfavorable conditions for germination and growth outside the vegetation period. Winter storage is thus an adaptation of species to the annual climatic cycle at middle and high latitudes (Runkle 1989). Extended periods of opportunistic dormancy (i.e., enforced and induced dormancy; see Harper 1977) can be regarded as an alternative to dispersal in space, sometimes referred to as dispersal in time (Willson 1993).

6.2.3.1 Orthodox and Recalcitrant Seeds

Seeds of many tree species are sensitive to desiccation, humidity, anoxic conditions, frost, fungal decay, or predation, and are therefore difficult to store. Seed storage patterns and conditions not only vary among species, but also within species and even provenances. The response of seeds to storage conditions and duration is generally expressed by the terms "orthodox" (desiccation-tolerant) and "recalcitrant" (desiccation-sensitive; see Roberts 1973). Orthodox seeds can be stored *in situ* for years, and also artificially under cool low-moisture conditions. Many forest tree species of important genera belong to this group, including all European coniferous trees. Sugars (e.g., sucrose and raffinose) play an important role as storage substances, accumulated in cells during maturation and degraded during the first 10–20 h of germination (Downie and Bewley 2000). In seed banks, seeds of species with dormancy adaptations (e.g., *Robinia pseudoacacia*, *Prunus serotina*) can survive dormancy periods of several years. In contrast, the seeds of oak (*Quercus* spp.), beech (*Fagus sylvatica*), or chestnut (*Castanea sativa*) are recalcitrant; they are intolerant of desiccation and cannot be stored in the soil for more than one winter without the loss of viability (Doody and O'Reilly 2008).

6.2.3.2 Seed Banks

Seed densities can increase rapidly following a major seed production event (e.g., masting in heavy-seeded oak and beech) or progressively by a combination of consecutive smaller dispersal events and seed storage (dormancy), thus building up seedling (ash, maple, and black cherry, *Prunus serotina*) or sapling banks (beech and fir) under closed canopies (Szwagrzyk et al. 2001). Most of these species produce fewer seeds per tree and lower seed densities on the ground over time than more light-demanding species (Ribbens et al. 1994; Clark et al. 1998). Most tree species in temperate forest ecosystems do not build up long-term soil seed banks or sapling banks (Halpern et al. 1999; Bossuyt and Honnay 2008). But existing seedling and sapling banks are strategies by which low seed production is compensated for by prolonged storage time. Underlying mechanisms include architectural adaptations such as the relative proportion of leaf and branch biomass (Kohyama 1987), high morphological plasticity such as opportunistic plagiotropy in beech (Brown 1951), and metabolic adaptations such as low respiration rates (Walters and Reich 2000). Individual seedlings are often capable of vigorous responses to sudden improvements in resource supply, and may thus occupy promising niches in advance of other less tolerant species.

Seed banks often serve as reservoirs of tree species diversity, which buffer the composition of plant populations and influence the postdisturbance dynamics of vegetation succession (Royo and Ristau 2012). Aerial seed banks, where seeds are retained in tree crowns, have been observed for some fire-adapted species outside Europe such as *Pinus contorta* (Richardson 1998). For forest restoration strategies, forest managers can use a range of regeneration methods relying on seed banks, but they must be familiar with the particular mechanisms needed to break the dormancy of the desired tree species. These mechanisms can be activated by light or heat (i.e., radiation), leading to particular cutting regimes or prescribed burning, as, for example, applied in eucalypts or serotinous pine and spruce species (Smith et al. 2007).

6.2.3.3 Widespread Susceptibility

Upon making ground contact (primary dispersal), seeds are exposed to new risks and mortality factors, among them are pathogens and predation. Postdispersal seed loss can cause an important bottleneck in the natural regeneration of many tree species. The relative importance of each mortality factor varies depending on seed quantity and microhabitat conditions, resulting in spatial discordance in the performance of each regeneration stage. In order to minimize postdispersal loss, some tree species produce seeds in irregular mast years. The intermittent mass production of seeds, for example, in beech stands, is often invoked as a strategy evolved by some plant taxa to overcome the capacity of seed predators to consume all seed (predator satiation; see Kelly and Sork 2002).

6.2.3.4 Predation by Rodents

Seed loss due to mice is common (Madsen 1995) and partly determined by storage duration. The ability of a rodent population to affect seed reserves at the scale of a forest stand depends on whether or not—and if so, how frequently—the overall rate of seed consumption exceeds the rate of seed production (Ruscoe et al. 2005). Significant effects of burial and microhabitat have been published for coexisting Mediterranean oak species in Spain (Pérez-Ramos and Marañón 2008). The highest predation rates occurred for acorns located on the ground surface (unburied), and in the most densely vegetated microhabitats, where rodents usually exhibit higher activity. The lowest predation rates were observed in years and at forest sites where the estimated seed production—and consequently resource availability—was higher than average, thus supporting the predator satiation hypothesis. This temporal pattern of higher seed predation in nonmast years has been also documented in other studies (Hulme and Borelli 1999).

Concerning restoration practices, an important recommendation for sowing is therefore to bury seeds 1–3 cm deep in the mineral soil (Birkedal et al. 2010, 2009). According to different studies (Pérez-Ramos and Marañón 2008), seed loss can thus be considerably reduced and the establishment of seedlings favored, as the leaf litter layer and mineral Ah horizon may improve seed performance via reduced soil temperature and water evaporation and increased local humidity levels (Rother et al. 2013). Moreover, it would be better for restoration sowing activities to focus on microhabitats with a low or moderate vegetation cover and to avoid microsites with dense shrub canopies, where rodent activity is usually high regardless of region and forest type (Fedriani et al. 2004).

6.2.3.5 Predation by other Vertebrates

Seeds of common tree species (e.g., oak and beech) are affected by vertebrate predation both pre and postdispersal, because they produce large seeds of high nutritional value (Nopp-Mayr et al. 2012). In return, birds disperse acorns away from the parent tree and create seed caches as food supply for the winter (Stimm and Böswald 1994). However, birds (as well as rodents) also predate on seeds such as acorns and beechnuts. Other omnipresent acorn and beechnut predators are wild boar, roe deer, red deer, and many beetles (e.g., weevil larvae), who can consume most of the primarily and secondarily dispersed seeds (González-Rodríguez and Villar 2012).

6.2.3.6 Scatter Hoarding

Seed dispersal by zoochory often results in highly clumped seed deposition, with high seed densities in some locations far away from the parent tree in areas used preferentially by dispersers (Schupp et al. 2010). Jensen (1985) found beechnuts scatter-hoarded with a mean number of five seeds per cache. The typical burial site is usually located a few cm below the soil surface close to the wall of a rodent runway. Predation on caches is generally high, but in experiments more caches survived if a surplus of seeds was offered or if rodent numbers were reduced. Age structure of saplings revealed that most individuals had germinated in the years following mast years. Thus, there is strong circumstantial evidence that scatter-hoarding animals influence the population biology and evolution of tree species by predation and dispersal of seeds. In turn, the synchronous production of seeds leads to prolonged reproduction periods in the rodent species, resulting in outbreaks.

Recent research results highlight the importance of clumped seed deposition for patterns of seedling survivorship and recruitment. By using spatially explicit simulation models, Beckman et al. (2012) showed that clumped seed deposition increased the probability of seedling establishment under both insect predation (host specific bruchid beetles) and pathogen attack (*Pythium* and *Phytophthora*), as it led to local satiation of insect seed predators and made it harder for pathogen distributions to track seeds.

6.2.3.7 Pathogens

High humidity levels at the forest floor generally favor micro-floral growth, which reduces the aeration of seeds and results in the production of toxins (Knudsen et al. 2004), thus causing seed rotting and reduced germination. Seeds are particularly susceptible to a variety of pathogens and suffer considerable loss of viability during storage.

The pathogen *Ciboria batschiana*, a fructicolous Discomycete, is a serious problem affecting acorn storage, which in combination with other native fungi is responsible for poor storability. *C. batschiana* can destroy up to 80% of an acorn crop and may also result in severe losses in oak seedling populations in forests (Schröder et al. 2004). Thus, oaks benefit from seed dispersal, as seedling recruitment is facilitated due to a decrease in pathogen infections following dispersal (Clark and Clark 1984). Beechnuts are affected by fungal infection to a similar degree. Massive mortality in stored beechnuts both in Europe and in the United States has been caused by several *Phytophthora* species. Mycological studies confirm the susceptibility of beech seedlings to *Phytophthora* spp. (Orlikowski and Szkuta 2004). Another relevant fungus in beech storage is *Rhizoctonia solani* (Hietala et al. 2005) displaying symptoms of cotyledon rot. The disease is characterized by decay resulting in reduced failure to sprout or death after emergence.

Pythium is a genus with high pathogenicity and causes the damping-off disease of germinating seeds and seedlings during storage, which is not restricted to a single species (Augspurger and Wilkinson 2007). As numerous species of this oomycetes have been described for nursery soils (Weiland et al. 2013), there is a potential risk of infecting restoration sites in the course of planting.

6.2.4 Germination in Stressful Surroundings: A Narrow Bottleneck

6.2.4.1 Germination and Safe Site

Occurrence and timing of germination play essential roles in subsequent plant establishment (Baskin and Baskin 2001; Manso et al. 2013) and are often discussed as a bottleneck within the demographic transition from seed to sapling (Rother et al. 2013). Seeds germinate only if certain conditions are met; including the breaking of dormancy. The species-specific maternal reserves within the seed endosperm must be mobilized to provide for the two autotrophic cotyledons, the primary leaves, and the proliferating roots. As the primary integrators of environmental signals (Farnsworth 2008), phytohormones enable the nascent seedling to adapt to its surrounding conditions. For many tree species, germination itself does not depend on light conditions (Nicolini et al. 2000), and radiation is thus not relevant until the reserves in the endosperm are depleted.

The results for seedling survivorship after germination have been widely studied (reviewed in Goulet 1995); the seedling environment is often described as extreme (Grime 1979) and conditions can be stressful and chronic, even in forest restoration. Factors inducing stressful surroundings for the initial seedling development can be abiotic (e.g., unfavorable light and temperature regime, drought, pH, and anoxia) or biotic (e.g., high intra- and interspecific competition, absence of mycorrhizae, herbivores, and pathogens). The latter often benefit from the same environmental conditions that are favorable for germinating seeds or succulent seedlings, and fungal attack and death by damping-off disease are thus common fates.

The microsite where the germination process can be successfully completed is termed a "safe site" (Harper 1977). Safe sites generally provide a seed with sufficient moisture, warmth, oxygen (Baier et al. 2007), and light of appropriate quality (Smith et al. 2007). Pathogens are absent or ineffective at safe sites. A lack of safe sites restricts germination even at many restoration sites, especially those with extensively exposed soils and, at the other extreme, those with a dense cover of competing vegetation (Urbanska 1997; Galatowitsch 2012).

6.2.4.2 Light

In many cases, seedling reserves are critical, especially under conditions of limited light availability (Ammer et al. 2008b). Even small reductions in biomass accumulation may lead to seedling mortality (Fenner and Thompson 2005). In some forest ecosystems, the light intensity at the forest floor is close to the photosynthetic compensation point, that is, the light intensity at which respiration is equal to photosynthesis (Modry et al. 2004; Facelli 2008). The successful establishment in such a stressful situation is only possible if the seedling achieves a sustained positive carbon balance; photosynthesis must exceed respirational carbon loss (Kitajima and Myers 2008). Seedlings produced by larger seeds (e.g., beech) are often more shade-tolerant (Leishman and Westoby 1994) due to more abundant reserves. Although seed size may not be an adaptation to limited light condition *per se* (e.g., *Abies alba*), it may assist in avoiding mortality as a result of herbivory, desiccation, or burial by litter.

The morphology of seedling organs is also affected by varying light availability. If growing in shaded conditions, leaves are generally broader and thinner, because this optimizes light capture (Bazzaz 1996). Most species also feature lower specific root length and increased leaf area under these conditions (Reich et al. 1998).

6.2.4.3 Litter and Soil Organic Matter

As a boundary layer, that is, a transition zone between the atmosphere and the soil, soil surface characteristics strongly determine microsite conditions. Over the range of a few cm, the environmental conditions can change dramatically. The boundary layer is primarily characterized by site, species-dependent litter quality and quantity, stand age, and stocking density. Existing vegetation can alter the boundary layer conditions by intercepting radiation, affecting moisture availability, and determining the thickness of the litter and humus layers and thus the humus form (George and Bazzaz 1999; Fischer et al. 2002).

The humus form, humus content in the mineral soil, and the thickness of the litter horizons are particularly crucial for overall moisture supply. The surface substrate and the vertical depth of litter and humus accumulation in relation to the relative size of seeds and germinants often determine whether the given site can serve as a safe site for a particular species or not. Exposed mineral soil is usually the best substrate for germination and initial growth because of its favorable moisture supply. Although litter has no effect on emergence (or even a positive one, see Facelli 2008), thick humus layers can considerably impede seedling emergence and germination. Apart from reducing light availability (Facelli and Pickett 1991), litter may release leachates with potentially allelopathic effects on seedling establishment (Olson and Wallander 2002). For example, Chrimes et al. (2004) and Mallik and Pellissier (2000) discussed these interactions based on allelopathy for *Picea abies* and *Vaccinium myrtillus* and other dominant ericaceous understory plants.

The fate of seedlings is thus affected by seed location at the time of germination. Plants are more successful when their seeds are near the mineral soil surface as roots growing in litter substrate can fail to obtain enough water for survival. Acorns and beechnuts often germinate best when buried 2–3 cm in the soil, regardless of the covering material (Millerón et al. 2013), whereas birch seeds require physical contact with fine humus material or the uppermost mineral soil (Ah) horizon to germinate successfully (Carlton and Bazzaz 1998; Karlsson 2001).

Thick humus layers (e.g., raw humus) may need to be reduced to promote survival of seeds and to facilitate germination, which can be done directly by prescribed fire or indirectly by preparatory cutting. However, if the thinning intensity is too high, advance regeneration of undesired species or competing ground vegetation may thus be promoted. Undesired species are often more shade-tolerant than desired ones, for example, Norway spruce (*Picea abies*) on wet soils where pedunculate oak (*Quercus robur*) is preferred or in beech stands on sites with good nutrient status where more valuable broadleaf tree species are preferred. In mixed stands, species composition can be altered by means of early thinning that selectively removes undesirable species.

Direct treatments aiming to make the physical environment of a site more suitable for germination are more common than preparatory cutting. These treatments are intended to modify the microclimate, improve access to water supply by exposing mineral soil, or eliminate competing vegetation (Morris et al. 1993). Although site preparation may interfere with natural succession (Nyland 1996), its benefits include the elimination of undesired tree species (Gordon et al. 1995; Lautenschlager 1995) as well as of grass and herbs which provide cover for seed predators such as mice. Measures are usually directed at species that have established advance regeneration and where control by weeding or precommercial thinning is not feasible.

Apart from mineral soil, the germination of many species is facilitated by downed deadwood, which consists of woody debris, stumps, and overgrown deadwood (LePage et al. 2000; Hagemann et al. 2009). This is particularly relevant at harsh sites, for example in mountain forests or in forests with a continuous grass cover (e.g., *Calamagrostis* spec., *Deschampsia flexuosa*, and *Carex brizoides*). The successful establishment of species such as Norway spruce, silver fir, sycamore, and rowan could be enhanced in high-elevation, grass-covered areas by increasing deadwood abundance (Santiago 2000; Motta et al. 2006). The role of downed deadwood is discussed in more detail in the next section.

6.2.4.4 Interaction with Mycorrhizae

Many case studies, both from laboratory experiments and field studies, show that specific mycorrhizal fungi promote seedling establishment through increased access to soil resources (Nara 2006), drought tolerance, and resistance to pathogens, among other benefits (St. John 1997). In most undisturbed European ecosystems, high root density and the ubiquity of long-lived mycorrhizal plants lead to conditions where inoculum (both spores and hyphae) is almost always available to newly germinated seedlings (e.g., Janos 1992; Börner et al. 1995). However, in restoration sites, especially where afforestation is required, conditions can be different as unfavorable soil status and/or artificial vegetation structures often prevail. After a site and its vegetation have been disturbed, its soil is more likely to have reduced and/or patchy mycorrhizal infectiveness (Janos 1992), especially for fungi that colonize new roots predominantly via hyphae rather than spores (Borner et al. 1996). Soils of early successional forests therefore typically show a low abundance and diversity of mycorrhizal fungi (Galatowitsch 2012).

Allen et al. (2002) noted that restoration measures typically attempt to establish late successional vegetation by planting late-seral species in early successional soils. The benefits of mycorrhizal symbiosis are therefore generally more readily apparent in forest restoration than in stands with high naturalness. Even so, plants intended for afforestation or enrichment planting in restoration should be inoculated in the nursery with appropriate mycorrhizal fungi, whereas an inoculation will be of little benefit for plants to be established in undisturbed forest soils.

6.2.4.5 Physical Damage to Seedlings

Even after emergence, seedlings and saplings can be subjected to substantial substrate movement, especially during forest restoration. As the result of wind erosion, young trees are often killed from partial root exposure, as often described for postmining landscapes (Hüttl and Weber 2001). Yet another phenomenon also creates unstable substrates for seedlings: in freezing air and low surface temperatures without snow cover, seedlings can be heaved out of the soil by ice crystals forming near the soil surface (Facelli 2008).

6.2.5 Seedling Survival and Establishment: The Fine Step to Maturity

As all autotrophic plants require the same resources (i.e., water, nutrients, and light), competition is the most important process during the postemergent phase of the young tree. Both overstory and ground vegetation reduce resource availability for seedlings, and mortality is thus common during subsequent seedling development.

Competition between individual plants can be differentiated into the effect of biomass production on resource availability and the response of individual fitness to resource limitation (Goldberg 1990). In recent years, knowledge of different strategies used by tree seedlings to acquire and allocate resources has improved considerably, particularly for photosynthetically active radiation (PAR; see von Lüpke 1987; Coates and Burton 1999; Hertel et al. 2012) and nutrients (Johansson et al. 2012; Guo et al. 2013). In contrast, knowledge of belowground mechanisms and their importance for seedling vitality is still rudimentary (Havranek and Benecke 1978; Flaig and Mohr 1990; Ammer 2003). Belowground resources in forests feature an extremely high variability at both micro- and macro-scales, which is difficult to control and separate from variability in PAR (Huss and Stephani 1978; Reed et al. 1994; Walters and Reich 1997; Finzi and Canham 2000). In particular, it is not well known how inter- and intraspecific root competition influence the survivorship of individual roots (Rust and Savill 2000; Beyer et al. 2013). Nevertheless, species-specific responses to the complex resource pool are the key to understand the coexistence of species (Tilmann 1982).

Restoration measures controlling competing vegetation during the seedling establishment stage can increase soil temperature, PAR availability, and nutrient availability, thereby improving seedling survival and growth particularly under unfavorable site conditions (Brand 1991; Madsen 1995; Groot 1999). The impacts of limited PAR availability due to competing vegetation are species specific (Küßner et al. 2000), but also depend on interactions with other environmental resources (Lautenschlager 1999; Küßner et al. 2000). Specific recommendations for vegetation control depend on the target tree species (Löf 2000), the dominant weed species (Lautenschlager 1995), and the site. Specifying appropriate treatment intensities is not easy (Tappeiner and Wagner 1987; Cain 1991), and should account for the difference between a competition threshold and a critical-period threshold (Wagner 1999). The former refers to the vegetation density at which yield loss occurs (Jobidon 1994), the latter to the time when vegetation control should begin to prevent yield loss (Wagner et al. 2010, 2011).

The main aim of this chapter was to illustrate the individual stages on the path to successful natural regeneration and to allow for a more accurate interpretation of the general recruitment probabilities. A clear focus was set on the identification of critical environmental and biological factors potentially affecting the recruitment of the progeny tree population. Although the young stand can mirror the seed rain distribution originating from the parental stand, its characteristics more likely will be modified by differential seed dispersal, seed storage and germination, mortality or predation, and growth (see Figure 6.1). This background information will help to understand the specific silvicultural strategies for forest restoration presented in the following two sections, discussing restoration strategies considering the overstory structure and below-canopy stand components.

6.3 Overstory Restoration Strategies Oriented toward the Natural Disturbance Regime

Forest restoration measures that take place in degraded forests can take advantage of existing canopies, which are to some extent degraded, that is, altered with regard to species composition, structure, density, and so on. The forest canopy also affects tree regeneration below, however, and these effects may be deliberately modified by treatments. This section describes some options of canopy treatments to facilitate forest restoration through regeneration.

6.3.1 Emulating Overstory Conditions Originating from Large-Scale Disturbances to Support Ecological Processes

Two main functions of large-scale disturbances (between 10 and 100 km²; Temperli 2012) in forest ecosystems have been identified as the gradual progression toward natural forest development stages, and consequently the possibility for adaptation to environmental changes after the destruction of the current forest system (Mitchell et al. 2004; Walker et al. 2007). The postdisturbance, more or less time-consuming successional process includes a broad range of subprocesses, among them mineralization, transpiration, regeneration, and recolonization. The progression and quality of each subprocess are influenced by the disturbance origin and the degree of destruction of different structural compartments. Possible abiotic causes for natural large-scale disturbances in temperate forest ecosystems are storms, drought, fires, floods, avalanches, and wet snow (Frelich 2002), which all change the structure of the previous overstory tree layer. The damage inflicted upon overstory trees and single stems strongly differs between broken, uprooted, and burned stands (Busing et al. 2009; Goldammer 2013; Mitchell 2013). The same applies to large-scale biotic disturbances, caused by bark beetle attacks or fungal infections (Seidl 2009; Netherer and Schopf 2010), because of differences in individual tree or tree species resilience. Although the overstory tree layer is often disturbed over a large, more or less contiguous area, the understory vegetation and the surface soil layer are characterized by heterogeneous disturbance patches of different size (Walker et al. 2007; Busing et al. 2009; Jonášová et al. 2010). The resulting mosaic of the previous understory is influenced by disturbances in a different way than the overstory, because natural large-scale disturbances never feature homogenous damage intensities throughout the entire disturbed area. This is mainly due to the heterogeneity of specific site conditions such as topography, relief, soil horizonation, or groundwater level. Moreover, the different traits of the present ground vegetation also contribute to spatial heterogeneity (Cater and Chapin 2000). Both categories of influential factors increase the site specific heterogeneity after large-scale disturbances, and thus support the formation of diverse ecological niches (Hutchinson 1978; Honnay et al. 2002). For example, small-scale patches and structures occur frequently following different large-scale disturbances, including broken branches, crowns and stumps, fallen trees, uprooted and leaning trees, as well as root plates, pits and mounds (Ulanova 2000; Brang 2005a,b). Overall, the original stand conditions are crucial for postdisturbance ecosystem development. It can be assumed that large-scale disturbances lead to more heterogeneous conditions in natural forests than in plantation forests (Keidel et al. 2008), where the overall variability is low due to mono-structured stand conditions and artificially homogenized site conditions. Moreover, extreme climatic conditions probably increase the frequency and risk of large-scale storm, drought, or flood events (Rowell and Moore 2000; Dale et al. 2001). Even though heterogeneity following large-scale disturbances will increase the diversity of plantation forest stands; it can be assumed that the next stand generation will still be less diverse than disturbed natural forest stands (Millar et al. 2007).

For large-scale restoration, efforts aimed at modifying disturbance regimes can result in the support of successional processes. According to Walker et al. (2007), *"succession and restoration are intrinsically linked because succession comprises species and substrate change over time and restoration is the purposeful manipulation of that change."* These authors argue that restoration strategies are usually geared toward short time scales, while natural succession processes need longer time frames to successfully develop (Thomasius and Schmidt 2004). Taking this into account, the practical implementation of natural processes and structures associated with large-scale disturbances can be divided into primary and secondary manipulations aiming to restore the system. Primary large-scale manipulations are associated with considerable technical efforts and create important structures for supporting specific successional processes. Restoration measures in this manipulation category primarily affect the natural regeneration indirectly via altering the overstory conditions as summarized in Table 6.1.

TABLE 6.1

Possible Primary Large-Scale Restoration Measures Associated with Altered Overstory Conditions

Primary Restoration Measures	Effects on the Regeneration Process
Complete clearing Cutting of overstory trees and removal of all woody material	Striving for a complete change in tree species composition due to high environmental risks and to prevent further seed dispersal and regeneration of undesired tree species. This procedure is typically used for controlling invasive/alien tree species (D'Antonio and Meyerson 2002). Usually, low amounts of deadwood are left, because of the nontarget tree species within the previous overstory
Simulated windthrow Pulling trees down or breaking stems using winch systems or excavators	Decoupling and reducing overstory tree competition in favor of tree regeneration. As a result, light, water, and nutrient availability at the forest floor increase significantly (Kliejunas et al. 2005; Koizumi et al. 2007). Particularly newly established light-demanding tree species, advance regeneration and early successional ground vegetation will benefit from this measure. Root plates create pits and mounds. High accumulations of different deadwood categories are present
Prescribed burning Implementation of low or high intensity burns by means of controlled fire lines	Low-intensity fires kill the existing ground vegetation and reduce the litter layer, but most of the soil seed bank (Hille and den Ouden 2005; Goldammer et al. 2013) and the overstory trees (Kozlowski 1974) will be unaffected. Only some trees will suffer cambium damage and subsequent die-back, but the degree of canopy closure is largely unchanged. It should be noted that, some tree species will be present in the next stand generation through their ability of resprouting. Therefore, favorable conditions for the regeneration of intermediate tree species exist. In contrast, crown fires with higher intensity destroy the overstory trees as well as the ground vegetation and soil seed bank. Thus, light availability at the forest floor will significantly increase. The following succession dynamics depend on the seed dispersal rate from surrounding tree species (Piha et al. 2013)
Water level regulation Technical implementation of different topographical structures within floodplains such as floodplain drainage, blocking of channels, restoration of meandering riverbeds, and rewetting of swamp areas	The regeneration of broadleaved tree species depends on the availability of seed trees which are well adapted to fluctuating water levels (Leyer et al. 2012). Seed production, seed dispersal adaptations, and stream velocity decide about the presence of different tree species in the regeneration layer (Peterken and Hughes 1995). The composition and the distribution of tree regeneration can be controlled by small-scale modification of soil conditions, elevation, water level, and duration of flooding (Deiller et al. 2003)

6.3.2 Manifold Harvesting as a Strategy for Managing Structural and Species Diversity

Two common restoration objectives—the development of natural conditions within protected areas without any further forest utilization, or the establishment of near-natural conditions as an integral part of silvicultural management strategies—require a combination of various structures to enhance species diversity. Restoration strategies within protected areas (e.g., reserves and national parks) are often stratified according to zoning categories, for example, "core zones" where forests are left to develop freely, "development zones" defined as protected areas managed mainly for ecosystem protection, and "recreation zones" subjected to near-natural management concepts (e.g., Federal Agency for Nature Conservation 2012). For national parks or reserves, restoration measures generally have only initializing character (Wojczulanis 2002; McComb 2007). Continuous restoration measures are therefore mainly required for application in managed forest areas, and

TABLE 6.2

Overview of Typical Stand Characteristics within (a) Plantations and (b) Near-natural Forests, and Restoration Measures Suitable for Initiating Change from (a) to (b) (marked by arrows).

(a) Plantation Forests	Suitable Restoration Measures	(b) Natural/Near-natural Forests
Single-species overstory	*Promotion and revitalization of admixed tree species via control of crown competition* -----→	Minimal tree species admixture
Single-layer overstory	*Establishment of vertical and horizontal stand structures via single-tree selection systems and combinations of uneven-aged tree species admixture* -----→	Single to multilayered overstory
Homogeneous canopy closure	*Combination of various differently-sized disturbances to implement heterogeneity in light, nutrient and water availability and influence inter- and intraspecific competition* -----→	Heterogeneous canopy closure
Homogeneous soil conditions	*Establishment of variable site conditions to promote the diversity of ecological niches available for different tree and ground cover species* -----→	Natural, unaffected soil conditions
Large-scale management blocks with homogenous treatment	*Increase of the variability of treatment sizes, methods and techniques and decrease of the size of management blocks* -----→	Natural mosaic of successional patches
Few selected, often not site-adapted tree species	*Selection of vital trees, well adapted to specific site conditions; approximation to natural tree species admixture with variable vertical and horizontal structure* -----→	Long-term site-adapted tree species

must be differentiated by spatial scale to successfully combine restoration measures with harvesting procedures. At the landscape scale, forest restoration practices should combine manifold harvesting systems instead of implementing homogenous harvesting methods over large areas (Lindenmayer and Franklin 2002). One of the main deficits of homogenous harvesting regimes with respect to restoration efforts and naturalness is the loss of small-scale diversity of species and structures. Species with a low activity density need transition zones or corridors across scales to maintain their populations (Hunter 1999; McComb 2007). At smaller scales, the selection of suitable harvesting systems for restoration activities aiming to establish complex natural processes is often restricted by existing stand structures, site conditions, and the available space. Usually, the selection of management strategies in forest enterprises is affected by numerous external factors that are not directly linked to the topic of restoration (Nyland 2002; Messier et al. 2013). In order to deliberately use specific harvesting systems in forest restoration, the resulting structures or processes must be identified (Fries et al. 1997; Kerr 1999). This section analyzes "classic" harvesting systems and their suitability to establish or maintain near-natural structures and regeneration conditions (Table 6.2).

6.3.2.1 Clearcut System

Although the use and the size of clearcuts have been restricted in most temperate forest regions, the system is still preferentially practiced in some forest regions because of the high degree of mechanization, low short-term costs, and simple demands with respect to management. The removal of the entire overstory leads to extreme changes of environmental conditions (i.e., light, temperature, water, and nutrient regimes). As a result, clearcut sites are often completely restocked using artificial regeneration, particularly plantations where stocking control is an objective. Clearcutting is also used to naturally regenerate forests where desired species are shade-intolerant. With respect to restoration objectives and near-natural processes, clearcuts only emulate those aspects of natural disturbances that are linked to the complete loss of overstory trees (Priewasser et al. 2013), while—in contrast to natural disturbance regimes—important structural elements such as deadwood, small-scale soil surface heterogeneity and islands of potential seed trees are lacking. If advance regeneration is already present prior to clearcutting, the tree species composition of the future overstory is determined by natural conditions, likely filtered by climate and interaction processes. Thus, clearcutting can be used as an initial event for forest restoration oriented toward establishing natural succession processes, as for example, practiced when converting late successional forests into non-forest ecosystems such as meadows or heathlands (Leuschner 1994; Härdtle et al. 2009). One relevant difference between regular clearcuts and naturally established open areas is the presence of deadwood at these sites (Priewasser et al. 2013). Moreover, if suitable seed trees are missing, small-scale supplementation with appropriate seeds or seedlings is possible.

6.3.2.2 Seed Tree System

The seed tree system can be described as a "soft version" of the clearcut system with respect to the creation of extreme environmental conditions and suitable seed trees (Kuuluvainen and Pukkala 1989; Nyland 2002). This system is mainly used to regenerate one light-demanding tree species throughout the entire site, but more natural modifications of the system can be easily made. Parts of or entire overstory trees can be left on site to increase deadwood stocks (Rosenvald et al. 2008). Areas without dense regeneration of target tree species can be left to regenerate naturally by other surrounding tree species, which will result in an admixture of different tree species with a broader range of age and development stages. For the seed tree system, the change from closed stand conditions to open-site conditions is less pronounced than for the clearcut system.

6.3.2.3 Retention Tree System

When both forest practitioners and forest scientists had realized the discontinuity of the complete clearcut system, they developed the idea of retaining a number of selected trees, so-called retention trees to compensate for the loss of key species important for forest ecosystems (Mitchell et al. 2007; Rosenvald and Lõhmus 2008; Lindenmayer et al. 2012). In the retention tree system, also called reserve tree system, retention trees can create a permanent link to original forest-adapted floral and faunal species of soil-dwelling organisms, insects, mycorrhizae, or specialized seed dispersers such as small mammals and birds (Matveinen-Huju et al. 2006; Gustafsson et al. 2010; Lindenmayer et al. 2010). This "link function" also applies to important processes associated with typical forest conditions. The shelterwood system—a special retention tree system—creates homogenous low-density canopy conditions, which promote the establishment of dense tree regeneration throughout the entire site. The regeneration of shade-tolerant tree species such as European beech using the shelterwood reserves system has thus led to large areas with dense homogenous beech regeneration (Wagner et al. 2010).

6.3.2.4 *Irregular Shelterwood*

Without supporting activities, it is therefore impossible to integrate admixtures of other tree species. Hence, manipulations which create irregular canopy openings to increase tree species heterogeneity within the regeneration layer are one approach to increase biodiversity by means of restoration measures (Raymond et al. 2009). Similarly, small-scale preparatory cuttings can be temporally staggered to stimulate seed production by overstory trees and to increase vertical structure. Additional cuttings within the regeneration layer can be used to reduce the competitive pressure on admixed tree species with lower shade tolerance and small admixture percentage. Without additional restoration measures, all harvesting systems outlined above produce even-aged tree regeneration (Fries et al. 1997; Nyland 2002). In contrast, the following silvicultural methods are oriented toward the establishment of uneven-aged tree regeneration. These also include the two general types of group-selection methods, which reduce the degree of canopy closure by means of gap and group-sized shelter cuttings (Sagheb-Talebi and Schütz 2002). Light demanding pioneer tree species receive greater support by the imitation of larger natural gaps. Small-scale heterogeneity within those gaps naturally results from the spatial within-gap heterogeneity of environmental conditions (e.g., light availability) and from edge effects created by the surrounding border trees. The stronger the environmental gradients, the more ecological niches can become colonized by different ground vegetation and tree species (Sagheb-Talebi and Schütz 2002; Poorbabaei and Poor-Rostam 2009). Therefore, the exclusive use of shelterwood-group selection cutting limits the establishment and growth of tree species requiring higher light availability. The adaptation of group selection to natural disturbance variability can increase the variability of small-scale environmental conditions as well as the dynamics of the ensuing regeneration processes. Although group-selection methods generally feature a high flexibility for use in forest restoration, single-tree selection is often considered the preferential uneven-aged reproduction method (Nyland 2002). However, single-tree selection systems are often criticized for excluding light-demanding tree species (Bauhus et al. 2013; Huth and Wagner 2013). Summarizing the possibilities for choosing harvesting systems aimed at promoting regeneration processes in the course of temperate forest ecosystem restoration, the large-scale combination of different harvesting systems, which at smaller scales are oriented toward emulating natural disturbance regimes, is a feasible compromise (Schütz 2002).

6.3.2.5 *Emulating Old-Growth Gap Conditions for Regeneration Processes*

It has long been known that canopy gaps are essential for the natural regeneration of forests (e.g., Watt 1925, 1947). Compared to those parts of the forest with a closed canopy, the resource availability at the forest floor in gaps is altered, that is, improved. This improvement in availability applies to radiation (Canham et al. 1990), soil water (Gray et al. 2002), and soil nutrients (Bauhus 1996). All of these resources are essential to tree regeneration; the high relevance of gaps to tree regeneration is hence understood as a response of the regeneration to increased resource availability (e.g., Beckage and Clark 2003). The occurrence of canopy gaps is linked to canopy "disturbance," that is, a canopy gap is "a patch created by the removal of the canopy" (Connell 1989). In the majority of events, the removal of parts of the canopy means that old trees or tree crowns are partly or completely damaged or destroyed. Thus, the decay of old trees that create canopy gaps is linked to the rejuvenation of the forest. This association between tree decay and regeneration is integrated in the framework of the so-called "gap-theory." A comprehensive overview about the gap-theory of forest regeneration is given by Yamamoto (1992). According to this theory, forests are perceived as spatial mosaics (e.g., Remmert 1991) of various compositional and structural phases that are changing cyclically over time, implying that canopy gaps are pivotal to tree regeneration. Many scientists working in natural forests have adopted the gap-theory explicitly (Zeibig et al. 2005) or implicitly (Leibundgut 1982; Korpel 1995; Christensen and Emborg 1996).

To date, the gap-theory has proved useful in analyzing a variety of forest ecosystems ranging from temperate deciduous forests dominated by shade-tolerant species (European beech: Wagner et al. 2010; mixed stands: Busing 1994) to subtropical forests (longleaf pine, *Pinus palustris*: Brockway and Outcalt 1998), and to tropical moist evergreen forests (Brokaw 1985).

Within natural forests, gap size is highly variable. Runkle (1982) noted that the gap size-frequency distribution of a forest may be approximated by a lognormal distribution, that is, that a forest features many small and few large gaps. In natural forests, the lognormal or near-exponential shape of the gap size-frequency distribution seems to be very similar regardless of forest type (Foster and Reiners 1986; Wagner et al. 2010). However, differences between forests have been observed regarding the total fraction of area in gaps as well as the average gap size (Runkle 1982; Denslow 1987).

In addition to merely providing space for tree regeneration as such, gaps are also structural elements relevant for the nontree species diversity of natural forests, for example, for butterflies (Hill et al. 2001), birds (Levey 1988), and ground flora (Hahn and Thomsen 2007). Not surprisingly, research on the regeneration ecology of forest tree species has also very intensively focused on the relevance of gaps for diversity, particularly in mixed stands. Two hypotheses are worth mentioning as they may help to understand the relevance of canopy gaps to tree species diversity in forests: (1) the "gap partitioning hypothesis" (Ricklefs 1977); and (2) the "intermediate disturbance hypothesis" (Connell 1978). The first hypothesis considers both the variability of the individual gap area and the gap size-frequency distribution. The variability in gap size may serve as a template for the niche specialization of forest tree species, thus leading to gap partitioning among species. The latter hypothesis claims that "diversity is highest when disturbances are intermediate in intensity or size and lower when disturbances are at either extreme" (Connell 1978). Over time, confirming evidence has been presented for both hypotheses: the gap partitioning hypothesis (e.g., Brokaw 1985; Denslow 1987; Abe et al. 1995; Sipe and Bazzaz 1995) and the intermediate disturbance hypothesis (Sheil 2001). However, there is also a remaining debate about the validity of either hypothesis (e.g., Brokaw and Busing 2000; Fox 2013).

While the gap-partitioning hypothesis seems to be most appropriate for temperate, low-diversity climax forests (Yamamoto 1992), the intermediate disturbance hypothesis is most suitable for transient successional forests at nonequilibrium (Sheil 2001). An important differentiation between disturbance regimes leading to succession and the disturbance regime in climax forests can be made based on disturbance intensity. Following Yamamoto (1995), we agree upon a somewhat artificial upper limit of 1000 m² for small-scale gap disturbances in climax (old-growth) forests. More importantly, "gap disturbance regimes in forest dynamics may not be important in the forest communities where the large-scale disturbance occurs at intervals shorter than the longevity of the trees" (Yamamoto 1995). Here, we will exclusively follow the small-scale disturbance regime and the gap-partitioning hypothesis. Moreover, we appreciate a differentiation of niches within a single gap reported for gaps of medium to large (but still ≤1000 m²) size (Sipe and Bazzaz 1994; Wagner et al. 2010).

The idea "to capitalize on natural forest dynamics and gap dependency" (Hartshorn 1989) in forest management may be especially appropriate in forest restoration efforts. In forest science, the utilization of gaps—created by either nature or man—was first advocated by Gayer (1886) for the promotion of mixed stands; and as early as 1927, Wiedemann reported about artificial gaps used in experimental forest research. Recent approaches to emulate the gap traits of natural forests in forest management have been described by Coates and Burton (1997). Single-tree-selection, group-selection, and irregular shelter-wood cutting come to mind as being appropriate silvicultural tools for implementing a gap-based approach at the stand level.

Emulating gap dynamics for the purpose of forest restoration, however, means more than simply cutting patches in forest canopies. This holds even more true if we agree on the objective of restoration as being "to reinstate ecological processes, which accelerate recovery of forest structure, ecological functioning, and biodiversity levels toward those typical of climax forest" (Elliott et al. 2013). Following this deterministic perspective, the emulation of gap dynamics is a measure which unavoidably becomes easier to successfully implement when the forest state is closer to climax conditions. In particular, this refers to the tree species and age distribution of the respective forest, because these are the structural components which would be most affected by the implementation or alteration of gap dynamics. Both mentioned aspects are well understood by forest science: the transformation of even-aged into uneven-aged forests aimed at prolonged and selective regeneration cuttings (Schütz 1999); and group-selection cuttings are designed for regenerating mixed forests stands (Vanselow 1949; Nyland 2002; Wagner et al. 2010). The feasibility of gap-based concepts in forestry is indicated by work of Hagemann et al. (2013), Trotsiuk et al. (2012), and Tabaku and Meyer (1999) who reported about the similarity of the gap size-frequency distributions in old-growth forests and in examples of practiced close-to-nature forestry with the same species. Even in plantation-like forests, a lognormal shape of the gap size-frequency distribution may be observed (e.g., for spruce: see Huth and Wagner 2006).

In an attempt to conclude on gap-based regeneration approaches in forest restoration, two critical aspects are especially important to be mentioned:

- In stands made up of species not adapted to the particular site, for example, in plantations of exotic species, small-scale gap-based approaches to convert by natural regeneration seem unlikely to add much to the naturalness of a stand and are therefore questionable. However, combined with artificial regeneration of native site-adapted species, gap-based approaches may make sense to implement uneven-aged structures.

- Small-scale gap-based approaches become easier to successfully implement the closer to climax conditions in terms of tree species assemblage and age structure. Following large-scale disturbances, for example, in the preclimax seral stages of forest succession, the gap approach is questionable because most preclimax species feature a poor ability to reduce resource availability. Gap effects with respect to altered resource availability may therefore be small in these communities, for example, in early successional pine forests (Bolte and Bilke 1998). When emulating gap conditions in such forests, a range of potential problems should be kept in mind (Table 6.3).

TABLE 6.3

Questions, Problems and Recommendations When Emulating Gap Conditions

Questions	Problems	Recommendations
Are there risks in implementing selection cuts for the first time in a stand with formerly closed canopy?	As canopy openings increase air turbulence and wind loading of the remaining trees (Panferov and Sogachev 2008), wind-throw disturbance may result, which would be quite different from natural events.	A high vitality and stability of the individual trees of the respective stand is required (e.g., Röhrig et al. 2006)
Are there differences between man-made and natural gaps (Schliemann and Bockheim 2011)?	Soil disturbance and biomass removal differ between artificial and natural gaps. The latter refers especially to 'gap-makers' in natural forests, which are those—often old and large—tree individuals that create canopy gaps as a result of damage or death. In managed forests, gap-makers are cut and harvested.	Gap-makers should not always be removed from the forest during restoration measures. Crown damage, girdling and pulling down are appreciable alternatives to true 'cutting'.
Which consequences for management result from shifting the cutting regime to a gap-based concept?	Periodic selection cuts may have to be repeated for a very long period of time, for example, for the next 100 years, and this may pose management obstacles (but will be cost-intensive). Extreme reductions of canopy density in a single cut tend to shift the forest to preclimax condition and should thus be reconsidered.	The intensity of a single cut, that is, the removal of canopy trees, should be based on the fraction of forest area in canopy gaps under old-growth conditions. As gap birth rates per year in old-growth forests may vary (e.g., 0.5%–2% per year for mesic forests of Eastern North-America; Runkle 1982) and the interval between consecutive cuts should be at least 10 years, the canopy fraction to be removed in one cut may also vary (e.g., a gap birth rate of 1% per year and a 10-year cutting cycle would lead to approx. 10% of canopy removal in a single cut).
Which trees should be cut?	The diameter and the spatial pattern of the cut trees determines the financial outcome of the restoration measures as well as the ecological consequences and may lead to undesired results.	Proposed selection criteria are: tree species (predominantly undesired species), vitality (preferentially dominant individuals), and spatial pattern (irregular distribution appreciated). In addition, where advance regeneration of native species occurs, canopy trees may be cut to vitalize the regeneration.

6.4 Restoration Strategies for the Active Manipulation of Below-Canopy Stand Components

Restoration measures altering the overstory layer of forest stands also influence all lower layers and associated ecological processes. In general, overstory changes caused by natural disturbances or harvesting increase the below-canopy availability of light, water, and nutrients due to the removal of dominant trees. Below-canopy restoration measures can be used to improve resource availability for target tree species and to directly promote their establishment. But considering that restoration measures within forest ecosystems strive to enhance diversity of different compartments, it is necessary to keep in mind that overstory manipulations indirectly affect tree regeneration. Particularly for the early tree regeneration stages, ground vegetation and its interactions with tree seedlings are important for successful establishment of tree regeneration. Further, the definition of target tree species for a given forest stand needs to consider the site-specific options for artificial restoration measures, which might be used to enrich the current system, to expedite the process of near-natural development, or simply to ensure regeneration success. For additionally supporting the regeneration of target tree species, individual measures at smaller scales can be used and underpinned with specific knowledge drawn from the regeneration cycle. Regeneration can also be facilitated through the availability of deadwood, a component whose ecological role in natural forest ecosystems has been described by numerous studies.

6.4.1 Emulating Large-Scale Disturbances to Support Ecological Below-Canopy Processes

Although primary and secondary large-scale restoration measures are interlinked, it is helpful to clearly differentiate between them to identify the key ecological processes that can be influenced through specific treatments. Secondary large-scale manipulations use existing below-canopy structures established by primary overstory manipulations and try to increase the quality or rate of specific processes. In this context, the recolonization of ground vegetation or tree species depends on the quality of the soil seed bank, which is strongly influenced by primary manipulations. While changes in the existing soil seed bank are low after wind-throws, the effect of wildfire depends on factors including rate of spread, intensity, and duration (Piha et al. 2013). Ground fires of low intensity can favor the regeneration process by reducing litter layer depth and competing ground vegetation (Hille and Ouden 2004; Balandier et al. 2006). This gives rise to an accelerated mineralization and exposes mineral soil, thus facilitating the establishment of pioneer species. But as Ryan (2002) pointed out, fires vary greatly in space and time, which makes their outcomes hard to predict. Further, in areas as densely populated as Europe, the use of large-scale fire restoration is problematic with respect to economic efficiency and social acceptance (Hille and Ouden 2004; Král et al. 2012). Moreover, a fire will likely destroy any desired advance regeneration—in contrast to simulated wind-throws—even if the resprouting potential of many tree species is often underestimated. On restoration sites where a wind-throw has occurred, only slight changes of former ground vegetation structures and regenerated tree species can be expected. The enrichment with pioneer tree species (such as birch, willow, poplar, and larch) and thus the succession process itself will be slow.

The restoration of temperate floodplain forest ecosystems calls for very specific measures aiming to reestablish natural water level fluctuations over large areas along rivers and streams (Leyer et al. 2012). According to Peterken and Huges (1995) restoration measures are connected with fluvial processes and geomorphologic features and linked to the different vegetation types of particular forest ecosystems. Combinations of natural reproductive and vegetative regeneration processes are also described as suitable restoration methods for such areas (Deiller et al. 2003).

Numerous different combinations of the primary and secondary restoration measures listed in Table 6.4 are possible and applied in restoration practice including typical features of possible secondary large-scale measures and their effects on the regeneration process.

The use of planting and sowing is well established in forest management and has also been an important restoration practice for several decades. Both techniques are characterized by a high degree of flexibility with respect to site conditions, tree species admixture, spatial heterogeneity, and growth potential. Depending on the primary manipulation, the

TABLE 6.4

Possible Secondary Large-Scale Restoration Measures Manipulating Below-Canopy Conditions

Secondary Restoration Measures	Effects on the Regeneration Process
Planting Introduction of site adapted and natural tree species. Creation of near-natural admixtures by the use of successful plant assortments and densities	Planting is possible after all primary restoration measures, but requires different efforts (Zerbe and Wiegleb 2009). Planting in wind-thrown sites among broken trunks is more complicated than in drained sites. Future tree species composition can be directly determined (Olsthoorn et al. 1999). The remaining overstory trees and their condition decide about the growth increment and the survival rate of the planted tree species. Small-scaled patches (e.g., pits and mounds) can be considered.
Direct sowing Introduction of site-adapted and natural tree species	For successful sowing, soil preparation is generally required to reduce above- and below-ground competitive pressure. The planning of species admixtures is possible, but hardly predictable as germination depends strongly on soil surface and microclimate conditions (Cole et al. 2011). The realization of species-specific growth edges is sometimes more complicated than for planted species (Löf et al. 2004; Birkedal et al. 2010). Sowing densities should be adapted to the surrounding site conditions (Kutscher et al. 2009).
Nurse plants Promoting plants that have positive effects for target tree species establishment and recovery	Following the stress-gradient hypothesis, the use of nurse plants after large-scale disturbances increases the survival, germination and growth rate of target tree species (Siles et al. 2010). The establishment of specific nurse plants (e.g., shrubs, grasses, or mosses) can buffer unfavorable postdisturbance climate or site conditions. It is important to note that benefits for tree species regeneration from nurse plants depend on site characteristics, species-specific interactions, individual distance from nurse plants and the regeneration stage of target tree species (Padilla and Pugnaire 2006).
Soil preparation Establishment of humus and mineral soil mixtures to reduce the inhibitory effect of ground vegetation	Large-scale soil disturbances diminish the competitive pressure of ground vegetation. The loosening of the surface layer also increases nutrient availability, water, and root permeability for tree seedlings (Löf et al. 2006). Large-scale soil preparation also influences the surface climate of the disturbed sites (Kubin and Kemppainen 1994). Following soil preparation (e.g., plowing), natural regeneration, sowing, and planting are options for establishing target tree species. The techniques are manifold: Strips, bands, and spots in some cases are preferred to full-areal treatment to minimize the surface area affected by the treatment. The degree of mechanization depends on overstory conditions and the amount of deadwood.
Mowing Cutting the above-ground biomass of competing ground vegetation	Mowing interrupts the above-ground competition of ground vegetation and thus increases the probability of successful tree species establishment by natural regeneration, sowing or planting. As root competition is retained, mowing measures usually have to be repeated (Willoughby et al. 2009). The degree of mechanization depends on overstory conditions and the amount of deadwood.
Grazing Reducing the above-ground biomass of ground vegetation	The reduction of above-ground competing biomass can lead to better seedling establishment, if the regenerated tree species are not preferentially browsed by the grazing animals. Low intensity grazing supports also the resprouting and flowering of specific ground vegetation species (Adams 1975; Kuiters et al. 1996), which makes this procedure suitable for the restoration of open areas within forests or for orchard meadows.
Mulching Application of organic materials at the soil surface	Forest restoration methods use different mulch materials originating from plants (e.g., leaves, wood, and bark chips). The mulch layer decreases the competition by ground vegetation, the loss of nutrients, and protects the soil surface layer against drought or erosion. A loose mulched soil surface can promote seed germination and the primary growth of tree seedlings (Haywood 2000).

applicability of planting or sowing must be weighed against the technical and economical effort (e.g., in the case of high deadwood abundance). Planting and direct sowing can be implemented as an important part of large-scale restoration even without overstory manipulation (Zerbe 2002; Birkedal et al. 2010). However, intense primary manipulations within the overstory tree layer increase light and water availability for planted or sown tree species as well as for ground vegetation (Ammer et al. 2002; Table 6.4). Taking this into account, planting can be adapted accordingly with respect to plant spacing and choice of assortment. Primary manipulations such as prescribed burning lead to a distinct temporal head start for the planted tree species and generally decrease competitive pressure. In most cases, successful direct sowing requires additional surface manipulation, for example, mowing or plowing, but seedlings may be better adapted to specific site conditions (Kutscher et al. 2009; Cole et al. 2011).

The degree of mechanization of large-scale planting or sowing depends on the remaining overstory density as well as on the amount and distribution of downed deadwood. This also applies to mowing or soil preparation (Kubin and Kemppainen 1994; Löf et al. 2006; Willoughby and Jinks 2009). Although this approach results in lower restoration costs for primary establishment, the renewed homogenization of site conditions is problematic. In temperate forest ecosystems, intensive soil preparation was often used for the restoration of extremely air-polluted forest sites (Kozlov et al. 2000).

The use of grazing as a continuous large-scale disturbance is mostly connected with the conversion of a forest ecosystem into nonforested ecosystems (Bengtsson et al. 2000), while periodic grazing under a light shelter of tree species favors flowering ground vegetation and can be used for restoring orchard meadows as a specific element of forest ecosystems (Adams 1975; Putman 1996; Tasker and Bradstock 2006). Mulching is also typically practiced in open landscapes, but successful examples for the establishment of tree regeneration are given by Haywood (2000) and Blanco-Garcia and Lindig-Cisneros (2005). A specific mulching strategy applying an admixture of mulch material and seeds was described by Krautzer and Klug (2009).

The use of nurse plants to restore forest ecosystems after large-scale disturbances is frequently used in harsh climates (Callaway 1995; Castro et al. 2002). However, even in the temperate zone, large-scale disturbances (and restoration measures such as complete clearings) can lead to extreme site conditions (e.g., slopes and dry sites), where the facilitative function of specific ground vegetation can be used to accelerate succession processes and tree species establishment (Padilla and Pugnaire 2006). For this purpose, it is necessary to have detailed information about species-specific interactions (Siles et al. 2010) and potential nurse plants must be established or exist within the soil seed bank. Such nurse plants need to have a higher natural recovery potential (e.g., vegetative or generative development strategies) than tree species regeneration (Rent et al. 2010).

6.4.2 The Function of Ground Vegetation: Facilitation Versus Competition

Compared to boreal and tropical forest ecosystems, the overall diversity of most temperate forests and their floral and faunal elements are described as medium and inherently limited by nature (Ricklefs 1977; Thomas and MacLellan 2004). Although the terms used to describe the lower layer of forest ecosystems vary (e.g., ground vegetation, ground cover, understory vegetation, herbaceous layer, ground layer, etc.), only two fundamental approaches have been used to describe and quantify its functions and processes in forest ecosystems. Both classifications are based on the hierarchy of vertical layers or strata (Gillam 2007; Owens 2007). The first approach involves all vascular plants with an absolute

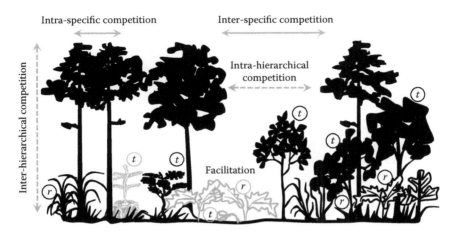

FIGURE 6.2
Overview of the hierarchical interaction levels in temperate forest ecosystems; and the differentiation between transient (*t*) and resident (*r*) plants. (From Gilliam, F.S. and M.R. Roberts. 2003. *The herbaceous layer in forests of eastern North America*. New York: Oxford University Press.)

height of 0.5–2.0 m (Gilliam and Roberts 2003), with the moss layer mostly separated (e.g., Barbier et al. 2008). The second approach combines the vertical hierarchy with a classification regarding future development. "Resident species" will never outgrow (i.e., become higher than) the defined lower stratum (see Figure 6.2). In comparison, "transient species" (i.e., tree regeneration) have a high potential to grow into the overstory layer (Gilliam and Roberts 2003). Even though this approach seems to be plausible within a development-oriented analysis of forest ecosystems, the term "resident species" can be confused with the way it is used in the context of invasive species (Pyšek et al. 2012).

6.4.2.1 The Role of Ground Vegetation in Natural Forests

Ground vegetation is influenced by local climate and abiotic conditions (e.g., temperature and precipitation, soil nutrient, and water availability). Without human impact, typical ground vegetation communities and the abundance of specific species are primarily determined by the site potential (Gracia et al. 2007). The biodiversity of ground vegetation in temperate forests is much higher than found in the overstory tree layer (Gillam 2007). Restoration measures aimed at the ground vegetation layer are frequently associated with the protection of a particular rare species, for example, orchids (Hermy et al. 1999; Honnay et al. 2002; Dorland and Willems 2006), which is deemed possible by means of modified forest management or manual regulation of ground vegetation. However, the preservation of complete plant communities requires complex measures at larger spatial scales (Honnay et al. 2002). As a result of increasing ground vegetation diversity in entire ecosystems, ecological niches for soil dwelling organisms, insects, and herbivores will also become more abundant (Hutchinson 1978; Gilbert and Lechowicz 2004; Silvertown 2004).

The ground vegetation can provide information about the potential natural vegetation (pnV) of a site and thus the degree of naturalness. Often, the locally defined pnV is used to determine the theoretical framework for restoration objectives and guide restoration activities (Zerbe 1998). Ground vegetation species can thus be regarded as the "historical memory of ecosystems" (Oheimb et al. 1999; Härdtle et al. 2003; Verheyen et al. 2003). Furthermore, ground vegetation responds quickly to changes in site conditions, for example, atmospheric depositions, habitat fragmentation, edge effects, and disturbances (Nabuurs 1996; Hermy et al. 1999; Bossuyt and Hermy 2000; Gaudio et al. 2008), causing large shifts in dominance of ground vegetation species as well as the abundance of alien species (Zerbe and Kreyer 2007).

In order to interpret ground vegetation characteristics and to distinguish the environmental factors driving its vitality, growth, and abundance; different theoretical indicators (e.g., light or soil humidity described by indicator values (Ellenberg 1996)) have been developed. Ground vegetation has been used as an indicator for classifying forest ecosystem status and guide for management based on decisions (Carignan and Villard 2002; LaPaix et al. 2009). Ferris and Humohreyn (1999) divided this indicator function into three main categories, differentiating (1) *compositional diversity* (number of plants within a defined area), (2) *structural diversity*, and (3) *functional diversity*. Applying these categories, numerous studies in natural and near-natural temperate forests have shown that site and stand structures can be described as spatial and temporal heterogeneous patches or mosaics (Korpel 1995; Bobiec et al. 2000; Emborg et al. 2000). A higher diversity of the overstory trees does not automatically lead to a higher diversity within the ground vegetation (Gilliam and Roberts 2003; Gillam 2007; Mölder et al. 2008). Ground vegetation has the important role of quickly recolonizing, covering, and protecting the exposed mineral soil (Pyšek 1993; Honnay et al. 2002; Royo and Carson 2006). Manifold combinations of ground vegetation and tree regeneration result from this early successional function (Bazzaz 1996; Brang 2005a; Walker et al. 2007).

6.4.2.2 *Interactions between the Different Hierarchical Strata of Forest Ecosystems*

For a long time, natural forests without any human impact have been rare or almost non-existent in the temperate climate zone. Therefore, species composition and abundance within the ground vegetation layer has also been shaped by human alteration of the tree layer (Økland et al. 2003; Paquette 2006; Denner 2007). Plant–plant interactions can have competing or facilitating characteristics (Bazzaz 1996; Wagner et al. 2011). Competition and facilitation effects across different hierarchical layers within one ecosystem are termed as *inter-hierarchical* (see Figure 6.2); those within the same hierarchical layer as *intra-hierarchical* (Goldberg and Landa 1991; Casper and Jackson 1997). Competition between individuals of the same species is defined as *intraspecific*; competition between individuals of different species as *interspecific* (Bazzaz 1996). These interactions between different strata and development stages lead to a complex system of competing and facilitating plant actors within forest ecosystems (Casper and Jackson 1997).

The dominance of overstory trees is obvious for those resources and environmental factors which are dependent on vertical gradients and growth space (Table 6.2). Under continuous forest cover, overstory trees thus determine the environmental conditions and resource availability for ground vegetation as well as for tree regeneration (Sydes and Grime 1981; Augusto et al. 2003; Økland et al. 2003; Penne et al. 2010). Impacts of overstory trees are mostly associated with light transmittance (van Oijen et al. 2005; van Couwenberghe et al. 2010, 2011; Wagner et al. 2011), precipitation interception, and throughfall (Bredemeier et al. 2011), litter accumulation (Carli and Drescher 2002; Augusto et al. 2003), fine root density in the humus layer (Ammer and Wagner 2005; Meinen et al. 2009), and mycorrhization (Hunter and Aarssen 1988; Luoma et al. 2006). Despite strongly reduced resource availability at the forest floor, ground vegetation species show special adaptations to these conditions, for example, underneath dense beech canopies (Nagaike et al. 1999; Gilbert and Lechowicz 2004; Silvertown 2004; Gaudio et al. 2008). According to Härdtle et al. (2003) and Nagaike et al. (1999), the diversity of ground vegetation in such systems is not hindered by low light availability. Besides the unfavorable effects of competition, some positive effects of overstory trees are also known, for example, the shelter function which mitigates extreme climate conditions such as late spring frosts (Hunter and Aarssen 1988; Nyland 2002). These positive influences caused by tree species or ground vegetation are defined as facilitation effects (Padilla and Pugnaire 2006; see Figure 6.2). The importance of facilitation effects generally increases with increasingly unfavorable or harsh environmental conditions (Freestone 2006; Paquette and Messier 2011). Within temperate forest ecosystems, facilitation has been particularly observed as a part of early tree development stages (Jensen 2011).

6.4.2.3 Using Ground Vegetation to Promote or Discriminate Against Specific Tree Regeneration

The previous sections have shown a complex system of linkages between overstory tree composition, ground vegetation, and site conditions. In order to integrate ground vegetation and tree species regeneration into restoration strategies, concise objectives need to be defined, such as the increase of ground vegetation diversity, specific ground vegetation assemblages (distributions), or the promotion of ground vegetation that facilitates tree species regeneration (Bakker et al. 2000). The dominance of certain ground vegetation groups is a typical phenomenon of intensively managed forest ecosystems (Paquette et al. 2006; Gaudio et al. 2008). It can be assumed that in managed forests or forest plantations (even-aged, single-layered stands; see Table 6.2), the homogenization of site and stand conditions also results in the homogenization of ground vegetation (Zerbe and Brande 2003; Bauhus and Schmerbeck 2010). Restrictions on the use of herbicides for competition control limit the number of active measures to manipulate ground vegetation. Willoughby and Jinks (2009) compiled a list of ground vegetation types (i.e., grasses or monocotyledons, dicotyledons, pteridophytes, and woody species) that are disproportionally abundant in managed European forests (i.e., *Deschampsia* sp., *Calamagrostis* sp., *Carex* sp., *Pteridium* sp., *Epilobium* sp., *Senecio* sp., *Urtica* sp., and *Rubus* agg.). All of these are considered potential inhibitors of successful tree regeneration (Löf and Welander 2004).

Local management strategies to reduce ground vegetation dominance mostly include the use of mechanical tools such as plows, mulchers, or cultivators (Löf et al. 2012; Ammer et al. 2009; Ammer et al. 2011). These activities primarily influence soil conditions and microtopography and reduce the above- and belowground competitive pressure on tree regeneration (Balandier et al. 2006; Metlen and Fiedler 2006). Other restoration activities focusing on aboveground conditions are mowing, trampling, grazing, and prescribed burning (Adams 1975; Vanha-Majamaa and Tuittila 1996; Hille and den Ouden 2005; Vandenberghe et al. 2006; Král et al. 2012). Subsequent sowing with near-natural herbaceous plant or tree species are other possible restoration methods (Roovers et al. 2005). Light-demanding seedlings (e.g., *Quercus* spp.) need a growth edge compared to the competing ground vegetation (Lorimer et al. 1994; Davis et al. 1998; Jensen 2011), while shade-tolerant tree species (e.g., *Abies alba* Mill.) can survive high aboveground competitive pressure by dominating ground vegetation for longer time periods (Leibundgut 1984; Grassi and Giannini 2005).

Compared to aboveground competition (Harmer et al. 2012), the regulation of belowground competition between tree species regeneration and ground vegetation by means of direct belowground measures or overstory management is more complicated. Tree species with fast growing taproots and dense fine root networks are competitive under water stress (Pallardy and Rhoads 1993; van Hees 1997) but most competitive ground vegetation species readily resprout and survive aboveground damage (Mallik and Gimingham 1985;

Fotelli et al. 2001; Klimešova and Klimeš 2007). Early successional pioneer tree species show a similar potential of vegetative response if environmental conditions are suitable (Koop 1987; Mallik 1997). However, even a short window post-cutting without competing ground vegetation can be sufficient to offer an advantage for tree regeneration. A more integrated restoration strategy based on silvicultural management options includes the utilization of pioneer nurse crops and the manipulation of overstory canopy closure (Augusto et al. 2003; Wagner et al. 2011; Paquette and Messier 2011).

6.4.3 Direct Control of Species Composition by Establishing Artificial Regeneration

Natural regeneration can be augmented with artificial regeneration when important species are lacking or when density or spatial distribution of the natural regeneration is unsatisfactory. These conditions are also of concern in managed forests and measures to remedy the conditions are termed enrichment or sowing, or reinforcement planting (Nyland 2002). There is plenty of experience with enrichment measures in managed forests as restoration plantings in Europe have involved beech (Ammer et al. 2008a), silver fir (Wiedemann 1927), and oak species (Mortzfeldt 1896). This section will focus on the effects of the old stand on the regeneration and the effects of the regeneration technique itself, that is, sowing versus planting.

The ideal planting material is adapted to the soil and site, and suitable for the chosen method and time of planting. Two types of nursery material are available: bareroot and container seedlings. Either type may provide good results given appropriate handling and cultural practices. In general, more intensive measures are necessary in more extreme situations, such as caused by competing vegetation, herbivores, drought, or high ground water table. Special preparation methods for the exact planting location and a careful choice of high-quality planting material might prove necessary.

The planting of wildlings is another way to establish desirable species. Wildlings may have fewer establishment problems under the canopy of an old stand than standard nursery stock, as they are already adapted to shade by physiological conditioning under an overstory (Nörr et al. 2002). Moreover, wildlings may be less prone to browsing damage as deer often preferentially browse well-watered and fertilized nursery stock (Suchant et al. 2000). Wildlings need careful selection and handling, however, to protect leader shoots and obtain a sufficient mass of fine roots.

In many cases, a moderate residual canopy can have positive effects on the height increment of the regeneration (Paquette et al. 2006; Balandier et al. 2007). Moderately dense canopies may favor tree regeneration over aggressive shade-intolerant graminoids or forbs, particularly for shade-tolerant and intermediate shade-tolerant tree species (Wagner et al. 2011). Successful enrichment measures have been reported for irregular shelterwood and gap-based approaches with predominantly small- to medium-sized canopy openings (Parker et al. 2001; von Lüpke and Hauskeller-Bullerjahn 2004). The effects of canopy density on ground vegetation were discussed above.

Sowing has some advantages compared to planting for restoration as the latter does not always guarantee site-adapted and near-natural root development (Ammer and Mosandl 2007). As successful sowing is also cheaper than planting, it may be worth trying. Sowing by broadcast, strip, or spot techniques (Nyland 2002) is generally recommended at the beginning of the growing season in order to reduce the risk of seed predation, fungal infection, or unfavorable environmental conditions such as excess moisture or heavy frost. The probability of success increases if seeds are pretreated (e.g., by stratification or incubation) to obtain prompt and vigorous germination (Stoehr and El-Kassaby 2011). Moreover, seeds may be treated with repellents to control predation. Soil scarification can be used to prepare a site for artificial sowing, which may be mechanized, such as practiced for beechnuts or acorns in drills or spots (Leder et al. 2003), as well as sowing of acorns in hoe strips (Preuhsler and Pinto da Costa, 1994). However, machine-sowing can cause injuries in shallow-rooted old stands, particularly in spruce, whereas this is rarely a problem in pine stands. As the germinants and the seedlings of small-seeded conifers are particularly tiny and prone to environmental stressors, a careful investigation of sowing techniques and appropriate stand conditions in advance of sowing measures is recommended (Hamm et al. 2014).

6.4.4 Manipulating the Small-Scale Seedling Environment

Almost all manipulation measures described above can be transferred to smaller spatial scales, but they differ regarding the application of the restoration techniques. The degree of mechanization decreases if manipulations aim for immediate environmental changes in the vicinity of individual or small groups of seedlings. For natural or near-natural forest conditions, typical gap size distributions resulting from natural disturbance regimes have been described (Runkle 1985; Mountford 2001) and have shown that small gaps caused by the loss of single or small groups of overstory trees dominate unmanaged forest conditions (Lorimer 1989). Manipulating the seedling environment at small scales starts with changing overstory density by establishing variable numbers of gap creators and gap shapes (Schütz 2002; Wagner et al. 2011). Particularly in the center of canopy gaps, competition with an overstory for resources is reduced, resulting in improved growth and vitality of tree seedlings. As light-demanding tree species will be disproportionately favored with increasing gap size (Huth and Wagner 2006), small parts of a forest stands can be transferred "back" to earlier successional stages, thus increasing overall species diversity. In contrast, intermediate and shade-tolerant tree regeneration shows a high competitive ability and potential for survival in smaller gaps or at the edges of larger gaps.

Gap creation for restoration can enrich tree species diversity and vertical structures at small scales (Brokaw and Busing 2000). However, periodical regulation is required to preserve intensive seedling admixtures and prevent highly competitive tree species from dominating over time (Petritan et al. 2007). The diversity of small-scale niches must therefore be continuously preserved by diverse restoration measures, if a maximum small-scale structural diversity is the objective (Grubb 1977).

Small-scale manipulations can be based on different development stages throughout the regeneration cycle (Figure 6.1). To begin with seed trees, seeds and the soil seed bank, the opportunities to increase the degree of naturalness within these development stages seem to be limited by natural processes. Promotion and revitalization of potentially rare, native target seed trees in forest ecosystems can be realized with low additional effort by using release cuttings or coppicing (Karlsson 2000). Most tree species in temperate forest ecosystems do not build up soil seed banks for long-term storage, but this can be different for ground vegetation species (Leck et al. 2008). Nevertheless, the overview of literature by Bossuyt and Honnay (2008) concluded that temperate forest soil seed banks fulfill the following functions with respect to restoration processes: (1) they provide information on the existing regeneration potential if the overstory and/or any competing ground vegetation were removed; (2) they give an impression about the discrepancy between the current forest conditions and the natural potential, including the dispersal input from surrounding areas; and (3) the spatial variability of the soil seed bank indicates the actual species-specific diversity and dominance potential without the influence of site conditions and interaction processes (Augusto et al. 2001). Restorative manipulations can thus aim to activate the soil seed bank by means of small-scale disturbances, for example, through the removal of ground vegetation, partial mixing of the humus and upper mineral soil layers, or the interruption of overstory canopy closure (Korb et al. 2005). Another option is to transfer humus and litter samples from natural forest sites to stands that are to be enriched (Bakker et al. 2000; Rodrigues et al. 2009). Direct sowing on small patches is also possible (Ren et al. 2012). Such secondary restoration measures as sporadic grazing have been shown to conserve the soil seed bank (Chaideftou et al. 2011).

Small-scale restoration measures focus on the creation of species-specific safe sites (Schupp 1995; Smit et al. 2006; Leck et al. 2008) by reducing the germination-hampering litter layer or ground vegetation, or by limiting the influence of overstory trees. Unfavorable soil conditions such as stony surfaces, mineral soil and litter accumulations, or acidic soils must be modified. Measures for soil improvement (e.g., fertilization or liming) are useful techniques to facilitate tree germination and to improve overall site conditions (Pabian et al. 2012). Further, small-scale surface heterogeneity can be manually established by creating specific microtopography. For example, microclimate within small pits is characterized by higher soil humidity and possibly lower light availability (Ulanova 2000), while mounds dry out faster but provide better light conditions (Carlton and Bazzaz 1998; Du et al. 2013). Germination on or near downed deadwood or stumps can be favorable for some species.

Light manipulation via overstory regulation represents the classical approach to protect advance regeneration. Approaches to manipulate competitors surrounding individual trees are trampling (manual) and cutting with trimmers or brushcutters; damage to individual target trees, such as caused by herbivory, can be reduced by fences or browsing repellents (Côté et al. 2004; Willoughby and Jinks 2009). Special nurse plants to directly influence the seedling environment and to buffer small-scale climate extremes have been used (Padilla and Pugnaire 2006). Highly specific restoration methods such as the inoculation with mycorrhizae (Allen 1991; Nara 2006; see Section 6.2.4) or the establishment of legumes (Carpenter et al. 2004; Siddique 2008) are mainly found at very small scales or used for single plant restoration activities.

6.4.5 Utilizing Deadwood to Improve Regeneration Survival

Deadwood has long been recognized as an integral component of many forest ecosystems with particular relevance for biodiversity (Jonsson et al. 2005; Stokland et al. 2012), carbon and nutrient cycling (Cornwell et al. 2009; Kahl et al. 2012), structural integrity (Franklin et al. 1987; Debeljak 2006), and forest regeneration. Regeneration is particularly facilitated by downed deadwood (DDW), which consists of woody debris, stumps and overgrown deadwood (Hagemann et al. 2009), and provides microsites for tree seedling germination and growth. The role of DDW as "nurse logs" for regeneration is relevant in many forest types (Figure 6.3a), including temperate (Harmon and Franklin 1989; McGee and Birmingham 1997; Kuuluvainen and Kalmari 2003), boreal (Hofgaard 1993), and tropical (Lack 1991; Sanchez et al. 2009) as well as submontane (Korpel 1995; Reif and Przybilla 1995; Zielonka and Niklasson 2001; Motta et al. 2006) and montane forests (Santiago 2000). Microsites created by DDW are especially common in natural forests where stocks are continuously replenished by gap creation and stand-replacing disturbances (Debeljak 2006; Stokland et al. 2012).

FIGURE 6.3
Spruce regeneration on decaying log in the Harz mountain, Germany (a) and yellow birch regenerating on stump in Quebec, Canada (b).

6.4.5.1 How Deadwood Facilitates Regeneration

In order for the tree regeneration to be successfully establish on or near DDW, seeds must be intercepted and retained by logs providing favorable conditions for germination, establishment, and growth. By modifying the microclimate, water, and nutrient availability, the presence of DDW thus affects several phases of the regeneration process (Figure 6.1), including seed interception and storage (Harmon 1989a), germination (Iijima and Shibuya 2010), and seedling survival (Szewczyk and Szwagrzyk 1996).

6.4.5.2 Seed Interception, Retention and Storage

Decaying wood generally covers <10% of the forest floor (Harmon 1989a; Zielonka 2006). Seed retention mainly depends on the size and characteristics of the log surface (Chambers 1991; Baier et al. 2007; Bače et al. 2012). Due to a smaller surface area, narrow logs trap significantly fewer seeds than large logs (Iijima et al. 2007). Seed retention on DDW generally increases with progressive decay (Bače et al. 2012), and is higher for moss- or litter-covered logs compared to logs with smooth bark or without bark (Harmon 1989a; Iijima et al. 2007). Tree species also influence seed retention, with better retention on logs with rough bark such as *Picea* or *Pseudotsuga* spp. (Harmon 1989a,b; Iijima and Shibuya 2010), and higher retention likelihood for smaller seeds (Szewczyk and Szwagrzyk 1996; Iijima et al. 2007).

6.4.5.3 Germination and Seedling Establishment

Deadwood modifies species-specific seed germination rates through its surface properties, decay status, and size. Germination rates increase with log size and progressive decay, because larger logs feature more stable moisture conditions than smaller logs (Stokland et al. 2012). Water retention capacity of DDW is generally higher than mineral soil and increases as decomposition progresses (Sollins et al. 1987; Zielonka 2006). The thickness of the moisture-retaining moss cover is higher on more decayed logs (Harmon 1989a; Dynesius et al. 2010). However, seedling establishment on DDW may be impeded by very thick moss layers preventing the extension of radicals into the substrate below (Harmon and Franklin 1989; LePage et al. 2000), potentially discriminating between species with different germinant size, for example, favoring *Abies* over *Picea* (Szewczyk and Szwagrzyk 1996).

6.4.5.4 Seedling and Sapling Survival

Particularly for conifers, the establishment and survival of seedlings on DDW is often better than on soil (Knapp and Smith 1982; Harmon and Franklin 1989; Szewczyk and Szwagrzyk 1996; Kupferschmid et al. 2006), and it is most pronounced for the genus *Picea* (Svoboda et al. 2010; Bače et al. 2012). However, higher seedling densities have also been observed for hardwoods such as yellow birch (*Betula alleghaniensis*; see McGee and Birmingham 1997) and rowan (Stöckli 1995). The increased survival rates of tree seedlings on logs is mainly an effect of decreased competition with ground vegetation compared to seedlings on the soil (Figure 6.3; see Ponge et al. 1998; Ran et al. 2010; Bače et al. 2012). Fallen logs can serve as a regeneration substrate until the moss mat becomes thick enough to smother seedlings (Harmon and Franklin 1989). However, differences in survival rates cannot be entirely explained by reduced competition, indicating that seedlings elevated on logs additionally have a lower risk of mortality from soil-borne pathogens (Harmon and Franklin 1989; Szewczyk and Szwagrzyk 1996; O'Hanlon-Manners and Kotanen 2004). Especially in harsh environments, DDW also promotes seedling growth and survival by offering better conditions for the growth of beneficial microorganisms and mycorrhizal fungi (Ponge et al. 1998; Zielonka 2006), increased nitrogen availability due to microbial fixation and transport from soil to DDW (Zimmerman et al. 1995; Brunner and Kimmins 2003) and a more favorable moisture regime than soil (Sollins et al. 1987; Ran et al. 2010). In contrast, DDW provides microsites with improved aeration on waterlogged soils (Santiago 2000).

6.4.5.5 Protection from Browsing Damage

In addition, DDW can protect tree regeneration against herbivore browsing via physical and visual obstruction (de Chantal and Granström 2007). Similar to other refuges, deadwood piles originating from wildfires, insect outbreaks, or harvesting can impede herbivore access to regeneration (Forester et al. 2007), thus reducing browsing damage particularly to hardwood saplings (Grisez 1960; Rumble et al. 1996; Ripple and Larsen 2001; de Chantal and Granström 2007; Smit et al. 2012). However, DDW refuges are not always efficient (Fredericksen et al. 1998; Kupferschmid and Bugmann 2005; Forester et al. 2007). The quality of protection likely depends on the dominating herbivore species as well as on log pile characteristics such as size or branchiness (de Chantal and Granström 2007), which is associated with disturbance type (Peterson and Pickett 1995).

6.4.6 Effects of Deadwood Properties

6.4.6.1 Abundance

As a result of climate and tree species, DDW abundance naturally differs among forest types (Harmon et al. 2004; Christensen et al. 2005). Regardless of forest type, DDW abundance in managed forests is typically drastically reduced compared to natural stands (Hodge and Peterken 1998; Debeljak 2006; Motta et al. 2006; Meyer and Schmidt 2011), as harvesting often removes most senescent and dead trees. Forest management also modifies DDW decay and size distribution (Stokland et al. 2012). As restoration aims to rehabilitate natural structures, processes, and species composition in modified forest ecosystems (Bradshaw 1997), restoration measures should aim at increasing DDW abundance particularly in coniferous and mixed forests where DDW provides an important regeneration substrate (Bače et al. 2012). In addition to DDW abundance, factors such as origin, type, size, and decay status also influence the relevance of DDW for tree regeneration and other ecological forest functions such as biodiversity (Harmon et al. 2004).

6.4.6.2 Origin and Type

Natural stand-level disturbances create large amounts of DDW immediately (storm) or several decades (insects, wildfire) following disturbance (Jonášová and Prach 2004; Hagemann et al. 2009; Jonášová et al. 2010), while small-scale tree mortality results in the more or less continuous creation of all sizes and types of DDW (Harmon et al. 2004). In contrast, harvesting leaves mostly crown material and stumps (Hagemann et al. 2009; Stokland et al. 2012). Like logs, stumps also offer microsites with reduced competition and favorable growing conditions (Motta et al. 2006; Bače et al. 2011). Regardless of origin, stumps may even be superior to logs as seedling establishment occurs earlier and at higher densities, likely due to better seed retention in surface depressions and faster decay (Nakagawa et al. 2001; Bače et al. 2011). Even standing deadwood can indirectly affect tree regeneration by accommodating birds which disperse the seeds of many tree species, such as spruce, rowan or beech (Jonášová and Prach 2004). The cause of tree death can also influence regeneration density, with higher seedling densities observed on logs uprooted or broken by wind compared to logs originating from bark beetle attacks (Bače et al. 2012). Bark beetles facilitate the entry of brown-rot fungi (Stokland et al. 2012), and DDW decayed by brown-rot fungi is less suitable for seedling establishment (Bače et al. 2012).

6.4.6.3 Size

It is well known that large-diameter DDW is particularly important for biodiversity, especially for saproxylic beetles, fungi and lichen (Jonsson et al. 2005; Müller and Bütler 2010; Stokland et al. 2012). As larger logs offer more surface area for seed retention and seedling establishment (Iijima et al. 2007; Bače et al. 2012), an increased abundance of large-diameter DDW also favors the natural regeneration of conifers. Moreover, the retention and creation of large DDW piles are recommended for protecting regeneration in areas with high browsing pressure (de Chantal and Granström 2007).

6.4.6.4 Decay Status

The suitability of DDW as a substrate for regeneration changes over time as decomposition alters its physical and chemical properties. This is reflected in changing seedling abundance, survival rates, and even physiological traits such as photosynthetic capacity (Ran et al. 2010; Bače et al. 2012). Seedling density generally increases as decay with progresses (Szewczyk and Szwagrzyk 1996; Zielonka 2006), but may slightly decreases in the last decay stages due to increased competition among seedlings and other vegetation (Zielonka and Niklasson 2001; Zielonka and Piatek 2004). For example, the highest number of spruce seedlings is typically found on 30–60-year-old logs, but seedling establishment may start as early 10 years after tree death (Zielonka 2006).

6.4.7 Deadwood and Forest Restoration

Although DDW is not the most important factor in forest regeneration (Szewczyk and Szwagrzyk 1996), it plays an extremely large role in coniferous forests where trees regenerate predominantly on logs (Franklin et al. 1987; Svoboda et al. 2010). In order to facilitate natural regeneration in these forest types, DDW abundance could be increased by a

TABLE 6.5

Restoration Measures for Increasing Deadwood Abundance

Restoration Measures	Effects on Deadwood Abundance and Quality
a. Passive measures	
Abandonment of forest activities	Gradual creation of snags and DDW by natural mortality (senescence, small-scale or large-scale disturbances)
Abandonment of salvage logging	Retention of disturbance-generated snags and DDW with given variability of decay, size, and species distribution
Retention of live (over-)mature trees	Gradual creation of snags and DDW as trees are left to die naturally over time, diversification of decay and size distribution
Retention of standing dead trees	Gradual creation of DDW as snags are left to fall naturally over time, diversification of decay and size distribution
Retention of logging debris	Immediate creation of fresh DDW of various size classes and species
b. Active measures	
Creation of snags and high stumps (girdling, topping, crown blow-up)	Immediate creation of fresh snags (and downed crown material) of various size classes, species and with desired spatial distribution
Creation of DDW (felling, pulling down, breaking, uprooting)	Immediate creation of fresh DDW of various size classes and species and with desired spatial distribution
Introduction of DDW (log sections)	Immediately increased availability of DDW with desired decay, size and species distribution as substrate for regeneration

range of passive or active restoration measures (Table 6.5; Stöckli 1995; Ježek 2004; Vanha-Majamaa et al. 2007; Svoboda et al. 2010; Meyer and Schmidt 2011; Stokland et al. 2012). However, the effects of such measures on tree regeneration will only become visible in the medium- or long-term because DDW accumulation following abandonment of forest activities is a very slow process (Bobiec 2002; Vandekerkhove et al. 2009; Meyer and Schmidt 2011) and it takes decades for newly generated DDW to become a suitable regeneration substrate (Zielonka 2006). Increased DDW abundance can also contribute to more natural stand structures, as the strong association between DDW and seedling establishment for some tree species results in nonrandom (i.e., linear or clumped) spatial patterns of regeneration (Svoboda et al. 2010). However, measures to increase DDW availability need to be supplemented by measures regulating canopy closure to ensure the successful growth of seedlings established on decaying wood (Zielonka 2006; Svoboda et al. 2010).

References

Abe, S., T. Masaki, and T. Nakashizuka. 1995. Factors influencing sapling composition in canopy gaps of a temperate deciduous forest. *Vegetatio* 120: 21–32.

Adams, S.N. 1975. Sheep and cattle grazing in forests: A review. *Journal of Applied Ecology* 12 (1): 143–52.

Allen, M.F. 1991. *The Ecology of Mycorrhizae.* Cambridge studies in ecology. Cambridge University Press, Cambridge, New York.

Allen, C.D., M. Savage, D.A. Falk, K.F. Suckling, T.W. Swetnam, T. Schulke, P.B. Stacey, P. Morgan, M. Hoffmann, and J. T. Klingel. 2002. Ecological restoration of ponderosa pine ecosystems: A broad perspective. *Ecological Applications* 12(5): 1418–33.

Ammer, C. 2003. Growth and biomass partitioning of *Fagus sylvatica* L. and *Quercus robur* L. seedlings in response to shading and small changes in the R/FR-ratio of radiation. *Annales of Forest Science* 60: 163–71.

Ammer, C., P. Balandier, N. Bentsen, L. Coll, and M. Löf. 2011. Forest vegetation management under debate: An introduction. *European Journal of Forest Research* 130: 1–5.

Ammer, C., E. Bickel, and C. Kölling. 2008a. Converting Norway spruce stands with beech—A review of arguments and techniques. *Austrian Journal of Forest Science* 125: 3–26.

Ammer, C., M. Blaschke, and P. Muck. 2009. Germany. In: Willoughby, I., P. Balandier, N.S. Bentsen, N. McCarthy, and J. Claridge (eds.), *Forest Vegetation Management in Europe—Current Practice and Future Requirements.* COST Office, Brussels, pp. 43–50.

Ammer, C. and R. Mosandl. 2007. Which grow better under the canopy of Norway spruce planted or sown seedlings of European beech? *Forestry* 80 (4): 385–95.

Ammer, C., R. Mosandl, H. El Kateb. 2002. Direct seeding of beech (*Fagus sylvatica* L.) in Norway spruce (*Picea abies* [L.] Karst.) stands—Effects of canopy density and fine root biomass on seed germination. *Forest Ecology and Management* 159: 59–72.

Ammer, C., B. Stimm, B., and R. Mosandl. 2008b. Ontogenetic variation in the relative influence of light and belowground resources on European beech seedling growth. *Tree Physiology* 28: 721–8.

Ammer, C. and S. Wagner. 2005. An approach for modelling the mean fine-root biomass of Norway spruce stands. *Trees* 19: 145–53.

Aronson, J. and S. Alexander. 2013. Ecosystem restoration is now a global priority: Time to roll up our sleeves. *Restoration Ecology* 21 (3): 293–96.

Ashley, M.V. 2010. Plant parentage, pollination, and dispersal: How DNA microsatellites have altered the landscape. *Critical Reviews in Plant Sciences* 29: 148–61.

Augspurger, C.K. and Wilkinson, H.T. 2007. Host specificity of pathogenic *Pythium* species: Implications for tree species diversity. *Biotropica* 39: 702–8.

Augusto, L., J.-L. Dupouey, J.-F. Picard, and J. Ranger. 2001. Potential contribution of the seed bank in coniferous plantations to the restoration of native deciduous forest vegetation. *Acta Oecologica* 22 (2): 87–98.

Augusto, L., J.-L Dupouey, and J. Ranger. 2003. Effects of tree species on understory vegetation and environmental conditions in temperate forests. *Annales of Forest Science* 60: 823–31.

Baasch, A., S. Tischew, and H. Bruelheide. 2009. Insights into succession processes by temporally repeated habitat models: Results from a long-term study in a post-mining landscape. *Journal of Vegetation Science* 20: 629–38.

Bače, R., M. Svoboda, and P. Janda. 2011. Density and height structure of seedlings in subalpine spruce forests of Central Europe: Logs vs. stumps as a favourable substrate. *Silva Fennica* 45 (5): 1065–78.

Bače, R., M. Svoboda, V. Pouska, P. Janda, and J. Červenka. 2012. Natural regeneration in Central-European subalpine spruce forests: Which logs are suitable for seedling recruitment? *Forest Ecology and Management* 266 (15): 254–62.

Bacilieri, R., T. Labbe, and A. Kremer 1994. Intraspecific genetic structure in a mixed population of *Quercus petraea* (Matt.) Liebl. and *Q. robur* L. *Heredity* 73: 130–41.

Baier, R., J. Meyer, and A. Göttlein. 2007. Regeneration niches of Norway spruce (*Picea abies* [L.] Karst.) saplings in small canopy gaps in mixed mountain forests of the Bavarian Limestone Alps. *European Journal of Forest Research* 126: 11–22.

Bakker, J.P., A.P. Grootjans, M. Hermy, and P. Poschlod. 2000. How to define targets for ecological restoration?—Introduction. *Applied Vegetation Science* 3 (1): 3–6.

Balandier, P., C. Collet, J.H. Miller, P.E. Reynolds, and S.M. Zedanker. 2006. Designing forest vegetation management strategies based on the mechanisms and dynamics of crop tree competition by neighbouring vegetation. *Forestry* 79 (1): 3–27.

Balandier, P., H. Sinoquet, E. Frak, R. Giuliani, M. Vandame, S. Descamps, L. Coll, B. Adam, B. Prevosto, and T. Curt. 2007. Six-year time course of light-use efficiency, carbon gain and growth of beech saplings (*Fagus sylvatica*) planted under a Scots pine (*Pinus sylvestris*) shelterwood. *Tree Physiology* 27: 1073–82.

Barbier, S., F. Gosselin, and P. Balandier. 2008. Influence of tree species on understory vegetation diversity and mechanisms involved—A critical review for temperate and boreal forests. *Forest Ecology and Management* 254 (1): 1–15.

Barnes, B.V., D.R. Zak, S.R. Denton, and S.H. Spurr. 1998. *Forest Ecology*, 4th edn. John Wiley & Sons, New York.

Baskin, C.C. and Baskin, J.M. 2001. Seeds. Ecology, biogeography, and evolution of dormancy and germination. *Nordic Journal of Botany* 20: 598.

Bauhus, J. 1996. C and N mineralization in an acid forest soil along a gap-stand gradient. *Soil Biology and Biochemistry* 28 (7): 923–32.

Bauhus, J. and J. Schmerbeck. 2010. Silvicultural options to enhance and use forest plantation biodiversity. In: Bauhus, P. van der Meer, and M. Kanninen (eds.), *Ecosystem Goods and Services from Plantation Forests*. Earthscan, London–Washington, DC, pp. 96–139.

Bayer, D., S. Seifert, and H. Pretzsch. 2013. Structural crown properties of Norway spruce (*Picea abies* [L.] Karst.) and European beech (*Fagus sylvatica* [L.]) in mixed versus pure stands revealed by terrestrial laser scanning. *Trees* 27 (4): 1035–47.

Bazzaz, F.A. 1996. *Plants in Changing Environments: Linking Physiological, Population, and Community Ecology*. Cambridge University Press, Cambridge, New York, Melbourne.

Beckage, B. and J.S. Clark. 2003. Seedling survival and growth of three forest tree species: The role of spatial heterogeneity. *Ecology* 84: 1849–61.

Beckman, N.G., C. Neuhauser, and H.C. Mueller-Landau. 2012. The interacting effects of clumped seed dispersal and distance- and density-dependent mortality on seedling recruitment patterns. *Journal of Ecology* 100: 862–73.

Bengtsson, J., S.G. Nilsson, A. Franc, and P. Menozzi. 2000. Biodiversity, disturbances, ecosystem function and management of European forests. *Forest Ecology and Management* 132 (1): 39–50.

Beyer, F., D. Hertel, and C. Leuschner. 2013. Fine root morphological and functional traits in *Fagus sylvatica* and *Fraxinus excelsior* saplings as dependent on species, root order and competition. *Plant Soil* 373: 143–56.

Birkedal, M., A. Fischer, M. Karlsson, M. Löf and P. Madsen 2009. Rodent impact on establishment of direct-seeded *Fagus sylvatica*, *Quercus robur* and *Quercus petraea* on forest land. *Scandinavian Journal of Forest Research* 24: 298–307.

Birkedal, M., M. Löf, G.E. Olsson, and U. Bergsten. 2010. Effects of granivorous rodents on direct seeding of oak and beech in relation to site preparation and sowing date. *Forest Ecology and Management* 259 (12): 2382–89.

Blanco-Garcia, A. and R. Lindig-Cisneros. 2005. Incorporating restoration in sustainable forestry management: Using pine-bark mulch to improve native species establishment on tephra deposits. *Restoration Ecology* 13 (4): 703–09.

Bobiec, A. 2002. Living stands and dead wood in the Białowieża forest: Suggestions for restoration management. *Forest Ecology and Management* 165 (1–3): 125–40.

Bobiec, A., H. van der Burgt, K. Meijer, C. Zuyderduyn, J. Haga, and B. Vlaanderen. 2000. Rich deciduous forests in Białowieża as a dynamic mosaic of developmental phases: Premises for nature conservation and restoration management. *Forest Ecology and Management* 130 (1–3): 159–75.

Boerner, R.E.J., G. Brent, P. DeMars, and P.N. Leicht. 1996. Spatial patterns of mycorrhizal infectiveness of soils long a successional chronosequence. *Mycorrhiza* 6: 79–90.

Bolte, A. and A. Bilke. 1998. Wirkung der bodenbelichtung auf die ausbreitung von *Calamagrostis epigejos* in den kiefernforsten Norddeutschlands. *Forst und Holz* 53: 232–36.

Bonnet-Masimbert, M. 1987. Floral induction in conifers: A review of available techniques. *Forest Ecology and Management* 19: 135–46.

Bossema, I. 1979. Jays and oaks: An eco-ethological study of a symbiosis. *Behaviour* 70 (1/2): 1–117.

Bossuyt, B. and M. Hermy. 2000. Restoration of the understorey layer of recent forest bordering ancient forest. *Applied Vegetation Science* 3 (1): 43–50.

Bossuyt, B. and O. Honnay. 2008. Can the seed bank be used for ecological restoration? An overview of seed bank characteristics in European communities. *Journal of Vegetation Science* 19 (6): 875–84.

Bradshaw, A.D. 1997. What do we mean by restoration? In: Urbanska, K.M., N.R. Webb, and P.J. Edwards (eds.), *Restoration Ecology and Sustainable Development*, Cambridge University Press, Cambridge, UK, pp. 8–14.

Bradshaw, A.D. 2002. Introduction and philosophy. In: Perrow, M.R., and Davy, A.J. (eds.), *Handbook of Ecological Restoration*, Vol. 1. Principles of Restoration, Cambridge University Press, Cambridge, UK, pp. 3–9.

Brand, D.G. 1991. The establishment of boreal and sub-boreal conifer plantations: An integrated analysis of environmental conditions and seedling growth. *Forest Science* 37: 68–100.

Brang, P. 2005a. Spatial distribution of natural regeneration on large windthrow areas created by the hurricane Lothar in 1999. *Schweizerische Zeitschrift für Forstwesen* 156 (12): 467–76.

Brang, P. 2005b. Virgin forests as a knowledge source for central European silviculture: Reality or myth? *Forest Snow and Landscape Research* 79 (1–2): 19–32.

Bredemeier, M., S. Cohen, D.L. Godbold, E. Lode, V. Pichler, and P. Schleppi. 2011. *Forest Management and the Water Cycle: An Ecosystem-Based Approach*. Ecological Studies 212. Dordrecht, New York: Springer.

Brockway, D.G. and K.W. Outcalt. 1998. Gap-phase regeneration in longleaf pine wiregrass ecosystems. *Forest Ecology and Management* 106: 125–39.

Brokaw, N.V.L. 1985. Gap-phase regeneration in a tropical forest. *Ecology* 66: 682–87.

Brokaw, N. and R.T. Busing. 2000. Niche versus chance and tree diversity in forest gaps. *Trends in Ecology and Evolution* 15 (5): 183–88.

Brown, J.M.B. 1951. Influence of shade on the height growth and habit of beech. Forestry Commission. *Report on forest research for the year ending*: 62–67.

Brunner, I., E. Graf Pannatier, B. Frey, A. Rigling, W. Landolt, S. Zimmermann, and M. Dobbertin. 2009. Morphological and physiological responses of Scots pine fine roots to water vegetation dynamics in northeastern USA. *Ecology* 85: 519–30.

Brunner, A. and J.P. Kimmins. 2003. Nitrogen fixation in coarse woody debris of *Thuja plicata* and *Tsuga heterophylla* forests on northern Vancouver Island. *Canadian Journal of Forest Research* 33 (9): 1670–82.

Buchert, G.P. 1992. Genetic diversity—An indicator of sustainability. *Advances Boreal Mixedwood Management* 10: 190–3.

Buiteveld, J., G.G. Vendramin, S. Leonardi, K. Kamer and T. Geburek. 2007. Genetic diversity and differentiation in European beech (*Fagus sylvatica* L.) stands varying in management history. *Forest Ecology and Management* 247: 98–106.

Bullock, J.M. and R.T. Clarke. 2000. Long distance seed dispersal by wind: Measuring and modelling the tail of the curve. *Oecologia* 124: 506–21.

Bullock, J.M. and R. Nathan. 2008. Plant dispersal across multiple scales: Linking models and reality. *Journal of Ecology* 96: 567–8.

Busing, R.T. 1994. Canopy cover and tree regeneration in old-growth cove forests of the Appalachian Mountains. *Vegetatio* 115: 19–27.

Busing, R.T., R.D. White, M.E. Harmon, and P.S. White. 2009. Hurricane disturbance in a temperate deciduous forest: Patch dynamics, tree mortality, and coarse woody detritus. *Plant Ecology* 201: 351–63.

Cain, M.D. 1991. The influence of woody and herbaceous competition on early growth of naturally regenerated loblolly and shortleaf pines. *Southern Journal of Applied Forestry* 15: 179–85.

Callaway, R.M. 1995. Positive interactions among plants. *Botanical Review* 61 (4): 306–49.

Cameron, A.D. 1996. Managing birch woodlands for the production of quality timber. *Forestry* 69: 357–71.

Canham, C.D., J.S. Denslow, W.J. Platt, J.R. Runkle, T.A. Spies, and P.S. White. 1990. Light regimes beneath closed canopies and tree-fall gaps in temperate and tropical forests. *Canadian Journal of Forest Research* 20 (5): 620–31.

Carignan, V. and M.A. Villard. 2002. Selecting indicator species to monitor ecological integrity: A review. *Environmental Monitoring and Assessment* 78 (1): 45–61.

Carli, A. and A. Drescher. 2002. Die verbesserung der humusauflage durch laubbäume—das beispiel sekundärer fichtenforste in der SE-Steiermark. *Mitteilungen des Naturwissenschaftlichen Vereines für Steiermark* 132: 153–68.

Carlton, G.C. and F.A. Bazzaz. 1998. Resource congruence and forest regeneration following an experimental hurricane blowdown. *Ecology* 79 (4): 1305–19.

Carpenter, F.L., J.D. Nichols, and E. Sandi. 2004. Early growth of native and exotic trees planted on degraded tropical pasture. *Forest Ecology and Management* 196 (2–3): 367–78.

Casper, B.B. and R.B. Jackson. 1997. Plant competition underground. *Annual Review of Ecology Systematics* 28 (1): 545–70.

Castro, J., R. Zamora, J.A. Hodar, and J.M. Gomez. 2002. Use of shrubs as nurse plants: A new technique for reforestation in Mediterranean mountains. *Restoration Ecology* 10 (2): 297–305.

Cater, T.C. and F.S. Chapin. 2000. Differential effects of competition or microenvironment on boreal tree seedling establishment after fire. *Ecology* 81 (4): 1086–99.

Cattarino, L., C. McAlpine, and J.R. Rhodes. 2013. The consequences of interactions between dispersal distance and resolution of habitat clustering for dispersal success. *Landscape Ecology* 28: 1321–34.

Chaideftou, E., C.A. Thanos, E. Bergmeier, A.S. Kallimanis, and P. Dimopoulos. 2011. The herb layer restoration potential of the soil seed bank in an overgrazed oak forest. *Journal of Biological Research* 15: 47–57.

Chambers, J.C., J.A. MacMahon, and J.H. Haefner. 1991. Seed entrapment in alpine ecosystems: Effects of soil particle size and diaspore morphology. *Ecology* 72 (5): 1668–77.

Chandler, L.M. and J.N. Owens. 2004. The pollination mechanism of *Abies amabilis. Canadian Journal of Forest Research* 34: 1071–80.

Chrimes, D., L. Lundqvist, and O. Atlegrim. 2004. *Picea abies* sapling height growth after cutting *Vaccinium myrtillus* in an uneven-aged forest in northern Sweden. *Forestry* 77: 61–66.

Christensen, M. and J. Emborg. 1996. Biodiversity in natural versus managed forest in Denmark. *Forest Ecology and Management* 85 (1–3): 47–51.

Christensen, M., K. Hahn, E.P. Mountford, P. Ódor, T. Standovár, D. Rozenbergar, and J. Diaci 2005. Dead wood in European beech (*Fagus sylvatica*) forest reserves. *Forest Ecology and Management* 210 (1–3): 267–82.

Ciccarese, L., A. Mattsson, and D. Pettenella. 2012. Ecosystem services from forest restoration. Thinking ahead. *New Forests* 43 (5–6): 543.

Clark, D.A. and D.B. Clark, 1984. Spacing dynamics of a tropical rain forest tree: Evaluation of the Janzen–Connell model. *The American Naturalist* 124: 769–88.

Clark, J.S., E. Macklin, and L. Wood, 1998. Stages and spatial scales of recruitment limitation in Southern Appalachian forests. *Ecological Monographs* 68: 213–35.

Clark, J.S., M. Silman, R. Kern, E. Macklin, and J. Hille Ris Lambers. 1999. Seed dispersal near and far: Patterns across temperate and tropical forests. *Ecology* 80: 1475–94.

Coates, K.D. and P.J. Burton. 1997. A gap-based approach for development of silvicultural systems to address ecosystem management objectives. *Forest Ecology and Management* 99 (3): 337–54.

Coates, K.D. and P.J. Burton, 1999. Growth of planted tree seedlings in response to ambient light levels in northwestern interior cedar–hemlock forests of British Columbia. *Canadian Journal of Forest Research* 29 (9): 1374–82.

Cole, R.J., K.D. Holl, C.L. Keene, and R.A. Zahawi. 2011. Direct seeding of late-successional trees to restore tropical montane forest. *Forest Ecology and Management* 261 (10): 1590–97.

Connell, J.H. 1971. On the role of natural enemies in preventing competitive exclusion in some marine animals and in rain forest trees. In: Den Boer, P.J., and G. Gradwell (eds.), *Dynamics of Populations.* Center for Agricultural Publication and Documentation, Wageningen, The Netherlands, pp. 298–312.

Connell, J.H. 1978. Diversity in tropical rain forests and coral reefs. *Science* 199 (4335): 1302–10.

Connell, J.H. 1989. Some processes affecting the species composition in forest gaps. *Ecology* 70 (3): 560.

Cornwell, W.K., Cornelissen, J.H.C., S.D. Allison, J. Bauhus, P. Eggleton, C.M. Preston, F. Scarff, J.T. Weedon, C. Wirth, and A.E. Zanne. 2009. Plant traits and wood fates across the globe: Rotted, burned, or consumed? *Global Change Biology* 15 (10): 2431–49.

Côté, SD., T.P. Rooney, J.-P. Tremblay, C. Dussault, and D.M. Waller. 2004. Ecological impacts of deer overabundance. *Annual Review of Ecology, Evolution, and Systematics* 35 (1): 113–47.

Crone, E.E., E.J.B. McIntire, and J. Brodie. 2011. What defines mast seeding? Spatio-temporal patterns of cone production by whitebark pine. *Journal of Ecology* 99: 438–44.

Curran, L.M. and M. Leighton. 2000. Vertebrate responses to spatiotemporal variation in seed production of mast-fruiting *Dipterocarpaceae*. *Ecological Monographs* 70 (1): 101–28.

Dale, V.H., L.A. Joyce, S. McNulty, R.P. Neilson, M.P. Ayres, M.D. Flannigan, P.J. Hanson et al. 2001. Climate change and forest disturbances. *BioScience* 51 (9): 723–34.

D'Antonio, C. and L.A. Meyerson. 2002. Exotic plant species as problems and solutions in ecological restoration: A synthesis. *Restoration Ecology* 10 (4): 703–13.

Davis, M.A., K.J. Wrage, and P.B. Reich. 1998. Competition between tree seedlings and herbaceous vegetation: Support for a theory of resource supply and demand. *Journal of Ecology* 86 (4): 652–61.

Debeljak, M. 2006. Coarse woody debris in virgin and managed forest. *Ecological Indicators* 6 (4): 733–42.

de Chantal, M. and A. Granström. 2007. Aggregations of dead wood after wildfire act as browsing refugia for seedlings of *Populus tremula* and *Salix caprea*. *Forest Ecology and Management* 250 (1–2): 3–8.

Deiller, A.-F., J.-M.N. Walter, and M. Trémolières. 2003. Regeneration strategies in a temperate hardwood floodplain forest of the Upper Rhine: Sexual versus vegetative reproduction of woody species. *Forest Ecology and Management* 180 (1–3): 215–25.

Denner, M. 2007. *Auswirkungen des ökologischen Waldumbaus in der Dübener Heide und im Erzgebirge auf die Bodenvegetation: Ermittlung phytozönotischer Indikatoren für naturschutzfachliche Bewertungen* (Effects of ecological forest conversion in Dübener Heide and Erzgebirge on the ground vegetation—Determination of phytocoenotic indicators for nature conservation evaluations). Forstwissenschaftliche Beiträge Tharandt 29. Stuttgart: Ulmer.

Denslow, J.S. 1987. Tropical rainforest gaps and tree species diversity. *Annual Review of Ecology and Systematics* 18 (1): 431–51.

Dobrowolska, D., S. Hein, A. Oosterbaan, S. Wagner, J. Clark, and J.P. Skovsgaard. 2011. A review of European ash (*Fraxinus excelsior* L.): Implications for silviculture. *Forestry* 84: 133–48.

Doody, C.N. and C. O'Reilly. 2008. Drying and soaking pretreatments affect germination in pedunculate oak. *Annales of Forest Science* 65 (509).

Dorland, E. and J.H. Willems. 2006. High light availability alleviates the costs of reproduction in *Ophrys insectifera* (Orchidaceae). *Journal Europäischer Orchideen* 38 (2): 501–18.

Downie, B. and J.D. Bewley. 2000. Soluble sugar content of white spruce (*Picea glauca*) seeds during and after germination. *Physiologia Plantarum* 110: 1–12.

Du, S., W. Duan, L. Wang, L. Chen, Q. Wie, M. Li, and L. Wang. 2013. Microsite characteristics of pit and mound and their effects on the vegetation regeneration in *Pinus koraiensis* dominated broadleaved mixed forest. *Chinese Journal of Applied Ecology* 24 (3): 633–38.

Dynesius, M., H. Gibb, and J. Hjältén. 2010. Surface covering of downed logs: Drivers of a neglected process in dead wood ecology. *PLoS ONE* 5 (10): e13237.

Ellenberg, H. 1996. *Vegetation Mitteleuropas mit den Alpen in ökologischer, dynamischer und historischer Sicht: 170 Tabellen.* 5., stark veränd. und verb. (Vegetation of Central Europe and the Alps with respect to ecology, dynamic and history: 170 tables. 5th Edition) Aufl. UTB 8104. Stuttgart: Ulmer.

Elliott, S.D., D. Blakesley, and K. Hardwick. 2013. *Restoring Tropical Forests: A Practical Guide.* Kew Publications, London.

Emborg, J., M. Christensen, and J. Heilmann-Clausen. 2000. The structural dynamics of Suserup Skov, a near-natural temperate deciduous forest in Denmark. *Forest Ecology and Management* 126 (2): 173–89.

Facelli, J.M. 2008. Specialized strategies I: Seedlings in stressful environments. In: M.A. Allessio Leck, V.T. Parker, and R.L. Simpson (eds.), *Seedling Ecology and Evolution.* Cambridge University Press, Cambridge.

Facelli, J.M. and S.T.A. Pickett. 1991. Plant litter: Light interception and effects on an oldfield plant community. *Ecology*, 72, 1024–31.

Farnsworth, E.J. 2008. Physiological and morphological changes during early seedling growth: Roles of phytohormones. In: Leck, M.A., V.T. Parker, and R.L. Simpson (eds.), *Seedling Ecology and Evolution*. Cambridge University Press, Cambridge, UK, 150–71.

Fedriani, J.M., P. Rey, J.L. Garrido, J. Guitian, C.M. Herrera, M. Mendrano, A. Sanchez-Lafuente, and X. Cerdá. 2004. Geographical variation in the potential of mice to constrain an ant-seed dispersal mutualism. *Oikos* 105: 181–91.

Fenner, M. and K. Thompson. 2005. *The Ecology of Seeds*, 2nd edn. Cambridge University Press, Cambridge, UK.

Ferris, R. and J.W. Humohreym. 1999. A review of potential biodiversity indicators for application in British forests. *Forestry* 72 (4): 313–28.

Finkeldey, R. and M. Ziehe. 2004. Genetic implications of silvicultural regimes. *Forest Ecology and Management* 197: 231–44.

Finzi, A.C. and C.D. Canham. 2000. Sapling growth in response to light and nitrogen availability in a southern New England forest. *Forest Ecology and Management* 131: 153–65.

Fischer, H., O. Bens, and R. Hüttl. 2002. Changes in humus form, humus stock and soil organic matter distribution caused by forest transformation in the north eastern lowlands of Germany. *Forstwissenschaftliches Centralblatt* 121: 322–34.

Fischer, A. and H. Fischer. 2012. Restoration of temperate forests: A European approach. In: van Andel, J. and J. Aronson (eds.), *Restoration Ecology: The New Frontier*, 2nd ed. Vol. 12. Blackwell Publishing Ltd., Chichester, UK, pp. 145–160.

Fischer, H. and S. Wagner. 2009. Silvicultural responses to predicted climate change scenarios. *Westnik—Journal of Forest Ecology and Forest Management*, Mari El. 2: 12–23.

Flaig, H. and H. Mohr. 1990. Auswirkungen eines erhöhten Ammoniumangebots auf die Keimpflanzen der gemeinen Kiefer (*Pinus sylvestris* L.). *Allgemeine Forst Jagdzeitung* 162: 35–42.

Forester, J.D., D.P. Anderson, and M.G. Turner. 2007. Do high-density patches of coarse wood and regenerating saplings create browsing refugia for aspen (*Populus tremuloides* Michx.) in Yellowstone National Park (USA)? *Forest Ecology and Management* 253 (1–3): 211–19.

Forget, P.M., J.E. Lambert, P.E. Hulme and S.B. Vander Wall (eds.) 2005. *Seed Fate: Predation, Dispersal and Seedling Establishment*. CAB International, Wallingford, UK.

Foster, J.R. and W.A. Reiners. 1986. Size distribution and expansion of canopy gaps in a northern Appalachian spruce-fir forest. *Vegetatio* 68: 109–14.

Fotelli, M.N., A. Gessler, A.D. Peuke, and H. Rennenberg. 2001. Drought affects the competitive interactions between *Fagus sylvatica* seedlings and an early successional species, *Rubus fruticosus*: Responses of growth, water status and delta13C composition. *New Phytologist* 151 (2): 427–35.

Fox, J.W. 2013. The intermediate disturbance hypothesis should be abandoned. *Trends in Ecology and Evolution (Amst.)* 28 (2): 86–92.

Franklin, J.F., H.H. Shugart, and M.E. Harmon. 1987. Tree death as an ecological process. *Bioscience* 37 (8): 550–6.

Fredericksen, T.S., B. Ross, W. Hoffmann, M. Lester, J. Beyea, M.L. Morrison, and B.N. Johnson. 1998. Adequacy of natural hardwood regeneration on forestlands in northeastern Pennsylvania. *Northern Journal of Applied Forestry* 15: 130–34.

Freestone, A.L. 2006. Facilitation drives local abundance and regional distribution of rare plant in a harsh environment. *Ecology* 87 (11): 2728–35.

Frelich, L.E. 2002. *Forest Dynamics and Disturbance Regimes*. Cambridge University Press, Cambridge.

Fries, C., O. Johansson, B. Petterson, and P. Simonsson. 1997. Silvicultural models to maintain and restore natural stand structures in Swedish boreal forests. *Forest Ecology and Management* 94: 89–103.

Galatowitsch, S.M. 2012. *Ecological restoration*. Sinauer Associates, Sunderland, Massachusetts, USA.

Gaudio, N., P. Balandier, and A. Marquier. 2008. Light-dependent development of two competitive species (*Rubus idaeus*, *Cytisus scoparius*) colonizing gaps in temperate forest. *Annales of Forest Science* 65 (1): 104.

Gayer, K. 1886. *The Mixed Forest.* Parey Verlag, Berlin.

Geburek Th. and J. Turok. 2005. *Conservation and Management of Forest Genetic Resources in Europe.* Arbora Publishers, Zvolen, Slovakia, 693p.

George, L.O. and F.A. Bazzaz. 1999. The fern understory as an ecological filter: Emergence and establishment of canopy-tree seedlings. *Ecology* 80: 833–45.

Gholz, H.L. and L. Boring. 1991. Characterizing the site: Environment, associated vegetation and site potential. In: Duryea, M.L. and P. Daugherty (eds.), *Regeneration Manual for the Southern Pines.* Springer, New York, pp. 163–82.

Gilbert, B. and M.J. Lechowicz. 2004. Neutrality, niches, and dispersal in a temperate forest understory. *Proceedings of the National Academy Sciences U.S.A.* 101 (20): 7651–56.

Gillam, F.S. 2007. The ecological significance of the herbaceous layer in temperate forest ecosystems. *BioScience* 57 (10): 845.

Gilliam, F.S. and M.R. Roberts. 2003. *The Herbaceous Layer in Forests of Eastern North America.* Oxford University Press, New York.

González-Rodríguez, V. and R. Villar. 2012. Post-dispersal seed removal in four Mediterranean oaks: Species and microhabitat selection differ depending on large herbivore activity. *Ecological Research* 27 (3): 587–94.

González-Varo, J.P., J.V. López-Bao, and J. Guitián. 2013. Functional diversity among seed dispersal kernels generated by carnivorous mammals. *Journal of Animal Ecology* 82: 562–71.

Goldammer, J.G. 2013. *Prescribed Burning in Russia and Neighbouring Temperate-Boreal Eurasia.* Kesssel Publishing House, Remagen-Oberwinter.

Goldberg, D.E. 1990. Components of resource competition in plant communities. In: Grace, J.B., and D. Tilman (eds.), *Perspectives of Plant Competition.* Academic Press, New York, NY, pp. 27–49.

Goldberg, D.E. and K. Landa. 1991. Competitive effect and response: Hierarchies and correlated traits in the early stages of competition. *Journal of Ecology* 79 (4): 1013–1030.

Gómez, J.M., D. García, and R. Zamora. 2003. Impact of vertebrate acorn- and seedling-predators on a Mediterranean *Quercus pyrenaica* forest. *Forest Ecology and Management* 180: 125–34.

Goodall-Copestake, W., M.L. Hollingsworth, P.M. Hollingsworth, G. Jenkins, and E. Collin. 2005. Molecular markers and ex situ conservation of the European elms (*Ulmus* spp.). *Biological Conservation* 122: 537–46.

Gordon, A.M., J.A. Simpson, and P.A. Williams. 1995. Six-year response of red oak seedlings planted under a shelterwood in central Ontario. *Canadian Journal of Forest Research* 25: 603–13.

Goulet, F. 1995. Frost heaving of forest tree seedlings: A review. *New Forests* 9 (1): 67–94.

Gracia, M., F. Montané, J. Piqué, and J. Retana. 2007. Overstory structure and topographic gradients determining diversity and abundance of understory shrub species in temperate forests in central Pyrenees (NE Spain). *Forest Ecology and Management* 242 (2–3): 391–97.

Grassi, G. and R. Giannini. 2005. Influence of light and competition on crown and shoot morphological parameters of Norway spruce and silver fir saplings. *Annales of Forest Science* 62 (3): 269–74.

Gray, A.N., T.A. Spies, and M.J. Easter. 2002. Microclimatic and soil moisture responses to gap formation in coastal Douglas-fir forests. *Canadian Journal of Forest Research* 32 (2): 332–43.

Greene, D.F. and E.A. Johnson. 1996. Wind dispersal of seeds from a forest into a clearing. *Ecology* 77: 595–609.

Grime, J.P. 1979. *Plant Strategies and Vegetation Processes.* John Wiley & Sons, Bath.

Grisez, T.J. 1960. Slash helps protect seedlings from deer browsing. *Journal of Forestry* 58: 385–87.

Groot, A. 1999. Effects of shelter and competition on the early growth of planted white spruce (*Picea glauca*). *Canadian Journal of Forest Research* 29: 1002–14.

Grubb, P.J. 1977. The maintenance of species-richness in plant communities: The importance of the regeneration niche. *Biological Review* 52 (1): 107–45.

Guitián, J. and I. Munilla. 2010. Responses of mammal dispersers to fruit availability: Rowan (Sorbus aucuparia) and carnivores in mountain habitats of northern Spain. *Acta Oecologica* 36 (2): 242.

Guo, L., J. Chen, X. Cui, B. Fan, and H. Lin. 2013. Application of ground penetrating radar for coarse root detection and quantification: A review. *Plant Soil* 362: 1–23.

Gustafsson, L., L. Kouki, and A. Sverdrup-Thygeson. 2010. Tree retention as a conservation measure in clear-cut forests of northern Europe: A review of ecological consequences. *Scandinavian Journal of Forest Research* 25: 295–308.

Hagemann, U., G. van der Kelen, and S. Wagner. 2013. Comparative assessment of natural regeneration quality in two northern hardwood stands. *Northern Journal of Applied Forestry* 30 (1): 5–15.

Hagemann, U., M. Moroni, and F. Makeschin. 2009. Deadwood abundance in Labrador high-boreal black spruce forests. *Canadian Journal of Forest Research* 39 (1): 131–42.

Hahn, K. and R.P. Thomsen. 2007. Ground flora in Suserup Skov: Characterized by forest continuity and natural gap dynamics or edge-effect and introduced species? *Ecological Bulletins* 52: 167–81.

Halpern, C.H.B., S.A. Evans, C.R. Nelson, D. McKenzie, D.A. Liguori, D.E. Hibbs, and M.G. Halaj. 1999. Response of forest vegetation to varying levels and patterns of green-tree retention: An overview of a long-term experiment. *Northwest Science (Special Issue)* 73: 27–44.

Hamm, T., J. Weidig, F. Huth, W. Kulisch, and S. Wagner. 2014. Wachstumsreaktionen junger weißtannen-voraussaaten auf begleitvegetation und strahlungskonkurrenz. *Allgemeine Forst- und Jagdzeitung* 185 (3/4): 45–59.

Härdtle, W., T. Aßmann, R. Diggelen, and G. Oheimb. 2009. Renaturierung und management von heiden. In: S. Zerbe and G. Wiegleb (eds.), *Renaturierung von Ökosystemen in Mitteleuropa*, Spektrum Akademischer Verlag, Heidelberg, pp. 317–47.

Härdtle, W., G. von Oheimb, and C. Westphal. 2003. The effects of light and soil conditions on the species richness of the ground vegetation of deciduous forests in northern Germany (Schleswig-Holstein). *Forest Ecology and Management* 182 (1–3): 327–38.

Harmer, R., A. Kiewitt, and G. Morgan. 2012. Can overstorey retention be used to control bramble (*Rubus fruticosus* L. agg.) during regeneration of forests? *Forestry* 85: 135–44.

Harmon, M.E. 1989a. Effects of bark fragmentation on plant succession on conifer logs in the *Picea-Tsuga* forests of Olympic National Park. *The American Midland Naturalist* 121 (1): 112–24.

Harmon, M.E. 1989b. Retention of needles and seeds on logs in *Picea sitchensis—Tsuga heterophylla* forests of coastal Oregon and Washington. *Canadian Journal of Botany* 67 (6): 1833–37.

Harmon, M.E. and J.F. Franklin. 1989. Tree seedlings on logs in *Picea-Tsuga* forests of Oregon and Washington. *Ecology* 70 (1): 48.

Harmon, M.E., J.F. Franklin, J.F. Swanson, P. Sollins, S.V. Gregory, J.D. Lattin, N.H. Anderson et al. 2004. Ecology of woody debris in temperate ecosystems. *Advances in Ecological Research Classic Papers* 34: 59–234.

Harper, J.L. 1977. *Population Biology of Plants*. Academic Press, London.

Hartshorn, G.S. 1989. Application of gap theory to tropical forest management: Natural regeneration on strip clear-cuts in the Peruvian Amazon. *Ecology* 70 (567–9).

Hattemer, H.H. 2005. Phenotypic and genetic variation. In: Geburek, T., Turok, J. (ed.), *Conservation and Management of Forest Genetic Resources in Europe*. Arbora Publishers, Zvolen, Slovakia, pp. 129–48.

Havranek, W.M. and V. Benecke. 1978. The influence of soil moisture on water potential, transpiration and photosynthesis of conifer seedlings. *Plant and Soil* 49: 91–103.

Haywood, J.D. 2000. Mulch and hexazinone herbicide shorten the time longleaf pine seedlings are in the grass stage and increase height growth. *New Forests* 19 (3): 279–90.

Hermy, M., O. Honnay, L. Firbank, C. Grashof-Bokdam, and J.E. Lawesson. 1999. An ecological comparison between ancient and other forest plant species of Europe, and the implications for forest conservation. *Biological Conservation* 91 (1): 9–22.

Hertel, O., C.A. Skjøth, S. Reis, A. Bleeker, R.M. Harrison, J.N. Cape, D. Fowler et al. 2012. Governing processes for reactive nitrogen compounds in the European atmosphere, *Biogeosciences*, 9: 4921–54.

Hietala, A.M., L. Mehli, N.E. Nagy, H. Kvaalen, and N. La Porta. 2005. *Rhizoctonia solani* AG 2-1 as a causative agent of cotyledon rot on European beech (*Fagus sylvatica*). *Forest Pathology* 35 (6): 397–410.

Hill, J.K., K.C. Hamer, J. Tangah, and M. Dawood. 2001. Ecology of tropical butterflies in rainforest gaps. *Oecologia* 128 (2): 294–302.

Hille, M. and J. den Ouden. 2005. Fuel load, humus consumption and humus moisture dynamics in Central European Scots pine stands. *International Journal of Wildland Fire* 14: 153–59.

Hille, M. and J. Ouden. 2004. Improved recruitment and early growth of Scots pine (*Pinus sylvestris* L.) seedlings after fire and soil scarification. *European Journal of Forest Research* 123 (3): 213–18.

Hilton G, J. Packham. 2003. Variation in the masting of common beech (*Fagus sylvatica* L.) in northern Europe over two centuries (1800–2001). *Forestry* 76: 319–28.

Hodge, S.J. and G.F. Peterken. 1998. Deadwood in British forests: Priorities and a strategy. *Forestry* 71 (2): 99–112.

Hofgaard, A. 1993. Structure and regeneration patterns in a virgin *Picea abies* forest in northern Sweden. *Journal of Vegetation Science* 4 (5): 601–08.

Holl, K.D. and T.M. Aide. 2011. When and where to actively restore ecosystems? *Forest Ecology and Management* 261: 1558–63.

Holmsgaard, E. and H.C. Olsen. 1966. Experimental induction of flowering in beech. *Forstliches Forsogsraes. Danemark* 30: 3–17.

Honnay, O., B. Bossuyt, K. Verheyen, J. Butaye, H. Jacquemyn, and M. Hermy. 2002. Ecological perspectives for the restoration of plant communities in European temperate forests. *Biodiversity and Conservation* 11: 213–42.

Hosius, B., L. Leinemann, M. Konnert, and F. Bergman. 2006. Genetic aspects of forestry in the Central Europe. *European Journal of Forest Research* 125: 407–17.

Hulme, P.E. and T. Borelli. 1999. Variability in post-dispersal predation in deciduous woodland: Relative importance of location, seed species, burial and density. *Plant Ecology* 145: 149–56.

Hulme, P.E. and M.K. Hunt. 1999. Rodent post-dispersal seed predation in deciduous woodland: Predator response to absolute and relative abundance of prey. *Journal of Animal Ecology* 68: 417–28.

Hüning, Chr., S. Tischew, and G. Karste. 2008. Erfolgskontrolle der renaturierungsmaßnahmen auf der brockenkuppe im Nationalpark Harz. *Hercynia N.F* 41: 201–17.

Hunter, M.L. 1999. *Maintaining Biodiversity in Forest Ecosystems*. Cambridge University Press, Cambridge, UK, New York, NY, USA.

Hunter, A.F. and L.W. Aarssen. 1988. Plants helping plants: New evidence indicates that beneficence is important in vegetation. *BioScience* 38 (1): 34–40.

Huss, J. and A. Stephani. 1978. Lassen sich angekommene buchennaturverjüngungen durch frühzeitige auflichtung, durch düngung oder unkrautbekämpfung rascher aus der gefahrenzone bringen? *Allgemeine Forst- und Jagdzeitung* 149 (8): 133–45.

Hutchinson, G.E. 1978. *An Introduction to Population Ecology*. Yale University Press, New Haven.

Huth, F. 2009. Untersuchungen zur verjüngungsökologie der sand-birke (*Betula pendula* Roth). phd. Technische Universität Dresden, Fakultät Forst-, Geo- und Hydrowissenschaften, 383p.

Huth, F. and S. Wagner. 2006. Gap structure and establishment of Silver birch regeneration (*Betula pendula* Roth.) in Norway spruce stands (*Picea abies* L. Karst.). *Forest Ecology and Management* 229 (1–3): 314–24.

Huth, F. and S. Wagner. 2013. Ecosystem services and continuous cover forests—A silvicultural analysis. *Schweizerische Zeitschrift für Forstwesen* 164 (2): 27–36.

Hüttl, R.F. and A.D. Bradshaw. 2001. Ecology of postmining landscapes. *Ecological Engineering Special Issue* 17 (2–3).

Hüttl, R.F.J. and E. Weber. 2001. Forest ecosystem development in post-mining landscapes: A case study of the Lusatian lignite district. *Naturwissenschaften* 88: 322–29.

Hynynen, J., P. Niemistö, A. Vihera-Aarnio, A. Brunner, S. Hein, and P. Velling. 2010. Silviculture of birch (*Betula pendula* Roth and *Betula pubescens* Ehrh.) in northern Europe. *Forestry* 83: 103–19.

Iijima, H. and M. Shibuya. 2010. Evaluation of suitable conditions for natural regeneration of *Picea jezoensis* on fallen logs. *Journal of Forest Research* 15 (1): 46–54.

Iijima, H., M. Shibuya, and H. Saito. 2007. Effects of surface and light conditions of fallen logs on the emergence and survival of coniferous seedlings and saplings. *Journal of Forest Research* 12 (4): 262–69.

Janos, D.P. 1992. Heterogeneity and scale in tropical vesicular-arbuscular mycorrhiza formation. In: Read, D.J., D.H. Lewis, A.H. Fitter, and I.J. Alexander (eds.), *Mycorrhizas in Ecosystems*. CAB, Wallingford, pp. 276–82.

Jansen, P.A., F. Bongers, and L. Hemerik, L. 2004. Seed mass and mast seeding enhance dispersal by a neotropical scatter-hoarding rodent. *Ecological Monographs* 74: 569–89.

Janzen, D.H. 1970. Herbivores and the number of tree species in tropical forests. *The American Naturalist* 104: 501–28.

Jensen, T.S. 1985. Seed-seed predator interactions of European beech (*Fagus sylvatica*) and forest rodents (*Clethrionomys glareolus* and *Apodemus flavicollis*). *Oikos* 44: 149–56.

Jensen, A.M., M. Löf, and E.S. Gardiner. 2011. Effects of above- and below-ground competition from shrubs on photosynthesis, transpiration and growth in *Quercus robur* L. seedlings. *Environmental and Experimental Botany* 71 (3): 367–75.

Ježek, K. 2004. Contribution of regeneration on dead wood to the spontaneous regeneration of a mountain forest. *Journal of Forest Science* 50: 405–14.

Jobidon, R. 1994. Light threshold for optimal black spruce (*Picea mariana*) seedling growth and development under brush competition. *Canadian Journal of Forest Research* 24: 1629–35.

Johansson, K., O. Langvall, and J. Bergh. 2012. Optimization of environmental factors affecting initial growth of Norway spruce seedlings. *Silva Fennica* 46 (1): 27–38.

Jonášová, M. and K. Prach. 2004. Central-European mountain spruce (*Picea abies* (L.) Karst.) forests: Regeneration of tree species after a bark beetle outbreak. *Ecological Engineering* 23 (1): 15–27.

Jonášová, M., E. Vávrová, and P. Cudlín. 2010. Western Carpathian mountain spruce forest after a windthrow: Natural regeneration in cleared and uncleared areas. *Forest Ecology and Management* 259 (6): 1127–34.

Jonsson, B.G., N. Kruys, and T. Ranius. 2005. Ecology of species living on dead wood—Lessons for dead wood management. *Silva Fennica* 39 (2): 289–309.

Josa, R., M. Jorba, and V. Ramon. 2012. Opencast mine restoration in a Mediterranean semi-arid environment: Failure of some common practices. *Ecological Engineering* 42: 183–91.

Jump, A. and J. Penuelas. 2007. Extensive spatial genetic structure revealed by AFLP but not SSR molecular markers in the wind-pollinated tree, *Fagus sylvatica*. *Molecular Ecology* 16: 925–36.

Kahl, T., M. Mund, J. Bauhus, and E.-D. Schulze. 2012. Dissolved organic carbon from European beech logs: Patterns of input to and retention by surface soil. *Ecoscience* 19 (4): 364–73.

Karlsson, C. 2000. Seed production of *Pinus sylvestris* after release cutting. *Canadian Journal of Forest Research* 30: 982–9.

Karlsson, M. 2001. Doctoral diss. Vol. 196. Southern Swedish Forest Research Centre, SLU. Acta Universitatis agriculturae Sueciae, Silvestria. Natural Regeneration of Broadleaved Tree Species in Southern Sweden; p. 1–44.

Kauppi, S., M. Romantschuk, R. Strömmer, and A. Sinkkonen. 2012. Natural attenuation is enhanced in previously contaminated and coniferous forest soils. *Environmental Science and Pollution Research* 19 (1): 53–63.

Keidel, S., P. Meyer, and N. Bartsch. 2008. Regeneration eines naturnahen Fichtenwaldökosystems im Harz nach großflächiger Störung. *Forstarchiv* 79: 187–96.

Kelly, D. and V.L. Sork. 2002. Mast seeding in perennial plants: Why, how, where? *Annual Review Ecology Systematics* 33: 427–47.

Kerr, G. 1999. The use of silvicultural systems to enhance the biological diversity of plantation forests in Britain. *Forestry* 72: 191–205.

Kimmins, J.P. 1987. *Forest Ecology*. Macmillan Publishing Company, New York, 531 pp.

Kitajima, K. and J.A. Myers. 2008. Seedling ecophysiology: Strategies towards achievement of positive carbon balance. In: Leck, M.A., V.T. Parker, and R.L. Simpson (eds.), *Seedling Ecology and Evolution*. Cambridge University Press, Cambridge, UK, pp. 172–188.

Kliejunas, J.T., W.J. Otrosina, and J.R. Allison. 2005. Uprooting and trenching to control annosus root disease in a developed recreation site: 12-year results. *Western Journal of Applied Forestry* 20 (3): 154–9.

Klimešova, J. and L. Klimeš. 2007. Bud banks and their role in vegetative regeneration—A literature review and proposal for simple classification and assessment. *Perspectives in Plant Ecology, Evolution and Systematics* 8 (3): 115–29.

Knapp, A.K. and W.K. Smith. 1982. Factors influencing understory seedling establishment of Engelmann spruce (*Picea engelmannii*) and subalpine fir (*Abies lasiocarpa*) in southeast Wyoming. *Canadian Journal of Botany* 60 (12): 2753–61.

Knudsen, I.M.B., K.A. Thomsen, B. Jensen, and K.M. Poulsen. 2004. Effects of hot water treatment, biocontrol agents, disinfectants and a fungicide on storability of English oak acorns and control of the pathogen, *Ciboria batschiana*. *Forest Pathology* 34: 47–64.

Kohyama, T. 1987. Stand dynamics in a primary warm temperate rain forest analyzed by the diffusion equation. *Botanical Magazine, Tokyo* 100, 305–317.

Koizumi, A., N. Oonuma, Y. Sasaki, and K. Takahashi. 2007. Difference in uprooting resistance among coniferous species planted in soils of volcanic origin. *Journal of Forest Research* 12: 237–242.

Koop, H. 1987. Vegetative reproduction of trees in some European natural forests. *Vegetatio* 72 (2): 103–10.

Korb, J.E., J.D. Springer, S.R. Powers, and M.M. Moore. 2005. Soil seed banks in *Pinus ponderosa* forests in Arizona: Clues to site history and restoration potential. *Applied Vegetation Science* 8 (1): 103–12.

Korpel, S. 1995. *Old growth Forests of the West Carpathians*. Gustav Fischer, Stuttgart, Jena, New York.

Kozlov, M.V., E. Haukioja, A.V. Bakhtiarov, D.N. Stroganov, and S.N. Zimina. 2000. Root versus canopy uptake of heavy metals by birch in an industrially polluted area: Contrasting behaviour of nickel and copper. *Environmental Pollution* 107 (3): 413–20.

Kozlowski, T.T. (ed.), 1974. *Fire and Ecosystems: Physiological Ecology*. Academic Press, New York.

Král, K., J. Trochta, and T. Vrška. 2012. Can fire and secondary succession assist in the regeneration of forests in a national park? In: Jongepierová, I., P. Pešout, J.W. Jongepier, and K. Prach (eds.), *Ecological Restoration in the Czech Republic*, Nature Conservation Agency of the Czech Republic, Prague, pp. 24–26.

Krautzer, B. and B. Klug. 2009. Renaturierung von subalpinen und alpinen Ökosystemen (Restoration of subalpine and alpine ecosystems). In: Zerbe, S. and G. Wiegleb, (eds.), *Renaturierung von Ökosystemen in Mitteleuropa* (Restoration of Ecosystems in Central Europe). Spektrum Akademischer Verlag, Heidelberg, pp. 209–34.

Kubin, E. and L. Kemppainen. 1994. Effect of soil preparation of boreal spruce forest on air and soil temperature conditions in forest regeneration areas. *Acta Forestalia Fennica* 244: 1–56.

Kuiters, A.T., G.M.J. Mohren, and S.E. Van Wieren. 1996. Ungulates in temperate forest ecosystems. *Forest Ecology and Management* 88: 1–5.

Kupferschmid, A.D., P. Brang, W. Schönenberger, and H. Bugmann. 2006. Predicting tree regeneration in *Picea abies* snag stands. *European Journal of Forest Research* 125: 163–179.

Kupferschmid, A.D. and H. Bugmann. 2005. Effect of microsites, logs and ungulate browsing on *Picea abies* regeneration in a mountain forest. *Forest Ecology and Management* 205 (1–3): 251–65.

Küßner, R., P. Reynolds, and F.W. Bell 2000. Growth response of *Picea mariana* seedlings to competition for radiation. *Scandinavian Journal of Forest Research* 15 (3): 334–342.

Kutscher, M., M. Bachmann, and A. Göttlein. 2009. Renaissance der Saat im Alpenraum? (Renaissance of sowing in the Alps)? *Waldbau—Planung, Pflege, Perspektiven, LWF aktuell* 68.

Kuuluvainen, T. and R. Kalmari. 2003. Regeneration microsites of *Picea abies* seedlings in a windthrow area of a boreal old-growth forest in southern Finland. *Annales Botanici Fennici* 40 (6): 401–13.

Kuuluvainen, T. and T. Pukkala. 1989. Effect of Scots pine seed trees on the density of ground vegetation and tree seedlings. *Silva Fennica* 23 (2): 159–67.

Lack, A.J. 1991. Dead logs as a substrate for rain forest trees in Dominica. *Journal of Tropical Ecology* 7 (03): 401.

LaMontagne, J.M. and S. Boutin. 2009. Quantitative methods for defining mast-seeding years across species and studies. *Journal of Vegetation Science* 20: 745–53.

LaPaix, R., B. Freedman, and D. Patriquin. 2009. Ground vegetation as an indicator of ecological integrity. *Environmental Review* 17: 249–65.

Lässig, R., S. Egli, O. Odermatt, W. Schönenberger, B. Stöckli, and T. Wohlgemuth. 1995. Beginn der Wiederbewaldung auf Windwurfflächen (Reforestation by natural regeneration after Windthrow). *Schweizerische Zeitschrift für Forstwesen* 146 (11): 893–911.

Lässig, R. and S.A. Mocalov. 2000. Frequency and characteristics of severe storms in the Urals and their influence on the development, structure and management of the boreal forests. *Forest Ecology and Management* 135, 1–3: 179–94.

Lautenschlager, R.A. 1995. Competition between forest brush and planted white spruce in north-central Maine. *Northern Journal of Applied Forestry* 12: 163–7.

Lautenschlager, R.A. 1999. Environmental resources interactions affect raspberry growth and its competition with white spruce. *Canadian Journal of Forest Research* 29: 906–16.

Leck, M.A., V.T. Parker, and R. Simpson. 2008. *Seedling Ecology and Evolution*. Cambridge University Press, Cambridge.

Leder, B., S. Wagner, J. Wollmerstadt, and C. Ammer. 2003. Bucheckern-Voraussaat unter Fichtenschirm—Ergebnisse eines Versuchs des Deutschen Verbandes Forstlicher Forschung-sanstalten/Sektion Waldbau. Direct Seeding of European Beech (*Fagus sylvatica* L.) in Pure Norway Spruce Stands (*Picea abies* [L.] Karst.)—Results of an Experiment by the German Union of Forest Research Organizations/Silviculture Division. *Forstwissenschafliches Centralblatt* 122 (3): 160–74.

Lefèvre, F., 2004. Human impacts on forest genetic resources in the temperate zone: An updated review. *Forest Ecology and Management* 197: 257–71.

Leibundgut, H. 1982. *European Old-Growth Forests*. P. Haupt, Bern, Stuttgart.

Leibundgut, H. 1984. *Die natürliche Waldverjüngung*. 2. überabeitete und erw. Aufl. Bern: P. Haupt.

Leishman, M.R. and M. Westoby. 1994. The role of large seeds in seedling establishment in dry soil conditions—Experimental evidence from semi-arid species. *Journal of Ecology* 82: 249–258.

LePage, P.T., C.D. Canham, K.D. Coates, and P. Bartemucci. 2000. Seed abundance versus substrate limitation of seedling recruitment in northern temperate forests of British Columbia. *Canadian Journal of Forest Research* 30: 415–27.

Leuschner, C. 1994. Walddynamik auf Sandböden in der Lüneburger Heide (NW-Deutschland). *Phytocoenologia* 22: 289–324.

Levey, D.J. 1988. Tropical wet forest treefall gaps and distributions of understory birds and plants. *Ecology* 69: 1076–89.

Leyer, I., E. Mosner, and B. Lehmann. 2012. Managing floodplain-forest restoration in European river landscapes combining ecological and flood-protection issues. *Ecological Applications* 22 (1): 240–49.

Li, H.J. and Z.B. Zhang. 2007. Effects of mast seeding and rodent abundance on seed predation and dispersal by rodents in *Prunus armeniaca* (Rosaceae). *Forest Ecology and Management* 242: 511–7.

Lindenmayer, D., E. Knight, L. McBurney, D. Michael, and S.C. Banks. 2010. Small mammals and retention islands: An experimental study of animal response to alternative logging practices. *Forest Ecology and Management* 260: 2070–78.

Lindenmayer, D. and J.F. Franklin. 2002. *Conserving Forest Biodiversity: A Comprehensive Multiscaled Approach*. Island Press, Washington.

Lindenmayer, D.B., J.F. Franklin, A. Lõhmus, S.C. Baker, J. Bauhus, W. Beese, A. Brodie et al. 2012. A major shift to the retention approach for forestry can help resolve some global forest sustainability issues. *Conservation Letters* 5 (6): 421–31.

Löf, M. 2000. Establishment and growth in seedlings of *Fagus sylvatica* and *Quercus robur*: Influence of interference from herbaceous vegetation. *Canadian Journal of Forest Research* 30: 855–64.

Löf, M., A. Thomsen, and P. Madsen. 2004. Sowing and transplanting of broadleaves (*Fagus sylvatica* L., *Quercus robur* L., *Prunus avium* L. and *Crataegus monogyna* Jacq.) for afforestation of farmland. *Forest Ecology and Management* 188 (1–3): 113–23.

Löf, M., D.C. Dey, R.M. Navarro, and D.F. Jacobs. 2012. Mechanical site preparation for forest restoration. *New Forests* 48: 825–48.

Löf, M., D. Rydberg, and A. Bolte. 2006. Mounding site preparation for forest restoration: Survival and short term growth response in *Quercus robur* L. seedlings. *Forest Ecology and Management* 232 (1–3): 19–25.

Löf, M. and N.T. Welander. 2004. Influence of herbaceous competitors on early growth in direct seeded *Fagus sylvatica* L. and *Quercus robur* L. *Annales of Forest Science* 61: 781–8.

Lorimer, C.G. 1989. Relative effects of small and large disturbances on temperate hardwood forest structure. *Ecology* 70 (3): 565.

Lorimer, C.G., J.W. Chapman, and W.D. Lambert. 1994. Tall understorey vegetation as a factor in the poor development of oak seedlings beneath mature stands. *Ecology* 82 (2): 227–37.

Lundqvist, L. 1995. Simulation of sapling population dynamics in uneven-aged *Picea abies* forests. *Annals of Botany* 76, 371–80.

Luoma, D.L., C.A. Stockdale, R. Molina, and J.L. Eberhart. 2006. The spatial influence of *Pseudotsuga menziesii* retention trees on ectomycorrhiza diversity. *Canadian Journal of Forest Research* 36 (10): 2561–73.

Madsen, P. 1995. Effects of soil water content, fertilization, light, weed competition and seed bed type on natural regeneration of beech (*Fagus sylvatica*). *Forest Ecology and Management* 72: 251–64.

Malanson, G.P. and M.P. Armstrong. 1996. Dispersal probability and forest diversity in a fragmented landscape. *Ecological Modelling* 87: 91–102.

Mallik, A.U. and C.H. Gimingham. 1985. Ecological effects of heather burning II. Effects on seed germination and vegetative regeneration. *Journal of Ecology* 73 (2): 633–44.

Mallik, A.U. and F. Pellissier. 2000. Effects of *Vaccinium myrtillus* on spruce regeneration: Testing the notion of coevolutionary significance of allelopathy. *Journal of Chemical Ecology* 26: 2197–209.

Mallik, A.U., F.W. Bell, and Y. Gong. 1997. Regeneration behavior of competing plants after clear cutting: Implications for vegetation management. *Forest Ecology and Management* 95 (1): 1–10.

Manso, R., M. Fortin, R. Calama, and M. Pardos. 2013. Modelling seed germination in forest tree species through survival analysis. The *Pinus pinea* L. case study. *Forest Ecology and Management* 289: 9–21.

Matveinen-Huju, K., J. Niemelä, H. Rita, and R.B. O'Hara. 2006. Retention-tree groups in clear-cuts: Do they constitute 'life-boats' for spiders and carabids? *Forest Ecology and Management* 230 (1–3): 119–35.

McComb, B.C. 2007. *Wildlife Habitat Management: Concepts and Applications in Forestry*. Boca Raton, New York: CRC Press/Taylor & Francis Group.

McGee, G.G. and J.P. Birmingham. 1997. Decaying logs as germination sites in northern hardwood forests. *Forest Ecology and Management* 14 (4): 178–82.

McWilliams, W.H., S.L. Stout, T.W. Bowersox, and L.H. McCormick. 1995. Adequacy of advance tree-seedling regeneration in Pennsylvania's forests. *Northern Journal of Applied Forestry* 12 (4): 187–91.

Meinen, C., D. Hertel, and C. Leuschner. 2009. Biomass and morphology of fine roots in temperate broad-leaved forests differing in tree species diversity: Is there evidence of below-ground overyielding? *Oecologia* 161 (1): 99–111.

Messier, C.C., K.J. Puettmann, and K.D. Coates. 2013. *Managing Forests as Complex Adaptive Systems: Building Resilience to the Challenge of Global Change*. The Earthscan forest library. Routledge Taylor & Francis Group, London and New York.

Metlen, K.L. and C.E. Fiedler. 2006. Restoration treatment effects on the understory of ponderosa pine/Douglas-fir forests in western Montana, USA. *Forest Ecology and Management* 222 (1–3): 355–69.

Meyer, P. and M. Schmidt. 2011. Accumulation of dead wood in abandoned beech (*Fagus sylvatica* L.) forests in northwestern Germany. *Forest Ecology and Management* 261 (3): 342–52.

Millar, C.I., N.L. Stephenson, and S.L. Stephens. 2007. Climate change and forests of the future: Managing in the face of uncertainty. *Ecological Applications* 17 (8): 2145–51.

Millerón, M., U. Lopez de Heredia, Z. Lorenzo, J. Alonso, and A. Dounavi. 2013. Assessment of spatial discordance of primary and effective seed dispersal of European beech (*Fagus sylvatica* L.) by ecological and genetic methods. *Molecular Ecology* 22: 1531–45.

Mitchell, S.J. 2013. Wind as a natural disturbance agent in forests: A synthesis (Review). *Forestry* 86 (2): 147–57.

Mitchell, R.J., J.F. Franklin, B.J. Palik, K.K. Kirkman, L.L. Smith, R.T. Engstrom, and M.L. Hunter, Jr. 2004. *Natural disturbance-based silviculture for restoration and maintenance of biological diversity. Final Report to the National Commission on Science for Sustainable Forestry*.

Mitchell, A.K., R. Koppenaal, G. Goodmanson, R. Benton, and T. Bown. 2007. Regenerating montane conifers with variable retention systems in a coastal British Columbia forest: 10-Year results. *Forest Ecology and Management* 246 (2–3): 240–50.

Modry, M., D. Hubeny, and K. Rejsek. 2004. Differential response of naturally regenerated European shade tolerant tree species to soil type and light availability. *Forest Ecology and Management* 188: 185–95.

Mölder, A., M. Bernhardt-Römermann, and W. Schmidt. 2008. Herb-layer diversity in deciduous forests: Raised by tree richness or beaten by beech? *Forest Ecology and Management* 256 (3): 272–81.

Moncur, M.W. and O. Hasan. 1994. Floral induction in *Eucalyptus nitens*. *Tree Physiology* 14: 1303–12.

Morris, L.A., S.A. Moss, and W.S. Garbett. 1993. Competitive interference between selected herbaceous and woody plants and *Pinus taeda* L. during two growing season following planting. *Forest Science* 39: 166–87.

Mortzfeldt, U. 1896. Über horstweisen Vorverjüngungsbetrieb. *Zeitschrift für Forst- und Jagdwesen* 28: 2–31.

Mosandl, R. and A. Kleinert. 1998. Development of oaks (*Quercus petraea* (Matt.) Liebl.) emerged from bird-dispersed seeds under old-growth pine (Pinus silvestris L.) stands. *Forest Ecology and Management* 106: 35–44.

Motta, R., R. Berretti, E. Lingua, and P. Piussi. 2006. Coarse woody debris, forest structure and regeneration in the Valbona Forest Reserve, Paneveggio, Italian Alps. *Forest Ecology and Management* 235 (1–3): 155–63.

Mountford, E.P. 2001. Natural Canopy Gap Characteristics in European Beech Forests. Nat-Man Project Report.

Mueller-Dombois, D. and H. Ellenberg. 1974. *Aims and Methods of Vegetation Ecology*. Wiley and Sons, New York, 547p.

Müller, J. and R. Bütler. 2010. A review of habitat thresholds for dead wood: A baseline for management recommendations in European forests. *European Journal of Forest Research* 129 (6): 981–92.

Müller-Starck, G., M. Ziehe, and R. Schubert. 2005. Genetic diversity parameters associated with viability selection, reproductive efficiency, and growth in forest tree species. In: Scherer-Lorenzen, M., Korner, C., Schulze, E.-D. (eds.), *Forest Diversity and Function: Temperate and Boreal Systems*. Springer, Berlin, pp. 87–108.

Nabuurs, G.J. 1996. Quantification of herb layer dynamics under tree canopy. *Forest Ecology and Management* 88 (1–2): 143–48.

Nagaike, T., T. Kamitani, and T. Nakashizuka. 1999. The effect of shelterwood logging on the diversity of plant species in a beech (*Fagus crenata*) forest in Japan. *Forest Ecology and Management* 118 (1–3): 161–71.

Nakagawa, M., A. Kurahashi, M. Kaji, and T. Hogetsu. 2001. The effects of selection cutting on regeneration of *Picea jezoensis* and *Abies sachalinensis* in the sub-boreal forests of Hokkaido, northern Japan. *Forest Ecology and Management* 146 (1–3): 15–23.

Nara, K. 2006. Ectomycorrhizal networks and seedling establishment during early primary succession. *New Phytologist* 169: 169–76.

Nathan, R. and H.C. Muller-Landau. 2000. Spatial patterns of seed dispersal, their determinants and consequences for recruitment. *Trends in Ecology and Evolution* 15: 278–85.

Netherer, S. and A. Schopf. 2010. Potential effects of climate change on insect herbivores in European forests—General aspects and the pine processionary moth as specific example. *Forest Ecology and Management* 259 (4): 831–8.

Nicolini, E., D. Barthélémy and P. Heuret. 2000. Influence de la densité du couvert forestier sur le développement architectural de jeunes chênes sessiles, *Quercus petraea* (Matt.) Liebl. (Fagaceae), en régénération forestière. *Canadian Journal of Botany* 78: 1–14.

Nopp-Mayr, U., I. Kempter, G. Muralt, and G. Gratzer. 2012. Seed survival on experimental dishes in a central European old-growth mixed-species forest—effects of predator guilds, tree masting and small mammal population dynamics. *Oikos* 121 (3): 337–46.

Nörr, R., M. Ganz, and A. Waechter. 2002. Wildlinge. *Allgemeine Forstzeitschrift/Der Wald* 5: 225–27.

Nyari, L. 2010. Genetic diverstiy, differentiation and spatial genetic structures in differently managed adult European beech (*Fagus sylvatica* L.) stands and their regeneration. *Forstarchiv* 81: 156–64.

Nyland, R.N. 1996. *Silviculture Concepts and Applications*. McGraw-Hill, New York, NY.

Nyland, R.D. 2002. *Silviculture: Concepts and Applications,* 2nd ed. McGraw-Hill, New York.

O'Hanlon-Manners, D.L. and P.M. Kotanen. 2004. Logs as refuges from fungal pathogens for seeds of Eastern hemlock (*Tsuga canadensis*). *Ecology* 85 (1): 284–89.

Oheimb, G.V., H.J. Ellenberg, J. Heuveldop, and W.-U. Kriebitzsch. 1999. Einfluß der Nutzung unterschiedlicher Waldökosysteme auf die Artenvielfalt und -zusammensetzung der Gefäßpflanzen in der Baum-, Strauch- und Krautschicht unter besonderer Berücksichtigung von Aspekten des Naturschutzes und des Verbißdruckes durch Wild. *Mitteilungen der Bundesforschungsanstalt für Forst- und Holzwirtschaft Hamburg* 195: 279–450.

Økland, T., K. Rydgren, R.H. Økland, K.O. Storaunet, and J. Rolstad. 2003. Variation in environmental conditions, understorey species number, abundance and composition among natural and managed *Picea abies* forest stands. *Forest Ecology and Management* 177 (1–3): 17–37.

Oliet, J. and D. Jacobs. 2012. Restoring forests: Advances in techniques and theory. *New Forests* 43 (5–6): 535–41.

Olson, B.E. and R.T. Wallander. 2002. Does ruminal retention time affect leafy spurge seed of varying maturity? *Journal Range Management* 55: 65–69.

Olsthoorn, A.F., H.H. Bartelink, J.J. Gardiner, H. Pretzsch, H.J. Hekhuis, and A. Franc. 1999. *Management of mixed-species forest:* Silviculture and economics. Wageningen, the Netherlands: IBN Scientific Contributions 15.

Orlikowski, L.B. and G. Szkuta. 2004. First notice of *Phytophthora ramorum* on *Calluna vulgaris, Photinia fraseri* and *Pieris japonica* in Polish container-grown ornamental nurseries. *Phytopathologia Polonica* 33: 87–92.

Otto, J. 1996. *Waldökologie,* UTB, Stuttgart, 1994.

Ouden, J., P.A. Jansen, R. Smit, P.M. Forget, J.E. Lambert, R.E. Hulme, and S.B. Vander Wall, S.B., 2005. Jays, mice and oaks: Predation and dispersal of *Quercus robur* and *Q. petraea* in North-western Europe. In: Forget, P.M., Lambert, J., Vander Wall, S.B. (eds.), *Seed Fate: Predation, Dispersal and Seedling Establishment.* CABI Publishing, Wallingford, pp. 223–40.

Övergaard, R., P. Gemmel, and M. Karlsson. 2007. Effects of weather conditions on mast year frequency in beech (*Fagus sylvatica*) in Sweden. *Forestry* 80 (5): 555–65.

Owen, M.D.K. 1994. Impact of crop tolerance to specific herbicides on weed management systems: Corn and soybeans. *Proceeding of the North Central Weed Science Society.* 49: 167–8.

Owens, J.N. 1995. Constraints to seed production: Temperate and tropical forest trees. *Tree Physiology* 15: 477–84.

Owens, J.N. and M.D. Blake. 1985. Forest tree seed production. A review of literature and recommendations for future research. *Environment Canada, Canadian Forest Service,* Information Report PI-X-53, 161 p.

Owens, J.N. and D.D. Fernando. 2007. Pollination and seed production in western white pine. *Canadian Journal of Forest Research* 37 (2): 260–75.

Pabian, S.E., N.M. Ermer, W.M. Tzilkowski, and M.C. Bittingham. 2012. Effects of liming on forage availability and nutrient content in a forest impacted by acid rain. *PLoSONE* 7 (6): e39755.

Pacala, S.W., C.D. Canham, J. Saponara, J.A. Silander, Jr., R.K. Kobe, and E. Ribbens. 1996. Forest models defined by field measurements: II. Estimation, error analysis and dynamics. *Ecological Monographs* 66: 1–43.

Padilla, F.M. and F.I. Pugnaire. 2006. The role of nurse plants in the restoration of degraded environments. *Frontiers in Ecology and Environment* 4 (4): 196–202.

Pairon, M.C., M. Jonard, and A.L. Jacquemart. 2006. Modelling seed dispersal of black cherry, an invasive forest tree: How microsatellites may help? *Canadian Journal of Forest Research* 36: 1385–94.

Pallardy, S.G. and J.L. Rhoads. 1993. Morphological adaptations to drought in seedlings of deciduous angiosperms. *Canadian Journal of Forest Research* 23 (9): 1766–74.

Palmer, M.A. and S. Filoso. 2009. Restoration of ecosystems services for environmental markets. *Science* 325 (5940): 575–76.

Panferov, O. and A. Sogachev. 2008. Influence of gap size on wind damage variables in a forest. *Agricultural and Forest Meteorology* 148 (11): 1869–81.

Paquette, A., A. Bouchard, and A. Cogliastro. 2006. Survival and growth of under-planted trees: A meta-analysis across four biomes. *Ecological Applications* 16 (4): 1575–89.

Paquette, A. and C. Messier. 2011. The effect of biodiversity on tree productivity: From temperate to boreal forests. *Global Ecology and Biogeography* 20 (1): 170–80.

Parker, W.C., D.C. Dey, S.G. Newmaster, K.A. Elliott, and E. Boysen. 2001. Managing succession in conifer plantations: Converting young red pine (*Pinus resinosa* Ait.) plantations to native forest types by thinning and underplanting. *The Forestry Chronicle* 77 (4): 721–34.

Paulsen, T. and G. Högstedt. 2002. Passage through bird guts increase germination and seedling growth of *Sorbus aucuparia*. *Functional Ecology* 16: 608–12.

Penne, C., B. Ahrends, M. Deurer, and J. Böttcher. 2010. The impact of the canopy structure on the spatial variability in forest floor carbon stocks. *Geoderma* 158 (3–4): 282–97.

Perala, D.A. and A.A. Alm. 1990. Regeneration silviculture of birch—A review. *Forest Ecology and Management* 32: 39–77.

Perea, R., A. San Miguel, and L. Gil. 2011. Leftovers in seed dispersal: Ecological implications of partial seed consumption for oak regeneration. *Journal of Ecology* 99: 194–201.

Pérez-Ramos, I.M. and T. Marañón. 2008. Factors affecting post-dispersal seed predation in two coexisting oak species: Microhabitat, burial and exclusion of large herbivores. *Forest Ecology and Management* 255 (8–9): 3506–14.

Peterken, G.F. and F.M.R. Huges. 1995. Restoration of floodplain forests in Britain. *Forestry* 68 (3): 187–202.

Peterson, C.J. and S.T. Pickett. 1995. Forest reorganization: A case study in an old-growth forest catastrophic blowdown. *Ecology* 76 (3): 763.

Petritan, A.M., B. von Lüpke, and I.C. Petritan. 2007. Effects of shade on growth and mortality of maple (*Acer pseudoplatanus*), ash (*Fraxinus excelsior*) and beech (*Fagus sylvatica*) saplings. *Forestry* 80 (4): 397–412.

Piha, A., T. Kuuluvainen, H. Lindberg, and I. Vanha-Majamaa. 2013. Can scar-based fire history reconstructions be biased? An experimental study in boreal Scots pine. *Canadian Journal of Forest Research* 43 (7): 669–75.

Ponder, F., Jr. 1997. Survival and growth of hardwood seedlings following preplanting-root treatments and treeshelters. In: Pallardy, S.G., R.A. Cecich, H.G. Garrett, and P.S. Johnson (eds.), *Proceedings of the 11th Central Hardwood Forest Conference*. General Technical Report. NC-188. USDA Forest Service North Central Forest Experiment Station, St. Paul., pp. 332–40.

Ponge, J.-F., O. Zackrisson, N. Bernier, M.-C. Nilsson, and C. Gallet. 1998. The forest regeneration puzzle. *BioScience* 48 (7): 523–30.

Pons, J. and J.G. Pausas. 2008. Modelling jay (*Garrulus glandarius*) abundance and distribution for oak regeneration assessment in Mediterranean landscapes. *Forest Ecology and Management* 256: 578–84.

Poorbabaei, H. and A. Poor-Rostam. 2009. The effect of shelterwood silvicultural method on the plant species diversity in a beech (*Fagus orientalis* Lipsky) forest in the north of Iran. *Journal of Forest Science* 55 (8): 387–94.

Preuhsler, T., P. da Costa, and E. Maria. 1994. Growth of mixed-species regeneration below Pinus shelter. Symposium of working groups of S4.01: Mensuration, Growth and Yield: MIXED STANDS—Research Plots.Measurement and Results. In *IUFRO 25–29 April*, 207–17. Lousá/Coimbra Portugal.

Priewasser, K., P. Brang, H. Bachofen, H. Bugmann, and T. Wohlgemuth. 2013. Impacts of salvage-logging on the status of deadwood after windthrow in Swiss forests. *European Journal of Forest Research* 132 (2): 231–40.

Putman, R.J. 1996. Ungulates in temperate forest ecosystems: Perspectives and recommendations for future research. *Forest Ecology and Management* 88 (1–2): 205–14.

Pyšek, P. 1993. What do we know about *Calamagrostis villosa*?—A review of the species behaviour in secondary habitats. *Preslia* 65: 1–20.

Pyšek, P., V. Jarošík, P.E. Hulme, J. Pergl, M. Hejda, U. Schaffner, and M. Vilà. 2012. A global assessment of invasive plant impacts on resident species, communities and ecosystems: The interaction of impact measures, invading species' traits and environment. *Global Change Biology* 18 (5): 1725–37.

Ran, F., C. Wu, G. Peng, H. Korpelainen, and C. Li. 2010. Physiological differences in Rhododendron calophytum seedlings regenerated in mineral soil or on fallen dead wood of different decaying stages. *Plant Soil* 337 (1–2): 205–15.

Rapp, J.M., E.J.B. McIntire, and E.E. Crone. 2013. Sex allocation, pollen limitation and masting in whitebark pine. *Journal of Ecology* 101: 1345–52.

Raymond, P., S. Bédard, V. Roy, C. Larouche, and R. Tremblay. 2009. The irregular shelterwood system: Review, classification, and potential application to forests affected by partial disturbances. *Journal of Forestry* 107 (8): 405–13.

Reed, B.C., J.F. Brown, and D. Vanderzee. 1994. Measuring phenological variability from satellite imagery, *Journal of Vegetation Science* 5 (5): 703–14.

Reich, P.B., D.S. Ellsworth and M.B. Walters. 1998. Leaf structure (specific leaf area) modulates photosynthesis–nitrogen relations: Evidence from within and across species and functional groups. *Functional Ecology* 12: 948–58.

Reif, A. and M. Przybilla. 1995. On the regeneration of spruce (*Picea abies*) in the montane zone of the Bavarian Forest National Park. *Hoppea* 56: 467–514.

Remmert, H. (ed.), 1991. *The Mosaic-Cycle Concept of Ecosystems*. Ecological Studies. Springer Berlin Heidelberg, Berlin, Heidelberg.

Ren, H., H. Lu, J. Wang, N. Liu, and Q. Guo. 2012. Forest restoration in China: Advances, obstacles, and perspectives. *Tree and Forestry Science and Biotechnology* 6 (1): 7–16.

Rey Benayas, J.M., A.C. Newton, A. Diaz, and J.M. Bullock. 2009. Enhancement of biodiversity and ecosystem services by ecological restoration: A meta-analysis. *Science* 325 (5944): 1121–24.

Ribbens, E., J.A. Silander, and S.W. Pacala. 1994. Seedling recruitment in forests: Calibrating models to predict patterns of tree seedling dispersion. *Ecology* 75: 1794–1806.

Richardson, D.M. 1998. Forestry trees as invasive aliens. *Conservation Biology* 12 (1): 18–26.

Ricklefs, R.E. 1977. Environmental heterogeneity and plant species diversity: A hypothesis. *The American Naturalist* 111 (978): 376–81.

Ripple, W.J. and E.J. Larsen. 2001. The role of postfire coarse woody debris in aspen regeneration. *Western Journal of Applied Forestry* 16: 61–4.

Roberts, E.H. 1973. Predicting the storage life of seeds. *Seed Science and Technology* 1: 499–514.

Rodrigues, R.R., R.A. Lima, S. Gandolfi, and A.G. Nave. 2009. On the restoration of high diversity forests: 30 years of experience in the Brazilian Atlantic Forest. *Biological Conservation* 142 (6): 1242–51.

Röhrig, E., N. Bartsch, A. Dengler, and B. von Lüpke. 2006. *Waldbau auf ökologischer Grundlage: 91 Tabellen*. 7., vollst. aktual. Aufl (Silviculture on an ecological foundation). UTB Forst- und Agrarwissenschaften, Ökologie, Biologie 8310. Stuttgart: UTB.

Roovers, P., H. Gulinck, and M. Hermy. 2005. Experimental assessment of initial revegetation on abandoned paths in temperate deciduous forest. *Applied Vegetation Science* 8 (2): 139–48.

Rosenvald, R. and A. Lõhmus. 2008. For what, when, and where is green-tree retention better than clear-cutting? A review of the biodiversity aspects. *Forest Ecology and Management* 255 (1): 1–15.

Rosenvald, R., A. Lõhmus, A. Kiviste, R. Rosenvald, A. Lõhmus, and A. Kiviste. 2008. Preadaptation and spatial effects on retention-tree survival in cut areas in Estonia. *Canadian Journal of Forest Research* 38 (10): 2616–25.

Ross, S.D. 1989. Temperature influences on reproductive development in conifers. In: MacIver, D.C., R.B. Street, and A.N. Auclair (eds.), *Forest Renewal and Forest Production: Forest Climate, '86 Symp.* Can. Gov. Printing Cent., Ottawa, Ont., pp. 40–43.

Ross, S.D. and R.P. Pharis. 1985. Promotion of flowering in tree crops: Different mechanisms and techniques, with special reference to conifers. In: Cannell M.G.R., and J.E. Jackson (eds.), *Attributes of Trees as Crop Plants*. Inst. Terrestrial Ecology, Monks Wood Exp. Stn., Abbots Ripton, Huntingdon, UK, pp. 383–97.

Rother, D.C., Jordanob, P., Rodriguesc, R.R., and M.A. Pizod. 2013. Demographic bottlenecks in tropical plant regeneration: A comparative analysis of causal influences. *Perspectives in Plant Ecology, Evolution and Systematics* 15: 86–96.

Rowell, A. and P.F. Moore. 2000. *Global Review of Forest Fires*. Gland: IUCN The World Conservation Union; Forests for Life Programme Unit WWF International.

Royo, A.A. and W.P. Carson. 2006. On the formation of dense understory layers in forests worldwide: Consequences and implications for forest dynamics, biodiversity, and succession. *Canadian Journal of Forest Research* 36 (6): 1345–62.

Royo, A.A. and T.E. Ristau. 2012. Stochastic and deterministic processes regulate spatio-temporal variation in seed bank diversity. *Journal of Vegetation Science* 24: 724–34.

Rüdinger, M.C.D., J. Glaeser, I. Hebel, and A. Dounavi 2008. Genetic structures of common ash (*Fraxinus excelsior*) populations in Germany at sites differing in water regimes. *Canadian Journal of Forest Research* 38: 1199–1210.

Rumble, M.A., T. Pella, J.C. Sharps, A.V. Carter, and J.B. Parrish. 1996. Effects of logging slash on aspen regeneration in grazed clearcuts. *Prairie Naturalist* 28: 199–210.

Runkle, J.R. 1982. Patterns of disturbance in some old-growth mesic forests of eastern North America. *Ecology* 63 (5): 1533.

Runkle, J.R. 1985. Disturbance regimes in temperate forests. In: Pickett, S.T.A. and P.S. White (eds.), *The Ecology of Natural Disturbance and Patch Dynamics*. Academic Press, New York, pp. 17–33.

Runkle, J.R. 1989. Synchrony of regeneration, gaps, and latitudinal differences in tree species diversity. *Ecology* 70: 546–7.

Ruscoe, W.A., J.S. Elkinton, D. Choquenot, and R.B. Allen. 2005. Predation of beech seed by mice: Effects of numerical and functional responses. *Journal of Animal Ecology* 74: 1005–19.

Rust, S. and P.S. Savill 2000. The root system of *Fraxinus excelsior* and *Fagus sylvatica* and their competitive relationships. *Forestry*. 73: 499–508.

Ryan, K.C. 2002. Dynamic interactions between forest structure and fire behavior in boreal ecosystems. *Silva Fennica* 36 (1): 13–39.

Sagheb-Talebi, K. and J.P. Schütz. 2002. The structure of natural oriental beech (*Fagus orientalis*) forests in the Caspian region of Iran and potential for the application of the group selection system. *Forestry* 75 (4): 465–72.

Sagnard, F., C. Pichot, G.G. Vendramin, and B. Fady. 2011. Effects of seed dispersal, adult tree and seedling density on the spatial genetic structure of regeneration at fine temporal and spatial scales. *Tree Genetics and Genomes* 7, 37–48.

Sanchez, E., R. Gallery, and J.W. Dalling. 2009. Importance of nurse logs as a substrate for the regeneration of pioneer tree species on Barro Colorado Island, Panama. *Journal of Tropical Ecology* 25 (04): 429.

Santiago, L.S. 2000. Use of coarse woody debris by the plant community of a Hawaiian montane cloud forest. *Biotropica* 32 (4a): 633–41.

Schirmer, W., T. Diehl and C. Ammer. 1999. Zur entwicklung junger eichen unter kiefernschirm. *Forstarchiv* 70: 57–65.

Schliemann, S.A. and J.G. Bockheim. 2011. Methods for studying treefall gaps: A review. *Forest Ecology and Management* 261 (7): 1143–51.

Schmidt, W., 2006. Temporal variation in beech masting (*Fagus sylvatica* L.) in a limestone beech forest (1981–2004). *Allgemeine Forst- und Jagdzeitung* 177: 9–19.

Schmiedel, D., F. Huth, and S. Wagner. 2013. Using data from seed dispersal modelling to manage invasive tree species: The example of *Fraxinus pennsylvanica* in Europe. *Environment Management* 52 (4): 851–60.

Schröder, T., R. Kehr, R., Z. Prochazkova, and J.R. Sutherland. 2004. Practical methods for estimating the infection rate of *Quercus robur* acorn seedlots by *Ciboria batschiana*. *Forest Pathology* 34: 187–96.

Schupp, E.W. 1995. Seed-seedling conflicts, habitat choice, and patterns of plant recruitment. *American Journal of Botany* 82 (3): 399.

Schupp, E.W., P. Jordano, and J.M. Gómez. 2010. Seed dispersal effectiveness revisited: A conceptual review. *New Phytologist* 188: 333–53.

Schurr, F.M., O. Steinitz, and R. Nathan. 2008. Plant fecundity and seed dispersal in spatially heterogeneous environments: Models, mechanisms and estimation. *Journal of Ecology* 96: 628–41.

Schütz, J.-P. 1999. Praktische bedeutung der überführung für die umsetzung der plenteridee. *Forst und Holz* 54 (104–8).

Schütz, J.-P. 2002. Silvicultural tools to develop irregular and diverse forest structures. *Forestry* 75 (4): 329–37.

Seidl, R., M.J. Schelhaas, M. Lindner, and M.J. Lexer. 2009. Modelling bark beetle disturbances in a large scale forest scenario model to assess climate change impacts and evaluate adaptive management strategies. *Regional Environmental Change* 9: 101–19.

Sheffer, E., C.D. Canham, J. Kigel, and A. Perevolotsky. 2013. Landscape-scale density-dependent recruitment of oaks in planted forests. More is not always better. *Ecology* 94: 1718–28.

Sheil, D. 2001. Long-term observations of rain forest succession, tree diversity and responses to disturbance. *Plant Ecology* 155 (2): 183–99.

Shibata, M. and T. Nakashizuka. 1995. Seed and seedling demography of four co-occurring *Carpinus* species in a temperate deciduous forest. *Ecology* 76: 1099–108.

Siddique, I., V.L. Engel, J.A. Parrotta, D. Lamb, G.B. Nardoto, J.P.H.B. Ometto, L.A. Martinelli, and S. Schmidt. 2008. Dominance of legume trees alters nutrient relations in mixed species forest restoration plantings within seven years. *Biogeochemistry* 88 (1): 89–101.

Siles, G., P.J. Rey, and J.M. Alcántara. 2010. Post-fire restoration of Mediterranean forests: Testing assembly rules mediated by facilitation. *Basic Applied Ecology* 11 (5): 422–31.

Silvertown, J. 2004. Plant coexistence and the niche. *Trends in Ecology and Evolution (Amst.)* 19 (11): 605–11.

Sipe, T.W. and F.A. Bazzaz. 1994. Gap partitioning among maples (*Acer*) in Central New England: Shoot architecture and photosynthesis. *Ecology* 75 (8): 2318.

Sipe, T.W. and F.A. Bazzaz. 1995. Gap partitioning among maples (*Acer*) in Central New England: Survival and growth. *Ecology* 76: 1587–1602.

Smit, C., D.P. Kuijper, D. Prentice, M.J. Wassen, and J.P. Cromsigt. 2012. Coarse woody debris facilitates oak recruitment in Białowieża Primeval Forest, Poland. *Forest Ecology and Management* 284: 133–41.

Smit, C., M. Gusberti, and H. Müller-Schärer. 2006. Safe for saplings; safe for seeds? *Forest Ecology and Management* 237 (1–3): 471–77.

Smith, K.T., W.C. Shortle, J. Jellison, J. Connolly, J., and J. Schilling. 2007. Concentrations of Ca and Mg in early stages of sapwood decay in red spruce, eastern hemlock, red maple, and paper birch. *Canadian Journal of Forest Research* 37: 957–65.

Sollins, P., S.P. Cline, T. Verhoeven, D. Sachs, and G. Spycher. 1987. Patterns of log decay in old-growth Douglas-fir forests. *Canadian Journal of Forest Research* 17 (12): 1585–95.

Sonohat, G., H. Sinoquet, V. Kulandaivelu, D. Combes, and F. Lescourret. 2006. Three-dimensional reconstruction of partially 3D-digitized peach tree canopies. *Tree Physiology* 26(3): 337–51.

Soons, M.B. 2006. Wind dispersal in freshwater wetlands: Knowledge for conservation and restoration. *Applied Vegetation Science* 9: 2721–278.

Sork, V.L., J. Bramble, and O. Sexton. 1993. Ecology of mast-fruiting in three species of North American deciduous oaks. *Ecology* 74: 528–41.

Stimm, B. and K. Böswald. 1994. Die häher im visier. Zur Ökologie und waldbaulichen Bedeutung der Samenausbreitung durch Vögel (The jay in focus: Ecology and silvicultural relevance of seed dispersal by birds). *Forstwissenschaftliches Centralblatt* 113: 204–23.

Stimm, B. and T. Knoke. 2004. Hähersaaten: ein Literaturüberblick zu waldbaulichen und ökonomischen Aspekten (Sowing by jays: A literature review to silviculture and forest economy). *Forst Holz* 59: 531–4.

St. John, T. 1997. Arbuscular mycorrhizal inoculation in nursery practice. In: Landis, T.D., South, D.B. (technical coordinators.), *National Proceedings, Forest and Conservation Nursery Associations*. US Department of Agriculture, Forest Service, Portland (OR), Pacific Northwest Research Station General Technical Report PNW-GTR-389. pp. 152–58.

Stöckli, B. 1995. Decaying wood for natural regeneration in montane forests. *Forest and Timber* 16: 8–15.

Stoehr, M.U. and Y.A. El-Kassaby. 2011. Challenges facing the forest industry in relation to seed dormancy and seed quality. *Methods in Molecular Biology* 773: 3–15.

Stokland, J.N., J. Siitonen, and B.G. Jonsson. 2012. *Biodiversity in Dead Wood*. Ecology, biodiversity, and conservation. Cambridge University Press, New York.

Stoyan, D. and S. Wagner. 2001. Estimating the fruit dispersion of anemochorous forest trees. *Ecological Modeling* 145: 35–47.

Streiff, R., T. Labbe, R. Bacilieri, H. Steinkellner, J. Glössl, and A. Kremer. 1998. Within population genetic structure in *Quercus robur* L. and *Quercus petraea* (Matt.) Liebl. assessed with isozymes and microsatellites. *Molecular Ecology* 7: 317–28.

Suchant, R., R. Baritz, and F. Armbruster. 2000. Werden Wildlinge weniger verbissen? *Allgemeine Forst- u.Jagd-Zeitung* 5: 251–54.

Svoboda, M., S. Fraver, P. Janda, R. Bače, and J. Zenáhlíková. 2010. Natural development and regeneration of a Central European montane spruce forest. *Forest Ecology and Management* 260 (5): 707–14.

Sydes, C. and J.P. Grime. 1981. Effects of tree leaf litter on herbaceous vegetation in deciduous woodland: I. Field investigations. *Journal of Ecology* 69 (1): 237–48.

Szewczyk, J. and J. Szwagrzyk. 1996. Tree regeneration on rotten wood and on soil in old-growth stand. *Vegetatio* 122 (1): 37–46.

Tabaku, V. and P. Meyer. 1999. Lückenmuster albanischer und mitteleuropäischer Buchenwälder unterschiedlicher Nutzungsintensität. *Forstarchiv* 70: 87–97.

Tappeiner, J.C. and R.G. Wagner. 1987. Principles of silvicultural prescriptions for vegetation management. In: J.D. Walstad and P.J. Kuch (eds.), *Forest Vegetation Management for Conifer Production*. John Wiley, New York, N.Y. pp. 399–429.

Tasker, E.M. and R.A. Bradstock. 2006. Influence of cattle grazing practices on forest understorey structure in north-eastern New South Wales. *Austral Ecology* 31 (4): 490–502.

Temperli, C.W. 2012. Climate change, large-scale disturbances and adaptive forest management. Diss. No. 20899, ETH Zürich.

Thomas, S.C. and J. MacLellan. 2004. Boreal and temperate forests. In: Owens J.N., and H.G. Lund (eds.), *Forests and Forest Plants*. Encyclopedia of Life Support Systems (EOLSS); UNESCO, Eolss Publishers, Oxford, UK, pp. 152–75.

Thomasius, H. and P.A. Schmidt. 2004. Forest, forest management and environment. In: Buchwald K., and W. Engelhardt (eds.), *Environmental Protection—Basics and Practice*. Economica 10, Bonn, 435p.

Thomsen, K.A., E.D. Kjær. 2002. Variation between single tree progenies of *Fagus sylvatica* in seed traits, and its implications for effective population numbers. *Silvae Genetica* 51: 183–90.

Tilmann, D. 1982. Resource competition and community structure. Princeton Monographs in Population Biology 17. Princeton University Press, Princeton, NJ.

Trabucchi, M., C. Puente, F.A. Comin, G. Olague, and S.V. Smith. 2012. Mapping erosion risk at the basin scale in a Mediterranean environment with opencast coal mines to target restoration actions. *Regional Environmental Change* 12 (4): 675–87.

Trotsiuk, V., M.L. Hobi, and B. Commarmot. 2012. Age structure and disturbance dynamics of the relic virgin beech forest Uholka (Ukrainian Carpathians). *Forest Ecology and Management* 265: 181–90.

Ulanova, N.G. 2000. The effects of windthrow on forests at different spatial scales: A review. *Forest Ecology and Management* 135 (1–3): 155–67.

Urbanska, K.M., N.R. Webb, and P.J. Edwards, (eds.) 1997. *Restoration Ecology and Sustainable Development*. Cambridge University Press, Cambridge, UK.

van Couwenberghe, R., 2011. Effets des facteurs environnementaux sur la distribution et l'abondance des espèces végétales forestières aux échelles locales et régionales. Ecosystems. AgroParisTech, French. Available at: https://pastel.archives-ouvertes.fr/pastel-00604628].

van Couwenberghe, R., C. Collet, E. Lacombe, J.-C. Pierrat, and J.-C. Gégout. 2010. Gap partitioning among temperate tree species across a regional soil gradient in windstorm-disturbed forests. *Forest Ecology and Management* 260 (1): 146–54.

Vandekerkhove, K., L. de Keersmaeker, N. Menke, P. Meyer, and P. Verschelde. 2009. When nature takes over from man: Dead wood accumulation in previously managed oak and beech woodlands in North-western and Central Europe. *Forest Ecology and Management* 258 (4): 425–35.

Vandenberghe, C., F. Freléchoux, F. Gadallah, and A. Buttler. 2006. Competitive effects of herbaceous vegetation on tree seedling emergence, growth and survival: Does gap size matter? *Journal of Vegetation Science* 17 (4): 481–88.

Van Der Meer, P.J., P. Dignan, and A.G. Saveneh. 1999. Effect of gap size on seedling establishment, growth and survival at three years in mountain ash (*Eucalyptus regnans* F. Muell.) forest in Victoria, Australia. *Forest Ecology and Management* 117 (1–3): 33–42.

Vanha-Majamaa, I. and E.-S. Tuittila. 1996. Seedling establishment after prescribed burning of a clear-cut and a partially cut mesic boreal forest in southern Finland. *Silva Fennica* 31 (1): 31–45.

Vanha-Majamaa, I., S. Lilja, R. Ryömä, J.S. Kotiaho, S. Laaka-Lindberg, H. Lindberg, P. Puttonen, P. Tamminen, T. Toivanen, and T. Kuuluvainen. 2007. Rehabilitating boreal forest structure and species composition in Finland through logging, dead wood creation and fire: The EVO experiment. *Forest Ecology and Management* 250 (1–2): 77–88.

van Hees, A.F. 1997. Growth and morphology of pedunculate oak (*Quercus robur* L.) and beech (*Fagus sylvatica* L.) seedlings in relation to shading and drought. *Annales of Forest Science* 54 (1): 9–18.

van Oijen, D., M. Feijen, P. Hommel, J. Ouden, and R. Waal. 2005. Effects of tree species composition on within-forest distribution of understorey species. *Applied Vegetation Science* 8 (2): 155–66.

Vanselow, K. 1949. *Theorie und Praxis der natürlichen Verjüngung im Wirtschaftswald* (Theory and Practice of Natural Regeneration in Managed Forests). 2nd ed., Neumann Verlag, Radebeul, Berlin.

Verheyen, K., G.R. Guntenspergen, B. Biesbrouck, and M. Hermy. 2003. An integrated analysis of the effects of past land use on forest herb colonization at the landscape scale. *Journal of Ecology* 91 (5): 731–42.

von Lüpke, B. 1987. Einflüsse von altholzüberschirmung und bodenvegetation auf das wachstum junger buchen und traubeneichen (Effect of canopy structure and ground vegetation on growth of young beeches and oaks). *Forstarchiv* 58: 18–24.

von Lüpke, B. and K. Hauskeller-Bullerjahn. 2004. Beitrag zur modellierung der jungwuchsentwicklung am beispiel von traubeneichen-buchen-mischverjüngung. *Allgemeine Forst- u.Jagd-Zeitung* 175: 61–69.

Wada, N. and E. Ribbens. 1997. Japanese maple (*Acer palmatum* var. Matsumurae, Aceraceae) recruitment patterns: Seeds, seedlings, and saplings in relation to conspecific adult neighbors. *American Journal of Botany* 84: 1294–300.

Wagner, S. 1999. Ökologische Untersuchungen zur Initialphase der Naturverjüngung in Eschen-Buchen-Mischbeständen. Schriftenreihe der Forstlichen Fakultät der Uni Göttingen und der Niedersächsischen Forstlichen Versuchsanstalt Göttingen, Sauerländer's Verlag.

Wagner, S., C. Collet, P. Madsen, T. Nakashizuka, R.D. Nyland, and K. Sagheb-Talebi. 2010. Beech regeneration research: From ecological to silvicultural aspects. *Forest Ecology and Management* 259: 2172–82.

Wagner, S., H. Fischer, and F. Huth. 2011. Canopy effects on vegetation caused by harvesting and regeneration treatments. *European Journal of Forest Research* 130 (1): 17–40.

Walker, L.R., J. Walker, and R.J. Hobbs. 2007. *Linking Restoration and Ecological succession.* Springer series on environmental management. New York, NY: Springer.

Walters, M.B. and P.B. Reich. 1997. Growth of *Acer saccharum* seedlings in deeply shaded understories of northern Wisconsin: Effects of nitrogen and water availability. *Canadian Journal of Forest Research* 27: 237–47.

Walters, M.B. and P.B. Reich. 2000. Seed size, nitrogen supply, and growth rate affect tree seedling survival in deep shade. *Ecology* 81: 1887–901.

Wang, C.W. and T.B. Smith. 2002. Closing the seed dispersal loop. *Trends in Ecology and Evolution* 17: 379–85.

Watt, A.S. 1923. On the ecology of British beechwoods with special reference to their regeneration. *Journal of Ecology* 11: 1–48.

Watt, A.S. 1925. On the Ecology of British Beechwoods with Special Reference to Their Regeneration: Part II, Sections II and III The development and structure of beech communities on the Sussex Downs (continued). *Journal of Ecology* 13: 27–73.

Watt, A.S. 1947. Pattern and process in the plant community. *Journal of Ecology* 35: 1–22.

Weiland, J.E., B.R. Beck, and A. Davis. 2013. Pathogenicity and virulence of *Pythium* species obtained from forest nursery soils on Douglas-fir seedlings. *Plant Disease* 2013. 97: 744–8.

Wenny, D.G. 2000. Seed dispersal, seed predation, and seedling recruitment of a Neotropical montane tree. *Ecological Monographs* 70: 331–51.

Wiedemann, E. 1927. Über den künstlichen gruppenweisen Voranbau von Tanne und Buche. *Allgemeine Forst- u.Jagd-Zeitung* 103: 433–52.

Willoughby, I. and R.L. Jinks. 2009. The effect of duration of vegetation management on broadleaved woodland creation by direct seeding. *Forestry* 82 (3): 343–59.

Willson, M.F. 1993. Dispersal mode, seed shadows, and colonization patterns. *Vegetatio* 107/108: 261–80.

Wojczulanis, B. 2002. *Forest Ecosystems of the Karkonosze National Park.* Agencja Fotograficzno-Wydawnicza, Mazury.

Wolf, H. 2003. EUFORGEN. Technical guidelines for genetic conservation and use for Silver fir (*Abies alba*). International plant Genetic resources institute, Rome, Italy. Available at: http://www.euforgen.org/publications/publication/silver-fir-emabies-albaem/.

Yamamoto, S.-I. 1992. The gap theory in forest dynamics. *The Botanical Magazine, Tokyo* 105: 375–83.

Yamamoto, S.-I. 1995. Gap characteristics and gap regeneration in subalpine old-growth coniferous forests, central Japan. *Ecological Research* 10: 31–39.

Zeibig, A., J. Diaci, and S. Wagner. 2005. Gap disturbance patterns of a Fagus sylvatica virgin forest remnant in the mountain vegetation belt of Slovenia. *Forest Snow and Landscape Research* 79: 69–80.

Zerbe, S. 1998. Potential natural vegetation: Validity and applicability in landscape planning and nature conservation. *Applied Vegetation Science* 1: 165–72.

Zerbe, S. 2002. Restoration of natural broad-leaved woodland in Central Europe on sites with coniferous forest plantations. *Forest Ecology and Management* 167 (1–3): 27–42.

Zerbe, S. and A. Brande. 2003. Woodland degradation and regeneration in Central Europe during the last 1,000 years—A case study in NE Germany. *Phytocoenologia* 33 (4): 683–700.

Zerbe, S. and D. Kreyer. 2007. Influence of different forest conversion strategies on ground vegetation and tree regeneration in pine (*Pinus sylvestris* L.) stands: A case study in NE Germany. *European Journal of Forest Research* 126 (2): 291–301.

Zerbe, S. and G. Wiegleb (eds.), 2009. *Renaturierung von Ökosystemen in Mitteleuropa.* Spektrum Akademischer Verlag, Heidelberg.

Zielonka, T. 2006. When does dead wood turn into a substrate for spruce replacement? *Journal of Vegetation Science* 17 (6): 739–46.

Zielonka, T. and M. Niklasson. 2001. Dynamics of dead wood and regeneration pattern in natural spruce forest in the Tatra Mountains, Poland. *Ecological Bulletin* 49: 159–63.

Zielonka, T. and G. Piątek. 2004. The herb and dwarf shrubs colonization of decaying logs in subalpine forest in the Polish Tatra Mountains. *Plant Ecology* 172 (1): 63–72.

Zimmerman, J.K., W.M. Pulliam, D.J. Lodge, V. Quiñones-Orfila, N. Fetcher, S. Guzmán-Grajales, J.A. Parrotta et al. 1995. Nitrogen immobilization by decomposing woody debris and the recovery of tropical wet forest from hurricane damage. *Oikos* 72 (3): 314.

Zywiec, M. and T. Zielonka. 2013. Does a heavy fruit crop reduce the tree ring increment? Results from a 12-year study in a subalpine zone. *Trees* 27: 1365–73.

7

Plantations: Forests: Wilderness: The Diversity of Forest Landscapes in Europe as a Consequence of Social Change, Technological Progress, and Disturbance

Norbert Weber and Sandra Liebal

CONTENTS

7.1 Introduction

The continent of Europe, ranging from the Iberian Peninsula to the European part of Russia, is subject to considerable climatic and orographic variety. According to Meeus (1995), landform, soil, and climate, but also regional culture, habits, and history, are decisive factors in delineating European landscape types. Forests cover 1.02 billion ha, equivalent to 45% of Europe's land area, and together with an additional area of 109 million ha of other wooded land, forests shape many European landscapes (Forest Europe et al. 2011). Enormous differences in the conditions for the growth of forest trees leads to a high diversity of forest landscapes, ranging from tundra, taiga, mountain forests, and semibocages to regional landscapes including montados and dehesas (Meeus 1995). Several typologies have been applied to split the continent into major regions with different forest characteristics; for example, Great Britain, the Nordic region, central Europe, and Iberia

(Edwards et al. 2012: 14); central-east, central-west, north, Russian Federation, southeast, and southwest (Forest Europe et al. 2011: 19); and Nordic, Atlantic, central, continental, and Mediterranean regions (Pröbstl 2007). For this reason, some of the findings presented in this chapter are of greater significance for specific regions, yet most have wide validity.

Forest landscapes have always been subject to change (see Hendinger 1960). Most of the landscape changes that have taken place in central Europe over the last few thousand years have been driven by material and technological advances, and also by social developments (Bastian and Bernhardt 1993). Before the initiation on a large scale of active measures for reforestation in the nineteenth century, the forest areas of Europe had been shrinking. The causes included permanent grazing by cattle, collection of litter for agricultural purposes, industrial uses in salt refineries, glass production, and mining, and harvesting wood for domestic fuel and material uses. Human influence had served to transform most of the closed forests into open landscapes, more reminiscent of parks than forests. Even 15 years after World War II, many forest ecosystems had not recovered from prior intensive agricultural exploitation (Welzholz and Johann 2007). Following the devastation wrought during the wars, and as a result of post-war overexploitation, the state of the forests gradually began improving again. Since the late twentieth century, forest area has expanded considerably, through afforestation and natural recolonization (Weber 2005), and through the establishment of plantations on disturbed land (Savill et al. 1997). In the area known as the Pan-European region, this increase has amounted to 17 million ha since the 1990s (Forest Europe et al. 2011). Individual tree growth has also accelerated. Growing stocks (Kahle et al. 2008) and carbon stocks (Nabuurs et al. 2003) have improved due to more favorable environmental conditions. The corresponding negative developments, mostly human induced, include the increased fragmentation of forests as a result of clearance for roads, railway tracks, and other infrastructural elements. This has resulted in smaller parcels of forest, and a reduction in biodiversity brought about by the establishment of barriers to dispersal affecting many species (Forest Europe et al. 2011).

High forests, managed for timber, dominate the current landscape, whereas coppice and coppice with standards are found only on a small scale. In 2010, the area of coppice maintained in some European countries such as France, Germany, Bulgaria, the Netherlands, Switzerland, the United Kingdom, Slovakia, and Belgium amounted to 2.9 million ha. However, the energy crisis of the 1970s led to a temporary renewal of interest in coppice crops managed in short rotations, "as alternative sources of wood, charcoal, and liquid fuels, a basis for chemical processes, wood pulp, and sometimes as a fodder supplement" (Savill et al. 1997). Mostly classified as agricultural land use, short rotation coppice (SRC) features on the political agenda again today, as a means to support the ambitious aims of renewable energy policies introduced at European, national, and subnational levels (see Bemmann and Knust 2010; Weih and Dimitriou 2012).

Landscape change is in the first instance an issue of individual perception, based primarily on the personal experiences people make in the early stages of their lives (Edwards et al. 2012). Using a landscape image sketching technique, Ueda et al. (2012) identified cultural differences between citizens of different countries with respect to their perception of forests. Perception is not the only measure, however, and "real" landscape change can be quantified, mapped and documented. There are well-developed tools to identify forest landscape change and to estimate future developments, such as forest inventory, aerial photography, remote sensing and laser scanning. Highly specialized applications include the use of remote sensing to estimate future bioenergy potentials in forests (Straub and Koch 2011), whereas forest canopy change research combines remote sensing and terrestrial inventories (Schleeweis et al. 2012). The Swedish target-tailored forest damage inventory

(TFDI) works in a similar manner (Wulff et al. 2012). Figures pertaining to the changes in the forest area as a consequence of afforestation and the abandonment of land are available (see Weber 2005). It is difficult, however, to compare data on changes to the forest area over a particular time series due to changing definitions of forest (Zanchi et al. 2007).

Our main hypothesis is that the appearance of the forest landscapes of Europe has changed fundamentally since World War II for several reasons. These include human-induced factors such as social and political changes, as well as global climate change; although these factors are in fact strongly linked to one another (Dale et al. 2001). The major drivers of change are assumed to be an increase in disturbance, changes in silvicultural paradigms, and tendencies toward land sparing for nature conservation purposes. In this chapter, we will outline some of the major characteristics of the large-scale changes to forest landscapes in Europe. We will concentrate on the period since World War II and focus on the forest area as a whole, irrespective of the types of ownership and the manifold national peculiarities. We will describe different patterns of forest landscape change in Europe, focusing on three stages along a continuum from plantations to production forests and wilderness (Liebal 2013). In a subsequent step, we will explore social change, technological development and large-scale disturbances as potential explanations for the changes to forest landscapes. Finally, we will develop upon our main findings in the discussion.

7.2 Characteristics of Forest Landscape Change

There are several factors that influence the landscape: climatic effects, fire (either wildfire or prescribed), and management being the decisive ones. As stated by Schumacher and Bugmann (2006), forest landscape dynamics are the result of complex interactions of driving forces as well as ecological processes at various scales. After the introduction of sustainable yield forestry in Europe at the beginning of the nineteenth century, the density of forest stands increased and many monospecies stands (mostly of *Picea* and *Pinus* species) were established. When it emerged that these forests were not only more prone to damage by insects and extreme meteorological events, but also with the growing need for protective and recreational services provided by forests, a preference for multifunctional forests took hold and a process of conversion to mixed forests and broadleaf stands began (e.g., Cameron et al. 2001; Dedrick et al. 2007). Another trend that can be observed today is a growing emphasis on "land sparing" (see Egan and Mortensen 2012), with forests segregated according to intensively managed areas and areas taken entirely out of production. While the amount of monocultural plantations is increasing, especially in Ireland, the United Kingdom, Denmark, Belgium, Luxembourg, Portugal, and Iceland (Forest Europe et al. 2011); many forests have been set aside and protected in national parks and other designated areas. The number of protected areas in the 32 member states covered by the European Environmental Agency (EEA) has increased from zero in 1938 to nearly 100,000 in 2009, covering an area of about 1,200,000 km² (EEA 2012).

Hendinger (1960) presented landscape-based indicators to assess and scrutinize change in forest landscapes. To these belong (1) a change of silvicultural system and changes to the distribution of the respective types of production forests; (2) changes to tree species distribution or a shift in tree species composition; (3) the percentage of afforested areas and areas of natural recolonization; (4) the type and intensity of regeneration and tending; (5) the percentage of cleared areas and the distribution of agricultural settlement in comparison to forest; (6) the state and scope of forest roads and skidding trails; and (7) relicts of agriculture, for example, woodland pasture and traces of former arable land in forests. More than 50 years later, this classification is still applicable; although, new trends and challenges are becoming apparent.

According to Bastian and Bernhardt (1993), anthropogenic changes to the landscape are characterized by an acceleration of the sequence of changes, a continual increase in the scope and complexity of ecological problems, a growing destabilization of the natural environment, and a rising proportion of irreversible changes. In a similar vein, (Vanhanen et al. 2010) referred to, "scenarios of rapid change that tend to push both social and ecological systems toward conditions where the historical relationships between the components can no longer be sustained." Unfortunately, "the speed and intensity of these changes leave little opportunity for learning" (Vanhanen et al. 2010). An indicator of more change to come, the global demand for food, feed, and fiber is expected to nearly double with an increase in the global population (Vanhanen et al. 2010).

Today, major changes to forest landscapes can be observed in the following categories:

- *Forest area* has increased in Europe, although this has mostly been due to afforestation and natural recolonization in remote areas. Larger losses have been documented on the fringes of urban agglomerations. New recreational forests around cities have been established especially in the United Kingdom, the Netherlands, Denmark, and Belgium (Elands et al. 2010). At the same time, a further fragmentation of existing forests has occurred. This fragmentation is a major source of alteration in the landscape with considerable impacts; that is, insufficient total forest habitat area, isolation of forest habitat patches, and edges where forest habitat areas adjoin modified ecosystems (Forest Europe et al. 2011: 84).

- *Silvicultural* patterns have changed in many parts of Europe; principally with the introduction of low impact silvicultural systems and more recently; adaptation measures implemented to address climate change (Hickler et al. 2014). A reduction in the number and size of clearcuts has been prescribed in silvicultural guidelines; in several countries in central Europe this has been fixed in legislation. Intensive site preparation (cultivation for the establishment of plantations) has been reduced considerably. Close-to-nature forestry and, more recently, retention forestry, are approaches gaining support amongst foresters, stakeholders, the general public, and politicians. This is mirrored in the conversion of many conifer stands to broadleaf or mixed species stands (Figure 7.1). Indeed, forest areas consisting of single tree species decreased by 0.6% annually during a 15 year period up to 2011. The majority of European forests (about 70%) are dominated by two or more tree species (Forest Europe et al. 2011). Further evidence of this is a greater use of natural regeneration (instead of planting and seeding), longer distances between skidding trails, etc. By contrast, at least in the long term and after realizing the sustainable biomass potential of existing forests, the increasing demand for wood and energy would appear to create a new need for SRC. Agroforestry is also growing in importance; and is in some cases supported by the forest legislation (e.g., amendment to the German Federal Forest Act, Bundeswaldgesetz 2010). Methods of enhancing biomass production from forests should also be mentioned; for example, the establishment of two-storied forests delivering an initial harvest for energy use and a second for material use, and pioneer cropping, sometimes also termed as energy nurse crops, for bioenergy.

FIGURE 7.1
Forest conversion visualized: Conifer forests are enriched and/or replaced by broadleaves. An example from the Ore Mountains, Saxony. (Photo: N. Weber.)

- *Disturbance*, either natural or human induced, caused by wildfire, wind, pests, and other agents, has increased considerably in recent decades (Schelhaas et al. 2003). There is empirical evidence that the frequency and severity of large wildfires have grown. Bark beetle outbreaks have also reached unexpectedly high levels in central Europe (Seidl et al. 2014). Due to climate change, disturbances to forest ecosystems will be even greater in the future (Hickler et al. 2014). In the case of forest fires, arson is the major human-induced cause. Examples of prescribed burning can also be found in Finland, France, and in *Pinus nigra* stands in northeastern Spain (Forest Europe et al. 2011; Jentsch 2013). Forests of the low mountain ranges in central Europe have recovered from severe damage caused by air pollution, whose effects were particularly severe in the 1980s and the early 1990s. The issue of disturbance will be dealt with in greater detail later in this chapter.

- *Political shaping of forest landscapes* (see Livingstone 2010) is a more recent phenomenon. Land use in Europe is strongly influenced by energy policy. While "yellow" agricultural landscapes are the result of the increased production of oilseed rape for bioenergy, forest landscapes have been impacted by energy policy through the establishment of wind power and solar power plants. Pumped-storage hydropower plants sometimes necessitate the clearance of forests for the construction of upper basins. Another political shaping of forest landscapes has resulted from the conversion of conifer forests into broadleaf and mixed forests. While in the past, this process was justified by policies aiming at close-to-nature forestry, today an important driver is to increase their capacity to adapt to the effects of climate change.

In the following section, three typical types of European forest landscape will be described. While production forests still comprise the majority of the forests, plantations and—at the opposite end of the continuum between intensive and extensive use—protected areas are attaining greater significance. Accordingly, about 87% of the forest areas have been classified as seminatural, 9% as plantations, and 4% as undisturbed (Forest Europe et al. 2011).

7.3 Production Forests, Single-Purpose Plantations, and Protected Areas

7.3.1 Production Forests

The majority of forests in the 27 countries of the European Union (EU), encompassing 88.6 million ha and covering more than half of the forest area in EU in 2010, primarily serve productive purposes. Even so, other goals play an important role in these forests as well (Figure 7.2). Around 20% are labeled multiple-use, although in the Netherlands and Germany this category is much higher at close to three quarters, whereas in Luxembourg, Slovakia, the United Kingdom, and Belgium multiple-use forests cover over half of the forest area (EUROSTAT 2011). Productive purposes refer to forests managed for wood and nonwood products, including cork, Christmas trees, mushrooms and truffles, fruits, berries and edible nuts, resins, decorative foliage, and ornamental plants. Regional differences are evident. Christmas tree production, for example, is more common in northern Europe, whereas management for mushrooms and truffles is mainly found in southern countries such as Italy and Spain (Forest Europe et al. 2011). In contrast to other regions of the world (e.g., North America, New Zealand, Brazil), clearcuts in most parts of Europe are limited in size and most have irregular shapes.

A large share of Europe's production forests consist of conifers (*Picea abies*, *Pinus* spp.), even on sites where broadleaves grew in the past. The cultivation of conifer species outside of their natural range can be explained by the "widespread over-exploitation and devastation of forests, as well as by the fear for timber shortage" in centuries past (cf. Figure 7.3; Dedrick et al. 2007).

The conversion of production forests to SRC is prevented by forest laws in most European countries. However, intensification of biomass production can be achieved within the scope

FIGURE 7.2
Multiple-purpose production forest in Saxony, Germany. (Photo: N. Weber.)

FIGURE 7.3
Single-purpose plantation, a *Picea sitchensis* stand in Ireland. (Photo: N. Weber.)

of current silvicultural systems by choosing more productive tree species, using improved genetic material, reducing standing volume without reducing the increment, optimizing productivity per unit area, increasing felling intensity, utilizing early stages of succession, and through mixed forests. While all of these options may serve to increase the added value of forestry and the incomes of forest owners, some of these options may have negative impacts on soil nutrient cycles, biodiversity, and nature conservation (cf. the considerations of Bauhus in Liebal 2013).

Pedroli et al. (2013) emphasized differences in the utilization of forests for bioenergy between Scandinavia (Finland and Sweden) and countries in Central and Western Europe (Slovakia, Belgium, and The Netherlands). They stated that the harvesting of wood plays an important role in the provision of energy and is practiced in many forests in Scandinavia. According to their data, crown biomass is harvested in 50% of the forests in southern Finland and stumps are utilized in 5% of the forests. Negative impacts on biodiversity are to be expected as a result of the loss of dead wood associated with the removal of stumps, roots, and logging residues; and increased fertilization with nitrogen and wood ash (Pedroli et al. 2013). Logging residues and stumps are not extracted from the forests in Belgium, The Netherlands and Slovakia, and the annual harvest amounts to less than 60% of the annual increment. Presumably the danger of biodiversity loss in Scandinavia as a consequence of increased biomass removal is greater than in those countries where harvest rates are much lower (Pedroli et al. 2013).

7.3.2 Plantations

Plantations are "forests of introduced species, and in some cases native species, established through planting and seeding" (FAO 2006), often with few species, regular spacing, and even-aged stands. These are divided into productive plantations and protective plantations (Kanninen 2010). Excluding the Russian Federation, intensively-managed plantations cover about 9% of the forest area in Europe (Forest Europe et al. 2011).

New forest plantations have mostly been established on former agricultural land. According to Kanowski and Savill (1992) in Savill et al. 1997),

"In its simplest form, plantation forestry describes the intensive management of a forest crop for a limited range of products. ... Complex plantation forestry also implies relatively intensive management which controls the origin, establishment and development of the forest crop, but which integrates other land uses within its boundaries, and which promotes the early and continuing production of a wide variety of goods, services and values." (Kanowski and Savill 1992, in Savill et al. 1997)

Countries such as Ireland, the United Kingdom, Denmark, and Iceland suffered a major loss of forest cover in the past, as a result of which the proportion of plantations relative to their total forest area today approaches 100% (Kanninen 2010).

Several tree species are preferred for plantations in Europe: *Pseudotsuga menziesii*, *Picea sitchensis*, *Pinus contorta*, *Pinus pinaster*, *Larix* ssp., *Populus* clones, and *Eucalyptus* spp. (Savill et al. 1997; FAO 2006; Forest Europe et al. 2011). Although occurring naturally in several regions in Europe, *Picea abies* stands are often managed as monocultural plantations. *Populus* cultivars from a controlled genetic base are used for veneer, matchsticks, and vegetable crates (Savill et al. 1997; Stanturf and van Oosten 2014). Apart from forest plantations cultivated for the production of woody biomass, some specialized plantations have been established for nonwood forest products. Some examples include oak woodlands planted exclusively for the production of truffles (*Tuber melanosporum*) in Spain and other Mediterranean countries.

The FAO estimates that about 18 million ha are currently covered by forest plantations in Europe. Within that category, in 2010, a total of 2,848,000 ha, equivalent to 2% of the forest area, were classified as coppice. These stands are situated in Belgium, France, Bulgaria, Germany, Netherlands, Slovakia, Switzerland, and the United Kingdom (Forest Europe et al. 2011). Plantation silviculture ensures product uniformity and sustained yields (Savill et al. 1997). Nevertheless, the expansion of intensive plantations may require justification and reasonable argumentation as the positive perceptions of forests, attributed on the basis of their nature, ecological diversity, social functions, and esthetics, might be damaged (Vanhanen et al. 2010). Problems with plantations can involve a range of factors: biological risks, instability, esthetics issues, loss of conservation value, timber quality, and increased soil and stream acidity as a result of the interception of atmospheric pollutants (Savill et al. 1997). Negative examples of the effects of single tree species plantations from England date back to the 1920s (Vanhanen et al. 2010). At the same time, plantations can have positive effects on landscapes and on communities. Plantations are expected to provide a resource base for developing wood-processing industries, a source of energy for industrial and nonindustrial production, and rural employment, where labor is available and affordable (Savill et al. 1997). Environmental benefits arise when plantations are used for the rehabilitation of industrial waste sites, shelterbelts, the stabilization of sand dunes, amenity purposes, recreation, and for the enhancement of landscape values; for the protection of natural forests from exploitation for commercial and subsistence needs, for carbon sequestration, and for soil and slope stabilization (Savill et al. 1997). Plantation forests are often not as valuable as natural forests in terms of biodiversity, but provide "valuable habitat, even for some threatened and endangered species, and may contribute to the conservation of biodiversity by various mechanisms" (Brockerhoff et al. 2008).

As SRC is often regarded as an instrument of energy policy (Faaij 2006; European Union 2009) and legally classified as agricultural land use, it is difficult to ascertain the exact area occupied by this land use from available statistics (Figure 7.4). While previous estimates were more optimistic, as in the case of Italy, increasing competition of low-priced imports

FIGURE 7.4
Short rotation coppice consisting of Populus species, Saxony. (Photo: S. Liebal.)

of wood from other regions, preferences for more profitable crops, and an unfavorable political framework have even reduced the area of SRCs. Notwithstanding, countries such as Hungary, Croatia, and Italy proposed to increase SRC areas in the next years mainly for energy purposes while Spain intends to compensate for a shortfall in roundwood production (Forest Europe et al. 2011). Environmental impacts associated with SRC have been assessed in detail in recent years (Weih and Dimitriou 2012). There is evidence that well-managed cultivation of SRC can prevent nitrate leaching in sensitive areas and the impact in terms of groundwater recharge can be mitigated through appropriate management (Schmidt-Walter and Lamersdorf 2012). By providing habitat for plants with different requirements, SRC plantations exhibit a significant share of γ-diversity. Therefore, they can affect species diversity positively on the landscape scale, especially in areas with lower habitat diversity (Baum et al. 2012).

7.3.3 Protected Areas

Protected forests and areas designated as wilderness are situated at the opposite end from plantations (Figure 7.3) of the intensification–extensification continuum of forest use (Figure 7.5). There has been a growth in the number and size of protected areas incorporated in national parks, Natura 2000 sites, reference areas according to certification systems such as FSC, and other categories. Out of a total of 120,000 nationally designated nature protection sites in 52 countries, over 105,000 of these are located in the 39 countries that come

FIGURE 7.5
Protected forest areas: the Koli National Park in Finland. (Photo: N. Weber.)

under the purview of the EEA. As these comprise 69% of the entries in the UNEP-WCMP world database of protected areas, Europe is host to more protected areas than any other part of the world (EEA 2012). Today, about 10% of Europe's forests (excluding those of the Russian Federation) are protected mainly for the purposes of conserving biodiversity, with a further 9% primarily targeting landscape protection; a total area of 39 million ha. During the last 10 years, these areas designated for biodiversity and landscape protection have increased by 500,000 ha annually (Forest Europe et al. 2011). The combination of biodiversity conservation and landscape protection is a European specialty; its regulatory basis is found in the Pan-European Biological and Landscape Diversity Strategy adopted in 1995 (www.peblds.org) and implemented within the framework of the annual Biodiversity for Europe conferences.

Forest protection for biodiversity conservation displays considerable variation among European regions in terms of the level of regulatory stringency. Whereas in northern Europe and in certain eastern European countries, a regime of nonintervention or only minimal intervention prevails; active management is allowed in protected areas in central and southern European countries. In the Russian Federation alone, there are 17 million ha in protected forests with no or minimal intervention allowed. The respective natural conditions, but also the traditions and population density give rise to these different policies across Europe (EEA 2012). "Forests undisturbed by man are those where the natural forest development cycle has remained or been restored, and show characteristics of natural tree species composition, natural age structure, deadwood component and natural regeneration and no visible sign of human activity" (Forest Europe et al. 2011). Undisturbed forests cover 8 million ha, which is 4% of the total forest area of the EEA region. Bulgaria, Estonia, Finland, Romania, Slovenia, Sweden, and Turkey are the countries with the largest undisturbed forest areas (each with areas over 100,000 ha), while the highest share of undisturbed forest as a percentage of total forest area is found in the Russian Federation with 32%, and in northern Europe (EEA 2012). The total number of sites designated for nature protection in Europe amounts to between 65,000 and 70,000 (Frank et al. 2007).

Around 20.4 million ha of forest within the EU, equivalent to 13% of the total area, had protected area status in 2010, with the largest areas located in Italy, Germany, and Spain (EUROSTAT 2011). There are different national categories of protected forest area and increasingly forests are subject to international conservation regulations. These include regulations stipulated by IUCN, UNESCO, and the Natura 2000 Network. The protected terrestrial area globally grew from about 200 to 240 million ha in the period from 1990 to 2007 (FAO 2009). In an earlier survey (MCPFE 2003), protected forests accounted for 11.7% of the total forest area of Europe, equivalent to about 127 million ha. Out of this, 85% was designated for the conservation of forest biodiversity and 15% was classified as protected landscape.

People do not value forests for their use functions alone. Often they are attributed a "great symbolic value as a constituent of rural identity or as a representation of nature" (Elands et al. 2004). Current conceptions of wilderness are diverse, ranging from "all good things are wild and free" to "threatening, hostile and wild" (Vanhanen et al. 2010). Forest wilderness is often considered to stimulate tourism, but strict definitions of wilderness can lead to reduced accessibility of these areas for visitors, for the reasons of biodiversity conservation (European Parliament 2009). Wilderness concepts potentially include the reintroduction of large predators (such as bear and wolf) and herbivores (e.g., bison and moose) to regions where they became extinct long ago. Debates between supporters and opponents of such reintroductions are often emotionally charged.

Landscape quality and esthetics are increasingly perceived as being important for human well being (Vanhanen et al. 2010). This might be another reason for the increasing interest in wilderness areas. People from rural areas mostly perceive forests from

> "the perspective of nature and landscape quality and less as an economic activity or carrier of services… forests are foremost valued in relation to their perceived contribution to the rural identity rather than in relation to their production and income generation capacity" (Elands et al. 2004).

Perceptions of forest landscapes can also be influenced by their recreational value. Standing and fallen dead wood is deemed a fundamentally important component of protected areas for its role in enhancing natural processes. It also plays a role in increasing biodiversity in production forests, although this can conflict with the obligation of forest owners to ensure the safety of visitors to the forests.

7.4 Reasons for Forest Landscape Change

Landscape transformation results from both natural and human-induced changes that continuously mold and remold landscapes. These changes vary in character and intensity from time to time, and from region to region. A temporal and a regional specification is required as landscapes differ in their responses to these influencing agents (Klijn 2003). Population increase, social change, and variations in spatial distribution of populations are particularly strong influences on the type and intensity of forest use (Vanhanen et al. 2010). Two paradigm shifts can be observed in this context. The first is a shift from sustainable raw material production to ecosystem management and the second is a switch from discussion merely at the national level to discussion at the international level (Ottitsch et al. 2005). Both of these developments serve to support efforts for the establishment of more protected forest areas and wilderness areas. Three drivers appear to be decisive for the diversification of forest landscapes: social change, technological progress, and disturbance.

7.4.1 Social Change

Although no clear proportional relationship between cause (intensity of social impact) and effect (landscape change) is in evidence (Bastian and Bernhardt 1993), social change influences landscapes in many respects. For instance, in European societies, the demand for wood for energy is increasing as a result of the debate over renewable energies and the resulting political statements (e.g., European Commission 2010 and the respective national regulations). Further influences arise from a growing concern over biodiversity issues, climate change, and emissions trading. The social significance and individual perceptions of "forest" are culturally founded, and general attitudes toward forests differ between societies (Vanhanen et al. 2010). Differences can be observed with regard to local climate, natural conditions, and also between individuals and groups. "Where the forest starts and what it consists of, which tree species make a forest, what is accepted as trees, are all open to different interpretations" (Vanhanen et al. 2010). Schanz stressed that although technological and ecological aspects are responsible for the specific format of forestry in a certain context, "it is obviously social aspects that constitute forestry as such" (Schanz 1999).

In many parts of Europe, social acceptance of wood production has diminished to a large extent. The reasons for this may be a "decline in the role of traditional forestry, and increasingly diverse views about some management practices" (Vanhanen et al. 2010). In a telephone survey of 11,106 citizens in 27 states of the EU, "conservation and protection" was cited as an important concern by those people surveyed. It was ranked first by 49% of respondents, and so was deemed much more important than climate change, which was considered most important by only 10% (European Commission 2009). However, 85.5% of all respondents suggested that there should be more or even much more active management of forests with a view to provide recreation opportunities, whereas 74% were of the opinion that active management is necessary to protect people from disasters. This means that a larger number of respondents called for more active management of forests to protect people from disasters and climate change than for the provision of wood (European Commission 2009).

Key social changes relevant for forestry involve the globalization of the timber industry, shifts in national policies dealing with forest management and, at least in a few cases, the devolution of power to the local community level (Vanhanen et al. 2010). It goes without saying that shifts in the demands placed on forests by society may eventually result in political discussion, amendments to forest laws and regulations, and altered funding schemes. A striking example of this is former Regulation 2080/92 of the EU, which provided subsidies for afforestation differentiated according to tree species. Whereas 2000 ECU were granted toward costs for *Eucalyptus* plantations, the sum for conifer plantations was 3000 ECU and 4000 ECU for broadleaf plantations and mixed plantations with not less than 75% broadleaf species (EC 1992). Even nonstate market-driven instruments such as forest certification can influence the appearance of forests. In European forestry, certification has developed as a kind of social license to operate. Most of the forests of the EU are certified under one or both of the major certification schemes. In 2005, forest areas certified under PEFC and FSC covered 42,577,747 and 22,239,070 ha, respectively (Commission of the European Communities 2005). Under FSC certification, landowners are obliged to set-aside at least 5% of the productive forest area for conservation purposes. This land is to serve as unused reference areas to allow long-term natural processes to take place. Given that this was applied to the entirety of the area certified by FSC in Europe in September 2013 (FSC Facts and Figures 2013), nearly 4 million ha of forest areas may currently be set-aside from production.

7.4.2 Technological Progress

Social change and technological progress are strongly linked. New technologies are either boosted or hindered by the views of the people affected. For example, social acceptance of new forms of bioenergy is dependent upon the acceptance of the corresponding technology and whether societies and individuals can benefit from it; if the price the consumer must pay for alternative energy is attractive and new jobs are created, contributions to the regional economy can be expected and incomes can be improved (Vanhanen et al. 2010). Irrespective of these potential benefits, urban visitors searching for tranquility in remote areas might feel disturbed by the harvesting and transport techniques, and by the damage that may be caused to forest roads and hiking trails (see Schaffner 2002).

Technological development is the application of scientific or other organized knowledge to practical tasks. This relates to tools, techniques, products, processes, methods, organizations, and systems (Hetemäki et al. 2010). From the perspective of the issue at hand, this includes progress in silvicultural techniques. Obviously, technological progress in the eighteenth and nineteenth centuries leading to the substitution of forest products previously put to material and energy uses served to reduce the pressure on forest areas, thereby changing their appearance from open to closed forests.

The introduction of harvesters and forwarders as the main harvesting tools did not change forest landscapes per se. Systematic thinning, however, leads to the increased uniformity of forest structures. Certain other approaches, such as whole-tree harvesting and slash harvesting, can influence forest appearance substantially. The technology employed for the latter is so effective at collecting small-dimension biomass (branches and twigs) that forests managed in this way may lose a large proportion of their small biomass components. Whereas the use of technology for site preparation after World War II to establish conifer plantations reflected a similar trend occurring in agricultural landscapes, close-to-nature-forestry minimizes intensive silvicultural interventions, for example by reduced impact logging and natural regeneration.

An approach employed in centuries passed, pioneer forests are today attracting attention as a means to increase biomass production. Fast growing trees such as poplar are planted simultaneously with longer rotation tree species, for example, oak or fir, with the former harvested after a period of about 10 years without causing damage to the remaining trees. The planting of fast growing tree species along ride lines and extraction tracks, to be harvested at the time of first thinning of the primary stand, is also being considered (Unseld and Bauhus 2012).

In the future, biotechnology applications might exert the strongest influence on the appearance of forest landscapes, although strict corresponding legal restrictions are likely to come into effect. Biotechnology to improve fast growing species (clonal propagation, marker-aided selection and breeding, and genetic engineering) and the specific preparation of trees for environmental restoration (phytoremediation using woody plants) are some examples of the new possibilities (Hetemäki et al. 2010).

7.4.3 Disturbance

Forests are strongly affected by human-induced and natural disturbances, which exert an influence on their composition, structure and functional processes (Dale et al. 2001). Disturbance events such as windthrow, fire, flooding, bark beetle outbreak, snow, frost, hail, drought, and heat waves influence forest landscapes and their respective dynamics (see Jentsch 2013). Complex interactions exist between forests, disturbances, climate change, and management strategies. "Climate change can affect forests by altering the frequency, intensity, duration, and timing of fire, drought, introduced species, insect and pathogen outbreaks, hurricanes, windstorms, ice storms or landslides" (Dale et al. 2001). Climate projections developed by the MOTIVE project (Zimmermann et al. 2013) pointed to different results for forests in the northern, central, and southern European regions. While the negative impacts for northern parts may be limited, and remain quite unclear for central Europe, the Mediterranean region and neighboring parts of southern Europe are expected to suffer from decreasing precipitation in summer and winter, and increasing temperatures in both seasons, with an even larger predicted increase for winter than for summer. The drier growing conditions are expected to have severe impacts on the forests in these regions, where even today water availability is limited (Zimmermann et al. 2013).

Millions of hectares of forest land are affected by either natural disturbances such as fire, wind or insects, or human disturbances in the form of harvesting and land conversion over large areas. However, there is little information available about the spatial and temporal patterns of these disturbances (Schleeweis et al. 2012). For this reason, we will concentrate on large-scale disturbances. On the European continent, the annual damage caused by disturbances in the period 1950–2000 amounted to an average of 35 million m^3 of wood. Storms were responsible for 53%, fire 16%, snow 3% and other biotic agents contributed 5%. This 35 million m^3 of wood is the equivalent of 8.1% of total annual fellings in Europe (Schelhaas et al. 2003). Apart from the economic damage, disturbances create irregular patterns of forest shape and forest use, and have no respect for long-term forest management plans, administrative boundaries, or national borders. This becomes obvious from the example of gale-force winds that have raged across large parts of Europe in recent years, known as "Vivien" and "Wiebke" that occurred in 1990, "Lothar" in 1999, "Gudrun" in 2005, "Kyrill" in 2007, "Klaus" in 2009, and "Xynthia" in 2010. Affecting southwestern France and northwestern Spain, the storm Klaus felled over 40 million m^3 of timber, mostly pine (UNECE/FAO 2011), while Gudrun caused heavy damage to forests in Sweden and other countries around the Baltic Sea. Localized extreme weather events also cause additional damage (Figure 7.6).

Although it would appear that most types of damage are on the rise, this may be an artifact arising from the greater availability of information than ever before. It should also be borne in mind that forest area, average standing volume, and average stand age are all increasing (Schelhaas et al. 2003). Affecting an average area of 213,000 ha of forest annually between the years 1961 and 2000, fire constitutes an important disturbance mechanism. Many of the wildfires that devastated these wooded areas were the result of arson (Gardiner et al. 2011). Large-scale disturbance caused by fire and social change are linked. For instance, in Mediterranean regions, the depopulation of rural areas is connected to the abandonment of land and a rising risk of forest fire. That is why an exponential increase in the extent of burnt areas was observed in southern Europe between the years 1970 and 2000. In Portugal, estimates from 2003 suggest that about 11% of the country's forests, reaching 350,000 ha, were burned, including a lot of protected areas (Niemelä et al. 2005). For a better adaptation of forests to a changing climate, several countries in Europe (Belgium, Ireland, and Spain) are addressing resilience questions in their policy objectives (Forest Europe et al. 2011).

FIGURE 7.6
Disturbances: A formerly closed forest near Großenhain, Saxony, struck by the Tornado of 2010. (Photo: N. Weber.)

7.5 Conclusions

7.5.1 Major Changes to Forest Landscapes

Major changes to the forest landscapes of Europe have undoubtedly occurred in the decades since 1945. *First,* forest area has been expanding in many parts of the continent due to afforestation and natural recolonization on abandoned land. There has also been an increase in individual tree growth and standing volume. *Second,* recent forest management in many productive forests has tended toward close-to-nature-forestry, low-impact silviculture, and ecosystem management. With the conversion of many forests from conifer to broadleaf species for ecological reasons and with the aim of climate adaptation, new questions have emerged as certain conifers such as Douglas-fir, although not a native species, may be better adapted to a changed climate than many native broadleaves (Spiecker et al. 2010). Another issue is that the forest-based sector is not yet prepared for the expected large-scale replacement of conifer timber by broadleaf wood. Although only applied on a limited scale as yet, retention forestry seeks to reintroduce "patches of nature" into production forests, thereby promoting biological processes, but also decreasing the productive area. *Third,* there is a tendency away from classical silvicultural planning toward ecosystem management approaches incorporating disturbance regimes. Strategies to cope with windthrow, fire, pests, etc., are being reconsidered in many places as disturbances are no longer necessarily classified as disasters. To an increasing degree, disturbances are being allowed to occur in protected areas and in some cases are even being artificially induced (natural disturbance emulation [NDE] and prescribed burning in protected areas) (OMNR 2001; Kraus and Zeppenfeld 2013). *Fourth,* trends toward segregation and "land sparing" have led to a rise in single-purpose plantations and a simultaneous increase in protected forests and areas designated for wilderness development. *Fifth,* some forest landscapes are changing as a result of European energy policy, which promotes biomass use and wind power stations. *Sixth,* scientific progress has made it easier to measure changes to forest area and forest condition, and to identify social preferences. In combination with landscape simulation techniques, planners are increasingly involving citizens and stakeholders in planning processes affecting changes to forest landscapes (Ruschkowski 2009; Edwards et al. 2012). It is not yet clear, however, how changes to forest management intensity, a shift toward the greater use of broadleaves, higher tree species diversity, and low impact silvicultural systems are really valued by the general public (Edwards et al. 2012).

7.5.2 Underlying Causes

An examination of the causes of the changes mentioned above reveals that societal shifts and technological advances are only partly adequate as explanatory factors. Large-scale and long-term modifications to the appearance of forest landscapes often depend on a combination of drivers. Economic factors such as the individual decisions of landowners, supply and demand in international trade, and major policy shifts in sectors such as energy should not be underestimated. As Meeus (1995) put it, "international confrontation with regional landscapes creates both inspiration and a lot of confusion." Similarly, "landscape conservation would involve the unrealistic aim of preserving past economic patterns (Meeus 1995). Current paradigm shifts are recognizable in concepts such as retention forestry and forest landscape restoration. The latter seems to replace the previous concept of afforestation with monoculture plantations, and "seeks to create a framework whereby both ecological integrity can be regained and human well-being enhanced in deforested or degraded forest landscapes" (Maginnis 2005). It centers on restoring the functionality of forests; that is, the provision of goods, services and ecological processes forests can provide at the broader landscape level (Maginnis 2005; Stanturf 2015).

7.5.3 Perspectives for the Future

The growing conditions to which the forests of Europe are subject are expected to change in the future, in terms of nutrient cycles, temperature, and the occurrence of extreme events such as heat, drought and precipitation. Phenology, tree growth and mortality patterns, will be influenced and the geographic distributions of certain tree species may alter. Changes might even become more rapid in the future (Dobbertin et al. 2006). Scenario simulations predict further increases in the damage caused by disturbances annually across all agents (wind, forest fire, and bark beetles), in the magnitude of $+0.91 \times 10^6 \text{ m}^3 \text{ yr}^{-1}$ as a median over several climate change scenarios. Uncertainties remain over the future trajectories of Europe's disturbance regimes, however, and over the effects of the predicted intensification of these regimes in different ecoregions. What seems certain is that many of the ecosystem services provided by European forests are likely to be adversely affected if the projected increase in natural disturbances comes to pass (Seidl et al. 2014: 807). The conflict between land-sparing and land-sharing approaches for reasons of biodiversity conservation is also likely to intensify. Today, according to estimates by Lindenmayer et al. (2012), only 3% of the forests in Europe are plantations and 4% are designated protected areas. According to Mantau (2010), however, "... the demand for energy wood will more than double by 2020, if the energy demand develops approximately according to the [EU] policy targets." Under this policy scenario, energy efficiency is estimated to increase by 20%, with biomass accounting for "only" 40% of the renewable supply of energy. This implies an imbalance between supply and demand with respect to the existing wood requirement for material applications and the extrapolated supply needed to meet the renewable energy targets (UNECE/FAO 2011). The growing energy needs could be satisfied by woody biofuel imports from other parts of the world, but this would merely displace competing land use demands elsewhere including the developing world, and raise serious concerns with respect to equity and sustainability. A wood deficit situation could also result in a "shift toward production of smaller-size wood grown in shorter rotations" (Söderberg and Eckerberg 2013). Strategies adopted on the supply side (increasing harvest levels, expanding forest areas and SRC) might influence the appearance of forest landscapes directly. At the same time, there is also likely to be a further intensification of the quest to set aside forests, both for reasons of biodiversity protection and the restoration of ecological systems (Söderberg and Eckerberg 2013). Certainly, the societal discourse and political struggle over the future of forestry in Europe, between intensification and extensification, integration *vs.* segregation, land-sharing *vs.* land-sparing, is set to continue. While "extensification through intensification" (Giessen 2013) might serve to solve the conflict between the increasing economic, ecological, and social demands placed upon the forests in part, multifunctional forestry would still appear to be the best model for most parts of the Europe (Pröbstl 2007).

Acknowledgments

We specially thank Dominik Graßhoff for his initial literature survey and a fruitful discussion during our graduate seminar. We also thank two anonymous reviewers for their substantial and constructive comments, which helped us to improve the manuscript.

References

Bastian, B. and Bernhardt, A. 1993. Anthropogenic landscape changes in Central Europe and the role of bioindication. *Landscape Ecology.* 8(2): 139–151.

Baum, S., Bolte, A., and Weih, M. 2012. Short Rotation Coppice (SRC) plantations provide additional habitats for vascular plant species in agricultural mosaic landscapes. *Bioenergy Research* 5: 573–583

Bemmann, A. and Knust, C. 2010. *AGROWOOD—Kurzumtriebsplantagen in Deutschland und europäische Perspektiven.* Weißensee-Verlag; Berlin; 342 p.

Brockerhoff, E.G., Jactel, H., Parrotta, J.A., Quine, C.P., and Sayer, J. 2008. Plantation forests and biodiversity: Oxymoron or opportunity? *Biodiversity Conservation* 17: 925–951.

Bundeswaldgesetz. 2010. German Federal Forest Act (Bundeswaldgesetz) of 2 May 1975 (BGBl. I S. 1037), amended by Article 1 of the Law of 31 July 2010 (BGBl. I S. 1050).

Cameron, A.D., Mason, W.L., and Malcolm, D.C. (Eds.). 2001. Transformation of plantation forests. *Forest Ecology and Management* 151: 1–224.

Commission of the European Communities. 2005. Anne to the Communication on the implementation of the EU Forestry Strategy. Commission Staff Working Document. SEC (2005) 333. Brussels, 10.3.2007

Dale, V.H., Joyce, L.A., McNulty, S., Nelson, R.P., Ayres, M.P., Flannigan, M.D. et al. B.M. 2001. Climate change and forest disturbances. *BioScience* 51(9): 723–734.

Dedrick, S., Spiecker, H., Orazio, C., Tome, M., and Martinez, I. (Eds.). 2007. Plantation or Conversion—The debate! Ideas presented and discussed at a joint EFI Project-Centre conference held 21–23 May 2006 in Freiburg, Germany. European Forest Institute, Discussion Paper 13, 2007.

Dobbertin, M., de Vries, W., and Sterba, H. 2006. Response of forest ecosystems to changing environmental conditions. *Presentation at the ICP Forests seminar "Forests in a Changing Environment"*, 25–28.10.2006, Göttingen, available at http://www.icp-forests.org/DocsCrown/Go2_Dobbertin.pdf, accessed 27 September 2013.

EC 1992: Council Regulation (EEC) No 2080/92 of June 1992 instituting a Community aid scheme for forestry measures in agriculture. *Official Journal of the European Communities* No. L 215/96, 30.07.1992.

Edwards, D., Jay, M., Jensen, F.S., Lucas, B., Marzano, M., Montagné, C. et al. 2012. Public preferences for structural attributes of forests: Towards a pan-European perspective. *Forest Policy and Economics* 19: 12–19.

EEA—European Environmental Agency. 2012. Protected areas in Europe—an overview. EEA Report No 5/2012. Copenhagen.

Egan, J.F. and Mortensen, D.A. 2012. A comparison of land-sharing and land-sparing strategies for plant richness conservation in agricultural landscapes. *Ecological Applications* 22: 459–471.

Elands, B.H.M., O'Leary, T., Boerwinkel, H., and Wiersum, F. 2004. Forests as a mirror of rural conditions; local views on the role of forests across Europe. *Forest Policy and Economics* 6: 469–482.

Elands, B.H.M., Bell, S., Blok, J., Colson, V., Curl, S, Kaae, B.C. et al. 2010. Atlantic Region. In: Pröbstl, U. et al. (Eds.): *Management of Recreation and Nature Based Tourism in European Forests*, DOI 10.1007/9783-642-03145-8_2, Springer, Berlin, Heidelberg.

European Commission. 2009. Shaping forest communication in the European Union: public perceptions of forests and forestry, Final Report. European Commission, DG Agriculture and Rural Development, Rotterdam, 17 September 2009.

European Commission. 2010. Analysis of options to move beyond 20% greenhouse gas emission reductions and assessing the risk of carbon leakage. COM (2010) 265 final.

European Parliament. 2009. European Parliament resolution of 3 February 2009 on Wilderness in Europe.

European Union. 2009. Directive 2009/28/EC. On the promotion of the use of energy from renewable sources and amending and subsequently repealing Directives 2001/77/EC and 2003/30/EC.2009.04.23. Off. J. Eur. Union 2009;L 140:16e62.

EUROSTAT. 2011. Forestry in the EU and the world: A statistical portrait. 2011 edition. Luxembourg: Publications Office of the European Union. KS-31-11-137-EN-C.

Faaij, A.P.C. 2006. Bio-energy in Europe: changing technology choices. *Energy Policy* 36: 322–342.

FAO. 2006. Global planted forests thematic study: results and analysis, by A. Del Lungo, J. Ball and J. Carle. Planted Forests and Trees Working Paper 38. Rome (also available at www.fao.org/forestry/site/10368/en).

FAO. 2009. State of the World's Forest 2009. Food and Agriculture Organization of the United Nations, Rome.

Forest Europe, UNECE and FAO. 2011. State of Europe's Forests: Status & Trends in Sustainable Forest Management in Europe. Oslo.

Frank, G., Parviainen, J., Vandekerhove, K., Latham, J, Schuck, A., and Little, D. (Eds.). 2007. Protected Forest Areas in Europe—Analysis and Harmonisation (PROFOR): Results, Conclusions and Recommendations. Vienna.

FSC Facts and Figures. September 2013. https://ic.fsc.org/facts-figures-2013.692.htm, accessed Oct. 23, 2014.

Gardiner, B., Aleksandrov, N., Blennow, K., Bouriaud, O., Bouriaud, L, Didion, M. et al. 2011. Models for Adaptive Forest Management. FP 7 Project no. 22564. Part D4.1 Report On Abiotic and Biotic Risks: Case Study Areas. Available online at: http://www.motive-project.net/files/DOWNLOAD2/MOTIVE_Deliverable_4.1a_1.pdf, accessed October 28, 2014.

Giessen, L. 2013. Extensivierung durch Intensivierung—Experten diskutieren politische Herausforderungen. *AFZ-DerWald* 14/2013, p. 33.

Hendinger, H. 1960. Der Wandel der Mittel- und Nordeuropäischen Waldlandschaft durch die Entwicklung der Forstwirtschaft im industriellen Zeitalter. *Geografiska Annaler*, 42(4): 294–305.

Hetemäki, L. and Mery, G. (convening lead authors), Holopainen, M., Hyyppä, J., Vaario, L.-M., Yrjälä, K. 2010. Implications of technological development to forestry. In: Mery, G., Katila, P., Galloway, G., Alfaro, R.I., Kanninen, M., Lobovikov, M., Varjo, J. (Eds.). *Forests and Society— Responding to Global Drivers of Change*. IUFRO World Series, International Union of Forest Research Organizations (IUFRO), Vienna, 25: 157–181. ISBN 978-3-901347-93-1.

Hickler, Th., Bolte, A., Hartard, B., Beierkuhnlein, C., Blaschke, M., Blick, Th. et al. 2014. Folgen des Klimawandels für die Biodiversität in Wald und Forst. In: Mosbrugger, V., Brasseur, G.P., Schaller, M., Stribrny, B. (Eds.). *Klimawandel und Biodiversität: Folgen für Deutschland*. 2. Aufl. Wiss. Buchges. Darmstadt: 164–220.

Jentsch, A. 2013. Störungsökologie—da kommt Bewegung auf! *AFZ-DerWald* 15/2013, 4–5.

Kahle, H.P., Karjalainen, T., Schuck, A., Ågren, G.I., Kellomäki, S., Mellert, K.H. et al. (Eds.). 2008. Causes and Consequences of Forest Growth Trends in Europe—Results of the Recognition Project. Brill. Leiden. 272 S.

Kanninen, M. 2010. Plantation forests: global perspectives. In: Bauhus, J., van der Meer, P., and Kanninen, M. (Eds.). *Ecosystem Goods and Services from Plantation Forests*. EarthScan, London, Washington, DC. pp. 16–42.

Kanowski, P.J. and P.S. Savill. 1992. Forest plantations: Towards sustainable practice. Chapter 6. In: Sargent, C. and Bass, S. (Eds). *Plantation Politics: Forest Plantations in Development*. Earthscan, London, pp. 121-151.

Klijn, J.A. 2003. Driving forces behind landscape transformation in Europe, from a conceptual approach to policy options: agriculture and climate change as examples. In: Jongman, R. (Ed.), *Proceedings of the Frontis Workshop*. Kluwer, Dordrecht, the Netherlands.

Kraus, D. and Zeppenfeld, T. 2013. Feuer als Störfaktor in Wäldern. *AFZ-DerWald* 15/2013, pp. 8–9.

Liebal, S. 2013. Plantage oder Wildnis? Nachhaltigkeits-Tagung Tharandt 2013. *AFZ-DerWald* 14/2013, p. 32–35.

Lindenmayer, D.B., Franklin, J.F., Löhmus, A., Baker, S.C., Bauhus, J., Beese, W. et al. 2012. A major shift to the retention approach for forestry can help resolve some global forest sustainability issues. *Conservation Letters* 5(6): 421–431.

Livingstone, D. 2010. Landscape of knowledge. In: Meusburger, P. et al. (Eds.). *Geographies of Science*, Knowledge and Space 3, DOI 10.1007/978-90-481-8611-2. Springer, Berlin, p. 247.

Maginnis, S. 2005. What is forest landscape restoration? In: Veltheim, T., Pajari, B. (Eds.): *Forest Landscape Restoration in Central and Northern Europe*. EFI Proceedings No. 53, 2005, European Forest Institute, Joensuu, p. 25–26.

Mantau, U. 2010: Is there enough wood for Europe? pp 19–34. in: EUwood—Real potential for changes in growth and use of EU forests, Final report. Hamburg/Germany, June 2010. 160 p.

MCPFE—Ministerial Conference on the Protection of Europe 2003. Protected Forests in Europe. Booklet based on "State of Europe's Forests 2003 – The MCPFE Report on Sustainable forest Management in Europe by the MCPFE Liaison Unit Vienna & UNECE/FAO, Vienna.

Meeus, J.H.A. 1995. Pan-European landscapes. *Landscape and Urban Planning* 31: 57–79.

Nabuurs, G.J., Schelhaas, M.J., Mohren, G.M.J., and Field, C.B. 2003. Temporal evolution of the European forest sector carbon sink from 1950 to 1999. *Global Change Biology* 9(2): 152–160.

Niemelä, J., Young, J., Alard, D., Askasibar, M., Henle, K., Johnson, R. et al. 2005. Identifying, managing and monitoring conflicts between forest biodiversity conservation and other human interests in Europe. *Forest Policy and Economics* 7: 877–890.

OMNR. 2001. Forest management guide for natural disturbance pattern emulation, Version 3.1. *Ont. Min. Nat. Res.*, Queen's Printer for Ontario, Toronto. 40 p.

Ottitsch, A., Michier, B., Palahi, M., Wardle, P., Janse, G., Moiseyev, A. et al. 2005. Changes in the Forest Sector in Europe and Russia. In: Gerardo, M., Rene, A., Markku, K., and Maxim, L., (Eds.), *Forests in the Global Balance—Changing Paradigms*. IUFRO, Helsinki, pp. 231–242.

Pedroli, B., Elbersen, B., Frederiksen, P., Grandin, U., Heikkilä, R., Henning Krogh, P. et al. 2013. Is energy cropping in Europe compatible with biodiversity?—Opportunities and threats to biodiversity from land-based production of biomass for bioenergy purposes. *Biomass and Bioenergy* 55: 73–86.

Pröbstl, U. 2007. Forests in Balance? Forest under the spell of economic, ecological and recreational requirements—Considerations about the European Model. *Allgemeine Forst und Jagdzeitung* 178(4): 68–73.

Ruschkowski, E. von. 2009. Causes and Potential Solutions for Conflicts between protected area management and local people in Germany; in: Weber, Samantha, ed. 2012. *Rethinking Protected Areas in a Changing World: Proceedings of the 2011 GWS Biennial Conference on Parks, Protected Areas, and Cultural Sites.* Hancock, Michigan: The George Wright Society, pp. 240–244.

Savill, P., Evans, J., Auclair, D., and Falck, J. 1997. *Plantation Silviculture in Europe.* Oxford University Press, Oxford, New York, Tokyo.

Schaffner, S. 2002. Hat modern Forsttechnik einen Platz im Waldbild der Gesellschaft? *AFZ-DerWald* 57: 1113–1117.

Schanz, H. 1999. Social changes and forestry. In: Pelkonen, P., Pitkänen, A., Schmidt, P., Oesten, G., Piussi, P., Rojas, E. (Eds.), *Forestry in Changing Societies in Europe*, Part 1, University Press, Joensuu, p. 59–81.

Schelhaas, M.-J., Nabuurs, G.-J., and Schuck, A. 2003. Natural disturbances in the European forests in the 19th and 20th centuries. *Global Change Biology* 9(11): 1620–1633.

Schleeweis, K., Goward, S.N., Huang, C., Masek, J., and Moisen, G.G. 2012. Understanding trends in observations of forest disturbance and their underlying causal processes. In: Morin, R.S., Liknes, G.C., comps. 2012. *Moving from Status to Trends: Forest Inventory and Analysis (FIA) Symposium 2012*; 2012 December 4–6, Baltimore, MD. Gen. Tech. Rep. NRS-P-105. Newtown Square, PA: U.S. Department of Agriculture, Forest Service, Northern Research Station. [CD-ROM], pp. 131–136.

Schmidt-Walter, P and Lamersdorf, N.P. 2012. Biomass production with willow and poplar short rotation coppices on sensitive areas—the impact on nitrate leaching and groundwater recharge in a drinking water catchment near hanover, Germany. *Bioenergy Research* 5: 546–562.

Schumacher, S. and Bugmann, H. 2006. The relative importance of climatic effects, wildfires and management for future landscape dynamics in the Swiss Alps. *Global Change Biology*, 12: 1435–1450.

Seidl, R., Schelhaas, M.-J., and Rammer, W., Verkerk, P.J. 2014. Increasing forest disturbances in Europe and their impact on carbon storage. *Nature Climate Change* 4: 806–810.

Söderberg, C. and Eckerberg, K. 2013. Rising policy conflicts in Europe over bioenergy and forestry. *Forest Policy and Economics* 33: 112–119.

Spiecker, H., Kohnle, U., Makkonen-Spiecker, K., and von Teuffel, K. (Eds.) 2010. Opportunities and risks for Douglas-fir in a changing climate. Abstracts. *Berichte Freiburger Forstliche Forschung*, Heft 85. Albert-Ludwigs-Universität Freiburg; Forstliche Versuchs- und Forschungsanstalt Baden-Württemberg. ISSN 1436–1566.

Stanturf, J.A. 2015. What is forest restoration? In: Stanturf, J.A. (Ed.), *Restoration of Boreal and Temperate Forests*, 2nd ed., CRC Press, Boca Raton; pp. 1–15.

Stanturf, J.A. and van Oosten, C. Operational poplar and willow culture., In: Isebrands, J.G., Richardson (Eds.): *Poplars and Willows: Trees for Society and the Environment*, J., CABI, UK, 200. 2014.

Straub, C. and Koch, B. 2011. Enhancement of bioenergy estimations within forests using airborne laser scanning and multispectral line scanner data. *Biomass and Bioenergy* 35: 3561–3574.

Ueda, H., Nakajima, T., Takayama, N., Petrova, E., Matsushima, H., Furuya, K. et al. 2012. Landscape image sketches of forests in Japan and Russia. *Forest Policy and Economics* 19: 20–30.

UNECE/FAO. 2011. Forest Products Annual Market Review, 2010–2011. Available at http://www.unece.org/fpamr2011.html (accessed 29 September 2013).

Unseld, R. and Bauhus, J. 2012. Energievorwälder—alternative Bewirtschaftungsformen zur Steigerung der energetisch nutzbaren Biomasse im Wald durch Integration von schnellwachsenden Baumarten. *Berichte Freiburger Forstliche Forschung*, Heft 91, Waldbau-Institut Albert-Ludwigs-Universität Freiburg; 213 p.

Vanhanen, A., Rayner, J., Yasmi, Y., Enters, T., Fabra-Crespo, M., Kanowski, P. et al. 2010. Forestry in changing social landscapes. In: Mery, G., Katila, P., Galloway, G., Alfaro, R.I., Kanninen, M., Lobovikov, M. et al. (Eds.). 2010. *Forests and Society—Responding to Global Drivers of Change.* IUFRO World Series Volume 25. International Union of Forest Research Organizations, Vienna. 509 p.

Weber, N. 2005. Afforestation in Europe: Lessons learnt, challenges ahead. In: Stanturf, J.A., Madsen, P. (Eds.): *Restoration of Boreal and Temperate Forests.* Boca Raton: CRC Press, pp. 121–135.

Weih, M. and Dimitriou, I. 2012. Environmental impacts of short rotation coppice (SRC) grown for biomass on agricultural land. *Bioenergy Research* 5: 535–536.

Welzholz, J.C. and Johann, E. 2007. History of protected areas in Europe. In: Frank, G., Parviainen, K., Vandekerhove, J., Latham, J., Schuck, A., and Little, D. (Eds.): *COST Action E27 – Protected Forest Areas in Europe—Analysis and Harmonisation (PROFOR): Results, Conclusions and Recommendations,* Federal Research and Training Centre for Forests, Natural Hazards and Landscape (BFW), Vienna, pp. 17–40.

Wulff, S., Lindelöw, Å., Lundin, L., Hansson, P., Axelsson, A-L., Barklund, P., Wijk, S., Stahl, G. 2012. Adapting forest health assessments to changing perspectives on threats—A case example from Sweden. *Environmental Monitoring and Assessment* 84(4): 2453–2464.

Zanchi, G., Thiel, D., Green, T., and Lindner, M. 2007. Forest area change and afforestation in Europe: Critical analysis of available data and the relevance for international environmental policies. EFI Technical Report 27. Joensuu. 45 p.

Zimmermann, N.E., Schmatz, D.R., and Psomas, A. 2013. Climate change scenarios to 2100 and implications for forest management. In: Fitzgerald, J., and Lindner, M. (Eds.): *Adaptation to Climate Change in European Forests—Results of the MOTIVE Project,* Pensoft Publishers, Sofia, pp. 9–14.

8

Afforestation and Land Use Dynamics in the Baltic States

Kalev Jõgiste, Marek Metslaid, and Veiko Uri

CONTENTS

8.1 Introduction

The Baltic States of Estonia, Latvia, and Lithuania have undergone drastic upheaval in their political systems during the last century. The Baltic States were independent nations between the two world wars (1918–1939); at other times in their history, they were first a part of Tsarist Russia and later the part of Soviet Union. Independence was regained in the 1990s. Changes in economic systems and land tenure, from the centrally planned economy of the Soviet Union to the free market and private ownership of today's newly independent states, have significantly affected land use, especially the balance between forestry and agriculture. In all the Baltic States, large areas of agricultural land were abandoned during the last decades and became available for afforestation. Mining operations also result in areas requiring reclamation, particularly the oil shale mined lands in Estonia (Toomik and Liblik 1998). Although afforestation was practiced in the Baltic States for nearly a century, the past several decades following independence have been the most active period. Many abandoned areas are undergoing old-field succession following recolonization by pioneer hardwood species such as birches (*Betula* spp.), alders (*Alnus* spp.), aspen (*Populus tremula*), and willows (*Salix* spp.).

The main aim of afforestation in the Baltic States until recently has been to establish commercially valuable, particularly conifer-dominated forests. Spruce (*Picea* spp.) and pine (*Pinus* spp.) plantations on abandoned farmland, however, are often plagued with root rot, and timber quality is reduced by the wide spacing of low density plantings that allows the development of large branches. Recently, afforestation has been set new objectives based on concern about environmental effects, and the desire to enhance biodiversity. Moreover, changing land use from agriculture to forests provides the opportunity to create carbon sinks with the proper tree species resistant to climatic fluctuations. Objectives of protecting soils against erosion and from pollution of water resources increasingly drive afforestation programs and restore traditional land uses. Our goal in this chapter is to provide a historical backdrop to the present effort to afforest public and private land in the Baltic States.

8.1.1 Natural Conditions

The Baltic States of Estonia, Latvia, and Lithuania are situated (Figure 8.1) in the transition zone between boreal and temperate forests (56 to 59°N; 24 to 26°E). The region is regarded as a hemiboreal zone; southern Lithuania belongs to the temperate zone (Ahti et al. 1968). Average temperatures range from +20°C in July to −6°C in February, and mean annual precipitation varies from 500 to 930 mm. Dominant soil types are podzols and gleyic podzols. In areas with calcareous parent material, podzolization processes in soil development are retarded.

FIGURE 8.1
Location of the Baltic States (d-maps.com; http://d-,aps.com/carte.php?num_car=5039&land=en).

Mixed-coniferous and deciduous forests are the characteristic natural vegetation in the Baltic region. Nearly half of the land area in Estonia and Latvia (Tullus and Uri 2002), and approximately one-third of the area in Lithuania (Gaizutis 1998) is covered by forest. The main tree species are Scots pine (*Pinus sylvestris* L.), Norway spruce (*Picea abies* [L.] Karst), silver and downy birch and hybrids (*Betula pendula* Roth. and *Betula pubescens* Ehrh.), grey and black alder (*Alnus incana* [L.] Moench and *Alnus glutinosa* [L.] Gaertn.), common aspen (*Populus tremula* L.), pedunculate oak (*Quercus robur* L.), and European ash (*Fraxinus excelsior* L.). Oak is more common in Lithuania than in the other Baltic States. Without human influence, the distribution of Norway spruce would probably be wider than at present in Estonia and elsewhere in the boreonemoral zone (Poska et al. 2004; Reitalu et al. 2013).

The variety of different forest habitats is high in the Baltic region. Dry sandy soils are predominantly occupied by Scots pine, whereas Norway spruce is found on more fertile soils. Mixtures of conifers and broadleaved trees are common. The proportion of semi-natural communities, such as grasslands, pastures, and abandoned agricultural land is considerable. Seminatural vegetation is typical for coastal areas (pastures covered with *Juniperus communis* L.) and also for inland woody meadows. Over time and without human influence, these communities will become woodlands.

During the last century, most attention was devoted to cultivating conifers for industrial needs, primarily Scots pine and Norway spruce. These species were mostly planted, but direct seeding has been used as well, mainly for Scots pine. A few exotic tree species were planted, for example *Larix* spp.

8.2 History and Land-Use Change

Forestry and agricultural land uses are closely linked in the Baltic States. For example in Estonia, during the period of the introduction and expansion of permanent agriculture (3950–1850 before the present), human impact became important for forest composition (Reitalu et al. 2013). Extensive use of forest areas started during the period from the tenth century of the Common Era until the beginning of the thirteenth century (Paal 1997). By the end of the nineteenth century, the forest resource was at its lowest ebb, exploited for construction timber and fuel wood (Anon. 2000). Deforestation was also due to peat extraction, which left areas without vegetation. A considerable area of peatland has been drained and afforested.

Clearing for agriculture was intense even through the early twentieth century. Agricultural land use predominated between the two world wars in the independent Baltic States, and forest cover decreased. At the end of the 1930s, for example, forest cover of Estonia was less than 30% of the land area (Figure 8.2). Nevertheless, during this same period, many of today's most valuable commercial forests were planted.

After World War II, the communist government repressed farmers, causing abandonment of large amounts of agricultural land as large, collective farms were established, and forest area increased in the Baltic States. For example, during the period 1950–1987 in Estonia, pine and spruce plantations on former agricultural fields increased forest area by about 65,000 ha (Tullus 2000). In Lithuania, during the period 1951 to 1985, afforestation of nonforest area was considerably larger: 279,000 ha (Table 8.1), mainly with Scots pine and Norway spruce. Afforestation in Latvia is documented from the nineteenth

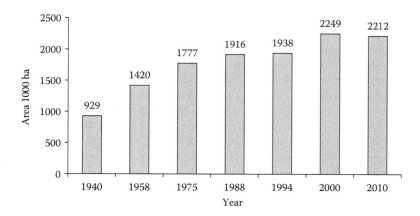

FIGURE 8.2
Forest land changes in Estonia during 1940–2010; the increase is mainly due to the abandonment of agricultural lands.

century (Daugaviete 2000). Between 1935 and 1983, forest area increased over one million ha through afforestation or natural recolonization in Latvia (Anon. 2002).

After regaining their independence in 1991, the Baltic States changed over to market economies. Land reform measures and restitution processes started to return public land to private ownership. Forest establishment on abandoned agricultural land, however, did not start instantly after regaining independence in Estonia. Massive economic restructuring often left the ownership status of land unclear and, in this unstable environment, financing for forest and agricultural management was unavailable. Consequently, natural colonization processes occurred on large-field areas that had been abandoned, including pioneer tree species such as grey alder, silver and downy birch, and willow sp.

Today, approximately half of the forests in all the Baltic States are privatized and with the other half remaining in state ownership. In Latvia, for example, the state owns 51.1% of the forest land in 2001, private forest owners 45.0%, and municipalities 3.9% (Anon. 2002). The situation is similar in Lithuania: 50% of the forest area remains in state ownership,

TABLE 8.1

Afforested Area Relative to Total Area Planted and Predominant Species in Lithuania over the Period 1951–1985

	Area (1000 ha)		Predominant Tree Species as a Percentage of Total Area Planted			
Period	Total	On Nonforest	Scots Pine	Norway Spruce	Oak	Other
1951–1955	97	64	66	21	9	4
1956–1960	80	53	65	25	5	5
1961–1965	75	63	58	34	3	5
1966–1970	67	49	35	58	1	6
1971–1975	50	16	34	62	1	3
1976–1980	55	16	26	68	—	6
1981–1985	52	16	21	71	—	8

Source: Adapted from Gabrilavicius, R., Review and Assessment of Lithuanian Experience with Afforestation of Abandoned Agricultural Land, Lithuanian Forest Research Institute, Project No. 055-17, Manuscript, http://miskai.gamta.lt/agro/reports.htm, 1999.

23% was privatized in the course of restitution, and 27% was reserved for later privatization. The proportion of the state owned forests is lower in Estonia, approximately 40%.

These changes in forest ownership have greatly influenced management, especially the afforestation methods used on the abandoned fields. In private forestry, both land use and forest management are more profit oriented and intensive than in the state forests. Timber prices have pushed afforestation toward softwoods; the market value of the timber in a 50-year-old Norway spruce stand is 4950 EUR/ha; corresponding values for birch and grey alder are 3410 and 1695 EUR/ha, respectively (Maamets 2003). The high afforestation and regeneration costs have meant less mechanical and more manual work is carried out in privately owned forests.

In the national forest policies of all the Baltic States, afforestation has a high priority in order to increase the efficiency of land management. In Estonia, a Swedish–Estonian cooperation program was initiated in 1999 to establish research and demonstration areas. The Forestry Development Plan 2002–2010 included afforestation; legislation to offer subsidies is being developed. Earlier calculations estimated a realistic support level for afforestation at 960 EUR/ha. The actual calculation, based on the current situation, indicates the effective cost to be around 1300 EUR/ha (Uus, personal communication). The total afforestation area in Estonia reached almost 9000 ha in the years 1991–2013 (Table 8.2).

A similar Lithuanian afforestation project was launched in 1999 as a cooperative program between the Danish Ministry of Environment and Energy and the Ministry of Environment of the Republic of Lithuania (Gabrilavicius 1999). Afforestation in Lithuania has focused on planting coniferous tree species (Danusevicius, personal communication). At least 228,000 ha of abandoned agricultural land is considered to be afforested. Mostly, afforestation has occurred passively as natural recolonization. Scarce funding limited active afforestation to 2500 ha in private and state forests between 1997 and 2001. After that, a considerable increase has resulted in almost 40,000 ha in the years 2002–2013 (Table 8.3).

In 1994, the State Forest Service of the Latvian Ministry of Agriculture commissioned a research program on afforestation. Several acts enacted between 1998 and 2000 included exemptions from the real-estate tax on young afforestation stands. Afforestation subsidies in Latvia are set at 150 EUR/ha (Daugaviete 2000). Surplus farmland in Latvia available

TABLE 8.2

The Main Trends of Afforestation in Estonia during 2005–2013

Year	Total area (ha)
2005	1324
2006	1592
2007	*
2008	1404
2009	1475
2010	1077
2011	*
2012	*
2013	*

Source: Adapted from State Forest Management Centre (RMK); Estonian Agricultural Registers and Information Board (PRIA); Foundation Private Forest Centre (PFC); Yearbook Forest 2013. Compiled and edited by the Estonian Environment Agency, Tartu, 2014. (in Estonian).

*Data not available.

TABLE 8.3

Afforestation in Lithuania during 1997–2013

Year	Private Forest (ha)	State Forest (ha)	Total area (ha)
1997	*	27	27
1998	*	562	562
1999	*	788	788
2000	*	822	822
2001	*	150	150
2002	*	964	964
2003	*	1164	1164
2004	*	1136	1136
2005	829	1337	2166
2006	2066	1272	3338
2007	1987	1397	3384
2008	2470	1789	4259
2009	2386	1022	3408
2010	2919	837	3756
2011	3974	736	4710
2012	3780	882	4662
2013	2198	742	2940
Total	*	15,629	38,238

Source: Adapted from Lithuanian State Forest Service.
*Data not available.

for afforestation exceeds 300,000 ha. Since 1994, 150 ha of experimental and demonstration stands were established (Daugaviete 2002). A rural development program began planting abandoned farmland in 1999; the goal was to plant 10,000 ha (2% of the total abandoned farmland) by 2006, mostly with birch (Anon. 2002). By 2012 the afforested area increased to 28,000 ha (Table 8.4).

TABLE 8.4

Afforestation in Latvia during 2001–2012

Year	Total Area (ha)
2001	700
2002	400
2003	1100
2004	2000
2005	1700
2006	2700
2007	2200
2008	1500
2009	3200
2010	5100
2011	4800
2012	3300
Total	28700

Source: Adapted from Valsts MežaDienesta 2012. Gada Publiskais Parskāts (APSTIPRINĀTS ar Valsts meža dienesta 2013. gada 12. jūnija rīkojumu Nr. 84) http://tpi.mk.gov.lv/ui/documentcontent.aspx?Type=attach&ID=266 (18.03.2014).

FIGURE 8.3
Opencast oil shale mine area reclaimed with Scots pine (*Pinus sylvestris*). The 3-year-old plantation was planted with 2-0 bareroot seedlings. (Photo by Veiko Uri.)

Oil shale is an important raw material for energy and chemicals and almost all (98%) of the electric energy produced in Estonia comes from oil shale (Toomik and Liblik 1998). The exhausted opencast mines of oil-shale extraction are intended to be afforested (Figure 8.3). In 2001, 77% of the abandoned mine land had been afforested (Kaar 2002). Afforestation of such areas is difficult because of high soil pH, low organic matter content, altered water regime, and stoniness. Scots pine is the primary tree species planted on these areas, accounting for 86% of the forest growing on the abandoned opencast mines (Figure 8.4). Mostly 2-year-old seedlings have been planted at a density of 5360 to 6667 seedlings per ha (Kaar 2002). Norway spruce forms 4% of the plantations on the opencast mines, but generally spruce is unsuitable because of its sensitivity to frost and infertile soil. Recent research suggests that the proportion of deciduous tree species should be higher on reclaimed opencast mines (Kaar 2002; Kaar and Kiviste 2010; Kuznetsova et al. 2011).

8.3 Afforestation Practice

8.3.1 Species Selection

The Baltic States are similar to the Nordic countries in the sense that the initial preference was to plant conifer forests due to their high production value and low establishment costs. Today, foresters understand that hardwood species were undervalued during the last century, when silvicultural recommendations aimed at totally cleaning young stands of hardwoods. Experimental hardwood plantations established at different locations in Estonia have shown that silver birch is the hardwood most promising economically for afforestation of abandoned agricultural lands (Vares et al. 2001). The cost of afforestation using seedlings of deciduous tree species such as silver birch was 857 EUR/ha (Kaimre 2001). One of the most expensive operations was planting: seedlings cost 195 EUR/ha, planting 116 EUR/ha, protection fencing 253 EUR/ha, and soil preparation 5% of the total costs. Polyethylene mulch was the most effective cover material in birch plantings,

FIGURE 8.4
Successful restoration: a reclaimed opencast oil shale mine with a 40-year-old Scots pine plantation. (Photo by Veiko Uri.)

reducing grass competition and conserving moisture. Estimated cleaning and precommercial thinning costs were 130 EUR/ha at the age of 15 years.

Latvian experiments on different soil types have looked at a large number of tree species, including birch, aspen, black alder, wild cherry, oak, ash, beech, larch, spruce, and pine (Daugaviete 2000, 2002). In general, ash, oak, and beech are slow-growing and more susceptible to late-season frosts. Based on these experiments and observations in older abandoned farmland, afforestation in Latvia will mostly favor birch, which reaches veneer size in 30 years (Anon. 2002).

Few exotic species have been widely used in the Baltic States. Several experimental stands were established after World War II, including exotic spruces, pines, larches, and Douglas-fir (*Pseudotsuga menziesii* [Mirb. Franco]). Douglas-fir and larches (*Larix decidua* Mill., *L. sibirica* Ledeb., and *L. kurilensis* Mayr) produced successful results, compared to Scots pine. Larch plantations form 2% of the plantations growing on the reclaimed opencast oil shale mine areas (Kaar 2002). Hybrid aspen (*Populus* x*wettsteinii Hämet-Ahti*) has become an important species during the last decades (Vares et al. 2003). Because it is grown on relatively short rotation, it is not susceptible to stem rot caused by *Phellinus tremulae* that commonly affects European aspen. Hybrid aspen achieves maturity for pulpwood production at 25 to 30 years in northern Europe. Today, there are more than 600 ha of hybrid aspen plantations in Estonia (Tamm 2000).

Biodiversity, or lack of it, is a concern when single-species plantations cover large areas. Thinning and pruning in young stands, which allow undergrowth to develop, may improve the structure of monoculture plantations (Fujimori 2001). Under Baltic conditions, Norway spruce is the most common shade-tolerant species creating undergrowth. Another approach is to increase landscape diversity by planting several site-adapted tree species. Thus, more fertile sites can be afforested with hardwood species. On low fertility sites, pioneer hardwoods (mainly birches and alders) and Scots pine can be used. Mixtures are a third approach. Various methods are available to create mixtures, but performance varies depending on species characteristics and environmental conditions (Vares et al. 2001). For example, pedunculate oak suffers from late frost and game damage in Estonia; similar problems have been reported from Latvian afforestation sites (Daugaviete 2002).

8.3.2 Site Preparation

Afforestation experience in Finland (e.g., Hynönen 2000) indicated that besides the choice of tree species, the next most important factor in planting is competition from ground vegetation. Site preparation reduces the competition and improves soil conditions, but it is expensive. Polyethylene and paper mulch materials reduce grass competition as well as conserve soil moisture (Vares et al. 2001). Moreover, mulch materials or herbicides can increase seedling growth (Table 8.5). The most common methods in Estonia for site preparation are ploughing and disk-trenching. If soil treatment is not used, afforestation success depends on former land use and time since the last cultivation. In Latvia, fertilization is considered essential for afforestation, especially in nutrient-poor old fields. Fertilization enhanced survival of pedunculate oak, ash, wild cherry (*Prunus avium* L.), Scots pine, and Norway spruce. Fertilizer prescriptions were estimated after observing deficiencies of N, P, and K; applying fertilizer in the ratio 5:8:11 as NH_4, P_2O_5, and K_2O targeted optimal fertility for conifers and hardwoods (Daugaviete 2000).

8.3.3 Spacing

Spacing in timber plantations is debated in all the Baltic States. Although low spacing increases individual stem growth; it is believed to reduce wood density in conifers. However, the wood density of deciduous tree species has not been shown to be affected by spacing. Experience in Finland indicated that the low density of silver birch was beneficial; pruning produced the same results with spacing of 1600 and 5000 seedlings per ha (Niemistö 1995). Tests of hardwoods in Latvia have included densities of 1111 stems/ha to 10,000 stems/ha (espacement of 3 m × 3 m to 1 m × 1 m) (Daugaviete 2000).

TABLE 8.5

Relative Height of Young Trees from different Experimental Treatments in Latvia, Expressed as Percentage of Height of Untreated Trees (Control)

Tree Species	Tending Applied				
	Hoeing	Mowing	Herbicide Application	Mulching	Control
Birch (*Betula pendula*)	108	100	156	132	100
Oak (*Quercus robur*)	108	134	125	100	
Ash (*Populus tremula*)	100	100	115	120	100
Pine (*Pinus sylvestris*)	125	130	144	100	
Spruce (*Picea abies*)	100	171	133	100	

Source: Adapted from Daugaviete, M., Afforestation of agricultural lands in Latvia, in NEWFOR—New Forest for Europe: Afforestation at the Turn of the Century, Weber, N., ed. EFI Proceedings No. 35, 175, 2000.

8.3.4 Planting and Direct Seeding

Although natural recolonization has proven the effectiveness of seed dispersal, planting is preferred in Estonia because it results in better stem quality. Planting with bare-root seedlings is the most common method. Although root suckers of grey alder have been planted experimentally, growth is inferior compared to seedlings (Uri 2001). Nevertheless, direct seeding is an inexpensive method for afforestation. Establishment success of direct seeding depends very much on weather conditions; the weather must not be too dry or too wet. Rodents, pine weevils, and competing vegetation are problems for direct seeding of large seeds or acorns. Lithuanian experience in afforestation highlights critical questions in technology development and species choice: failures are ascribed to weed competition and poor quality of planting stock (Gabrilavicius 1999).

8.3.5 Natural Recolonization

Natural regeneration is a tool for regenerating harvested hardwood forests and this experience can be applied to passive restoration. The main problems with reliance on natural invasion are: (1) the long regeneration time on harsh sites, (2) uneven spacing that varies from a few individuals to thousands of trees per hectare, (3) lowered wood quality, and (4) the time that is needed for dense stands to differentiate through self-thinning (Jõgiste et al. 2003). The main benefit is that its natural regeneration is essentially free and may be the only option for some private landowners. However, it is necessary to apply precommercial thinning when relaying on natural regeneration.

Old-field succession on abandoned agricultural lands usually leads to the development of forests (Bazzaz 1998). Low-intensity afforestation can rely upon natural recolonization, which has proven effective under the growing conditions of the Baltic region. Early successional tree species such as black alder and grey alder are common colonizers on abandoned arable fields and pastures. Moreover, birches (Figure 8.5) and willows are common colonizers of abandoned arable fields (Reitalu et al. 2013) in fact, silver birch is often planted on abandoned agricultural land (Figure 8.6).

FIGURE 8.5
Natural invasion of birches (*Betula* spp.) on abandoned agricultural land. (Photo by Veiko Uri.)

FIGURE 8.6
Planted silver birch (*Betula pendula* Roth.) stand on abandoned agricultural land. (Photo by Veiko Uri.)

8.3.6 Protection and Tending

Browsing by wildlife significantly damages forest regeneration in all the Baltic States. Moose (*Alces alces* L.) are a serious threat to both deciduous and coniferous species in plantations (e.g., Metslaid et al. 2013). Even though fencing is important for establishment success, it is very expensive and can account for as much as 30% of the total afforestation costs (Kaimre 2001). Different individual seedling protectors have been tested and proven effective against small mammals (Daugaviete 2000; Vares et al. 2003). Nevertheless, long-term problems may arise due to poor light quality in tubes (light absorption by tube walls) and instability of the sapling if the tube is not removed (Madsen and Löf 2005).

Today, the planning of management measures (tending, thinning, and supplemental planting for diversity) is an important part of forest restoration (Tullus 2001). Optimal tending of stands has been stressed in afforestation strategies (Tullus 2000; Daugaviete 2000). The EU programs in Latvia clearly stated the need for tending the plantations three years after establishment (Anon. 2002). Restoration for purposes other than timber production, however, is just beginning to receive attention in the Baltic States and guidelines for tending restored stands are needed.

8.4 Research Needs

In the Baltic States, the greatest needs for successful afforestation are to develop (1) management capacity, (2) methods for the establishment of forests with high biodiversity, and (3) a system for funding landowners. The use of alternative species to the traditional conifers has resulted in highly stable and sustainable stands. Silver and pubescent birch species, in particular form commercially valuable forests under the right conditions. Nevertheless, research is needed to increase the variety of suitable species and affordable afforestation methods. Species adaptations to sites, spacing, and stand development of both planted and direct-seeded stands need to be better understood. Research on tending methods is needed in order to affect stand density and age structure in order to produce resilient forests.

Natural recolonization can be included in afforestation schemes as a low-cost alternative. Tending and thinning, however, may be needed to overcome poorly stocked or overly dense stands. Supplemental planting may be needed to increase diversity or to fill in gaps. Research is needed to: (1) reveal the conditions where natural recolonization is reliable, (2) estimate stand dynamics, and (3) to understand development pathways. However, the necessary research is complex because of the variety of stand establishment conditions that must be examined, suggesting that a combination of experimental sites and modelling would be the most efficient approach.

Afforested stands will increasingly make important contributions to meeting national obligations related to biodiversity conservation, carbon sequestration, and provision of ecosystem services. Research is needed to better understand these processes and to quantify the benefits flowing from afforestation stands. One of the primary tasks should be the collection of existing information into a database; adequate documentation of afforestation results can improve the basis for decision making.

In all the Baltic States, conifer monocultures, such as pure Norway spruce forests, have been recognized problematic from the viewpoint of biological diversity, and the need to restore forests for the increase of biodiversity has been understood. For the present, there is only a little experience to convert conifer monocultures to hardwood forests with the exception of using natural regeneration of birch. Perhaps the experience gained within afforestation will help us to make more informed regeneration decisions in forest regeneration.

References

Ahti, T., Hämet-Ahti, L., and Jalas, J., Vegetation zones and their sections in Northwestern Europe, *Ann. Bot. Fennici*, 5, 169, 1968.

Anon., Forest Sector in Latvia, Ministry of Agriculture of the Republic of Latvia, http://www.zm.gov.lv/ data/forest_sector_2002_web.pdf, 2002.

Anon., Majanduslicult väheväärtuslike lehtpuualade hooldamise ja rekonstrueerimise vajaduse analüüs (Economic Analysis of Reconstruction of Hardwood Areas of Low Commercial Value), Project report, Estonian Forest Survey Centre, Tallinn, Manuscript, 2000 (in Estonian).

Bazzaz, F.A., *Plants in Changing Environments. Linking Physiological, Population, and Community Ecology*, Cambridge University Press, Cambridge, 1998.

Danusevicius, D., personal communication, 2003.

Daugaviete, M., Afforestation of agricultural lands in Latvia, in NEWFOR—New Forest for Europe: Afforestation at the Turn of the Century, Weber, N., Ed., EFI Proceedings No. 35, 175, 2000.

Daugaviete, M., Research Results on the Afforestation of Surplus Farmland in Latvia, Finnish Forest Research Institute, Research Paper 847, 96, 2002.

Fujimori, T., *Ecological and Silvicultural Strategies for Sustainable Forest Management*, Elsevier, Amsterdam, 2001.

Gabrilavicius, R., Review and Assessment of Lithuanian Experience with Afforestation of Abandoned Agricultural Land. Lithuanian Forest Research Institute, Project No. 055-17, Manuscript, http:// miskai.gamta.lt/agro/reports.htm, 1999.

Gaizutis, A., The role of forestry in the economy of Lithuania, in Social Sustainability of Forestry in the Baltic Sea Region, Hytönen, M., Ed., The Finnish Forest Research Institute, Research Paper 704, 111, 1998.

Hynönen, T., Pellometsitysten onnistuminen Itä-Suomess (Field Afforestation Results in East-Finland), The Finnish Forest Research Institute, Research Paper 765, 2000 (in Finnish).

Jõgiste, K., Vares, A., and Sendrós, M., Restoration of former agricultural fields in Estonia: Comparative growth of planted and naturally regenerated birch, *Forestry*, 76, 209, 2003.

Kaar, E., Coniferous trees on exhausted oil shale opencast mines, Metsanduslikud Uurimused (Forestry Studies), XXXVI, 125, 2002.

Kaar, E. and Kiviste, K. (Eds.), Maavarade kaevandamine ja puistangute rekultiveerimine Eestis (Mining and rehabilitation in Estonia), Eesti Maaülikool, Tartu, 2010 (in Estonian; English summary).

Kaimre, P., Väheväärtuslike põllumajandusmaade metsastamise ökonoomilised aspektid, Summary: Economic aspects of afforestation on abandoned agricultural lands, in *Proceedings of the Estonian Academical Forestry Society*, 14, 68, 2001.

Kuznetsova, T., Lukjanova, A., Mandre, M., and K. Lõhmus, Aboveground biomass and nutrient accumulation dynamics in young black alder, silver birch and Scots pine plantations on reclaimed oil shale mining areas in Estonia, *Forest Ecology and Management*, 262, 56–64, 2011.

Maamets, L., Metsastamistoetuste põhjenduse finantsmajanduslik analüüs (The Financial and Economic Analysis as a Basis for Subsidies of Afforestation), Forest Inventory Bureau Ltd., Tallinn, Manuscript, 2003 (in Estonian).

Madsen, P. and Löf, M., Reforestation in southern Scandinavia using direct seeding of oak (*Quercus robur* L.), *Forestry*, 78, 55–64, 2005.

Metslaid, M., Palli, T., Randveer, T., Sims, A., Jõgiste, K., and Stanturf, J.A., The condition of Scots pine stands in Lahemaa National Park, Estonia 25 years after damage by moose (*Alces alces*), *Boreal Environment Research* 18(suppl. A), 25–34, 2013

Niemistö, P., Influence of initial spacing and row-to-row distance on the crown and branch properties and taper of silver birch (*Betula pendula*), *Scand. J. Forest Res.*, 10, 235, 1995.

Paal, J., Eesti taimkatte kasvukohatüüpide klassifikatsioon (Classification of Estonian vegetation site types), Keskkonnaministeeriumi Info- ja Tehnokeskus, Tallinn, 1997 (in Estonian).

Poska, A., Saarse, L., and Veski, S., Reflections of preand early-agrarian human impact in the pollen diagrams of Estonia, Palaeogeography, Palaeoclimatology, *Palaeoecology*, 209, 37–50, 2004.

Reitalu, T., Seppä, H., Sugita, S., Kangur, M., Koff, T., and Avel, E. et al. Long-term drivers of forest composition in a boreonemoral region: The relative importance of climate and human impact, *Journal of Biogeography*, 40, 1524–1534, 2013.

Tamm, Ü., Haab Eestis (Aspen in Estonia), Eesti Loodusfoto, Tartu, 2000 (in Estonian; English summary).

Toomik, A. and Liblik V., Oil shale mining and processing impact on landscapes in north-east Estonia, *Landscape and Urban Planning*, 41, 285–292, 1998.

Tullus, H. Lehtpuupuitute kasvatamine (Growing broadleaves), in *Proceedings of the Academical Forestry Society*, XIV, 5, 2001 (in Estonian; English summary).

Tullus, H., Põllumajandusmaade metsastamine ja metsastumine (Natural and artificial afforestation of abandoned agricultural lands), Agraarteadus (*J. Agric. Sci.,*), 11, 22, 2000 (in Estonian; English summary).

Tullus, H. and Uri, V., Baltic afforestation, in *Proceedings of the IUFRO Conference on Restoration of Boreal and Temperate Forests—Documenting Forest Regeneration Knowledge and Practices in Boreal and Temperate Ecosystems*, Gardiner, E.S. and Breland, L.J., Eds., Danish Centre for Forest, Landscape and Planning, 11, 2002.

Uri, V., The Dynamics of Biomass Production and Nutrient Status of Grey Alder and Hybrid Alder Plantations on Abandoned Agricultural Lands, Dissertationes Scientarium Naturalium Universitatis Agriculturae Estoniae IX, 2001.

Vares, A., Jõgiste, K., and Kull, E., Early growth of some deciduous tree species on abandoned agricultural lands in Estonia, *Baltic Forestry*, 7, 52, 2001.

Vares, A., Tullus, A., and Raudoja, A., Hübriidhaab. Ökoloogia ja majandamine. (Hybrid aspen. Ecology and Management), Estonian Agricultural University, Tartu, 2003 (in Estonian; English summary).

Yearbook Forest 2013. Compiled and edited by the Estonian Environment Agency, Tartu, 2014. (in Estonian).

9

Afforestation in Denmark

Palle Madsen, Finn A. Jensen, and Søren Fodgaard

CONTENTS

9.1 Landscape History

Denmark is a small (43,000 km²) Scandinavian country characterized by its 7000 km coastline and almost 500 islands. Approximately 11% (486,000 ha) of the country is forest land. Denmark shares its only land border (50 km) with Germany on the Jutland peninsula (near 55°N latitude), which is the largest part of the country. Other interesting characteristics of this nation, with a population of 5.3 million, are the lack of fully exposed bedrock in the landscape (only on the island of Bornholm in the Baltic Sea) and the low relief; the highest elevation is only 170 m above sea level. The last ice age (the Weichselian) lasted 110,000 years, and it ended 12,000 years before present (ybp). Ice advances and retreats considerably shaped and rearranged the landscape, coastline, and distribution of soils. The land surface was changed, not only by the force of moving glaciers, but also from melt water in front of the glacier running in rivers, floating in deltas, or passing though lakes. As the ice retreated, the windy periglacial climate also affected the landscape.

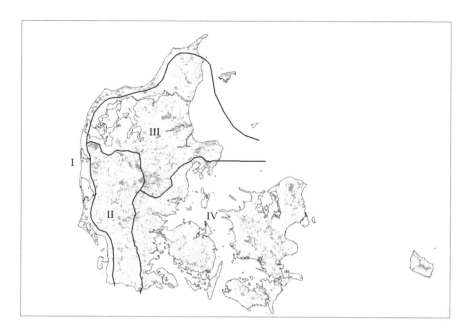

FIGURE 9.1
Map of Denmark showing Forest Regions (see text) and the present forest cover. Used with permission from the Danish Forest and Nature Agency.

Glacial ice came from Norway and Sweden, covering northeastern Jutland and leaving sandy deposits of limited fertility (Figure 9.1, Forest Region III). Another ice lobe left more loamy deposits containing lime in the southeastern part of Jutland, as well as at the southeastern islands (Figure 9.1, Forest Region IV). The southwestern part of Jutland was never covered with ice (Figure 9.1, Forest Region II); it was a tundra landscape with sparse vegetation during the whole glacial period. Soils in the west were not renewed by materials left by the glaciers; deposits here were sand and gravel from glacier meltwater that created sandy outwash plains between slightly more elevated structures, which are deposits from a previous ice age (the Saalian). The soils of western Jutland are generally sandy and poor, mostly classified as Spodosols or Entisols, whereas Alfisols and Inceptisols are the most common soil types in the rest of the country (Vejre et al. 2003). Sands moving along the west coast of Jutland carried by the near-coastal and sea currents in the sea formed dunes (Figure 9.1, Forest Region I). The glacial ice was probably more than 2 km thick, and when it melted, the land rebounded from the released weight. Simultaneously, sea level rose and the relative speed of the two processes has changed the coastlines several times. At times, sea level was lower than now and sometimes it was higher. This change is illustrated by Viking inlet harbors (fjords), which were in cities that today are impossible to reach by boat.

9.2 Climate

The mild and windy Atlantic climate supports annual mean temperatures between 7.5°C and 8.5°C, with January and July means close to 0°C and 16°C, respectively, and annual mean precipitation between 500 and 900 mm. Despite the overall mild climate, weather can be harsh for forests, in particular for regeneration. Atlantic weather is unpredictable and different from year to year. Late spring frost is a major problem for the regeneration of many species such as beech (*Fagus sylvatica* L.), Sitka spruce (*Picea sitchensis* [Bong.] Carr.), and European silver fir (*Abies alba* Mill.). Warm weather in early spring, from late March to early April, may stimulate an early flush of growth but the risk of late spring frost is high, depending on the site. Near-coastal areas are much less exposed than interior sites in Jutland or on Zealand. In fact, late July to early August may be the only time in central Jutland when there is no risk of frost. Site-specific features such as local terrain, grass cover, and shelterwood may influence the risk of late spring frost.

Denmark is a windy country; low-pressure fronts travel east from the Atlantic across the country, creating the predominant westerly or northwesterly winds. The almost perpetual west wind mainly influences the trees in the western, northern, and central regions of Jutland. Wind can stress trees in Denmark, not only when the low pressures develop into hurricanes and blow trees and forests down but also due to long-term effects. Young trees may be killed or severely damaged in the first several years after planting. A worst case occurs on exposed sites in winter, when there may be no snow cover, intense frost, and a dry and cold east wind.

9.3 Land-Use History: Loss and Gain of Forestland

9.3.1 From Ice Age to Black Death

Humans arrived shortly after the ice left, hunting reindeer in the open arctic landscape. As the temperature gradually increased, tree species such as aspen (*Populus tremula* L.) and birch (*Betula pendula* Roth. and *B. pubescens* Ehrh.) arrived, followed later by Scots pine (*Pinus sylvestris* L.). Hazel (*Corylus avelana* L.) was one of the first shade-tolerant species to invade these open forests and became dominant. Later, oak (*Quercus robur* L. and *Q. petraea* [Matt.] Liebl.), lime (*Tilia cordata* Mill. and *T. platyphyllos* Scop.), elm (*Ulmus glabra* Huds.), and ash (*Fraxinus excelsior* L.) dominated and formed a dense forest. Many large animals, now extinct in Denmark (e.g., wild boar, moose, aurochs, lynx, and wolf), inhabited the virgin forest. Our hunting and fishing ancestors mainly settled at the coasts. The average temperature reached a climax (2 to 3°C higher than now) about 5000 ybp; the first steps toward agriculture about 6000 ybp were the beginning of a long-lasting decline in forest cover. Beech did not arrive until 3500 ybp in the Bronze Age, when the forests were already strongly influenced by human activity.

Gradually increasing population and developing technology put more pressure on the forests. Too-short rotations of shifting cultivation and heavy livestock grazing gave ericaceous plants, such as heather (*Calluna vulgaris* [L.] Hull), the opportunity to establish on the poorest soils. These developments were sometimes interrupted by periods of war or epidemic diseases such as the Black Death, which arrived in Denmark in 1348 and killed about one-third of the population. Forest vegetation invaded the abandoned fields, but when the Black Death was over and human population recovered, forest exploitation began anew. Wood was used for many purposes, mainly construction timbers for houses and ships, as well as firewood and charcoal to manufacture steel, glass, and bricks.

9.3.2 Cutting, Grazing, and Fire

Wood consumption was by itself a serious threat to the forest, but intensive grazing by domestic stock also significantly contributed to the rapid forest decline. Until the late eighteenth century, peasants had rights to coppice the understory trees and to graze the forest with domestic stock. The trees of the overstory belonged to the landowner, often the local landlord. Consequently, trees were cut for timber and firewood, but the regeneration was removed or excluded by grazing. If the landowner had an interest in growing high forest for good-quality timber, there would be a conflict with the peasants over the rights to graze the forest. A closed canopy would not support good grazing and a fence would have been necessary during the regeneration phase. Stories of early attempts to plant trees are among the more amusing anecdotes of Danish forest history. The peasants, who did all the work of planting, made all kinds of "mistakes" that killed the trees and thereby made sure the planting failed.

Through the eighteenth century, the population faced increasingly severe shortages of wood and agricultural products. Almost complete deforestation was reached by 1800, when forest cover reached a low of only 4%. Jutland, except for the southeast, was largely covered by heathland (Figure 9.2), which had no trees and supported only sheep grazing. The heather was regularly burned so that it regenerated and retained its grazing value. In many locations, overgrazing or fires destroyed the vegetation and exposed the underlying sand to serious wind erosion. Dunes developed primarily on the west coast of Jutland but also in the interior, and in many places the drifting sand threatened the populace. Blowing and drifting sand covered villages, churches, farms, and fields.

9.3.3 The Complete Ecological Disaster

Today, such natural disasters and their impact on communities are difficult for modern people to envision. The now revegetated dunes and remaining heathland are seen as

FIGURE 9.2
Historic landscape of western Jutland, showing a peasant herding sheep on heathland. (Painted by F. Vermeren in 1855.)

interesting landscape elements or habitat for rare species. However, these past events and the landscapes they created were truly ecological disasters caused by exploitation of the forests and landscape. Old travel accounts from interior Jutland described the harsh conditions and the poverty of the poor people (Kousgaard 2000). Many locations with names that indicate a past forest cover were barren heathland in the mid-nineteenth century. This level of forest and landscape degradation and accompanying sand drift and heathland development, however, did not develop in the eastern parts of Denmark to the same extent it did in western Jutland.

From the mid-eighteenth century, the government tried to alter this unfortunate situation by various means. Reforms were launched over decades to change farmland ownership and tenure from large estates to independent peasant owners. Different governmental initiatives were launched over nearly 50 years, including assistance from German foresters who introduced basic forestry knowledge such as planning, testing species, regeneration methods, silvicultural systems, and forest education. In 1805, new forest legislation separated forest from farmland by fences in order to exclude grazing livestock from the forests. Fencing and the gradually changing ownership structure of farmland and forest, including a new social class of independent peasants/farmers, were important elements for the restoration of forests and landscapes. This became a crucial step forward for forestry, particularly management of high forests.

Late in the 18th and early nineteenth centuries, the state forest service established plantations at several sites on the heathland of western Jutland. Afforestation was slow; only a few thousand hectares were established as plantations by the mid-nineteenth century. Regeneration was difficult at these exposed sites, with infertile, acid soils containing a dense subsoil pan (B-horizon) that perched water and impeded root penetration. Severe competition from heather as well as the risk of severe frost added to the difficulty. There was hardly any silvicultural knowledge or experience to guide afforestation, and no rich private association or government agency to support the movement in the beginning. Other elements of the initial failures were choosing the wrong species and provenances, as well as inadequate cultivation methods (direct seeding).

9.3.4 An Aid Project of the Nineteenth Century

As a result of war, Denmark lost southern Jutland in 1864 to Prussian Germany, led by Bismarck. The northern third of this region became Danish again in 1920 by plebiscite, after Germany's defeat in World War I. The southern two-thirds of the region in question had a clear German majority and remained Germany. A national movement arose, in response to this loss of southern Jutland which boosted reclamation of bare heathland and other low-productivity areas in Denmark, particularly western Jutland. This revegetation movement (Stanturf et al. 2014) had three main objectives: timber production, soil erosion control, and, in the twentieth century, job creation in times of high unemployment. Moreover, tree shelterbelts have been regularly distributed in the landscape to create windbreaks to shelter agricultural crops.

In 1866, an independent organization (Hedeselskabet) was established to support reclamation by public relations, coordination, and knowledge. The goal was to persuade wealthy people to invest, government to subsidize, and local people to work, invest, and believe that reclamation would improve their future. The symbol of this land reclamation movement became E.M. Dalgas, an army engineer and officer. The timing was undoubtedly right for him and his vision, advanced by his organizational skills combined with his persuasiveness and speaking gifts. This seemed to be the necessary blend of elements to catalyze the whole process. Dalgas and his organization (Hedeselskabet) supported the reclamation by a range of activities, including

- Find relevant sites.
- Agitate at meetings for the reclamation movement.
- Negotiate with the landowners, often poor peasants.
- Find investors willing to invest in land, planning, amelioration, and planting.
- Obtain government subsidies for planting.
- Establish experiments to develop and demonstrate planting methods.
- Transfer knowledge through an extension service, lectures, and newspaper articles and other publications.
- Plan, manage, and control specific projects.

Hedeselskabet worked very much like a modern nongovernmental organization (NGO), but within the same country where it was formed. Today, the approach seems very modern: local people and cooperative organizations were freely involved, knowledge that was based on research and development was utilized, documented, and distributed, practical planning and implementation was properly guided, and (last but not the least) proper financial support was obtained.

Getting the finances in place is an interesting story on its own. Rich landlords, nobility, merchants, and industrialists were willing to invest in reclamation (Kousgaard 2000; Jensen 2001). Fortunes developed due to the growth of industry and the economy. It became fashionable and popular for such people to invest in reclamation projects, perhaps for a good reputation and image in society as a patriotic individual, while becoming a kind of landlord on their own estate, where they could recreate with their family. Local people and communities also invested in reclamation and considerably contributed to afforestation. Because these people were generally not wealthy, their plantations were financially constructed like corporations. Even relatively poor people could participate by buying one or a few shares, and thereby contribute to the movement.

The reclamation of the barren heathland was almost completed by 1950; by then, there were 180,000 ha in plantations and the forestland cover in Jutland increased from 2.7% in 1860 to 10% in 1950 (Figure 9.3). Thus, much of the reclamation led to farmland. Hedeselskabet established 59% of the new plantations from 1866 to 1950 and the government was responsible for 41%. Local people and rich investors were the main actors in this nongovernmental afforestation (unpublished data, Hedeselskabet). The local people (45%) were mainly farmers; rich landlords and businessmen from other parts of the country accounted for 38% of the investment. Hedeselskabet itself (9%) and local authorities such as counties and communities (8%) were relatively minor investors. Today it may seem that there was complete agreement on the advantage of reclamation and on the distribution between afforestation and reclamation for agriculture, but such overall agreement never existed (Jensen 2001). Voices of disagreement have always been present, but only in recent decades have nature conservation, biodiversity, or environmental protection interests gained sufficient support to influence afforestation efforts.

This first wave of Danish afforestation in most of Jutland undoubtedly involved hard work and sometimes even hunger and disappointment for at least some of the workers and peasants. Undoubtedly, the present farmland and conifer plantations appear more successful and productive than even the visionary pioneers and rich urban investors imagined some 70 to 140 years ago.

(a)   (b)

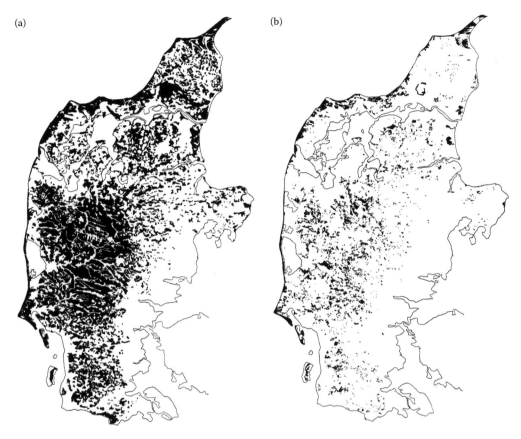

FIGURE 9.3
Maps showing the heathland distribution in 1800 (a) and 1950 (b). (Jensen, F.A., *Aktieselskabet Sønder Omme Plantage år 2001*, Aktieselskabet Sønder Omme Plantage, 96 pp., 2001. With permission from Hedeselskabet.)

9.3.5 How Was the Reclamation Done?

Generally, the best sites were reclaimed to farmland and the poor sites to forestry. When E.M. Dalgas and Hedeselskabet increased the speed of reclamation, there was still much to learn about it. However, the knowledge was soon greatly improved and reliable planting methods were developed by the late 1870s (Kousgaard 2000). From today's perspective, the approach was very intensive; planning might involve careful surveying, land leveling, and mapping site types. Often it was a practical question of how to divide the land between "poor" and "very poor" sites. Access to fertilizers, lime, and heavy machinery was not an option then, as it is today. Instead, labor was inexpensive and horses were, apart from the workers, the main source of power.

Dalgas soon realized that mountain pine (*Pinus mugo* Turra) was extremely tolerant of the harsh conditions of frost, drought, and infertile soil, and served as an effective nurse crop that could shade out heather and assist the establishment of the main species such as Norway spruce (*Picea abies* [L.] Karst.) (Kousgaard 2000). Mountain pine became the only suitable species for the first-generation plantation on the poorest sites. However, mountain pine does not tolerate shallow rooting, such as from a dense subsoil pan that blocks root penetration. A single-stem subspecies of mountain pine (*P. mugo* ssp. *uncinata* (Ramond) Domin) that gained popularity in the twentieth century has similar nurse tree effects as the regular mountain pine. Generally, the species choice was simple:

- The very poor sites were planted with approximately 3500 mountain pine seedlings per ha.
- Medium sites were planted with 50% mountain pine and 50% Norway spruce (approximately 3500 per ha in total).
- Good sites were planted with approximately 3500 Norway spruce seedlings per ha.

Various soil preparation methods were employed, depending on the site; on very sandy sites, mountain pine was just planted in holes prepared by hand and spades. Plowing was intensively employed on sites where heather dominated because it was important to control heather and stimulate decomposition and humus mineralization. Such superficial but intensive soil preparation also overcame some of the problems created by pans. Deep plowing to approximately 45 cm depth was used to some extent, but it was expensive and slow. Deep plowing was mainly used to establish plantations on the most exposed sites in the west, where white spruce (*Picea glauca* Moench) was planted instead of Norway spruce.

Norway spruce became the main timber species because it also tolerated frost, drought, and infertile soil, although not as well as mountain pine. It was reliable, relatively inexpensive to plant, and produced high-quality timber, which until recently brought high stumpage prices. Norway spruce is not without problems; it suffers from root rots, windthrow, and poor stability (Hahn et al. 2005). Windthrow problems became apparent early on, and silviculturists soon attempted to create more diverse and stable forests. Japanese larch (*Larix kaempferi* [Lamp.] Carr.) is a fast-growing pioneer species, which serves well as a nurse crop except that it is sensitive to drought. Consequently, larch somewhat replaced mountain pine as a nurse crop on many sites. Microclimate and soil fertility improved as the plantations developed, and so did the conditions which are necessary to establish a number of other species such as Sitka spruce, European silver fir, Douglas fir (*Pseudotsuga menziesii* Mirb. Franco), and grand fir (*Abies grandis* Dougl. Lindley). Great effort was dedicated toward establishing European silver fir, often mixed with Norway spruce. European silver fir resists root rots and was thought to be more stable in older late spring frost. Moreover, it is very attractive to deer. Consequently, it is easily suppressed if mixed with Norway spruce, and many expensively planted mixtures have ended up today as pure Norway spruce stands. Improved microclimate and soil fertility led to shorter regeneration periods and faster growth. Norway spruce productivity has increased; the typical average annual production (100-year rotation) has increased from 7 to 11 m³ ha⁻¹.

The leading role of Norway spruce is uncontested: 94% of the area was in conifer species and, on average, 67% of this was Norway spruce (Neckelmann 1986). No other conifer species exceeded 5% on average, although some species reached, exceeded, or came close to a 10% share as they were popular for a period (e.g., lodgepole pine [*Pinus contorta* Dougl.], European silver fir, and Sitka spruce). Broadleaves only covered 6% of the heathland afforestation, with oak as the most important species, covering 58% of the broadleaf area. Scots pine never attained the leading role expected at the beginning of the afforestation program (Neckelmann 1986). Although it was the main species planted during the first 50 to 60 years in the early nineteenth century, the German provenances used were poor choices and the species gained a bad reputation. Today, we know that it is definitely a tolerant species and as such suitable as a nurse crop species, but not as highly productive as Norway spruce.

9.3.6 Along the West Coast

Close to the west coast of Jutland, site conditions are quite different from the interior. The North Sea deposits high levels of salt, carried by the predominant westerly winds, which is not tolerated by Norway spruce. However, frost problems are less frequent along the coast. Besides the "blanket of mountain pine" that covered the Danish west coast, the relatively mild coastal climate called for species that are sensitive to frost but tolerant of the wind and salt deposition such as Sitka spruce and to a lesser extent European silver fir.

9.3.7 Afforestation in Eastern Denmark, 1800 to 1950

In the rest of the country (southeastern Jutland and the islands of Funen and Zealand), there were still forest remnants in the early nineteenth century. Between one-third and one-half of the forestland was given up and cleared for farmland, and land-use reforms initially reduced the forestland. Future land-use (farmland and forestry) was separated; the forests were fenced against livestock and hay harvest ended. It was generally easier to regenerate the forests and plant trees in eastern Denmark, and this part of the afforestation history appears less spectacular than reclamation of the heath plains. Afforestation in eastern Denmark included expansion of existing forests and afforestation on sites unsuitable for agriculture because of steep terrain or poor soil. The forest land area gradually increased in this part of the country as well, and the better growing conditions made it possible to use a wider range of tree species of both broadleaves and conifers.

Growing conditions in the southeastern part of the country (Figure 9.1, Forest Region IV) were better than on the heath plains. The soils offered good nutrient and water supply, hilly terrain protected against late spring frost, and often regeneration was present, although browsed. Particularly, beech regeneration sprouted in many places. Beech, oak, and ash were the main species planted, mixed with alder (*Alnus glutinosa* [L.] Gaertn.) and elm. New species were introduced, such as European silver fir, Norway spruce, larch, and sycamore maple (*Acer pseudoplatanus* L.). Interestingly, oaks were planted or sown by the state forest service (e.g., in the royal forests) in the early nineteenth century to meet the demand for naval timbers. The motivation for this planting increased after Denmark lost its navy ships in the Napoleonic wars to Great Britain in 1807. Nevertheless, conifers gradually became more frequently used for afforestation in eastern Denmark, particularly in the twentieth century until the 1980s.

9.4 Recent Afforestation

9.4.1 Changing Goals: Like a Moving Target

The first wave of afforestation (1800–1950) was very successful. Forest cover increased to the present 11% and two thirds are exotic conifers. Although 11% forest cover may seem a limited success from an international perspective, Danish culture is deeply rooted in agriculture, which has served its economy well and remains one of our largest exports. Nevertheless, the success of the early afforestation effort depends on the viewer, and today there are alternative views on the success or failure of what foresters accomplished decades ago. Even as we harvest what they planted, our modern affluent society questions whether the result of their efforts is optimal after all. Prices for softwood timber are at historically low levels, which has reduced or even eliminated the formerly important source of forestry income. Considerable windthrow risk in conifer stands before the trees reach target diameter further stresses the poor economics of spruce timber production. Such plantations often show a poor resilience and consequently both aspects of stability (resistance and resilience) (Larsen 1995) suffer in these plantations (Hahn et al. 2005).

Yet, afforestation is still under way. After two decades (1960s and 1970s) with limited activity, afforestation became an issue again from the mid-1980s. The goals differ now from those of 150 years ago, and even from the goals of 25 years ago. No doubt, they will probably change again. Subsidized production that produced a surplus of agricultural products fueled new governmental interest in afforestation, in order to reduce farm production and related subsidies. Another attractive goal was increased timber production, as the annual wood consumption substantially exceeded harvest (8–9 million m^3 wood consumed but only 2–3 million m^3 harvested).

Additionally, the potential for more multifunctional forests came into focus. Pure conifer stands were still planted, but in 1989, the Danish government initiated a new afforestation program aiming at doubling forest cover within one tree generation (80–100 years). Timber production was still an issue, but other values such as nature conservation, biodiversity, recreation, carbon sequestration, bioenergy, and protection of environment and groundwater increasingly gained support (Skov- og Naturstyrelsen 2000; Danish Forest and Nature Agency 2003). These are typical national goals, well known in many countries with afforestation programs.

The regeneration stages challenge foresters to anticipate management goals 50 to 120 years into the future, at least until the end of a rotation. Goals may well change because of the long production time, and the longer the rotation length, the greater the difficulty in anticipating preferences. The typical rotation age in Denmark is between 50 and 120 years so that stable and flexible forests, which may be subject to changing preferences, seems to be a reasonable strategy and should allow even for periods of no management. In Denmark, this includes using site-adapted and stable species, which on a certain site may be both native species and exotic species such as Douglas fir, Sitka spruce, or European silver fir. Norway spruce is still planted but not to the same extent as before.

9.4.2 Public Afforestation

The central government and some communities may buy farmland from farmers and plant trees. Such afforestation has been done by the Danish Forest and Nature Agency (Skov- og Naturstyrelsen), in cooperation with local community authorities or water companies responsible for drinking water supply. Public afforestation usually creates new and relatively large forests (greater than 200 ha). These are located close to urban areas to protect important groundwater resources and provide recreational opportunities. Participatory planning and design involving local people encourages use of the forest from the earliest stages of development. Such young forests provide valuable recreation for people living nearby (walking the dog, jogging, etc.); the closer the forest located to the neighborhood, the better it is (Jensen and Koch 1997). Moreover, proximity of residential housing to forests affects house prices positively, up to 500 m, where the effect fades out (Anthon and Thorsen 2002; Præstholm et al. 2002).

Afforestation was not planned only for government land; private landowners are subsidized to participate in the afforestation program. Subsidies seem necessary to reach afforestation goals, because the most common alternative land use is agriculture, which is heavily subsidized. Private landowners require subsidies for establishment costs as well

TABLE 9.1

Subsidy for Afforestation on Private Land, 2004

	Afforestation Areas (US$ ha^{-1})	Outside Afforestation Areas (US$ ha^{-1})
(A) Establishment		
Planting broadleaved forest or forest edge	3.33	2.13
Planting conifer forest	2.13	1.33
Direct seeding	2.13	1.33
(B) No pesticides used	400	
(C) "Gentle" soil preparation—no deep penetration	400	
(D) Fence	US$2 m^{-1}	
(E) Income compensation for 10 years	320 year^{-1}	
(F) Preparatory investigations		
Site mapping	US$133 + US$27 ha^{-1a}	
Mapping and survey	67$ + US$6.7^a	

Source: Modified after Skov- og Naturstyrelsen, Tilskud til skovrejsning, http://www.sns.dk/skov/tilskud/ skovrejs/, 2004.

^{a} Example: Subsidy for 5 ha site mapping: US$133 + (5 × 27) = US$ 268.

as additional compensation for the lost agricultural income. Subsidies for afforestation on private land have been as high as US$4800 ha^{-1} (US$1 = 7 DKR) plus an annual income compensation of US$320 ha^{-1} year^{-1} for the first 10 years (Table 9.1).

9.4.3 How Is Afforestation Done Today?

The high subsidy rates reflect the high cost for regeneration under Danish conditions. Stock density is high (2500 to 3500 conifer seedlings ha^{-1} or 4000 to 6000 broadleaf seedlings ha^{-1}; Figure 9.4). Bare-root seedlings presently are the most common stock type, usually 2–0 or 2–1 seedlings (2 to 3 years in the nursery). Price negotiations for materials and planting make it difficult to obtain a clear picture of costs. The price per planted seedling, including machine planting, ranges from US$0.30 to US$0.55, depending on stock type, species, and area to be planted. The total regeneration cost usually ranges between US$2600 and 5000 ha^{-1}, depending on material and planting costs plus the need for fencing, weeding, or other measures.

Soil preparation and weeding before planting is common, followed by weeding the first years after establishment. Fencing, mainly against roe deer (*Capreolus capreolus* L.), is also common practice. Depending on species, from 2500 to 4000 saplings must reach an average height of 1 m within 8 years after planting in order to obtain the full subsidy. Because faster growth means a quicker subsidy payment, intensive regeneration methods including weeding, fencing, and deep plowing are favored.

Deep plowing (up to 70 to 80 cm depth), in particular, has large and positive effects on survival and growth in poor sandy soils, due to several factors. The topsoil contains many weed seeds and has a relatively high water-holding capacity due to higher humus content than the subsoil. Deep plowing buries the topsoil, which eliminates most of the weed competition for the first growing season, protects the buried topsoil from evaporation as the hydraulic connectivity is broken, and plow pans, that may have developed in soils with higher clay content, are broken.

However, deep plowing is not popular among people who value cultural heritage, which may be severely disturbed or destroyed. Landowners must consult archeological experts

(a)

(b)

FIGURE 9.4
Typical afforestation in eastern Denmark. (a) Beech with larch as nurse trees. The beech is approximately 2 m tall. An old broadleaved forest is in the background. (b) Ash planted at the same location but on a moister site, showing the dense stocking; the lowest branches are approximately 1 m above the forest floor. These ash trees have since died due to ash dieback. (Photos by P. Madsen.)

before a site is deeply plowed. Further, establishment without deep plowing received a higher subsidy (Table 9.1). Similarly, afforestation without using herbicides obtained additional subsidies. The Danish Forest and Nature Agency does not use pesticides or herbicides, as the agency has converted to close-to-nature silviculture. Afforestation by water companies is probably done without using herbicides for competition control; harrowing or rotary cultivation between the rows is a common alternative. These mechanical methods are more expensive, may be difficult in a rainy summer, and may increase nutrient leaching, which is undesirable for groundwater quality (Pedersen et al. 2000). Increased nutrient leakage may also result from deep plowing.

Preparatory survey of soil conditions, mapping, and survey may also release additional subsidies (Table 9.1). Site classification is supposed to guide the choice of tree species, for example, planting ash and alder in moister areas and oaks primarily on sites with hard pans, pseudogley, or on relatively dry sites. The ultimate goals are better stability of the future forest and a higher probability of successful natural regeneration. Such attention to matching site and species was absent during most of the twentieth century, when conifers were planted almost everywhere.

9.4.4 Restrictions, Priorities, and Landscape Planning

The Danish landscape is intensively used and managed and the administration of afforestation subsidies is a good example of the intensity of land-use. Several restrictions and priorities seek to regulate how and where afforestation is carried out on private land in order to reflect the interests of the society. Such restrictions include the following ones:

- The new forest must be more than 2 ha and forests between 2 and 5 ha are only subsidized close to urban areas, and the public must have access.
- Habitats such as lakes, meadows, heathland, and moors must not be drained or otherwise changed and planted with trees or shrubs; they may, however, become integrated open areas within a new forest.
- New forests need at least 20 m wide forest edges with minimum 20% shrubs to the north and the west, and a minimum of 10 m forest edge is needed to the south and the east.
- Broadleaved stands must have minimum 75% broadleaved species.
- Both broadleaved and conifer stands must be intermixed with minimum 10% other species than the main species.

Budget limits prevent the government from supporting all afforestation projects on private land. The highest priority, and therefore the greatest probability to successfully gain subsidies, is reached under the following conditions:

- In areas classified as afforestation areas by the regional planning authority.
- In areas where groundwater is extracted for drinking water.
- When native species are more than 75% of the planted seedlings.
- When no deep plowing or other deep soil preparation is used.
- When the new forest area is more than 5 ha, better when it is more than 10 ha.
- When the new forest is located close to an urban area or in an area with low forest cover.
- When the afforestation project is coordinated with other landscape restoration and afforestation projects such as environmentally friendly farming, habitat restoration, or improved public access to forest and nature.

9.4.5 Regional Planning: Where Is Afforestation Wanted?

These restrictions and priorities offer government a detailed control of afforestation on private land, which can be changed as new needs are identified. Moreover, regional authorities are responsible for general land-use planning, which is reviewed and adjusted at regular intervals. The landscape is divided into three main categories: suitable, unsuitable, or neutral for afforestation. At present, 6% of the land area is classified as suitable for afforestation, whereas 25% is classified as unsuitable. The most recent classification guidelines stress the importance of groundwater and recreational interest near urban areas for the areas suitable for afforestation. The guidelines also encourage regional authorities to favor "green corridors" in the landscape for biodiversity, connectivity, and to show consideration for agricultural interests. Generally, afforestation areas should not be located in areas where it conflicts with the interests of geology, cultural heritage (e.g., churches and castles), and landscape. New forests should not cover or hide buildings, geological formations, or landscape vistas. Such areas may instead be classified as unsuitable, along with the areas where afforestation conflicts with future urban development or technical installations such as windmills.

9.4.6 Why Do Private Landowners Enter into Afforestation?

Private landowners undertaking afforestation are of two kinds. One group includes part-time farmers, who earn their main income elsewhere and own relatively small properties. Their average afforestation project includes 7 ha of new forest, primarily for recreation and nature conservation. The other group includes professional and full-time farmers who want to create a hunting property out of part of their farm, perhaps the least suitable part for farming.

Professional full-time farmers seem to need increasingly more farmland to stay in business. In addition to arable land, a farmer must have access to a certain amount of land per cow or pig to spread manure in order to avoid concentrated loss of nutrients to the groundwater or surface water. Consequently, farmland is a limited and expensive resource; a price for farmland of US$15,000 ha^{-1} would be usual. Forestland is less subsidized, and thus less expensive, approximately US$7000 to 17,000 ha^{-1} including standing volume. Thus, subsidies for afforestation are in competition with subsidies for continued farming, which explains the importance of income compensation. By planting a minimum 35 ha without subsidies, the landowner acquires a property with a very attractive feature: it can be sold to anybody who can afford to buy it, without having to live there year-round, as with ordinary farms. This new status of forestland makes such properties very valuable and earns the farmer who planted the 35 ha a good return.

Wood production is not a driving force for afforestation of private land (Skov- og Naturstyrelsen 2000). The value of timber in the distant future is theoretical, not real, for most individuals. Private landowners may simply afforest land to enjoy ownership of forestland for recreation, wildlife watching, and hunting. Hunting rights on private property belong to the landowner and may generate recreation and pleasure for the owner or rental income of US$50 to 100 ha^{-1} or more annually. Projects may call for tree planting on about half of the afforestation area, with the rest left open for livestock grazing or small lakes or ponds, which may be designed for duck hunting. The most important economical aspect of afforestation for the landowner may be its influence on overall property value, which typically includes the transformation of a traditional farm dominated by bare fields to an attractive property with amenity values. Consultants and extension services offer assistance to private landowners for designing, planning, and implementing afforestation, as well as applying for subsidies.

9.4.7 The Afforestation Program Is behind Schedule

Both public afforestation and subsidized private afforestation are expensive. Expenses for public afforestation include purchasing land and regeneration costs, whereas the subsidies for private landowners include compensation for agricultural income lost over 10 years. Costs are US$17,000 ha^{-1} for public and US$9500 ha^{-1} for private afforestation (Skov- og Naturstyrelsen 2003). Not surprisingly, afforestation fell behind schedule (Skov- og Naturstyrelsen 2000; Danmarks Statistik 2002; Skov- og Naturstyrelsen 2003). Over the period 1991 to 2001, the program resulted on average in 1650 ha of new forest annually, far behind the target set in 1989 of 4000 to 5000 ha annually. Public afforestation in particular is behind schedule; only 3000 ha (11%) were established in the 10-year period, whereas 8500 ha of subsidized private and 5000 ha of private afforestation without subsidies were established.

Competition with subsidized farming for land is clearly an obstacle for the afforestation program. An additional subsidy to compensate for the loss of annual income, launched in 1998, boosted subsidized private afforestation in 1999–2001 to 1300–3300 ha annually (Skov- og Naturstyrelsen 2003). Private afforestation without subsidies was probably reduced. When first introduced, the income-compensation period was 20 years but private landowners are now supported for only 10 years. Subsidized private forests must not be cleared again. The future will show how agricultural subsidies and earnings will influence land prices and the need for income compensation.

The development of new and inexpensive regeneration methods may provide some help for the afforestation program. New methods need to be independent of intensive weeding, either by herbicides or machinery, and deep soil preparation in order to meet the requirements of public afforestation, and for private afforestation to receive the highest subsidies and priorities. Direct seeding and small container stock, in combination with cover crops and nurse trees; have shown promising results. These regeneration techniques better match the principles of close-to-nature silviculture than traditional, more expensive and intensive means. These methods continue to be developed and there has been limited operational use. Natural recolonization is an even cheaper means of restoration that will probably gain importance. Although it is indeed very close to nature, landowners and the agencies administering subsidies will need more patience than is needed with other methods.

References

Anthon, S. and Thorsen, B.J., Værdisætning af statslig skovrejsning—en husprisanalyse, *Dansk Skovbrugs Tidsskrift*, 87, 73, 2002.

Danish Forest and Nature Agency, The Danish National Forest Programme in an International Perspective, Ministry of the Environment, Danish Forest and Nature Agency, Haraldsgade 53, DK-2100 Copenhagen Ø, available at http://www.sns.dk/internat/dnf-eng.pdf, 2003.

Danmarks Statistik, Skove og Plantager 2000, Danmarks Statistik, 171 pp., available at http://www.dst.dk, 2002.

Hahn, K., Emborg, J., Larsen, J.B., and Madsen, P., Forest rehabilitation in Denmark using nature-based forestry. In: *Restoration of Boreal and Temperate Forests*, Stanturf, J. and Madsen, P. (Eds.). CRC Press, Boca Raton, FL, 299–317, 2005.

Jensen, F.A., De jyske heders opdyrkning, in *Aktieselskabet Sønder Omme Plantage år 2001*, Jensen, F.A., Ed., Aktieselskabet Sønder Omme Plantage, 96 pp., 2001.

Jensen, F.S. and Koch, N.E., Friluftsliv i skovene 1976/77–1993/94, Forskningsserien nr. 20, Forskningscentret for Skov & Landskab, Hørsholm, 1997.

Kousgaard, P., Hvor heden før var gold nu skov og læhegn findes, St. Hjøllund Plantage A/S og Palle Kousgaard, 128 pp., 2000.

Larsen, J.B., Ecological stability of forests and sustainable silviculture, *For. Ecol. Manage.*, 73, 85, 1995.

Neckelmann, J., Træartsvalg og dyrkningserfaringer i hedeplantagerne, *Vækst 107, 20 and Vækst* 107, 12, 1986.

Pedersen, L.B., Riis-Nielsen, T., Ravn, H.P., Dreyer, T., Krag, M., Nielsen, A.O. et al. Alternativer til pesticisprøjtning i skovkulturer, *Skoven*, 32, 355, 2000.

Præstholm, S., Jensen, F.S., Hasler, B., Damgaard, C., and Erichsen, E., Forest and afforestation in the neighbourhood: Attractiveness and value of local areas in Denmark, *Proceedings from the Conference "The Changing Role of Forestry in Europe; Between Urbanisation and Rural Development,"* *11–14 November 2001, Proceedings 2002-02*, Forest and Nature Conservation Policy Group, Wageningen University, The Netherlands, 2002.

Skov- og Naturstyrelsen, Evaluering af den gennemførte skovrejsning 1989–1998, Skov- og Naturstyrelsen, Haraldsgade 53, DK-2100 Copenhagen Ø, http://www.sns.dk/skov/net-pub/evaluering, 2000.

Skov- og Naturstyrelsen, Skov og Natur i tal 2003. Miljøministeriet, Skov- og Naturstyrelsen, http://www.skovognatur.dk, 2003.

Skov- og Naturstyrelsen, Tilskud til skovrejsning, http://www.sns.dk/skov/tilskud/skovrejs/, 2004.

Stanturf, J.A., Palik, B.J., Williams, M.I., Dumroese, R.K., and Madsen, P., Forest restoration paradigms, *J. Sust. For.*, 33 (suppl. 1), S161–S194, 2014.

Vejre, H., Callesen, I., Vesterdal, L., and Raulund-Rasmussen, K., Carbon and nitrogen in Danish forest soils—contents and distribution determined by soil order, *Soil Sci. Soc. Am. J.*, 67, 143, 2003.

10

Forest Restoration and Rehabilitation in the Republic of Korea

Don Koo Lee, Pil Sun Park, and Yeong Dae Park

CONTENTS

10.1 Introduction

Following more than 50 years of social and political upheaval in the early twentieth century, the Republic of Korea (ROK) faced massive environmental problems of forest degradation and severe soil erosion. Moreover, rural populations looked to the remaining forests to supply fuelwood for domestic heating and cooking needs. The government responded with a series of 10-year Forest Development Plans that attempted to meet the most pressing immediate needs and begin the long-term restoration of the country's forests. Beginning with the first and second Forest Development Plans, fuelwood plantations were successfully established and most of the degraded forests rehabilitated. The first functional improvements in the environment have already appeared, such as erosion control, improvement of soil quality, and increasing biodiversity. Nevertheless, the full effects of the forest restoration will not be seen for at least another 50 years. Increasingly, rehabilitation has shifted from an emphasis on plantations and exotic species toward native species and natural regeneration.

10.1.1 The Environment

The Korean Peninsula is located in the eastern edge of the Eurasian Continent and in the western coast of the Pacific Ocean. Climatically, Korea lies in the East-Asian Monsoon belt.

The Korean Peninsula encompasses 221,000 km², of which 45% (99,600 km²) makes up the ROK. The topography is complex with two major mountain ranges. The location and topography of the Korean Peninsula bring four distinct seasons in Korea. Spring and autumn are relatively short and summer and winter are longer. Summer is hot, humid, and wet. From June to September, summer monsoons hit the Peninsula, bringing more than 60% of the annual precipitation. Heavy showers with thunder and lightning are common, with periodic flooding. Winter is dry and freezing cold due to northwesterly Siberian air masses that sweep down from the north. The average temperature of Korea is hotter in summer and colder in winter than that of the countries located in the similar latitudes on the continent. Mean temperatures are 12–14°C in the middle and south region, and 3–10°C in the north. Annual mean precipitation ranges from 600 to 1600 mm. The rainy season beginning in late June lasts approximately 30 days and commonly between June and October, more than two typhoons strike the Peninsula every year, resulting in concentrated rainfall (Shin 2002).

Because of the variation in seasonal patterns of temperature and rainfall, there are three major forest zones on the Korean Peninsula (Figure 10.1): warm temperate forest, cool temperate forest, and subfrigid (or subalpine) forest. The warm-temperate forest covers the area south of 35°N, a part of the southern coastal regions, including Jeju Island and many smaller islands where the annual mean temperature is above 14°C. Natural vegetation in this zone is dominated by evergreen broadleaved trees including *Quercus acuta, Camellia japonica, Castanopsis cuspidata, Cinnamomum japonicum,* and various bamboo species. However, most of the natural forests were destroyed by forest fire and overexploitation, and changed to deciduous broadleaf, pine, or mixed forests. The cool temperate forest zone covers most of Korea between 35° and 43° N except the mountainous highlands and southern coastal zone. The annual mean temperature ranges from 6°C to 13°C. Species predominantly grown in this region are deciduous broadleaved trees, with pine forests established after destruction by natural or human causes. Major tree species in this zone are *Quercus* spp., *Acer* spp., *Zelkova* spp., *Carpinus* spp., *Prunus* spp., *Fraxinus* spp., *Betula* spp., *Pinus densiflora,* and *P. koraiensis.* This zone can be divided into three subzones, namely the southern, central, and northern temperate forest subzones.

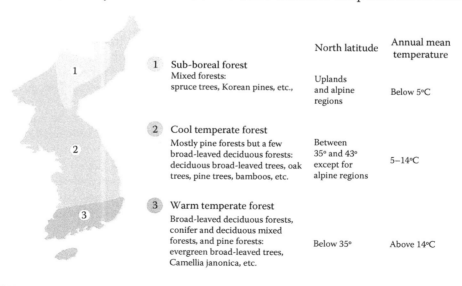

		North latitude	Annual mean temperature
1 Sub-boreal forest Mixed forests: spruce trees, Korean pines, etc.,		Uplands and alpine regions	Below 5°C
2 Cool temperate forest Mostly pine forests but a few broad-leaved deciduous forests: deciduous broad-leaved trees, oak trees, pine trees, bamboos, etc.		Between 35° and 43° except for alpine regions	5–14°C
3 Warm temperate forest Broad-leaved deciduous forests, conifer and deciduous mixed forests, and pine forests: evergreen broad-leaved trees, Camellia janonica, etc.		Below 35°	Above 14°C

FIGURE 10.1
Major forest zones in the Republic of Korea.

The subfrigid temperate forest covers the northern end of Korea and high mountainous area where the average annual temperature is 5°C and lower. Dominant tree species are *Abies nephrolepis, Picea jezoensis, Larix gmelinii, Juglans mandshurica,* and *Betula platyphylla* (Korea Forest Service 2010).

10.1.2 Historical Vegetation Change

Glacial activity is not evident in Korea except in the highest mountains (Nelson 1993). Because of the lack of glaciers, which at least temporarily eliminated flora in other locations, plant diversity is great (4662 known vascular plant species). Information about the vegetation history of Korea during the postglacial period was obtained from pollen analyses (Yamazaki 1940; Matsushima 1941; Oh 1971; Jo 1979; Yasuda et al. 1980; Choi 2001). The beginning of the postglacial period in Korea is not certain but assumed to be around 10,000 BCE. The period between about 10,000 and 6000 BCE was characterized by the dominance of deciduous broadleaved forests (e.g., *Quercus*) with rapid decline in subarctic conifers across the peninsula (Choi 2002). Around 6000 BCE, these dominant species clearly occupied the lowlands of the eastern and western coasts. According to pollen analysis (Choi 2002), the western region changed from *Quercus* to *Alnus* followed by *Pinus;* while the eastern region changed from *Quercus* to *Pinus*. In around 2000 BCE, the distribution of deciduous broadleaved forests dominated by *Quercus* and *Alnus,* and of evergreen broadleaved forests dominated by *Cyclobalanopsis,* became narrower while that of coniferous forests dominated by *Pinus (Diploxylon)* expanded.

Pine forests were protected for a long time, particularly during the Chosun Dynasty (1392–1910 CE) and affected the cultural life of the people. The national level forest control system during Chosun Dynasty was initiated by removing the provisions of the "Shiji" (lands for collecting fuelwood) and the "Gumsan system" that regulated the forests for specific national uses (e.g., coffins of the royal family). In the nineteenth century, the forests of the ROK were extensive with many old-forest characteristics. These forests were totally destroyed, however, by overharvesting and illegal cutting for fuel and building material throughout the chaotic periods of the Japanese occupation (1910–1945) and the Korean War (1950–1953). During this time of over-exploitation, the average growing stock volume decreased from about 100 m³ ha⁻¹ in 1900 to only 10 m³ ha⁻¹ in 1960 (Table 10.1).

The shortage of fuelwood needed for domestic cooking and heating continued even after this wartime turmoil. In 1960, there were about 2.4 million households in the country, each requiring about 0.5 ha of forest to meet fuel needs. One of the first priorities of the ROK government after the Korean War (1950–1953) was to establish fuelwood plantations, which could meet demands for fuelwood as well as beginning the rehabilitation of devasted

TABLE 10.1

Forestland Area and Growing Stock by Year

Year	Area (10^3 ha)	Growing Stock (10^3 m³)	Growing Stock (m³ha⁻¹)
1960	6700	63,995	9.55
1970	6611	68,772	10.40
1980	6567	145,694	22.18
1990	6476	248,426	38.36
2000	6422	407,576	63.46
2010	6369	800,025	125.62

Source: Adaped from Korea Forest Service. 2013. *Statistical Yearbook of Forestry,* 486 pp.

forestland. During the last 40 years of rapid economic development and urbanization of Korea, there has been an unprecedented demand for new land. Clearance of forests for the construction of various types of social infrastructure, industrial estates, and new towns has been the primary cause of deforestation and forest fragmentation. As a result of these anthropogenic activities and biotic disturbances (e.g., pine gall midges), the amount of *Pinus densiflora* in forests has decreased while *Quercus mongolica*, *Robinia pseudoacacia*, and *Styrax japonicum* have increased in density.

10.2 Large-Scale Rehabilitation in ROK

Since the end of the Korean War in 1953, the ROK government has encouraged replanting forests and has promoted the search for alternative fuels, especially for domestic use. To restore deforested areas, the government initiated massive reforestation programs with fast-growing, mostly nonnative trees such as *L. kaempferi*, *Pinus rigida*, and *Populus* species; nurse trees, such as *Robinia pseudoacacia* and *Alnus* species, were also used. Reforestation was implemented from 1959 to improve soil conditions and control erosion, which by 1999 resulted rehabilitation of about 97.4% of the deforested areas (Lee et al. 2004). Reforestation was initiated on a large scale in 1959 mainly by planting fast-growing trees to meet the fuelwood demand.

Over the first 31 years of the restoration program, approximately, 12 billion trees were planted, mostly fast-growing trees for fuelwood (Korea Forest Research Institute; KFRI, 1997). The plantation area is now estimated to cover 70% of the total forest land area. The major species planted in degraded forest areas in 1960s were *Robinia pseudoacacia*, *Pinus rigida*, and *Alnus*, *Quercus*, and *Lespedeza* species. The choice of species shifted in the 1970s and 1980s to *L. kaempferi* and *Pinus koraiensis* in order to increase the commercial value of forest resources (Table 10.2). During the period from 1962 to 1992, reforestation has occurred on more than 4.5 million ha. As a result, the growing stock increased to 60.3 m^3 ha^{-1} as of 1999 and annual growth reached 2 m^3 ha^{-1} (Korea Forest Service 2000). The large-scale planting effort was successful in terms of the original aims but it is now recognized that certain problems exist, including simple stand structures and mostly nonnative species composition. The uniform structure and reduced biodiversity has led to lower stability and declining productivity of the ecosystem (Lee et al. 2004).

The emphasis of forest rehabilitation in Korea is changing. The focus on revegetation, fuelwood supply, and erosion control using exotic species has shifted toward native species and natural regeneration. The restoration approach using native species is becoming one of the main issues in forest ecosystem management in Korea and may contribute to the improvement of forest environmental conditions (Jordan et al. 1997; Urbanska et al. 1997; and Sayer et al. 2001).

10.2.1 Lessons of the Korean Experience

The massive reforestation program since 1960s of the Korean Government has achieved impressive results. The focus on revegetation, without regard to composition or structure (Stanturf et al. 2014), is now recognized as problematic but the Korean experience may provide some useful insights for developing countries that are faced with conditions similar

TABLE 10.2

List of Native Species, Exotic Species, Hybrid Pine, Poplars and Aspens, and Nut Trees Used in Reforestation during 1962–1992

Species	1962–1972 Trees		1973–1992 Trees	
	Planted (1000)	Area (ha)	Planted (1000)	Area (ha)
Native species				
Pinus densiflora	42,275	14,902	51,150	17,050
Pinus thunbergii	204,364	68,121	90,761	30,254
Pinus koraiensis	159,998	53,333	831,093	277,031
Paulownia coreana	940	2350	12,746	31,865
Rhus verniciflua	4,404	4404	141	141
Quercus acutissima	78,232	26,077	18,819	6,273
Exotic species and hybrids				
Pinus rigida	1,252,387	417,462	634,483	211,494
Pinus taeda	19,052	6351	12,787	4262
P. rigida × P. taeda	31,458	10,486	83,124	27,708
Larix kaempferi	797,320	265,773	1,179,918	393,306
Cryptomeria japonica	106,262	35,421	160,459	53,486
Chamaecyparis obtusa	98,234	32,745	243,679	81,226
Robinia pseudoacacia	1,244,070	414,690	333,250	111,083
Alnus species	372,081	124,027	481,564	160,521
Populus alba × P. glandulosa	—	—	142,438	142,438
Populus nigra × P. maximowiczii	—	—	75,725	151,450
Nut trees				
Castanea creanata	94,359	235,898	60,588	151,470
Juglans sinensis	1956	4890	1830	4575
Ginkgo biloba	667	1668	3857	9643
Zizyphus jujuba var. inermis	122	305	1024	2560
Diospyros kaki	1325	3313	1230	3075
Diospyros lotus	1353	3383	170	425
Others	69,161	23,054	803,841	267,947
Total	4,580,020	1,748,650	5,224,677	2,139,283

Source: Adaped from Korea Forest Service. 1993. *Statistical Yearbook of Forestry.*

to Korea in the 1960s. The most important lesson from the Korean experience may be that the restoration of denuded forest lands and forest protection are not obstacles to economic development but instead may be a catalyst to sustainable development. The Korean experience was that through the forest restoration policy, other sectors such as agriculture and industries were simultaneously stabilized. Before adopting the Korean model, or expecting similar success, developing countries should recognize that the unique Korean social system played a critical role. National identity is very strong; the Korean people believe they are one ethnic group. Furthermore, the Confucian tradition influenced social norms; at that time, political leaders and the government were held in high esteem by the people. As shown in the Korean experience, effective local and national leadership is very important for the success of forest restoration projects.

10.3 Case Study: Rehabilitation of Eroded Land in Yeongil District

Land degradation and forest exploitation in the Yeongil district began with the invasion by Japan and other countries in 1800. Forest degradation continued through the 1900s, especially during World War II and the Korean War. Economic difficulty, political, and social disorder, including fuel shortages and illegal harvesting, accelerated the degradation of forests and caused severe soil erosion. Numerous (approximately 50) small-scale erosion control plans were implemented in Yeongil district. These efforts were unsuccessful, however, and large areas of degraded land were abandoned for a long time. In September 1971, President Jung-Hee Park visited the district and declared that "since this site has been the root location of droughts and floods to Yeongil district, which is also a flying route for international airlines, it is recommended to establish fundamental measures for the complete restoration of the devastated lands." The President's recommendation caused large-scale erosion control activities to be started (Gyeongsangbuk-do 1977).

The Yeongil district is in the temperate forest zone. Original tree species included *Carpinus tschonoskii, Zanthoxylum schinifolium, Lindera glauca, Celtis sinensis, Platycarpa strobilacea, Pinus thunbergii,* and *Pinus densiflora.* These species had disappeared due to illegal logging and devastation caused during the Korean War; only dwarfed pine trees remained on eroded slopes. The area for the erosion control project extended from 129°5′ to 129° 26′ E and from 35° 54′ to 36°16′ N, which includes Yeongil, Weolseong County and Pohang City. Annual precipitation is 1000–1100 mm, with half (500–600 mm) concentrated as rainfall during June to August. Parent rocks are shale and mud stone, which have low infiltration and low water holding capacity. Thus, the topsoil is thin and highly erosive because of much overland flow caused by heavy rainfall. Annual average temperature is 13°C, ranging from 37°C to –14°C. Because of the dry westerly winds, evaporation exceeds rainfall by more than 560 mm (Gyeongsangbuk-do 1977).

10.3.1 Application of Erosion Control Techniques

Harsh site conditions, especially little or no topsoil, continuing erosion, and moisture stress were the challenges to be overcome by the forest restoration program in Yeongil District. Four different techniques were used for gently sloping areas; these included sod-patching channel (discharges rain water safely and is used to stabilize small channels), terrace-sodding structures (controls rill erosion on slopes), contour-trenching (improves water holding capacity and promotes vegetation growth), and underground-laying structures (Gyeongsangbuk-do 1977).

Five different techniques were required for steep slopes caused by slope failure or cutting. These methods included stone-patching channels (safe discharge of rain water and stabilization of small channels), stone soil-arresting structures (protection from rill erosion on slopes), stone terracing structures (controls overland flow and provides rooting depth for tree growth), latticed block works (controls slope failure), and mulching structures (effectively covers eroded land surfaces). In the uplands valleys, revetments were employed for the stabilization of foot slopes and protection of steep slopes. Other methods used included erosion check dams to stabilize foot slopes and control horizontal and vertical slope erosions and erosion control dams.

Seeding and planting were used for soil improvement (supplying good soil, i.e., sandy loam to improve tree growth), sod patching (control of sedimentation and fast reforestation by covering completely bare land), and planting of fast-growing species as nurse trees (*Alnus* spp. and *Robinia pseudoacacia*) to supplement fertilization for a strong acid soils, and pioneer species such as *Pinus thunbergii* (Japanese black pine) and *Pinus rigida* (pitch pine) for coastal areas and grasslands.

The erosion control project at Yeongil area was planned between 1971 and 1972. The project was conducted over the 5-year period from 1973 to 1977. The total size of the completed area was 4538 ha (general erosion control on 3291 ha, special erosion control on 479 ha, and reforested bare land of 768 ha). The project cost US$ 3,190,000 (labor cost US$ 2,700,000; material cost US$ 390,000; and additional costs of US$ 100,000). The number of participants in the project was 3,556,000 persons. The materials used were 23,890,000 seedlings, 4,161 tons of fertilizers, 101 tons of seeds, 552,000 tons of stones, 20,800,000 m^3 of sods, 114,259,000 m^3 of turfs, 727,000 m^3 of transported soil, and 16,000 m^3 of rubble.

The amount of seeds sown was over 20 kg ha^{-1} of *Robinia pseudoacacia*, *Lespedeza bicolor* and *Cyperaceae* spp. The mixing ratio of tree species was: one-half grass seed, a fourth *Lespedeza bicolor*, and a fourth *Robinia pseudoacacia*. Height growth in one year after sowing was 150 cm for *Robinia pseudoacacia*, 70–100 cm for *Lespedeza bicolor*, and 80–90 cm for grass. After two years, *Robinia pseudoacacia* occupied the overstory in the planted area (Gyeongsangbuk-do 2002).

10.3.2 Ecological Effects of Rehabilitation and Erosion Control

Surface soil on the hillside had accumulated over 24 years to 11 cm in depth, which was far greater than the 1–2 cm depth at the outset when trees were planted. Surface soil depth ranged from 4 cm the near summits to 112 cm at the base of the hill. This was induced by the surface soil stabilization through the root system development, rainfall interception, and a windbreak effect after crown closure. Even though the surface soil was considered soft by penetrometer measurements (0.7 kg m^{-2}) the surface depth was so shallow that root system development was affected except on the base of the hill. The root systems of jack pine was distributed roughly within the top 10 cm of surface soil, while pitch pine roots developed vertically, which was better for planting in mudstone areas. Soil texture was changed from sandy-loam or silt-loam to fine sandy silt-loam as the ratio of clay increased. Soil acidity was changed from pH 4.1–4.5 to pH 4.6–5.5. Nitrogen and phosphorus were greatly deficient in this area but fertility increased after planting. Available calcium and the cation exchange capacity remained below that of average forest soils. Natural vegetation, *Pinus thunbergii,* and *P. densiflora*, were in the upper stratum of 5 to 20 trees ha^{-1}. Middle and lower story vegetation is still variable even after 12 years of planting, ranging from 500 to 2000 trees ha^{-1} (Lee 2001).

Several positive results have been generated from the erosion control work at the site of Yeongil. Serious damage from both drought and flood has been prevented due to the restoration of diverse forests. Trees have reduced soil erosion and protected arable land and the coastline, allowing for better management. Forest environment and productivity have improved and timber production can be expected in the future and the trees can provide a fuel resource where fuel is scarce. Overall biodiversity has gradually improved and wildlife that was almost extinct has recovered, including *Parus majore* (great tit), *Phasianus colchicus karpowi* (pheasant), dove, *Sus scrofa creanus* (wild boar), *Lepus brachyurus* (hare), and *Capreolus capreolus* (roe deer). In addition, one of the planted species, *Robinia pseudoacacia* became a good source to produce honey, which generates income for the local people.

10.4 Case Study: Natural Regeneration by Indigenous Species in Exotic Species Plantations

In Korea, some plantations of exotic species are being replaced by indigenous species, such as *P. densiflora* and *Quercus* species (Lee et al. 2004). In our study area, *L. kaempferi*, an exotic fast-growing species was planted in the 1960s and *P. koraiensis* was planted in 1970s for timber production after logging the native *P. densiflora* and *Quercus* species. However, *P. densiflora* has regenerated naturally from a few mother trees that remained in the planted *P. koraiensis* and *L. kaempferi*. *Quercus* species have regenerated mainly by sprouting. *Pinus koraiensis* showed either good or poor growth in southeast and ridge-slope areas. When tending works on *P. koraiensis,* such as weed control and thinning, were conducted the plantation showed good growth rates but when tending works were not conducted, the plantation changed into *P. densiflora* or *Quercus* dominated stands (Table 10.2). Thus, tending works are necessary for *P. koraiensis* especially at early stand stages to avoid competition with native species on these sites (Shin, 1989).

L. kaempferi showed good survival and growth in fertile soils but survival and growth were poor on infertile soils (Kim, 1999). In our study, *L. kaempferi* showed good survival and growth only in northeast and valley areas with fertile soils and available moisture, but poor survival and growth in other sites where it was replaced by *P. densiflora* or *Quercus* spp. (Table 10.3). Apparently, *L. kaempferi* is more demanding of good site conditions because it needs much moisture and nutrients as a fast-growing species (Mackensen and Fölster, 2000).

Soil conditions seemed to be the main factors in determining vegetation growth and distribution (Shin et al. 2001). In the *P. koraiensis* and *L. kaempferi* plantations that exhibited good growth, the soil conditions in terms of depth of the A horizon, moisture content, organic matter, and total nitrogen, were better than soil conditions in plantations with suppressed growth and replacement by *P. densiflora* (Table 10.4).

The dominant species in the study plots are shown in Figure 10.2, as indicated by the importance value index (IVI). In plots where *P. koraiensis* was growing well, it accounted for more than 70% of the IVI because invading *Quercus* species were removed by continuous silvicultural treatments. In plots where silvicultural management was neglected, the IVI of *P. koraiensis* was only 20% and the plantation was a failure. Other species associated with the plantation of *P. koraiensis* were *Prunus sargentii, Zanthoxylum schinifolium, Morus bombycis, Lindera obtusiloba,* and *Styrax obassia,* in mid- and understory layers with little density.

Even where *L. kaempferi* grew well, it was less dominant than *P. koraiensis* with about 40% IVI. In plantations without tending treatments, other species such as *Carpinus cordata, Acer pseudosieboldianum, Cornus controversa, Fraxinus rhynchophylla, P. sargenti, Lindera obtusiloba,* and *Styrax obassia* accounted for about 40% of IVI. Where *L. kaempferi* was suppressed by indigenous species and replaced by *P. densiflora* or *Quercus* species, IVI of *L. kaempferi* was less than 10%. *L. kaempferi* is more shade-intolerant than *P. koraiensis* species, thus it had lower IVI and is more susceptible to suppression by native species.

10.4.1 Growth Pattern

Stem analysis is a useful tool for determining growth patterns of tree species and contributes to developing site index and height growth models (Wang et al. 1994). The total height growth of *P. koraiensis* in the better growing conditions was estimated to be 8 m at age 20; on the poor sites where growth of *P. koraiensis* was suppressed by *P. densiflora* or *Quercus* spp., height growth at the same age was only 2.94 or 2.71, respectively (Figure 10.3). The

TABLE 10.3

Description of the Study Plots in the Plantations of *P. koraiensis* and *L. kaempferi*

P. koraiensis Plantations

	Good Growth Areas			Poor Growth Areas					
				Replaced by *P. densiflora*			Replaced by *Quercus* Species		
Altitude (m)	320	250	320	230	290	375	260	265	275
Topography	Ridge-slope	Ridge-slope	Ridge-slope	Ridge	Ridge	Ridge-slope	Ridge-slope	Ridge-slope	Ridge-slope
Aspect (°)	S70E	S84E	S10E	S20W	S35E	S25W	S60E	S80E	N79E
Slope (°)	23	18	18	23	23	18	22	33	17
Stand age (years)		27–30			26–35			26–30	

L. kaempferi plantations

	Good Growth Areas			Poor Growth Areas					
				Replaced by *P. densiflora*			Replaced by *Quercus* Species		
Altitude (m)	290	240	300	320	200	270	240	270	220
Topography	Valley	Valley	Valley	Ridge	Ridge	Ridge-slope	Valley-slope	Ridge-slope	Valley-slope
Aspect (°)	N10E	N20E	N35E	S70E	N70E	S8E	S10E	S30E	S60E
Slope (°)	22	27	15	12	25	21	25	27	17
Stand age (years)		30–40			30–45			20–30	

TABLE 10.4

Soil Characteristics of the Study Plots in the Plantations of *P. koraiensis* and *L. kaempferi*

Properties / Study Plots	Soil Depth in A Layer (cm)	Texture	pH (1:5)	Moisture (%)	Organic Matter (%)	Total N (%)
Pk (A)	28	Silty loam	6.65	30.4	7.08	0.134
Pk (B)	—	Silty loam	6.70	28.1	4.10	0.066
Pd-P (A)	12	Sandy loam	5.88	17.8	4.46	0.057
Pd-P (B)	—	Sandy loam	6.56	20.3	3.65	0.039
Qs-P (A)	22	Sandy loam	6.44	28.8	6.22	0.099
Qs-P (B)	—	Sandy loam	6.86	24.3	4.29	0.045
Lk (A)	36	Silty loam	6.03	36.4	9.20	0.210
Lk (B)	—	Loam	6.75	37.4	6.77	0.120
Pd-L (A)	11	Sandy loam	5.04	17.5	5.02	0.065
Pd-L (B)	—	Sandy loam	6.93	12.1	2.56	0.021
Qs-L (A)	24	Sandy loam	6.03	27.8	6.48	0.189
Qs-L (B)	—	Silty loam	6.89	29.0	3.81	0.050

Note: Pk: *P. koraiensis* plantation on good sites; Pd-P: *P. koraiensis* plantation with suppressed growth and replacement by *P. densiflora;* Qs-P: *P. koraiensis* plantation with suppressed growth and replacement by *Quercus* species; Lk: *L. kaempferi* plantation on good sites; Pd-L: *L. kaempferi* plantation with suppressed growth and replacement by *P. densiflora;* Qs-L: *L. kaempferi* plantation with suppressed growth and replacement by *Quercus* species.

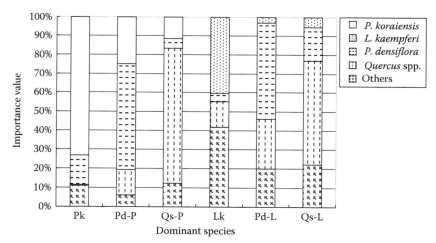

Pk: *P. koraiensis* plantation on good sites

Pd-P: *P. koraiensis* plantation with suppressed growth and replacement by *P. densiflora*

Qs-P: *P. koraiensis* plantation with suppressed growth and replacement by *Quercus* species

Lk: *L. kaempferi* plantation on good sites

Pd-L: *L. kaempferi* plantation with suppressed growth and replacement by *P. densiflora*

Qs-L: *L. kaempferi* plantation with suppressed growth and replacement by *Quercus* species

FIGURE 10.2

Importance value index (IVI) of major species in the *P. koraiensis* and *L. kaempferi* plantations where they showed well or suppressed growth.

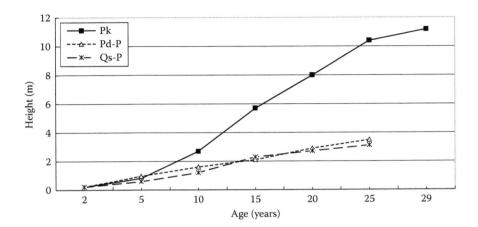

FIGURE 10.3
Total height growth of *P. koraiensis* at the plots where good (Pk) or suppressed growth by *P. densiflora* (Pd-P) or *Quercus* species (Qs-P) prevails.

total DBH growth of *P. koraiensis* was similarly greater on the better sites (14.05 cm at age 20) compared to the poor sites (2.75 or 2.15 cm).

The faster-growing *L. kaempferi* was taller than the *P. koraiensis* (17.20 m at age 30) on good sites. On poor sites with growth suppressed by *P. densiflora* or *Quercus*, height growth was 5.19 or 7.79 m, respectively (Figure 10.4). Comparing DBH growth of *L. kaempferi* in a similar fashion, trees on good sites were 16.25 cm at age 30 and only 4.2 cm and 7.85 cm on poor sites where growth was suppressed. Few trees of *L. kaempferi* remained at the plots where *P. densiflora* or *Quercus* species invaded.

Total volume growth of several species is shown in Figure 10.5. The volume of *P. densiflora* was greater than the *Quercus* species. Among the *Quercus* species, the volume was highest in *Q. mongolica*, then in descending order, *Q. variabilis* and *Q. serrata*. The low soil

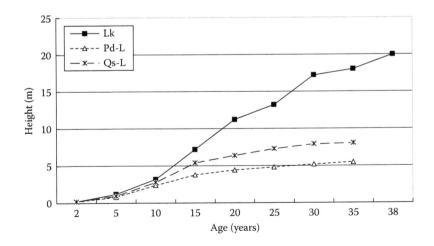

FIGURE 10.4
Total height growth of *L. kaempferi* at the plots where good (Lk) or suppressed growth by *P. densiflora* (Pd-P) or *Quercus* species (Qs-P) prevails.

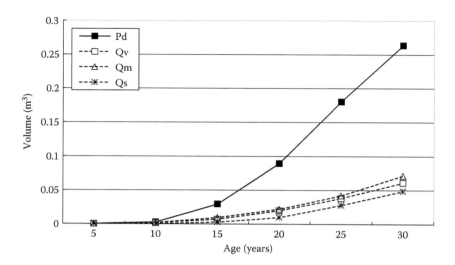

FIGURE 10.5
Total volume growth of *P. densiflora* (Pd) and *Quercus* species (*Q. variabilis*; Qv, *Q. mongolica*; Qm, and *Q. serrata*; Qs) in the plantations of *P. koraiensis* and *L. kaempferi*.

fertility in these sites favored *P. densiflora* (Park and Lee 1996) and the pine mushroom (*Tricholoma matsutake*), which only grows naturally in *P. densiflora* stands. The pine mushroom is one of the most valuable forest products due to difficulties in artificial cultivation, therefore we recommend maintaining naturally regenerated *P. densiflora* stands because of their ecological and economic value.

The *Quercus* species on this site were found to have a severe heartwood rot, probably due to their sprout origin. The *Quercus* has low commercial value for timber and needs effective silvicultural treatments to produce any value (Lee et al. 1997). The *Quercus* stands can be managed on short rotation to produce logs for cultivating oak mushrooms.

Figure 10.6 shows the site indices, which represent the relationship between height growth and age for each species in Korea. The dotted line in each of the figures indicates the site index of the major species occurring on this study site. The site index of *P. koraiensis* on this site was lower than found on other sites in Korea even though the measured tree was selected from the better site with continuous tending. This indicates that *P. koraiensis* is not a proper species for this area, which is affected by warm oceanic temperatures, and the high-cost of the need for continuous silvicultural treatments to produce timber renders it inefficient (Chon et al. 1999).

However, the site index of *P. densiflora* was higher than found elsewhere in Korea and indicates that *P. densiflora* is an appropriate native species for restoration in this area.

10.5 Conclusion

The shift of emphasis in restoration from plantations and exotic species toward native species and natural regeneration has been seen elsewhere (Healey and Robert 2003; Lee et al. 2005). The planting of *P. koraiensis* and *L. kaempferi* for timber production was conducted in this area in the 1960s and 1970s to restore forests degraded by half a century of conflict

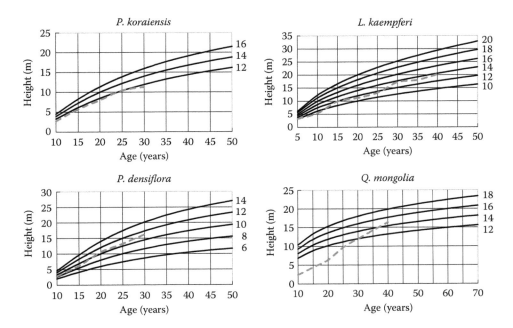

FIGURE 10.6
Site indices of major species in Korea (solid line) and in the study site (dotted line).

and exploitation. However, many *P. koraiensis* plantations failed due to the lack of tending works especially at early stages of stand development. Most of the *L. kaempferi* plantations, except in valley sites, also failed because of ignorance of detailed consideration of its site characteristics. Thus, indigenous species such as *P. densiflora* and *Quercus* species started to reoccupy the failed plantations and these areas have been replaced by the native species.

Although *P. koraiensis* showed suppressed growth where *P. densiflora* or *Quercus* species invaded in the plantation, it still remained with high density. However, few trees of *L. kaempferi* survived where *P. densiflora* or *Quercus* species invaded. The recommendation is to manage the native *P. densiflora* or *Quercus* species that have established in the failed exotic plantation of *P. koraiensis* and *L. kaempferi*. Where the *Quercus* species have developed severe heartwood rot, they should be managed on short rotation to produce logs for oak mushrooms. *Pinus densiflora* has exhibited a better growth rate than any of the *Quercus* species and a better than average site index for Korea. This explains why *P. densiflora* became the dominant species and occupied the site. It is an important native species for restoration in this area, and elsewhere in Korea.

References

Choi, K.R. 2001. Vegetation and climate history of the lowland on the Korean peninsula *Journal of the Korean Physical Society* 39: 762–765.

Choi, K.R. 2002. *Vegetational and Climatic History of the Lowland in Korea*, pp. 61–70, In Lee, D.W. (ed.) Ecology of Korea. Bumwoo Publishing Company, Seoul, 406pp.

Chon, S.K., M.Y. Shin, D.J. Chung, Y.S. Jang, and M.S. Kim. 1999. Characteristics of the early growth for Korean white pine (*Pinus koraiensis* Sieg. et Zucc.) and effects of local climatic conditions on the growth. *Journal of Korean Forestry Society* 88(1): 73–85.

Gyeongsangbuk-do. 2002. Erosion control of Gyeongsangbuk-do Province. 140pp. (in Korean).

Gyeongsangbuk-do. 1977. History of Yeongil district erosion control project. 286pp. (in Korean).

Healey, S.P., and I.G. Robert. 2003. The effect of a teak (*TeSSSctona grandis*) plantation on the establishment of native species in an abandoned pasture in Costa Rica. *Forest Ecology and Management* 176: 497–507.

Jordan, III., W.R. M.E. Gilpin, and J.D. Aber. 1997. *Restoration Ecology–A Synthetic Approach to Ecological Research-*. Cambridge University Press. Cambridge, UK, 342pp.

Jo, W.R. 1979. Palynological studies on postglacial region, Korea peninsula. *Annuals of the Tohoku Geographical Association* 31: 23–35. (in Japanese).

Kim, C.S. 1999. Aboveground nutrient distribution in pitch pine (*Pinus rigida*) and Japanese larch (*Larix leptolepis*) plantations. *Journal of Korean Forestry Society* 88(2): 266–272.

Korea Forest Research Institute. 1997. In-depth country study in the republic of Korea—status, trends and prospects to 2010. Asia-Pacific Forestry Sector Outlook Study Working Paper Series APFSOS/WP/06. FAO.

Korea Forest Service. 1993. *Statistical Yearbook of Forestry.*

Korea Forest Service. 2000. *Green Korea.* 29pp.

Korea Forest Service. 2010. *Vision for Green Korea.* 47pp.

Korea Forest Service. 2013. *Statistical Yearbook of Forestry.* 486pp.

Lee, C.Y. 2001. Traditional Knowledge for Soil Erosion Control in the Republic of Korea. Korea Forest Research Institute, No. 184. 110pp.

Lee, D.K., H.S. Kang, and Y.D. Park. 2004. Natural restoration of deforested woodlots in South Korea. *Forest Ecology and Management* 201(1): 23–32.

Lee, E.W.S., B.C.H. Hau, and R.T. Corlett. 2005. Natural regeneration in exotic tree plantations in Hong Kong, China. *Forest Ecology and Management* 212: 358–366.

Lee, K.J., D.K. Lee, and I.Y. Park. 1997. Development of low input sustainable forestry system using natural oak resources. Pages 149–162 in *Proceedings of the Strategies for Development of Plant Environment and Growth.* 28 November 1997, College of Agriculture and Life Sciences, Seoul National University, Suwon, Korea.

Mackensen, J. and H. Fölster. 2000. Cost-analysis for a sustainable nutrient management of fast growing-tree plantations in East-Kalimantan, Indonesia. *Forest Ecology and Management* 131: 239–253.

Matsushima, S. 1941. Betrachtung zur Waldentwicklung in Korea auf grund von pollenststistik. *Journal of the Japanese Forestry Society* 23: 441–450 (in Japanese).

Nelson, S.M., *The Archaeology of Korea.* Cambridge University Press, Cambridge, 1993.

Oh, C.Y. 1971. A pollen analysis in the peat sediments from Pyungtack country, Korea. *Korean Journal of Botany* 14, 126–133. (in Korean).

Park, P.S. and D.K. Lee. 1996. Factors affecting the early natural regeneration of *Pinus densiflora* S. et Z. after forest works at Mt. Joongwang located in Pyungchang-Gun, Kangwon-Do. *Journal of Korean Forestry Society* 85(3): 524–531.

Sayer, J., U. Chokkalingam, and J. Poulsen. 2001. The restoration of forest biodiversity and ecological values. pp. 1–12 in *Proceeding of International Seminar on Restoration Research on Degraded Forest Ecosystems.* 13–14 April 2001, Seoul National University, Seoul, Korea.

Shin, J.H. 1989. Crown architecture and differentiation in tree classes, their growth strategies and growth model in *Pinus koraiensis* S. et Z. plantations. Ph.D. Dissertation, Department of Forest Resources, Seoul National University, Seoul, Korea.

Shin, J.H. 2002. Ecosystem geography of Korea. *Pages 19–46, in* D.W. Lee et al. (*eds.*) *Ecology of Korea.* Bumwoo Publ. Co., Seoul. 406pp.

Shin, M.Y., S.Y. Chung, and D.K. Lee. 2001. Estimation of microclimate by site types in natural deciduous forest and relation between periodic annual increment of diameter and the microclimatic estimates: A case study on the National Forest in Pyungchung, Kangwon Province. *Korean Journal of Agricultural and Forest Meteorology* 3(1): 44–54.

Stanturf, J.A., B.J. Palik, and R.K. Dumroese. 2014. Contemporary forest restoration: A review emphasizing function. *Forest Ecology and Management* 331: 292–323.

Urbanska, K.M., N.R. Webb, and P.J. Edwards. 1997. *Restoration Ecology and Sustainable Development.* Cambridge University Press, Cambridge, 397pp.

Wang, G.G., P.L. Marshall and K. Klinka. 1994. Height growth pattern of white spruce in relation to site quality. *Forest Ecology and Management* 68: 137–147.

Yamazaki, T. 1940. Beitrage zur verwandlung der baunarten im sudlichen teile von Korea durch die pollenanalyse. *Journal of the Japanese Forestry Society* 22: 73–85. (in Japanese).

Yasuda, Y., M. Tsukada, C.M. Kim, S.T. Lee, and Yim, Y. J. 1980. History of environmental changes and origin of agricultural in Korea: History of environmental changes in Korea. Nippon Monbushyo Kaigai Gakujutsu Report No. 404332, 1–19. (in Japanese)

11

Forest Landscape Restoration in China: A Case Study in the Minjiang River Watershed, Southwest China

Shirong Liu, Jiangming Ma, and Ning Miao

CONTENTS

11.1 Introduction

China is a vast country with a diverse physical environment. With a few exceptions, China contains all of the main forest vegetation types of the northern hemisphere. Widespread ecological degradation has constrained sustainable socioeconomic development in recent decades, particularly in the period before the end of twentieth century (Li 2004). For instance, in the early 1980s and 2000s, 23% of the land area (on which approximately 35% of the Chinese population depended for ecosystem services) suffered ecological degradation (Lü et al. 2012), including a reduced capacity for carbon sequestration. The estimated economic costs of interrelated problems associated with this degradation, including resource depletion, environmental pollution, and ecological damage, have amounted to over 13% of the national gross domestic product (Shi et al. 2011).

Timber production was the main goal of forest management until recently. Increasing economic pressures and over-exploitation resulted in ecologically sensitive natural forests being harvested with little or no regeneration and significant loss of ecosystem services of the remnant forests. In the wake of the 1998 floods in the Yangtze River basin, focus shifted from timber production to environmental protection, with policy redirected toward the rehabilitation of damaged forest ecosystems, reforestation in cleared or degraded areas, and a ban on logging in natural forests (Liu 2011). The Chinese government implemented a series of policies toward ecological restoration that included formal approval and implementation of six key State Forestry development programs: (1) Natural Forest Protection Program (NFPP), (2) Key Shelterbelt Construction Program, (3) Grain to Green Program (GTGP), (4) Desertification Control Program, (5) Conservation of Biodiversity and Nature Conservation Construction Program, and (6) Establishment of Fast-growing and High-yielding Timber Plantations (Wu et al. 2008). The idea of ecological restoration of degraded ecosystems runs through these actions.

GTGP, launched in 1999, is the largest land retirement program in the developing world (Delang and Yuan 2015). This program uses a public payment scheme that directly engages millions of rural households as core agents of project implementation. This is distinct from China's other soil and water conservation and forestry programs because it is one of the first, and certainly the most ambitious "payment for ecosystem services" program in China (Li 2009). From 1998 to 2008, the Chinese Central Government directly invested 191.8 billion RMB (approximately 28.8 billion USD) to implement the GTGP and involved 120 million farmers in retiring and restoring 9.27 million ha of sloping croplands (Yin et al. 2009).

More needs to be fulfilled in order to improve our understanding of program implementation and impacts, and future work should pay close attention to the NFPP and other programs, especially concerning their environmental impacts and implementation effectiveness. To these ends, ecologists and foresters must gather more data on changing ecosystem conditions and socioeconomic circumstances across scales and disciplines, applying geospatial technology and more effective modeling (Yin et al. 2009).

Forested ecosystems are complex dynamic systems characterized by multiscale heterogeneity that highlights the need for a holistic ecosystem-level approach to landscape restoration (Naveh 2000). The upper reaches of the Minjiang River watershed in Southeastern China are important for environmental protection and social-economic well being in both the watershed and the entire Yangtze River basin. Due to its strategic significance, severe deforestation in the past, and high sensitivity to climate change, the watershed has been recognized as one of the highest priority watersheds in China for scientific research and resource management (Cui et al. 2012). Forest restoration in this area is a critical element for sustainable forest land management and provision of watershed ecosystem services. In this chapter, we illustrate an eco-hydrological approach to ecological restoration as it is being applied toward the Minjiang River watershed.

11.2 Key Issues and Challenges Facing Ecological Restoration in China

At present, little research has been done in China on the mechanisms underlying the restoration processes of degraded natural forest ecosystems. Likewise, there has been little research on how to accelerate the recovery of ecosystem structure and function from secondary forest to old-growth forest features. Although much is known of the ecological benefits of plantations, our overall understanding of the roles and benefits of planted forests has changed from a focus on timber outputs toward multiple ecosystem goods and services. Therefore, more research is needed on how to induce the natural transformation of plantations to "close-to-nature" forestry (Green 2001; Bauhus et al. 2013; Larsen 2013). For the successful ecological restoration of degraded natural forests, a number of important questions need to be addressed (Liu 2011):

1. Where the forest has been fragmented and old-growth forest has been lost, how to protect and restore genetic diversity?
2. Where ineffective natural or artificial regeneration has resulted in an unsatisfactory secondary forest, how to accelerate restoration of ecological function and productivity?
3. Where the original dominant community has been replaced with plantations, how to transform these plantations to the original dominant community?
4. How to conduct forest restoration within multiobjective landscape planning in order to achieve the optimal allocation of landscape pattern in the process of land use and land cover change in a natural forest area?
5. Where there has been serious degradation of the forest and loss of the soil seed bank, how to restore soil biological community and function and subsequently the natural forest community?

Different approaches to forest restoration raise a number of concerns, such as the effectiveness of a simple revegetation approach; restoration only focused on a stand-level or at a site-level; lack of systematic approaches targeting ecosystem structure and function; unbalanced consideration of ecological, social and economic benefits from restoration; and lack of an integrated landscape restoration approach carried out at multiple scales (Stanturf et al. 2014b). Many artificially established monoculture forests have low biodiversity and are unstable. Nonnative species have been widely imported and planted. Features required by healthy ecosystems, such as heterogeneity and mutual biotic interactions among different species, are often ignored. Rare and endangered forest climax species are seldom considered in reforestation. The ecological function of vegetation is often neglected in cities and urban landscapes (Ren et al. 2004). It is uncertain how ecological restoration should be applied over the following wide range of final objectives: ecological restoration driven fully by a natural succession process, or natural ecological restoration driven by human-aided processes, or totally human-designed restoration with a goal to establish a new artificial ecosystem that has not existed historically but that provides a better fit for human needs or expectations.

A remaining challenge is how to couple environmental objectives with poverty reduction objectives in programs launched by the Chinese government. Almost every ecological restoration program is treated as a mechanism to improve environmental quality and provide financial aid to the poor. Achieving an optimal solution to a problem with multiple goals needs to use multiple instruments. This suggests that it may be necessary to investigate the extent to which these goals are compatible with each other. It is still unclear how the poverty alleviation objective has influenced the environmental objective, and whether and how the outcome can be improved by using multiple instruments (Yin et al. 2009).

11.3 Integrated Landscape Restoration Planning

In many parts of the world there is an urgent need for landscape restoration to conserve biodiversity (Lindenmayer et al. 2002; Stanturf et al. 2014a; Grunewald et al. 2014). Landscape restoration is not straightforward, however, because many issues and processes must be understood for effective action to take place (Lindenmayer et al. 2002). Understanding the structure and function of the natural forested ecosystems forms the necessary basis for all forest restoration activities. Knowledge of the composition, structure and function of natural forests—both the average values and historical range of variation—is needed to set goals for restoration and to evaluate the success of particular restoration actions (Baskerville 1995; Holling 1995). However, defining the natural forest is not a simple task. Especially in ecosystems that are frequently disturbed such as the boreal forest, we often lack knowledge about the range of natural variability of the forest structure in the area to be restored. Even if we have this knowledge, a natural forest may be so variable over time that it does not provide any static targets for restoration (Baskerville 1995).

To overcome the uncertainty of what constitutes the natural forest target for restoration, goals can and should vary to cover the natural range of variability, which in turn can be defined using existing information from multiple sources. Potential sources of information include (1) local analyses of biological archives by ecological methods, (2) retrospective analyses of forest structure based on historical materials, (3) research on ecologically similar but more natural forests, and (4) modeling.

Successful restoration of forested landscapes requires long-term planning and multiple interventions. However, as a result of the long periods of time involved, the occurrence of unexpected events becomes an inevitable companion of restoration. This is partly due to our ignorance of ecosystem functioning, which restricts the ability to precisely predict the outcomes of restoration activities (Baskerville 1995). Another reason, even more fundamental, is that the occurrence of unexpected events is related to stochastic features of the environment and to the many nonlinear relationships among processes that regulate ecosystem dynamics. Therefore, all restoration planning should acknowledge that surprises can occur and try to buffer the ecosystem to be restored against surprises, such as abrupt environmental changes. One way to do this is to restore larger ecosystem complexes instead of small areas.

Restoration ecology is a science that is closely connected to practice, but practice should also be closely connected to science (Holling 1995). Practical restoration projects should be closely linked with monitoring and research whenever possible. Monitoring enables us to adjust what we do in order to better achieve our goals (that is the key for adaptive management). Incorporation of research into management generates synergy benefits, for example, by making it possible to set up experiments on scales that are relevant, both ecologically and managerially. It also helps in ensuring the formation of a basis of knowledge of the long-term effects of restoration, which in turn can be used in planning future restoration efforts (Holling 1995; Gunderson and Holling 2002).

11.4 Forest Landscape Restoration in Minjiang Watershed

Restoration of the forests in the upper reaches of the Minjiang River watershed aims at rehabilitating forest structures and functions for natural forest ecosystems. Restoration actions are typically discrete events in time, but they aim at enhancing long-term developmental processes, such as forest succession. Restoration of forest ecosystems may be focused on species, structures, or dynamics. However, as all these aspects of forest ecosystems are closely interrelated, it is not feasible to focus only on one aspect, such as species, without simultaneously considering other aspects of the ecosystem. This is particularly true for the entire Minjiang River watershed as different ecosystems or landscape components are ecologically and physically connected by eco-hydrological processes (e.g., evaporation and transpiration, evaporation-related rainfall, soil water infiltration, and movement) among different ecosystems (e.g., grasslands, meadows, shrublands, and forests). Situated at headwater areas, these alpine meadows and subalpine forests in western Sichuan play an important role in regulating the hydrological balance and conserving soil and water (Zhang et al. 2007; Cui et al. 2012).

11.4.1 General Situation of Minjiang River Watershed

The watershed of the Minjiang River, the largest tributary of the Yangtze River, is located at an extension of the Qinghai-Tibetan Plateau (31°–33°N, 102°–104°E) (Figure 11.1). The Minjiang watershed is an important ecological zone of Sichuan Province and the west of China for biodiversity and water resource conservation, which includes five counties: Songpan, Heishui, Maoxian, Lixian, and Wenchuan. The area is about 24,000 km². The climate is characterized by cool summers and cold winters. Annual precipitation in this region is approximately 491–836 mm. Mean annual temperature is 3.0°C. The mean temperature during the warmest month (July) is 12.6°C, dropping to −8.0°C during the coldest month (January). Evaporation varies from 1100 to 1600 mm per year, and the relative humidity is in the range of 62%–72%. The elevation of the region is 2200 to 5500 m above sea level (masl). The terrain is steep, complex, and deeply dissected and the soils are predominantly brown forest soils.

Vegetation types vary with elevation. Forests occur in the valleys and on mountain slopes (from 2600 m to 4000 masl). The dark coniferous forest type is identified and managed as a water conservation forest. Forests in the region were extensively logged beginning in the 1950s, changing dramatically over the past 60 years. In the early 1950s, forests were dominated by old growth *Abies* species, with associated species of *Picea*, *Betula*, and others. Large-scale logging occurred in the region between 1954 and 1965. Clearcutting was the predominant harvesting method. After 1965, the scale of harvesting and annual yields of timber dropped sharply, and harvesting ceased in 1998 when the NFPP and GTGP began

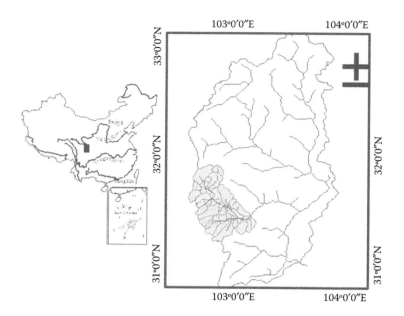

FIGURE 11.1
The location of the upper reaches Minjiang River watershed in China. The dark color area is Miyaluo, which is the study area.

in the region (Zhang et al. 2013). The long period of extensive timber harvesting led to great changes in forest cover, landscape pattern, and microclimate that seriously influenced ecosystem services (e.g., soil eco-hydrological conditions, biodiversity conservation, wildlife habitat, and timber production).

11.4.2 Ecosystem Degradation

Following harvesting, old-growth fir forests were replaced by naturally regenerated broadleaved forests, croplands, grasslands, shrublands, or monoculture plantations. The result was increasing problems of floods, drought, degradation of aquatic habitats, and extinction of fish species, soil erosion and inadequacy of water resources.

The current forests show a mosaic landscape pattern comprised of pure plantations (e.g., spruce plantation forest, SPF), naturally regenerated forest (e.g., natural birch forest, NBF; fir and birch mixed forest, FBM), mixed man-made and naturally regenerated forest (spruce and birch mixed forest, SBM), and remaining old-growth fir forest (OGF), with different ages, densities, and species composition. After clearing and burning of logging debris, 3- to 5-year-old seedlings of *Picea asperata* were planted on many clear-cut sites. The plantations were periodically treated to maintain them as monoculture stands. Top soil was harrowed twice a year, and grasses and shrubs were cleared once a year during the first 8 years after planting. After that, shrubs were cleared up once every 3–4 years. Most SPFs were thinned to stimulate tree growth for increasing stand volume.

Since 1998, the term "ecological public-welfare forest" has been gradually recognized and advocated in China, and, therefore, most SPFs in the upper reaches of the Minjiang River watershed, which were originally aimed at timber production, have been reclassified as ecological public-welfare forest. However, since SPFs have low biodiversity, all silvicultural practices aimed at timber production with SPFs have been stopped in order to increase biodiversity. In the 1960s and 1970s, many SPFs failed to maintain monoculture structure due to inadequate management; SBMs gradually developed as a result of the invasion by *Betula albosinensis*, *B. utilis*, and *Acer* species (Zhang et al. 2013).

11.4.3 Eco-Hydrological Characteristics of Typical Water-Conservation Forests

Water yield and eco-hydrological characteristics of degraded forests and different restoration approaches can be fundamental to guide strategies of water conservation forests and watershed restoration. Studies of the eco-hydrological functions of forest ecosystems have been conducted, including old-growth forests dominated by *Abies* spp. (Ma 1963, 1987; Jiang 1981; Zhang et al. 2009, 2011), secondary broad-leaved forests (Zhang et al. 2004, 2009), shrubs (Zhang et al. 2006, 2009), and *Picea* spp. plantations (Hu and Liu 2001; Pang et al. 2003; Zhang et al. 2011).

11.4.3.1 Old-Growth Forests and Secondary Naturally Regenerated Forests

Alpine old-growth forests dominated by fir (*Abies* spp.) are adapted to the cold and wet conditions in the upper watershed, and play an important role in regulating watershed hydrology, including base flow and peak flows. Due to relatively lower evapotranspiration (ET) rates and higher water holding capacity, mature forests (over 200 years old) have large catchment water yield even though they have high leaf area index and accumulate high levels of biomass. The annual ET of fir forests is quite low compared with shrubs, secondary broad-leaved forests and spruce plantations, accounting for only 5% of the annual precipitation (Zhang et al. 2004).

After logging of old-growth forests, the secondary forests comprised mainly of birch (*Betula* spp.) regenerated naturally. The maximum water holding capacity (MWHC, %) value of litter and moss in the secondary forests was 945% for litter and 573% for moss. The MWHC values of the surface soil (0–40 cm) were not significantly different among stands with different ages, however, significant differences were found among the stands located at different elevations. The moss in the secondary forests normally needs a long period to recover the original status of the old-growth forests after a clear cut.

11.4.3.2 Spruce Plantations and Shrubs

The standing mass of litter in spruce plantations from the 1960s was one of the highest among different aged plantations, and higher than litter mass in older plantations from the 1940s (60 a) (Hu and Liu 2001). Soil water content and soil porosity was also lower in 1940s spruce plantations (60 a). The water-holding capacity of moss and litter layers increased with age in the order of 30 < 40 < 60 < 50 a of the plantations. Water-holding capacity of the litter layer is greater than that of the moss layer (Lin et al. 2002; Table 11.1).

TABLE 11.1

Water-Holding Capacity of Vegetation Types in the Minjiang River Watershed

Vegetation Type	Water-Holding Capacity of the Moss Layer (t/ha)	Water-Holding Capacity of the Litter Layer (t/ha)
Spruce plantation	44.8	182
Rhododendron scrub	46.7	140
Oak scrub	1.64	72

Alpine and subalpine shrubs, distributed above tree line or on south-facing slopes, are the major vegetation components in the watershed. The accumulated mass (CM, t/ha) of moss and litter layers, along with their maximum water holding capacities (MWHC, t/ha) and maximum water holding rate (MWHR, %) of the three main types of shrub (*Rhododendron przewalskii*, *Quercus aquifolioides*, and *Quercus cocciferoides*) at different elevations in western Sichuan were conducted (Zhang et al. 2006). The water-holding capacity of *R. przewalskii* shrub is highest among the three shrub types (*R. przewalskii*, *Q. aquifolioides*, and *Q. cocciferoides*). The CM and MWHC of moss decreased significantly with increasing elevation in the *R. przewalskii* scrub type, and reached the maximum at 3400 m in the *Q. aquifolioides* shrub type.

11.4.3.3 Water Yield of Different Forests

There are large variations in ET, and consequently in water production among different types of forests. Mature or old-growth natural coniferous forests had the lowest ET (39.5%–43.8%), followed by shrubs, broadleaved forests, mixed coniferous and broadleaved forests, and coniferous plantations (Zhang et al. 2011). The low ET may also be related to the foggy and moist climate in the high-elevation topography (Liu et al. 2001). The lowest ET in areas of the watershed dominated by old coniferous forests had the highest water yield potentials. The highest ET in spruce plantations suggests that increasing plantation area could reduce catchment water yield. An isotope-based study to identify water sources in seven subwatersheds of the Minjiang River showed that reducing the cover of total vegetation in mixed forests and subalpine coniferous forests would increase surface and subsurface water yields; water yield was higher, however, with the alpine shrub and meadow cover (Liu et al. 2006).

11.4.3.4 Landscape Hydrology

Landscape hydrology, which is largely dependent on climate and vegetation, and topography across the whole watershed, contributes to our understanding of the water budget and water cycle at watershed scale. It covers not only forest-related hydrological processes but also includes those influenced by other landscape components. Natural shrub and meadow vegetation types comprise a large proportion of vegetation cover in the watershed, exceeding even the coniferous or broad-leaved forest cover. The annual watershed-scale ET is higher in fast growing shrubs and grasses than in the coniferous or broad-leaved forests in the watershed. Also, glaciers and permanent snow cover are patchily distributed at the higher elevations, acting as an important water source in the Minjiang Valley. The isotope-based source study in one subwatershed (Heishui) showed that glaciers and snow meltwater contributed a large proportion of base flow in the river network, ranging from 63.8% to 92.6%; while, rain contributed from 7.4% to 36.2% in the wet season (from May to June) (Liu et al. 2008a,b). The isotope source analysis confirmed a close hydrological connection between subalpine forests and alpine meadows. A large proportion of the rainfall in the alpine meadows derived from secondarily evaporated water. Fog derived from the evaporated water was produced shortly after rainfall events. This suggests that evaporated water falling as precipitation in the alpine meadows largely originated from subalpine forests at lower elevation, constituting a unique natural hydrological process in the landscape system. These ecohydrological linkages between subalpine and alpine highlight the important role of the alpine meadows in regulating watershed hydrology and should be taken into consideration for ecological restoration and watershed management at the landscape level.

11.5 Ecological Restoration Practices to Restore Degraded Ecosystems

Much is known about the upper reaches Minjiang River watershed. For example, effects of logging on soil and water conservation of alpine dark coniferous forest (Ma 1963; Shi et al. 1988), width of edge influence of the agricultural-forest landscape boundary (Li et al. 2007), landscape changes (He et al. 2006), impacts of land-use change (Zhang et al. 2007, 2008), spatial pattern analysis on the structure of secondary forests (Miao et al. 2008, 2009), dynamics of stand biomass and volume of the tree layer (Zhang et al. 2012), effects of different restoration practices on water conservation and water yield (Zhang et al. 2004, 2006, 2011), the roles of remnant trees (Miao et al. 2014), and aboveground carbon stock evaluation (Liu et al. 2012; Zhang et al. 2013). Based on current and past research, taking into consideration disturbance regime, degradation, and restoration status, we analyzed key issues relating to ecological restoration of the degraded natural forests in the upper reaches of the Minjiang River watershed. By integrating ecological theory, restoration practice, and the traditional Chinese concept of "restricting access to hills for forest conservation," we developed an approach based on the following principles:

- Use the species composition and community structure of old-growth stands as the reference target.
- Use spontaneous succession (natural regeneration) as much as possible.
- Introduce appropriate key species at critical successional stages.
- Use indigenous species as much as possible.

The degradation process can be reversed, or a given succession stage can be skipped or shortened, thereby accelerating the restoration process. This approach resulted in five ecological restoration practices appropriate to different objectives. Common to all five practices were measures to protect forest sites by restricting access and prohibiting tree-cutting, logging, hunting, and grazing. These measures included (1) confirming the forest types and delimiting their area; (2) marking the area with flags placed on the ridge, at intersections with the river, and at the main traffic intersections near the restricted area boundary; (3) fencing areas frequented by people and livestock; (4) constructing enclosures around sites vulnerable to disturbance with iron wire, bamboo, or wood fencing; (5) and employing forest rangers to patrol, with the number of rangers needed based on the size of the restricted area and the level of disturbance by people or livestock. The extent of restrictions on access should be varied as needed for relevant objectives.

11.5.1 Conservation of Natural Forests

Conservation of natural forests, old-growth forest and forest with completely natural regeneration is accomplished by restricting access and limiting disturbances (Figure 11.2). Absolute restrictions on access and activities in these areas are enforced for 10 years, or longer as needed. Through strictly restricted access (along with prohibiting harmful activities), ecosystem structure and function can be conserved or passively restored.

11.5.2 Restore Degraded Land

In severely degraded sites, such as shrub land and grassland (Figure 11.3), where natural forest has been completely lost, human-assisted measures must be taken to restore the

FIGURE 11.2
Natural forests, old-growth forest and forest with completely natural regeneration in Miyaluo in the upper reaches of Minjiang River watershed.

forest environment in terms of soil function and reintroducing the pioneer species needed to reinitiate a natural forest succession process. Restoration procedures include the following activities: (1) Strictly enforce restricted access for 7–9 years. (2) Based on the conditions of sites suitable for reforestation, select suitable tree species to initiate forest succession. Plant seedlings or saplings of the required tree species in rectangular holes (length of 0.4–0.5 m, width of 0.4–0.5 m, and depth of 0.3–0.4 m). For needle-leaved trees, use 3- to 5-year-old seedlings and for broad-leaved trees, plant 2- to 3-year-old seedlings in spring. (3) Monitor progress of the planted trees and take special measures if needed, such as providing shade or water or changing the microhabitat conditions of the seedlings.

FIGURE 11.3
The shrubland and grassland in Miyaluo in the upper reaches of the Minjiang River watershed.

11.5.3 Regulate Species Composition of Naturally Regenerated Secondary Forest

For moderately degraded secondary naturally regenerated forest, appropriate measures may be needed to facilitate its natural succession process and to enhance provision of ecosystem service by regulating the species composition at early successional stages. This applies to forests with inappropriate species composition, density, or nonforest land with good conditions for natural seeding.

In addition to restricting access, the following activities are undertaken: (1) The duration of restricted access is based on the height of the shrubs and the height increments of the planted trees. Restrict access for broad-leaved forest 5–7 years; conifers for 7–9 years. (2) In young forests, remove shrubs and herbs, and then loosen soil to ensure that the tree seedlings and saplings have sufficient nutrients and space to grow. (3) For nonforest land with good conditions for natural seeding, seeds may not reach the soil because of heavy shrub and herb cover or thick litter layer; therefore, dig holes and plant seedlings. (4) For the sites with an uneven distribution of seedlings and saplings or poor regeneration, dig rectangular holes and clear away shrubs and weeds within 1 m² around the hole. Plant 3- to 5-year-old seedlings of needle-leaved trees (spruce and fir), or with 2- to 3-year-old seedlings of broad-leaved trees (birch and poplar). (5) In land appropriate for reseeding, plant seeds in small holes to a depth of 0.2 m, and clear away shrubs and weeds near the hole within 1 m². Replanting and reseeding should be done in the spring.

11.5.4 Regulate Structure in Overly Dense Naturally Regenerated Secondary Forests

The main objective for regulating structure is to reduce unusually high density, especially of even-aged trees, or undesirable natural regeneration of target species. For example, this includes naturally regenerated birch forests with unusually high densities that can hinder the growth of the birch and block the regeneration of later successional species, such as spruce and fir (Figure 11.4). The main aim is to reduce the competition among individuals and promote undergrowth vegetation to form a natural forest community with high biodiversity and a multilayer structure.

Restoration activities include the following: (1) Strictly restrict access for 4–6 years for young forests. The duration of the restricted access depends on the growth of the trees and the threat of the disturbance. (2) Determine target stand density based on the stand canopy cover. Under the criterion of stand within 50%–70% of canopy cover, calculate the

FIGURE 11.4
The natural secondary forests in Miyaluo in the upper reaches of the Minjiang River watershed.

FIGURE 11.5
Pure plantations in Miyaluo in the upper reaches of the upper reaches of the Minjiang River watershed.

numbers of trees that need to be regulated. Remove suppressed and weak trees; keep the stand density of target species above 200 stems per ha.

11.5.5 Rehabilitate Pure Plantations

For pure plantations (Figure 11.5), in particular pure conifer plantations, dense canopy cover and simplified structure should be changed in order to encourage natural regeneration processes to adjust species composition toward close-to-nature reference stands. It is strongly suggested that landscape pattern and heterogeneity be taken into consideration for restoring degraded natural forests, that is, optimal land use pattern and land cover configuration by forest landscape planning (Liu et al. 2009).

Procedures include these activities: (a) Restricted access is enforced for 7–9 years for young forests. (b) Use thinning to decrease stand density and increase light conditions inside the stands, so as to promote the growth of the understory. (c) Regulate stand canopy cover through hemispherical photographs at the criterion of 70%. The objects to be regulated are the suppressed and weak trees. (d) Seed or plant shrubs and herbs based on the specific local circumstances and, at the same time, replant and tend the target species to improve the species composition and reduce stand density in order to develop a more natural community with high biodiversity and multilayer stand structure. (e) Select mainly indigenous species for replanting and keep the replanting density above 200 individuals per ha.

11.6 Research Needs

Research on ecological restoration on mountain forest landscapes is needed on two broad fronts: obtaining a better understanding of the mechanisms underlying forest landscape dynamic and improving the effectiveness of restoration approaches. Process-based landscape models address the interactions of landscape pattern and landscape processes in the context of a heterogeneous land surface. Well-structured landscape models appropriate to targeted landscape restoration are needed. Because of nonlinear characteristics and scaling effects of terrestrial ecological processes across the forest landscape, process-based ecosystem models based on slope profile or small catchments cannot make explicit predictions of landscape processes over a large area. At the same time, landscape functions cannot be quantitatively evaluated if landscape patterns and ecological processes are not adequately coupled. Several issues should be explored in future forest landscape restoration research: developing process-based landscape models with comparable scaling of spatial scale and temporal features and integrated with geographic information systems. Further, high resolution data sets are needed for land use, land-use change, and vegetation cover, along with other environmental factors.

The need remains for both basic and applied research to set general restoration strategies and goals and for developing monitoring and research to document the responses of ecosystems to restoration activities. Monitoring and research should be essential components of long-term restoration projects. At the moment, there is a lack of research results dealing with different aspects of restoration in subalpine forest ecosystems. Restoration projects should be organized so as to enable experimental testing of methods. This would allow the continuous accumulation of knowledge that can be used to direct the restoration efforts more efficiently in the future. From the ecological restoration point of view, four important areas of research emerge.

11.6.1 Structure, Dynamics, and Species Composition of Natural Forest Mosaic Landscapes

Knowledge of the natural variability of the structure and dynamics of natural forests forms the necessary reference and background for all restoration activities. At present, our limited understanding of the natural variability of the structure and dynamics of natural forest landscapes makes it difficult to set restoration goals and assess restoration results. Above all, we lack a full understanding of interactions between different disturbance agents and the long-term cumulative effects of disturbance dynamics in natural forest ecosystems. Ultimately, we should be able to define landscape-specific targets for restoration, since each forested landscape is likely to be a special case. However, in many cases the direction of restoration actions is evident, and, during long-term restoration projects, the restoration methods and goals can be modified based on monitoring and new research results (adaptive management).

11.6.2 Conservation of Old-Growth Remnant Trees after Logging

Disturbances are integral features of ecosystems, and most ecosystems are subject to several disturbance regimes that take place at different temporal and spatial scales (Turner et al. 1998). Biological legacies have been defined as "the organisms, organic materials, and organically generated patterns that persist through a disturbance and are incorporated into the recovering ecosystem" (Franklin et al. 2000). Remnant trees, the trees that survived from large-scale disturbances (hurricane, fire, volcano, earthquake, logging, grazing, etc.), act as biological legacies, since they are persisting representatives of organically-generated patterns which may act to influence the recovery of ecosystem composition and function (Franklin et al. 2000; Keeton and Franklin 2005; Herrera and García 2009). Therefore, remnant trees play not only an important role at early stages of forest regeneration but also may have a profound impact on the structure and dynamics of the secondary forest that becomes established (Keeton and Franklin 2005; Miao et al. 2014).

The deliberate retention of remnant trees can be incorporated in practical forest management, that is, the application of green-tree retention (Lõhmus et al. 2006; Rosenvald and Lõhmus 2008) in forest harvesting in order to reduce the damages to ecosystem services and functions caused by extensive tree removal management. While leaving remnant trees may appear to reduce short-term commercial profits, the application of "green-tree retention" serves to enhance wildlife habitat, biodiversity, and ecosystem function, including soil protection and nutrient retention, and, ultimately, to foster the natural regeneration of the forest ecosystem (Franklin et al. 2002).

11.6.3 Aboveground Carbon Stock Evaluation Using Tree Ring Chronosequences

As an alternative to short-term growth data from plots and complex growth models, dendrochronology is a tool to evaluate carbon stock dynamics. Tree rings provide reliable data on lifetime growth rates, thus showing realistic change of diameter at breast height (DBH) for each tree (Brienen and Zuidema 2006). Moreover, these data can be obtained without long-term and labor-intensive measurements in permanent sample plots. The acquired information from tree rings on ages and long-term growth rates can be directly applied to estimate carbon stocks at stand level with allometric models. Hence, tree ring analysis could be a valuable and reliable tool to detect growth trends and carbon sequestration. Tree ring analysis may be useful for retrospective study assessing long-term growth changes and carbon sequestration in forest stands. As indicators of tree growth, they provide an estimate of the quantity of biomass produced. Tree rings have already been used to assess aboveground net primary productivity trends at the stand level (Graumlich et al. 1989), but only recently they have been used to assess carbon sequestration trends (Mund et al. 2002).

Forests often recover rapidly from less severe, small-scale disturbance such as tree-fall gaps and small-scale shifting cultivation. However, the impact of large-scale clear cutting on forest growth and carbon stock is significant. Different restoration approaches developed from the cutover areas have a remarkable effect on forest structure, biodiversity, hydrology, and carbon sequestration (Zhang et al. 2004). Few studies in subalpine forest had used tree rings to determine stand age or study historical growth patterns (Dang et al. 2009), and none had used tree rings in the context of different restoration approaches to evaluate the variance of carbon sequestration. The evaluation of various growth patterns may provide useful insight into whether restoration approaches are actually inducing variations in carbon stocks.

11.6.4 Monitoring Methods for Restoration

Cost-efficient methods of monitoring restoration success should be developed, because ecosystems do not always respond to restoration as expected. In addition, the fact that environmental changes are occurring emphasizes the importance of monitoring in restoration projects. Well-designed experiments and field measurements across landscapes at multiple scales are needed to integrate landscape dynamics with restoration. Inter-species relationship, disturbance, and succession all play a large role in the formation of landscape pattern (Turner et al. 2001) and ecosystem components need to be linked to forest landscape dynamics.

Acknowledgments

We are grateful to Timothy Moermond for helpful comments and revisions that improved this chapter. This study was jointly funded by the Ministry of Science and Technology (2012BAD22B01 and 2011CB403205), the Ministry of Finance (201104006 and 200804001), and China's National Natural Science Foundation (31290223, 31100380, and 31200477).

References

Baskerville, G.L. 1995. The forestry problem: Adaptive lurches of renewal. In *Barriers and Bridges to the Renewal of Ecosystems and Institutions*, L.H. Gunderson, C.S. Holling, and S.S. Light, (Eds.), pp. 37–102. Columbia University Press, New York.

Bauhus, J., Puettmann, K.J., and Kühne, C. 2013. Close-to-nature forest management in Europe: Does it support complexity and adaptability of forest ecosystems? In *Managing Forests as Complex Adaptive Systems: Building Resilience to the Challenge of Global Change*, C. Messier, K.J. Puettmann, and K.D. Coates, (Eds.), pp. 187–213. The Earthscan forest library, Routledge, London.

Brienen, R.J.W. and Zuidema, P.A. 2006. The use of tree rings in tropical forest management: Projecting timber yields of four Bolivian tree species. *Forest Ecology and Management* 226(1–3): 256–267.

Cui, X., Liu, S., and Wei, X. 2012. Impacts of forest changes on hydrology: A case study of large watersheds in the upper reaches of the Minjiang River watershed in China. *Hydrology & Earth System Sciences* 16(11): 4279–4290.

Dang, H., Jiang, M., Zhang, Y., Dang, G., and Zhang, Q. 2009. Dendroecological study of a subalpine fir (*Abies fargesii*) forest in the Qinling Mountains, China. In *Forest ecology*, A.G. Valk, (Ed.), pp. 67–75. Springer, Netherlands.

Delang, C.O. and Yuan, Z. 2015. *China's Grain for Green Program: A Review of the Largest Ecological Restoration and Rural Development Program in the World*. Springer International Publishing, Switzerland, ISBN: 978-3-319-11504-7.

Franklin, J.F., Lindenmayer, D., MacMahon, J.A., McKee, A., Magnuson, J., Perry, D.A et al. 2000. Threads of continuity. *Conservation in Practice* 1(1): 8–17.

Franklin, J.F., Spies, T.A., Pelt, R.V., Carey, A.B., Thornburgh, D.A., Berg, D.R et al. 2002. Disturbances and structural development of natural forest ecosystems with silvicultural implications, using Douglas-fir forests as an example. *Forest Ecology and Management* 155(1–3): 399–423.

Graumlich, L.J., Brubaker, L.B., and Grier, C.C. 1989. Long-term trends in forest net primary productivity: Cascade Mountains, Washington. *Ecology* 70(2): 405–410.

Green, T. (ed.) 2001. *Ecological and Socio-Economic Impacts of Close-to-Nature Forestry and Plantation Forestry: A Comparative Analysis*. EFI Proceedings No. 37, European Forest Institute, Joensuu, Finland.

Grunewald, K., Syrbe, R.U., and Bastian, O. 2014. Landscape management accounting as a tool for indicating the need of action for ecosystem maintenance and restoration–Exemplified for Saxony. *Ecological Indicators* 37(Part A): 241–251.

Gunderson, L.H. and Holling, C.S. (eds.) 2002. *Panarchy: Understanding Transformations in Human and Natural Systems*. Island Press, Washington, DC.

He, X.Y., Zhao, Y.H., Hu, Y.M., Chang, Y., and Zhou, Q.X. 2006. Landscape changes from 1974 to 1995 in the upper Minjiang River Basin, China. *Pedosphere* 16: 398–405.

Herrera, J.M. and García, D. 2009. The role of remnant trees in seed dispersal through the matrix: Being alone is not always so sad. *Biological Conservation* 142(1): 149–158.

Holling, C.S. 1995. What barriers? What bridges? In *Barriers and Bridges to the Renewal of Ecosystems and Institutions*, L.H. Gunderson, C.S. Holling, and S.S. Light, (Eds.), pp. 3–34. Columbia University Press, New York.

Hu, H. and Liu, S.Q. 2001. Changes of soil properties during artificial recovery of subalpine conifer-ous forests in western Sichuan. *Chinese Journal of Applied & Environmental Biology* 7(4): 308–314 (in Chinese, with English abstract).

Jiang, Y.X. 1981. Phytoecological role of forest floor in subalpine fir forests in western Sichuan prov-ince. *Acta Phytoecologica et Geobotanica Sinica* 5(2): 89–98 (in Chinese, with English abstract).

Keeton, W.S. and Franklin, J.F. 2005. Do remnant old-growth trees accelerate rates of succession in mature Douglas-fir forest? *Ecological Monographs* 75(1): 103–118.

Larsen, J.B. 2013. Close-to-nature forest management: The Danish approach to sustainable forestry. In *Sustainable Forest Management–Current Research*, Martin Garcia, J., and Díez Casero, J.J., (Eds.), pp. 201–218. InTech, Rijeka, Croatia, and Shanghai, China.

Li, W.H. 2004. Degradation and restoration of forest ecosystems in China. *Forest Ecology and Management* 201(1): 33–41.

Li, Y.C. 2009. Grain for green program is a great concrete action towards the ecological civilization in China: Summary of a decade program implementation. *Forestry Construction* 27(5): 3–13 (in Chinese, with English abstract).

Li, L., He, X., Li, X., Wen, Q., and He, H. 2007. Depth of edge influence of the agricultural-forest land-scape boundary, Southwestern China. *Ecological Research* 22: 774–783.

Lin, B., Liu, Q., Wu, Y., He, H., and Pang, X.Y. 2002. Water-holding capacity of moss and litter layers of subalpine coniferous plantations in western Sichuan, China. *Chinese Journal of Applied and Environmental Biology*, 8(3): 234–238 (in Chinese, with English abstract).

Lindenmayer, D.B., Manning, A.D., Smith, P.L., Possingham, H.P., Fischer, J., Oliver, I et al. 2002. The focal-species approach and landscape restoration: A critique. *Conservation Biology* 16(2): 338–345.

Liu, S.R. 2011. *Ecological Restoration Principle and Techniques of Natural Forests*. Vol. 4. Chinese Forestry Press, Beijing.

Liu, Y., An, S., Deng, Z., Fan, N., Yang, H., Wang, A et al. 2006. Effects of vegetation patterns on yields of the surface and subsurface waters in the Heishui Alpine Valley in west China. *Hydrol. Earth Syst. Sci. Discuss.* 3: 1–23.

Liu, Y., An, S., Xu, Z., Fan, N., Cui, J., Wang, Z et al. G. 2008a. Spatio–temporal variation of stable iso-topes of river waters, water source identification and water security in the Heishui Valley (China) during the dry–season. *Hydrogeol. J.* 16: 311–319, doi:10.1007/s10040–007–0260–3.

Liu, Y., Fan, N., An, S., Bai, X., Liu, F., Xu, Z et al. 2008b. Characteristics of water isotopes and hydro-graph separation during the wet season in the Heishui River, China. *J. Hydrol.* 353: 314–321, doi:10.1016/j.jhydrol.2008.02.017.

Liu, S.R., Shi, Z.M., Ma, J.M., Zhao, C.M., and Liu, X.L. 2009. Ecological strategies for restoration and reconstruction of degraded natural forests on the upper reaches of the Yangtze River. *Scientia Silvae Sincae* 45(2): 120–124 (in Chinese, with English abstract).

Liu, S., Sun, P., Wang, X., and Chen, L. 2001. Hydrological functions of forest vegetation in upper reaches of the Yangze River. *J. Nat. Resour.* 16(5): 451–456.

Liu, Y., Zhang, Y., and Liu, S. 2012. Aboveground carbon stock evaluation with different restoration approaches using tree ring chronosequences in southwest China. *Forest Ecology and Management* 263: 39–46.

Lõhmus, P., Rosenvald, R., and Lõhmus, A. 2006. Effectiveness of solitary retention trees for conserv-ing epiphytes: Differential short-term responses of bryophytes and lichens. *Canadian Journal of Forest Research* 36(5): 1319–1330.

Lü, Y., Fu, B., Feng, X., Zeng, Y., Liu, Y., Chang, R et al. 2012. A policy-driven large scale ecological restoration: Quantifying ecosystem services changes in the Loess Plateau of china. *PLoS One* 7(2): e31782.

Ma, X.H. 1963. Cutting and soil & water conservation of alpine dark coniferous forest in west Sichuan. *Scientia Silvae Sinicae* 8(2): 149–158.

Ma, X.H. 1987. Preliminary study on hydrologic function of fir forest in Miyaluo region of Sichuan. *Scientia Silvae Sincae* 23(3): 253–264.

Miao, N., Liu, S.R., Shi, Z.M., Yu, H., and Liu, X.L. 2009. Spatial pattern of dominant tree species in sub-alpine *Betula-Abies* forest in west Sichuan of China. *Chinese Journal of Applied Ecology* 20(6): 1263–1270 (in Chinese, with English abstract).

Miao, N., Liu, S.R., Yu, H., Shi, Z.M., Moermond, T., and Liu, Y. 2014. Spatial analysis of remnant tree effects in a secondary *Abies-Betula* forest on the eastern edge of the Qinghai–Tibetan Plateau, China. *Forest Ecology and Management* 313: 104–111.

Miao, N., Shi, Z.M., Feng, Q.H., Liu, X.L., and He, F. 2008. Spatial pattern analysis of *Abies faxoniana* population in sub-alpine area in western Sichuan. *Scientia Silvae Sinicae* 44(12): 1–6 (in Chinese, with English abstract).

Mund, M., Kummetz, E., Hein, M., Bauer, G.A., and Schulze, E.D. 2002. Growth and carbon stocks of a spruce forest chronosequence in central Europe. *Forest Ecology and Management* 171(3): 275–296.

Naveh, Z. 2000. What is holistic landscape ecology? A conceptual introduction. *Landscape and Urban Planning* 50(1–3): 7–26.

Pang, X.Y., Liu, S.Q., Liu, Q., Wu, Y., Lin, B., He, H et al. 2003. Influence of plant community succession on soil physical properties during subalpine coniferous plantation rehabilitation in western Sichuan. *Journal of Soil and Water Conservation* 17(4): 42–45 (in Chinese, with English abstract).

Ren, H., Peng, S., and Lu, H. 2004. The restoration of degraded ecosystems and restoration ecology. *Acta Ecologica Sinica* 24(8): 1756–1764 (in Chinese, with English abstract).

Rosenvald, R. and Lõhmus, A. 2008. For what, when, and where is green-tree retention better than clear-cutting? A review of the biodiversity aspects. *Forest Ecology and Management* 255(1): 1–15.

Shi, L.X., Wang, J.X., Su, Y.M., and Hou, G.W. 1988. The early succession process of vegetation at cut-over area of dark coniferous forest in Miyaluo, western Sichuan. *Acta Phytoecologica et Geobotanica Sinica* 12(4): 306–313 (in Chinese, with English abstract).

Shi, M., Ma, G., and Shi, Y. 2011. How much real cost has China paid for its economic growth? *Sustainability Science* 6(2): 135–149.

Stanturf, J.A., Palik, B.J., and Dumroese, R.K. 2014a. Contemporary forest restoration: A review emphasizing function. *Forest Ecology and Management* 331: 292–323.

Stanturf, J.A., Palik, B.J., Williams, M.I., Dumroese, R.K., and Madsen, P. 2014b. Forest restoration paradigms. *Journal of Sustainable Forestry* 33(sup. 1): S161–S194.

Turner, M.G., Barker, W.L., Peterson, C.J., and Peet, R.K. 1998. Factors influencing succession: Lessons from large, infrequent natural disturbances. *Ecosystems* 1(6): 511–523.

Turner, M.G., Gardner, R.H., and O'Neill, R.V. 2001. *Landscape Eology in Theory and Practice: Pattern and Process*. Springer, New York.

Wu, Q.B., Wang, X.K., Duan, X.N., Deng, L.B., Lu, F., Ouyang, Z.Y et al. 2008. Carbon sequestration and its potential by forest ecosystems in China. *Acta Ecologica Sinaca* 28(2): 0517–0524.

Yin, R., Yin, G., and Li, L. 2009. Assessing China's ecological restoration programs: What's been done and what remains to be done? In *An Integrated Assessment of China's Ecological Restoration Programs*, R. Yin, (ed.), pp. 21–38. Springer, Netherlands.

Zhang, Y.D., Gu, F.X., Liu, S.R., Liu, Y., and Li, C. 2013. Variations of carbon stock with forest types in subalpine region of Southwestern China. *Forest Ecology and Management* 300: 88–95 (in Chinese, with English abstract).

Zhang, W.G., Hu, Y.M., Hu, J.C., Chang, Y., Zhang, J., and Liu, M. 2008. Impacts of land-use change on mammal diversity in the upper reaches of Minjiang River, China: Implications for biodiversity conservation planning. *Landscape and Urban Planning* 85: 195–204.

Zhang, W.G., Hu, Y.M., Zhang, J., Liu, M., and Yang, Z.P. 2007. Assessment of land use change and potential eco-service value in the upper reaches of Minjiang River, China. *Journal of Forestry Research* 18: 97–102.

Zhang, Y.D., Liu, S.R., and Gu, F.X. 2011. The impact of forest vegetation change on water yield in the subalpine region of Southwestern China. *Acta Ecologica Sinica* 31(24): 7601–7608 (in Chinese, with English abstract).

Zhang, Y.D., Liu, Y.C., Liu, S.R., and Zhang, X.H. 2012. Dynamics of stand biomass and volume of the tree layer in forests with different restoration approaches based on tree-ring analysis. *Chinese Journal of Plant Ecology* 36(2): 117–125 (in Chinese, with English abstract)

Zhang, Y.D., Liu, S.R., Luo, C.W., Zhang, G.B., and Ma, J.M. 2009. Water holding capacity of ground covers and soils in different land uses and land covers in subalpine region of Western Sichuan, China. *Acta Ecologica Sinica* 29(2): 627–635 (in Chinese, with English abstract).

Zhang, Y.D., Liu, S.R., and Ma, J. 2006. Water-holding capacity of ground covers and soils in alpine and sub-alpine shrubs in Western Sichuan, China. *Acta Ecologica Sinica* 26(9): 2775–2781 (in Chinese, with English abstract).

Zhang, Y.D., Zhao, C.M., and Liu, S.R. 2004. Woodland hydrological effects of spruce plantations and natural secondary series in sub-alpine region of western Sichuan. *Journal of Natural Resources* 19(6): 761–768 (in Chinese, with English abstract).

12

Restoration and Ecosystem-Based Management in the Circumboreal Forest: Background, Challenges, and Opportunities

Timo Kuuluvainen, Yves Bergeron, and K. David Coates

CONTENTS

> Any serious attempt at ecological forestry ... in this region must confront the simplified age structures and altered compositions of repeatedly harvested stands using a patient restoration approach.
>
> **Robert S. Seymour (2005)**

12.1 Introduction

Boreal forestry is confronted with increasing demands to restore structurally simplified managed forest toward their natural state of complexity (Burton et al. 2010; Messier et al. 2013). Since the rapid expansion of industrial forestry across the circumboreal zone after World War II (WWII), forest management has predominantly been based on the even-aged forest management, where timber is periodically extracted by clear-cutting. This practice, together with short cutting rotations relative to the biological age of the main tree species or to the natural stand development, has created landscapes which are patchworks of relatively young structurally homogeneous stands. This contrasts with primeval boreal forests, which are characterized by more uneven-aged forest, old trees, abundant dead wood, and higher structural variability. These more variable conditions are common in unmanaged forests because high severity disturbances are infrequent and/or of limited scale, or because partial disturbances are prevalent in the forest. The strong contrast between managed and unmanaged forest is due to their divergent disturbance regimes (Cyr et al. 2009; Kuuluvainen 2009). The disturbance regime of the managed forest differs fundamentally compared to the historical disturbance regime under which the forest biota has evolved. This poses a threat to both biodiversity and long-term maintenance of important ecosystem services. To reconcile timber management with maintenance of other ecosystem values and services, boreal forest management should restore important natural ecosystem structures and dynamics.

The circumboreal forest biome is an important part of the global ecosystem. The boreal forest accounts for about one quarter of the global area of closed canopy forest, almost 1 billion ha (Burton et al. 2010, Figure 12.1). The boreal forests and associated peatlands are estimated to contain more than 35% of total terrestrial carbon. This is more than five times the amount of carbon found in the world's temperate forests, and almost double the carbon found in tropical forests (Kasischke 2000; Burton et al. 2010).

Although heavily exploited for a long time in its southern parts, the boreal zone still contains about half of the world's unexploited frontier forests (Bradshaw et al. 2009; Burton et al. 2010). This makes the boreal forest equal to tropical forests in terms of hosting the last large intact forest landscapes on earth (Aksenov et al. 2002). However, increased human pressure through logging and various other kinds of resource extraction is rapidly diminishing the area of intact boreal forests and changing the structure and dynamics of exploited forests. This development is rapidly weakening the ability of the boreal forest to

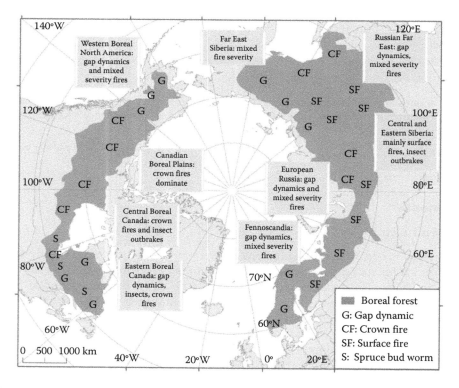

FIGURE 12.1
A coarse-scale illustration of the circumboreal distribution of the main forest dynamics types that are driven by specific disturbance regimes. Note that in all regions, all these types are present in variable proportions depending on the variability of disturbance regimes. In many regions human influence has significantly altered or even completely changed the inherent forest dynamics. (Modified from Shorohova, E. et al., 2011. *Silva Fenn.* 45(5), 785–806.)

act as a carbon sink, maintain biological diversity and, possibly, ecosystem resiliency in the face of global change (Bradshaw et al. 2009).

Despite such worrisome threats, the boreal forests are largely missing from the global considerations and efforts of climate change mitigation and biodiversity conservation (Warkentin and Bradshaw 2012; Moen et al. 2014). Instead, international attention has been mainly focused on the tropical forests and their role in carbon emissions (e.g., REDD) and biodiversity conservation. More recently, however, the uniqueness of the boreal forests has amply been emphasized for their importance to biodiversity and the global carbon cycle (e.g., Bradshaw et al. 2009; Moen et al. 2014). Accordingly, there is a need to critically analyze the current forest management practices in the boreal region and to assess whether and to what extent they are compatible with the modern sustainable management principles, including restoration and the ecosystem approach highlighted since 1990s (CBD 2004; Halme et al. 2013).

Historically human utilization of the boreal forest has taken many forms ranging from hunting and gathering by stone-age humans to extensive slash-and-burn cultivation in the seventeenth to nineteenth century in parts of northern Europe. Timber harvesting in the boreal forest has increased with increasing human population, in concert with the overall intensification of natural resource utilization. At the outset of the sawmill industry in the boreal zone in the nineteenth and early twentieth centuries, harvesting of timber was first mostly based on opportunistic cutting of the largest best quality trees on easily accessible sites (e.g., Siiskonen 2007; Keto-Tokoi and Kuuluvainen 2014). Lower quality and smaller trees became valuable with the expansion of the pulp and paper industry after WWII. Advances in harvesting technology together with cheap fossil fuels made operation over much larger areas feasible. When all timber could be utilized there was a shift toward clear-cutting regimes. This was an obvious choice when the primary objective was low logging costs per harvested timber volume.

Starting from the 1990s onwards, the goals of forest management set by governments, institutions and forest industry have greatly diversified to include various aspects of ecological and sociocultural sustainability. These trends have paralleled our improved understanding of the ecological dynamics and biodiversity of unmanaged boreal forests (Angelstam and Kuuluvainen 2004; Bergeron et al. 2004; Gauthier et al. 2009; Shorohova et al. 2009; Kuuluvainen and Aakala 2011). From the accumulated body of research, two conclusions are particularly important from the forest restoration and management points of view. First, it has become evident that the unmanaged boreal forest displays high variability in structure and dynamics, which is important for biodiversity and ecosystem functioning. Secondly, unmanaged boreal forest typically contains a high share of old forests (Cyr et al. 2009; Kuuluvainen 2009). These ecological findings are of fundamental importance because such intrinsic ecosystem properties define ecological constraints on how the forest should be restored or managed without losing its defining properties and the ecological services that it provides.

The forest management approach based on singularly dominant use of clear-cut harvesting and growing of even-aged stands does not provide an adequate strategy for sustainable forest management, where the aim is to reconcile timber production goals with maintenance of important ecosystem values and services. To demonstrate this we first review the current state of understanding of the structure and dynamics of unmanaged boreal forest in the circumboreal zone. Secondly, we contrast this knowledge with that of current managed forest landscapes using examples from both North America and North Europe. Finally, we discuss possible approaches and strategies to better reconcile timber management objectives with restoration and maintenance of other important ecosystem values and services. Our main point of view is ecological, but economic and social aspects of sustainability are also briefly considered.

12.2 Intrinsic Dynamics of the Circumboreal Forest: More Diverse Than Expected

The conventional view of the boreal forest holds that across the boreal zone, stand-replacing fires over short return intervals are the most important and overall dominant natural disturbance factor driving forest structure and development (Sirén 1955; Zackrisson 1977; Johnson 1992). However, this generalization has been challenged and reassessed by recent research (Bergeron 2004; Kuuluvainen 2009; Kneeshaw et al. 2011; Bergeron and Fenton 2012). The accumulating body of scientific evidence indicates that the natural disturbance regimes of boreal forests (i.e., forests with only negligible human impact) exhibit substantial variability across the vast circumboreal zone (Bartemucci et al. 2002; Wooster and Zhang 2004; Shorohova et al. 2009; Kneeshaw et al. 2011; Kuuluvainen and Aakala 2011). For example, boreal fire regimes can range from regions with frequent stand-replacing fires in continental Canada (Johnson 1992; Payette 1992) to regions of nonpyrogenic forests that possibly have avoided fire altogether since the last glaciation (Zackrisson et al. 1995; Pitkänen et al. 2003).

The high variability in boreal forest fire regimes is a reflection of the variability in factors influencing fire ignition and spread, such as climatic conditions, fuels, tree species traits, and stand structural and landscape characteristics (Ryan 2002; Kneeshaw et al. 2011). The ecological outcome of fire is strongly dependent on tree species traits that have evolved under specific historical disturbance regimes in different parts of the boreal zone. For example, in continental regions where short return intervals of stand-replacing fires are common, the forest is dominated by fire-adapted pyrogenic tree species. In boreal North America these species include jack pine (*Pinus banksiana*), black spruce (*Picea mariana*), or trembling aspen (*Populus tremuloides*) (Kneeshaw et al. 2011, Figure 12.2a). These species all have adaptations to crown fire, such as serotinous cones, or the ability to resprout from root suckers. The Eurasian species such as Scots pine (*Pinus sylvestris*) and larches (*Larix* spp.) have adapted to fires differently, by having a thick heat-insulating bark that helps especially large trees to survive low- and medium-intensity fires (Kuuluvainen 2002; Shorohova et al. 2009, Figure 12.2b).

In areas where fire return intervals are longer than the longevity of the main tree species, or where low-intensity surface fires dominate, autogenic disturbances such as tree senescence-related insect outbreaks, fungi, and wind storms become important drivers of forest dynamics (Kuuluvainen 1994; McCarthy 2001). It is common that a diverse set of different allogenic and autogenic disturbance agents are jointly in action in boreal forest landscapes. As a consequence, the natural forests often "self-generate" structures and dynamics that exhibit fine-scale variability (Kuuluvainen 1994; McCarthy and Weetman 2006; Bergeron and Fenton 2012).

Despite geographic, regional, and landscape scale variability in boreal forest disturbance regimes, some broad differences can be distinguished between the North American and Eurasian continents (Kneeshaw et al. 2011). In particular, the disturbance regimes in the continental North American boreal forest are more often characterized by severe stand-replacing fires (Johnson 1992; Payette 1992), with the exception of some pine stands, for which a regime of nonlethal surface fires are reported (Bergeron and Brisson 1990; Smirnova et al. 2008).

In Eurasian boreal forests, nonstand replacing disturbances appear more prevalent (Shorohova et al. 2009; Kuuluvainen and Aakala 2011). Typical forest dynamics types are

FIGURE 12.2
(a) Stand-replacing disturbances are typical in boreal North American forests under continental climate. A burned lodgepole pine-black spruce stand in the Yukon, Canada. (b) Partial disturbances, where only some trees die, are common in many Eurasian boreal forests, as illustrated by this Scots pine stand in Eastern Fennoscandia after a surface fire. (Photo (a) by Dave Coates, (b) by Timo Kuuluvainen.)

the so-called cohort-dynamics, driven by low- and medium-severity fires and windstorms, and fine-scale gap dynamics in late-successional forests (Figure 12.2b, Wein and MacLean 1983; Gromtsev 2002; Kuuluvainen 2009; Kuuluvainen and Aakala 2011). Extensive highly devastating fires and windstorms do occur but infrequently (Syrjänen et al. 1994; Niklasson and Granström 2000). Although rare, such disturbances have a pronounced and long-lasting impact on landscape structure and dynamics (Aakala et al. 2009). In Eurasia such events are most common in the continental parts of Eurasia, such as Komi Republic and western Siberia (Syrjänen et al. 1994; Shorohova et al. 2009).

The prevalence of low-intensity nonstand-replacing fires in boreal Eurasia is evidently related to factors such as climate, type of fuels, landscape characteristics, and the fire-resistant character of important Eurasian tree species, particularly pines (e.g., *P. sylves-tris, P. sibirica*) and larches (e.g., *Larix gmelinii, L. sibirica*) (Helmisaari and Nikolov 1992). It is possible, however, that increased fire frequency due to human activity, and the associated reduction in burnable fuels, also has contributed to the observed prevalence of ground fires in Eurasia.

From the accumulating body of scientific evidence two conclusions can be drawn that are important for forest restoration and sustainable management. First, natural boreal forests in the circumboreal zone are structurally and developmentally diverse; second, they contain more old forests than previously assumed (see further discussion below for North America and Eurasia). These conclusions bear important ramifications for forest restoration and ecosystem-based forest management in the boreal region.

12.3 Comparison of Unmanaged and Managed Boreal Forest Landscapes

12.3.1 Boreal North America

In boreal North America, forest fires can generally be characterized as severe and stand-replacing, with the exception of coastal maritime regions with humid climate conditions. However, and importantly, the fire return interval is generally much longer than the rotation length in industrial forests. This is illustrated in Table 12.1, which summarizes results from published fire history reconstructions along an east–west gradient in Canada's boreal forests. It includes data from Labrador (Foster 1983), Central and Western Quebec (Bergeron et al. 2001), Eastern (Bergeron et al. 2001) and Western (Suffling et al. 1982) Ontario, Saskatchewan (Weir et al. 2000) and Alberta (Larsen 1997) (A more extensive dataset is presented in Bergeron et al. (2004)). Average age of the forest (time since fire), or if age was not available, the fire cycle before large-scale clear-cutting began was used to estimate historic burn intervals. The average age of the forest was preferred to the historic fire cycle, because it integrates climatically induced changes in fire frequency over a long period, and because it is easier to evaluate than a specific fire cycle (Bergeron et al. 2001).

Although there are variations in the mean age of the forests, probably caused by changes in climate conditions along the east–west gradient in Canada, in all cases there is a significant proportion of forest that is older than 100 years. The reported historical mean ages are mostly well above the mean age of similar landscapes under a normal even-aged rotation.

TABLE 12.1

Geographic Location and Proportion of Forests That Are over 100 Years Old According to Reported Forest Age for Each Study Area

Study Area	Study Area (km²)	Time Period	Mean Age	% >100 Years Old
Wood Buffalo Park	44,807	1750–1989	71	24
Prince Albert	3461	<1890	97	36
Northern Ontario	24,000	~1870–1974	52	15
LAMF	8245	1740–1998	178	78
Western Quebec	15,793	~1750–1998	139	57
Central Quebec	3844	1720–1998	102	35
Southeastern Labrador	48,500	1870–1975	500	81

In fact, under a 100-year forest rotation, the mean age of the *fully regulated forest* would be 50 years while a natural landscape with a 100-year cycle of stand-replacing fire would have a mean age of 100 years (or mean fire interval). This is because fire can be assumed to occur more or less randomly in the landscape while in forest management, harvesting always targets the oldest economically mature stands. Because of this it can be shown theoretically that with a 100-year cycle fire in a landscape subject only to fire disturbance, 37% of stands are older than 100 years, while no stands in a fully regulated managed landscape are older than 100 years (van Wagner 1978, Figure 12.3a). This means that a large proportion of the preindustrial landscape was composed of forests older than the 100-year commercial forest rotation.

Forest structural diversity tends to increase with succession (Bergeron and Fenton 2012), thus the distribution of forest age classes over the landscape can be regarded as a good indicator of habitat conditions and variability, which in turn controls forest biodiversity. Cyr et al. (2009) reconstructed the fluctuation of the fire interval during the Holocene (during the last 10,000 years) in northwestern Quebec using a paleoecological approach. They showed that the mean fire interval has fluctuated roughly between 100 and 250 years with an extended range of variability from 75 to 400 years (Figure 12.3b). The distribution of the stand ages associated with this range of mean fire return intervals clearly demonstrates that at no time during the Holocene was the landscape composed of less than 30% of forests over 100 years of age. Current forest management has already decreased older forest levels well below this percentage (Figures 12.3c and 12.4). This means that intensive management has driven the landscape outside its historical range of variability (Cyr et al. 2009).

In the mixedwood forest located in the southern portion of the eastern boreal forest in North America, we generally observe a postfire invasion of shade-intolerant deciduous trees (*Betula* spp. and *Populus* spp.) that are gradually replaced in the canopy by shade-tolerant conifers (Bergeron 2000). Thus, successive replacement of deciduous tree stands by mixed stands and then by coniferous stands occurs over a 200-year period. Further north, in the coniferous boreal forest dominated by black spruce, stand establishment following fire is often dominated by an initial cohort of spruce that gives rise to a dense forest principally of seed origin. As the stand matures, this even-aged structure is gradually replaced by a more structurally complex forest containing trees that regenerated from seed after the fire and trees that developed vegetatively by layering. In the prolonged absence of fire, these stands develop a very heterogeneous uneven-aged structure (Boucher et al. 2006; Lecomte et al. 2006). The presence and abundance of birds, insects, and vascular and non-vascular plants changes gradually along the time-since-fire gradient and many species could be threatened by a decrease in the abundance of old forest stands and associated habitats (Drapeau et al. 2009).

In light of these results it becomes clear that in comparison with unmanaged landscapes, even-aged forest management under relatively short rotations will lead to a significant decrease in old and overmature stands and habitats crucial for the maintenance of biodiversity (Kneeshaw and Gauthier 2003).

12.3.2 Boreal Eurasia

In contrast to boreal North America where severe fires prevail, the Eurasian fire regime is more diverse and may perhaps be best characterized as "mixed-severity" (Gromtsev 2002; Shorohova et al. 2009). Factors contributing to high variability in fire regimes include, in addition to variation in climate conditions and patchiness of landscape characteristics, high fire resistance of important tree species such as *P. sylvestris* and *Larix* spp. Under these

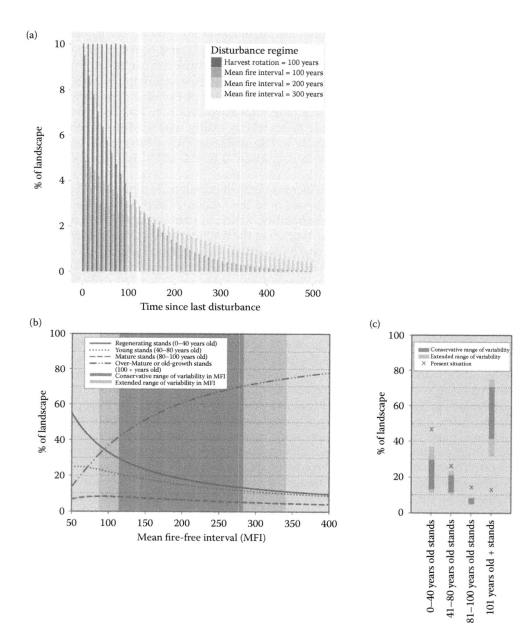

FIGURE 12.3
Illustrations of forest management impacts on forest age class distribution across the landscape. (a) Distribution of age classes according to different mean fire return intervals compared with a 100-year harvest rotation. (b) Proportion in the landscape of stand ages according to the mean fire return intervals. Dark grey shaded area represents the observed variation in mean fire return intervals during the Holocene in northwestern Quebec, while light grey is a 95% confidence interval around the natural range of variability. (c) Estimated proportion of the landscape covered by different age classes during the Holocene compared with current proportion, which is indicated by the x-mark. (Modified from Cyr, D. et al., 2009. *Front. Ecol. Environ.* 10, 519–524.)

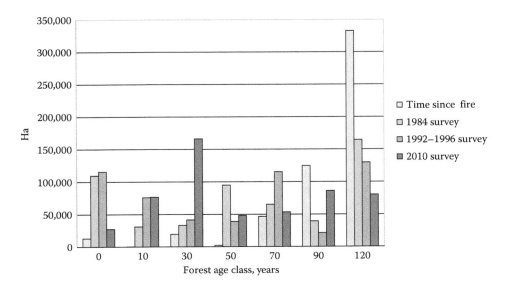

FIGURE 12.4
Change in the forest age class composition in northwestern Quebec, Canada, from 1980 to 2010. Prelogging conditions are represented by time since fire distribution in 1997. (Data provided by A. Belleau, Quebec Ministry of Forest, Wildlife and Park).

conditions significant sections of a landscape can remain unburned for very long periods of time (hundreds of years), or they are affected only by low-intensity ground fires (e.g., Gromtsev 2002; Lampainen et al. 2004; Wallenius et al. 2004; Shorohova et al. 2009).

In addition, recent research suggests that past fire cycles (i.e., the time in which an area equivalent to the landscape in question has burned) at the time of low human impact have been significantly longer than previously assumed (Kuuluvainen 2009 and references therein). For example, studies in Fennoscandia, based on dendrochronological samples and charcoal samples from peat layers of small hollows, have shown that prior to significant human influence, fire cycles often have ranged from 200 to 500 years or even more. Fire cycles were shorter in dryer forests dominated by *P. sylvestris* and longer in moister *Picea abies* dominated forests (Pitkänen et al. 2002, 2003; Wallenius et al. 2010). These values contrast with the well-documented and much shorter fire cycles of 30–60 years that prevailed from the seventeenth to twentieth centuries, for example, in southern Finland, which were caused by slash-and-burn cultivation and other human activity in forests. This strongly human-impacted fire regime has sometimes erroneously been regarded as "natural."

The prevalence of either long fire cycles or low-intensity surface fires means that disturbance agents other than fire play a major role in driving Eurasian boreal forest dynamics (Shorohova et al. 2009). In the absence of severe disturbances, trees are eventually weakened by senescence, fungi, insect attacks, and finally broken down by wind or heavy snow loads thereby creating gaps (e.g., Rouvinen et al. 2002; Lännenpää et al. 2008). Such gap-phase disturbance regimes seem to be particularly prevalent in those parts of Eurasia with maritime or semi-maritime climate, such as western boreal Fennoscandia and the Russian Far East (Shorohova et al. 2009).

Although partial disturbances are intrinsic to Eurasian boreal forests, also large stand-replacing disturbances may occur but infrequently (Gromtsev 2002; Kuuluvainen 2002,

2009; Shorohova et al. 2009; Kuuluvainen and Aakala 2011). Stand-replacing fires are most common in more continental climates, such as the Komi Republic and the western Siberian plateau (Syrjänen et al. 1994; Schulze et al. 2005). Although such large fires are relatively infrequent (return intervals ranging from several hundreds to thousands of years), they can leave legacy structures and effects, which last for hundreds of years and may prevent the landscape from approaching any steady-state structure and composition (e.g., Aakala et al. 2009).

The rarity of large stand-replacing disturbances suggests that even-aged forest conditions play a minor role in the dynamics of unmanaged forests in boreal Eurasia (Kuuluvainen 2002; Shorohova et al. 2009). On moist spruce- and fir-dominated sites this is due to intrinsically long fire return intervals (Pitkänen et al. 2002, 2003; Shorohova et al. 2009; Ohlson et al. 2011). On dry sites, where fires are more common, the fire-resistant character of the dominant tree species *P. sylvestris* and *Larix* spp. results in survival of large trees. As a result, this type of forest ends up consisting of several age cohorts of trees that have either survived or regenerated after fires (Kuuluvainen 2002).

To summarize, disturbance regimes that do not replace stands are naturally prevalent in boreal Eurasia. Such disturbance regimes are characterized by various partial (cohort) and/or small-scale (gap) disturbances driven by multiple disturbance agents. The high number of contributing factors results in highly diversified successional pathways and dynamics that contribute to the maintenance of heterogeneous habitat structures (e.g., Kuuluvainen 1994; Pennanen 2002; Rouvinen et al. 2002).

The landscape-level forest age structure is an outcome of the prevailing disturbance types and their impact on stand dynamics (e.g., Wallenius et al. 2004). Based on empirical studies of fire ecology and stand dynamics, and synthesized by a spatial landscape simulation model, Pennanen (2002) concluded for eastern Fennoscandia that uneven-aged stands with a continuous presence of old big trees, living up to their biological lifespan of 200–400 years, would be a prevalent feature of natural forest landscapes (Figure 12.5a, Pennanen 2002; Kuuluvainen 2009). This conclusion is supported by both historical studies from Fennoscandia and recent empirical studies carried out in unmanaged landscapes in Russian Karelia in eastern Fennoscandia (Kuuluvainen 2002, 2009 and references therein).

The extensive use of clear-cutting introduced since WWII brought about drastic changes in forest landscape structures. Unmanaged forest landscapes, dominated by late-successional forests, faced a strong decline in old age classes (Figure 12.5a and b). The situation was similar to that documented in Canada (see Figures 12.3 and 12.4): management has driven the forest outside its historical range of variability (Kuuluvainen 2009).

12.4 Challenges of Boreal Forest Restoration and Ecosystem-Based Management

Boreal forest management is currently facing increasing requirements to shift from fulfilling solely timber extraction and associated economic goals toward incorporating a broader set of values and providing for a variety of ecosystem services (Burton et al. 2010). It is generally acknowledged that to accomplish these broader goals, management practices have to conform to the general principles of an ecosystem approach (CBD 1995, 2004). This means sustaining the intrinsic structures, functions, dynamics, and biodiversity of ecosystems under management. This is an ambitious goal and it requires that old management

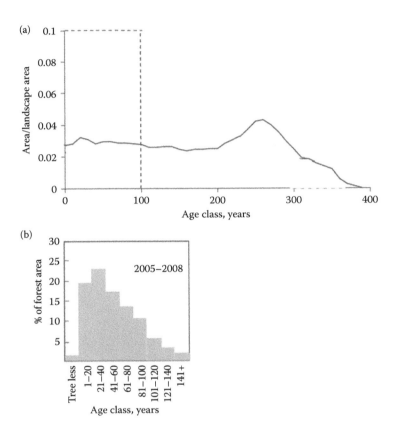

FIGURE 12.5
Management impacts on forest age distribution in eastern Fennoscandian conditions. (a) Solid line: stand age distribution of an unmanaged forest landscape with a 150-year fire cycle. (Redrawn from Pennanen, J., 2002. *Silva. Fenn.* 36, 213–231.) Stands are classified according to the oldest tree cohort, but in most cases they display uneven-aged structure. Dotted line: theoretical distribution of stand age classes when a landscape is managed as a "fully regulated forest" with a 100-year cutting cycle. (b) The stand age distribution of forests in southern Finland in 2005–2008 according to national forest inventory, showing the scarcity of old forests. (From Metsätilastollinen vuosikirja 1996–2009. Available at http://www.metla.fi/metinfo/tilasto/julkaisut/vsk/2002/index.html.)

approaches are reevaluated and new ones developed (Bergeron et al. 2002; Bauhus et al. 2009; Puettmann et al. 2009; Kuuluvainen and Grenfell 2012).

Ecological research carried out over the last few decades has greatly increased and diversified our understanding of the structure, function, and dynamics of boreal forest ecosystems. In particular, the previous monolithic conception that the circumboreal forest was dominated by stand-replacing fires is now put into history (Messier et al. 2013). Yet the management approach for these forests has changed little; the emphasis continues on clear-cutting or low retention systems in uniform stands. Our review has shown that this kind of management, and the disturbance regime it imposes on ecosystems, contradicts the historical disturbance regime of boreal forests (Cyr et al. 2009; Kuuluvainen 2009). The widespread enforcement of the even-aged, single species management approach interferes with fundamental ecological and evolutionary processes that are currently understood as critical for the maintenance of biodiversity of forest ecosystems and their ability to withstand and adapt to environmental change (resilience) (Bengtsson et al. 2003; Drever et al. 2006; Rist and Moen 2013).

The new knowledge from unmanaged "wild" forests stresses that for restoring and sustaining crucial ecosystem characteristics, greater attention must be paid to maintaining variability in stand structures and tree species compositions at different spatial scales and across all temporal (or successional) stages of forest development (Bergeron et al. 2002; Bengtsson et al. 2003, Figures 12.6 and 12.7). This is because the ecological processes that regulate the availability of ecosystem services occur at characteristic spatial scales ranging from the microscale (e.g., microhabitats for small immobile creatures), to the stand-scale (e.g., timber production), to the watershed- or landscape-scale (e.g., visual quality, water purification, carbon sequestration, or species migration) (Table 12.2). The maintenance of ecological processes interacting across scales is important for ecological resilience, and subsequently for the long-term provision of the "services" that the ecosystem provides (Bengtsson et al. 2003; Rist and Moen 2013).

Favoring green tree retention and deadwood abundance in low retention systems is one avenue that has gained acceptance and used at the stand-scale in boreal forestry

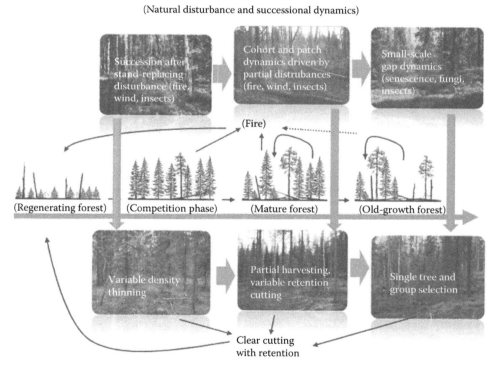

FIGURE 12.6
A silvicultural system that emulates natural disturbances and stand structures to restore and maintain important structural features of the unmanaged forest; indicated here by dividing the forest area into three types of disturbance dynamics (even-aged, cohort, and gap). Different cutting methods and intensities are shown, with variable harvest rotations applied to maintain landscape-level structure and age distributions similar to that existing under a natural disturbance regime. (From Bergeron, Y. et al., 2002. *Silva Fenn.* 36, 81–95; Kuuluvainen, T., 2009. *Ambio* 38, 309–315; Kuuluvainen, T., Grenfell, R., 2012. *Can. J. Forest Res.* 42, 1185–1203, Drawing by J. Karsisto.)

FIGURE 12.7
In most boreal regions, emulating natural disturbances in forest restoration and management requires application of versatile and small-scale timber harvesting techniques. (a) Sufficient retention on clear-cuts increases structural variability and sustains dead wood availability and continuity. (b) Small-gap cuttings are used to imitate gap-phase dynamics and to create fine scale heterogeneity in stand structure and composition. (c) Single tree selection can be used to create multilayered canopies and to open growing space for understory seedlings. (Photos by Timo Kuuluvainen.)

TABLE 12.2

Structural Features and Ecological Processes Requiring Restoration Attention in Boreal Forest Stands and Landscapes

	Structural Features to be Restored	Processes to be Restored
Stand-Level	• Old trees, especially deciduous trees • Broken, leaning, damaged, and cavity trees • Trees with abundant epiphytic lichen flora • Various kinds of fallen deadwood, especially large logs • Standing dead trees (snags) • Burned living and dead trees • Mounds and pits caused by uprooting • Mixtures of coniferous and deciduous tree species • Multiaged and multisized tree stand structures • Structurally and compositionally diverse understory canopies • Diverse microhabitat mosaics in relation to water table in peatland forests	• Small-scale gap disturbances • Fine-scale soil disturbances • Postfire successions • Successions following other disturbances besides fire • Natural tree stand succession and self-thinning • Multiple pathways of wood decay successions • Natural successions of peatland forests
Landscape-Level	• Natural variability of postfire and other successional stages • Natural-like landscape connectivity • Natural-like ecotone structures	• Natural variability of fire regime • Natural variability and distribution and spatial pattern of young deciduous successional stages • Natural variability and distribution and spatial pattern of old successional stages • Natural variability and distribution and spatial pattern of fire-free areas (fire refugia) • Natural variability of dynamics and spatial distribution of deadwood • Natural-like hydrology of peatland forests

Note: Structural features describe the state of forest ecosystems at a given point in time, while processes refer to the dynamics of disturbance and successional processes in time. Structures and processes are closely interlinked.

(Vanha-Majamaa and Jalonen 2001; Gustafsson et al. 2010). However, this approach may not succeed in reproducing the structural features of late-successional forests at the stand- and landscape-scales (Burton et al. 1999). Therefore, a more diverse array of management methods applied over a range of scales is required for restoration and ecosystem management (Kuuluvainen et al. 2002; Bauhus et al. 2009; Gauthier et al. 2009; Kuuluvainen and Grenfell 2012; Halme et al. 2013). For example, greater use of irregular shelterwood systems could be employed in forests naturally subjected to intermediate-severity partial disturbances (Seymour 2005; Raymond et al. 2009, see Figure 12.6).

Global change, especially the projected climate change in northern latitudes, presents a further challenge for boreal forest restoration and sustainable management (Bradshaw et al. 2009; Burton et al. 2010). The potential for major ecosystem changes as well as uncertainties about the degree and rate of climate change can be seen as another incentive to promote a shift in our approach to forest management (Drever et al. 2006). In the future, forest management will probably need to operate in a rapidly changing and uncertain environment. In such circumstances, forest and land managers must pay more attention to restoring ecosystem components crucial for system resiliency. This will minimize the risk of undesirable outcomes or "surprises" that could arise from boreal forests being ill-adapted to change to future conditions (e.g., Costanza 2000; Millar et al. 2007).

Management of boreal forests in the future may involve deliberate restorative practices that aim at maintaining ecosystem complexity and species diversity across multiple scales of time and space, and facilitate ecosystem adaptation in response to climate change (Drever et al. 2006; Burton et al. 2010). Forest harvest, regeneration, and stand-tending activities need to be developed to maintain or restore ecosystem complexity and the resultant diverse responses to environmental changes (Messier et al. 2013). Such a management and restoration strategy can be realized through diversifying silvicultural and management practices to create and maintain a range of forest structures inspired by knowledge of unmanaged ecosystems (Angelstam 1998; Kneeshaw and Bergeron 1998; Burton et al. 1999; Bartemucci et al. 2002; Harvey et al. 2002; Kuuluvainen 2009; Gauthier et al. 2009; Shorohova et al. 2009; Bauhus et al. 2009; Raymond et al. 2009).

In practice this means that multicohort and uneven-aged management practices should be dramatically increased in most boreal forest landscapes (Figures 12.6 and 12.7). A common argument against this is that uneven-aged forestry is characterized by low economic performance. However, recent studies indicate that this may not be the case. In fact, several studies suggest that uneven-aged management can be fully competitive with existing even-aged management (Tahvonen 2009; Pukkala et al. 2010; Tahvonen et al. 2010; Kuuluvainen et al. 2012; Ruel et al. 2013). This promising development suggests that reconciliation of timber production and maintenance of other ecosystem values and services is feasible in management that restores important intrinsic natural ecosystem structures and dynamics.

12.5 Conclusion

During the past decades substantial changes have taken place in the scientific understanding of forests and in the societal values about what kind of forest management is desirable and sustainable (Burton et al. 2003, 2010; Lindenmayer et al. 2006). It is widely acknowledged that forest managers need to take into account and accommodate a much wider range of ecological and social values than focusing only on timber production. Together with new ecological findings, this broader perspective leads to a critical reassessment of long-standing management practices in boreal forestry (Burton et al. 2003, 2010). In particular, the exclusive use of clear-cut harvesting in boreal forests, and other forms of low retention harvesting systems with short rotations, create forest structures that lie far outside the natural or historical range of variability of most boreal forests. The simplification of forest structures to fulfill short-term timber management goals has resulted in declines in forest biodiversity and associated ecosystem services. To address this problem, diversifying silviculture to restore tree species and structural diversity at multiple scales, as inspired by unmanaged ecosystems, is a promising avenue toward a new more balanced and ecosystem-based restoration and management strategy for the circumboreal forest.

References

Aakala, T., Kuuluvainen, T., Wallenius, T., Kauhanen, H., 2009. Contrasting patterns of tree mortality in late-successional *Picea abies* stands in two areas of northern Fennoscandia. *J. Veg. Sci.* 20, 1016–1026.

Aksenov, D., Dobrynin, D., Dubinin, M., Egorov, A., Isaev, A., Karpachevskiy, M., Laestadius, L., Potapov, P., Purekhovskiy, A., Turubanova, S., Yaroshenko, A., 2002. *Atlas of Russia's Intact Forest Landscapes. Global Forest Watch Russia, Moscow.* Available at: http://www.forest.ru/eng/publications/intact/[Cited 16 Mar 2010].

Angelstam, P., 1998. Maintaining and restoring biodiversity in European boreal forests by developing natural disturbance regimes. *J. Veg. Sci.* 9, 593–692.

Angelstam, P., Kuuluvainen, T., 2004. Boreal forest disturbance regimes, successional dynamics and landscape structures—A European perspective. *Ecol. Bull.* 51, 117–136.

Bartemucci, P., Coates, K.D., Harper, K.A., Wright, E.F., 2002. Gap disturbances in northern old-growth forests of British Columbia, Canada. *J. Veg. Sci.* 13, 685–696.

Bauhus, J., Puettmann, C., Messier, C., 2009. Silviculture for old-growth attributes. *For. Ecol. Manage.* 258, 525–537.

Bengtsson, J., Angelstam, P., Elmqvist, T., Emanuelsson, U., Folke, C., Ihse, M., Moberg, F., Nyström, M. 2003. Reserves, resilience and dynamics landscapes. *Ambio* 32, 389–396.

Bergeron, Y., 2000. Species and stand dynamics in the mixed woods of Quebec's southern boreal forest. *Ecology* 81, 1500–1516.

Bergeron, Y., 2004. Is regulated even-aged management the right strategy for the Canadian boreal forest? *For. Chron.* 80, 458–462.

Bergeron, Y., Brisson, J., 1990. Fire regime in red pine stands at the northern limit of the species' range. *Ecology* 71, 1352–1364.

Bergeron, Y., Fenton, N., 2012. Boreal forests of eastern Canada revisited: Old growth, nonfire disturbances, forest succession and biodiversity. *Botany* 90, 509–523.

Bergeron, Y., Flannigan, M., Gauthier, S., Leduc, A., Lefort, P., 2004. Past, current and future fire frequency in the Canadian boreal forest: Implications for sustainable forest management. *Ambio* 33, 356–360.

Bergeron, Y., Gauthier, S., Kafka, V., Lefort, P., Lesieur, D., 2001. Natural fire frequency for the eastern Canadian boreal forest: Consequences for sustainable forestry. *Can. J. For. Res.* 31, 384–391.

Bergeron, Y., Leduc, A., Harvey, B.D., Gauthier, S., 2002. Natural fire regime: A guide for sustainable management of the Canadian boreal forest. *Silva Fenn.* 36, 81–95.

Boucher, D., Gauthier, S., De Grandpré, L., 2006. Structural changes in coniferous stands along a chronosequence and a productivity gradient in the northeastern boreal forest of Québec. *Ecoscience* 13, 172–180.

Bradshaw, C.J.A., Warkentin, I.G., Sodhi, N.S., 2009. Urgent preservation of boreal carbon stocks and biodiversity. *Trends. Ecol. Evol.* 24, 541–548. doi 10.1016/j. tree.2009.03.019.

Burton, P., Messier, C., Smith, D.W., Adamowicz, W.L., (eds.), 2003. *Towards Sustainable Management of the Boreal Forest*. NRC Research Press, Canada.

Burton, P.J., Bergeron, Y., Bogdanski, B.E.C., Juday, G.P., Kuuluvainen, T., McAfee, B.J., Ogden, A., et al. 2010. Sustainability of boreal forests and forestry in a changing environment. In: G. Mery, P. Katila, G. Galloway, R.I. Alfaro, M. Kanninen, M. Lobovikov, J. Varjo, (eds.), *Forests and Society—Responding to Global Drivers of Change*. International Union of Forest Research Organizations (IUFRO), Vienna, Austria, pp. 249–282.

Burton, P.J., Kneeshaw, D.D., Coates, K.D., 1999. Managing forest harvesting to maintain old growth in boreal and sub-boreal forests. *For. Chron.* 75, 623–631.

CBD, 1995. *Malawi Principles*. Convention on Biological Diversity, United Nations Environment Programme, Nairobi, Kenya. http://www.cbd.int/ecosystem/principles.shtml.

CBD, 2004. The Ecosystem Approach, (CBD Guidelines) Montreal: Secretariat of the Convention on Biological Diversity. http://www.cbd.int/doc/publications/ea-text-en.pdf.

Costanza, R., 2000. Environmental sustainability, indicators, and climate change. In: M. Munasinghe, R. Swart, (eds.), *Climate Change and its Linkages with Development, Equity, and Sustainability.* Intergovernmental Panel on Climate Change (IPCC) Geneva, Switzerland, pp. 109–142.

Cyr, D., Gauthier, S., Bergeron, Y., Carcaillet, C., 2009. Forest management is driving the eastern North American boreal forest outside its natural range of variability. *Front. Ecol. Environ.* 10, 519–524.

Drapeau, P., Leduc, A., Bergeron, Y., 2009. Bridging ecosystem and multiple-species approaches for setting conservation targets in managed boreal landscapes. In: M.-A. Villard, B.G. Jonsson, (eds.), *Setting Conservation Targets for Managed Forest Landscapes.* Cambridge University Press, Cambridge, pp. 129–160.

Drever, C.R., Peterson, G., Messier, C., Bergeron, Y., Flannigan, M., 2006. Can forest management based on natural disturbances maintain ecological resilience? *Can. J. For. Res.* 36, 2285–2299.

Foster, D.R., 1983. The history and pattern of fire in the boreal forest of southeastern Labrador. *Can. J. Bot.* 61, 2459–2471.

Gauthier, S., Vaillancourt, M.-A., Leduc, A., De Grandpre, L., Kneeshaw, D., Morin, H., Drapeau, P., Bergeron, Y., 2009. *Ecosystem Management in the Boreal Forest.* Les Presses de l'Université du Québec, Québec, 568p.

Gromtsev, A., 2002. Natural disturbance dynamics in the boreal forests of European Russia: A review. *Silva Fenn.* 36, 41–55.

Gustafsson, L., Kouki, J., Svedrup-Thygeson, A. 2010. Tree retention as a conservation measure in clear-cut forests of northern Europe: A review of ecological consequences. *Scand. J. For. Res.* 25(4), 295–308.

Halme, P., Allen, K.A., Aunins, A., Bradshaw, R.H.W., Brumelis, G., Cada, V., Clear, J.L., et al. 2013. Challenges of ecological restoration: Lessons from forests in northern Europe. *Biol. Conserv.* 167, 248–256.

Harvey, B., Leduc, A., Gauthier, S., Bergeron, Y., 2002. Stand-landscape integration in natural disturbance-based management of the southern boreal forest. *For. Ecol. Manage.* 155, 369–385.

Helmisaari, H., Nikolov, N., 1992. Silvics of the circumboreal forest tree species. In: H.H. Shugart, R. Leemans, G. Bonan, (eds.), *A Systems Analysis of the Global Boreal Forest.* Cambridge University Press, Cambridge, pp. 13–84.

Johnson, E.A., 1992. *Fire and Vegetation Dynamics: Studies from the North American Boreal Forest.* Cambridge Studies in Ecology, Cambridge University Press, Cambridge.

Kasischke, E.S., 2000. Boreal ecosystems in the global carbon cycle. In: E.S. Kasischke, B.J. Stocks, (eds.), *Fire, Climate Change and Carbon Cycling in the Boreal Forest.* Ecological Studies Series, Springer-Verlag, New York.

Keto-Tokoi, P., Kuuluvainen, T., 2014. *Primeval Forests of Finland. Cultural History, Ecology and Conservation.* Maahenki, Finland, 302p.

Kneeshaw, D., Bergeron, Y., Kuuluvainen, T., 2011. Forest ecosystem structure and disturbance dynamics across the circumboreal forest. In: Millington, A., Blumler, M., Schickhoff, U., (eds.), *The SAGE Handbook of Biogeography.* Texas A&M. SAGE, London.

Kneeshaw, D.D., Bergeron, Y., 1998. Canopy gap characteristics and tree replacement in the southeastern boreal forest. *Ecology* 79, 783–794.

Kneeshaw, D.D., Gauthier, S., 2003. Old-growth in the boreal forest at stand and landscape levels. *Environ. Rev.* 11, S99–S114.

Kuuluvainen, T., 1994. Gap disturbance, ground microtopography, and the regeneration dynamics of boreal coniferous forests in Finland: A review. *Ann. Zool. Fenn.* 31, 35–51.

Kuuluvainen, T., 2002. Natural variability of forests as a reference for restoring and managing biological diversity in boreal Fennoscandia. *Silva Fenn.* 36, 97–125.

Kuuluvainen, T., 2009. Forest management and biodiversity conservation based on natural ecosystem dynamics in northern Europe: The complexity challenge. *Ambio* 38, 309–315.

Kuuluvainen, T., Aakala, T., 2011. Natural forest dynamics in boreal Fennoscandia: A review and classification. *Silva Fenn.* 45(5), 823–841.

Kuuluvainen, T., Aapala, K., Ahlroth, P., Kuusinen, M., Lindholm, T., Sallantaus, T., Siitonen, J., Tukia, H., 2002. Principles of ecological restoration of boreal forested ecosystems: Finland as an example. *Silva Fenn.* 36(1), 409–422.

Kuuluvainen, T., Grenfell, R., 2012. Natural disturbance emulation in boreal forest ecosystem management: Theories, strategies and a comparison with conventional even-aged management. *Can. J. For. Res.* 42, 1185–1203.

Kuuluvainen, T., Tahvonen, O., Aakala, T., 2012. Even-aged and uneven-aged forest management in boreal Fennoscandia: A review. *Ambio* 41(7), 720–737.

Lampainen, J., Kuuluvainen, T., Wallenius, T.H., Karjalainen, L., Vanha-Majamaa, I., 2004. Long-term forest structure and regeneration after wildfire in Russian Karelia. *J. Veg. Sci.* 15, 245–256.

Larsen, C.P.S. 1997. Spatial and temporal variations in boreal forest fire frequency in northern Alberta. *J. Biogeogr.* 24, 663–673.

Lännenpää, A., Aakala, T., Kauhanen, H., Kuuluvainen, T., 2008. Tree mortality agents in pristine Norway spruce forests in northern Fennoscandia. *Silva Fenn.* 42, 151–163.

Lecomte, N., Simard, M., Bergeron, Y., 2006. Effects of fire severity and initial tree composition on stand structural development in the coniferous boreal forest of northwestern Québec, Canada. *Ecoscience* 13, 152–163.

Lindenmayer, D.B., Franklin, J.F., Fisher, J., 2006. General management principles and a checklist of strategies to guide forest biodiversity conservation. *Biol. Cons.* 131, 433–445.

McCarthy, J., 2001. Gap dynamics of forest trees: A review with particular attention to boreal forests. *Environ. Rev.* 9, 1–59.

McCarthy, J.W., Weetman, G., 2006. Age and size structure of gap-dynamic, old-growth boreal forest stands in Newfoundland. *Silva Fenn.* 40, 209–230.

Messier, C., Puettmann, K.J., Coates, K.D., 2013. *Managing Forests as Complex Adaptive Systems. Building Resilience to the Challenge of Global Change.* Routledge, London.

Metsätilastollinen vuosikirja 1996–2009. Available at http://www.metla.fi/metinfo/tilasto/julkaisut/vsk/2002/index.html.

Millar, C.I., Stephenson, N.L., Stephens, S.L., 2007. Climate change and forests of the future: Managing in the face of uncertainty. *Ecol. Appl.* 17, 2145–2151.

Moen, J., Rist, L., Bishop, K., Chapin III, F.S., Ellison, D., Kuuluvainen, T., Petersson, H., Puettmann, K.J., Rayner, J., Warkentin, I.G., Bradshaw, C.J.A., 2014. Eye on the taiga: Removing global policy impediments to safeguard the boreal forest. *Conserv. Lett.* 7(4), 408–418.

Niklasson, M., Granström, A., 2000. Numbers and sizes of fires: Long-term spatially explicit fire history in a Swedish boreal landscape. *Ecology* 81, 1484–1499.

Ohlson, M., Kendrick, J.B., Birks, J.B., Grytnes, J.-A., Hörnberg, G., Niklasson, M., Seppä, H., Bradshaw, R.H.W. 2011. Invasion of Norway spruce diversifies the fire regime in boreal European forests. *J. Ecol.* 99, 395–403.

Payette, S., 1992. Fire as a controlling process in the North American boreal forest. *A Systems Analysis of the Global Boreal Forest.* In: H.H. Shugart, R. Leemans, G.B. Bonan, (eds.), Cambridge University Press, New-York, pp. 144–169.

Pennanen, J., 2002. Forest age distribution under mixed-severity fire regimes—a simulation-based analysis for middle boreal Fennoscandia. *Silva Fenn* 36, 213–231.

Pitkänen, A., Huttunen, P., Jugner, K., Tolonen, K., 2002. 10,000 year local forest fire history in a dry heath forest site in eastern Finland, reconstructed from charcoal layer records of a small mire. *Can. J. For. Res.* 32, 1875–1880.

Pitkänen, A., Huttunen, P., Tolonen, K., Jugner, K., 2003. Long-term fire frequency in the spruce dominated forests of Ulvinsalo strict nature reserve, Finland. *For. Ecol. Manage.* 176, 305–319.

Puettmann, K.J., Coates, K.D., Messier, C., 2009. *A Critique of Silviculture: Managing For Complexity.* Island Press, Washington, DC.

Pukkala, T., Lähde, E., Laiho, O., 2010. Optimizing the structure and management of uneven-sized stands in Finland. *Forestry* 83, 129–142.

Raymond, P., Bédard, S., Roy, V., Larouche, C., Tremblay, S., 2009. The irregular shelterwood system: Review, classification, and potential application to forests affected by partial disturbances. *J. For.* 107, 405–413.

Rist, L., Moen, J., 2013. Sustainability in forest management and the new role of resilience thinking. *For. Ecol. Manage.* 310, 416–427.

Rouvinen, S., Kuuluvainen, T., Siitonen, J., 2002. Tree mortality in a *Pinus sylvestris* dominated boreal forest landscape in Vienansalo wilderness, eastern Fennoscandia. *Silva Fenn.* 36, 127–145.

Ruel, J.-C., Fortin, D., Pothier, D., 2013. Partial cutting in old-growth boreal stands: An integrated experiment. *For. Chron.* 89(03), 360–369.

Ryan, K., 2002. Dynamic interactions between forest structure and fire behaviour in boreal ecosystems. *Silva Fenn.* 36, 13–39.

Schulze, E.D., Wirth, C., Mollicone, D., Ziegler, W., 2005. Succession after stand replacing disturbances by fire, wind throw and insects in the dark Taiga of central Siberia. *Oecologia* 146, 77–88.

Seymour, R.S., 2005. Integrating natural disturbance parameters into conventional silvicultural systems: Experience from the Acadian forest on northeastern North America. In: C.E. Peterson, D.A. Maguire, (eds.), *Balancing Ecosystem Values: Innovative Experiments for Sustainable Forestry*. USDA Forest Service Pacific Northwest Research Station, Gen. Tech. Rep. 635: Portland, Oregon, pp. 41–48.

Shorohova, E., Kneeshaw, D., Kuuluvainen, T., Gauthier, S., 2011. Variability and dynamics of old-growth forests in the circumboreal zone: Implications for conservation, restoration and management. *Silva Fenn.* 45(5), 785–806.

Shorohova, E., Kuuluvainen, T., Kangur, A., Jogiste, K., 2009. Natural stand structures, disturbance regimes and successional dynamics in the Eurasian boreal forests: A review with special reference to Russian studies. *Ann. For. Sci.* 66(2), 1–20.

Siiskonen, H., 2007. The conflict between traditional and scientific forest management in the 20th century Finland. *For. Ecol. Manage.* 249, 125–133.

Sirén, G., 1955. The development of spruce forest on raw humus sites and its ecology. *Acta. For. Fenn.* 62, 363.

Smirnova, E., Bergeron, Y., Brais, S., 2008. Influence of fire intensity on structure and composition of jack pine stands in the boreal forest of Québec: Live trees, understory vegetation and dead wood dynamics. *For. Ecol. Manage.* 255, 2916–2927.

Suffling, R., Smith, B., Dal Molin, J., 1982. Estimating past forest age distributions and disturbance rates in North-western Ontario: A demographic approach. *J. Environ. Manage.* 14, 45–56.

Syrjänen, K., Kalliola, R., Puolasmaa, A., Mattson, J., 1994. Landscape structure and forest dynamics in subcontinental Russian European taiga. *Ann. Zool. Fenn.* 31, 19–36.

Tahvonen, O., 2009. Optimal choice between even-and uneven-aged forestry. *Natural Resource Modeling* 22(2), 289–321.

Tahvonen, O., Pukkala, T., Laiho, O., Lähde, E., Niinimäki, S., 2010. Optimal management of uneven-aged Norway spruce stands. *For. Ecol. Manage.* 260, 106–115.

Vanha-Majamaa, I., Jalonen, J., 2001. Green tree retention in Fennoscandian forestry. *Scand. J. For. Res.* 16 (Suppl. 3), 79–90.

Van Wagner, C.E. 1978. Age-class distribution and the forest fire cycle. *Can. J. For. Res.* 8, 220–227.

Wallenius, T., Kauhanen, H., Herva, H., Pennanen, J., 2010. Long fire cycles in northern boreal Pinus forests in Finnish Lapland. *Can. J. For. Res.* 40, 2027–2035.

Wallenius, T., Kuuluvainen, T., Vanha-Majamaa, I., 2004. Fire history in relation to site type and vegetation in eastern Fennoscandia, Russia. *Can. J. For. Res.* 34, 1400–1409.

Warkentin, I.G. and Bradshaw, C.J.A. 2012. A tropical perspective on conserving the boreal 'lung of the planet'. *Biol. Conserv.* 151, 50–52.

Wein, R.W., MacLean, D.A., 1983. An overview of fires in northern ecosystems. In: R.W. Wein, D.A. MacLean (eds.), *The Role of Fire in Northern Circumpolar Ecosystems*. John Wiley, Chichester, pp. 1–18.

Weir, J.M.H., Johnson, E.A., Miyanishi, K., 2000. Fire frequency and the spatial age mosaic of the mixed-wood boreal forest in western Canada. *Ecol. App.* 10, 1162–1177.

Wooster, M.J., Zhang, Y.H., 2004. Boreal forest fires burn less intensely in Russia than in North America. *Geo. Res. Lett.* 31, 1–3.

Zackrisson, O., 1977. Influence of forest fires on the north Swedish boreal forest. *Oikos* 29, 22–32.

Zackrisson, O., Nilsson, M., Steijlen, I., Hörnberg, G., 1995. Regeneration pulses and climate-vegetation interactions in nonpyrogenic boreal Scots pine stands. *J. Ecol.* 83, 469–483.

13

Integrating Forest Restoration into Mainstream Land Management in British Columbia, Canada

Tanis L. Gower, Philip J. Burton, and Mike Fenger

CONTENTS

13.1 Introduction

Most land in the Province of British Columbia, Canada, has never been settled or cleared, and 94% of it remains in public hands (Government of British Columbia 2010a). The Ministry of Forests, Lands and Natural Resource Operations manages or administers approximately 80% of the land in British Columbia (BC), and an additional 14%—half of which is forested—is protected in federal and provincial parks (Government of British Columbia 2010a, b). Of the 60% of BC that is forested (55 million ha), 40% is managed for timber and fiber, and the remaining is considered uneconomic for commercial exploitation or is in protected areas (Government of British Columbia 2010b).

British Columbia is ecologically diverse and large, occupying an area of 950,000 km², greater than the area of France and Germany combined, and stretching from 48.3° to 60.0°N latitude. Its ecosystems range from offshore areas along its sinuous coastline with the Pacific Ocean, to alpine habitats greater than 4000 m above sea level. People come from around the world to visit BC's fog-enshrouded coastal rain forests, view its salmon runs, bears and whales, hunt big game in its northern wilderness, and to fish for steelhead trout in its glacier-fed rivers. These facts and the impressions they engender would suggest that there is little need for ecological restoration in such a natural corner of the world. Yet 4.6 million people and 150 years of agricultural expansion have replaced or degraded most of the province's rare southern, low-elevation ecosystems. A highly mechanized forest products industry logs an average of 190,000 ha of mostly wild (previously uncut) forest every year, and domestic livestock roam over hundreds of thousands of hectares of public land.

A 2001 review of the state of provincial ecosystems (Holt 2001) documented examples of ecosystem degradation in every part of the province (Table 13.1). At that time, government-sponsored programs were in place to fund riparian and stream habitat restoration projects, as well as to stabilize sediment sources and develop other approaches to forest restoration. In the period from 1995 to 2002, these programs, which were funded by royalties on timber production, supported a large and unprecedented effort to address damage done by past forestry practices. A great deal was learned, and this chapter presents case studies of projects from that era, highlighting experience gained.

Since 2002, large-scale government funding is no longer in place. Some smaller restoration efforts are ongoing with support from various funders, including a Foundation supported by hunters and anglers, and a compensation program to address the loss of fish and wildlife habitat associated with publicly owned hydroelectric dams. Private funding for ecological restoration work has also been available through various nonprofit foundations, and is commonly associated with community-based, volunteer-led initiatives. The most active restoration program in BC, the Rocky Mountain Trench Ecosystem Restoration Program, has acquired funding from some 30 different sources to fund over CAD$14 million in activities over 15 years (Rocky Mountain Trench Ecosystem Restoration Program 2014). These activities have been addressing forest in-growth due to fire suppression, and have wide social support as the restoration activities benefit ranchers, forest products companies, hunters, and human settlements affected by wildfire, as well as rare species and ecosystem types. This restoration program demonstrates that projects that provide clear benefits to people, including economic benefits, will have the widest support and the greatest chance of implementation, even during lean economic times.

The provincial government is currently coordinating the assessment of fish passage at road crossings in public forests. There is also a small amount of funding for restoring fish passage and for other restoration activities carried out by private forest products companies. However, at the time of publication, public forest management efforts in BC are focused mainly on timber and fiber production at the expense of other values. In this chapter we highlight forest restoration needs and solutions in BC, including the need to prevent and restore damage to forest attributes that support biodiversity, aquatic and wildlife habitats, and forest ecosystem resilience. The case studies highlighted here have been selected to share some lessons learned and to promote the inclusion of innovative restoration techniques and habitat and biodiversity management techniques in mainstream forestry management in BC, and in other similar forests worldwide.

TABLE 13.1

Causes of Ecological Degradation and Associated Restoration Needs in the Forested Ecosystems of British Columbia

Ecosystem Type	Ecosystem Characteristics	Restoration Challenges			
		Loss of Biodiversity Features	Liquidation of Old Growth	Invasive Species	Loss of Wildfire
Interior Cedar-Hemlock	Interior wet belt, highly productive and diverse forests, some very old forests	***	***	**	*
Interior Douglas-Fir and Ponderosa Pine	Low-elevation open forests of dry interior; historical regime of frequent surface fires	***	*	***	***
Montane Spruce	Interior transition between high and low elevations; lodgepole pine, stand-replacing wildfire, and insect outbreaks prevail	***	**		***
Engelmann Spruce-Subalpine Fir	High elevation interior forests; whitebark pine decimated by blister rust	***	***	**	
Sub-Boreal Pine-Spruce	Interior dry, cold, low-productivity zone dominated by lodgepole pine and wetlands; important habitat and lichen forage for caribou	**	**		**
Sub-Boreal Spruce	Productive interior forests with distinct tree species mix and extensive wetlands	***	***	**	***
Coastal Douglas-Fir	Dry coastal forests of limited extent; high species diversity and high urban encroachment	***	***	***	*
Coastal Western Hemlock	Extremely productive, complex, and old coastal forests with very high biodiversity values	***	***	*	
Mountain Hemlock	High-elevation coastal forests, now being logged as low-elevation forests have been depleted	*	***		
Boreal White and Black Spruce	Northern forest and muskeg east of the Rockies; heavy oil and gas exploration	**	**	*	**
Spruce-Willow-Birch	Sparse northern forests and shrublands west of the Rockies; primary land uses are mining and big-game hunting				**

Note: The number of stars indicates the degree to which the restoration challenge applies to the ecosystem type, with three stars the highest degree.

13.2 A Strategy for Forest Restoration

Ecological restoration is the process of assisting the recovery of an ecosystem that has been degraded, damaged, or destroyed (Society for Ecological Restoration 2004). As indicated in previous chapters, ecosystem restoration is a rapidly growing endeavor worldwide, and represents the widespread recognition that (a) ecological systems have been negatively impacted by human actions in the past and (b) that well-conceived human actions can subsequently reverse some of that damage (Burton and Macdonald 2011). In many ways, the roots of modern forestry and silviculture owe their origins to the same admissions in the eighteenth and nineteenth centuries, when controls on the rate of harvest and the planned regeneration of forests started to be instituted in Europe and subsequently in North America (Apsey et al. 2000).

Despite the site- and problem-specific nature of all restoration projects, some general themes and approaches in BC and other jurisdictions (e.g., Nuzzo and Howell 1990) often include the following components:

- *Prioritization and triage*—identifying those ecosystems (especially rare ones) and sites where intervention is urgently needed to save species at risk, or where restoration is expected to be comparatively effective for the least amount of effort; this implies the necessary postponement of work on more common ecosystems, those suffering from less severe problems, and those in which even heroic efforts may be unable to repair damage.

- *Prevention*—identifying ongoing threats to ecological integrity so that the agents of ecological degradation at the project site and elsewhere can be stopped before undertaking restoration work, to avoid the need for still more restoration.

- *Goal setting*—clearly stating measurable, meaningful, and achievable objectives, including social as well as ecological benefits where possible, to help recruit participation from the necessary agencies, corporations, and members of the public, and to keep projects "on track."

- *Continuity*—following through with effectiveness monitoring and maintenance activities, as most restoration efforts are multistep processes that typically need adjustment and repeat treatment to achieve their objectives, and to inform future projects.

Examples of past and current forest restoration activities in BC include the reintroduction of surface fires in dry-forest ecosystems, incorporating wildlife trees, coarse woody debris (CWD) or canopy gaps in homogenous secondary forests, the rehabilitation of compacted landings and access roads, the control of up-hill sediment sources to salmon-bearing streams, the promotion of late-successional vegetation in riparian areas, and the control of invasive species such as knapweed (*Centaurea* spp.), thistles (*Cirsium* spp.), and broom (*Cytisus scoparius*). Fuel management has been another driver of restoration after several destructive forest fires and the loss of homes in the rural–urban forest interface in 2003 (Filmon 2004). Since then, the reestablishment of more open, fire-proof ecosystems (as had historically prevailed) has been actively promoted (Rocky Mountain Trench Ecosystem Restoration Program 2014), and is eligible for funding under BC's Strategic Wildfire Prevention Initiative (UBCM 2014).

13.3 Overview of British Columbia's Forested Ecosystems

British Columbia's diversity has been classified into 16 biogeoclimatic zones that identify potential climax vegetation and prevailing climatic constraints, and serve as the basis for most natural resources planning in the province (Pojar et al. 1987; Government of British Columbia 2012). Undisturbed forests in each of these zones and the variation they exhibit over topographic gradients provide templates for ecological restoration and help identify opportunities and constraints to forest manipulation. Zones dominated by coastal influences include wet rain forests of the Coastal Western Hemlock (CWH) zone, the drier and sometimes more open forests of the Coastal Douglas-Fir (CDF) zone (situated in the rain shadow of Washington state's Olympic Mountains and the Vancouver Island mountains), and the high-elevation Mountain Hemlock (MH) zone in the coastal mountains. The southern interior of the province, much of which is in the rain shadow of the Coast and Cascade Mountains, includes open Bunchgrass (BG) grasslands at the lowest elevations, above which are found Ponderosa Pine (PP) savannahs, and both uneven-aged and even-aged forests of the Interior Douglas-Fir (IDF) zone. The Montane Spruce (MS) zone is found on high plateaus and mountain slopes above the IDF where the climate is dry, while the windward slopes of the Cariboo, Selkirk, and Rocky Mountains support an interior rain forest or "wet belt" known as the Interior Cedar-Hemlock (ICH) zone. Subalpine forests in both the southern interior and northern interior are classified as the Engelmann Spruce—Subalpine Fir (ESSF) zone. The central plateaus of the northern interior are dominated by the Sub-Boreal Spruce (SBS) zone and the drier, colder Sub-Boreal Pine-Spruce (SBPS) zone. Colder and transitional subzones of the ICH zone are also found in northern BC. The extensive plains of northeastern BC (on the leeward side of the Rocky Mountains) and the northwestern valleys are part of Canada's boreal biome, and are referred to as the Boreal White and Black Spruce (BWBS) zone. In the northern one-third of the province, subalpine tree cover is discontinuous and often dwarfed, so these high-elevation areas are classified as the Spruce-Willow-Birch (SWB) zone. Alpine tundra is found at the highest elevations throughout the province, encompassing alpine meadows as well as large areas of bare rock and permanent snow and ice fields. The alpine is differentiated into Coastal Mountain-heather Alpine (CMA), Interior Mountain-heather Alpine (IMA), and Boreal Altai Fescue Alpine (BAFA) zones, reflecting major climatic and geographic divisions in this high-elevation treeless environment (Government of British Columbia 2012). More detailed descriptions of these zones are provided by Meidinger and Pojar (1991) and the Government of British Columbia,* and differences among ecosystem restoration priorities and activities in BC (Table 13.1) must be considered in this biogeoclimatic context.

13.4 Forest Management in British Columbia

Forest products industries, especially those associated with logging, silviculture, the milling of dimensional lumber, panelboard manufacturing, and the production of pulp and

---

* See an interactive map of biogeoclimatic zones at: http://www.for.gov.bc.ca/hfd/library/documents/treebook/biogeo/biogeo.htm.

paper, employ 56,400 BC residents directly. Together with associated economic spin-offs, this sector accounts for an estimated 7% of the employment and 6% of the gross domestic product (GDP) of the province. These numbers are down from 90,000 employed (14% of employment) and 13% of GDP as recently as a decade ago.

Plans for road development, logging, and silviculture are prepared by government foresters on part of the forest land base, while most commercial forest lands are directly managed by company foresters responsible for Tree Farm Licenses (area-based tenures) or for particular operating areas within broader Timber Supply Areas (under a timber volume-based tenure) that are covered by Forest Stewardship Plans subject to government approval.

From a global perspective, many forest management policies and silvicultural practices prevalent in BC are sustainable, environmentally sound, and represent implementation of some principles of conservation biology and restoration ecology (Fenger 1996a; Burton 1998; Burton et al. 2006). For example, BC's forests remain almost exclusively composed of indigenous species (unlike the widespread practice of establishing exotic tree crops in the British Isles, Chile, New Zealand, and elsewhere), and all cutover forests are likewise regenerated to native tree species of suitable genetic origin. These practices reflect BC's good fortune in being home to some of the world's most productive temperate tree species, and in being characterized by sharp differences in climate where the necessity of genetic adaptation is readily apparent. The basic elements of reforestation and the establishment of native species, key to the restoration of many other temperate forest ecosystems (Stanturf and Madsen 2002), have been a responsibility of forest products companies operating in BC since 1987. Ecologically based site stratification, diagnosis and regeneration planning are now standard components of all silvicultural prescriptions. A mixture of tree species is frequently planted, and the tolerance for noncrop hardwood species is increasing, though the economic and regulatory bias favoring conifers remains (Fenger 1996b). Efforts at landscape planning and the designation of riparian buffers, wildlife tree patches, forested ecosystem networks, and old-growth management areas initiated in the 1990s were making progress in retaining more diversity on the landscape than when trees were managed under the pretense that they were an agricultural crop (Fenger 1996b). Unfortunately, political and market changes since 2001, coupled with the drive to salvage as much timber as possible after a widespread mountain pine beetle (*Dendroctonus ponderosae*) outbreak (Burton 2010; Burton et al. 2015), has led to significantly accelerated harvest rates with no commensurate increase in conservation measures.

Through decades of protest, land-use negotiations, and changes of government, the forest products industry has remained the primary agent of landscape change, and is a key player in the economy and politics of the province. Consequently, any serious effort at reforming forest management practices or introducing forest restoration techniques must effectively engage land managers who come from a professional culture that is driven by commercial imperatives and a fiber-based emphasis. There has been a growing receptivity to the concepts of ecosystem management, sustainable forest management, and ecological restoration as younger foresters make their way up the management hierarchy, but implementation of ecological stewardship is limited by political priorities, global competition, and commercial imperatives.

Clear-cut logging and even-aged stand management remain the norm, although partial cutting and variable retention harvesting (Franklin et al. 1997) are on the increase (Beese et al. 2003). In many northern and interior forests, even-aged management is often appropriate, as even-aged stands of lodgepole pine (*Pinus contorta*) historically experienced wildfire approximately every 80–120 years. It is widely recognized that understanding and mimicking natural disturbance regimes is the most effective approach to conserving native species and ecological processes at a landscape and stand level (Government of British Columbia 1995a; Seymour and Hunter 1999; Perera et al. 2004). For coastal forests (and wetter interior forest types), this would mean landscape level conservation planning, and using other options in addition to clear-cut harvesting and even-aged management, as these ecosystems experience wildfire and other large-scale natural disturbances only rarely, and are historically multiaged (McClellan et al. 2000; Beese et al. 2003; Deal 2007). Unfortunately, provincial policy has long been to liquidate old-growth forests to establish more productive second-growth stands. In addition, the rate of cut in many land management units is greater than the long-run sustained yield. As a result, important structures, functions, and composition associated with older and naturally disturbed forests are not sufficiently protected or generated. So despite low human populations and wide open spaces, much of the managed forest land in BC would need restoration, if the biodiversity and ecosystem services of old and diverse forests are desired in some landscapes.

Since the mid-1980s, and particularly in the 1990s, the provincial government introduced requirements to manage important habitats and biodiversity within areas to be logged. However, economic pressures and changing political priorities have limited or reduced the implementation of measures to manage nontimber values. Also, there has been a shift in the way that BC's forests are managed, giving more discretion to the private sector to meet broad government objectives. The government's role is now primarily limited to assessing conditions after harvest and stand establishment, and determining whether the results have achieved or will achieve documented objectives. There are no consequences to licensees if their results and strategies do not meet the government objectives, which are also very difficult to measure.

Some tenure holders also maintain third-party certification that provides greater transparency and standards than that set by government. As a result of certification, there is some interest in incorporating ecosystem restoration practices into day-to-day forest management in BC, for example, to protect biodiversity, rebuild soil productivity, aid in the recovery of salmonid populations, or to bolster the credibility of sustainable forest management plans needed under the Sustainable Forestry Initiative. Ecosystem restoration may also be required for corporate applications for forest product certification under the more stringent Forest Stewardship Council standards.

It has been argued that forest restoration can make good business sense for forest products companies, aiding in certification and market access, minimizing costs associated with the liability for untreated ecological damage, and as a component of a risk management strategy (Douglas and Burton 2005). Nevertheless, large-scale forest restoration is not currently happening in BC, with two or three exceptions. There is a program in the Rocky Mountain Trench to create more open habitats and open forests, to the benefit of rare ecosystems and species as well as forest products companies, hunters, ranchers, and rural human communities. There is also a process underway in the mid-coast region known as the "Great Bear Rainforest," where forest products companies have been cooperating directly with environmental groups and First Nations to carry out conservation planning and ecosystem-based management plans. In the first example, forest restoration is clearly providing benefits to people, allowing its implementation even through lean economic times (Rocky Mountain Trench Ecosystem Restoration Program 2014). In the second example, direct action by environmental groups affected markets for forest products from the Great Bear Rainforest, bringing industry to the table to discuss new ways of managing the area that would give them greater social license (Price et al. 2009).

13.5 Forest Restoration Priorities in British Columbia

We here review some of the priorities for ecological restoration in different forest zones of the province, and then showcase a number of recurrent ecological issues and some cases of on-the-ground forest restoration activities that have addressed them. Eight broad ecological issues are identified, with general statements of their impact and options for remediation and prevention; five case studies provide a more detailed description of some on-the-ground restoration activities. The authors of this chapter participated in the evaluation of the restoration activities described below, but did not design or implement most of them, nor monitor their effectiveness.

While BC is predominantly a forested jurisdiction, it is the comparatively rare grasslands, shrublands, and open meadows that receive most of the attention for conservation and restoration. The most degraded ecosystems in BC are those southern-most habitats in which European settlement has been most pervasive over the last 150 years. Most of the land formerly occupied by Garry oak (*Quercus garryana*) savannahs and meadows is now dominated by privately held residential and commercial land in the greater Victoria area on southern Vancouver Island and on the Gulf Islands. The restoration of Garry oak ecosystems is well informed by the resources and energy of an active Garry Oak Ecosystems Recovery Team, with annual symposia or research colloquia, and widely embraced recovery plans that address both private and public lands (Fuchs 2001; Douglas and Smith 2006). Likewise, much of the land formerly occupied by antelope brush (*Purshia tridentata*) scrub and bluebunch wheatgrass (*Pseudoroegneria spicata*) grasslands in the southern Okanagan Valley is now occupied by homes, orchards, vineyards, and recreational property. Restoration activities in these ecosystems have required a cooperative effort spanning all levels of governments and a large degree of public education and volunteer involvement. It is fortunate, yet ironic, that the greatest resources for ecological restoration (high public awareness, large pools of volunteers) are found in the very population centers that were responsible for the degradation of rare ecosystems in the first place.

In the mainly forested regions of the provinces, ecological degradation is mostly related to the impacts of industrial forestry, mining, and associated road development, rather than as a direct consequence of human population pressures and land use conversion. On the other hand, the role of provincial forest land in providing forage for livestock has also prompted some interest in protecting and restoring the open habitats that provide the best grazing land and habitat for many of the rarer species in BC (Neal and Anderson 2009). Four sources of degradation are noted throughout many of BC's public forest lands (Holt 2001; Neal and Anderson 2009; SERN 2013):

1. The loss of structural and compositional diversity, in terms of
   a. Physical habitat elements such as large trees, snags and fallen logs, canopy layering, and mixed tree species composition at the stand level.
   b. Natural variation in age- and size-class distributions at the landscape level.
2. Liquidation of old-growth forests and stands of the largest and most productive trees as a result of commercial logging.
3. Invasive species, especially shrubby and herbaceous plants of Eurasian origins, which tend to establish and persist in open habitats.

4. The loss of wildfire (whether burning at ground level or through tree crowns) as a generator of biocomplexity and an agent of stand renewal and compositional diversity.

More information on causes of ecological degradation and associated restoration challenges are described by ecosystem type in Table 13.1. Each of these forested ecosystem types have high-priority restoration needs in certain components, subzones, or areas. Many of these restoration needs will be shared with the biogeographically similar US Pacific Northwest and Inland Empire (O'Hara and Waring 2005; Jain and Graham 2015). Likewise, lessons applicable in BC are on offer from the dominance of industrial forestry and efforts to restore its ecological damage in Fennoscandia (Angelstam et al. 2005; Kuuluvainen et al. 2015).

13.6 British Columbia Experience in Addressing Key Ecological Issues

Integrating ecosystem approaches into management—such as mimicking natural disturbance patterns during forest harvesting (Seymour and Hunter 1999; Perera et al. 2004)—can minimize the need for forest restoration in the future. However, there is an extensive backlog of land suffering from some degree of ecological degradation and simplification, calling for the application of a variety of restoration techniques. In this section, we explore some of the more widespread issues and priorities described above, with emphasis on those that have been addressed through restoration projects in BC.

Ecological restoration includes a very broad range of activities—from reconstructing a local plant community such as a riparian fringe, to restoring ecological processes such as fire in a large area of forest. Impacts and implications are outlined for different types of restoration needs in BC, followed by suggested preventative and restorative measures. Case studies are presented to detail how government agencies, forest products companies, and community groups have been undertaking these types of restoration projects.

13.6.1 Ecological Issue: Lack of Natural Structural Elements or Species in Second Growth Stands

Extensive areas of dense, young, closed canopy stands provide little habitat for many forest species. This loss of natural stand structure (and species and genetic mixes) leads to a loss of biodiversity, wildlife and aquatic habitat, and ecological resilience. The loss of natural stand structure, species, or genetic mix in BC is due to

- Highly uniform harvesting practices, particularly clear-cutting without appropriate retention of wildlife trees, live and dead large wood pieces in riparian areas, and wildlife tree patches.
- Silvicultural practices resulting in high stand uniformity—development of dense monocultures, highly uniform tree spacing, and very high levels of crown closure.
- Short-rotation forestry resulting in lack of recruitment of important stand structural attributes.

Plantations can be of limited habitat value due to high uniformity and lack of stand structural elements like large wildlife trees and CWD. Shrubs and herbs can be virtually eliminated due to high crown closure, and highly uniform stands also are at higher risk of pest
and disease damage. These issues can be prevented through proactive design of harvest
activities to retain structural elements. In addition, species and genetic diversity can be maintained during reforestation. Reforestation can also be designed to produce more diversity in
densities through management of stocking levels and minimum spacing between trees.

In areas where complex stand structure and tree species diversity have been lost, restoration approaches can include patchy precommercial or commercial thinning to reduce
crown closure and stand uniformity. A more complex stand structure can be introduced
in selected areas through wildlife tree and CWD addition techniques. Restoring the abundance of structural elements (such as wildlife trees, CWD, and canopy gaps) is a common
feature of many forest restoration activities. Such activities are also used for the fine-filter
management of species dependent on old-growth forests (such as the northern spotted
owl, *Strix occidentalis caurina*) in areas where most old-growth forest has already been
logged, but second-growth forest is well developed. Northern spotted owls require relatively open habitats with large-diameter trees and snags, a multilayered canopy, and relatively high amounts of shrub and CWD cover—conditions typical in old forests (Meyer
2007). Commercial thinning is being used in southwestern BC to improve habitat for spotted owls in 60-year-old Douglas-fir (*Pseudotsuga menziesii* var. *menziesii*) stands, enhancing
their future value from Type C ("suitable") to Type B ("moderate") spotted owl habitat,
while at the same time making a profit. Although these habitat improvements have generally been deemed successful, the continued decline of spotted owl numbers in BC have
prompted an emphasis on control of the competing barred owl (*Strix varia*) and captive
breeding instead (Fenger et al. 2007).

13.6.1.1 Case Study 1: Fungal Inoculation to Create Wildlife Trees

Large, old trees—both dead and alive—with cavities of various sizes and in different stages
of decomposition are key habitat features for wildlife in all forests (Fenger et al. 2006). These
trees are commonly referred to as "wildlife trees." Fungal inoculation is a relatively new and
promising technique used to create wildlife trees. Methods are still under development, but
appear to be a highly efficient and effective means for recruiting one of the most valuable
structural elements of wildlife habitat—a tree that contains heart rot. Trees in this condition
are excavated for use by wildlife and eventually become hollow trees, further increasing their
habitat value. Eventually they break apart to become fallen woody debris and hollow logs.

Fungal inoculation is used on live, healthy trees to initiate heart rot. The fungus does
not usually kill the tree; instead a compartmentalized decay column is produced in the
live tree within 3–6 years. The inoculated tree is able to maintain its foliage and growth
form, and continues to put on new incremental growth. Wildlife trees in this condition (as
opposed to a dead snag) will provide habitat for a longer period of time, and will provide
few worker safety or operational concerns. They will also continue to function as a seed
source, and are less likely to be felled by firewood cutters.

Two inoculation procedures have been used in BC and results have been compared. In
one method, spores of a native heart rot fungus (*Phellinus pini*) are injected into the tree by
climbing it, drilling a hole, and inserting a wooden dowel that is cultured with a locally
collected strain of the fungus (Figure 13.1). The second technique uses a rifle to shoot the
tree trunk with a bullet that consisting of a smaller wooden dowel cultured with the same
fungus. Both techniques result in fungal decay spreading within the tree above and below

FIGURE 13.1
This ponderosa pine at Fort Shepherd (Kootenay Region, B.C.) was partially topped and full-ring girdled, and inoculated above the girdle in the remaining upper section. A tree with a dead, decaying top will result, providing a combination of short and medium-term habitat supply. Note the artificial scarring which was added to simulate stem damage, this provides a visual stimulus to cavity excavators (Photo by Todd Manning).

the point of inoculation (Manning 2014). The decayed (and eventually hollow) bole center that results can provide cavity nesting and feeding habitats that are critical for many forest species. There is virtually no risk of unwanted spread to nontarget trees using these techniques because of the natural reproductive history of the fungi.

Data from the US Pacific Northwest show that fungal inoculation is less expensive and faster than techniques such as topping trees with a chainsaw or explosives, top girdling, or cavity creation using chainsaws. However, all these techniques have shown some success at providing tree creation techniques have the advantage of creating or enhancing habitat in a relatively short period of time (10–15 years), as opposed to recruiting similar stand structure through natural cycles. In contrast, it usually requires ≥100 years to naturally recruit trees of sufficient size and condition to function as useful wildlife trees.

In British Columbia, various operational trials as well as ecosystem restoration treatments involving fungal inoculation have been conducted since 2002. These have included work in second growth Douglas-fir forests on Vancouver Island (Manning 2014), and dry interior Douglas-fir—ponderosa pine forests in the East Kootenay region of British Columbia. The results to date, particularly from the Kootenay region, are very promising (Manning and Manley 2014), as these latter projects involved significant improvements to the inoculation techniques which were not applied in earlier projects.

Fungal inoculation can be used in a variety of wooded habitats that require restoration or enhancement. For example, the method can be used to

- Create more diverse stand structure and suitable wildlife habitat in riparian management areas.
- Accelerate the production of mature forest elements (e.g., wildlife trees, snags and CWD) in large areas of relatively homogeneous immature forest; and directly increase the habitat supply for species that require mature forest elements, either directly or indirectly (e.g., including woodpeckers, owls, fishers and goshawks).

13.6.2 Ecological Issue: Invasive Species Replacing Native Species

Human disturbance and some kinds of management activities have favored invasive plant species, which can replace native species and degrade native ecosystems. Invasive species can also reduce production of economically valuable native species, or make it more difficult to reestablish native species. In BC, an estimated 200,000 ha of grassland and open forest are infested with a variety of invasive plant species, and at least another 20 million ha are susceptible to invasion (Forest Practices Board 2010).

Problems with invasive plant species can be prevented by minimizing site disturbance and by prompt revegetation and reforestation. Areas of bare ground should be seeded or planted with native species. Plant materials used for erosion control should also be carefully selected to ensure that invasive species are not introduced. Once invasive plants are discovered, they should be proactively removed, particularly those species which are relatively recent arrivals and whose spread can be contained. Current forestry law in BC contains obligations to control invasive plants, but it is unclear whether there are any repercussions if this does not occur.

13.6.2.1 Case Study 2: Scotch Broom Control

Numerous groups and volunteers use manual pulling to reduce populations of the invasive leguminous shrub Scotch broom (*Cytisus scoparius*) throughout the CDF zone on Vancouver Island. This includes agency and nongovernmental programs (Garry Oak Ecosystems Recovery Team 2014; Parks Canada 2014) which are in place for broom removal in rare Garry Oak meadows that are part of the CDF zone.

Like most invasive species, broom is exceptionally difficult to control due to its lack of natural enemies, its ability to outcompete native plants, and its profuse production of seeds that persist in seed banks. Broom can significantly reduce conifer regeneration in clearcuts, while in Garry oak meadows it threatens endangered plant communities (Fuchs 2001; Burton 2002). New invasions of broom have been found on Haida Gwaii (formerly known as the Queen Charlotte Islands), on the Lower Mainland, and around Castlegar and Kootenay Lake (Prasad 2003). Scotch broom is just one of several invasive species being battled by the Coastal Invasive Species Committee, which in turn is just one of 17 regional weed committees in the province that are addressing the threat of invasive species. For example, the Japanese knotweed (*Fallopia japonica*) has also become a problem on BC's south coast.

The most effective method of broom removal is by manual cutting or pulling, with repeated treatments usually required at any given site. Care needs to be taken to minimize soil disturbance to ensure that conditions are not recreated that are suitable for additional inseeding by broom or other exotic weeds. Preventing the spread of broom through these manual methods must continue until an acceptable biological control agent is developed. The Canadian Forest Service recommends the following actions to forest managers (Prasad 2000, 2003):

- Carefully inspect road ballast and materials brought from other areas for broom seed.
- Bring new invasions to the attention of researchers, forest managers, and the local British Columbia Resource District office.
- Cut broom stems as close to the ground as possible before the seed matures, taking care not to disturb the surrounding soil; hand-pulling is preferable to prevent resprouting, but this is not always practical.

- Remove broom before it has a chance to flower.
- Inspect conifer plantations on a regular basis.
- Support urban efforts to remove Scotch broom in parks.
- Where broom invasion is likely, selective timber extraction should be considered over clear-cutting.
- On warm, well-drained sites, avoid excessive soil disturbance and exposed mineral soil.

13.6.3 Ecological Issue: Reduction and Changes to Open Habitat Types

In BC and other jurisdictions (e.g., O'Hara and Waring 2005; Jain and Graham 2015), fire control has had a major impact on open habitat types such as grasslands, savannahs, and open forests. Forests in the drier areas of BC (i.e., the IDF, PP, and BG biogeoclimatic zones, see Table 13.1) are often ingrown at densities much higher than historical levels, due to decreased fire frequencies since European settlement (Figure 13.2). The result is a loss of species richness and diversity, and a loss of habitat for those species that depend on open habitats or that depend on fire for some part of their life history (Klenner et al. 2008). There are also economic impacts due to the loss of grazing lands for cattle, and the lack of vigorous trees that would occur at lower densities. The increased likelihood of catastrophic wildfire also increases the risk of serious economic losses for forestry interests and for rural settlements adjacent to forest lands.

Restoration and management approaches to maintain open habitats include commercial and precommercial thinning to reduce stand densities, clearing of trees from historic

FIGURE 13.2
Treatment of densely ingrown Douglas-fir stands in the IDFxm subzone of the Cariboo Forest Region, British Columbia. (a) Prior to any treatment; (b) after logging (commercial thinning); (c) after logging slashing (non-commercial thinning); and (d) during a light underburn treatment. (Photos by Ordell Steen.)

grassland areas, and controlled reintroduction of ecosystem-maintaining fire. As noted above, addressing forest ingrowth and reintroducing fire is currently the most active forest restoration activity occurring in BC today, and is addressed through multiple programs in western states of the United States as well (see Jain and Graham 2015).

13.6.4 Ecological Issue: Extensive Uniform Stands

Forest stands that are even-aged with uniform stand characteristics can occur across extensive areas. This is usually a result of harvesting large areas over a short period of time, or can be due to fire control increasing the dominance of a single age class. Having such uniform stands over such extensive areas is a concern, as it results in loss of essential habitats for those species dependent on certain forests structures and stand ages. The uniformity also increases the risk of large-scale pest outbreaks, disease, or wildfires.

More diverse and ecologically healthy stands can be created by designing management activities to mimic natural disturbance patterns. This can be done by planning landscape-scale harvesting that is more like natural disturbance (windthrow, windstorms, fire, pests, and disease), in that it leaves a great variety of biological legacies (large logs, snags, clusters of healthy and lightly burned trees) which many forest-dwelling species are dependent on. For instance, where logging now replaces wildfire as the dominant agent of ecological disturbance, timber harvesting activities can be done in a manner to create irregular stand structures and patterns (Seymour and Hunter 1999). Careful harvest planning will also avoid the problem of having a large amount of forest at similar ages or with similar stand structures, and regeneration plans can be used to increase the mix of tree species and ages within a given stand. When salvage harvesting is done after a major fire or pest outbreak, it is important to follow ecological best management practices (see Lindenmayer et al. 2008) so that a diverse stand structure and landscape pattern can be maintained.

When a more diverse stand is desired, stand heterogeneity can be significantly increased through precommercial thinning and commercial thinning to produce a variety of stand densities and within-canopy openings. Stand structure can also be introduced in selected areas of the landscape through wildlife tree and CWD addition techniques.

13.6.4.1 Case Study 3: Restoring Structural Diversity to the Kitimat Valley

The Kitimat Valley in northwestern BC has been extensively logged, with less than 3% of its original CWH forest remaining. Since harvest, even-density spacing and intensive pruning have created second-growth habitat with little structural diversity in horizontal or vertical dimensions. This extensive uniformity of secondary forest habitat limits opportunities for mammals, birds, and other elements of biodiversity dependent on snags, CWD, or canopy variability associated with old forests. The Kitimat Valley has approximately 19,400 ha of these heavily managed pole-sapling stands on a productive, but heavily roaded valley bottom. Coarse-filter biodiversity conservation measures such as stream-side buffers and wildlife tree patches are no longer options in this landscape because little mature forest remains. Improving biodiversity in this landscape requires stand-level intervention, to introduce stand elements upon which many species depend. In 2001, a pilot project was initiated to reintroduce old-growth stand characteristics through stand management techniques. The long-term goal was to encourage the return of bird species temporarily lost from the area.

West Fraser Timber Co. Ltd. (Skeena Sawmills Division) and the BC Ministry of Forests worked together to complete year one of the project. Variable spacing, girdling, and installation of snags and CWD piles were used to create islands of old-growth characteristics within the mosaic of homogenous (young, second-growth, conifer-dominated) commercial forest. An important part of this project was determining the feasibility of incorporating these techniques into mainstream, volume-oriented stand management. Various baseline data were also collected to allow future monitoring of the success of this project.

The treated stands were primarily planted spruce (*Picea sitchensis* and its natural hybrids with *P. engelmannii* and *P. glauca*) affected by spruce leader weevil (*Pissodes sitchensis*), along with lower densities of naturally established western hemlock (*Tsuga heterophylla*), fir (*Abies lasiocarpa* and *A. amabilis*), and western redcedar (*Thuja plicata*). These stands were chosen because of their high densities (greater than 4000 stems/ha), their lack of previous stand treatments, and because they were relatively young (20–30 years old). Spacing brought stand densities down to 800 stems/ha in many areas, 400 stems/ha in others, with no treatment of control areas. A proportion of trees were girdled as part of the treatments. Once spacing was complete, CWD piles were created, and imported snags were installed. The CWD piles were created by piling slashed material and placing cast-off cedar logs on top. The snags were logs purchased from Skeena Sawmills and were planted in upright positions using two backhoes. The Terrace Rod and Gun Club provided machine time and qualified operators at reduced cost.

Adding CWD and snags to second-growth stands to the levels naturally found in old-growth forests (e.g., average levels of 300–500 m^3 ha^{-1} in the CWH zone; Stevens 1997) would be extremely costly. Instead, these structural elements were obtained and placed at lower densities on the basis of the availability of suitable logs. As a result, the snags and CWD piles were concentrated in two areas. However, it is expected that their addition will significantly improve habitat values over existing conditions.

Forest Renewal BC funded this project, and project costs were tracked to determine operational feasibility. If one-time costs are removed and anticipated efficiencies achieved, costs were approximately CAD \$3200 ha^{-1} (2014 dollars) for the package treatment of spacing, and snag and CWD addition. This compares very favorably to restoration silviculture projects elsewhere in the province, but is approximately twice the cost of standard pre-commercial thinning treatments at this stage of stand development. Breeding bird surveys and other monitoring activities conducted by volunteers continue to assess its success over time, and these informal evaluations have confirmed their effectiveness (Brad Pollard, pers. comm., 2014). The obvious difficulty in projects of this type is to treat enough area to have a landscape-scale effect.

13.6.5 Ecological Issue: Under-Represented Stand Types in the Landscape

Some landscapes do not contain the full natural range of potential forest stand types. This can be due to preferential harvesting of particular species, age classes, or stand structures at unsustainable levels. It can also happen as a result of fire control that affects the rate of replacement of specific stand types, or tree mortality due to extreme fire or insect outbreak events, or the effects of exotic pests such as fungal diseases.

If certain forest habitats, stand structures, tree species, or age classes are lost, it will affect the species dependent on them. Loss of stand types can result in low or endangered population levels of dependent species. Future economic opportunities can also be lost. There is also an increased risk of catastrophic damage (e.g., from herbivorous insects that have preferred species and sizes of host trees) in over-represented forest age or species types.

As a preventative measure, harvesting can be designed to retain representative amounts of all stand types (based on compositional and structural criteria). Restoration approaches include the following:

- Thinning and planting to increase the future supply of under-represented stand types.
- Use of slashing and prescribed fire to increase supply of open forest stand types, where ecologically appropriate.
- Reintroduction or recovery, through breeding and nursery research, of tree species that have been devastated by exotic pests or have been overharvested.

Under-represented stand types can be defined compositionally (e.g., those dominated by western white pine, *Pinus monticola*), structurally (e.g., uneven-aged, multistoried stands), or through combinations of composition, structure, and site type (e.g., stands of large, old white spruce, *Picea glauca*, on boreal floodplains).

Western five-needle pines have been devastated by the arrival from Asia a century ago of white pine blister rust, *Cronartium ribicola*. Efforts to restore western white pine stands in BC and neighboring US states have consisted of a two-pronged approach: (1) reducing mortality in regenerating stands by pruning lower branches to a height of three meters; and (2) selecting rust-resistant lines in tree breeding programs. Earlier efforts to control the spread of the disease through eradication of alternate hosts (shrubs of the genus *Ribes*) were largely ineffectual (Government of British Columbia 1996; Maloy 1997). As the range of blister rust continues to expand northward and to higher elevations, the challenge is now to save stands of whitebark pine (*Pinus albicaulis*), a high-elevation species recently declared to be federally endangered in Canada. Often found at alpine treeline, this species has also suffered from mountain pine beetle attack, produces large seeds that are dependent on Clark's nutcracker (*Nucifraga columbiana*) for dispersal, and benefits from open (typically fire generated) habitats for regeneration. Genetic research is now focusing on screening populations for blister rust resistance, and seedlings propagated from such populations now constitute the focus of whitebark pine restoration efforts (McLane and Aitken 2012; Haeussler et al. in press).

BC Parks undertook restoration efforts after removing a campground from a rare old-growth Sitka Spruce—Salmonberry ecosystem on the Exchamsiks River in northwestern BC (Burton and Burton 2002). "Old-growth restoration" is certainly a long-term prospect, but it is expected that various structural and compositional goals can be accelerated through the placement of large CWD, and the planting of salmonberry (*Rubus spectabilis*) and redosier dogwood (*Cornus stolonifera*) shrubs as well as Sitka spruce (*Picea sitchensis*) and other conifer seedlings (Figure 13.3). A decade later, much of the area formerly used for road access and parking recreational vehicles is now dominated by thickets of red alder (*Alnus rubra*) with scattered Sitka spruce seedlings underneath (pers. obs.). Old-growth restoration is increasingly needed to achieve landscape-level goals for forest seral class distributions and connectivity. Many restoration efforts targeting particular stand types are often coupled with other objectives, such as maintaining critical wildlife habitat or riparian zone integrity, as discussed next.

13.6.6 Ecological Issue: Riparian Habitat Integrity

Riparian habitat includes forested areas alongside streams and rivers. Historically these areas were harvested to the stream bank, or only a narrow buffer was left. This has resulted

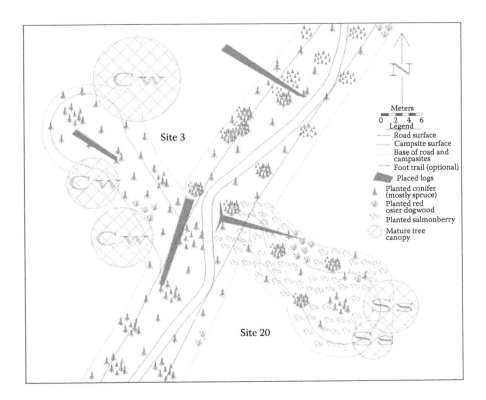

FIGURE 13.3
The first steps in restoring a rare old-growth Sitka Spruce—Salmonberry ecosystem after the closing of a campground require not only the planting of trees, but also the placement of very large pieces of coarse woody debris, and the planting of shrubs and ferns. (From Burton, P.J., and C.M. Burton. 2002. *Phase II Campground Restoration Prescriptions for Exchamsiks River Provincial Park*. Prepared for BC Parks, Terrace Field Office, BC Ministry of Water, Land and Air Protection. Symbios Research & Restoration, Smithers, BC.) Cw and Ss denote crowns of mature western redcedar and Sitka spruce trees, respectively.

in limited habitat value in formerly biodiverse areas—riparian areas frequently contain the highest number of plant and animal species found in forests, and provide critical habitats, home ranges, and travel corridors for wildlife (Décamps et al. 2010). Healthy riparian areas are also critical for the proper functioning of the stream ecosystems they surround (Government of British Columbia 1995b).

When riparian buffers are not retained during harvest, the result is a loss of stand structural elements (wildlife trees, CWD) and older age classes in areas highly important for biodiversity conservation. There is a critical lack of large woody debris input to stream channels, and subsequent impacts to fish habitat. In some cases, formerly coniferous or mixed forests are converted to deciduous stands or brush. These changes can mean a loss or reduction in animal movement corridors. Stream shading, nutrient inputs, and water quality filtering may also be affected.

The most important proactive action is to design harvest to include sufficient buffer widths, taking windthrow probability into account. Silvicultural standards and techniques can also be designed to maximize riparian habitat value—for instance, use of more diverse species and spacing as well as retaining some structural diversity. In areas affected by past timber harvesting, restoration approaches include silvicultural techniques to restore stand composition (species and stocking levels). Important stand structural elements such as wildlife trees and CWD can also be reintroduced. An important restoration technique for streams is to add large woody debris, often in conjunction with restoration thinning. In cases where stream banks or bars need stabilizing, bioengineering methods can be used to rapidly establish cuttings of cottonwood (*Populus balsamifera* ssp. *trichocarpa*) and willow (*Salix* spp.)

13.6.6.1 Case Study 4: Western Forest Products and Weyerhaeuser Riparian Silviculture

Historic logging practices around streams have contributed to a loss of fish and riparian habitat in some important salmon-producing rivers in forest lands managed by Western Forest Products Limited on Vancouver Island. To deal with these issues, Western Forest Products (WFP) surveyed candidate areas to determine priorities for restoration silviculture. In the early 2000s, three watersheds were selected for restoration on northern Vancouver Island, and a total of 171 ha were treated in high-priority riparian areas. The creation of stands of larger, well-spaced conifer trees and wildlife trees was an objective of the treatments. This work was planned to speed the recovery of fish habitat and channel stability at the same time that wildlife habitat was improved.

Many coastal riparian zones are dominated by red alder (*Alnus rubra*) after disturbance, but these fast-growing deciduous trees cannot attain the structural size and durability of streamside conifers. Consequently, the objective of many riparian silviculture projects is to enhance the establishment and growth of long-lived conifers. Other typical objectives of riparian restoration projects (Peters 1999; Bancroft and Zielke 2002) include the restoration of riparian zone ecosystem functioning by managing for the provision of

- Large future wildlife trees
- Large future coarse woody debris on the ground and as large woody debris in-stream
- Open areas and canopy gaps for maintenance of shrub communities
- Ungulate and bear forage
- Habitat for birds, amphibians, small mammals, insects, plants, and other organisms
- Shade and nutrients to the stream
- Filtering of overland flow (run-off)

All the watersheds in the WFP license area on northern Vancouver Island are in long-term forestry licenses. Most riparian stands were logged 25–45 years previous to treatment, with the exception of the retained mature or old-growth timber that occupies 18%–52% of the total riparian area. The harvested stands are often dominated by red alder, with an understory of scattered and suppressed conifers. Restoration treatments were designed to increase and promote the conifer components of these stands, with a secondary objective of managing for large alder trees. Conifer growth was improved by removing all alder stems within a radius of 3–10 m of each conifer tree. This resulted in the removal of approximately 80% of the alder stems in the stand—the minimum necessary to achieve a target of 40% full sunlight to the understory trees. Posttreatment conditions depended on the site; alder stands with a conifer understory were thinned to 100–300 stems ha^{-1} of overstory alder. Pockets of pure alder were also thinned, and up to 600 stems ha⁻¹ were retained depending on the age and height of the stand.

The drier riparian sites were generally overstocked with conifers, particularly western hemlock. These sites were also thinned with the goal of improving conifer growth and establishment. Both uniform and variable-density thinning was used to take advantage of productive growing sites, while allowing natural gaps and clusters of trees to prevail. Thinning took densities down to 300–600 stems ha⁻¹, with preferential retention of western redcedar and Sitka spruce where these species were present. Cluster planting was also done in selected areas of both the alder- and conifer-dominated stands. Spruce and cedar were companion-planted with black cottonwood to increase the species and structural diversity of the riparian forests.

Thinning provided excellent opportunities to further enhance the biodiversity value of these stands. Many of the cull trees were topped or top-girdled to create snags, while others were damaged to initiate heart rot. Habitat features such as bat slits and cavity starts were also created in some trees. Over 900 such features have been introduced into WFP's riparian stands.

Thinning also provided the opportunity to add large woody debris to the stream. Approximately 80 in-stream structures were created using thinnings from both alder and conifer stands. The structures typically consisted of 2–10 trees, depending on the availability of thinnings and site suitability. Directional felling allowed the creation of self-locking log jams that mimicked natural windthrow. Jams such as these are more likely to persist during floods.

Costs to complete the WFP work described above ranged from CAD\$3000 to \$4200 ha⁻¹ (expressed in 2014 dollars). These costs included thinning, girdling, planting, and modifications for biodiversity and in-stream structures. Forest Renewal BC provided project funding. These treatments are based in part on experience garnered from similar ecosystem issues in the Pacific Northwest (Poulin et al. 2000), though project effectiveness in restoring fish habitat is as yet undetermined. The alder log jams are experimental and are being monitored over time for their durability.

Weyerhaeuser Company Limited (in forests now owned by Island Timberlands LP) similarly treated 730 ha (approximately 70 linear km) in riparian reserve zones over a 5-year period at the turn of the millennium. This work was part of a larger policy of variable-density spacing to meet a number of management goals in its coastal British Columbia Tree Farm Licenses. In addition to providing for long-term stream stability and fish habitat, these density reduction treatments were undertaken with the objectives of (1) creating uneven stocking and distribution, and horizontal and vertical diversity; (2) retaining or promoting shrub and herb layers (particularly for use as ungulate browse); and (3) promoting some conifer growth (through release) on hardwood-dominated areas while retaining a hardwood component on conifer-dominated areas (Perry and Muller 2002). The goal was spatial variability, often achieved by creating open space around key ecological attributes to enhance horizontal and vertical complexity.

Innovative features of the Weyerhaeuser program included the interspersion of 5-m-wide untreated leave strips within the treatment blocks, in addition to 3 m "no touch zones" left along all stream banks. Up to 50 stems ha⁻¹ of western hemlock were partly screefed or girdled to develop trunks with a rotten core and a live hard shell for cavity-nesting birds and small mammals. Red elderberry (*Sambucus racemosa*) was randomly cut back to a height of 1 m to stimulate growth for browsing by black-tailed deer (*Odocoileus hemionus columbianus*) and Roosevelt elk (*Cervus canadensis roosevelti*).

Weyerhaeuser's riparian restoration treatments with an experienced crew ranged in cost from CAD$2600 to $3300 ha⁻¹ (2014 dollars). The long-term success of these projects will not be clear for decades, and subsequent monitoring has proceeded on an ad hoc basis. Short-term indications of success include the fact that stand diversity has increased via the promotion of previously suppressed species such as cedar, and that shrub growth (including browse for elk) has increased considerably in some areas.

13.6.7 Ecological Issue: Access Impacts to Sensitive Ecosystems and Populations

Forestry roads and other access roads (e.g., used for mining and hydropower) can have serious impacts to sensitive forest ecosystems and populations. Some sensitive species are killed by encounters with road traffic, while others such as grizzly bear (*Ursus arctos horribilis*) and bull trout (*Salmo confluentus*) are more generally sensitive to human access and disturbance. Important habitats can be fragmented, while the habitat that remains is of reduced value. Roads are also a major vector for the spread of invasive plant species. Lastly, sediment generation and slope failures (e.g., landslides) as a result of poor road construction or maintenance can occur, and can have significant negative effects on streams and aquatic ecosystems.

In BC it has been found that closed-bottom road culverts often result in complete or partial barriers to fish passage, and it is estimated that there are approximately 200,000 closed-bottom culverts on fish-bearing streams province-wide (Eastman et al. 2014). This statistic guarantees that there are thousands of kilometers of fish streams that have limited or no migratory (e.g., salmon) fish populations, due to fish passage barriers.

The best proactive measures are to minimize roaded area through long-term operational planning, and to put open-bottom culverts or bridges on stream crossings on new and upgraded roads. When roads are no longer required they can be decommissioned to deter recreational use, and to avoid slope failures and sediment inputs into fish streams. Disturbed road-side areas and deactivated roads should be revegetated promptly with native plant seed, to preclude establishment of invasive species.

Roads created to facilitate timber harvesting and other development in BC have increased public access to the backcountry, mainly for recreation. Limiting public access through preplanned road closure (physical barriers) and signage can be effective and important for certain sensitive species. Educating the public about sensitive areas and species is a crucial part of any successful access management plan. Although access management and public education are not usually considered elements of forest ecosystem restoration, these activities are crucial to the recovery and health of forest wildlife populations. In addition, the decommissioning and restoration of old forest access roads, as financed in BC between 1995 and 2002 under the Forest Renewal BC Watershed Restoration Program, is an important element of landscape restoration. Various degrees of road deactivation (from access and sediment control to full deconstruction and reforestation) can help restore aquatic values and populations of hunted wildlife.

13.6.8 Ecological Issue: Landscape Fragmentation due to Timber Harvesting

Forested landscapes are commonly fragmented by timber harvesting as well as other land uses. Timber harvesting in BC is done mainly on public land, where a lack of planning for connectivity between important habitats has been the norm. In landscapes without planned connectivity, there is a loss of habitat for some large animal species and species dependent on forest interior habitat and connected habitats. Depending on how well forested buffers and wildlife tree patches are designed, there is a potential for increased losses of timber due to blow-down. In addition, regenerated stands can be at increased risk for pests and disease if buffers between stands of similar age and species mix are not maintained.

Planning for connectivity and maintaining connections between habitat types is the best and most economical approach, as described in Case Study 5. In areas where connectivity is lost, it is possible to restore connective corridors through silvicultural techniques that accelerate recovery of mature forest attributes in areas designated as landscape corridors and for old-growth recruitment.

13.6.8.1 Case Study 5: Planning for Landscape Connectivity

Restoring landscape connectivity once it is lost is a difficult, costly, and long-term endeavor. As a result, most examples of addressing landscape fragmentation are planning examples in areas where options still exist to maintain connections among important habitats on a landscape scale. It is a fundamental tenet of ecological restoration that avoidance of degradation is always to be preferred over repair, on the grounds of effectiveness as well as cost. One example of planning for connectivity was conducted under the Morice and Lakes Innovative Forest Practices Agreement (IFPA), where program partners worked cooperatively on pilot projects to minimize landscape fragmentation. Both "fine-filter" and "coarse-filter" biodiversity conservation approaches were applied to maximize connectivity between important habitats. The incentive for licensees in this pilot project was to maintain environmental standards while increasing access to timber.

Another example, involving land use decisions (e.g., protection of one-third of the land base from resource extraction) and greater aboriginal control over a 64,000 km² area was the Great Bear Rainforest ecosystem-based management (EBM) planning exercise undertaken for the mid-coast and north-coast regions of western BC (Price et al. 2009, and Nanwakolas Council et al. 2012).

When planning for landscape connectivity, ecosystems that are particularly rare in the region or are important to the conservation of particular species are identified on the landscape prior to harvest. Often these are ecosystems that form critical habitat for one or more rare species. Using grizzly bears as an example, critical habitats would include avalanche chutes, seasonal foraging areas, denning areas, and traditional trails. Critical habitats are then mapped and an analysis is done to determine how they can be connected using coarse-filter biodiversity measures.

Coarse-filter conservation measures generally entail managing a proportion of the landscape in keeping with the natural disturbance regime. In areas with infrequent natural disturbance, this will require keeping a certain amount of the landscape in a condition that is less disturbed, consisting primarily of mature and older seral stages, under the assumption that this will provide sufficient habitat to sustain the majority of forest-dependent species that use the area. For the grizzly bear, coarse-filter measures that pertain to connectivity include unharvested riparian corridors and old-growth management areas. In the Morice and Lakes IFPA area, important grizzly feeding, denning, and trail areas were connected using riparian corridors and old-growth patches to provide for all the bear's habitat needs on the landscape. Where inadequate continuity of riparian and old-growth habitats can be identified, restoration activities (as described in the previous sections) were prescribed to accelerate the attainment of large tree and log sizes and complex canopy structures. In recent years, implementation of these landscape plans has been challenged by the recent mountain pine beetle outbreak, accelerated (salvage) logging activities, and proposed pipeline construction, making the forest restoration increasingly important.

13.7 Conclusions

The 1990s marked a growing acceptance of the need to take nontimber values into account in the management of BC's forests. This awareness, often linked to market demand for more environmentally sensitive practices, resulted in the initiation of various restoration projects throughout the province. To date, forest restoration has been conducted most commonly in interior dry forests and coastal riparian forests, and was more prevalent in the 1990s and early 2000s, when public funds were available for these purposes. At that time, the deterioration of salmon stocks and spawning beds in many of BC's rivers and streams prompted a major effort at restoration work in upland areas (sediment sources), riparian areas, and within stream channels. The scope of restoration work then gradually expanded to include a variety of broader terrestrial ecosystem restoration and forest productivity maintenance goals as well, although available funding for all restoration work has since declined. In recent years, concerns about forest health and the need to restore habitat for open forest species has fortunately coincided with an emphasis on fuel reduction treatments at the forest–urban interface.

There is a diversity of restoration needs and solutions in BC, and a high degree of public support and professional expertise to carry them out (Gayton 2001; Ritchlin 2001; Neal and Anderson 2009; Rocky Mountain Trench Ecosystem Restoration Program 2014). Many innovative and effective techniques have been developed to restore ecological features and processes to BC's forests. Interagency cooperation, public participation through volunteer organizations, and the identification of mutually supportive objectives have characterized successful restoration efforts to date.

Many of the issues of ecological degradation only emerge at the landscape scale. Homogenous stand structures and compositions are not without natural precedent, and do not constitute a problem in themselves, but are associated with reduced biodiversity and less ecological resilience when they dominate large areas. Nevertheless, the remedial solutions to landscape-scale problems can only be implemented on a stand-by-stand basis, or even in pockets within stands. This bottom-up approach consumes considerable time and resources, and its effectiveness remains to be demonstrated. Top-down approaches of preventing the drastic alteration of landscape pattern and disturbance processes are clearly more practical and more effective. Land use planning has addressed this need to some extent in BC, but these plans are now dated and do not cover the province as a whole. A need for strategic planning remains, to address landscape-scale issues of habitat conservation, connectivity, cumulative effects, and biodiversity and species management.

British Columbia is a jurisdiction blessed with exceptional landscapes and rich biodiversity that attracts visitors from around the world. The province has a wealth of forest restoration expertise and experience to draw from, as well as excellent precedents for land management practices to provide connectivity, conserve biodiversity and manage sensitive species. A great deal has been learned about planning practices and effective restoration of degraded ecosystems, and about how to prevent ecosystems from crossing thresholds from which there is no return. More needs to be done to change forestry policies and practices as resource development proceeds apace so that restoration intervention is no longer needed in Canada's most biodiverse and productive province.

Acknowledgments

We are indebted to the contributions of the many original researchers and practitioners who provided the information here, including the case study contributors Glen Dunsworth, Reinhard Muller, Eric Gagné, Vince Poulin, Todd Manning, Brad Pollard, Art Moi, Morice & Lakes IFPA participants, Jan Jonker, and Raj Prasad. We thank Colene Wood, Robert Seaton, Rachel Holt and the other members of the Technical Advisory Committee to FRBC's Terrestrial Ecosystem Restoration Program for identifying and structuring ecological issues in the manner reported above. The constructive comments of two anonymous reviewers are gratefully appreciated.

References

Angelstam, P., L. Laestadius, and J.M. Roberge. 2005. Data and tools for conservation, management and restoration of northern forest ecosystems at multiple scales. In *Restoration of Boreal and Temperate Forests*, J.A. Stanturf, and P. Madsen (eds.), pp. 269–283. Boca Raton, FL: CRC Press.

Apsey, M., D. Laishley, V. Nordin, and G. Paillé. 2000. The perpetual forest: Using lessons from the past to sustain Canada's forests in the future. *The Forestry Chronicle* 76(1): 29–53.

Bancroft, B. and K. Zielke. 2002. *Guidelines for Riparian Restoration in British Columbia: Recommended Riparian Silviculture Treatments, 1st Approximation*. Victoria, BC: BC Ministry of Forests, http://www.for.gov.bc.ca/hfp/publications/00077/riparian_guidelines.pdf [viewed February 26, 2014].

Beese, W.J., B.G. Dunsworth, K. Zielke, and B. Bancroft. 2003. Maintaining attributes of old-growth forests in coastal BC through variable retention. *The Forestry Chronicle* 79(3): 570–578.

Burton, P., C. Messier, W. Adamowicz, and T. Kuuluvainen. 2006. Sustainable management of Canada's boreal forests: Progress and prospects. *EcoScience* 13(2): 234–248.

Burton, P.J. 1998. *An Assessment of Silvicultural Practices and Forest Policy in British Columbia from the Perspective of Restoration Ecology*. Paper presented at Helping the Land Heal: Ecological Restoration in British Columbia Conference, November 5–7, 1998, Victoria, BC Conference hosted by the BC Environmental Network Educational Foundation, Vancouver BC.

Burton, P.J. 2010. Striving for sustainability and resilience in the face of unprecedented change: The case of the mountain pine beetle outbreak in British Columbia. *Sustainability* 2: 2403–2423. DOI: 10.3390/su2082403.

Burton, P.J. (ed.). 2002. *Garry Oak Ecosystem Restoration: Progress and Prognosis*. Proceedings of the Third Annual Meeting of the BC Chapter of the Society for Ecological Restoration, April 27–28, 2002, University of Victoria. BC. Chapter of the Society for Ecological Restoration, Victoria, British Columbia.

Burton, P.J. and C.M. Burton. 2002. *Phase II Campground Restoration Prescriptions for Exchamsiks River Provincial Park*. Prepared for BC Parks, Terrace Field Office, BC Ministry of Water, Land and Air Protection. Symbios Research & Restoration, Smithers, BC.

Burton, P.J. and S.E. Macdonald. 2011. The restorative imperative: Assessing objectives, approaches and challenges to restoring naturalness in forests. *Silva Fennica* 45(5): 843–863.

Burton, P.J., M. Svoboda, D. Kneeshaw, and K.W. Gottschalk. 2015. Options for promoting the recovery and rehabilitation of forests affected by severe insect outbreaks. In *Restoration of Boreal and Temperate Forests*, Stanturf, J.A. (Ed.), pp. 495. Boca Raton, FL: CRC Press.

Deal, R.L. 2007. Management strategies to increase stand structural diversity and enhance biodiversity in coastal rainforests of Alaska. *Biological Conservation* 137: 520–532.

Décamps, H., R.J. Naiman, and M.E. McClain. 2010. Riparian zones. In *River Ecosystem Ecology: A Global Perspective*, G.E. Likens, (ed.), pp. 182–189. New York: Elsevier.

Douglas, T. and P.J. Burton. 2005. Integrating ecosystem restoration into forest management in British Columbia, Canada. In *Restoration of Boreal and Temperate Forests*, J.A. Stanturf and P. Madsen, (eds.), pp. 423–444. Boca Raton, FL: CRC Press.

Douglas, G.W. and S.J. Smith. 2006. Recovery Strategy for Multi-Species at Risk in Garry Oak Woodlands in Canada. In *Species at Risk Act* Recovery Strategy Series. Ottawa: Parks Canada Agency. 58 p. Available on-line at http://www.sararegistry.gc.ca/default.asp?lang=En&n=A4BFCE40-1 [viewed May 11, 2015].

Eastman, D.S., R. Archibald, R. Ellis, and B. Nyberg. 2014. Trends in renewable resource management in British Columbia. *Journal of Ecosystems and Management* 14(3): 1–10. http://jem.forrex.org/index.php/jem/article/view/556/498

Fenger, M. 1996a. Implementing biodiversity conservation through the British Columbia Forest Practices Code. *Forest Ecology and Management* 85(1–3): 67–77.

Fenger, M. 1996b. Wildlife, harvesting and hardwoods. In *Ecology and Management of BC Hardwoods. FRDA Report 255*, P.G. Comeau, G.J. Harper, M.E. Blache, J.O. Boateng, and K.D. Thomas, (eds.), pp. 19–29. Victoria, BC: Canadian Forest Service and BC Ministry of Forests.

Fenger M., T. Manning, J. Cooper, S. Guy, and P. Bradford. 2006. *Wildlife and Trees in British Columbia*. Edmonton: Lone Pine Publishing.

Filmon, G. 2004. *Firestorm 2003: Provincial Review*. Government of British Columbia, Victoria, BC. http://bcwildfire.ca/History/ReportsandReviews/2003/FirestormReport.pdf (accessed March 24, 2014).

Forest Practices Board. 2010. *Follow Up Report: Update on Control of Invasive Plants*. Special Report 38, November 2010. https://www.bcfpb.ca/sites/default/files/reports/SR38%20-%20Follow%20Up%20Report%20-%20Update%20on%20Control%20of%20Invasive%20Plants%20-%20WEB.pdf (accessed November 4, 2014).

Franklin, J.F., D.R. Berg, D.A. Thornburgh, and J.C. Tappeiner. 1997. Alternative silvicultural approaches to timber harvesting: Variable retention harvest systems. In *Creating a Forestry for the 21st Century: The Science of Ecosystem Management*, K.A. Kohm and J.F. Franklin, (eds.), pp. 111–139. Washington, DC: Island Press.

Fuchs, M.A. 2001. *Towards a Recovery Strategy for Garry Oak and Associated Ecosystems in Canada: Ecological Assessment and Literature Review*. Technical Report GBEI/EC-00-030. Canadian Wildlife Service, Pacific and Yukon Region. Environment Canada, Vancouver, BC.

Garry Oak Ecosystems Recovery Team. 2014. *Restoration Projects and Places to Visit*. http://www.goert.ca/support/places_to_visit.php [accessed November 4, 2014].

Gayton, D. 2001. *Ground Work: Basic Concepts of Ecological Restoration in British Columbia*. FORREX Series No. 3. Southern Interior Forest Extension and Research Partnership, Kamloops, BC http://www.forrex.org/publications/forrexseries/ss3.pdf (accessed February 26, 2014).

Government of British Columbia. 1995a. *Forest Practices Code of British Columbia Biodiversity Guidebook*. Victoria, BC: BCMinistry of Forests and BC Environment. https://www.for.gov.bc.ca/tasb/legsregs/fpc/fpcguide/biodiv/biotoc.htm (accessed October 26, 2014).

Government of British Columbia. 1995b. *Forest Practices Code of British Columbia Riparian Management Area Guidebook BC*. Victoria, BC: Ministry of Forests and BC Environment. https://www.for.gov.bc.ca/tasb/legsregs/fpc/fpcguide/riparian/rip-toc.htm (accessed October 21, 2014).

Government of British Columbia. 1996. *Forest Practices Code of British Columbia Pine Stem Rust Management Guidebook*. Victoria, BC: BC Ministry of Forests and BC Environment. http://www.for.gov.bc.ca/tasb/legsregs/fpc/fpcguide/pinestem/pine-toc.htm (accessed February 26, 2014).

Government of British Columbia. 2010a. *Crown Land Indicators and Statistics Report*. Victoria, BC. BC Ministry of Forests, Lands and Natural Resource Operations. http://www.for.gov.bc.ca/land_tenures/documents/publications/Crown_Land_Indicators_&_Statistics_Report.pdf (accessed November 1, 2014).

Government of British Columbia. 2010b. *The State of British Columbia's Forests*. Third edition. BC Ministry of Forests, Lands and Mines, Victoria, BC. http://www.for.gov.bc.ca/hfp/sof/2010/SOF_2010_Web.pdf (accessed November 4, 2014).

Government of British Columbia. 2012. *Biogeoclimatic Zones of British Columbia (Map)*. Victoria, BC: BC Ministry of Forests, Lands and Natural Resource Operations. ftp://ftp.for.gov.bc.ca/HRE/external/!publish/becmaps/papermaps/BGCmap.2014.05.pdf (accessed February 22, 2014).

Haeussler, S., P.J. Burton, and A.J. Clason. in press. Combating decline of Whitebark pine ecosystems in West-Central British Columbia. In *The Integration Imperative—Cumulative Environmental, Community and Health Effects of Multiple Natural Resource Developments*, M. Gillingham, G. Halseth, M. Parkes, and C. Johnson, (eds.), pp. xx–yy New York: Springer.

Holt, R.F. 2001. *A Systematic Ecological Restoration Assessment in the Forest Regions of British Columbia, The Results of Six Workshops—Summary: Ecological Restoration Priorities by Region*. Prepared for Forest Renewal BC and BC Ministry of Environment Habitat Branch. Pandion Ecological Research Ltd., Nelson, BC. http://www.env.gov.bc.ca/wld/documents/fia_docs/sera_terp_summary.pdf (accessed March 23, 2014).

Jain, T.B. and R.T. Graham. 2015. Restoring dry and moist forests of the inland northwestern United States. *In Restoration of Boreal and Temperate Forests*, Stanturf, J.A. (ed.), pp. 467. Boca Raton: CRC Press.

Klenner, W., R. Walton, A. Arsenault, and L. Kremsater. 2008. Dry forests in the southern interior of British Columbia: Historic disturbances and implications for restoration and management. *Forest Ecology and Management* 256 (10): 1711–1722.

Kuuluvainen, T., Y. Bergeron, and K. David Coates. 2015. Restoration ecosystem-based management in the circumboreal forest: Background, challenges, and opportunities. In: *Restoration of Boreal and Temperate Forests*, Stanturf, J.A. (Ed.), pp. 251. Boca Raton, FL: CRC Press.

Lindenmayer, D.B., P.J. Burton, and J.F. Franklin. 2008. *Salvage Logging and its Ecological Consequences*. Washington, DC: Island Press.

Maloy, O.C. 1997. White pine blister rust control in North America: A case history. *Annual Review of Phytopathology* 35: 87–109.

Manning, T. 2014. *Wildlife Tree Creation—Fungal Inoculation Effectiveness Evaluation*. Prepared for BC Ministry of Forests, Lands and Natural Resource Operations, Coast Area Research Section, West Coast Region. Strategic Resource Solutions, Victoria, BC.

Manning, E.T. and I.A. Manley. 2014. Results of fungal inoculation treatments as a habitat enhancement tool in the East Kootenay region of British Columbia: 2007–2013. British Columbia Ministry of Forests, Land and Natural Resource Operations Extension Note 112. Victoria, B.C. http://www.for.gov.bc.ca/hfd/pubs/Docs/En/En112.htm

McClellan, M.H., D.N. Swanston, P.E. Hennon, R.L. Deal, T.L. De Santo, and M.S. Wipfli. 2000. *Alternatives to Clearcutting in the Old-Growth Forests of Southeast Alaska: Study Plan and Establishment Report*. General Technical Report PNW-GTR-494. Portland, OR: USDA Forest Service.

McLane, S.C. and S.N. Aitken. 2012. Whitebark pine (*Pinus albicaulis*) assisted migration potential: Testing establishment north of the species range. *Ecological Applications* 22(1): 142–153.

Meidinger, D. and J. Pojar, (eds.), 1991. *Ecosystems of British Columbia*. Special Report Series 6. Victoria, BC: BC Ministry of Forests, http://www.for.gov.bc.ca/hfd/pubs/Docs/Srs/SRseries.htm (accessed February 26, 2014).

Meyer, R. 2007. *Strix occidentalis. Fire Effects Information System*, US Department of Agriculture, Forest Service, Rocky Mountain Research Station, Fire Sciences Laboratory. Missoula, MT. http://www.fs.fed.us/database/feis/animals/bird/stoc/all.html (accessed May 5, 2015).

Nanwakolas Council, Coastal First Nations and BC and Ministry of Forests, Lands and Natural Resource Operations. 2012. Ecosystem Based Management on B.C.'s Central and North Coast (Great Bear Rainforest) Implementation Update Report July 2012. https://www.for.gov.bc.ca/tasb/slrp/lrmp/nanaimo/central_north_coast/docs/EBM_Implementation%20Update_report_July%2031_2012.pdf

Neal, A. and G.C. Anderson. 2009. *Ecosystem Restoration Provincial Strategic Plan—DRAFT*. Kamloops, BC: BC Ministry of Forests and Range, Range Branch, http://www.for.gov.bc.ca/hra/Restoration/ (accessed February 23, 2014).

Nuzzo, V.A. and E.A. Howell. 1990. Natural area restoration planning. *Natural Areas Journal* 10(4): 201–209.

O'Hara, K. and K.M. Waring. 20005. Forest restoration practices in the Pacific Northwest and California. In *Restoration of Boreal and Temperate Forests*, J.A. Stanturf and P. Madsen, (eds.), pp. 445–461. Boca Raton, FL: CRC Press.

Parks Canada. 2014. *National Parks Restoration Case Studies: Garry Oak Ecosystems Restoration* (Fort Rodd Hill National Historic Site). http://www.pc.gc.ca/progs/np-pn/re-er/ec-cs/ec-cs05.aspx (accessed November 4, 2014).

Perera, A.H., L.J. Buse, and M.G. Weber, (eds.), 2004. *Emulating Natural Forest Landscape Disturbances: Concepts and Applications*. New York: Columbia University Press.

Perry, J. and R. Muller. 2002. *Forest Project–Technical Project Summary Report #3, July 2002. Riparian Restoration Program*. Nanaimo, BC: BC Coastal Group, Weyerhauser Inc.

Peters, C.M. 1999. *Riparian Zone Habitat Rehabilitation of the Eve River, Vancouver Island, BC—a Preliminary Report. Work Term Report for Weyerhaeuser, Nanaimo Woodlands*. Victoria, BC: Department of Biology, University of Victoria.

Pojar, J., K. Klinka, and D.V. Meidinger. 1987. Biogeoclimatic ecosystem classification in British Columbia. *Forest Ecology and Management* 22: 119–154.

Poulin, V.A., B. Simmons, and C. Harris. 2000. *Riparian Silviculture: An Annotated Bibliography for Practitioners of Riparian Restoration. Prepared for the BC Ministry of Forests, Victoria, BC* Vancouver, BC: V.A. Poulin & Associates Ltd., http://www.for.gov.bc.ca/hfd/library/ffip/Poulin_VA2000_a.pdf (accessed November 6, 2014).

Prasad, R. 2000. Some aspects of the impact and management of the exotic weed, Scotch broom (*Cytisus scoparius* [L.] Link) in British Columbia, Canada. *Journal of Sustainable Forestry* 10: 341–347.

Prasad, R. 2003. *Scotch Broom, Cytisus scoparius L. in British Columbia*. Victoria, BC: Pacific Forestry Centre, Canadian Forest Service. http://cfs.nrcan.gc.ca/publications?id=31653.

Price, K., A. Roburn, and A. MacKinnon. 2009. Ecosystem-based management in the Great Bear Rainforest. *Forest Ecology and Management* 258(4): 495–503.

Ritchlin, J. 2001. *Healing the Land…Healing Ourselves: A Guide to Ecological Restoration Resources for British Columbia*. Vancouver, BC: BC Environmental Network Foundation.

Rocky Mountain Ecosystem Trench Ecosystem Restoration Program. 2014. *Blueprint for Action 2013: Progress and Learnings 1997–2013*. Rocky Mountain Trench Ecosystem Restoration Program, Cranbrook, BC. http://trench-er.com/our_blueprint (accessed February 26, 2014).

Seymour, R.S. and M.L. Hunter. 1999. Principles of ecological forestry. In: *Maintaining Biodiversity in Forest Ecosystems*, M.L. Hunter, (ed.), pp. 22–61. Cambridge: Cambridge University Press.

Society for Ecological Restoration. 2004. *SER International Primer on Ecological Restoration, Version 2. Society for Ecological Restoration International, Science & Policy Working Group*. Tucson, AZ: Society for Ecological Restoration. http://www.ser.org/resources/ (accessed February 23, 2014).

Society for Ecological Restoration in North Central British Columbia (SERN). 2013. *About Ecological Restoration.* http://www.sernbc.ca/sern_ecorestore.html (accessed March 23, 2014).

Stanturf, J.A. and P. Madsen. 2002. Restoration concepts for temperate and boreal forests of North America and Western Europe. *Plant Biosystems* 136(2): 143–158.

Stevens, V. 1997. *The Ecological Role of Coarse Woody Debris: An Overview of the Ecological Importance of CWD in BC Forests.* Working Paper 30. Research Branch, Victoria, BC: BC Ministry of Forests. http://www.for.gov.bc.ca/hfd/pubs/docs/Wp/Wp30.pdf (viewed February 26, 2014).

UBCM. 2014. *Strategic Wildfire Prevention Initiative.* Union of British Columbia Municipalities. http://www.ubcm.ca/EN/main/funding/lgps/current-lgps-programs/strategic-wildfire-prevention.html (viewed November 4, 2014).

14

A Hundred Years of Woodland Restoration in Great Britain: Changes in the Drivers That Influenced the Increase in Woodland Cover

Ralph Harmer, Kevin Watts, and Duncan Ray

CONTENTS

14.1 Introduction

Since the last ice age, Great Britain has lost most of its forest cover but a determined restoration effort over the last 100 years has managed to increase both the cover and productivity of woodlands. Most of the woodland restoration that took place during the twentieth century, which can be described as providing replacement woodland (Stanturf, 2005), was driven by the vision that it was necessary to create a strategic supply of homegrown timber (Anon, 1918). Much of this new woodland was established before the ecological aspects of restoration became generally well-established toward the end of the twentieth century (Newton and Kapos, 2003; Stanturf, 2005). This initial phase of woodland restoration was focused on the widespread afforestation of poor land where fast-growing non-native conifers would grow satisfactorily and losses to agricultural production would be minimal.

Modern concepts that it could be beneficial to create new woodland near to other wood-land and in areas rich in biodiversity (Newton and Kapos, 2003) have not been widely used as criteria for selecting the site of new woodland. Consequently, despite the establishment of large areas of new woodland, especially in upland areas of the country, woodland cover in most of Great Britain remains fragmented.

The requirement to improve the productivity of woodland also resulted in the degrada-tion of large areas of remaining seminatural broadleaved woodland. At the time when this occurred, these native woodlands were regarded as slow growing and uneconomic; about 40% of their area was converted to faster growing, more-productive species—primarily non-native conifers, although native broadleaves were also used (Pryor and Smith, 2002; Pryor, 2003). The biodiversity value of these native woodland fragments was belatedly recognized at the end of the twentieth century and restoration (= rehabilitation (Stanturf, 2005)) of these plantations on ancient woodland sites (PAWS) to woodland with a more native character is taking place. However, there have been few detailed studies to verify whether the restoration procedures recommended are suitable (Thompson et al., 2003) or the expected outcomes are being achieved (Harmer et al., 2012a,b). As there has been little overall change in practice or policy since the restoration of PAWS were described in the first edition of this book (Harmer et al., 2005), they are not considered in this chapter.

Factors influencing woodland restoration in Great Britain have changed considerably since it began. Initially, the objective was simple—to increase the amount and productivity of woodland. Over time, objectives have become more demanding with the need for new woodland to provide a much greater range of benefits including: recreational space; areas for wildlife; and to mitigate the vagaries of potential climate change. We briefly describe how the forest cover of Great Britain has changed over time and how a number of drivers have affected the restoration and creation of new woodland; these include: the need for timber and where trees can be grown; multipurpose forestry, sustainable forestry, and ecosystem services; a change in the policy for broadleaved trees; and the requirements for improved biodiversity and the demands of climate change. As in the previous edition of this chapter, a number of case studies are included to illustrate some of the topics.

14.2 Historical Context: Change in Forest Cover before Twentieth Century

The following description of changes in forest cover is presented to provide context for the woodland restoration activities which have occurred in Great Britain during the last century. It is a brief general summary, for more details see, for example, Anderson (1967a,b), James (1981), Linnard (2000), Rackham (1986), and Smout (2003).

As the ice retreated at the end of the last glaciations, Britain was colonized by plants with woodland becoming the predominant vegetation type and there may have been 90% cover in England and 50% or more in Scotland. However, recent estimates using pollen data indicate significant regional variation in woodland cover and suggest that at the broad scale, some areas of the country may have been less wooded than previously thought (Fyfe et al., 2013). While it has been widely accepted that these woodlands were largely closed canopy, Vera (2000) has proposed an alternative structure with a mixture of open areas and woodland of varying ages and densities in which large herbivores influence the regeneration cycle.

In prehistoric times, most management of woodland in Britain was related to its use as shelter for livestock, or to its destruction and conversion to agricultural use. Significant forest clearance began during the Bronze Age, becoming more intensive during the early Iron Age (2450–2750 BP) when axes and plows became commonly available. Woodland cover at this period has been estimated at about 5%–50% depending on location (Fyfe et al., 2013). Nevertheless, about 2000 years ago, the Roman geographer Strabo described Great Britain as a densely wooded country (Fernow, 1913). Extensive management of woodlands resulted in the creation of open, grazed woodland, and wood pasture that remain common features of the landscape throughout Scotland and other upland areas. There are few estimates of woodland cover before 1900 but those available suggest that the area of woodland continued to decline. In England, the area of land covered by woodland reduced from approximately 15% during the survey for the Domesday Book in 1086 to about 5% by 1900. The overall cover in Wales has been estimated to be 10% between the fifteenth and seventeenth centuries, but may have been much higher in remote valleys; in Scotland, cover had been reduced to 4% by the start of the eighteenth century. The supply of timber for strategic purposes has been a recurring theme of forest management for several centuries, but efforts to increase the area of woodland to provide a secure supply of timber were generally unsuccessful. Use of direct sowing and planting, both to improve existing woodland and create new woodlands did not become common until the eighteenth century, and at some periods during the nineteenth century, the rates of planting have been estimated at ~2500 ha yr^{-1} in Scotland and 500–1000 ha yr^{-1} in Wales. However at the beginning of the twentieth century, woodland cover in Great Britain was only ca. 5% of land area (Figure 14.1) and Fernow (1913) strongly implied that direct activity by the state would be necessary to improve the prospects for British forestry.

14.3 Afforestation and Timber Production

The primary focus for forestry restoration during much of the twentieth century was on the production of timber. The shortages of timber during the First World War highlighted the poor state of Great Britain's forests and home-grown timber supply, and in 1919, the Forestry Commission, a newly established government department, was given the task of developing a strategic reserve of timber (Pringle, 1994). Large areas of land were purchased or leased, and during the remainder of the twentieth century, many new plantations of conifers were established and managed by the state. Britain has only three native conifer species and of these only, Scots pine (*Pinus sylvestris*) is suitable for large-scale plantation forestry. Consequently, in order to use the range of site types available for many of the new plantations comprised of fast growing non-native species. At about the same time as state forestry was developing, a variety of incentive schemes were also introduced to encourage planting by private owners (James, 1981; Aldhous, 1997). These schemes changed throughout the century and during some periods, they included grant aid and fiscal incentives. The ability to claim tax relief on expenditure for new woodland plantations was a significant inducement for many individuals to invest in forestry.

Since the creation of the Forestry Commission and the introduction of planting grants for private owners woodland cover has increased significantly: between 1905 and 2014, total woodland cover in Great Britain has increased from about 1.1 million ha (4.9% of land area) to over 3.0 million ha (13.3% of land area). At the beginning of the twentieth century, percentage woodland cover was greatest in England, but since 1947, percentage woodland cover has been greater in both Wales and Scotland (Figure 14.1a). The proportion of land

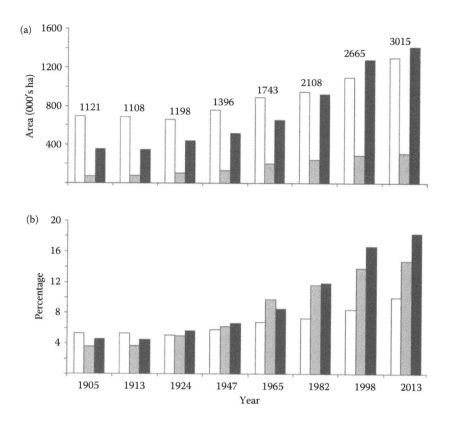

FIGURE 14.1
Development of woodland cover in Great Britain between 1905 and 2013 in England, Wales and Scotland. (a) Area of woodland in thousands of hectares; (b) Woodland cover as a percentage of land area. England = unshaded; Wales = gray; Scotland = black. Numbers in (a) are overall totals for all countries. (Information adapted from census data: Anon. 1928. *Report on Census of Woodlands and Census of Production of Home-Grown Timber 1924.* London: HMSO; Anon. 1952. *Census of Woodlands 1947–1949.* London: HMSO; Locke, G.M.L. 1979. *Census of Woodlands 1965–1967.* London:HMSO; Locke, G.M.L. 1987. *Census of Woodlands and Trees 1979–1982.* Forestry Commission Bulletin 63. London:HMSO; Forestry Commission. 2001. *National Inventory of Woodland and Trees— England.* Edinburgh: Forestry Commission; Forestry Commission. 2002a. *National Inventory of Woodland and Trees—Scotland.* Edinburgh: Forestry Commission; Forestry Commission. 2002b. *National Inventory of Woodland and Trees—Wales.* Edinburgh: Forestry Commission; Forestry Commission. 2003. *National Inventory of Woodland and Trees—Great Britain.* Edinburgh: Forestry Commission.)

area covered by woodland increased about fourfold in Wales and Scotland between 1905 and 2013, but did not double in England (Figure 14.1b). A large part of the increase in forest area resulted from the afforestation of upland grassland, moors, and bogs; these form a greater proportion of the land area of Scotland and Wales, and explain much of the change in the distribution of forest cover.

14.3.1 Influence of Soils and Climate on Species Choice

The increase in woodland cover during the twentieth century was accompanied by a change in overall species composition for Great Britain: in the early years of the twentieth century, about 25% of cover comprised coniferous species, but today, the figure is about

50% with much of this being non-native species (Forestry Commission, 2013a,b, 2014). However, there is significant variation between countries with conifers comprising about 75%, 50%, and 25% of cover in Scotland, Wales, and England, respectively. Overall, the proportion of conifers increases toward the north and west of the country, and the proportion of broadleaves to the south and east. These regional variations are largely due to differences in climate and soils across Great Britain.

The climate of Scotland and Wales is generally cool and wet, and the land which has been made available for forest expansion has tended to be of low agricultural value. Such conditions have resulted in the use of coniferous species, particularly those from the Pacific Northwest of North America (e.g., Sitka spruce—*Picea sitchensis* and lodgepole pine—*Pinus contorta*) which are well-suited to a moist climate and tolerant of wet soil conditions (Savill, 2013). The climate and species used are similar in the north and west of England, but central and southern areas are warmer and drier, and more suitable for broadleaved species with most of those planted being native.

The regional differences in forest soils across England, Scotland, and Wales, as shown in Table 14.1, reflects the general occurrence of woodland on land of low agricultural value and the distribution of species in different countries. In England and Wales, the majority of woodland is on brown earth, whereas in Scotland, it is on peaty gleys, surface water gleys, upland brown earths, and podzols. In Scotland, the forest areas can be regionally and edaphically divided approximately into two major types. The Sitka spruce forests in the south and west are largely associated with the wetter gley and peat soils, whereas the drier and less fertile ironpans and podzols generally support pine forest. Similarly in Wales and northern England, large areas of spruce forest have been planted on gley and peat soils with brown earths supporting broadleaved woodland. In central and southern England, woodland expansion has concentrated on thin rankers and rendzinas, and infertile sandy podzols.

Predicted changes in future climate suggest that some species will become incompatible with the combination of soil and climate at sites where they are currently used, and the choice of which species to use to restock existing sites and create new woodland is currently of significant interest to forest managers (Broadmeadow et al., 2005; Ray, 2008a,b; Read et al., 2009; Ray et al., 2010, 2014)

14.4 The Advent of Multipurpose Forestry and Sustainable Forest Management

During the latter half of the twentieth century, a series of legislative and policy changes moved the focus of forestry practices away from the simple provision of timber (Richards, 2003) toward multipurpose forestry. Although this had taken place for many years (e.g., timber/fuel wood production, feeding/shelter for livestock, protection of water catchments, hunting, and general recreation), it was not necessarily obvious. This shift in emphasis was partly in response to concerns over impacts of large-scale afforestation on cultural landscapes, important habitats and species; the questionable relevance of a strategic timber reserve in a nuclear age; and opportunities to enhance the wider public benefits from forestry (Mason, 2007). As a result, various regulations and incentive schemes were used to increase the planting of broadleaved trees, restrict the expansion of forestry on valued open landscapes, and enhance the protection, management and restoration of the remaining native woodland fragments (Watts et al., 2008). Consequently, woodland

TABLE 14.1

Major Soil Groups by Land Area (km2) and Forest Area (km2) in Great Britain

Soil Group	England				Scotland				Wales				Great Britain			
	Area (ha)	Forest (ha)	%Soil	%For	Area (ha)	Forest (ha)	%Soil	%For	Area (ha)	Forest (ha)	%Soil	%For	Area (ha)	Forest (ha)	%Soil	%For
Brown earths	46,537	4173	9.0	39.6	13,385	2165	16.2	16.2	10,987	1482	13.5	54.3	70,909	7820	11.0	29.4
Podzols and Iron pans	3840	953	24.8	9.1	8495	2164	25.5	16.2	2013	440	21.9	16.1	14,348	3557	24.8	13.4
Surface water gleys	30,975	2448	7.9	23.3	10,096	1399	13.9	10.5	3476	275	7.9	10.1	44,547	4122	9.3	15.5
Ground water gleys	11,273	382	3.4	3.6	41	0	0	0	605	29	4.8	1.1	11,919	411	3.4	1.5
Peaty gleys/ podzols	4208	631	15.0	6.0	30,094	5909	19.6	44.2	1624	269	16.6	9.9	35,926	6809	19.0	25.6
Deep peats	3942	318	8.1	3.0	8818	1492	16.9	11.2	697	123	17.6	4.5	13,457	1933	14.4	7.2
Rankers and rendzinas	7811	1607	20.6	15.3	4989	239	4.8	1.8	21	2	9.5	<1	12,821	1955	15.2	7.3
Others	21,215				2861				1195				25,271			
Total area	129,803	10,513			78,779	13,370			20,618	2727			229,198	26,607		

Source: Adapted from Morison, J. et al., 2012. *Understanding the Carbon and Greenhouse Gas Balance of Forests in Britain.* Edinburgh: Forestry Commission.

Note: Area = total area of each soil type; Forest = area of forest on each soil type; %Soil = percentage of each soil type supporting forest; %For = percentage of total forest area on each type of soil.

creation began to take place in areas which are not traditionally associated with forestry including sites next to large centers of urban population (Anon, 2005).

The move toward multipurpose forestry led to the emergence of sustainable forest management which explicitly considers the environmental, economic, social, and cultural objectives of forestry (Quine et al., 2013). The United Kingdom Forestry Standard (UKFS) (Forestry Commission 2011) articulates the basis for sustainable forest management and provides a framework for the pursuit of multiple benefits, while reducing adverse impacts. For example, the land-use focus to native woodland forest management surrounding Loch Katrine in Loch Lomond and the Trossachs National Park aims to improve water quality, provide small amounts of timber from native woodlands, and improve recreation and biodiversity in an area previously intensively managed for sheep (GPFLR, 2013; Lamont, 2006).

14.4.1 Case Study: The National Forest—New Woodland with Multipurpose Objectives

The idea to create a new forest in the center of England developed during the late 1980s when there was increasing concern about the quality and appearance of the environment, a greater interest in multipurpose forestry, and the perception that farmers could diversify their activities and use less land for agriculture. The broad objectives of the forest would be to (1) improve the ecological, visual, and cultural environment; (2) provide a major opportunity for agricultural diversification and also develop forestry and wood processing industries, (3) be a catalyst for economic development, and (4) be an important demonstration of the benefits of multipurpose forestry (Countryside Commission, 1989; Bell, 1992; Countryside Commission 1994).

The site selected for the National Forest covered an area of about 500 km² to the northeast of Birmingham. Although land-use was predominantly agricultural, especially dairy and livestock, there were significant areas of land degraded by sand and gravel extraction, and coal mining: reclamation of this derelict land was a prime target for woodland creation. Forest cover was ca. 6% with the majority being broadleaved/conifer mixtures. About 40% of the existing woodlands were small and fragments within the range of 0.25–2.0 ha in size. The aim was to increase woodland cover to about one-third of the landscape with the majority of new woodland comprising broadleaves. This would require planting of about 13,500 ha with 70% of new woodland being created during the first 10 years. The first trees were planted in December 1990 and by the mid-1990s, several hundred ha were being planted each year (Figure 14.2) which compares with <10 ha yr^{-1} prior to the formation of the National Forest (Kerr and Williams, 1999). Rates of woodland planting have declined during the last decade due to a variety of factors including: rising agricultural land prices and tenancy constraints, changes in agricultural and forestry incentive schemes, decisions made by mineral extraction companies, and budget cuts. Nevertheless, by 2013, woodland cover in the National Forest had increased to ca. 20% by the creation of about 7000 ha of woodland. The increase in woodland cover is clearly visible from the maps as shown in Figure 14.3.

14.5 Ecosystem Services

The latest manifestation of multipurpose forestry is the concept of ecosystem services which describes the goods and services provided to society by forests, woodland, and trees (Quine et al., 2013). The UK National Ecosystem Assessment (UKNEA, 2011) has

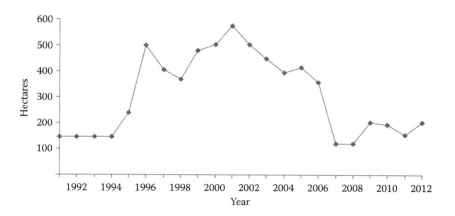

FIGURE 14.2
Area of woodland established in the National Forest between 1991 and 2013. Figures are from annual reports, data for individual years 1991–1994 are not available; values for these years are the average of total woodland area established during the first four years.

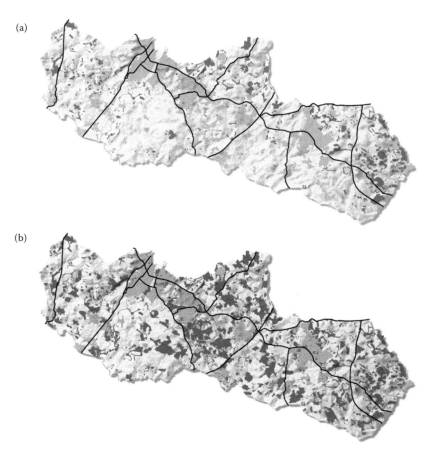

FIGURE 14.3
Woodland cover (area of darkest gray shading) in National Forest in (a) 1995 and (b) 2013.

identified the need to assess the impact of land use change and existing land management practice on the goods and services provided to society, and the value of the goods and services supplied by the natural environment to people. Consequently, ecosystem services are now helping to define the priorities within sustainable forest management, and in determining the targeting of woodland restoration where more benefits can be provided. While this is not a new idea, it is being applied to a wider range of services. For example, forests were often created in catchments to protect public water supplies by maintaining a high standard of drinking water. Nowadays, the visual, recreational, and biodiversity values that accrued as these forests developed, would also be regarded as services.

14.5.1 Case Study: Ecosystem Services—Potential Location of New Woodland to Alleviate Flooding

The upper catchment of the Derwent River covers about 54,000 ha of the Lake District in northwest England which is some of the highest and wettest land in Great Britain. There is a long history of flooding in the catchment which is unlikely to decrease: the intensity of rainstorms has doubled in the last 40 years and the number of days of heavy rainfall is predicted to increase 3- to 4-fold due to climate change. Soils range from peat and humic rankers on high ground to stagnogleys in the lowlands; they saturate quickly after heavy or prolonged rainfall generating surface runoff. About 11% of the catchment is currently wooded. Although investment in flood defenses will be needed in urban areas, the preferred method in rural areas is to manage risk by working with natural processes to create wetland habitats in order to reduce flood flows.

Forestry offers a number of options for flood alleviation, the most important of which is the ability of floodplain woodland to slow flood flows and enhance flood storage. Woodland can also attenuate flooding due to greater water use by trees and by the ability of woodland soils to intercept and delay movement of rainwater to streams and rivers. Moreover, the use of woodland for flood management may provide other services, including improvement of water quality, fisheries, nature conservation, recreation, and landscape diversity.

A GIS-based opportunity mapping procedure was used to identify sites where three different classes of woodland could be planted to alleviate flooding (Figure 14.4):

1. Flood plain woodland: at present only 5% of the total area of flood plain is wooded. The opportunity mapping procedure identified a total of 70 locations with a combined area of 440 ha (24% of the flood plain) which are suitable for new woodland planting.
2. Riparian woodland: an area of 6000 ha was identified as having high priority for potential new riparian woodland.
3. Woodland in the wider catchment: more than 35,000 ha of the catchment were classified as a high priority for new woodland planting which reflects the soil types present and their susceptibility to erosion. (For more information see Broadmeadow and Nisbet 2010).

14.5.2 Renewal of Interest in Broadleaves

For much of the twentieth century, broadleaved species were generally regarded as a less valuable crop than conifers, and large areas of broadleaved woodland were

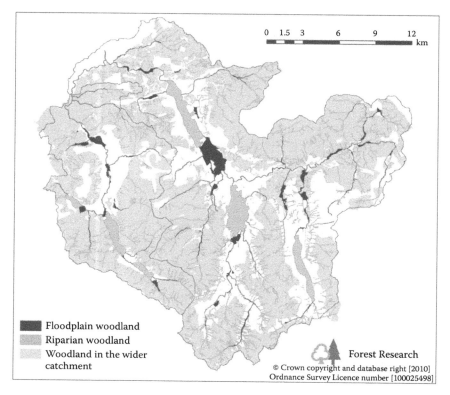

FIGURE 14.4
The Derwent River catchment showing areas where woodland creation has the potential to alleviate flooding.

converted to conifer plantations. However, by the 1980s, this practice was falling into disrepute, and there was a revival of interest in growing good quality broadleaved trees and a belated acknowledgment of the biodiversity value of seminatural broadleaved woodland. A new policy on broadleaved woodland and guidelines for management were introduced (Forestry Commission, 1985a,b). This policy marked a notable shift in woodland creation and although the estimated area of new woodland created during the last 40 years varied between countries, the trends were similar. Throughout the period, there was a general decline in the area of new coniferous woodland created with most being established in Scotland; the steep decline for this country at the end of the 1980s coincided with changes to tax laws (Figure 14.5). Initially, the area of conifer woodland created was greater than that of broadleaves, but the area of these began to increase during the late 1980s and has subsequently exceeded that for conifers in all the countries. The proportion of broadleaved woodland created was greatest for England. These data reflect the general aims of the policy to give more emphasis to broadleaves and the increased incentives for establishing native broadleaved trees. Much of this new woodland creation was undertaken by private owners supported by the Woodland Grant Scheme which favored broadleaved trees and additional schemes for woodland creation on farmland (Aldhous, 1997).

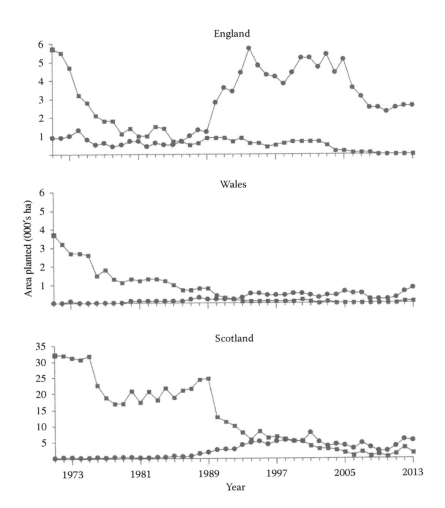

FIGURE 14.5
Area of new woodland planted in the three countries of Great Britain between 1971 and 2013. Area includes woodland established by natural recolonization. Broadleaves = ●; Conifers = ■. (Adapted from Forestry Commission. 2013b. *New Planting and Restocking 1976–2013*. http://www.forestry.gov.uk/pdf/planting1976-2013. xls/$FILE/planting1976-2013.xls (accessed October 14, 2013).)

14.6 Ecological Aspects of Woodland Creation

The renewed interest and increased use of broadleaves, which began in the 1980s, was also accompanied by changes in practice that aimed to improve the biodiversity value of new woodland. These recommendations were based on ecological ideas and included changes in species mixtures, methods of establishment, and the location of new woodland. For example, it became normal when establishing new woodland by planting to use mixtures of tree and shrub species described by the National Vegetation Classification as being suitable for the site (Rodwell, 1991; Rodwell and Patterson, 1994). While in the short-term, this will create woodlands different to those comprising monocultures or simple mixtures of one or two species; but in the long-term, the differences may depend on management of the stand.

14.6.1 Woodland Biodiversity and Fragmentation

This move toward planting of broadleaved woodland was partly in recognition that despite a long history of woodland creation in the twentieth century, much biodiversity was still restricted to small, isolated broadleaved woodland fragments. Although policies and grants were successful in increasing the cover of broadleaved woodland, many of these newly created woodlands were small and isolated from existing woodland fragments.

In an attempt to overcome this fragmentation of broadleaved woodlands, some grant schemes were later revised to favor the establishment of new woodlands beside existing woodlands and planting between woodlands to join them together. Such revisions were influenced by ecological ideas which suggest that larger woodlands linked together in the landscape are likely to be better for biodiversity than many small isolated woodlands scattered throughout an intensively managed landscape.

14.6.2 Case Study: Effectiveness of Grant-Aid to Reduce Fragmentation and Increase the Size of Individual Woods

Although Great Britain has about 13% forest cover, this is highly fragmented; around 75% of all woodlands are under 2 ha in size, with large nonnative conifer plantations accounting for the few larger forests (Figure 14.6). In general terms, fragmentation causes a decline

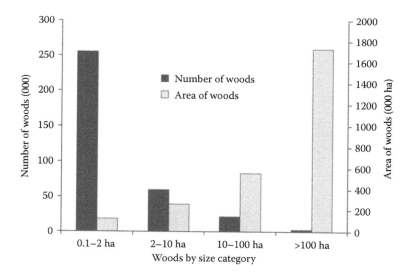

FIGURE 14.6
Number of woodlands in different size classes and total area of woodlands in each size class. (Data adapted from Forestry Commission. 2001. *National Inventory of Woodland and Trees—England*. Edinburgh: Forestry Commission; Forestry Commission. 2002a. *National Inventory of Woodland and Trees—Scotland*. Edinburgh: Forestry Commission; Forestry Commission. 2002b. *National Inventory of Woodland and Trees—Wales*. Edinburgh: Forestry Commission; Forestry Commission. 2003. *National Inventory of Woodland and Trees—Great Britain*. Edinburgh: Forestry Commission.)

in the size, and often the quality of habitat patches, and increases the physical isolation of the remaining habitat fragments (Fischer and Lindenmayer, 2007). This can have the biological impacts of reducing population sizes, reducing genetic diversity, and increasing the risk of local extinctions; while the increase in isolation may reduce the exchange of individuals and genes between fragmented populations, which may otherwise ameliorate these negative effects at a patch level (Fahrig, 2003).

In recognition of the problem of woodland fragmentation, the Forestry Commission introduced a spatially-targeted grant scheme that offered extra incentives for woodland creation that would expand, buffer, or join existing woodland habitats. This was known as Joining and Increasing Grant Scheme for Ancient Woodland (JIGSAW) and contrasted with previous WGS (Woodland Grant Schemes), which offered little spatial targeting (Forestry Commission, 1999).

A study on the Isle of Wight in southern England revealed that the targeted JIGSAW scheme reduced fragmentation compared to the nontargeted WGS scheme (Quine and Watts, 2009). This study found that woodlands planted under the JIGSAW schemes decreased the overall number of individual woodlands and increased their average size (Figure 14.7). In contrast, WGS schemes, which created a comparable area of woodland, increased the number of woodland patches and decreased the mean woodland size. This was due to the creation of greater number of smaller and more isolated woodlands. The study confirmed the merits of spatially-targeted woodland creation to combat fragmentation and potentially provide greater biodiversity benefits.

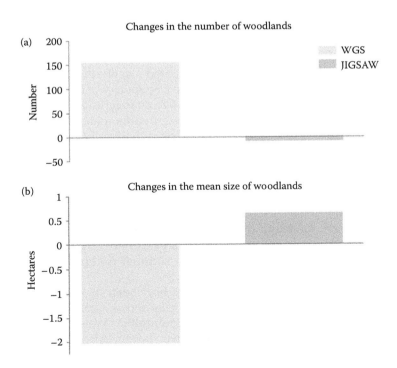

FIGURE 14.7
Comparison of the number (a) and size (b) of new woodlands created by spatially targeted JIGSAW grant scheme and nontargeted Woodland Grant Scheme (WGS) between 1998 and 2005.

14.7 Climate Change

In recent years, the prospect of significant climate change has become an important consideration in forestry policy, planning, and management due to its influence on species choice for different locations across the country. In addition, woodland creation also provides a cost-effective and achievable method of abating green house gas emissions (Matthews and Broadmeadow, 2009). Although the two most cost-effective abatement options were use of fast growing conifer plantations and other rapidly growing energy crops, the creation of mixed woodlands managed for multiple objectives will deliver substantial mitigation benefits, coupled with recreation and biodiversity gains (Read et al., 2009; Ray et al., 2014).

14.7.1 Species Choice and Site Suitability

In Great Britain, foresters increasingly use the Ecological Site Classification (ESC) decision support system (Pyatt et al., 2001; Ray, 2001) to help in classifying sites in order to choose species well suited to soil and climatic conditions. This system uses ground vegetation and soil type to estimate site fertility and water availability (Wilson et al., 2001; Wilson et al., 2005). Climatic suitability is estimated from accumulated temperature (mean annual degree-days above 5°C) and moisture deficit (mean annual excess of monthly potential evapotranspiration over rainfall). The accumulated temperature and moisture deficit climatic indices (Figure 14.8) can be used in ESC to evaluate the current and projected future climatic constraints on species suitability. A retrospective assessment of species choice in relation to site conditions for state forests found that across the state forests, species choice had been very good for the current climatic conditions (Moffat, et al., 2012). However, further assessment of suitability and yield, incorporating projected changes in climatic conditions, highlighted a spatially dependent and variable degree of risk to continued growth of some of the tree species. An example of the regional vulnerability of tree species resulting from climate change is shown in Figure 14.9. The three species shown (Sitka spruce—*Picea sitchensis*, pedunculate oak—*Quercus robur*, and Scots pine—*Pinus sylvestris*) exhibit different regional changes between the degree of suitability (Very suitable, Suitable, and Unsuitable) in the current climate and the suitability modeled by ESC for the climatic conditions projected in 2050. Sites identified as Suitable for Sitka spruce shift northwards and westward as those sites are currently Suitable in central England, whereas eastern Wales and Eastern Scotland become Unsuitable. In contrast, there is a general trend for a northwards shift in the suitability of sites for oak and pine. These changes are the result of a warmer climate with less change in the seasonal distribution of rainfall in the west and north of Great Britain (Figure 14.8) than in eastern and southern regions where moisture deficits in summer will be higher due to a reduction in rainfall.

A range of strategies to adapt to climate change have been suggested (Ray, 2008a,b; Ray et al., 2010) and are promoted by the forest policies across Great Britain (DEFRA, 2013; Forestry Commission Scotland, 2009; Forestry Commission Wales, 2009). At present, forest managers rely on a small range of productive species (Mason and Perks, 2011), but to allow for climate change, a range of measures are recommended; these include: an increase in tree species diversity in forests to spread the risk of losses, use of appropriate species for sites and the projected climate conditions that will develop, and better choice of provenance.

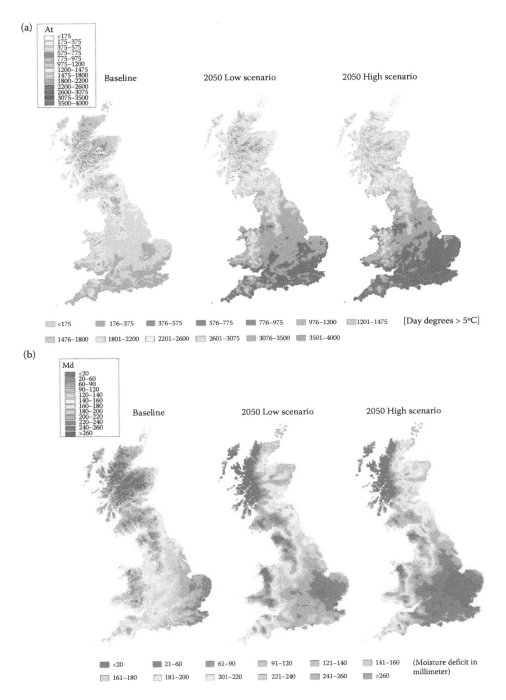

FIGURE 14.8
Climatic warmth defined by (a) accumulated temperature in degree-day over 5°C, and (b) climatic wetness defined by moisture deficit, and projections for the climatic period 2031–2060 (2050s). (Based on the B2 and A1Fi SRES scenario. (From Hulme, M. et al. 2002. *Climate Change Scenarios for the United Kingdom: The UKCIP02 Scientific Report.* Norwich: Tyndall Centre for Climate Change Research, University of East Anglia.))

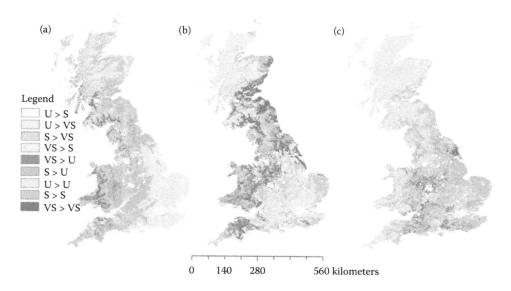

FIGURE 14.9

Modeled changes in the production suitability in Britain of (a) Sitka spruce, (b) Scots pine and, (c) pedunculate oak, between the baseline climatic period and the projected climate for the 30 year period 2041–2070. V = Very suitable, S = Suitable, U = Unsuitable. (Based on the SRES scenario A1Fi. (From Hulme, M. et al. 2002. *Climate Change Scenarios for the United Kingdom: The UKCIP02 Scientific Report.* Norwich: Tyndall Centre for Climate Change Research, University of East Anglia.))

14.8 Conclusion

At the beginning of the twentieth century, Great Britain had no national forestry policy, and more than 95% of woodlands were privately owned and managed as their owners wished. Woodlands were generally poorly-stocked producing an annual increment of 25%–50% of that expected from well-managed woods on similar soils (Anon, 1918). Following a report to government, a forestry policy for Great Britain was introduced and a national forestry department established in 1919 (Anon, 1918; Pringle, 1994). The report contained both short-term and long-term targets for woodland creation indicating that ca. 81,000 ha would be created in a decade and 720,000 after 80 years. The short-term target was not achieved and the long-term target was superseded within 25 years. The revised target is ca. 1.2 million ha of new woodland as part of a total of 2.0 million ha of woodland overall within 50 years (Forestry Commissioners, 1943). Although Figure 14.1 shows that woodland cover was greater than 2 million ha by 1982, the new target specified productive forest is the area which did not exceed the target until 1996 (Aldhous, 1997). While the policy introduced early in the twentieth century to increase the area of woodland has been very successful in delivering the desired outcome within the predicted timescale, it is more difficult to assess whether the policy changes relating to woodland creation using broadleaves will be effective.

The expansion of forest cover was based almost entirely on planting and there was little understanding or practical skill in the use of natural regeneration, even to restock existing woodland (Harmer, 1994). Despite this lack of knowledge, the use of natural regeneration to create new woodland became an acceptable practice that could receive grant aid; it was recommended for a variety of reasons including the establishment of woodlands with more natural characteristics comprising trees of local genotype. However available information suggests that success is site-dependent and on fertile soils, woodland may take many years to establish (Hodge and Harmer, 1996; Harmer et al., 2000). Consequently, use of natural regeneration may be more appropriate on sites with well drained, infertile soils than those with heavy soils that have been improved by agriculture. Although practices such as these have become embedded in the UKFS, most of the practices have not been systematically examined to determine whether they deliver the expected outcomes.

Although ecological principles relating to the creation of new woodland (Ferris-Kaan, 1995) have been incorporated into incentive schemes, the JIGSAW scheme described is one of few studied to have determined whether expected benefits have been achieved. However, this scheme aimed only to increase the size of woodland blocks and the study was an accounting exercise that measured the size of woodlands but not the actual ecological benefit achieved. Some studies of plant and animal species in newly created woodland have shown changes which meet expectations (e.g., Moore et al., 2003; DEFRA, 2003), but these were short-term and in very young woodland. Other ecologically inspired criteria of incentive schemes are largely unexamined. For example, the idea that new woodlands comprising local provenances, which are created by either natural regeneration or using transplants grown from seeds collected in the locality, remains untested and may be inappropriate if the predictions of future climate change are correct. While many of these criteria were logical, practical extensions of existing ecological ideas, examination of their effectiveness in delivering the expected benefits is unlikely in the current economic conditions.

The concept of multipurpose forestry became widespread during the 1980s and use of woodlands has changed noticeably since this time. Newly created woodlands, especially those near large centers of population have embraced the ideals of multipurpose forestry and are managed for a wide range of uses. For example, in the ca. 20 years since the creation of the National Forest: about 80% of the forest created has open public access; ca. 0.5 million children have taken part in environmental education sessions; about 50 new recreational facilities and 100 km of cycleways have been created; there were 7.8 million visitors in 2011 and about 4500 tourism related jobs; more than 2000 ha of wildlife habitat have been created or brought back into management and 150 wildlife ponds created. Most of these could be considered as environmental, economic, and social/cultural aspects of sustainable forest management delivering these important ecosystem services (Quine et al., 2013).

New methods of assessing the resilience of Britain's forests under the combined effects of climate and socio-economic change are being developed (Petr et al., 2014a,b; Ray et al., 2014). The aim is to help forest managers and planners to assess risk, reduce uncertainty, and choose robust forest plans into the future. This implies that the delivery of a broad range of ecosystem services from the natural capital of forested land will continue to be required. The dynamic coupling of different forestry models such as ESC, Forest Yield (Matthews, 2008), ASORT and BSORT (Matthews and Duckworth, 2005), and ForestGALES (Gardiner and Quine, 2000), and running these models with probabilistic climate projections (Petr et al., 2014b), demonstrates how ecosystem services are likely to change into the future under different combinations of management and environmental drivers. The methods look very promising for helping to direct strategic forest planning on the public forest estate in Britain.

Nevertheless, the effectiveness of some current schemes to improve sustainable forest management, provide ecosystem services and overcome the anticipated effects of climate change, are more difficult to assess. Targeted woodland creation, to provide specific ecosystem services such as flood management and improvements to water quality (Thomas and Nisbet, 2006; Nisbet et al., 2011), are likely to be successful if they are implemented. However, in forestry time scales, any established schemes are in their infancy, and it is likely to be many years before success can be judged. The likely effects of climate change are topical and of significant importance for those planning the restocking and creation of new woodlands to be resilient to future changes. The current expectations are that by careful selection of species and provenances (Hubert and Cottrell, 2007) appropriate for the future climate at the site, combined with appropriate silviculture (Meason and Mason, 2013), forests will be more resilient to climate change. Whether this approach will be effective is unknown but at present it seems to be the best available.

References

Aldhous, J.R. 1997. British Forestry: 70 years of achievement. *Forestry* 70(4):283–291.

Anon. 1918. *Reconstruction Committee. Forestry Sub-Committee.* Final report. Cd. 8881. London: HMSO.

Anon. 1928. *Report on Census of Woodlands and Census of Production of Home-Grown Timber 1924.* London: HMSO.

Anon. 1952. *Census of Woodlands 1947–1949.* London: HMSO.

Anon. 2005. *WIAT: Woodlands in and Around Towns.* Edinburgh: Forestry Commission and Scottish Executive.

Anderson, M.L. 1967a. *A History of Scottish Forestry. Volume 1: From the Ice Age to the French Revolution.* London: Nelson.

Anderson, M.L. 1967b. *A History of Scottish Forestry. Volume 2 From the Industrial Revolution to Modern Times.* London: Nelson.

Bell, S. 1992. Planting the New National Forest. In: *Report of the 9th Meeting of the National Hardwoods Programme at the Oxford Forestry Institute* (ed.). P.S. Savill, OFI Occasional papers 41, 13–16.

Broadmeadow, S. and Nisbet, T. 2010. *Opportunity Mapping for Woodland to Reduce Flooding in the River Derwent, Cumbria.* Forestry Commission, Edinburgh. http://www.forestry.gov.uk/pdf/Derwent_flooding_final_report_2010.pdf/$FILE/Derwent_flooding_final_report_2010.pdf (last accessed September 30, 2013).

Broadmeadow, M., Ray, D., and Samuel, C., 2005. Climate change and the future for broadleaved tree species in Britain. *Forestry* 78:145–167.

Countryside Commission. 1989. *A New National Forest in the Midlands: A Consultation Document.* Countryside Commission Report CCP 278. Cheltenham: Countryside Commission.

Countryside Commission. 1994. *The National Forest: The Strategy. The forest vision.* Countryside Commission Report CCP 468. Cheltenham: Countryside Commission.

DEFRA. 2003. Environmental impacts of farm woodland planting on an intensive arable farm: changes in biodiversity on the site of the 'Boxworth Project'. Project WD0131 http://sciencesearch.defra.gov.uk/Default.aspx?Menu=Menu&Module=More&Location=None&Completed=0&ProjectID=9504 (accessed December 11, 2014).

DEFRA. 2013. *Government Forestry and Woodlands Policy Statement: Incorporating the Government's Response to the Independent Panel on Forestry's Final Report.* London: Department for Environment Food and Rural Affairs.

Fahrig, L. 2003. Effects of habitat fragmentation on biodiversity. *Annual review of ecology, evolution, and systematics* 34: 487–515.

Ferris-Kaan, R. (Ed.). 1995. *The Ecology of Woodland Creation.* Chichester: John Wiley and Sons.

Fernow, B.E. 1913. *History of Forestry.* Toronto: Toronto University Press.

Fischer, J. and Lindenmayer, D. B. 2007. Landscape modification and habitat fragmentation: A synthesis. *Global Ecology and Biogeography* 16(3): 265–280.

Forestry Commission. 1985a. *The Policy for Broadleaved Woodland*. Policy and Procedure Paper 5, Edinburgh: Forestry Commission.

Forestry Commission. 1985b. *Guidelines for the Management of Broadleaved Woodland*. Edinburgh: Forestry Commission.

Forestry Commission. 1999. *England Forestry Strategy: A New Focus for England's Woodlands*. Cambridge: Forestry Commission.

Forestry Commission. 2001. *National Inventory of Woodland and Trees—England*. Edinburgh: Forestry Commission.

Forestry Commission. 2002a. *National Inventory of Woodland and Trees—Scotland*. Edinburgh: Forestry Commission.

Forestry Commission. 2002b. *National Inventory of Woodland and Trees—Wales*. Edinburgh: Forestry Commission.

Forestry Commission. 2003. *National Inventory of Woodland and Trees—Great Britain*. Edinburgh: Forestry Commission.

Forestry Commission. 2011. *The UK Forestry Standard*. Edinburgh: Forestry Commission.

Forestry Commission. 2013a. *Forestry Facts and Figures 2013*. Edinburgh: Forestry Commission. http://www.forestry.gov.uk/forestry/INFD-7AQF6J (accessed November 18, 2014).

Forestry Commission. 2013b. *New Planting and Restocking 1976–2013*. Edinburgh: Forestry Commission. http://www.forestry.gov.uk/pdf/planting1976-2013.xls/$FILE/planting1976-2013.xls (accessed October 14, 2013).

Forestry Commission. 2014. *Forestry Facts and Figures*. http://www.forestry.gov.uk/pdf/FFF2014.xls/$FILE/FFF2014.xls (accessed November 18, 2014).

Forestry Commissioners. 1943. *Post-War Forest Policy*. London: HMSO.

Forestry Commission Scotland. 2009. *Climate Change Action Plan 2009–2011*. Edinburgh: Forestry Commission Scotland.

Forestry Commission Wales. 2009. *Woodlands for Wales—The Welsh Assembly Government's Strategy for Woodlands and Trees*. Cardiff: Forestry Commission Wales.

Fyfe, R.M., Twiddle, C., Sugita, S. et al. 2013. The Holocene vegetation cover of Britain and Ireland: Overcoming problems of scale and discerning patterns of openness. *Quaternary Science Reviews* 73:132–148.

Gardiner, B.A. and Quine, C.P. 2000. Management of forests to reduce the risk of abiotic damage—a review with particular reference to the effects of strong winds. *Forest Ecology and Management* 135:261–277.

GPFLR 2013. *Learning Site—Loch Katrine*. Global Partnership for Forest Landscape Restoration, IUCN, Gland, Switzerland. http://www.forestlandscaperestoration.org/learning-site-loch-katrine (last accessed October 16, 2013).

Harmer, R. 1994. Natural regeneration of broadleaved trees in Britain: Historical aspects. *Forestry* 67:179–188.

Harmer, R., Kiewitt, A., and Morgan, G. 2012a. Can overstorey retention be used to control bramble (*Rubus fruticosus* L. agg.) during regeneration of forests? *Forestry* 85:135–144.

Harmer, R., Kiewitt, A., and Morgan, G. 2012b. Effects of overstorey retention on ash regeneration and bramble growth during conversion of a pine plantation to native broadleaved woodland. *European Journal of Forest Research* 131:1833–1843.

Harmer, R., Peterken, G., Kerr, G., and Poulton, P. 2000. Vegetation changes during 100 years of development of two secondary woodlands on abandoned arable land. *Biological Conservation* 101:291–304.

Harmer, R., Thompson, R., and Humphrey, J. 2005. Great Britain—Conifers to broadleaves. In: Stanturf, J.A. and Madsen, P. (eds.). *Restoration of Boreal and Temperate Forests*. Roca Baton: CRC Press.

Hodge, S. and Harmer, R. 1996. Woody colonization on unmanaged urban and ex-industrial sites. *Forestry* 69:245–261.

Hubert, J. and Cottrell, J. 2007. *The Role of Forest Genetic Resources in Helping British Forests Respond to Climate Change*. Information Note 86, Edinburgh: Forestry Commission.

Hulme, M., Jenkins, G.J., Lu, X. et al., 2002. *Climate Change Scenarios for the United Kingdom: The UKCIP02 Scientific Report*. Norwich: Tyndall Centre for Climate Change Research, University of East Anglia.

James, N.D.G. 1981. *A History of English Forestry*. Oxford: Basil Blackwell.

Kerr, G. and Williams H.V. 1999. *Woodland Creation: Experience from the National Forest*. Forestry Commission Technical Paper 27. Edinburgh: Forestry Commission.

Lamont, R., 2006. *Loch Katrine Forest Landscape Restoration*. United Nations Environment Programme, World Conservation Monitoring Centre, Cambridge, UK. http://www.unep-wcmc.org/medi alibrary/2011/05/24/4f1385fb/UK%20Loch%20Katrine%20highres.pdf (accessed October 16, 2013).

Locke, G.M.L. 1979. *Census of Woodlands 1965–1967*. London: HMSO.

Locke, G.M.L. 1987. *Census of Woodlands and Trees 1979–1982*. Forestry Commission Bulletin 63. London: HMSO.

Linnard, W. 2000. *Welsh Woods and Forests: A History*. Llandysul: Gomer Press.

Mason, B. and Perks, M.P. 2011. Sitka spruce (*Picea sitchensis*) forests in Atlantic Europe: Changes in forest management and possible consequences for carbon sequestration. *Scandinavian Journal of Forest Research* 26:72–81.

Mason, W.L. 2007. Changes in the management of British forests between 1945 and 2000 and possible future trends. *Ibis* 149:41–52.

Matthews, R.W. 2008. *Forest Yield a Software Framework for Accessing Forest Growth and Yield Information*. Edinburgh: Forestry Commission.

Matthews, R.W. and Broadmeadow, M. 2009. The potential of UK forestry to contribute to government's emissions reduction commitments. In: Read, D.J., Morison, J.I.L, Hanley N, West, C.C. and Snowdon, P. (Eds.). *Combating Climate Change—A Role for UK Forests. An Assessment of the Potential of the UK's Trees and Woodlands to Mitigate and Adapt to Climate Change*. Edinburgh: The Stationery Office.

Matthews, R.W. and Duckworth, R.R. 2005. *BSORT: A model of tree and stand biomass development and production in Great Britain*. In: M.S. Imbabi and C.P. Mitchell (Eds.). *Proceedings of the World Renewable Energy Congress (WREC 2005)*. Elsevier: Oxford, Aberdeen, UK, pp. 404–409.

Meason, D.F. and Mason, W.L. 2013. Evaluating the deployment of alternative species in planted conifer forests as a means of adaptation to climate change—Case studies in New Zealand and Scotland. *Annals of Forest Science*, 71:239–253.

Moffat, A.J., Morison, J.I.L., Nicoll, B et al. (Eds.). 2012. Climate change risk assessment for the forestry sector. In: *UK Climate Change Risk Assessment*. London: DEFRA.

Moore, N. P., Askew, N., and Bishop, J.D. 2003. Small mammals in new farm woodlands. *Mammal Review* 33:101–104.

Morison, J., Matthews, R., Miller, G et al. 2012. *Understanding the Carbon and Greenhouse Gas Balance of Forests in Britain*. Edinburgh: Forestry Commission.

Newton, A. and Kapos, V. 2003. Restoration of wooded landscapes: placing UK initiatives in a global context. In: Humphrey, J. Newton, A., Latham, J., Gray, H., Kirby, K., Poulsom, E, and Quine, C. (Eds.). *The Restoration of Wooded Landscapes*. Edinburgh: Forestry Commission.

Nisbet, T., Silgram, M., Shah, N., Morrow, K., and Broadmeadow, S. 2011. *Woodland for Water: Woodland Measures for Meeting Water Framework Directive Objectives*. Forest Research Monograph, 4. Edinburgh: Forestry Commission. http://www.forestry.gov.uk/pdf/FRMG004_ Woodland4Water.pdf/$FILE/FRMG004_Woodland4Water.pdf (accessed December 11, 2014).

Petr, M., Boerboom, L.G.J., Ray, D., and van-der-Veen, A. 2014a. An Uncertainty Assessment Framework for Forest Planning Adaptation to Climate Change. *Forest Policy and Economics* 41:1–11.

Petr, M., Boerboom, L.G.J., van-der-Veen, A., and Ray, D. 2014b. A spatial and temporal drought risk assessment of three major tree species in Britain using probabilistic climate change projections. *Climatic Change* 124:791–803.

Pringle, D. 1994. *The Forestry Commission—The First 75 Years*. Edinburgh: Forestry Commission.

Pryor, S. 2003. The costs and benefits of restoring plantations versus creating new native woodland. In: Humphrey, J. Newton, A., Latham, J., Gray, H., Kirby, K., Poulsom, E, and Quine, C. (Eds.). *The Restoration of Wooded Landscapes*. Edinburgh: Forestry Commission.

Pryor, S. and Smith, S. 2002. *The Area and Composition of Plantations on Ancient Woodland Sites*. Grantham: The Woodland Trust, 32 pp.

Pyatt, D.G., Ray, D., and Fletcher, J. 2001. *An Ecological Site Classification for Forestry in Great Britain*. Forestry Commission Bulletin 124. Edinburgh: Forestry Commission.

Quine, C.P., Bailey, S.A., and Watts, K. 2013. Sustainable forest management in a time of ecosystem services frameworks: common ground and consequences. *Journal of Applied Ecology* 50:863–867.

Quine, C.P. and Watts, K. 2009. Successful de-fragmentation of woodland by planting in an agricultural landscape? An assessment based on landscape indicators. *Journal of Environmental Management* 90: 251–259.

Rackham, O. 1986. *The History of the Countryside*. London: Dent.

Ray, D. 2001. *Ecological Site Classification Decision Support System 1.7*. Edinburgh: Forestry Commission. https://www.eforestry.gov.uk/forestdss/ (accessed November 18, 2014).

Ray, D. 2008a. *Impacts of Climate Change on Forestry in Wales*. Forestry Commission Wales Research Note 301, Aberystwyth: Forestry Commission Wales.

Ray, D. 2008b. *Impacts of Climate Change on Forests in Scotland—A Preliminary Synopsis of Spatial Modelling Research*. Forestry Commission Research Note 101, Edinburgh: Forestry Commission Scotland.

Ray, D., Bathgate, S., Moseley, D., Taylor, P., Nicoll B., Pizzirani, S., and Gardiner, B. 2014. Comparing the provision of ecosystem services in plantation forests under alternative climate change adaptation management options in Wales. *Regional Environmental Change*. DOI 10.1007/s10113-014-0644-6.

Ray, D., Morison, J., and Broadmeadow, M. 2010. *Climate Change: Impacts and Adaptation in England's Woodlands*. Forestry Commission Research Note 201. Edinburgh: Forestry Commission.

Read, D.J., Morison, J.I.L, Hanley N, West, C.C., and Snowdon, P. (Eds.). 2009. *Combating Climate Change—A Role for UK Forests. An Assessment of the Potential of the UK's Trees and Woodlands to Mitigate and Adapt to Climate Change*. Edinburgh: The Stationery Office.

Richards, E.G. 2003. *British Forestry in the Twentieth Century, Policy and Achievements*. Leiden: Koninklijke Brill.

Rodwell, J. (ed.). 1991. *British Plant Communities. Volume 1 Woodlands and Scrub*. Cambridge: Cambridge University Press..

Rodwell, J. and Patterson, G. 1994. *Creating New Native Woodlands*. Forestry Commission Bulletin 112. London: HMSO.

Savill, P. 2013. *The Silviculture of Trees Used in British Forestry*. Wallingford: CABI.

Smout, T.C. (Ed.). 2003. *People and Woods in Scotland. A History*. Edinburgh: Edinburgh University Press.

Stanturf, J.A. 2005. What is forest restoration? In: Stanturf, J.A. and Madsen, P. (eds.). *Restoration of Boreal and Temperate Forests*. Roca Baton: CRC Press.

Thomas, H. and Nisbet, T.R. 2006. An assessment of the impact of floodplain woodland on flood flows. *Water and Environment Journal* 21:114–126.

Thompson, R.N., Humphrey, J.W., Harmer, R., and Ferris, R. 2003. *Restoration of Native Woodland on Ancient Woodland Sites*. Edinburgh: Forestry Commission.

UKNEA. 2011. *The National Ecosystem Assessment: Synthesis of Key Findings*. Cambridge: UNEP-WCMC.

Vera, F.W.M. 2000. *Grazing Ecology and Forest History*. Wallingford: CABI.

Watts, K., Quine, C.P., Eycott, A.E., Moseley, D., and Humphrey, J.W. 2008. Conserving Forest Biodiversity: Recent Approaches in UK Forest Planning and Management. In: Lafortezza, R., Sanesi, G., Chen, J., and Crow, T.R. (Eds.). *Patterns and Processes in Forest Landscapes*. Berlin: Springer, 375–400.

Wilson, S.M., Pyatt, D.G., Malcolm, D.C., and Connolly, T. 2001. The use of ground vegetation and humus type as indicators of soil nutrient regime for an ecological site classification of British forests. *Forest Ecology and Management*, 140:101–116.

Wilson, S.M., Pyatt, D.G., Ray, D., Malcolm, D.C., and Connolly, T. 2005. Indices of soil nitrogen availability for an ecological site classification of British forests. *Forest Ecology and Management*, 220:51–65.

15

Irish Peatland Forests: Lessons from the Past and Pathways to a Sustainable Future

Florence Renou-Wilson and Kenneth A. Byrne

CONTENTS

15.1 Introduction

Lying between the 51° and 55° northern latitudes, the island of Ireland covers an area of 84,000 km² and is made up of a central plain enclosed by coastal highlands. The 4.6 million inhabitants living in the Republic of Ireland (Ireland hereafter) enjoy a mild temperate climate due to the influence of the Gulf Stream, characterized by cool summers, mild winters, and significant amounts of cloudiness, high atmospheric humidity and windiness. With these climatic conditions, it is not surprising therefore that one-fifth of the country is covered by peat soils. Peatlands are a significant element in the Irish landscape with peat soils estimated to cover 20% (1.46 million ha, see Figure 15.1) of the land area (Connolly and Holden 2009) and have been widely utilized over the centuries for energy and horticultural peat production, agriculture, and forestry. Of the Holocene extent of fens and bogs in Ireland, only 15% of the original peatland cover is in near-intact condition (i.e., low level of degradation) with an even smaller proportion being fully functioning mires (i.e., active peat forming ecosystem) (Wilson et al. 2013). The area of undisturbed fens is very small as they have long been drained and reclaimed for agricultural use. Two bog morphologies are found in Ireland: raised and blanket bogs, the latter being the most extensive and is subdivided into Atlantic blanket bog and mountain blanket bog. Peat soils, by definition,

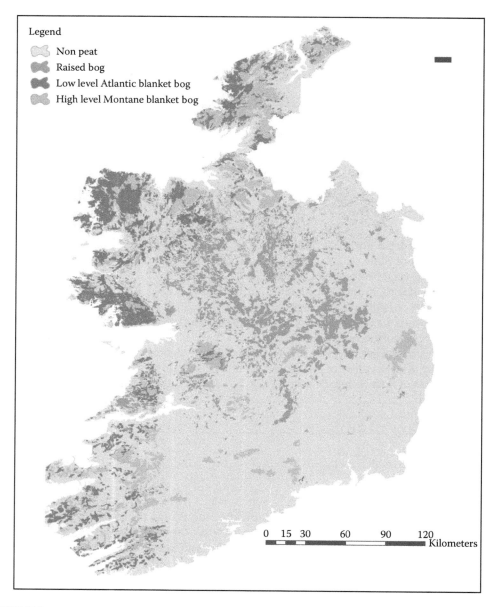

FIGURE 15.1
Map of Ireland showing the distribution of the three main bog types.

contains peat over a depth of at least 45 cm on undrained land and 30 cm deep on drained land; the depth requirement does not apply in the event that the peat layer is directly over bedrock (Renou-Wilson et al. 2011). However the range of biogeochemical characteristics of peat soils encountered in Ireland is wide because of the various processes leading to the formation of peat landforms.

Raised bogs are found mainly in the Irish midlands under a rainfall regime between 750 and 1000 mm annually. The surface is strongly acidic, but beneath the peat are alkaline moraine soils of limestone origin. Raised bogs are dominated by *Sphagnum* mosses, with heather (*Calluna vulgaris* [L.] Hull), bog cotton (*Eriophorum angustifolium* Honckeny), and several species of sundew (*Drosera* spp.) and orchids (e.g., *Dactylorhiza incarnata*) also present. They started to develop at the end of the last glacial period 10,000 years ago, when shallow lakes left by the retreating ice covered much of central Ireland. Nutrient-rich groundwater derived from calcareous glacial drift fed these lakes (Mitchell and Ryan 1997). Reeds and sedges encroached around the lake edges; their remains fell into the water where they were only partly decomposed, in time forming a thick layer of reed peat. The upward growth of sedges and other plants (brown moss spp. as well *Sphagnum* mosses) and the accumulation of their undecayed remains "raised" the peat surface above the influence of groundwater, becoming "rain-fed" and therefore ombrotrophic. This gave rise to raised bogs up to 14 m deep (averaging 6–7 m), a feature that made them particularly suitable for peat exploitation.

Blanket bogs developed about 4000 years ago; they are found chiefly along the western seaboard and on mountaintops throughout the rest of the country. Most of the blanket bogs were initiated through paludification as a result of changed climatic conditions but peat initiation can also be attributed to human activity in other areas. Blanket bogs are most widespread in areas where annual rainfall levels are greater than 1200 mm and the number of rain days exceeds 225. These bogs are shallow and form a blanket-like layer averaging 2.5 m in depth over an underlying acidic mineral soil. In their natural condition, these areas are dominated by *Eriophorum* species, black bog rush (*Schoenus nigricans* L.), and purple moor- grass (*Molinia caerulea* [L.] Moench). A particular characteristic of both Irish raised and blanket bog is their natural tree-less state.

15.2 Overview of Irish Forestry

15.2.1 Current State

Ireland belongs to the temperate deciduous forest biome and the whole island was originally covered with these types of woodlands (without the lime, beech and hornbeam which did not get as far as Ireland) together with pine forests on the poorer soils. By the end of the nineteenth century, following centuries of exploitation, the forests covered only 1% of the land area. The twentieth century witnessed the gradual restoration of this lost resource. Despite this, Ireland has the lowest proportion of total land under forest in the European Union (EU), at 11.1% or 731,652 ha (John Redmond, personal communication.) and it is the policy of the government to increase the area of forest to 1.2 million ha (i.e., 17% of the land area) by 2030 (Department of Agriculture Food and Forestry 1996). The main impetus for afforestation is the financial incentives provided by the EU and the Irish Government.

Ireland has a young, intensively managed forest estate, of which 57% is in public ownership (Redmond et al. 2007), reflecting the dominant role of the State in forestry development. The balance of ownership is changing however, as the private sector (farmers in particular) becomes increasingly involved (Gillmor 1998). The forestry estate contributes significantly to the Irish economy supporting direct employment in the forestry and wood processing sectors (Ní Dhubháin et al. 2013). Ireland has moved from a position of timber deficit to one of surplus in less than 100 years.

The national forest estate is dominated by exotic conifer species, which accounts for 74% of the total forest area. Broadleaf species account for the balance. The dominant coniferous species is Sitka spruce (*Picea sitchensis* [Bong] Carr.), which accounts for 53% of the forest estate. It is widely favored as a commercial species because it is capable of higher yields over a wider range of site types compared to other species (Joyce and O Caroll 2002). The emphasis on quick-growing softwoods reflects the favorable conditions of the Irish climate for these species. Timber production is the principal business of Irish forestry. Plantations are managed through planting and thinning; stands are dense and rotations are short (the crop is clear-felled from 35 years of age, depending on species and growth rate). As a result, Irish coniferous forests rarely develop to the open character of mature forests.

The level of broadleaf afforestation was very low up to the early 1990s, averaging less than 500 ha (3%–4%) of total annual afforestation. This rate was still only at 12% in the early 2000s but the first decade of the twenty-first century saw a dramatic increase to reach an unprecedented 30% in 2012. This coincides with the move away from planting unenclosed land (peatland), which in 2012 was reduced nationwide to cover barely 300 ha in total. The total forest area planted on organic soils accounts for 60% (Duffy et al. 2013), but the proportion on wet mineral soil has steadily increased. Pockets of semi-natural woodland survive around the country and these receive some protection under different schemes.

15.2.2 Historical Framework

Deciduous forests as well as Caledonian-style pine forests once covered a large part of the Irish landscape. Their demise was the result of a combination of paludification due to changed climatic conditions, which turned the landscape of the west of Ireland in particular into blanket bog, and anthropogenic activities. From earliest times, the forests were exploited. Early migrants came by sea and were both boat builders and craftsmen of some skill. A Viking ship discovered at Roskilde, Denmark in the 1960s, was shown by tree-ring analysis to have been constructed in 1059 of oak from Ireland (probably Dublin) (Olsen and Crumlin-Pedersen 1978). In the seventeenth century, Sir Arthur Chichester wrote that "the Irish build very good ships and many English merchants choose to build there. Their oak is very good and they have a very good store of it." Writing about the same era McCracken (1971) claimed that "the forests of Cork and Kerry in the southwest of Ireland were used to cask the wine that France and Spain produced." Exploitation became destruction under the Tudors in the sixteenth century as the resource was decimated to build ships for the English Navy. Ireland's remarkable demographic and settlement history from the seventeenth century onwards ensured that by the mid-nineteenth Century, it was among the countries most denuded of timber. According to one eighteenth century French observer (de la Latocnaye 1985) "they had not left wood enough to make a toothpick." At the beginning of the twentieth century, the total forest area of Ireland was under 50,000 ha or less than 1% of the land area (Neeson 1991).

Historians may one day view the twentieth century as the restoration phase of forest cover in Ireland. Paradoxically, because of the earlier devastation, experimentation was feasible and new techniques could be used without disrupting established traditions. Hence the early interest in conifers. Afforestation began with the establishment of the Forest Service in 1903 and planting increased with the passage of the Forestry Act of 1919 (Figure 15.2). Afforestation policy had only one objective: to produce timber. It relied almost totally on conifers, on the grounds that short-rotation softwood forests would have a greater commercial appeal than the slow growing, native hardwood, for which demand was weak at that time.

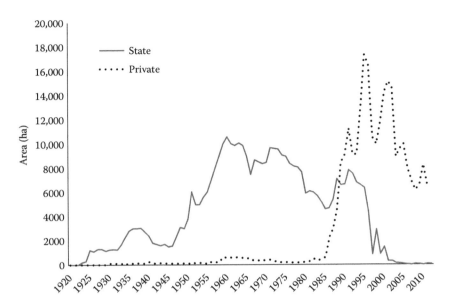

FIGURE 15.2
Annual planting rates on public and private land in the Republic of Ireland. (Annual data from the Department of Food, Agriculture and the Marine, 2013.)

Expansion of forestry was limited by the scarcity of suitable and affordable land. With the Great Famine of the 1840s still haunting memories, priority was given to food production and agricultural crops. Forestry was not perceived to have a role in this scheme and it rapidly became accepted policy that no land fit for agriculture should be afforested. Thus, in the absence of incentives, private landowners refrained from planting forests. The restoration of Irish forest cover became synonymous with state forestry, which from the early 1950s extended onto very poor soils described as agriculturally "marginal." Peatlands represented the largest proportion within this category.

15.3 Role of Peatlands in the Afforestation of Ireland

15.3.1 Birth of Peatland Forestry in Ireland

Irish bogs are mainly open, windswept, wilderness areas virtually devoid of tree cover due to the wet conditions that have prevailed (in particular in recent centuries) whereby high precipitation did not favor the spread of trees and a stagnant water table favored wetland communities over woody shrubs and trees (Charman 2002). It is to the past, to periods when bog growth was slower and the surface drier, that one must look to find some Irish peatlands covered with pine. Dendrochronology and radiocarbon dating have shown that this forest resource existed for a period of at least 500 years sometime between 2500 and 1800 BCE (Feehan et al. 2008). Over time, bogs grew and engulfed these forests under several meters of peat. None of these ancient forests, which principally included pine, birch, and alder, exist anymore. These buried trees were a valuable resource in the seventeenth century, when standing timber became scarce. The remnants of pine forests can still be seen on industrial cutaway peatlands (bogs where peat has been extracted for many years) and are sometimes used for firewood, construction, and wood sculpture.

Unlike Fenno-Scandinavian countries, wetland forestry was not popular in Ireland and little had been done to afforest peatlands prior to the 1940s. Fuel production was still seen as the priority for the raised bogs of the Midlands. Despite the fact that trees had not grown naturally on Irish bogs for several millennia, the government decided to increase the rate of afforestation significantly in the early 1950s. Blanket peatland was seen to offer the possibility of large-scale plantations on treeless land that was virtually unused, apart from some extensive grazing. The increased availability of inexpensive phosphatic fertilizers and the introduction of tracked vehicles that could pull large, heavy-duty plows made blanket peatland forestry possible.

Forestry moved onto blanket peatlands with some difficulty; blanket peat is extremely acidic and has very low fertility. Particularly, phosphorus is in short supply and nitrogen reserves are often much smaller than in mineral soils. Blanket peat has a high resistance to water movement (saturated hydraulic conductivity of less than 1 cm per day), an obstacle to drainage (Galvin 1976; Gleeson 1985). The subdivision of blanket bogs by altitude into low-level and high-level (montane) bog (Hammond 1981) is relevant to silviculture. Plantations on higher elevation sites are more prone than low-level plantations to acute nutritional problems, exposure, and have a higher risk of windthrow (Farrell and McAleese 1972; Dickson and Savill 1974; Gallagher 1974; Farrell and Mullen 1979; Schaible 1992). Peatland afforestation was primarily focused on the blanket bogs but significant areas of raised bogs were also afforested. This new planting regime was to last for at least three decades. In 2008, 43% or 300,070 ha of the total forest estate was located on peat soils with the majority located on blanket bog (218,850 ha) and the remainder on raised peat (74,080 ha) and cutaway peatlands (8840 ha) (Black et al. 2008). The principal species planted have been Sitka spruce (*Picea sitchensis* (Bong.) Carr.), and lodgepole pine (*Pinus contorta* Dougl.) with Norway spruce (*Picea abies* (L.) Karst) and Scot's pine (*Pinus sylvestris* L.), also having been used on raised bog sites. Lodgepole pine was deemed a viable option on Irish peatlands as it grows along most of the United States/Canadian Pacific coast with the subspecies *Pinus contorta var. contorta* found in the more cool and moist climate of British Columbia in bogs, on sand dunes and on the margins of pools and lakes (O'Driscoll 1980). While the productivity of plantations on some high-level bogs and most low-level blanket bogs has highly exceeded expectations, with yield class[*] estimates for Sitka spruce at 13 and for lodgepole pine at 10 (Farrell and Boyle 1990), peatland forests were deemed of low quality. Peatland afforestation rates peaked in 1995 with over 6000 ha planted in that year. However, since 2006, there has been a steep decline with a mere 265 ha planted in 2012 over the whole country (Department of Food, Agriculture and the Marine 2014).

15.3.2 Lessons Learned from Afforestation of Blanket Bogs

Due to the waterlogged and nutrient deficient condition of peat, in its unmodified state, it is unsuitable for any tree crop production and therefore management techniques were developed to overcome site preparation and establishment problems but many more issues arose such as nutrient deficiencies prior to canopy closure as well as high levels of nutrients and carbon leaching following clearfelling. Table 15.1 summarizes these difficulties and in some cases the solutions deployed at the time.

---

[*] Yield class is an index of potential maximum mean annual volume growth increment expressed in m^3 ha^{-1}.

TABLE 15.1

Afforestation of Peatlands in Ireland: Features, Practices, and Issues

Peatland Characteristics	Management Techniques	Issues
Waterlogging	• Blanket bog and raised bogs: drainage through single or double moldboard plowing, tunnel plowing or mounding combined with ditching • Cutaway bogs: cleaning and deepening of existing ditches	• Increased peat bulk density • Subsidence of bog surface • Shrinkage and cracking • Increased microbial oxidation leading to increased carbon dioxide (CO_2) fluxes to atmosphere and loss of carbon to waterways • Increased sediment losses • Dewatering of the peat mass at depths deeper than the root zone • Decreased pH
Low bearing capacity	• Wide-tracked machines and manual planting with bare root transplants	• Compaction during machine operations • Runoff
Nutrient poor	• Fertilization	• Leaching of phosphorus (P) due to low iron (Fe) content of peat • Additional leaching of P due to operations such as clearfelling with windrowing • Competition with vegetation such as *Juncus effusus*
Exposure (high rainfall, wind, and sea salt)	• Selection of adapted species • No thinning	• Increased dissolved organic carbon (DOC) from high rainfall passing through forest canopy • Shallow rooting and windthrow • Marine ions deposition leading to high acidity level in runoff water
Late spring frost in the Midlands	• Selection of species provenance with later bud burst dates • Nurse species	• Competition with nurse species requires additional management

15.3.3 Afforestation of Cutaway Peatlands

In Ireland, some 100,000 ha of peatlands (mainly raised bogs) have been utilized for industrial peat extraction and some 70,000 ha are currently in production, mainly in the Midlands. A state-funded research programme (BOGFOR) was initiated in 1998 in an attempt to develop new techniques to successfully establish forests on industrial cutaway peatlands (Renou et al. 2007a). This required a novel forestry approach which was only investigated in Finland at the time where peat deposits were very shallow, compared to their Irish counterparts (Aro and Kaunisto 1998). The BOGFOR research program established that with good planning and the application of site-specific establishment procedures, satisfactory results could be obtained (Renou et al. 2007b; Renou-Wilson et al. 2008a, 2009). The results from over 200 ha of experimental and demonstration plantations concluded that the successful afforestation of cutaway peatlands is possible but requires (1) a sound plan with specific objectives, (2) a careful selection of sites with suitable characteristics, and (3) the use of specific operational methods tailored to the site conditions and species requirements. Commercial forest crops have been successfully established on certain cutaway site types, each requiring a combination of necessary actions pertaining to site assessment, site preparation, species performance, tree establishment, nutrition and fertilization, late spring frost, pests, and vegetation management. Norway spruce may be the most suitable commercial forest species for planting on cutaways (Renou-Wilson et al. 2008a). Irish coniferous forestry plantations are typically densely planted (2×2 m) leaving no space for understory vegetation or indeed other tree species except where a shelter species such as birch is planted to mitigate late spring frost damage. Results from several field trials showed that various conifer and broadleaved species can also be established successfully (Renou-Wilson et al. 2008b, 2010). While there is still little information on the long-term performance of most of the species on such sites, the relatively wide range of suitable species affords the forester the opportunity to create multispecies landscapes and the potential for providing other options (e.g., diversity of products for market) at a later stage. The variation in site conditions encountered in any given cutaway peatland means that, rather than a single monoculture, several species might flourish within a given area, thus enhancing the sustainability of these new forests. More importantly, the investigation recognized the heterogeneity of cutaway peat soils in terms of their physical and biogeochemical characteristics and thus, the difficulties associated with their rehabilitation in general.

In contrast to both Finland and Sweden (Päivänen and Hånell 2012) where cutaway peatland forests would always remain a small proportion of the total forest area, Irish cutaway peatlands could have a role to play in national wood production, even if it is only for biomass (Renou-Wilson 2011). However, the future of the industrial cutaway peatlands is now fully encompassed in the objectives of the National Peatlands Strategy which will give direction to Ireland's approach to the management of its national peatland resource (Peatlands Council 2013). In this context, large-scale commercial plantations many not represent a major after-use of industrial cutaway peatlands in the future. Instead, restored wetland habitats and high-biodiversity drier habitats where birch and willow have naturally recolonized are likely to be the sustainable and acceptable options (Renou-Wilson et al. 2010), along with greener energy production systems, such as biomass burning and wind farms (Renou-Wilson et al. 2011).

15.4 Managing Irish Peatland Forests in the Twentieth Century

The principal focus of peatland forestry during the twentieth century was the establishment and management of forest plantations for timber production and as with all plantation forests in the country, this principally involves a combination of site preparation, fertilization, thinning on windfirm sites, and clearfelling. This focus has changed, as the goals of forest management have broadened to embrace sustainability and the provision of ecosystem services. Ireland has realized the importance of sustainable management of peatland forests in particular, similar to some of the Nordic countries which are already applying progressive management approaches, for example single tree selection on wetter sites (Päivänen and Hånell 2012) or using soil preparation techniques that maximize success at the lowest cost to the atmosphere (Pearson et al. 2012).

While conventional management practices will continue in the majority of the peatland forests in the near future, Tiernan (2008) estimated that slightly more than one-fifth of peatland forests (64,548 ha) are uneconomic and unsustainable. These forests will require immediate action in the form of alternative management approaches if these peatland ecosystems are to be placed on a more sustainable footing. This means considering not just issues regarding potential timber yield but also adopting an ecosystem management

TABLE 15.2

Alternative Management Approaches to Peatland Forests in the West of Ireland

Management Approach	Definition
1. Bog rewetting/restoration	• The restoration of peatland habitat following felling, and/or retention of existing unplanted areas
2. Natural regeneration	• Renewal by natural seeding
3. Water protection	• The creation of planted or unplanted buffer zones along watercourses, which may or not include areas suitable for riparian native woodland
4. Long-term retention	• Stands retained beyond the normal economic felling age (40–80 years)
5. Low impact regeneration with native species	• The establishment of a low-density scrub native forest using minimal site preparation techniques
6. Low impact regeneration with lodgepole pine	• The establishment of a low-density lodgepole pine forest using minimal cultivation and fertilizer inputs where feasible
7. No replanting (visual enhancement)	• Leaving strategic forest areas unplanted following clear-felling, as a means of improving the overall visual esthetics of the landscape
8. Retain existing unplanted areas	• The retention of existing unplanted areas in situ
9. No replanting following felling	• Clear-felling follows best practices
10. Replant with possible phased felling	• Restore to commercial forest

approach that incorporates environmental and social issues alongside timber production. Tiernan (2008) put forward 10 such options, most of them decreasing the input costs and therefore likely decreasing economic losses (Table 15.2).

Realizing that half of these "poor" peatlands forests should not have been planted in the first place, some of the aforementioned options have already been put in place with estimates suggesting that by 2015, more than 3100 ha of afforested blanket bogs and raised bogs have been restored to wetland ecosystems (Delaney and Murphy 2012). Latest estimates show that 10% of the western peatland forest estate (including unplanted areas) can be restored to functioning peatland ecosystems (Tierney, personal communication). However, several legal and logistical barriers have been highlighted during recent efforts to implement such innovative sustainable management options. For instance, due to an outdated national forest policy, a cumbersome limited felling license procedure is required to remove trees without replanting the same land. Finding replacement lands to fulfill felling conditions are also mostly unrealistic, given the shortage of land available for forestry in a small country like Ireland. In short, Ireland needs to regularize procedures to be followed when not restocking a forested peatland while also preventing any adverse impacts of deforestation, such as illegal land use (e.g., grazing and trespassing), fire hazard and invasion by nonnative species such as *Rhododendron ponticum*.

Irish blanket bogs are becoming increasingly appreciated as a unique natural and global resource albeit mostly disturbed to various extents, as well as an exceptional cultural landscape (Figure 15.3) contributing to a catchment where the streams are of particularly high quality, for example, for salmonid populations. Biodiversity enrichment incentives have been established to screen against planting in large open areas and on deep peat especially in the west of Ireland where the battle of the bogs versus the forests will be now fought as to which can provide more ecosystem services.

FIGURE 15.3
Forestry on blanket bog in the west of Ireland, Ox Mountains, County Sligo. (Photo by Florence Renou-Wilson.)

15.5 The Ecosystem Services of Peatland Forestry: Issues for the Future

Recent developments such as the Millennium Ecosystem Assessment (2005) have empha-sized the dependence of humankind on natural ecosystems. Furthermore, the value of ecosystem services has been central to the series of European Ministerial Conferences on the Protection of Forests and the Oslo Ministerial Decision: European Forests 2020 (www. foresteurope.org), which included a specific target for estimating "the full value of forest ecosystem services across Europe" by 2020. This will require the development of robust, scientifically based practices to guide the management of these ecosystems.

This creates specific challenges for peatland forestry, not least of which is their economic viability. As discussed above, these forests are typically of low productivity (Farrell and Boyle 1990) and significant areas are uneconomic (Tiernan 2008). In addition, these forests have been linked to problems with surface water quality (Cummins and Farrell 2003), although changes in forest practices through the introduction of guidelines in relation to water quality, harvesting, and aerial fertilization may have helped in reducing these impacts (O'Driscoll et al. 2011; Rodgers et al. 2010, 2011).

Carbon (C) sequestration is one of the key ecosystem services provided by forests (Byrne 2010) and forests are widely considered to be net sinks for carbon. However, peatland for-ests present a complex series of challenges in assessing their contribution to greenhouse gas (GHG) mitigation. This is because undrained peatlands are a long-term carbon store, being net sinks of carbon dioxide (CO_2) and sources of methane (CH_4). Drainage moves the ecosystem away from being a functional peatland with water level drawdown leading to a reduction or cessation of CH_4 emissions and increased CO_2 losses due to the acceler-ated rate of organic matter oxidation. Moreover, the vegetation communities typical of an undisturbed peatland are replaced by a monocultural forest cover with almost no under-story vegetation to sequester carbon. The fundamental issue to be addressed is whether the C losses from the soil are compensated by C inputs through litterfall and belowground C turnover. Other crucial issues are the rate C sequestration in growing biomass and ulti-mately, the impact of afforestation on the net GHG balance of peatlands.

Despite the importance of peatland forests in Ireland, knowledge of their C balance remains limited. Byrne and Farrell (2005) found that afforestation of blanket peats lowered the water table and promoted CO_2 loss. In a modeling study, Byrne and Milne (2006) suggested that peatland forests are a net sink for CO_2 with biomass C uptake compensating for soil CO_2 loss. Studies from the UK provide relevant findings. Hargreaves et al. (2003) reported that the forest crop takes 4–8 years to recoup the C losses due to drainage or felling. Given the low productivity of peatland forests, the rate of C sequestration in biomass may be low. The wide variation in C sequestration rates reported for peatland forests may reflect differences in site conditions and study methods. For example, Byrne et al. (2004) used standing crop figures to estimate a net sink of 0.1 t CO_2 eq ha^{-1} yr^{-1} while Worrall et al. (2011) report a net loss of 2.49 t CO_2 eq ha^{-1} yr^{-1}. In a study of a range of sites in Finland, Ojanen et al. (2013) found that the soil was a CO_2 source on fertile sites and a CO_2 sink on nutrient poorer sites. When the CO_2 sink of the tree stand and soil emissions of nitrous oxide (N_2O) and CH_4 were considered, fertile and nutrient poor sites were found to have a net cooling impact on the global climate. Further studies from Sweden (Meyer et al. 2013) and Germany (Hommeltenberg et al. 2014) confirm that the overall sink potential of afforested nutrient-rich organic soils are probably limited to only a short period when C accumulation from the biomass is at its peak.

Given the evidence to date regarding the impact of afforestation on the C balance of peatlands, as well as the recognition that large areas of peatland forests are not suitable for conventional forest management, it is clear that peatland forestry is not sustainable on many sites. There is a need for research to facilitate the identification of such areas and guide their management toward a more sustainable future. Firstly, there is a need for comprehensive assessment of the impact of afforestation, and subsequent forest management, on the soil C balance. This requires quantifying net ecosystem GHG balances as well as fluvial C fluxes from these peatland ecosystems. Secondly, there is a need for a site-based system which can rate the ecosystem services of both peatland forests and the range of management alternatives on an equivalent basis. Within a climate change mitigation context, additional options to maximize the C sequestration potential of such ecosystems should be investigated by promoting the return of wetland species; for example, a *Sphagnum* dominated ground cover would additionally fulfill biodiversity objectives. This would provide a scientific basis for identifying the most sustainable option for the future management of these forests.

15.6 Restoration of Afforested Peatlands

Despite the potential offered by the availability of sites for afforestation after the cessation of industrial peat extraction, peatland afforestation in Ireland has no more impetus. Peatland forestry can now be identified as the twentieth century exercise of planting naturally tree-less blanket peatlands in the west of Ireland. In contrast, the area of previously afforested bogs that are being restored is increasing, but remains modest and experimental (Figure 15.4). The next required step is to operationalize this restoration by drawing on the experience to date and to develop practices to guide future restoration techniques. For example, while rewetting forested peat soils may reduce soil CO_2 emissions, it will also remove or reduce the CO_2 uptake by the forest stand while adding CO_2 emissions from brash. Moreover, a return of CH_4 emissions can be expected (Komulainen et al. 1998).

FIGURE 15.4
Restored raised bog following fell-to-waste and rewetting by damming the drains, Sopwell, County Tipperary.
(Photo by Florence Renou-Wilson.)

Therefore, when considering the environmental sustainability of potential future restoration of unproductive peatland forests, it is necessary to consider their net radiative forcing impact (Ojanen et al. 2013). In addition, the impact of restoration on other ecosystem services should also be considered, especially with regard to water quality.

Restoration also has implications for climate change policy. Restoration of low productivity peatland forests involves deforestation and the associated C emissions are captured under Article 3.3 of the Kyoto Protocol and added to the national GHG emissions reporting. Hendrick and Black (2009) estimate that an annual deforestation rate of 1000 ha would release 500,000 t CO_2. Successful reestablishment of the soil C sink following rewetting could recoup some of these losses.

Further incentive for rewetting and restoration of peatlands will be provided by the decision of United Nations Framework Convention on Climate Change to recognize a new elective activity "Wetland drainage and rewetting (WDR)" under Article 3.4 of the Kyoto Protocol (Document FCCC/KP/AWG/2010/CRP.4/Rev.4). This will be supported by the *2013 Supplement to the 2006 IPCC Guidelines for National Greenhouse Gas Inventories: Wetlands* (*Wetlands Supplement*) which provides new guidance on estimating and reporting GHG emissions from organic soils and from wetlands, in so far as they are (directly) impacted by human activities ("managed"). Taken together, both of these developments pave the way for the inclusion of rewetting of peatlands in GHG offsetting schemes.

15.7 Conclusion

While it took the most part of the twentieth century to see the return of forests in Ireland, the expanses of peatlands onto which the trees marched are no longer fit for this purpose. The turn of the twenty-first century has seen the end of the afforestation of Irish bogs and peat soils in general, except for pockets of industrial cutaway peatlands. Despite the majority of the forest estate being on peat soils, the status of peatland forestry in Ireland remains marginal and its national economic significance has yet to be fully evaluated. Ultimately, the success of peatland forestry in Ireland will be measured not only by how these forests can meet their financial objectives but also by how well they meet social, esthetic, and environmental needs.

Acknowledgments

The authors wish to thank Dr. David Wilson for reviewing the text.

References

Aro, L. and Kaunisto, S., Nutrition and development of 7–17 year old Scots pine and silver birch plantations in cutaway peatlands, in *Peatland Restoration and Reclamation: Techniques and Regulatory Considerations; Proceedings of the 1998 International Peat Symposium*, Malterer, T., Johnson, K., and Stewart, J., Eds., Duluth, Minnesota, 109, 1998.

Black K., O'Brien P., Redmond J., Barrett F., and Twomey M., The extent of recent peatland afforestation in Ireland, *Ir. For.*, 65(1–2), 71–81, 2008.

Byrne, K.A., The role of plantation forestry in Ireland in the mitigation of greenhouse gas emissions, *Ir. For.* 67, 86–96, 2010.

Byrne, K.A. and Farrell, E.P. The effect of afforestation on soil carbon dioxide emissions in blanket peatland in Ireland, *Forestry*, 78(3), 217–227, 2005.

Byrne, K.A. and Milne, R., Carbon stocks and sequestration in plantation forests of the Republic of Ireland, *Forestry*, 79, 361–369, 2006.

Byrne, K.A., Chojnicki, B., Christensen, T.R., Droesler, M., Freibauer, A., Fribourg, T. et al. EU peatlands: Current carbon stocks and trace gas fluxes. *Carbo-Europe Discussion Paper for Concerted Action CarboEurope-GHG*, Lund, Sweden, 2004.

Charman, D., *Peatlands and Environmental Change*. John Wiley and Sons, Chichester, 2002.

Connolly, J. and Holden, N.M., Mapping peat soils in Ireland: Updating the derived Irish peat map, *Ir. Geo.*, 42(3), 343–352, 2009.

Cummins, T. and Farrell, E.P., Biogeochemical impacts of clearfelling and reforestation on blanket peatland streams. I. Phosphorus, *For. Ecol. Manage.*, 180, 545, 2003.

de la Latocnaye, C. [1798], *A Frenchman's Walk through Ireland*, Blackstaff Press, Belfast, 1985.

Delaney, M. and Murphy, P., Coillte and the EU Life Programme: 10 years of restoration works on afforested peatlands in Ireland. In *Peatlands in Balance, 14th International Peat Congress Stockholm, Sweden*, International Peat Society, Abstract No 296, 1–5, 2012.

Department of Agriculture, Food and Forestry, Growing for the Future—A Strategic Plan for the Development of the Forestry Sector in Ireland, Department of Agriculture, Food and Forestry, The Stationery Office, Dublin, Ireland, 1996.

Department of Food, Agriculture and the Marine, http://www.agriculture.gov.ie/forestservice/forestservicegeneralinformation/foreststatisticsandmapping/afforestationstatistics/ accessed September 2014.

Dickson, D.A. and Savill, P.S., Early growth of Picea sitchensis on deep oligotrophic peat in Northern Ireland, *Forestry*, 47, 57, 1974.

Duffy, P., Hanley, E., Hyde, B., O'Brien, P., Ponzi, J., Cotter, E. et al., *National Inventory Report 2013*. Greenhouse gas emissions 1990–2011 reported to the United Nations Framework Convention on Climate Change, 2013.

Farrell, E.P. and Boyle, G., Peatland Forestry in the 1990s. 1. Low-level blanket bog, *Ir. For.*, 47, 69, 1990.

Farrell, E.P. and McAleese, D.M., The response of Sitka spruce to sulphate of ammonia and ground rock phosphate on peat, *Ir. For.*, 29, 14, 1972.

Farrell, E.P. and Mullen, G.J., Rooting characteristics of Picea sitchensis and Pinus contorta on blanket peat in Ireland, *J. Life Sci.*, Royal Dublin Society, 1, 1, 1979.

Feehan, J., O'Donovan, G., Renou-Wilson, F., and Wilson, D. *The Bogs of Ireland—An Introduction to the Natural, Cultural and Industrial Heritage of Irish Peatlands. 2nd Edition, Digital Format*. University College Dublin, Ireland, 2008.

Gallagher, G.J., Windthrow in State forests in the Republic of Ireland, *Ir. For.*, 31, 154, 1974.

Galvin, L.F., Physical properties of Irish peats, *Ir. J. Agric. Res.*, 15, 207, 1976.

Gillmor, D.A., Trends and spatial patterns in private afforestation in the Republic of Ireland, *Ir. For.*, 55, 10, 1998.

Gleeson, T.N., Drainage of lowland peats and wet mineral soils, in Peatland Production Seminar, Lullymore, Co. Kildare, Ireland, 10, 1985.

Hammond, R.F., *The Peatlands of Ireland*, An Foras Taluntais, Dublin, Ireland, 1981.

Hargreaves, K.J., Milne, R., and Cannell, M.G.R., Carbon balance of afforested peatland in Scotland, *Forestry*, 76, 299, 2003.

Hendrick, E. and Black, K., *Climate Change and Irish Forestry*. COFORD Connects Environment No. 9. Forest Service. Dublin, 2009.

Hommeltenberg, J., Schmid, H.P., Drösler, M., and Werle, P., Can a bog drained for forestry be a stronger carbon sink than a natural bog forest? *Biogeosciences*, 11, 3477–3493, 2014.

Joyce, P. and O Caroll, N., *Sitka Spruce in Ireland*, COFORD, Dublin, 2002.

Komulainen, V.-M., Nykänen, H., Martikainen, P.J., and Laine, J., Short-term effect of restoration on vegetation change and methane emissions from peatlands drained for forestry in southern Finland, *Can. J. For. Res.* 28, 402–411, 1998.

McCracken, E., *The Irish Woods Since Tudor Times*, David and Charles, Newton Abbot, 1971.

Meyer, A., Tarvainen, L., Nousratpour, A., Bjork, R.G., Ernfors, M., Kasimir Klemedtsson, A. et al., A fertile peatland forest does not constitute a major greenhouse gas sink, *Biogeosciences*, 10, 7739–7758, 2013.

Millennium Ecosystem Assessment. *Ecosystems and Human Well Being: Synthesis Report*. Island Press, Washington, 2005.

Mitchell, F. and Ryan, M., *Reading the Irish Landscape*, Town House and Country House, Dublin, Ireland, 1997.

Neeson, E., *A History of Irish Forestry*, Lilliput Press and the Department of Energy, Dublin, Ireland, 1991.

Ní Dhubháin, A., Bullock, C., Moloney, R., and Upton, V., *Economic Evaluation of the Market and Non-Market Functions of Forestry*. COFORD, Dublin, Ireland, 2013.

Olsen, O. and Crumlin-Pedersen, O., *Five Viking Ships from Roskilde Fjord*, The National Museum, Copenhagen, 1978.

O'Driscoll C., Rodgers M., O'Connor M., Asam Z., de Eyto E., Poole R. et al. A potential solution to mitigate phosphorus release following clearfelling in peatland forest catchments, *Wat., Air, & Soil Poll.*, 221, 2011.

O'Driscoll, J., The importance of lodgepole line in Irish forestry, *Ir. For.*, 37, 7–22, 1980.

Ojanen, P., Minkkinen, K., and Penttilä, T., The current greenhouse gas impact of forestry-drained boreal peatlands, *For. Ecol. Manage.*, 289, 201–208, 2013.

Päivänen, J. and Hånell, B., *Peatland Ecology and Forestry—a Sound Approach*. University of Helsinki Department of Forest Sciences Publications, Helsinki, 3, 1–267, 2012.

Pearson, M., Saarinen, M., Minkkinen, K., Silvan, N., and Laine, J. Short-term impacts of soil preparation on greenhouse gas fluxes: a case study in nutrient-poor, clearcut peatland forest, *For. Ecol. Manage.*, 283, 10–26, 2012.

Peatlands Council, http://www.npws.ie/peatlandsturf-cutting/peatlandscouncil/, 2013.

Redmond, J., Gallagher, G., Černý, M., and Russ, R. Ireland's nation forest inventory—Results. In National Forest Inventory Republic of Ireland. *Proceedings of NFI conference,* M. Nieuwenhuis, J. Redmond and C. O'Donovan (eds.), Forest Service, Johnstown Castle Estate, Co. Wexford, Ireland, 2007.

Renou, F., Keane, M., McNally, G., O'Sullivan, J., and Farrell, E.P. *BOGFOR Project Final Report: A Research Programme to Develop a Forest Resource on Industrial Cutaway Peatlands in the Irish Midlands,* COFORD, Dublin, 2007a.

Renou, F., Scallan, Ú., and Keane, M., Early performance of native birch (*Betula* spp.) planted on cutaway peatlands: Influence of species, stock types and seedlings size, *Eur. J. For. Res.,* 126, 545–554, 2007b.

Renou-Wilson, F. *Optimal Practices in the Afforestation of Cutaway Peatlands.* Lambert Academic Publishing, Saarbrücken, Germany, 2011.

Renou-Wilson, F., Bolger, T., Bullock, C., Convery, F., Curry, J. P., Ward, S. et al., *BOGLAND—Sustainable Management of Peatlands in Ireland.* STRIVE Report No 75 prepared for the Environmental Protection Agency (EPA), Johnstown Castle, Co. Wexford, 2011.

Renou-Wilson, F., Keane, M., and Farrell, E.P., Effect of stocktype and cultivation treatment on the survival, morphology and physiology of Norway spruce on cutaway peatlands. *New Forests,* 36, 307–330, 2008a.

Renou-Wilson, F., Keane, M., and Farrell, E.P., Establishing oak woodland on cutaway peatlands: Effects of soil preparation and fertilization, *For. Ecol. Manage.,* 255, 728–737, 2008b.

Renou-Wilson, F., Keane, M., and Farrell, E.P., Afforestation of industrial cutaway peatlands in the Irish midlands: Site selection and species performance. *Ir. For.,* 66, 85–100, 2009.

Renou-Wilson, F., Pollanen, M., Byrne, K.A., Wilson, D., and Farrell, E.P., The potential of birch afforestation as an after-use option for industrial cutaway peatlands, *Suo* 61(3–4), 59–76, 2010.

Rodgers, M., O'Connor, M., Healy, M., O'Driscoll, C., Asama, Z., Nieminen, M. et al. Phosphorus release from forest harvesting on an upland blanket peat catchment, *For. Ecol. Manage.,* 260(12), 2241–2248, 2010.

Rodgers, M., O'Connor, M., Robinson, M., Muller, M., Poole, R., and Xiao, L.W., Suspended solid yield from forest harvesting on upland blanket peat, *Hydrol. Process* 25(2), 207–216, 2011.

Schaible, R., Sitka spruce in the 21st century, establishment and nutrition, *Ir. For.,* 49, 10, 1992.

Tiernan, D., Redesigning afforested western peatlands in Ireland, 13th International Peat Congress: *After Wise-Use: The Future of Peatlands,* in Farrell C.A. and Feehan J. (eds.), Tullamore, Co. Offaly, Ireland, IPS, pp. 520–523, 2008.

Wilson, D., Müller, C., and Renou-Wilson, F., Carbon emissions and removals from Irish peatlands: current trends and future mitigation measures, *Ir. Geo.* 46(1–2), 1–23, 2013.

Worrall, F., Chapman, P., Holden, J., Evans, C., Artz, R., Smith, P. et al. *A review of current evidence on carbon fluxes and greenhouse gas emissions from UK peatlands,* JNCC report, no. 442, 2011.

16

Forest Restoration in the French Massif Central Mountains

Philippe Balandier and Bernard Prévosto

CONTENTS

16.1 Introduction

16.1.1 Background

The Massif Central is an upland area of igneous origin in southern France (Figure 16.1) with a mean elevation of 800 m, from 300 m in the Limagne plain to 1886 m in the Sancy Mountains and covers about 70,000 km^2 (12.7% of France). Only the higher elevations will be considered here, as the low fertile plains are farmed (Figure 16.2). The climate is semi-continental, with low winter temperatures ($\geq$–20°C), late spring frosts (until June), and high summer temperatures ($\leq$30°C). A striking feature of the climate is the high amplitude of daily temperature variation, sometimes more than 25°C. Temperatures vary with elevation, about –0.6°C per +100 m elevation (Guitton 1986). Rainfall is about 800 mm per year but varies widely with exposure, from less than 600 mm in the rain shadow of hills to more than 1500 mm in the west slopes of the mountains. Rainfall is fairly evenly distributed throughout the year (Figure 16.3), especially in areas benefiting from oceanic influences, but summer droughts occur about one year in every five (Balandier et al. 2003a).

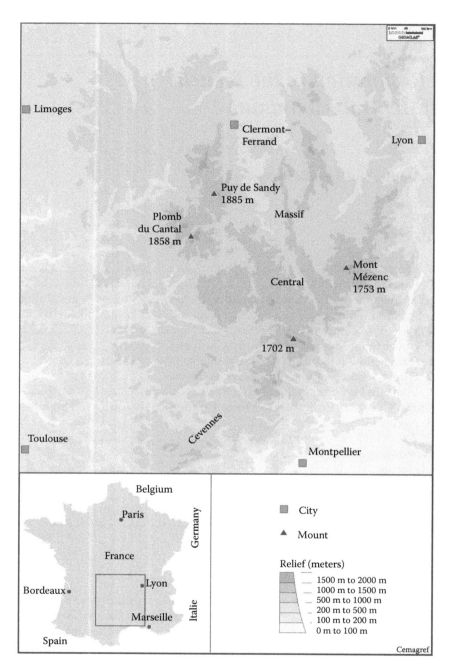

FIGURE 16.1
Geographical location, relief, and elevation of the Massif Central in France.

FIGURE 16.2
Massif Central landscape, with forested higher elevations and farmed lowlands. Sancy Mountain is in the background. (Photo: P. Balandier).

Soils are derived from granite, basaltic flows, and ash-fall deposits. The most frequently occurring soils are acid brown soils, podzolic podzols, and andosols on volcanic material. Soils derived from granite are often infertile: low pH (5–5.5), low cation exchange capacity (15–20 m-equiv 100 g^{-1}), and low organic matter content (5%–10%). In contrast, soils in volcanic areas are more favorable to forest production due to higher pH (about 6), higher cation exchange capacity (30–35 m-equiv 100 g^{-1}), and higher organic matter content (10%–15%).

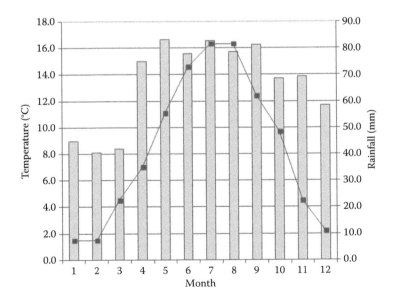

FIGURE 16.3
Mean monthly temperature (black squares) and mean monthly rainfall (Vertical bars) for the period 1993–2013 at Saint-Genès-Champanelle (890 m a.s.l.), France.

Vegetation history of the Massif Central is well documented due to numerous pollen analyses of peat bogs, swamps, and lakes (De Beaulieu et al. 1988). During the postglacial period, the area was successively dominated by *Betula* and *Pinus* forests (9900–9500 ybp), *Corylus* (9500–8000 ybp), and *Quercus* forests (8000–4200 ybp). From 4200 ybp, *Abies alba* and *Fagus sylvatica* forests were the dominant type until human disturbances occurred on a large scale. At the beginning of the twentieth century, forest cover was minimal and beech and fir forests remained only in some mountainous areas.

At the beginning of the eighteenth century, most of the land was in subsistence farms producing cereal and typically possessing a few cattle and a horse. In 1800, less than 10% of the total area was wooded (Michelin 1995). Agricultural specialization came with the arrival of the railway at the end of the nineteenth century, and with it began a slow and continuous decline of agriculture. Production systems are now mostly restricted to forage crops and ruminant livestock (cattle and sheep) for dairy products and meat. After World War II, further specialization and intensification of agriculture accelerated rural depopulation. The onset of common European agricultural policies (which favored the most productive farming systems) and natural hindrances (such as harsh climate and steep slopes) led to the abandonment of the most difficult upland areas (Balandier et al. 2002). Thus, old grasslands, heathlands, and alpine pastures gradually have been taken over, first by brushwood and later, or directly, by pioneering tree species, among which *Pinus sylvestris* and *Betula* spp. have played a major role (Prévosto et al. 2004). Alongside this natural colonization of former upland agricultural areas by trees, national aid programs have favored planting conifer stands.

16.1.2 Forests Today

Natural forests are very rare, as most disappeared or were drastically modified by human activities. Natural forests at the highest elevations were dominated by *F. sylvatica—A. alba* communities; oak (generally *Quercus pubescens*) takes the place of fir with decreasing altitude or decreasing latitude and becomes the dominant species in the south of the Massif Central. Often mixed with those dominant species, trees or shrubs such as *Prunus avium, Acer pseudoplatanus, Fraxinus excelsior,* and *Sorbus torminalis* can enrich these forests. Plantations, mainly conifers (*P. sylvestris, Picea abies,* and *Pseudotsuga douglasii*), were the main species of afforestation and, consequently, today are the most abundant forests. Upland areas today have greater than 60% forest cover, but with wide differences between regions (Balandier et al. 1997). In the Auvergne region, coniferous plantations account for 33% of private woods; for the entire Massif Central, conifer plantations account for about 25% of the forested area, that is, 500,000 ha. Afforestation initially responded to a shortage of wood for industrial purposes, such as mining and coal, beginning in the nineteenth century and continuing into the twentieth century (RTM: restoration of mountainous forests, FFN: national forest fund, Bianco 1998). Coniferous species (successively *P. sylvestris, P. abies,* and until 1980 *P. douglasii*) were the main species used in these plantations because of their fast growth and high productivity. A recent current trend (and particularly after the dramatic windstorm of 1999) is more and more to plant *Larix* sp. (*L. deciduas, L. kaempferi* or the hybrid *L. eurolepis*) in place of *P. sylvestris* and *P. abies*. It is the single coniferous species that drops its needles in winter, and therefore less sensitive to strong winds in winter.

Forests arising from spontaneous regeneration by pioneer trees such as *P. sylvestris* are also widespread, comprising as much as 45% of the forested area in some parts of the Massif Central (Lifran et al. 1997). This woodland is generally very fragmented, mostly privately owned (80%) with a mean holding of 2.2–2.5 ha per landowner in Auvergne (ONF 1993; DRAF 1999).

Utilization developed steadily as postwar plantations matured. Timber production is expected to double in the next 10 years from 3 million m³ at present to 6 million m³ (AFOMAC 2001). Salvage logging after the last catastrophic windstorm in 1999 changed the face of the wood industry toward mechanization; harvesting machines have replaced individual chain saws, except in areas with large trees on steep slopes. Small sawmills have closed and larger mills are expanding and modernizing. The wood pulp industry, however, is small and logs have to be transported to remote factories. Pulpwood demand has fallen owing to waste paper recycling and the direct supply of paper mills with saw-mills waste. This low demand for pulpwood reduces income from first thinnings and stands consequently lose their stability to wind and heavy snow (De Champ 1997).

16.1.3 Forest Restoration

Timber production is no longer the sole function expected from forests; management has to provide other goods and services demanded by the public, such as recreation, biodiversity restoration and conservation, and landscape value. For instance, coniferous afforestation is often disliked by local people (Balandier et al. 1997), because of its adverse effect on the landscape (screening, and dark-colored vegetation) or because it symbolizes the abandonment of earlier agriculture (Bouvarel and Larrere 1981). In addition, management of these generally small-sized plantations is often desultory and wood of low economic value accumulates (Lifran et al. 1997). The severe windstorm of December 1999 destroyed more than 20% of the standing volume in some parts of the Massif Central, representing seven times the annual harvest (140 million m³ for the whole of France) (Rérat 2000) and coniferous stands were the most frequently injured (mainly *P. abies* and *P. sylvestris*, although *Pseudostuga douglasii* and *Larix* sp. were less affected).

The public, diverse institutions and nongovernmental organization also favor sound and sustainable management practices and dislikes mechanical and chemical competition control in plantations (Frochot et al. 2002, 2009). They welcome the use of local tree species and a return to, or a creation of, mixed woodland with broadleaved species (Bianco 1998; Collective Report 1998). In the context of acceleration of problems linked to increasing climate changes, mixed woodlands are also advocated as an interesting management option with a higher resilience to different damages and a risk reduction strategy. Whether it is true or not, or supported by experimental results is still in debate.

The aim of forest restoration in the Massif Central in France is to find sustainable solutions for forest management that are environmentally friendly and responsive to public demands, whereas remaining economically profitable (eco-efficient), especially for the small owners who manage most of the forested area. In light of these considerations, we shall look at three main issues for forest restoration in the Massif Central:

- How can we treat competing ground vegetation in the most environmentally friendly way (i.e., with limited use of herbicides or mechanical operations) in afforestation and rehabilitation?
- How can we accelerate stand development in spontaneous forests of pine or birch toward near-climax forests (such as beech and silver fir)?
- Is it possible to convert monospecific coniferous stands into plurispecific mixed stands combining both softwood and broadleaved species?

16.2 Sound Management of Ground Vegetation in Afforestation or Regeneration of Forest Stands

Unless ground vegetation (herbs or shrubs) that competes with young seedlings is controlled, tree establishment fails in plantations on former agricultural land or in forest regeneration (Nambiar and Sands 1993; Balandier et al. 2006; Balandier and De Montard 2008; Mc Carthy et al. 2011). Cutting back competing vegetation has only a temporary effect; it grows again, often more vigorously. Mechanical and chemical applications are effective in controlling this ground vegetation (Davies 1987) but may harm the environment and are sometimes dangerous for human health.

In a search for more ecologically sound approaches to help in reducing the use of herbicides; two techniques are currently being tested: using accompanying woody species to control the herbaceous stratum, and using a cover of herbaceous plants that competes only weakly with tree seedlings.

16.2.1 Woody Accompanying Species

Ground vegetation has long been considered unfavorable to forest tree establishment. Nevertheless, the herbaceous stratum is often more damaging than the shrub stratum (Perrin 1963) and microclimate of a large open area is often unfavorable to forest tree establishment (Friedrich and Dawson 1984; Aussenac 1986). Certain woody species may be more beneficial than detrimental by providing protection against wind, late spring frost, and other hazards. Lateral shelter may even favor young tree growth (Collet et al. 1992). These accompanying woody species, also termed as nurse trees, nurse vegetation, or nurse crops, also improve forest tree shape (Hubert 1992). Other benefits of accompanying species include limiting the development of competing herbaceous species, mainly grasses (Frochot et al. 1986; Davies 1987; Hubert 1992), financial return from harvesting intermediate wood products from the accompanying species (e.g., pulpwood, and firewood), and improving tree growth by adding nitrogen by using N-fixing species such as *Alnus* sp. However, to manage accompanying species is difficult in practice. Decisions include choice of species, plantation density, and plantation design. The accompanying vegetation is beneficial only if it is sufficiently close to the crop trees and has a comparable height (Hubert 1992). However, increasing the density of the accompanying species around crop trees or using species with a fast development also intensifies the competition for light and soil water resources (Prévosto and Balandier 2007; Prévosto et al. 2012). Although the highest densities of the woody accompanying species accelerate height growth of crop trees, diameter growth tends to decrease correspondingly (Figure 16.4). At the highest densities (more than 2000 stem ha⁻¹), crop tree stems are slender, with a height/diameter ratio greater than 120, and thus vulnerable to strong winds (Balandier and Marquier 1998). The windstorm in France in December 1999 showed that this effect cannot safely be ignored.

Using woody accompanying species generally benefits tree stem shape (straightness), but these effects may only be appreciable at very high densities of accompanying species. For example, in a plantation of ash (*F. excelsior*) with alder (*Alnus glutinosa*) as an accompanying species at densities up to 1665 alder ha⁻¹, the ash trees showed no difference in the number of forks or trunk deformations (Balandier and Marquier 1998). However, the size and number of lateral shoots were strongly influenced by the density of the accompanying species, thereby increasing the density producing fewer, thinner lateral shoots. Despite the reduction in lateral shoot vigor with increasing alder density, the natural pruning of lateral

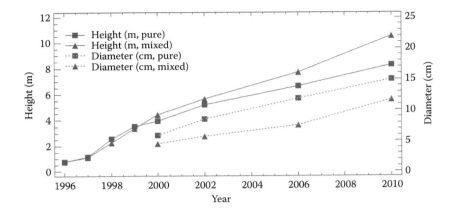

FIGURE 16.4
Evolution of stem height and diameter at breast height of hybrid walnut (*Juglans regia × nigra*) planted at 160 stem ha^{-1} near Clermont-Ferrand in the Massif Central in a pure stand or with *Eleagnus angustifolia* as an accompanying species (three *Eleagnus* around each walnut at a distance of 50 cm from the walnut trunk). The figure only gives the growth of walnut.

shoots is only obtained at very high densities (more than 5000 stem ha^{-1}). Therefore, artificial pruning is necessary, in most cases, to obtain a knot-free bole (Table 16.1, Balandier et al. 2008).

Plantations with accompanying species are difficult to proportion (Balandier and Marquier 1998; Prévosto and Balandier 2007; Balandier et al. 2008). Desirable effects on tree shape, size, and number of lateral shoots and control of the ground herbaceous species are only obtained at high densities. They lead to decreased trunk diameters due to higher competition but can also improve tree architecture. However, too intense competition can result in slender trees, easily broken by wind. Hence the use of neighboring tree vegetation is possible but it requires active management such as the regular harvest of the accompanying species to limit competition although these operations may cause incidental damage to crop trees because of their proximity. Improved tree growth can also be simply achieved using low densities of woody accompanying species as windbreaks, planted in lines regularly distributed in the stand. However, windbreaks neither have any effect on tree shape and nor do they provide control of ground herbaceous vegetation. Therefore, forest managers must compromise between different effects according to management

TABLE 16.1

Knot-Free Bole Height and Straightness of Hybrid Walnuts (*Juglans regia × nigra*) Planted at 160 stem ha^{-1} near Clermont-Ferrand in the Massif Central in a Pure Stand or with *Eleagnus angustifolia* as an Accompanying Species

	Knot-Free Bole Height of Unpruned Walnuts (cm)	Knot-Free Bole Height of Pruned Walnuts (cm)	Percentage of Unpruned Walnuts Perfectly Straight (%)	Percentage of Pruned Walnuts Perfectly Straight (%)
Pure	87	392	87	86
Mixed	171	452	83	87

Source: Adapted from Balandier, P., Allegrini, C., and Jay, D. 2008. *Forêt-entreprise* 178: 21–25.
Note: Three *Eleagnus* around each walnut at a distance of 50 cm from the walnut trunk.

objectives and should probably avoid dense plantings of accompanying species in areas vulnerable to windthrow.

16.2.2 Cover Plant Mixtures to Control Competition

Using herbaceous vegetation that competes weakly against young seedlings in order to control highly competitive species (e.g., *Agropyrum repens, Molinia caerulea, Deschampsia flexuosa, Agrostis stolonifera*) through competitive exclusion is not a new idea. Cereals, especially rye (*Secale cereale*), were used in France in direct seeding mixes with oak and pine (Cotta 1822 in Dimkic 1997); blue lupine (*Lupinus perennis*) was tested in forest plantations in Germany in the 1950s.

One recent innovation is to use a mixture rather than a single species to control the spontaneous competitive species (mainly grasses, see Balandier et al. 2006) and thus favor young tree establishment (Reinecke et al. 2002). The choice of the plants composing the mixture must have the following characteristics (Balandier et al. 2003b):

- Not be invasive species and must have low water and nutrient requirements to limit competition with tree seedlings.
- Have different phenological developments and be capable of replacing one another in order to keep the soil continuously covered in space and time until the trees establish.
- Disappear rapidly to give way to the natural vegetation.
- Be available in seed form, because the cover vegetation is sown before trees are planted or sown.

The sown vegetative cover, with its multilayered structure, is expected to exclude the most competitive species such as *A. repens*, and create a favorable microclimate that protects young trees from wind and late-spring or early-autumn frosts (Balandier et al. 2009a). In summer, the cover creates favorable shade. Using a mixture of cover plants rather than a single species also favors biodiversity because fauna and insects (notably butterflies) proliferate (Reinecke et al. 2002). The technique is still regarded as experimental, although more than 1000 ha of forests have been established in Germany over the last 20 years with apparent success (Frochot et al. 2002). Tree mortality has been limited, tree growth seems better, and the sown vegetation disappeared in 3 to 5 years (Balandier et al. 2003b). In the Massif Central, however, only a few experimental plantations have been established (Provendier and Balandier 2002; Balandier et al. 2009a). The composition of the plant mixture used by Reinecke will probably need adjusting to fit local soil and climate. Two experiments, one at high elevation (about 800 m, Charensat) and the other in the plain (about 300 m, Montoldre), were established to assess different plant mixtures and compare their efficiency to woody accompanying species in terms of tree establishment (Provendier and Balandier 2002; Balandier et al. 2009a). Despite a long list of drawbacks, cover plants can improve tree establishment, mostly in unfertile sites, in comparison with doing nothing. However, the use of herbicides still gives better results. The improvement of the microclimate (buffering of extreme events) and resource availability (mainly water) are by far the main positive effect of the cover plants (Balandier et al. 2009a). They can also bring an added value, such as an increase of biodiversity, an improvement of the landscape if flower species are used (Figure 16.5), or bring other productions as honey if melliferous plants are used. These different aspects have not been really investigated to actually assess the ecological footprint of the technique (Balandier et al. 2009b).

FIGURE 16.5
Mixture of cover plants with flowering species to favor tree establishment by sowing. (Montoldre, France, Photo P. Balandier)

16.3 Converting Fallow Land into More Sustainable Forest Stands

The forest cover of France has doubled over the last 150 years (DERF 1995) mainly by natural invasion of abandoned cropland or pasture (Figure 16.6). Natural invasion in the Massif Central has been primarily pine (*P. sylvestris*, *Pinus nigra*) and birch (*Betula* sp.) (Koerner et al. 2000). The resulting woodlands are usually of low economic value because of poor stem form and low wood quality. Over time, these natural secondary woodlands give way to forests of more shade-tolerant species such as oak (*Quercus* sp.) in lowlands and beech

FIGURE 16.6
Progressive colonization of old pastures by natural invasion of woody species following grazing decrease in the French Massif Central Mountains. (Photo P. Balandier).

(*F. sylvatica*) and silver fir (*A. alba*) at higher altitudes. Forest managers increasingly favor late-successional species, because they have more commercial value, are well-suited to site conditions, are better adapted than conifers to major disturbances such as windstorms; and native forests are usually more suitable as habitat for a wider range of native forest species than plantation forests (Hermy et al. 1999; Singleton et al. 2001; Brockerhoff et al. 2008). Late-successional species will establish naturally under pioneer species; for example, *F. sylvatica* under *P. sylvestris* or *P. nigra* overstory. This apparent facilitation model (Connell and Slatyer 1977) is a slow process with variable success controlled by numerous factors, such as distance to seed sources, abundance and persistence of the seed bank, site conditions, competition, and ungulate herbivory. Managers can accelerate these processes by direct seeding or planting.

16.3.1 Direct Seeding of Late-Successional Species

Direct seeding can be an economical method for establishing deciduous species below conifer canopies, provided some prerequisites are met: high germination capacity of seeds, appropriate sowing technique, and absence of excessively competitive vegetation (Ammer et al. 2002; Balandier et al. 2009a).

Little is known about the environmental conditions that influence germination and survival of direct seeded species, but secondary woodlands certainly present unfavorable conditions because ground vegetation is abundant due to good light penetration. Ground vegetation competes with seedlings and provides favorable habitat for small mammals, leading to higher predation. We tried direct seeding of beech (*F. sylvatica*) in pine (*P. sylvestris*) stands and meadows in the Chaîne des Puys, a volcanic mid-elevation mountain of the Massif Central. We used three Scots pine stands of a range of density (4000, 1500, and 500 stem ha^{-1}) that created a light gradient from 9% to 46% of full light and an open meadow with full light. Conditions in the stands were comparable: same altitude (≅900 m) and topography (flat area), and similar volcanic soils. Annual rainfall of 820 mm is evenly spread over the year and soil is well watered in the upper horizon. Nevertheless, there is a risk of rapid summer dryness on these soils (Coll et al. 2003). A total of 1200 seeds, with an 85% germination capacity (Suska et al. 1994), were sown in the different stands and meadow, either directly on the ground without modifying the existent vegetation, or on bare soil after removing vegetation and litter. Seedling emergence (cotyledons visible) was lower in open habitats (meadow and pine at 500 stem ha^{-1}) even in the absence of ground vegetation (Figure 16.7). After 20 weeks, survival was 23% in the high- and medium-density stands, 15% in the less crowded stand, and 0% in the meadow. One possible explanation is that small mammal predation was higher in open habitats owing to a more developed herb cover (Gill and Marks 1991). Another explanation could be that microclimatic conditions and variations in soil moisture may have affected seed germination and early survival (Ammer et al. 2002; Balandier et al. 2009a). Another explanation was that the ground vegetation, which followed the light gradient, may have negatively affected germination and young seedling survival.

16.3.2 Planting Late-Successional Species

Planting beneath the canopy of pioneer species can favor late-successional species. Studies with planting 2-year-old beech seedlings were carried out using the same experimental design as described above, where a light gradient was developed by thinning Scots pine stands and using an open meadow (either weeded or undisturbed). This study focused

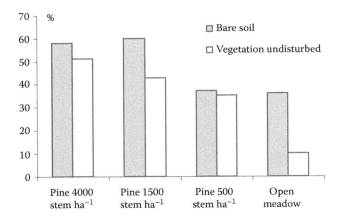

FIGURE 16.7
Cumulated percentage of beech (*Fagus sylvatica*) seedlings emergence (cotyledons visible) 20 weeks after seeding in Scots pine (*Pinus sylvestris*) stands of different densities and in an open meadow. Two seeding modalities were compared: in bare soil and in undisturbed vegetation. Plants that died during this period were counted.

particularly on the effect of light level on beech seedling growth and on the competitive interactions between beech and the ground vegetation.

Beech seedling growth was mainly driven by light availability, with higher increment for higher transmittance (Figure 16.8). The ranking was perfectly respected for stem diameter (Figure 16.8b), whereas stem height was sometimes more disordered due to stem damages by insects (Figure 16.8a). Water was a limiting factor only in the open meadow where the grass vegetation exerted a strong competition (Provendier and Balandier 2008). Predawn leaf water potential measured in summer in the beech seedlings was comparable in all treatments, except for that of the open, unweeded meadow, which exhibited a significantly lower value (Figure 16.9), indicating a higher water demand for beech seedlings in this treatment. Ground vegetation was absent in the most dense Scots pine plots (4000 stem ha^{-1}), weakly developed in the moderate density treatment, or well developed in the low-density treatment but composed of dicotyledonous species, which compete less for water (Coll et al. 2003) than the open meadow grasses.

Beech, the major late-successional species in this area, in the process of old field succession, establishes itself naturally under pioneer species but seldom on open sites. The dominant early-successional tree species develop both positive and negative interactions with young beech seedlings and saplings: direct competition for light, reduced water stress by control of the ground vegetation, and further modification of microclimatic conditions. Control of the shelterwood density of the established secondary woodlands (Scots pine or birch) can thus be a powerful tool to enhance beech establishment and growth.

16.4 From Monospecific Coniferous to Mixed Coniferous–Broadleaved Stands

In certain parts of the Massif Central, plantations of a single conifer species (*P. abies* or *P. douglasii*) cover large areas. Forest managers are seeking to replace these stands with mixed forests in order to reach a new equilibrium that combines the economic value of

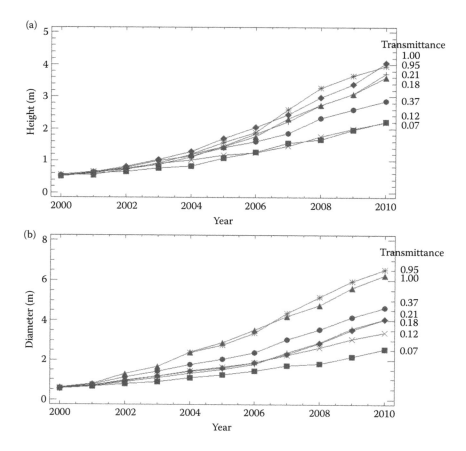

FIGURE 16.8
Mean growth of two-year-old beech (*Fagus sylvatica*) seedlings planted in Scots pine stands of different densities and in an open meadow. (a) Stem height and (b) basal stem diameter. Each curve corresponds to a plot with a given light availability, which is indicated by the transmittance value in the right of the figure. Transmittance is the ratio between the light measured below and above pine canopy; therefore transmittance equals 1 in the open meadow. (For the design see Coll, L. et al. 2003. *Ann. For. Sci.* 60, 7: 593–600; Curt, T. et al. 2005. *Ann. For. Sci.* 62: 51–60.)

exotic species with the ecological value of native trees to ensure sustainable, highly productive woodlands. They are attempting to mix exotic species such as Douglas fir, larch, which are very productive, with native species (*F. sylvatica, P. sylvestris, A. alba*).

Conifer plantations have always had high densities (Figure 16.10), with more than 3000 stems ha^{-1} with pines, 2500 with spruce, and 1500 with Douglas fir. High density was intended to ensure rapid occupancy of the site to forestall weed competition, allow a high selection ratio, and produce maximum wood volume. Before the first thinning at age 40, the understory and the ground vegetation were nearly absent due to severe light limitation. In fact, relative irradiance in these conifer plantations was very low, less than 2.5% of full sunlight. Irradiance levels of 5% of available sunlight are necessary for understory plants to develop, and maximum plant biodiversity is obtained at about 20% (Balandier et al. 2006). An average tree density of about 800 stems ha^{-1} will produce adequate light levels for understory vegetation to develop, but this density is rarely obtained after the first thinning. Plant and overall biodiversity are generally reduced and only a few sources

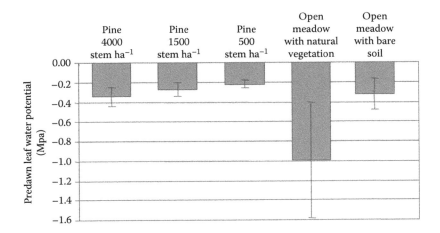

FIGURE 16.9
Predawn leaf water potential on 29 June 2000 of two-year-old beech seedlings planted in Scots pine stands of different densities and in an open meadow of which one part was weeded. (For the design see Coll, L. et al. 2003. *Ann. For. Sci.* 60, 7: 593–600; Curt, T. et al. 2005. *Ann. For. Sci.* 62: 51–60.)

FIGURE 16.10
Young exotic Douglas-fir (*Pseudostuga douglasii*) plantation at high density in the French Massif Central Mountains. (Photo by P. Balandier.)

of biodiversity can be found in such large plantation areas; mainly in gaps due to rocky or wet areas, deer damage, or uncleared spots. Even in areas where land ownership is more fragmentary, biodiversity inside woods is low.

Two approaches have been used to restore biodiversity in these plantations: natural or artificial regeneration.

16.4.1 Natural Regeneration

Biodiversity can be increased in plantations by changing to natural regeneration. Natural regeneration relies on existing adult trees to produce seeds and obtain new tree populations. Favoring natural recruitment needs to apply a careful silviculture in young plantations to gradually reduce tree density to 150 to 200 stems ha^{-1} and to select the best-formed trees as seed sources. According to the plantation's vitality (a function of soil, climate, and species adaptation), active management can obtain a final stand with a mean of 45 cm in 45 years for Douglas fir and 60 cm for Norway spruce. Longer rotations are required if thinnings are delayed; ideally, stands should be thinned as soon as the live crown height is less than half the total tree height. Stem taper remains high and trees are resistant to wind damage.

Gaps in the canopy allow native species (e.g., beech, and fir, Landré and Balandier 2006) to establish and be present, when the stand is regenerated. Natural regeneration of the conifers is obtained by opening the stand and harvesting the less valuable timber; thus, residual trees will produce seed in abundance, and seedlings will find enough light to develop. Fir and beech produce abundant and fertile seed crops every 2 to 3 years in the Massif Central and natural regeneration occurs readily if the soil is weed free. It must be emphasized that successful recruitment, early growth and survival can be strongly influenced by soil preparation and ground vegetation control treatments (see reviews by Balandier et al. 2006 and Wiensczyk et al. 2011). Seeds from secondary species can also establish, adding to diversity. Presently, beech is establishing in many stands throughout the Massif Central, adding color and diversity to the landscape.

16.4.2 Artificial Regeneration

Introducing biodiversity into new plantations is possible by planting at fairly low densities. There are few opportunities for afforestation in some areas due to agriculture subsidies, whereas in other areas with more than 70% high forest cover, there is pressure to preserve farmland and forbid new plantations. Nevertheless, planting is always used for reforestation after gale damage or clearcuts. Clearcuts are necessary in plantations where species were planted off-site or where upkeep such as thinnings was delayed. In these conditions, thinning is no longer possible because of lack of wind firmness and the stand cannot reach full maturity. About a quarter of the spruce plantations established in the 1960s at densities greater than 2500, and unthinned, will not tolerate thinning. The only solution is to clearcut these plantations as soon as they reach an average of 30 cm, before they are fallen by wind or snowstorms.

Mixed-species plantations have rarely been used because of the difficulty of managing species with different growth rates. Attempts in the 1950s and 1960s with "half-and-half" mixtures always failed dismally as one of the two species took over and displaced the second one (Bernard 1987). The French Forest Administration (DERF 2000) does not allow step-by-step mixtures and recommends planting single species in bunches, clumps, or screens, where the second species covers less than 20% of the area and is scattered inside the main stand. When mature, the stand will include natural seed sources for future cohorts developing following regeneration (for instance, fir or beech planted with spruce). In monospecific plantations, densities around 1000 stem ha^{-1} are commonly used with rectangular spacing such as 4 × 2.5 m. These patterns allow the natural establishment of other tree species inside the plantation. The forester must safeguard these "volunteers" when clearing the plantation and should release these stems when carrying out the first commercial or precommercial thinning. Because of the proliferation of roe deer (*Capreolus capreolus*), clearing is often done only in alternate rows to limit accessibility for deer and allow natural vegetation to grow in these areas. When pulpwood demand is low, precommercial thinning (cut stems are left on the ground) is used when stands are about 5 to

6 m high to reduce tree density. Precommercial thinning gives indigenous tree species a greater chance to establish and become quality crop trees.

The enrichment of existing stands, whether established by natural or artificial regeneration, with a planting of a third species inside gaps under the tree canopy can give good results. In young stages, it is advisable to use individual tree tubes that distinguish the planted seedlings from the general vegetation and so that cleaning and weeding the stand can promote tree growth. Fast-growing species can be used, such as larch or maple in fertile soils, mixed for instance with spruce. In older forest stands with taller trees, shade-tolerant species must be used for supplemental planting. In the center of the Massif Central, fir has often been planted under Scots pine stands promoting nowadays a transition toward fir stands.

16.5 Conclusion

Changing economic, environmental, and social conditions have challenged foresters to modify their silvicultural practices. Forests in the Massif Central begin the twenty-first century with new management strategies, including restoring diversity and transforming existing stands and plantations into diversified stands that are storm- and pest-resistant and probably more adapted to face climate change consequences (e.g., drought occurrence), using environmentally friendly soil preparation and weed control, and taking advantage of natural processes to favor mixtures of species. Experience to date suggests that more research should be devoted to fundamental exploration of four areas: competition between trees and ground vegetation, changes of biodiversity in relation to the main environmental variables, especially light level, ecology and uses of exotic species, and natural dynamics of forest stands in diverse contexts.

Acknowledgments

We are thankful to André Marquier, Fabrice Landré, Gilles Agrech, and René Jouvie for their field assistance. Experiments described in this chapter were partly funded by the Auvergne Region, the FNADT (National Fund for Territorial Development), and the IFB (French Institute for Biodiversity).

References

AFOMAC, 2001. Organiser l'offre de bois résineux dans l'espace central, compte-rendu de mission. France.

Ammer, C., Mosandl, R., and Kateb H.E. 2002. Direct seeding of beech (Fagus sylvatica L.) in Norway spruce (Picea abies [L.] Karst.) stands – effects of canopy density and fine root biomass on seed germination. *For. Ecol. Manage.* 159: 59–72.

Aussenac, G. 1986. La maîtrise du microclimat en plantation. *Rev. For. Fr.* 38: 285–292.

Balandier, P., Allegrini, C., and Jay, D. 2008. Des réactions similaires à l'accompagnement ligneux pour le frêne et le noyer hybride. *Forêt-entreprise* 178: 21–25.

Balandier, P., Bergez, J.E., and Etienne, M. 2003a. Use of the management-oriented silvopastoral model ALWAYS: Calibration and evaluation. *Agrofor. Syst.* 57: 159–171.

Balandier, P., Collet, C., Miller, J.H., Reynolds, P.E., and Zedacker, S.M. 2006. Designing forest vegetation management strategies based on the mechanisms and dynamics of crop tree competition by neighbouring vegetation. *Forestry* 79, 1: 3–27.

Balandier, P. and De Montard, F.X. 2008. Root competition for water between trees and grass in a silvopastoral plot of ten-year-old *Prunus avium*. In : *Ecological Basis of Agroforestry*, ed. D.R. Batish, R.K. Kohli, S. Jose, and H.P. Singh, 253–270. Boca Raton, FL, USA: CRC Press.

Balandier, P., Frochot, H., Charnet, F., Reinecke, H., Ningre, F., Koerner, W et al. 2003b. *Restauration de la Biodiversité Floristique lors des Operations de Boisements*. Paris, France: Cemagref, INRA, IDF, Reinecke Forstingenieurbüro, IFB.

Balandier, P., Frochot, H., and Sourisseau, A. 2009a. Improvement of direct tree seeding with cover crops in afforestation: microclimate and resource availability induced by vegetation composition. *For. Ecol. Manage.* 257: 1716–1724.

Balandier, P., Frochot, H., and Sourisseau, A. 2009b. Foundations of the cover crop technique and potential use in afforestation. In *Forest Vegetation Management—Towards Environmental Sustainability*, ed. N.S. Bentsen, 60–61. Vejle, Denmark: Forest and Landscape Working Papers no. 35.

Balandier, P. and Marquier, A. 1998. Vers une remise en question des avantages d'une plantation frêne—Aulne. *Rev. For. Fr.* 50: 231–243.

Balandier, P., Rapey, H., and Guitton, J.L. 1997. Improvement and sustainable development of moderate altitude areas through agroforestry: tree-grass-animal association. In *Proceedings of the XI World Forestry Congress*, vol. 1, topic 2. Antalya, Turkey.

Balandier, P., Rapey, H., and Ruchaud, F. 2002. Agroforesterie en Europe de l'Ouest: pratiques et experimentations sylvopastorales des montagnes de la zone tempérée. *Cahiers Agricultures* 11: 103–113.

Bernard, C. 1987. *Etude sur les Potentialités de Production de Plantations Mélangées dans le Département du Puy-de-Dôme*. Clermont-Ferrand, France: Cemagref.

Bianco, J.L. 1998. *La forêt: une Chance pour la France, Rapport de Propositions D'orientation d'un Projet de loi de Modernisation Forestière*. Paris, France.

Bouvarel, P. and Larrere, G. 1981. Les forêts de montagne. In *L'INRA et la montagne*, 59–65. Versailles, France: INRA.

Brockerhoff, E.G., Jactel, H., Parrotta, J.A., Quine, C.P., and Sayer, J. 2008. Plantation forests and biodiversity: Oxymoron or opportunity? *Biodiv. Conserv.* 17: 925–951.

Coll, L., Balandier, P., Picon-Cochard, C., Prévosto, B., and Curt, T. 2003. Competition for water between beech seedlings and surrounding vegetation in stands differing in light availability and vegetation composition. *Ann. For. Sci.* 60, 7: 593–600.

Collective report, 1998. *Prospective: La Forêt, sa Filière et Leurs Liens au Territoire*. Paris, France: INRA, Délégation permanente à l'agriculture, au développement et à la prospective.

Collet, C., Frochot, H., Pitsch, M., and Wehrlen, L. 1992. Effet d'un abri latéral artificiel sur le développement de jeunes merisiers (Prunus avium L.) installés en pépinière. *Rev. For. Fr.* 44: 85–90.

Connell, J.H. and Slatyer, R.O. 1977. Mechanisms of succession in natural communities and their role in community stability and organization. *Am. Nat.* 111: 1119–1144.

Curt, T., Coll, L., Prévosto, B., and Balandier, P. 2005. Growth, allocational flexibility and root morphological plasticity of beech seedlings in relation to light and herbaceous competition. *Ann. For. Sci.* 62: 51–60.

Davies, R.J. 1987. Trees and weeds. In *Weed Control for Successful Tree Establishment*. London, England: Forestry Commission Handbook, HMSO Publications.

De Beaulieu, J.L., Pons, A., and Reille, M. 1988. Histoire de la flore et de la végétation du Massif Central (France) depuis la fin de la dernière glaciation. *Cahiers de Micropaléontologie* 3: 5–36.

De Champ, J. 1997. Eclaircies et stabilité des peuplements. In *Le Douglas*, ed. J. De Champ, 206–210. France: AFOCEL.

DERF (French Directorate of Rural Areas and Forests) 2000. *Actualisation des Conditions de Financement par le Budget de l'Etat des Projets de Boisement-Reboisement, circulaire DERF/SDF/C2000-3021 du 18 août 2000*. Paris, France: Ministère de l'Agriculture et de la Pêche.

DERF (French Directorate of Rural Areas and Forests) 1995. *Les Indicateurs de Gestion Durable des Forêts Françaises*. Paris, France: Ministère de l'Agriculture et de la Pêche, Direction de l'Espace Rural et de la Forêt.

Dimkic, C. 1997. *L'enherbement des Plantations Forestières*. Paris, France: IDF.

DRAF 1999. *Orientations Régionales Forestières*. Clermont-Ferrand, France: DRAF Auvergne.

Friedrich, J.M. and Dawson, J.O. 1984. Soil nitrogen concentration and Juglans nigra growth in mixed plots with nitrogen-fixing *Alnus, Eleagnus, Lespedeza* and *Robinia* species. *Can. J. For. Res.* 14: 864–868.

Frochot, H., Balandier, P., Michalet, R., and Van Lerberghes, P. 2009. France. In *Forest Vegetation Management in Europe: Current Practice and Future Requirements*, ed. I. Willoughby, P. Balandier, N.S. Bentsen, N. McCarthy, and J. Claridge, 33–42. Brussels, Belgium: COST Office.

Frochot, H., Balandier, P., Reinecke, H., Boulet-Gercourt, B., Ningre, F., Lefèvre, Y et al. 2002. Using cover plants mixtures to favour tree establishment in afforestation: an alternative to repeated herbicides or mechanical vegetation controls? In *Popular Summaries of the Fourth International Conference on Forest Vegetation Management*, ed. H. Frochot, C. Collet, and P. Balandier, 233–235. Nancy, France: INRA.

Frochot, H., Picard, J.F., and Dreyfus, P. 1986. La végétation herbacée obstacle aux plantations. *Rev. For. Fr.* 37: 271–278.

Gill, D.S. and Marks, P.L. 1991. Tree and shrub seedling colonization of old fields in Central New York. *Ecol. Monogr.* 61: 183–205.

Guitton, J.L. 1986. *Problèmes des Peuplements Forestiers de Montagne en Auvergne*. Riom, France: Cemagref.

Hermy, M., Honnay, O., Firbank, L., Grashof-Bodkam, C., and Lawesson, J.E. 1999. An ecological comparison between ancient and other forest plant species of Europe, and the implications for forest conservation. *Biol. Conserv.* 91: 9–22.

Hubert, M. 1992. La vegetation d'accompagnement: un auxiliaire sylvicole à surveiller de près. *Forêt- Entreprise* 82: 39–47.

Koerner, W., Cinotti, B., Jussy, J.-H., and Benoît, M. 2000. Evolution des surfaces boisées en France depuis le début du XXe siècle: identification et localisation des territoires abandonnés. *Rev. For. Fr.* 52: 249–269.

Landré, F. and Balandier, P. 2006. Potentialités de régénération naturelle des feuillus sous épicéas ou douglas. *Forêt-entreprise* 168: 56–59.

Lifran, R., Rapey, H., and Valadier, A. 1997. Nouveau regard sur la tradition sylvo-pastorale en Lozère. *Agreste—les cahiers*, 21: 17–22.

Mc Carthy, N., Bentsen, N.S., Willoughby, I., and Balandier, P. 2011. The state of forest vegetation management in Europe in the 21st century. *Eur. J. For. Res.* 130: 7–16.

Michelin, Y. 1995. *Les jardins de Vulcain*. Paris, France: Maison des sciences de l'homme.

Nambiar, E K.S. and Sands, R. 1993. Competition for water and nutrients in forests. *Can J. For. Res.* 23: 1955–1968.

ONF, 1993. L'Auvergne, Pays de hautes terres au coeur du Massif Central. *Arborescences* 43: 25–.

Perrin, H. 1963. *Sylviculture*. Nancy, France: Ecole Nationale des Eaux et des Forêts.

Prévosto, B. and Balandier, P. 2007. Influence of nurse birch and Scots pine seedlings on early aerial development of European beech seedlings in an open-field plantation of central France. *Forestry* 80, 3: 253–264.

Prévosto, B., Dambrine, E., Moares, M., and Curt, T. 2004. Effects of volcanic ash chemistry and former agricultural use on the soils and vegetation of naturally regenerated woodlands in the Massif Central, France. *Catena* 56: 239–261.

Prévosto, B., Monnier, Y., Ripert, C., and Fernandez, C. 2012. To what extent do time, species identity and selected plant response variables influence woody plant interactions? *J. Appl. Ecol.* 49: 1344–1355.

Provendier, D. and Balandier, P. 2002. *Comparaison de Différentes Modalités de Reconstitution de la Forêt après Tempête en Auvergne dans une Perspective de Gestion Durable. Gestion de la Vegetation et Alternatives à L'utilisation D'herbicides en Plantation Forestière.* Clermont-Ferrand, France: Cemagref.

Provendier, D. and Balandier, P. 2008. Compared effects of competition by grasses (Graminoids) and broom (*Cytisus scoparius*) on growth and functional traits of beech saplings (*Fagus sylvatica*). *Ann. For. Sci.* 65: 510, 9p.

Reinecke, H., Koerner, W., Frochot, H., Balandier, P., Boulet-Gercourt, B., and Ningre, F. 2002. Sowing cover plants mixtures as a mean to control competitive vegetation in new forest plantations. In *Popular Summaries of the Fourth International Conference on Forest Vegetation Management*, ed. H. Frochot, C. Collet, and P. Balandier, 421–423. Nancy, France: INRA.

Rérat, B. 2000. Les dégâts des ouragans Lothar et Martin—France 1999. *Forêt-entreprise* 131: 8-.

Singleton, R., Gardescu, S., Marks, P.L., and Geber, M. A. 2001. Forest herb colonization of postagricultural forests in central New York State, USA. *J. Ecol.* 89: 325–338.

Suska, B., Muller C., and Bonnet-Masimbert, M. 1994. *Graines des feuillus forestiers: de la récolte au semis*. Paris, France: INRA.

Wiensczyk, A., Swift, K., Morneault, A., Thiffault, N., Szuba, K., Bell, W. 2011. An overview of the efficacy of vegetation management alternatives for conifer regeneration in boreal forests. *For. Chron.* 87: 175–200.

17

Conversion of Norway Spruce (Picea abies [L.] Karst.) Forests in Europe

Jörg Hansen and Heinrich Spiecker

CONTENTS

17.1 Introduction

Traditional forestry, in much of northern and central Europe, has concentrated on growing conifers in plantations for timber and pulp. Increased concern over ecological stability, nature conservation, possible climate change (Ulrich and Puhe 1994), and damage from major storms, with the resulting periods of reduced prices for softwood timber, however, has increased interest in alternative management approaches such as close-to-nature forestry (Dafis 2001). As a result, the widespread reliance on conifer monocultures has come into question and more stable alternatives, such as site-adapted mixtures of species, often dominated by broadleaves, are under consideration or already employed (Hansen et al. 2003; Spiecker et al. 2004).

Norway spruce (*Picea abies* [L.] Karst.) is found in almost every European country. It is affected by severe calamities caused by storms, snow, or ice (Ebert 2002; Kuhn 1995), and the corresponding secondary damage, such as decline of damaged trees or bark beetle attacks. As there are potentially unstable Norway spruce forests in many European countries, this species is the primary candidate for conversion to structured or mixed broadleaf-dominated forests (Teuffel et al. 2004; Gardiner and Breland 2002; Hasenauer 2000; Klimo et al. 2000). The stability of forests can be expressed in terms of the resistance and resilience of the forest ecosystem (Larsen 1995). Poor resistance can be manifested as a higher susceptibility than would be expected of a natural forest to damage or destruction by strong winds, drought, fire, or a complex of factors. Poor resilience may entail considerable difficulty in regaining a closed-forest condition after a catastrophe, because of regeneration problems.

Norway spruce is an example of a species that can regenerate successfully on sites where it is not adapted in the long run, or where browsing or management intervention has favored it over other more site-adapted species (Johann et al. 2004). A large proportion of potentially unstable Norway spruce stands exist today, which were artificially established by direct seeding or planting. A critical question facing managers is which Norway spruce stands should be converted. Selection of conversion stands cannot be based solely on the natural range of Norway spruce, but also needs to be compatible with the objectives of forest management.

17.2 Distribution of Norway Spruce

Norway spruce is one of the most profitable tree species in Europe and has been widely planted outside its natural range. The highest relative proportion of Norway spruce, as compared to the total national territory, is found in Sweden (~34%) followed by Austria with over 22% and the Czech Republic with around 17%. Germany, Luxemburg, Slovakia, and Switzerland have Norway spruce proportions of approximately 10% and therefore correspond to the European average (Teuffel et al. 2004). For Finland and Norway, there have been reports of Norway spruce being between 10% and 25% (Spiecker 2000). Although these countries are located mostly within the natural distribution of the species, Norway spruce was introduced to new sites (Koch 2000; Heinzel and Peters 2001) where it had been absent or, at most, was codominant or admixed.

Norway spruce's natural range is somewhat contentious. Because it was little affected by human interference until the third or fourth century AD (Hasel 1985), its range at that time provides a starting point for what can be considered "natural." At the end of the last glacial maximum, Norway spruce probably spread into central Europe from the east at the same time as other species, such as European silver fir (*Abies alba* Mill.), returned to central Europe from southwestern refugia. With its wide ecological amplitude and ability to compete against other species (Mayer 1984), Norway spruce might have migrated further into the northwest of continental Europe even without human intervention. Thus, the determination of "which stands are not site-adapted" is not a simple matter of referring to a map of its natural range (Figure 17.1).

Based on its third–fourth century AD distribution, Norway spruce can be described as a species of a cool and humid climate. The focus of its natural distribution in Europe lies in the boreal coniferous forest of Nordic countries and Russia, as well as at higher elevations in the temperate zone forests of Switzerland, Austria, the Czech Republic, Poland, Byelorussia, and the Baltic countries of Lithuania, Latvia, and Estonia (Figure 17.1). Water availability in years with low precipitation levels is thought to limit the distribution of Norway spruce in the south. In the west and southwest of central Europe, natural and pure Norway spruce stands occur in mountainous areas where short growing seasons, low annual mean temperatures ($\leq 5°C$), and high annual precipitation (≥ 800 mm) are typical (Schmidt-Vogt 1977). With the increasing competitiveness of European silver fir or European beech (*Fagus sylvatica* L.), Norway spruce is usually a codominant or admixed tree species.

Forest management has not only extended the geographical range of Norway spruce, but even within its natural range, Norway spruce has been planted on sites naturally dominated by broadleaves. This change in growing stock took place from the Middle Ages up

FIGURE 17.1
Natural distribution of Norway spruce (*Picea abies* [L.]Karst.) in Europe. (Simplified from Schmidt-Vogt, H., *Die Fichte. EinHandbuch in zweiBänden, Taxonomie, Verbreitung, Morphologie, Ökologie, Waldgesellschaften*, Vol. I, Paul Parey, Hamburg, Berlin, 647 pp., 1977.)

until the twentieth century and was mostly motivated by widespread overexploitation and devastation of forests, as well as the fear of timber shortage (Hasel 1985). From the eighteenth century onward, large-scale litter raking, which removed nutrients and diminished the humus layers, causing regeneration problems and lowering yield, degraded the forests. Removals for charcoal and potash production, plus intensive timber consumption for shipbuilding and mining resulted in extensive difficulty to obtain adequate regeneration. Increasingly, managers turned to coniferous trees, notably Norway spruce and Scots pine (*Pinus sylvestris* L.), for growing stock. Norway spruce is better adapted to the microclimate on clearcut sites than European beech, silver fir, or oak (*Quercus* spp.), which are more susceptible to damage by late frosts. Furthermore, Norway spruce exhibits high initial growth, an advantage in competing with other ground vegetation. The most recent wave of reforestation that increased the proportion of Norway spruce took place after World Wars I and II.

17.3 Management of Norway Spruce

Industrialization in the nineteenth century dramatically increased the demand for timber and special timber assortments, including small dimensions. Economic models for organizing forest management and evaluating silvicultural decisions were developed at the same time. Norway spruce was ideal for these developments; it was easier and less expensive to regenerate than broadleaves, grew faster, and reached an economic size quicker. These economic decision models postulated profit maximization as the main goal and favored dense planting, heavy thinning, and relatively short rotations. As the focus of analysis shifted to the stand or harvesting unit level, forests were organized into even-aged cutting units. This form of profitable and efficient forest management became widespread in Europe and provided industry with a continuous and even volume of timber in typical and uniform assortments of consistent quality. The demand for small-dimension spruce timber was met by "Dunkelwirtschaft" (German for dark management), which is characterized by high density and low thinning to produce uniform material with little or no taper. Although appreciated by timber processors, such management increases the susceptibility of stands to storm and snow damage. The introduction of alternative materials such as steel and plastic in the twentieth century reduced the demand for small-dimension timber. Without the demand for these timber assortments, required thinning became precommercial and was often postponed, increasing the instability of dense spruce stands (Figure 17.2)

At present, forests in Europe are managed to fulfill a range of functions: economic (timber and employment), nonconsumptive recreation, and ecological and societal services (soil, air, and water protection, nature conservation, noise absorption, and avalanche prevention). In central Europe, Norway spruce is faced with the risk of various biotic and abiotic damages (Tomiczek 2000; Nopp et al. 2001; Kulikova 2002). Storm and snow are the most detrimental (Spiecker 2000). Further negative factors include bark beetles (*Ips typographus* L., *Pityogenes chalcographus* L.), fire, ice, fungal diseases (*Heterobasidion annosum* [Fr.] Bref., *Fomes annosus* [Fr.] P. Karst., *Armillariella* spp.) or, in exceptional cases, damage by wildlife. The risk of damage to a stand increases if two or more of these detrimental factors coincide, or if they combine with other predisposing site or stand conditions such as poorly drained soils, few admixed species or nonsite-adapted provenances of spruce, or great tree heights.

In some parts of Europe, high volumes of Norway spruce are harvested as incidental exploitations (Hanewinkel 2002), which are entirely unintended or not performed for

FIGURE 17.2
Storm—damaged Norway spruce stand in central Germany.

economic motives (Kusché 2000; Spiecker 2000). Although implying an impaired stability and lack of resilience, general conclusions should not be drawn from such harvests without first analyzing the underlying causes. Damage review after the storms of the previous decades has shown that Norway spruce growing on physiologically shallow soil, such as on dual-layered soil profiles or poorly drained soils, develops a very shallow root system and bears an extreme risk for windthrow. This risk increases with high height-to-diameter ratios of individual trees, great tree height, and high stocking. Additional factors such as stand age and size, the proportion of other species, or trees impaired by fungi or cambial damage from wildlife also contribute to the risk of windthrow or windbreak.

17.4 The Question of Conversion

In order to achieve or safeguard the long-term stability of forests, foresters in many European countries are confronted with questions of whether to convert Norway spruce stands, which stands will benefit from conversion to other species or mixtures with Norway spruce, and how to accomplish conversion. A multifunctional approach implies that such decisions will be based on site factors, location, and potential effects on ecological processes (e.g., Duncker et al. 2012). In order to answer all questions related to the option of conversion, further aspects also need to be taken into account, including the conservation status of a stand, the form of ownership, and the level of local infrastructure development. All these factors affect the approach and intensity of silvicultural measures to be taken. With increasing labor costs and stagnant timber prices, new models of forest management are required. Close-to-nature forestry, with species mixtures and uneven-aged structure, as opposed to pure stands and forest management in even-aged cutting units, is commonly regarded as the most effective option (e.g., Hahn et al. 2005).

17.4.1 Is Conversion Needed?

Each forest owner determines the most dominant forest function(s) pursued in accordance with regulatory background, economic opportunity, and personal preference. In most instances, the objective will not be to achieve maximum fulfillment of all forest functions, but to have an optimized combination for implementing the forest owner's target system. The need for conversion will be manifested if a substantial disruption in one or several of the targeted forest functions occurs. Thus, the criteria to decide on conversion needs are always dependent on the objectives of forest management (Yousefpour and Hanewinkel 2014).

Management objectives differ greatly between forest ownerships. In several European countries, the conversion of Norway spruce stands in public ownership has progressed further than in privately owned forests (Schmid 1998). Timber from Norway spruce forests has been easily marketed due to its many uses and uniform assortments of standard quality. Unless excess supply, such as experienced after calamities due to storm, heavy snow, or bark beetle outbreaks, causes a slump in timber prices, there is little short-term incentive for private owners to convert. Timber certification and its influence on marketability of forest products may be an incentive for conversion in the approximately 40% of the European forest area in private ownership. However, it is difficult to make general statements for private forest owners, as this group has many different attitudes toward management (Schraml and Härdter 2002; Schraml and Volz 2004). Nevertheless, reduced forestry subsidies in many European countries may enhance the appeal of low-input techniques such as close-to-nature silviculture.

17.4.2 Which Stands Should Be Converted?

A decision to convert a stand should only be taken if superior options for achieving desired forest functions can be identified. The timing of conversion will be determined in a second step, where the conversion priority for each stand is established on the basis of the overall situation in the forest enterprise. The strategy chosen for conversion, including the prescribed silvicultural measures, needs to be selected in accordance with the priority for conversion. Large areas of Norway spruce blown down by a storm require immediate conversion. Neighboring stands unaffected by blowdown should also be converted if similar site and stand conditions apply. Adequate measures should be taken in the short-to-medium term in order to prevent replicating identical stand structures and species composition, for example, by introducing other species to the regenerating stand. Norway spruce stands have repeatedly suffered small-scale damage due to abiotic or biotic factors. Such warnings of potential catastrophe should motivate stand conversion measures in the medium to long term. The conversion process can be delayed, however, if the risk is deemed low or if conversion will be undertaken later to fulfill a new policy direction.

17.4.3 How Should Conversion Be Accomplished?

For each stand with an established need for conversion, the selection of a conversion strategy should be guided by the widest possible use of economic principles, both for conversion and for management of the target stand management. The objectives for conversion are to optimize stability in the target forests, which is achieved by a high resilience against detrimental factors, and the greatest flexibility in the management system, so as to allow easy adaptation to future shifts in objectives or changed ecological conditions (Nyland 2003). As a practical matter, conversion strategies requiring a minimum of long-term intervention are preferable. Nevertheless, the long-term nature of some processes in forests (e.g., soil degradation) hinders or completely disables a direct and immediate analysis of the ecological impact after human intervention. The time lag between cause and effect, or between management measures and their long-term consequences, may be as long as a rotation length and therefore it can also be as long as several generations of foresters.

There are two basic strategies for conversion: introduction of more complex structure within the same species, such as by creation of multiaged stands of Norway spruce, and expansion of species variety, as achieved by insertion of broadleaves (beech or maple [*Acer* spp.]) or other conifers (silver fir). To address a specific conversion task, these two strategies can be combined in varying degrees, which provide the forest manager with a broad range of conversion options. The silvicultural measures for conversion of pure secondary Norway spruce stands to mixed-species stands range from clearcutting Norway spruce and planting broadleaves that correspond to the potentially natural vegetation, to the less-intensive measure of tending broadleaves that naturally enter or are artificially inserted into pure Norway spruce stands. This range of measures reflects the large number of alternatives that can be chosen to fit site conditions, stand characteristics, management intensity, timing preference of the forest owner, and applicable laws and regulations. Each approach, however, will be based on a silvicultural strategy that must be implemented consistently over the long term. Some temporary destabilization of the stand may accompany intensive measures taken to convert Norway spruce.

Managing the light environment in the regeneration layer is the silvicultural challenge in conversion. Controlling light transmittance in the stand prevents suppression of broadleaves from dense Norway spruce regeneration or grass proliferation. Shade-tolerant broadleaves, which can be introduced into Norway spruce stands by underplanting or seeding, can be favored by these measures over Norway spruce regeneration (Lüpke et al. 2004). Knowledge and utilization of the shade tolerance of young broadleaves, such as European beech or oak, as well as utilization of gap development in Norway spruce forests, is the key component of a successful conversion strategy. Manipulating light levels in the stand requires more entries per planning period, which increases the risk of damage to residual trees and soil compaction from machinery traffic. Target-oriented (diameter limit) felling of individual trees or small groups to create small gaps usually constitutes the first and most important silvicultural measure in the conversion process. Harvesting will be spread across a larger area, as compared to conventional even-aged management. The initial phase of conversion presents severe economic challenges, including lower revenues and higher costs. As the conversion process may extend over several rotation lengths, the most effective strategies will include the fewest interventions with the greatest potential for revenue (Figure 17.3).

In addition to choosing an effective silvicultural technique for stand and site conditions, conversion decisions will be constrained by the socioeconomic differences within Europe and between forms of forest ownership. These differences not only cause a huge

FIGURE 17.3
Shelterwood treatments in Norway spruce stands result in seedlings of broadleaves and spruce.

disparity in available infrastructure and equipment, as well as in available means for purchasing new technology, but are also manifest as large differences between countries in labor costs and supply. As a result, low-tech and mostly manual implementation may be highly superior to the use of machinery and high-tech methods in some countries (Nordfjell et al. 2004). Seasonal investment of the forest owner's labor in small privately owned forests, common in many farming families, may be a practical scenario to achieve low labor costs.

References

Dafis, S., Ecological impacts of close-to-nature forestry on biodiversity and genetic diversity, in *Ecological and Socio-Economic Impacts of Close-to-Nature Forestry and Plantation Forestry: A Comparative Analysis*, Proceedings of the Scientific Seminar of the 7th Annual EFI Conference, Green, T., Ed., 3 September 2000, EFI Proceedings 37, 21–25, 2001.

Duncker, P.S., Barreiro, S.M., Hengeveld, G.M., Lind, T., Mason, W.L., Ambrozy, S., and Spiecker, H., 2012. Classification of forest management approaches: A new conceptual framework and its applicability to European forestry. *Ecology and Society* 17(4): 51. Available at: http://dx.doi.org/10.5751/ES-05262-170451

Ebert, H.P., Schädendurch"Lothar"imLehrrevier Rottenburg, *Allgemeine Forstzeitschrift/Der Wald*, 24, 1261, 2002.

Gardiner, E.S. and Breland, L.J., Eds., *Proceedings of the IUFRO Conference on Restoration of Boreal and Temperate Forests—Documenting Forest Restoration*, Vejle, Denmark, 28 April–2 May 2002, Skov ob Landskab, Report 11, 238 pp., 2002.

Hahn, K., Emborg, J., Larsen, J.B., and Madsen, P., Forest rehabilitation in Denmark using nature-based forestry, in *Restoration of Boreal and Temperate Forests*, Stanturf, J., and Madsen, P., Eds., CRC Press, Boca Raton, FL, 299, 2005.

Hanewinkel, M., Climatic hazards and their consequences for forest management—an analysis of traditional methodological approaches of risk assessment and alternatives towards the development of a risk control system, in *Risk Management and Sustainable Forestry*, Arbez, M., Birot, Y., and Carnus, J.M., Eds., EFI Proceedings, European Forest Institute, Joensuu, Finland, 45, 21, 2002.

Hansen, J. Spiecker, H., and Teuffel, K. von, Eds., *The Question of Conversion of Coniferous Forests*, Abstracts of the International Conference, 27 September–2 October 2003, Freiburg im Breisgau, Germany, Freiburger Forstliche Forschung, Berichte, 47, 85, 2003.

Hasel, K., *Forstgeschichte. Ein Grundrißür Studium und Praxis*, Paul Parey, Hamburg, Berlin, *Pareys Studientexte*, 48, 258 pp., 1985.

Hasenauer, H., Ed., Ecological and economic impacts of restoration processes in secondary coniferous forests, in *Proceedingsof the International Conference on Forest Ecosystem Restoration*, Vienna, Austria, 10–12 April 2000, 418 pp., 2000.

Heinzel, K.U. and Peters, S., Umbau von Fichten- und Kiefernreinbeständen zu naturnahen Mischwäldern., *Allgemeine Forstzeitschrift/Der Wald*, 56, 467, 2001.

Johann, E., Agnoletti, M., Axelsson, A.L., Bürgi, M., Östlund, L., Rochel, X. et al., History of secondary Norway spruce forests in Europe, in Spiecker, H., Hansen, J., Klimo, E. et al., Eds., *Norway Spruce Conversion—Options and Consequences*, EFI Research Report 18, S. Brill Academic Publishers, Leiden, 25, 2004.

Klimo, E., Hager, H., and Kulhavy, J., Eds., *Spruce Monocultures in Central Europe: Problems and Prospects*, International Workshop held in Brno, Czech Republic, 22–25 June 1998, EFI Proceedings, 33, 208, 2000.

Koch, G., Vergleich potenzieller natürlicher und aktuellerBaumartenverteilung in forstlichenProblemgebieten, in Mariabrunner—*Waldbautage 1999: UmbausekundärerNadelwälder*, Müller, F., Ed., Berichte der Forstlichen Bundesversuchsanstalt in Wien, 111, 31, 2000.

Kuhn, N., Die standörtliche Abhängigkeit der Vivian-Windwürfe, *Informationsblatt des Forschungsbereichs Landschaftsökologie*, 28, 5, 1995.

Kulikova, E.G., Phytosanitary risks, in *Risk Management and Sustainable Forestry*, Arbez, M., Birot, Y., and Carnus, J. M., Eds., EFI Proceedings, European Forest Institute, Joensuu, Finland, 45, 43, 2002.

Kusché, W., Geschichtliche Entwicklung, Umfang und Verteilungsekundärer Fichtenwälderim Alpenvorland, in *Mariabrunner—Waldbautage 1999: Umbausekundärer Nadelwälder*, Müller, F., Ed., Berichte der Forstlichen Bundesversuchsanstalt in Wien, 111, 45, 2000.

Larsen, J.B., Ecological stability of forests and sustainable silviculture, *For. Ecol. Manage.*, 73, 85, 1995.

Lüpke, B., von, Ammer, C., Bruciamacchie, M., Brunner, A., Ceitel, J., Collet, C. et al., Silvicultural strategies for conversion, in Spiecker, H., Hansen, J., Klimo, E. et al., Eds., *Norway Spruce Conversion—Options and Consequences*, EFI Research Report 18, S. Brill Academic Publishers, Leiden, 121, 2004.

Mayer, H., *Waldbau auf soziologisch-ökologischer Grundlage*, 3rd ed., Gustav Fischer, Stuttgart, 514 pp., 1984.

Nopp, U., Netherer, S., Eckmüller, O., and Führer, E., Parameters for the assessment of the predisposition of spruce-dominated forests to various disturbing factors with special regard to the 8-toothed Spruce Bark Beetle (*Ips typographus* L.), in *Criteria and Indicators for Sustainable Management at the Forest Management Unit Level*, Franc, A., Laroussinie, O., and Karjalainen, T., Eds., EFI Proceedings, European Forest Institute, Joensuu, Finland, 38, 99, 2001.

Nordfjell, T., Bacher, M., Eriksson, L., Kadlec, J., Stampfer, K., Suadicani, K. et al., Operational factors influencing the efficiency in conversion, in Spiecker, H., Hansen, J., Klimo, E. et al., Eds., *Norway Spruce Conversion—Options and Consequences*, EFI Research Report 18, S. Brill Academic Publishers, Leiden, 197, 2004.

Nyland, R.D., Even-to uneven-aged: the challenges of conversion, *For. Ecol. Manage.* 172, 291, 2003.

Schmid, S., Struktur und Entwicklung des Kleinprivatwaldes in Baden-Württemberg, *Forst und Holz*, 7, 199, 1998.

Schmidt-Vogt, H., *Die Fichte. Ein Handbuch in zwei Bänden, Taxonomie, Verbreitung, Morphologie, Ökologie, Waldgesellschaften*, Vol.I, Paul Parey, Hamburg, Berlin, 647 pp., 1977.

Schraml, U. and Härdter, U., Urbanität von Waldbesitzern und von Personen ohne Waldeigentum—Folgerungen aus einer Bevölkerungsbefragung in Deutschland, *Allgemeine Forst- und Jagdzeitung*, 7–8, 140, 2002.

Schraml, U. and Volz, K.-R., Conversion of coniferous forests—social and political perspectives. Findings from selected countries with special focus on Germany, in Spiecker, H., Hansen, J., Klimo, E. et al., Eds., *Norway Spruce Conversion—Options and Consequences*, EFI Research Report 18, S. Brill Academic Publishers, Leiden, 97, 2004.

Spiecker, H., Growth of Norway spruce (*Picea abies* [L.] Karst.) under changing environmental conditions in Europe, in *Spruce Monocultures in Central Europe: Problems and Prospects*, Klimo, E., Hager, H., and Kulhavy, J., Eds., International Workshop held in Brno, Czech Republic, 22–25 June 1998, EFI Proceedings, European Forest Institute, Joensuu, Finland, 33, 11, 2000.

Spiecker, H., Hansen, J., Klimo, E., Skovsgaard, J.P., Sterba, H., and Teuffel, K. von, Eds., *Norway Spruce Conversion—Options and Consequences*, EFI Research Report 18, S. Brill Academic Publishers, Leiden, 269 pp., 2004.

Teuffel, K. von, Heinrich, B., and Baumgarten, M., Present distribution of secondary Norway spruce in Europe, in Spiecker, H., Hansen, J., Klimo, E., Skovsgaard, J.P., Sterba, H., and Teuffel, K. von, Eds., *Norway Spruce Conversion—Options and Consequences*, EFI Research Report 18, S. Brill Academic Publishers, Leiden, 63, 2004.

Tomiczek, C., Überblick über die Forstschutzprobleme in sekundärenNadelwäldern, in *Mariabrunner—Waldbautage 1999: UmbausekundärerNadelwälder*, Müller, F., Ed., Berichte der Forstlichen Bundesversuchsanstalt in Wien, 111, 81, 2000.

Ulrich, B. and Puhe, J., Auswirkungen der zukünftigen Klimaveränderung auf mitteleuropäische Waldökosysteme und deren Rückkopplungen auf den Treibhauseffekt, in *Enquete- Kommission 'Schutz der Erdatmosphäre' des Deutschen Bundestages*, Studienprogramm, Volume 2, Wälder, Deutscher Bundestag, Ed., Economica Verlag, Bonn., III+ 208, 1994.

Yousefpour, R. and Hanewinkel, M., Balancing decisions for adaptive and multipurpose conversion of Norway spruce (*Picea abies* L. Karst) monocultures in the Black Forest area of Germany, *Forest Science*, 60, 73, 2014.

18

Restoration of Conifer Plantations in Japan: Perspectives for Stand and Landscape Management and for Enabling Social Participation

Takuo Nagaike

CONTENTS

18.1 Introduction

The primary role of plantations is timber production, and their secondary role is to reduce pressure on natural forests for timber production (Bauhus et al. 2010; Bremer and Farley 2010; Pawson et al. 2013). Moreover, this role has changed from serving a single purpose (i.e., timber production) to serving many (e.g., providing habitat for wild animals and plants) (Bauhus et al. 2010; Felton et al. 2010; Calviño-Cancela et al. 2012). Thus, the species composition and stand structure of some plantations have changed from pure simple to mixed complex forests (Nagaike 2010; Harmer et al. 2011).

In Japan, forests cover 66% of the land area, and plantations account for ~40% of these forests. *Cryptomeria japonica* (Japanese cedar) and *Chamaecyparis obtusa* (hinoki cypress), both evergreen conifers, account for 58% and 21% of trees planted, respectively. According to classification of the Food and Agriculture Organization of the United Nations, Japanese plantations are classed as "protective plantations" that are planted to protect the soil and water, to rehabilitate degraded land, and so on (Kanninen 2010), but the actual object of most plantations is to produce timber. Although the area covered by forests in Japan has remained stable over the last 50 years, there was a sharp increase in plantation area just after World War II due to the demand for timber for house building (Nagaike et al. 2005). Most of the new plantations were created by replacing existing natural forests and semi-natural grasslands. Consequently, some of these plantations were established on unsuitable sites (e.g., in areas of heavy snowfall and at high altitude) and were not productive (Masaki et al. 2004; Wada et al. 2009). Because these plantations are not able to produce high-quality timber for the construction industry, there has been a call for a program to restore them to their natural state (Nagaike et al. 2005, 2010; Nagaike 2012a).

Proper plantation management includes a schedule of tending (e.g., weeding and thinning) to produce high-quality timber. Because of low price for timber, however, most Japanese plantations have lacked appropriate tending, particularly thinning (Nagaike et al. 2005). Implementing such extensive tending would lead to issues about diminished ecological functioning (e.g., ensuring a stable water supply, habitat for wildlife, and prevention of landslides). Probably the most serious issues would be soil erosion. Because Japan has high rainfall (mean annual precipitation of 1700 mm), maintenance of the understory vegetation in plantations is vital to prevent soil erosion. Conversely, a lack of thinning creates a very dark understory in evergreen conifer plantations, resulting in sparse vegetation and increased risk of soil erosion (Miura et al. 2003; Razafindrabe et al. 2010). Over the last 10 years, to ensure appropriate plantation management, by 2013 fully 33 of 47 Japanese prefectures tax local residents to pay for ecosystem services, in a bid to maintain the forest ecosystem. Administrative bodies are using these funds to manage the untended plantations.

In 2012, Japan was only 28% self-sufficient in timber despite the fact that forests cover a high percentage (66%) of the nation's land area. An underdeveloped infrastructure (e.g., forest roads), challenging environmental factors (e.g., steep slopes) and so on underlie this low timber self-sufficiency rate. Furthermore, each individual landowner only owns a small area of private forest (Figure 18.1), which increases the cost of harvesting. For instance, if a small forest owner wants to harvest, constructing a forest road would require consultation with many neighboring forest owners to gain approval for the road. These obstacles to active forest management led to the establishment in 2011 of the Forestry Revival Plan, led by the Forestry Agency of Japan, which was established to consolidate the small private forests into a single aggregated management area, thus enabling forestry activities (i.e., effective timber production) to be intensified by reducing logging and extraction costs. The application of these measures is intended to increase the national timber self-sufficiency rate to 50% by 2020.

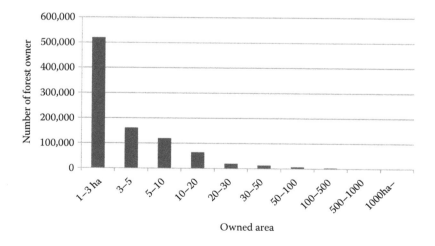

FIGURE 18.1
Frequency distribution of the area of private forests in Japan. Source: Data from Ministry of Agriculture, Forestry, and Fisheries, Japan, "2010 World Census of Agriculture and Forestry," http://www.rinya.maff.go.jp/j/kikaku/hakusyo/25hakusyo_h/all/other/a38_02.xls.

People and social interaction have become involved in forestry, with ecosystems and social systems becoming more closely linked (Emborg et al. 2012). In Japan, social interest in forest management has increased, along with forest preservation, particularly from the perspective of corporate social responsibility that affects private companies. This not only includes planting trees but also a host of forest management issues (e.g., tending and preserving the natural fauna) and social participation in drawing up forestry management plans.

This chapter is focused on ecological restoration activities in Japanese plantations. It first offers a review of the restoration of the stand- and landscape-level in conifer plantations. Forest restoration is considered with particular attention to social participation in the restoration process. To illustrate the issues mentioned, two projects are highlighted: (1) the AKAYA project, which aims to restore biodiversity in national forests through collaboration among a wide range of stakeholders and (2) the Mt. Fuji reforestation project, a collaborative forest restoration and management scheme formed by various related sectors (e.g., environmental nongovernmental organization, forest owners, and supporting firms).

18.2 Restoration of the Stand- and Landscape-Level in Conifer Plantations

Approaches to restoration have focused on productivity and biodiversity and have included considerations of secondary forest management, enrichment planting, agroforestry, monoculture plantations with buffers, mosaics of monocultures, and mixed-species plantations (Wilson et al. 2012). In Japan, the emphasis has been on increasing the development of the understory because of the large areas of untended evergreen conifer plantations. Some landowners seek to maximize timber production; others want to maximize the provision of ecosystem services; and still others want to achieve both (Lamb et al. 2012). Japanese forest restoration activities have focused primarily on enhancing ecological functions in plantations.

To establish naturally regenerated mixed-forest stands, a comprehensive research project was undertaken during 2007–2011 entitled "Developing of regeneration technology for leading artificial forests to broadleaf forests." It was led by the Forestry and Forest Products Research Institute of Japan. Based on research outcomes, three recommendations for the successful ecological restoration of plantations were offered: (1) the importance of advance regeneration, (2) consideration of past land use, and (3) spatial arrangement. They recommended first identifying stands that were easy to restore. It was important to identify the factors related to the abundant number of mature trees in the targeted plantations. The key factors included distance from mother tree seed sources and trees growing within the expected seed dispersal distances from *C. obtusa* (Kondo et al. 2007) and *C. japonica* (Kodani 2006; Gonzalez and Nakashizuka 2010) plantations. The distance from the seed source was associated with management intensity. Nagaike et al. (2012) examined the interactive effects of management and distance from seed source (i.e., broadleaf natural forests) on species composition of naturally regenerated trees species in *C. japonica* plantations (i.e., more intensive management done closer to seed source). The interactive effects of the distance from the seed source and thinning on species diversity and hardwood abundance in *C. japonica* plantations in northern Japan did not change the species diversity of the seedlings and saplings, probably due to distance-independent improvements in environmental conditions as a result of thinning (Utsugi et al. 2006).

In Japan, stands are usually replanted after clear-cutting. However, since the cost of replanting is high (e.g., site preparation, cost of saplings, labor costs), natural regeneration or failure to replant occurs. Consequently, studies on natural regeneration in clear-cut sites have increased (e.g., Noguchi and Okuda 2012; Takahashi et al. 2013). According to Yamagawa and Ito (2006), naturally regenerated trees in a clear-cut *C. obtusa* plantation were mostly the result of sprouting from stumps and not from seeds dispersed after cutting. Smaller sites with adjacent evergreen and broadleaved forests and a longer interval after clear-cutting were factors that enabled successful regeneration of broadleaved trees (Nagashima et al. 2009). Soil seed banks could contribute to the recovery of vegetation after clear-cutting in *C. obtusa* plantations, but did not assist the recovery of late-successional species (Sakai et al. 2010). After clear-cutting of conifer plantations, tree species found it difficult to regenerate (Saito et al. 2006). Thus, if only late-successional tree species were the targets for restored stands, it would be difficult to monitor and check the successful regeneration of these stands (Nagaike 2010). Differences in past land use would also significantly affect the composition of the naturally regenerated tree species in *C. japonica* plantations (Ito et al. 2004; Sugita et al. 2008).

Thinning in plantations is mainly to reduce resource competition among the overstory trees. Nevertheless, thinned stands with improved light conditions offer the possibility of regenerated tree species as a second-order effect. In Japan, low or crown thinning have traditionally been carried out; however, more recently mechanical thinning (particularly line or strip thinning) have become popular as they are more cost-effective. Low or crown thinning provides an opportunity for the regeneration of other tree species in plantations of *Larix kaempferi* (Hanada et al. 2006), *C. obtusa* (Noguchi et al. 2011), *Abies sachalinensis* (Nonoda et al. 2008), and *C. japonica* (Seiwa et al. 2009; Hirata et al. 2011). Seiwa et al. (2012) showed that intensive thinning in *C. japonica* plantations resulted in regeneration of more early- and mid-successional tree species but no late-successional tree species. Line or strip thinning usually opens up sites more than does low or crown thinning and was more effective in establishing natural regeneration in *C. japonica* (Ito et al. 2006; Ishii et al. 2008; Hirata et al. 2011), *A. sachalinensis* (Kon et al. 2007), and *C. obtusa* (Sakuta et al. 2009) plantations. However, abundant advance regeneration is a precondition for successful natural regeneration.

Extensive management of plantations on long rotation has become popular in Japan; by delaying the need to regenerate stands, owners are seeking to avoid the high cost of replanting after harvesting (Masaki et al. 2006; Nishizono et al. 2008). A long-rotation *C. japonica* plantation has a multilayered structure, with planted trees in the upper layer and naturally regenerated broadleaves in the lower layers (Suzuki et al. 2005). These long-rotation plantations could play an important role in developing complex stand structure and diverse species composition (Busing and Garman 2002; Nagaike et al. 2010). Furthermore, forest restoration should focus not only on stands but also on the surrounding landscape (Fischer et al. 2006; Lindenmayer et al. 2006), and an appropriate distribution of long-rotation plantations in a forested landscape is one strategy for forest landscape restoration (Stanturf et al. 2012). To date, however, Japanese authorities have only focused on landscape management for timber production in their forestry revival plans that consolidate small private holdings and that offer the potential for landscape management planning (Figure 18.2). Therefore, restoration targets should focus not only on timber production but also on biodiversity conservation (Ciccarese et al. 2012). By combining many small forest parcels into an aggregated management area, the triad approach using functional zoning could be realized, producing zones of intensive management, complex adaptive system management, and protected areas (Seymour and Hunter 1999; MacLean et al. 2009). However, such an approach is lacking in current forest planning.

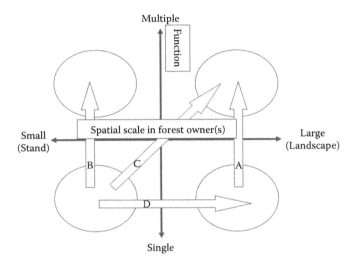

FIGURE 18.2
Schematic diagram of relationship between size of forest ownership and function. Large forest owner: Landscape management which demonstrated that multiple functions could be attainable by appropriate spatial arrangement of composed stands (Arrow A). Small forest owner: Owners could select from the following three alternatives for function. Arrow B: Since small forest owners do not consolidate surrounding stands, multifunction should be effective even in small stands, but its feasibility might be low. Arrow C: Many small forest owners consolidate surrounding stands. As a result, landscape management, which has demonstrated multiple functions from ecological forestry, such as the triad approach, could be attained by appropriate spatial arrangement of stands. Arrow D: Many small forest owners consolidate surrounding stands. Landscape management with multiple functions was not attainable, although cost-effectiveness of timber production could be improved.

18.3 Social Participation in Forest Restoration in Japan

Social participation is an increasing concern for forest management, particularly tree planting, in Japan. Interest in protective functions of coastal forests was heightened after the tragic tsunami disaster on March 11, 2011, in eastern Japan. Restoring plantations for protection from blowing sand near the coastline is of great interest to local people and urban residents. Recently, such activity has broadened from only planting to all aspects of forest management including collaborative planning between forest owners and other stakeholders. I would like to illustrate these developments with two case studies. First, a collaborative forest management and planning scheme is presented that involved the Forestry Agency of Japan, local residents, and an environmental nongovernmental organization, the Nature Conservation Society of Japan. This project focused mainly on using natural processes for ecological restoration of plantations in the project area that was inhabited by an endangered raptor. The second case study is about plantations damaged by an insect outbreak as the starting point. Ecological restoration and collaborative forest management was undertaken among Yamanashi Prefectural Forest, an interested private firm, and The Organization for Industrial, Spiritual and Cultural Advancement (OISCA), an environmental nongovernmental organization.

18.3.1 The AKAYA Project

The AKAYA project takes its name from its location in the Akaya district, Minakami town, Gunma Prefecture, central Japan (Figure 18.3). The project aimed to restore biodiversity in a model system representative of the Japanese national forests (~10,000 ha) through the collaboration of stakeholders. The project entailed collaboration among three core organizations: the Akaya Project Regional Conference organized by local citizens, the Nature Conservation Society of Japan, and the Kanto Regional Forest Office of the Forestry Agency of Japan. Resource management would only be successful if it was ecologically possible, technically feasible, and socially acceptable (Oliver et al. 2012). Specific objectives of the project included drawing up a forest management plan with stakeholders (e.g., local residents) and accomplishing ecological landscape management that encompassed conserving biological diversity, timber production, and the needs of the local residents. The Japanese golden eagle (*Aquila chrysaetos*) and the Japanese mountain hawk eagle (*Nisaetus nipalensis orientalis*), both endangered species as designated on the Red List of Japan, inhabit the project area. Since the reproduction rate of these species is low, it has been necessary to improve their foraging conditions. This area is in a region of high snowfall (>2 m winter) and mostly unsuitable for timber production. An experimental thinning of the *C. japonica* plantations was installed to enable the understory to grow and consequently increase numbers of prey (e.g., hares). Before logging, a large number of plots were surveyed, and assessments were made to identify stands that would be easier to convert (Nagaike et al. 2012). Experimental logging was undertaken determine effective dispersal distance from seed sources of tree species (Figure 18.4). The experimental stands were strip-cut (20 × 100 m, 40 × 250 m) in 2011 and have been studied and monitored for distance from seed sources for the following for species and conditions: natural regeneration of tree seedlings, vascular plants, mammals, bats, carabid beetles, birds, and light conditions.

The AKAYA project

The Mt. Fuji reforestation project

FIGURE 18.3
Map showing the location of the two case studies.

FIGURE 18.4
Experimental strip-cut site of the AKAYA project. Forest owner (national forests) and scientific advisory committee discussed ways to maintain low density of *Cervus nippon* (sika deer) that favors such open sites. The project anticipates obtaining natural regeneration and consequently recovering biological diversity in the *Cryptomeria japonica* plantations.

18.3.2 The Mt. Fuji Reforestation Project

Abies veitchii plantations are widely distributed on the northern slopes of Mt. Fuji in Narusawa Village, Yamanashi Prefecture, central Japan (Figure 18.3). In 2002, an outbreak of the *Epinotia piceae* moth caused severe damage to >100 ha of forest. Since these outbreaks occur every 8 years, plantations surrounding the severely damaged areas were converted from single-species to mixed stands by strip thinning and soil scarifying, as a precaution (Figure 18.5). Natural regeneration of broadleaved trees in the strip-cuts was expected but failure was the initial outcome, due mainly to of deer browsing on the young saplings and a lack of sufficient seed trees.

The OISCA took the initiative to plant trees in the strip-cuts (width of about 10 m), in collaboration with private companies, including large enterprises. Forests in the surrounding area were surveyed before planting, to determine the appropriate species for planting. The dominant species in that area, *Fagus crenata*, *Quercus crispula*, *Acer palmatum*, *Cerasus jamasakura*, and *Alnus japonica*, were selected for planting. To restore the forests and to avoid contamination of genetic diversity, seedlings were produced from seeds taken from local sources. Planting started in 2007; over the next 4 years, approximately 5100 people (mostly enterprise employees) planted 38,900 seedlings over 38.4 ha. Because of a high sika deer (*Cervus nippon*) population in the area, each seedling was protected by a biodegradable plastic tube, which would decompose within 5 years. Planting density was low (1000 seedlings ha^{-1}) compared to conventional conifer planting density (2300–3000 seedlings ha^{-1}). We expected this planted area to act as a hub for the restoration of the forests with these

FIGURE 18.5
Aerial photo of the Mt. Fuji reforestation project area. Center of the photo was clear-cut because of heavy damage by insect outbreak. Surrounding the strip-cut site was the main project area for planting and subsequent collaborative management.

species (Corbin and Holle 2012). However, the mortality rate of the planted seedlings was relatively high (24%–34%) because of strong wind from Mt. Fuji and poor soil conditions, making future enrichment planting necessary (Gómez-Aparicio et al. 2009). Also, in some planted sites, naturally regenerated *Larix kaempferi* and *Alnus hirsute* have been successful. How to accommodate the planted and naturally regenerated trees to compensate for high mortality is one of the future challenges.

18.4 Future Research Needs

Forestry operations affect herbivore dynamics (e.g., Edenius et al. 2011). In Japan, the numbers of sika deer have increased sharply (Takatsuki 2009; Iijima et al. 2013), resulting in substantial damage to forests, agricultural land, and natural vegetation (e.g., Nagaike and Hayashi 2003; Takatsuki 2009; Nagaike 2012b). Open sites created by forestry operations (e.g., clear-cuts) are suitable foraging sites for sika deer. Thinned stands and clear-cuts provide temporary open spaces that create a more suitable habitat for early successional or light-demanding species, which include vulnerable and endangered species (Yamaura et al. 2012). But these open spaces could also increase deer density. Shimada and Nonoda (2009) showed that natural regeneration establishing after thinning in *C. japonica* and *C. obtusa* plantations were severely browsed by sika deer. Thus, careful consideration is needed when thinning or harvesting where there is a high sika deer density. The vegetation in clear-cuts with abundant sika deer is biased towards grassland species rather than forest species (Nagashima et al. 2011). Where deer density is low, abundant understory vegetation initially would compete with woody regeneration. Thinning in particular would facilitate the growth of the understory, especially *Sasa* spp., and compete with new seedlings (Nagai and Yoshida 2006).

Forest restoration could contribute to mitigating the effects of climate change (Ciccarese et al. 2012) and further adaptation. The "Socio-ecological Model" aims at defining the coupled relationship between humans and nature. Restoration at stand- and landscape-levels should expand the scope of actions available to landowners, managers, and policymakers. Because these collaborative methods have not yet been studied in Japan, they should be the subject of essential future research.

References

Bauhus, J., P. van der Meer, and M. Kanninen, eds, 2010. *Ecosystem Goods and Services from Plantations Forests*. Earthscan, London.

Busing, R. T. and S. L. Garman. 2002. Promoting old-growth characteristics and long-term wood production in Douglas-fir forests. *For Ecol Manage* 160:161–175.

Bremer, L. L. and K. A. Farley. 2010. Does plantation forestry restore biodiversity or create green deserts? A synthesis of the effects of land-use transition on plant species richness? *Biodiv Cons* 19:3893–3915.

Calviño-Cancela, M., M. Rubido-Bará, and E. J. B. van Etten. 2012. Do eucalypt plantations provide habitat for native forest biodiversity? *For Ecol Manage* 270:153–162.

Ciccarese, L., A. Mattsson, and D. Pettenella. 2012. Ecosystem services from forest restoration: Thinking ahead. *New For* 43:543–560.

Corbin, J.D. and K. D. Holl. 2012. Applied nucleation as a forest restoration strategy. *For Ecol Manage* 265:37–46.

Edenius, L., G. Ericsson, G. Kempe, R. Bergström, and K. Danell. 2011. The effects of changing land use and browsing on aspen abundance and regeneration: A 50-year perspective from Sweden. *J Appl Ecol* 48:301–309.

Emborg, J., G. Walker, and S. Daniels. 2012. Forest landscape restoration decision-making and conflict management: Applying discourse-based approaches. In *Forest Landscape Restoration: Integrating Natural and Social Science*. J. Stanturf, D. Lamb, and P. Madsen, eds., 131–153, Springer, Berlin.

Felton, A., M. Lindbladh, J. Brunet, and Ö. Fritz. 2010. Replacing coniferous monocultures with mixed-species production stands: An assessment of the potential benefits for forest biodiversity in northern Europe. *For Ecol Manage* 260:939–947.

Fischer, J., D. B. Lindenmayer, and A. D. Manning. 2006. Biodiversity, ecosystem function, and resilience: Ten guiding principles for commodity production landscapes. *Front Ecol Environ* 4:80–86.

Gómez-Aparicio, L., M. A. Zaval, F. J. Bonet, and R. Zamora. 2009. Are pine plantations valid tools for restoring Mediterranean forests? An assessment along abiotic and biotic gradients. *Ecol Appl* 19:2124–2141.

Gonzales, R. S. and T. Nakashizuka. 2010. Broad-leaf species composition in *Cryptomeria japonica* plantations with respect to distance from natural forest. *For Ecol Manage* 259:2133–2140.

Hanada, N., M. Shibuya, H. Saito, and K. Takahashi. 2006. Regeneration process of broadleaved trees in planted *Larix kaempferi* forests. *J Jpn For Soc* 88:1–7 (In Japanese with English summary).

Harmer, R., G. Morgan, and K. Beauscamp. 2011. Restocking with broadleaved species during the conversion of *Tsuga heterophylla* plantations to native woodland using natural regeneration. *Eur J For Res* 130:161–171.

Hirata, A., T. Sakai, K. Takahashi, T. Sato, H. Tanouchi, H. Sugita, and H. Tanaka. 2011. Effects of management, environment and landscape conditions on establishment of hardwood seedlings and saplings in central Japanese coniferous plantations. *For Ecol Manage* 262:1280–1288.

Iijima, H., T. Nagaike, and T. Honda. 2013. Estimation of deer population dynamics using a Bayesian state-space model with multiple abundance indices. *J Wildlife Manage* 77:1038–1047.

Ishii, H. T., M. A. Maleque, and S. Taniguchi. 2008. Line thinning promotes stand growth and under-story diversity in Japanese cedar (*Cryptomeria japonica* D. Don) plantations. *J For Res* 13:73–78.

Ito, S., S. Ishigami, N. Mizoue, and G. P. Buckley. 2006. Maintaining plant species composition and diversity of understory vegetation under strip-clearcutting forestry in conifer plantations in Kyusyu, southern Japan. *For Ecol Manage* 231:234–241.

Ito, S., R. Nakayama, and G. P. Buckley. 2004. Effects of previous land-use on plant species diversity in semi-natural and plantations forests in a warm-temperate region in southwestern Kyusyu, Japan. *For Ecol Manage* 196:213–225.

Kanninen, M. 2010. Plantation forests: Global perspectives. In *Ecosystem Goods and Services from Plantations Forests*. J. Bauhus, P. van der Meer, and M. Kanninen, eds., 1–15, Earthscan, London.

Kodani, J. 2006. Species diversity of broad-leaved trees in *Cryptomeria japonica* plantaitons in relation to the distance from adjacent broad-leaved forests. *J For Res* 11:267–274.

Kon, H., I. Watanabe, and M. Yasaka. 2007. Effect of thinning on the natural regeneration of broad-leaved trees in *Abies sachalinensis* plantations. *J Jpn For Soc* 89:395–400 (In Japanese with English summary).

Kondo, M., Y. Mitsuda, S. Yoshida, N. Mizoue, and T. Murakami. 2007. Prediction of existence prob-ability of Abies firma seedlings in cypress (*Chamaecyparis obtusa*) plantations in Kirishima. *J Jpn For Soc* 89:407–411 (In Japanese with English summary).

Lamb, D., J. Stanturf, and P. Madsen. 2012. What is forest landscape restoration? In *Forest Landscape Restoration: Integrating Natural and Social Science*. J. Stanturf, D. Lamb, and P. Madsen, eds., 3–23, Springer, Berlin.

Lindenmayer, D. B., J. F. Franklin, and J. Fischer. 2006. General management principles and a check-list of strategies to guide forest biodiversity conservation. *Biol Cons* 131:433–445.

Masaki, T., T. Ohta, H. Sugita, H. Oohara, T. Otani, T. Nagaike, and S. Nakamura. 2004. Structure and dynamics of tree populations within unsuccessful conifer plantations near the Shirakami Mountains, a snowy region of Japan. *For Ecol Manage* 194:389–401.

Masaki, T., S. Mori, T. Kajimoto, G. Hitsuma, S. Sawata, M. Mori, K. Osumi, S. Sakurai, and T. Seki. 2006. Long-term growth analyses of Japanese cedar trees in a plantation: Neighborhood com-petition and persistence of initial growth deviations. *J For Res* 11:217–225.

MacLean, D. A., R. S. Seymour, M. K. Montigny, and C. Messier. 2009. Allocation of conservation efforts over the landscape: The TRIAD approach. In *Setting Conservation Targets for Managed Forest Landscapes*. M. A. Villard and B. G. Jonsson, eds., 283–303, Cambridge University Press, Cambridge.

Miura, S., S. Yoshinaga, and T. Yamada. 2003. Protective effect of floor cover against soil erosion on steep slopes forested with *Chamaecyparis obtusa* (hinoki) and other species. *J For Res* 8:27–35.

Noguchi, M., S. Okuda, K. Miyamoto, T. Itou, and Y. Inagaki. 2011. Composition, size structure and local variation of naturally regenerated broadleaved tree species in hinoki cypress plantations: A case study in Shikoku, south-western Japan. *Forestry* 84:493–504.

Nonoda, S., H. Shibuya, H. Saito, S. Ishibashi, and M. Takahashi. 2008. Invasion and growth pro-cesses of natural broadleaved trees and influences of thinning on the processes in an *Abies sachalinensis* plantation. *J Jpn For Soc* 90:103–110 (In Japanese with English summary).

Nagai, M. and T. Yoshida. 2006. Variation in understory structure and plant species diversity influ-enced by silvicultural treatments among 21- to 26-year-old *Picea glehnii* plantations. *J For Res* 11:1–10.

Nagaike, T. 2010. Effects of altitudinal gradient on species composition of naturally regenerated trees in *Larix kaempferi* plantations in central Japan. *J For Res* 15:65–70.

Nagaike, T., T. Yoshida, H. Miguchi, T. Nakashizuka, and T. Kamitani. 2005. Rehabilitation for spe-cies enrichment in abandoned coppice forests in Japan. In *Restoration of Boreal and Temperate Forests*. J.A. Stanturf and P. Madsen, eds., 371–381, CRC Press, FL.

Nagaike, T., A. Hayashi, and M. Kubo. 2010. Diversity of naturally regenerating tree species in the overstory layer of *Larix kaempferi* plantations and abandoned broadleaf coppice stands in central Japan. *Forestry* 83:285–291.

Nagaike, T. 2012a. Review of plant species diversity in managed forests in Japan. *ISRN Forestry* 2012: Article ID 629523, 7 pages. doi:10.5402/2012/629523.

Nagaike, T. 2012b. Effects of browsing by sika deer (*Cervus nippon*) on subalpine vegetation at Mt. Kita, central Japan. *Ecol Res* 27:467–473.

Nagaike, T., T. Fujita, S. Dejima, T. Chino, S. Matsuzaki, Y. Takanose, and K. Takahashi. 2012. Interactive influences of distance from seed source and management practices on tree species composition in conifer plantations. *For Ecol Manage* 283:48–55.

Nagaike, T. and A. Hayashi. 2003. Bark-stripping by Sika deer (*Cervus nippon*) in *Larix kaempferi* plantations in central Japan. *For Ecol Manage* 175:563–572.

Nagaike, T., E. Ohkubo, and K. Hirose. 2014. Vegetation recovery in response to the exclusion of grazing by sika deer (*Cervus nippon*) in seminatural grassland on Mt. Kushigata, Japan. *ISRN Biodiversity* 2014: Article ID 493495, 6 pageshttp://dx.doi.org/10.1155/2014/493495.

Nagashima, K., K. Omoto, and S. Yoshida. 2011. The patterns and factors of vegetation recovery at abandoned plantation clearcut sites in Kyushu region: Implication for management. *J Jpn For Soc* 93:294–302 (In Japanese with English summary).

Nagashima, K., S. Yoshida, and T. Hosaka. 2009. Patterns and factors in early-stage vegetation recovery at abandoned plantation clearcut sites in Oita, Japan: Possible indicators for evaluating vegetation status. *J For Res* 14:135–146.

Nishizono, T., K. Tanaka, K. Hosoda, Y, Awaya, and Y. Oishi. 2008. Effects of thinning and site productivity on culmination of stand growth: Results from long-term monitoring experiments in Japanese cedar (*Cryptomeria japonica* D. Don) forests in northeastern Japan. *J For Res* 13:264–274.

Noguchi, M. and S. Okuda. 2012. Changes in stand structure and species composition from 5 to 11 years after clear-cutting of a sugi plantation in the warm-temperate zone in Shikoku, Japan. *J Jpn For Soc* 94:192–195 (In Japanese with English summary).

Oliver, C. D., K. Covey, A. Hohl, D. Larsen, J. B. McCater, A. Niccolai, and J. Wilson. 2012. Landscape management. In *Forest Landscape Restoration: Integrating Natural and Social Science*. J. Stanturf, D. Lamb, and P. Madsen, eds., 39–65, Springer, Berlin.

Pawson, S. M., A. Brin, E. G. Brockerhoff, D. Lamb, T. W. Payn, A. Paquette, and J. A. Parrotta. 2013. Plantation forests, climate change and biodiversity. *Biodiv Cons* 22:1203–1227.

Razafindrabe, B. H. N., H. Bin, S. Inoue, T. Ezaki, and R. Shaw. 2010. The role of forest stand density in controlling soil erosion: Implications to sediment-related disasters in Japan. *Env Monit Assess* 160:337–354.

Saito, S., N. Inoue, R. Noda, Y. Yamada, K. Saho, T. Takamiya, K. Yokoo et al. 2006. Tree densities and prediction models for plantations and reforestation-abandoned sites after clear cutting in Kyushu Island, Japan. *J Jpn For Soc* 88:482–488 (In Japanese with English summary).

Sakai, A., T. Sakai, S. Kuramoto, and S. Sato. 2010. Soil seed banks in a mature hinoki (*Chamaecyparis obtusa* Endl.) plantation and initial process of secondary succession after clearcutting in southwestern Japan. *J For Res* 15:316–328.

Sakuta, K., S. Taniguchi, A. Inoue A, and N. Mizoue. 2009. Effects of strip-cutting on stand floor micro climate and tree-species diversity in a Japanese cypress plantation. *J Jpn For Soc* 91:86–93 (In Japanese with English summary).

Seiwa, K., M. Ando, A. Imaji, M. Tomita, and K. Kanou. 2009. Spatio-temporal variation of environmental signals inducing seed germination in temperate conifer plantations and natural hardwood forests in northern Japan. *For Ecol Manage* 257:361–369.

Seiwa, K., Y. Etoh, M. Hisita, K. Masaka, K. Imaji, N. Ueno, Y. Hasegawa et al. 2012. Roles of thinning intensity in hardwood recruitment and diversity in a conifer, *Cryptomeria japonica* plantation: A 5-year demographic study. *For Ecol Manage* 269:177–187.

Seymour, R. S. and M. L. Hunter, Jr. 1999. Principles of ecological forestry. In *Managing Biodiversity in Forest Ecosystems*. M. L. Hunter, Jr., ed., 22–61, Cambridge University Press, Cambridge.

Shimada, H. and T. Nonoda. 2009. Effects of deer browsing on broad-leaved tree invasion after heavy thinning in conifer plantations. *J Jpn For Soc* 91:46–50 (In Japanese with English summary).

Stanturf, J., D. Lamb, and P. Madsen, eds. 2012. *Forest Landscape Restoration: Integrating Natural and Social Science*. Springer, Berlin.

Sugita, H., T. Kunisaki, T. Takahashi, and R. Takahashi. 2008. Effects of previous forest types and site conditions on species composition and abundance of naturally regenerated trees in young *Cryptomeria japonica* plantations in northern Japan. *J For Res* 13:155–164.

Suzuki, W., T. Suzaki, T. Okumura, and S. Ikeda. 2005. Aging-induced development patterns of *Chamaecyparis obtusa* plantations. *J Jpn For Soc* 87:27–35 (In Japanese with English summary).

Takahashi, Y., M. Hasegawa, K. Zushi, and H. Aiura. 2013. Dynamics of the seedlings in the early phase of natural regeneration after clear cutting in *Cryptomeria japonica* plantations in Toyama prefecture. *J Jpn For Soc* 95:182–188 (In Japanese with English summary).

Takatsuki, S. 2009. Effects of sika deer on vegetation in Japan: A review. *Biol Cons* 142:1922–1929.

Utsugi, E., H. Kanno, N. Ueno, M. Tomita, T. Saitoh, M. Kimura, K. Kanou, and K. Seiwa. 2006. Hardwood recruitment into conifer plantations in Japan: Effects of thinning and distance from neighboring hardwood forests. *For Ecol Manage* 237:15–28.

Wada, S., T. Kaneko, T. Yagihashi, and H. Sugita. 2009. Factors affecting the success of *Cryptomeria japonica* plantations and regeneration of hardwoods in plantations in a snowy region of northern Japan. *J Jpn For Soc* 91:79–85 (In Japanese with English summary).

Wilson, K. A., M. Lulow, J. Burger, and M. F. McBride. 2012. The economics of restoration. In *Forest Landscape Restoration: Integrating Natural and Social Science*. J. Stanturf, D. Lamb, and P. Madsen, eds., 215–231, Springer, Berlin.

Yamagawa, H. and S. Ito. 2006. The role of different sources of tree regeneration in the initial stages of natural forest recovery after logging of conifer plantation in a warm-temperate region. *J For Res* 11:455–460.

Yamaura, Y., H. Oka, H. Taki, K. Ozaki, and H. Tanaka. 2012. Sustainable management of planted landscapes: Lessons from Japan. *Biodiv Cons* 21:3107–3129.

19

Restoration of Open Oak Woodlands in Mediterranean Ecosystems of Western Iberia and California

Mariola Sánchez-González, Guillermo Gea-Izquierdo, Fernando Pulido, Vanda Acácio, Doug McCreary, and Isabel Cañellas

CONTENTS

19.1 Introduction: Mediterranean Open Oak Woodlands

Mediterranean ecosystems are characterized by a period of water deficit in summer, which determines survival and reproduction of the plants living in this environment. Open woodlands dominated by an overstory of native *Quercus* species (oaks) are common landscape features in both California and the Iberian Peninsula. These woodlands typically consist of one to several species of native oaks and an understory of predominantly annual plants, principally grasses. Shrublands may or may not be present depending on past management history and environmental conditions. These oak woodlands are biodiversity-rich habitats (Díaz 2013), and priority ecoregions for global conservation (Olson 2002), as well as protected ecosystems under the Pan-European network "Natura 2000."

In California, more than half of 600 species of terrestrial vertebrates utilize oak woodlands at some time during the year (CIWTG 2005). In the Iberian Peninsula, Mediterranean open oak woodlands reach levels of 60–100 flowering plant species per 0.1 ha (Díaz 2003). The tree layer is important for specialized tree-dwelling species in one or more phases of their life cycles, the removal of trees resulting in significant species losses (Aragón 2010). Oaks are also critical in protecting watersheds and ensuring the quality of water resources; they anchor the soil, preventing erosion and sedimentation. Oak woodlands also provide the majority of forage that supports the local livestock industry, as well as acorns, an important food source for wildlife and livestock in the Iberian Peninsula. Finally, oak trees are also very desirable for recreation, including hunting and fishing, as well as growing recreational activities.

Open oak woodlands in Iberia and California are threatened by the lack of natural tree regeneration, which is strongly related to management practices, especially stocking rates of a variety of livestock species (Pulido 2005; Pausas 2009; Plieninger 2009 for Iberia; McCreary 2001 for California). Indeed, these oak woodlands have been transformed throughout history, with prolonged use often resulting in tree loss. Hence, oak woodlands require some intervention for ensuring tree maintenance in the long term. Such concern has led to efforts for developing oak restoration practices both in Iberia and California.

Different approaches have been used in California and Iberia to address the "regeneration problem," as oak woodlands have been managed more intensively in the latter. First, Iberian open woodlands provide highly valuable products, including forage for livestock and game, acorns for fattening hogs, and cork from cork oak trees (*Quercus suber* L.). Concer due to inadequate regeneration have triggered efforts to increase woodland cover through government-funded planting schemes in Spain and Portugal. However, in California marketable products derived from oak woodlands, such as extensive grazing and firewood, are less valued. As a result, California ranches have been the subject of limited tree restoration projects which have been promoted mainly by conservation trusts. In this chapter, we will discuss restoration techniques in relation to historical management and environmental threats of managed open oak woodlands in western Iberia and California.

19.2 California's Open Oak Woodlands

19.2.1 Distribution

The approximately 3 million ha of woodlands in California (Bolsinger 1988) are slow-growing hardwoods unsuitable for timber production. Open oak woodlands distribution is highly dispersed along the foothills of the Sierra Nevada, and Coast Ranges, interspersed with shrublands and dense forest ecosystems (Elena-Roselló 2013) (Figure 19.1, modified from (Standiford 1992). All these woodlands have a Mediterranean climate. The 20 native species of oak in California (Nixon 2002) are from three subgenera, including white oaks

FIGURE 19.1
Distribution of open oak woodlands in (a) California and (b) Iberia. (Modified from Standiford, R.B. and R.E. Howitt. 1992. *American Journal of Agricultural Economics* 74:421–433.) *(Continued)*

(Section *Quercus*), black or red oaks (Section *"Lobatae"*), and intermediate oaks (Section *"Protobalanous"*). The average rainfall within California's oak woodlands varies from 300 mm in the southern interior regions to over 1000 mm in northern coastal woodlands. The wetter portions of these woodlands have higher tree density, and in lower rainfall areas, oak woodlands grade into savannahs with tree density often less than 12 trees ha^{-1}.

19.2.2 Ownership, Historic Uses, and Values

More than 70% of California's oak woodlands are privately owned (Bolsinger 1988), in contrast to higher elevation coniferous forests that are mostly publicly owned. Many of the oak woodlands are unsuitable for intensive agriculture due to uneven terrain, poor soils, and inadequate moisture. Livestock grazing has been the main economic use of California's oak woodlands since the introduction of domestic herd animals in the 1700s. Livestock have impacted the oak woodlands by reducing organic matter, compacting soil and dispersing exotic annual plants from seeds brought from Europe in hay bales

(b)

FIGURE 19.1 (*Continued*)
Distribution of open oak woodlands in (a) California and (b) Iberia. (Modified from Standiford, R.B. and R.E. Howitt. 1992. *American Journal of Agricultural Economics* 74:421–433.)

and grain sacks that supplied Spanish settlements (Pavlik 1991). These introduced plants probably displaced many native perennials. Introduced annual grasses such as wild oats (*Avena fatua* L.), ripgut brome (*Bromus diandrus* Roth.), and Italian rye (*Lolium multiflorum* Lam.) dominate the understory of most of the oak woodlands in California today.

Indigenous people populated California's woodlands for approximately 10,000 years prior to the arrival of Europeans. These woodlands may have supported the highest population densities on the North American continent. Native Americans actively managed woodlands principally by their use of fire (Blackburn and Anderson 1993). Oaks were revered by California's native people and figured prominently in their culture. Acorns were a main food source for many tribal groups and are still consumed today.

Oaks continued to be important to later residents. The deep and endearing value of oaks in the psyche of early-settlers is apparent in many cities and landmarks that carry oak or the Spanish equivalent "encina" and "roble" in their names. There are even towns named "Dehesa" and "Bellota" the latter meaning acorn in Spanish. Today, golden brown hills dotted with gnarled oak trees epitomize the California landscape, and native oaks symbolize values that are held dear: strength, beauty, adaptability, and longevity.

The value of oaks in California is more than esthetic; oaks and oak woodlands support a diverse fauna. More than half of the 600-plus species of terrestrial vertebrates in California utilize oak woodlands during the year for essential food and shelter. Oaks are critical in protecting watersheds and ensuring the quality of water resources Oak trees anchor the soil, preventing erosion and sedimentation. More than 70% of California's oak woodlands are privately owned (Bolsinger 1988). Livestock grazing has been the main economic use of California's oak woodlands since the introduction of domestic herd animals in the 1700s. Livestock have impacted the oak woodlands by reducing organic matter, compacting soil, and dispersing exotic annual plants from seeds brought from Europe (Pavlik 1991). These introduced plants probably displaced many native perennials.

19.2.3 Oak Woodland Losses

Although there are few records to indicate the total area of oak woodlands before European settlement, the best estimate is that, roughly, half of the presettlement area has been converted to other uses (Burcham 1981). The original settlers converted substantial areas of oak woodland to agriculture, especially in the Central Valley where deep, fertile, alluvial soils that supported valley oak (*Q. lobata* Née) forests were converted to orchards and other crops. During the Gold Rush, oaks and other trees were important for firewood and for mining timbers, being also harvested for fuel for steamships, railroads, and industrial purposes. A large charcoal business that utilized native oaks flourished in the nineteenth century.

Loss of oak woodlands continued into the twentieth century. From the end of World War II until 1975, approximately 800,000 ha of woodlands and savannah were converted to tree-less pastures in order to promote greater forage production for livestock (Bolsinger 1988), often supported by government grants. In the 1970s, the sharp increase in the cost of petroleum increased the demand for oak firewood for home heating. Thousands of hectares were cut, primarily in the upper Sacramento Valley in north-central California. Today, pressures on oak woodlands continue. Approximately 3000 ha were converted annually for residential and commercial development in the 1970s and 1980s (Bolsinger 1988), and development pressure continues. In the 1990s the area in vineyards increased, largely at the expense of oak woodlands. In the same period, a new disease, sudden oak death (*Phytophthora ramorum Werres, de Cock & Man in't Veld*), killed thousands of oak trees in the coastal regions (Garbelotto 2003).

19.2.4 Poor Natural Regeneration

Three of California's native oak species have been reported to have insufficient natural regeneration to replace mortality. These include blue oak, valley oak, and Engelmann oak (*Q. engelmannii*) (Muick 1987; Bolsinger 1988), which are all deciduous white oaks. The determination that these species have insufficient natural regeneration to replace mortality has relied on inventories of the size-class distribution of oaks. The most commonly used practice to assess the success of regeneration for California oaks has been to identify trees in three general size classes: seedlings (Bachou 1987), saplings (>50 cm tall and <10 cm dbh—diameter at breast height), and mature trees (>50 cm tall and >10 cm dbh; Lawson 1993). While not perfect, this approach does provide a good general description of how a stand is progressing over time in that we can observe if seedlings are becoming saplings, and saplings growing into mature trees. Since saplings must be recruited into the mature size class when the older trees die, insufficient sapling numbers suggest that current population densities will decline.

However, it is important to note that, even for those oak species in California that have been shown to have poor regeneration, the spatial pattern can be highly variable and the success of regeneration can be vastly different. Examples include better regeneration on north slopes versus south slopes, and in swales (low-lying areas, dips) versus ridges. Hence, generalizations about regeneration must be made cautiously and underscored by the point that in some locales there is not a "regeneration problem." These patterns, as well as the previously difficulties in determining the true age of oak trees, led Tyler et al. (2006) to conclude that there was not enough information on mortality rates to support the conclusion of a regeneration problem after an extensive review of the literature, though others have formed different conclusions (Muick 1987; Bolsinger 1988; Mensing 1991; Swiecki 1998).

Until the 1980s, there was little interest in oak conservation or restoration of degraded woodlands in California. Because native oaks were widely distributed and had little economic value; they were often considered weeds and there was no incentive to learn how to propagate or establish them. The recognition in the mid-1980s that some oak species were not adequately regenerating naturally and that oak woodlands were being lost caused the state to support a research and education program on oak management (Passof 1987). In part, this program funded oak regeneration research addressing such fundamental questions as the best time to collect acorn, how they should be stored, how to grow seedlings in nurseries, and how to plant and maintain oak seedlings in the field. These efforts demonstrated that successful artificial regeneration of oaks is possible, but often requires considerable maintenance and follow-up (McCreary 2001).

19.2.5 Recruitment Limitations

Regeneration of oak woodlands in California has been extensively studied in the last two decades and, as a result, many different hypotheses have been proposed to explain recruitment failure. These can be grouped into three categories: effects of introduced Mediterranean annuals, heavy livestock grazing, and microhabitat modification due to fuel buildup after fire suppression. These limitations can be exacerbated in marginal oak populations facing climate warming (Bayer 1999). In addition, some authors suggest that gaps in age structure of oak populations are because recruitment occurs in infrequent pulses. At present, there is not much evidence to support this theory, since aging studies of blue oak stands (White 1966; McClaran 1986; Mensing 1991; Kertis 1993) tend to indicate that seedling recruitment has occurred over long intervals.

The change in the surface vegetation in oak woodlands, from predominantly perennial bunch grasses to introduced Mediterranean annuals, in California has created environmental conditions that hamper natural regeneration (Welker 1987; Gordon 1989). These plants often deplete soil moisture at a more rapid rate than perennials, especially in the early-spring when acorns are sending down their roots (Welker 1990). Livestock grazing is also suspected as being a primary cause of poor oak regeneration. Both cattle and sheep eat and destroy oak seedlings, as well as acorns and the foliage from tree branches. Browsed seedlings can remain stunted, repeatedly clipped back, for several decades before dying or becoming saplings (White 1966; McClaran 1989). Heavy grazing in woodlands, especially over many years, can also indirectly affect oak recruitment by increasing plant density and soil compaction and reducing organic matter, all of which can make it more difficult for oak roots to penetrate downward and obtain moisture (Welker 1987). Another theory of poor regeneration has to do with fire. Certainly historical fire frequency rates are very different today than they were when Native Americans regularly burned the oak woodlands and there were no efforts to put out naturally occurring fires (McClaran 1989). It has been suggested that since oaks clearly have evolved with, and adapted to naturally occurring fire, the change in fire regimes may adversely affect the ability of oaks to successfully recruit. Since postfire sprout growth can be rapid, fires in the past may have contributed to oak establishment (Plumb 1981; McClaran 1989). Also, fuel buildup as a result of fire suppression may have created conditions unfavorable for recruitment (Mensing 1992).

19.3 Iberian Open Woodlands

19.3.1 Defining Dehesa and Montado

The Spanish term "dehesa" comes from the Latin word "deffesa," the name for a fenced pasture or enclosed area that allows controlled grazing by local rangers, thus preventing invasion from foreign livestock (especially animals involved in transhumance systems). The most widely accepted definition today for dehesas and montados (the Spanish and Portuguese terms, respectively) is that of an agro-silvo-pastoral system consisting of an open overstory of oak trees of varying densities (20–80 trees ha^{-1}). Especially in Portugal, patches of different tree cover form a large-scale landscape mosaic (Pinto-Correia 2011). The dominant tree species are Mediterranean evergreen oaks—holm oak (*Quercus ilex* sbsp. *ballota* L.) (Section *Quercus*) and cork oak (*Q. suber* L.) (Section *Cerris*)—and, to a lesser extent, the Portuguese oak (*Q. faginea* Lam. sbsp. *broteroi*) (Section *Quercus*) in the areas of higher soil moisture. The understory herbaceous vegetation is dominated by winter annuals. Cork oak dominates in the coastal areas where the oceanic influence is stronger and thus Portugal has a larger area of cork oak than Spain, whereas the holm oak is characteristic of the driest continental areas (Castro 2009).

Iberian open oak woodlands are cultural landscapes created and maintained by human management and characterized by their associated biodiversity. Management can include cultivation of the understory, mostly with cereal crops. Besides producing food resources, cultivation limits the invasion of grasslands by shrubby vegetation. Indeed, grazing, ploughing, and shrub clearing are necessary to maintain open wood pastures (Huntsinger 1992; Díaz 2003). Oak density is determined by the need for space for pasture or crop cultivation in the understorey, the density being lower in intercropped areas (Gómez-Gutiérrez 1996).

From an economic point of view, the Iberian open woodland is considered the most efficient forest system to obtain diverse products under limiting environmental conditions (San Miguel 1994). With adequate management, it is in theory possible to optimize economic production while reducing the impact on natural regeneration, biodiversity and ecosystem functioning, though all of these components are currently threatened by intensification of agricultural practices (Moreno 2009).

According to the Spanish National Inventory, the dehesa agroforestry systems occupy an area of 3.6 M ha which means 13.3% of the Spanish forestry surface (Torres-Quevedo 2012) and in the Portuguese region of Alentejo about 800,000 ha (Costa 2009).

19.3.2 Ownership and Historic Uses

Most oak woodlands in Spain and Portugal are privately owned. For centuries, they have been intensively managed to maximize direct products in the form of grazing, browse, acorns, cork, cereals, firewood, and charcoal. Human influence has been intense for centuries, and what we now see are merely partial remnants of the original dense forests and shrublands (Manuel 1996; Ezquerra 2009; Gea-Izquierdo 2011).

For the dehesa of southwestern Spain, historical records show that wooded pastures were developed as a complex management system in early medieval times, at least near existing urban settlements (Linares 2003; Ezquerra 2009). Hereafter, the increase in the area covered by dehesas paralleled the growth of the human population, especially from the eighteenth century onward, as a growing population required more and more arable and grazing lands (Linares 2003). This process is considered to have been completed by the middle of the twentieth century, when almost all natural forest and shrublands in flat areas had been converted into open dehesas. In Portugal, early land changes were similar to Spain: between the fifteenth and the seventeenth centuries, oak forests were gradually transformed into open woodlands because wood was increasingly needed for ship building and the land was used for growing crops and for pastures. After the seventeenth century, cork began to be used for wine bottle stoppers and acquired an increasing commercial value and from the eighteenth century onward, cork oak montados expanded rapidly.

During the period 1940–1970, an intensification of agricultural practices and a number of socioeconomic changes led to a crisis in the traditional dehesa system (Díaz 1997). Thus, few dehesas remained untilled and tree cover was also reduced to increase arable land. Consequently, the dehesa suffered a sharp contraction due to tree cutting and lack of tree regeneration, a process that ceased during the 1980s as a result of new regulations and rising environmental awareness. Similar changes took place in southern Portugal where from 1929 until the early 1960s, most of the land was cultivated for wheat due to a government policy known as the Wheat Campaign that strongly subsidized wheat production, which resulted in depleted soil fertility and increased erosion in montado areas (Puig de Fábregas 1998; Pinto-Correia 2004). Wide-ranging international socio-economic changes (industrialization, the exodus to cities and emigration) during the 1960s and 1970s, brought about an intensive development of cropland and mechanization and massive rural depopulation. Hence dehesas and montados followed a dual process: (1) abandonment of understory use in the least productive land and smaller properties; (2) maintenance of the agro-silvopastoral system in flat areas and on larger properties, with intensified production within more fertile regions (Pinto-Correia 1999).

The area covered by wooded dehesa has been stable during the last three decades, which probably reflects that tree mortality due to ageing in some areas is compensated by natural tree regeneration in abandoned farms and planting in marginal agricultural lands (García del Barrio 2004; Plieninger 2006). In Portugal, the area covered by cork oak montados increased from 657,000 ha in 1975 to 737,000 ha in 2006, according to national forest inventories, mainly due to oak plantations and agri-environmental measures subsidized by Common Agriculture Policy (Pinto-Correia 1999; Caparros 2010). However, the area occupied by holm oak in Portugal has been decreasing since 1990.

19.3.3 Poor Natural Regeneration of Oaks

The main problem affecting Iberian open woodlands today is insufficient natural oak regeneration (Montero 2000; Pausas 2009). Livestock overgrazing, including depletion of acorns, plowing with heavy machinery, intensive pruning, shrub clearance, and the loss of animal dispersers of acorns have been pointed out as factors limiting oak regeneration in dehesas. Many stands with insufficient tree density are further limited by unfavorable climatic conditions and compacted soils. Unless the practices hampering natural regeneration and seedling establishment are stopped or reversed, large areas of oak woodland will not regenerate. Unfortunately, most landowners show little interest in oak regeneration, considering it as a very long-term problem without immediate consequences (Montero 2001; Gea-Izquierdo 2011, 2013).

The calls for adequate tree replacement in dehesas date back to at least the middle of the twentieth century. Recent lines of evidence have been crucial to attract the interest of land managers and policy makers. Recent research has demonstrated an almost complete lack of juvenile age classes in the population structure of most dehesa holm oak stands (Pulido 2001; Plieninger 2003; Ramírez 2008; MARM 2008). More interestingly, a positive correlation between "dehesa age" (the time elapsed from first forest clearance and grazing in historic times) and current mean tree age in the stands has been observed. Thus, older dehesas are formed by older trees and show a bell-shaped size structure, as compared with more recent dehesas with an inverse J-shaped distribution (Plieninger 2003). In Portugal, tree density of cork oak dehesas has fallen over the last decades: stands with less than 40 trees per hectare increased from 10% of the cork oak area in 1995 to 30% in 2005 (Vallejo 2009). These results indicate that the lack of oak regeneration is an inherent problem of these grazed woodlands. Though this is the currently prevailing view, some authors have pointed that old dehesas could have been self-regenerating by means of pulses of vegetative propagation (Martín 2006).

Silvicultural practices related to tree production such as pruning or cork extraction can seriously affect flowering and fruiting success (Cañellas 2002; Alejano 2008). At the understory level, stocking density, shrub control or cereal cropping greatly determine rates of dispersal and the postdispersal fate of seeds, seedlings and saplings (Plieninger 2004; Pulido 2005, 2010). The inability to direct acorns to safe (shaded) sites by means of efficient dispersers has also been shown to be one of the main recruitment limitations (Cañellas 2003; Pulido 2005; Pardos 2005). Accordingly, various studies from farm-level to regional scale showed that shrub encroachment generally results in higher oak recruitment rates by means of facilitation (Ramírez 2008; Smit 2008; Plieninger 2010; Pulido 2010). Nevertheless, seedling recruitment may be hampered beneath certain shrub species that act as competitors instead of facilitators (Acacio et al. 2007).

Another major concern about the sustainability of Iberian open woodlands is a complex disease known as oak decline, oak dieback, or oak mortality, depending on the area and the particular case taken into consideration. In Spain, it has been named "la seca" and it refers to the decline and death of individuals of the *Quercus* species, especially around the Mediterranean region (González Alonso 2008). In Portugal, cork oak decline has been reported in the south-west region since the 1890s (Cabral 1992) and from beginning of the 1980s, the Iberian Peninsula has experienced an evident increase in the spread and frequency of this disease and consequently, in the damages caused by it (Brasier 1996; Navarro Cerrillo 2004; González Alonso 2008; Costa 2010). This situation has led to efforts for determining the causes and possible treatments to slow down this process (González Alonso 2008; Camilo-Alves 2013).

19.4 Oak Restoration Practices

The following sections provide a description of the main restoration practices used in California and Iberian open woodlands. Table 19.1 shows a resume of these practices.

19.4.1 Acorn Collection, Storage, and Planting

Although there is little information about geographic variation in germination, establishment, and growth of oaks planted from acorn collected from different regions in Spain, Portugal, or California, it is generally agreed that local seed sources should be used, or at least seed from a region with similar environmental conditions whenever is possible

TABLE 19.1

General Oak Regeneration Practices in California and Iberian Dehesas and Montados

Act	California Oak Woodlands	Dehesa Region of Spain and Portugal
Acorn collection	Local genotypes recommended. Acorns considered ripe when caps easily dislodge. Acorns should be collected directly from tree branches if possible. Acorn crops vary widely from year to year.	Provenance regions established in Spain and Portugal. Acorns collected when ripe, fat, light-brown, and smooth. Generally collected from the ground from late October to December. Good acorn crops generally follow wet springs.
Acorn storage	Stored in a refrigerator or cooler preferably near, but above, freezing, in plastic bags that are left loosely sealed at the top. White oaks cannot be stored for longer than 3 months.	Kept under controlled conditions at −1°C to −3°C, and 40%–45% humidity, in nonhermetic containers with adequate ventilation. Mixing acorns with hygrophilous material (peat) sometimes used to lengthen the storage period.
Acorn planting	Better performance for acorns sown in winter, rather than in spring.	Acorns sown in autumn should be treated with a rodent repellent. Sowing in uneven terrain should be done by hand. When sown in furrows, 1–2 acorns are usually sown every 5 m.
Seedling production	Most seedlings are produced in containers of varying sizes. Limited numbers of bareroot seedlings are produced, but root pruning in nursery beds is critical.	Most seedlings are grown for 1 year in containers larger than 300 cm^3 that promote wide, well-developed root systems and prevent root circling.
Seedling storage	Seedlings generally kept in refrigerated units prior to planting.	Seedlings should be kept cool. Seedlings taken to the field should be placed in a sheltered place to ensure that the roots stay sufficiently cool and moist. Heating up and drying should be avoided.
Seedling planting	Seedlings should be planted in winter. Auguring planting spots can greatly reduce the soil bulk density and make it much easier for the oaks roots to grow.	Autumn is almost always the recommended time for planting seedlings. In areas that have had complete plowing, a furrow with a ripper is often created for planting.
Weed and shrub control	Weeds control in areas around individual seedlings using herbicides, mulch, or mechanical removal is usually essential.	Preparing sites using plowing is essential in areas that have shrub cover. In grassy areas, weeds should be controlled by hoeing, cultivation or with herbicides.
Seedling Protection	Tree shelters have proven very effective in protecting oak seedlings from a variety of damaging animals	Both fencing and individual protectors are used to keep grazing animals out of planted pastures for up to 20 years to prevent oak damage.

taking into account that seed sources should also consider possible climate change. To facilitate the choice of seeds, provenance regions have been defined for the main Spanish Mediterranean *Quercus* species based on environmental uniformity, which is presumed to be related to genetic homogeneity. Provenance regions have also been defined in Portugal for cork oak and holm oak. In California, the need for planting local genotypes has been recognized, but provenance regions have not been established. Nurseries are encouraged to keep records of where seeds were collected. No doubt, there have been instances where nonlocal stock has been planted, but the possible contribution to regeneration problems is unknown.

Acorn yield is highly variable in both California and Iberia, and typically there are no more than two or three abundant crops over a 10-year period. Acorns cannot be stored longer than 3–4 months (Cañellas 1992; Montero 1999). In California, acorns should be collected shortly after they are physiologically mature. While there are various indicators, such as moisture content, carbohydrate percent, or acorn color that have been used to predict ripeness for oak species in other locales in the U.S. (Bonner 1987), the most useful indicator for many oak species in California is the ease of dislodging acorns from the cupule or cap.

After collection, acorns are especially sensitive to drying and their ability to germinate can decrease rapidly with even small losses in moisture. For example, a 10% reduction in moisture content caused a 50% drop in germinating blue oak. Because acorns are so sensitive to drying and the weather condition in California when they drop can be quite hut and dry; acorns should be collected directly from tree branches (McCreary 1990, 2001). Once acorns fall to the ground, their quality can decline quickly. Sometimes, collecting from branches is impossible because they are so high that they are beyond reach, even with long poles. In these cases, frequent collections from the ground minimize exposure.

In Iberian oaks healthy acorns are generally large and dark brown when ripe. Acorns are collected from the ground from late-October to December. Insect-infested acorns, germinated acorns with broken radicles, and dry acorns must be rejected. Insect-damaged acorns rarely produce seedlings with the quality and vigor needed to adequately root and fully develop. Some parasites, specially the larvae of weevils (*Curculio*) and moths (*Cydia*), may be present inside the acorn, but infested acorns can be eliminated by float-testing in water. After discarding infested acorns, the healthy ones are collected from the bottom of the container and air-dried to prevent fungal development during storage. After collection, acorns should be stored in plastic bags that are loosely sealed at the top, in a refrigerator or cooler, just above freezing temperature. The cap should be removed before storage. Because respiration continues during storage, some gas exchange with the atmosphere is necessary and airtight containers should be avoided. Keeping acorns cool during storage slows down respiration, which can deplete carbohydrate reserves. Cool temperatures also reduce the tendency for acorns to germinate, which is especially common for white oaks. Finally, lower temperatures reduce the damage from harmful microorganisms that can rapidly degrade acorn quality in storage. When the acorns are storage properly, they can conserve their quality for a long period (Merouani 2001; Almeida 2009).

In Spain and Portugal, acorns should be sown immediately after collection, though this is not the common practice. Usually acorns collected from dehesas are stored until February or March. They are then directly planted in the field or in the nursery for outplanting the subsequent autumn. The transportation from the site of collection to the storage facility must be done rapidly in coarse-weave bags that allow ventilation. At the storage facility, bags should be emptied into trays or onto the floor in layers less than 10 cm deep and stirred frequently to allow ventilation (Montero 1999). Acorns should be stored at cool temperature, between $-1°C$ and $-3°C$ but not below $-5°C$. They should be kept under controlled conditions at 40%–45% humidity in containers with adequate ventilation. Because it is easier and cheaper, reducing temperature and humidity is the usual method for storage. Mixing seeds with hygrophilous material such as peat can help to lengthen the storage period.

The best time for directly sowing of acorns depends mainly on climate. In areas with very cold winters, it is advisable to sow acorns at the end of winter. But in areas with mild winters, sowing in autumn is preferable. This allows plants to enter the summer dry period with a more developed root system. A disadvantage of autumn sowing is that the seed usually takes longer to germinate and is longer exposed to attacks by mice and others seed predators. Hence, acorns sown in autumn should be treated with rodent repellent (Torres 1995; Cañellas 2003).

In California, acorns generally ripen in late-summer to mid-fall before the first heavy fall rains, when soils can still be extremely dry. Although even fairly dry soil can have relatively high humidity under the surface, the hard dry surface will restrict root penetration. Therefore, acorns should not be directly sown until there has been sufficient rainfall to soak the soil to a depth of at least several centimeters. Acorns should be planted as early as possible after the soil is sufficiently wet (McCreary 1990); as a rule of thumb, acorns should not be sown later than the end of January. Even this may be too late in areas with low rainfall and short winters. In general, acorns should be planted horizontally and placed 1- to 3-cm deep; however, it may be better to plant them even deeper if rodents are present, which could dig them up. Planting depth will also depend on soil texture, increasing in sandy soils (until 8–12 cm). Even though acorns tend to naturally fall on the soil surface, and some are able to germinate and became established, exposed acorns are more likely to suffer from predation and desiccation damage than buried ones. Burying acorns may not eliminate rodent damage, but it will reduce losses (Griffin 1971). Indeed, germination of buried acorns may be twice that of surface-sown ones (Borchert 1989).

19.4.2 Seedling Production

The majority of native oaks produced by nurseries in California are grown in containers rather than as bare-root seedlings. Seedlings of most oak species invest a tremendous amount of energy into producing roots, which initially have a taproot or carrot-like configuration. Such rapid root growth means that seedlings can quickly exceed the volume of the container root systems and become pot-bound. In general, better quality oak seedlings are produced in deep containers than in wide, shallow containers. Oak taproots will often grow to the bottom of a container before the shoots even emerge from the soil surface. At the bottom, roots will tend to circle round the container unless checked. Such root deformation can persist for decades after field planting and create problems of weakness, root growth, and lack of stability years later. Many container production systems employ air pruning to thwart root circling which results in a more fibrous root system.

Historically, a limited number of bareroot oak seedlings have been produced in California and were sold by the California Department of Forestry and Fire Protection (CDF). Seedlings must be root-pruned with an oscillating bar towed behind a tractor while they are in the nursery beds in order to produce acceptable plants (Krelle 1992). The timing of this pruning is also critical; if done too early in the season before the roots have grown down at least 20 cm, pruning has little effect on root form. If pruning occurs after seedlings have produced fairly thick, deep, roots, too much of the roots are lost during pruning and seedlings usually die. Bareroot seedlings can be lifted from early-December to early-February without seriously affecting seedling quality. They can be left in cold storage for 2 months without damage, as long as roots do not dry out. However, the combination of late lifting (February or later) and long storage can result in poor field survival and growth, because seedlings are not able to initiate sufficient root growth before conditions become harsh (McCreary 1994).

Plant quality is one of the most important factors governing the success of seedling-planting projects in Iberian oak woodlands (Cortina 2006). Genetic, morphological and physiological characteristics of seedlings are critical, and must be within certain established limits so that a good-quality seedling roots well and grows quickly. Most seedling stocks used in restoration are l-year-old seedlings with well-developed root systems grown under conditions as natural as possible (Montero 1999). Seedlings are moved outdoors from greenhouses as soon as possible, where they are exposed to higher light intensities and can become hardened-off before outplanting.

Forest trays are commonly used in reforestation programs in Iberian open woodlands (Peñuelas 1996). Different types of containers are continuously being developed to reduce handling costs and improve seedling quality (Almeida 2009). (Chirino 2008) demonstrates that the nursery cultivation of *Q. suber* in deep containers improves the morpho-functional attributes and seedling quality.

Both direct seeding with acorns and planting of young seedlings in Iberian oak woodlands have advantages and disadvantages. Seeding is simpler and cheaper, especially transportation to the field. Also, sowing will produce more plants per ha, enabling selection of the most vigorous trees; finally, oaks growing from directly sown acorns may be better adapted without root deformation. The main advantages of planting seedlings over direct seeding are that they are one or three years older when planted, which potentially reduces the time it takes to develop into trees and increase their resistance to adverse conditions. Also the initial planting density for seedlings can be lower, thereby reducing the costs of later operations such as thinning. Finally, fewer seeds are needed for growing seedlings in the nursery than for direct seeding in the field. González-Rodríguez et al. (2011), when comparing direct seeding and planting of one year and three years seedling, conclude that any of them can be similarly effective if appropriate nursery cultivation conditions are used and seeds are protected against predators. Other nursery techniques for increasing the success of the seedling establishment have been studied although the results are not concluding, such as the fertilization or preconditioning treatments (Almeida 2009).

19.4.3 Seeding, Planting, and Maintenance

19.4.3.1 *Planting Date and Site Preparation*

Sowing or planting date can influence subsequent field performance. The greatest problems in California arise from planting seedlings too late in the season. Since seedlings in the white oak subgenera are able to grow roots during winter, late-fall or early winter, planting allows them to develop well-established root systems while the soil is still cool and moist. Planting after March can result in poor establishment since soils generally become too dry (McCreary 1995).

In Iberian woodlands, the time period suitable for planting is not very long and is even shorter for direct seeding because the seeds must be collected in advance. Autumn is usually recommended for planting seedlings since subsequent rain fixes the soil around the roots and create favorable conditions for the root system (Cañellas 2002). Dehesa sites usually have intense competition from herbs or shrubs, and should be cleared before planting. The intensity of plowing (complete or in strips) largely depends on the characteristics of the site. Areas with high densities of shrubs need more intense tillage. Partial strip-plowing is more suitable for hills, and complete plowing is better for flatter, low-lying land. Complete plowing improves water storage by reducing plant competition (Montero 2001). Complete plowing is accompanied by the opening of furrows with a ripper to provide seedling roots access to deep soil moisture, which enhances survival and growth through the driest months of the summer.

Recently, new works have been presented other planting techniques to improve the success of restoration plans. Leiva et al. (2013) recommend the use of evergreen shrub as myrtle (*Myrtus communis* L.), as nurse plants for enhancing seedling recruitment in open oak forests. Jiménez et al. (2013) conclude that organic and stone mulching improved survival and growth of holm oak seedling at long term growing under Mediterranean conditions.

For direct seeding, 1–2 acorns should be sown every 5 m along the furrow opened by the ripper, for a density of 1000 acorns ha^{-1}. Sowing more acorns in the furrows to reduce the risk of failure due to seed predation could imply that culling is required before seedlings are 3-years old if all the acorns germinate. High densities, nevertheless, have the advantage of allowing the removal of poorly shaped trees or trees producing poor-quality cork, increasing stand yield and cork quality (Montero 2001). In uneven terrain, manual sowing is recommended as mechanized sowing may bury the acorns too deep. Navarro et al. (2006) when analyzing the effect of soil preparation on direct sowing of holm oak acorns report that preparing seed beds with a backhoe negatively affected survival.

In California, many of the hardwood rangelands have been grazed almost continuously for the past two centuries and a half, and soil compaction is commonplace. Augering planting spots can greatly reduce the bulk density of the soil and make it much easier for the oak's roots to grow downward, especially shallow-planted acorns. The depth of augering apparently makes little difference because most of the compaction is in the upper portion of the soil. As long as this area was broken up, augering, as shallow as 30 cm, is sufficient (McCreary 1995).

Other soil preparation techniques such as runoff harvesting that aims to intercept runoff to the planted seedlings by subsoiling or by microcatchments (Vallejo 2012) has been proved in arid areas (Whisenant 1995; Fuentes 2004; Bainbridge 2007) and could be taking into account as an affordable technique in open oak woodlands (Cortina 2009).

19.4.3.2 Transporting Seedlings

As a general rule, transportation of seedlings should not last more than 24 h and both roots and shoots must be protected. On long trips drying-out can be prevented by the use of special packing or, preferably, by using refrigerated trucks. Typically, seedlings are not planted immediately after arrival, but rather over periods of days, or even weeks. Therefore, it is necessary to protect seedlings against cold, heat, and desiccation by preserving them in a sheltered place to ensure that the roots stay sufficiently cool and moist, and away from domestic or wild animals.

19.4.3.3 Planting Densities

The initial planting density in the Iberian woodlands depends on management objectives. The minimum initial density should be around 625 plants ha^{-1} (Montero 1999). If cork is the main product, higher initial densities can be used. The disadvantage of high densities is that both holm oak and cork oaks sprout vigorously from stumps, which have to be removed to prevent overcrowding. However, the disadvantage of wide spacing (5 m × 6 m or 6 m × 6 m) is that large unproductive areas result if several adjacent seedlings die.

19.4.3.4 Controlling Competing Vegetation

Competition from grasses and other ground vegetation at the planting site is a primary obstacle to successful artificial regeneration of oaks in California. Competing vegetation can greatly reduce soil moisture and may severely limit survival and growth, therefore controlling competing plants around planted acorns or seedlings is essential. These plants also compete with oak seedlings for nutrients and light, and provide cover for populations pests such as voles (*Microtus californicus scirpensis*) and grasshoppers (*Melanoplus* spp.) that can seriously damage young seedlings. There are a variety of methods that can be used to eradicate competing plants, including herbicides, mulches, and physical removal. The actual procedure or technique chosen may depend on a host of variables including soil, topography, equipment or materials available, costs regulation, and the oak species planted (deciduous or evergreen).

In Iberian woodlands afforestation on abandoned farmland usually implies that a heavy grass cover will compete with oaks for moisture. Grasses can interfere with seedling growth, increases fire risk in summer, and provide shelter for acorn and seedling predators. Grasses should be removed before flowering to prevent seed dissemination. This may be done mechanically with a cultivator or a disk harrow, or chemically.

Although the competition of new seedlings with native vegetation is a major constraint for survival and growth, plant–plant interactions are not always negative and the importance of facilitative processes on Mediterranean restoration plans has been increasingly recognized (Moreno 2007; Pérez-Devesa 2008; Cortina 2009; Marañón 2009; Gómez-Aparicio 2009; Vallejo 2012). However, the shift from net positive to net negative plant–plant interactions cannot be easily defined for a particular set of interacting species because of the time-spatial variability of the outcomes of the interactions (Maestre 2004). This is the main reason for not having a common practice in restoration plans (Cortina 2009).

19.4.3.5 Protecting Seedlings from Animals

Animal damage is another common obstacle to successful regeneration in both California and the dehesa region. Without any animal protection, oak plantings often stand little chance of survival. Tree shelters have proven particularly effective in California (Tecklin 1997). They not only protect seedlings from a variety of animals, but also stimulate aboveground growth by elevating temperature, humidity, and CO_2 concentrations (Tuley 1983). In Iberia, wire netting tree shelters (2 m high) are the most suitable way to protect plantations from domestic and wild animals. Seedlings inside these enclosures should also be protected from rodents with short (60–70 cm high) plastic shelters. Although their cost is high, these devices allow livestock to graze after planting with no seedling damage. Planting densities depend on the cost of the shelters in such a way that large wire shelters are used in low densities.

Fencing against big game is desirable in estates with high deer populations. Livestock fencing (about 1.2 m tall) is cheaper and it is used to exclude large areas from livestock during the first five years after planting. Sheep grazing can be allowed at that point; however, cattle, goats and deer should be excluded from the planted area for at least 15 years.

19.5 Oak Restoration Efforts to Date

19.5.1 California

Regional and national governments are usually responsible for implementing rules or programs addressing oak woodland conservation and these approaches vary widely, depending on local threats to the resource (i.e., firewood harvesting, agricultural conversions, and development pressures) and the political climate. In some locales, there are ordinances, but these often focus on individual tree removal and rarely address oak woodland habitat conservation. In other places, counties have adopted language in their General Plans that promotes retention of woodlands and the values associated with them. Still other jurisdictions have voluntary oak conservation guidelines that are endorsed by County Boards of Supervisors and promoted by local oak conservation committees. These voluntary approaches seem to work best in rural areas where there are not large financial incentives to convert oak woodlands to other uses (i.e., vineyards or housing).

The total oak planted area in California in the last 20 years has been relatively low— probably less than 5000 ha. One reason for this is that there are few financial incentives for landowners, unlike Spain and Portugal, to expand hardwood forests through planting. Much of the planting that has occurred has been the result of requirements for mitigation by local jurisdictions to plant trees to make up for those that were removed, especially accompanying development. These plantings rarely try to replace habitat and often there is inadequate monitoring to ensure that the seedlings survive longer than three years. As a result, they were rarely successful in restoring the many ecological values lost when the original trees were removed.

Probably the largest oak plantings in California have been undertaken by The Nature Conservancy, an international conservation organization that has considerable holdings in the State. There have been restoration plantings of several hundred hectares at reserves along the Sacramento and Cosumnes Rivers in Northern California. These have focused on restoring riparian forests with valley oaks and other hardwoods in areas where oaks historically grew, but were eliminated as a result of agricultural conversions, flood control, and/or fuel-wood harvest. Many of these plantings have been aided greatly by volunteers who are enthusiastic about helping to restore oaks when given the opportunity (Ballard et al. 2001). Some of The Nature Conservancy's plantings are now 20 years old and have produced young riparian forests Spain and Portugal.

19.5.2 Spain and Portugal

Most of the restoration programs in dehesas and montados have been implemented in the form of reforestation actions (Cortina 1999; Vallejo 2003). In Spain, planting of nursery seedlings in treeless areas has been promoted with subsidies since 1994 under EU regulations within the framework of the Common Agricultural Policy (Caparros 2010). According to official figures, in the period 1996–2002, when most afforestation took place, over 186,000 ha have been planted, mostly with holm oak and cork oak (MAPA 2004; Ovando 2007). In these plantations, landowners were initially committed to exclude livestock for 20 years after planting, with the resulting loss of income compensated by EU subsidies. The goal of oak planting was the reduction of cropped area to reduce a cereal surplus, with no planned effect on dehesa regeneration and little control over the long-term consequences (Campos 2003). The program did increase oak woodland area and decrease cereal production in the 1990s but participation has dropped off in the last decade due to changes in the regulations of subsidies

Unfortunately, data on the long-term success of afforestation plans are very scarce, as no systematic monitoring effort has been conducted by regional agencies. Considering the three main regions included in the Spanish dehesa in the period 1994–2006, 78% of the CAP targeted land (125,669 ha) was planted in Andalucía, 67% in Castilla-La Mancha (75,021 ha), and 70% in Extremadura (53,855 ha). Within these programs, cork oak was planted mostly in pure stands, but also mixed with several pine species. Holm oak was planted in monospecific stands or in coexistence with other broadleaved species. According to the annual reports on forest management from the Spanish Government, in the period 2007–2011, the plantations subsidized by CAP have been reduced in the three aforementioned regions: Andalucía (1617.2 ha), Castilla-La Mancha (5800.6 ha), and Extremadura (11,478.7 ha).

Recent restoration actions in cork oak areas in Portugal have been carried out by WWF within its "Green Belt" ecoregional work in the Mediterranean region, which represents a series of functional landscape units in the Mediterranean seen as necessary to achieve ecoregional conservation priorities (Mansourian 2012). In southern Portugal, small-scale technical restoration projects have been developed at priority landscapes of the WWF Green Belt territory. Restoration activities included planting indigenous species, pruning, removing fireprone shrubs, and stabilizing slopes. Restoration activities have also been indirectly developed through payment of ecosystem services schemes (Bugalho 2011), particularly within the WWF "Green heart of cork project" (http://www.wwf.pt/corporativo/green_heart_of_cork/).

19.6 Identification of Need for Further Research

Although researchers and practitioners in California, Portugal, and Spain have come a long way in understanding the factors contributing to poor regeneration and developing successful approaches for artificial regeneration in oak woodlands, more information would help us to establish a greater degree of accuracy in this matter. Therefore, it is recommended that further research be undertaken, mainly in the following areas:

Climate change and oak decline. Precipitation has a critical effect on acorn development, so increasing drought in Mediterranean regions could compromise oak fecundity and regeneration (Pérez-Ramos 2010). In their dendroecological study of Spanish holm oak stands, Gea-Izquierdo et al. (2011) concluded that trees at warmer sites showed symptoms of growth decline, most likely explained by the increase in water stress over recent decades. Stands at colder locations did not show any negative growth trend and may benefit from the current increase in winter temperatures. Further growth projections under different climatic scenarios appear to confirm this conclusion (Gea-Izquierdo 2013). These results suggest that stands at warmer sites may be more threatened by climate change, as also suggested for oak populations in California (Bayer 1999).

Mixed stand plantations: More information is needed about plantations of mixed stands, which can achieve higher biodiversity levels, more resilience and resistance to disturbance and better adaptation strategies to global change. The use of certain oak species in conjunction with pines and other broadleaf species can provide these reforestations with greater capability to respond to different threats.

It is necessary to gain a more deep knowledge as regards the processes related to regeneration of oak woodlands to ensure that these stands are conserved into the future. Moreover, it is important that California, Portugal, and Spain join their efforts to find solutions to current problems. The new knowledge acquired through this joint research effort will help to enhance the conservation of Mediterranean oak woodlands both in California and in the Iberian Peninsula.

References

Acacio, V., M. Holmgren, P.A. Jansen, and O. Schrotter. 2007. Multiple recruitment limitation causes arrested succession in Mediterranean cork oak systems. *Ecosystems* 10:1220–1230.

Alejano, R., R. Tapias, M. Fernández, E. Torres, J. Alaejos, and J. Domingo. 2008. The influence of pruning and climatic conditions on acorn production in holm oak (*Quercus ilex* L.) dehesas in SW Spain. *Ann For Sci* 65:209–215.

Almeida, M.H., H. Merouani, F. Costa e Silva, J. Cortina, R. Trubat, E. Chirino, A. Vilagrosa et al. 2009. Germplasm selection and nursery techniques. In *Cork Oak Woodlands in Transition: Ecology, Management, and Restoration of an Ancient Mediterranean Ecosystem*, eds. J. Aronson, J.S. Pereira and J.G. Pausas. Washington DC: Island Press.

Aragón, G., R. López, and I. Martínez. 2010. Effects of Mediterranean dehesa management on epiphytic lichens. *Sci Tot Environ* 409:116–122.

Bachou, A. and Ch. Voreux. 1987. L'aménagement de la tétraclinaie de l'Amsittène (Maroc). *Forêt Privée* 174:57–68.

Bainbridge, D.A. 2007. *A Guide for Desert and Dryland Restoration: New Hope for Arid Lands*. Washington, USA: Island Press.

Ballard, H., R. Kraetch, and L. Huntsinger 2001. How collaboration can improve a monitoring program. In *Proceedings of the Fifth Symposium on Oak Woodlands: Oaks in California's Changing Landscape*, eds. R.B. Standiford, D.D. McCreary, and K.L. Purcell. Gen.Tech. Rep. PSW-GTR-184. Albany, CA: Pacific Southwest Research Station, Forest Service, U.S. Department of Agriculture, 846 p.

Bayer, R., D. Schrom, and J. Schwan. 1999. Global climate change and California oaks. In *Proceedings of the Second Conference of the International Oak Society*, ed. D.D. McCreary. San Marino, CA: International Oak Society.

Blackburn, T.C. and K. Anderson. 1993. Introduction: Managing the domesticated environment. In *Before the Wilderness: Environmental Management by Native Californians*. Menlo Park, CA: Ballena Press.

Bolsinger, C.L. 1988. The hardwoods of California's timberlands, woodlands and savannas. *USDA Forest Service Pacific NW Research Sta. Res. Bull*: PNW-148. 148 p.

Bonner, F.T. and J.A. Vozzo. 1987. Seed Biology and technology of Quercus. *USDA Forest Service SO. Forest Experiment Station Gen. Tech. Rep. SO-66*:22 p.

Borchert, M.I., F.W. Davis, J. Michaelson, and L.D. Oyler. 1989. Interactions of factors affecting seedling recruitment of blue oak (*Quercus douglasii*) in California. *Ecology* 70:389–404.

Brasier, C.M. 1996. Phytophthora cinnamomi and oak decline in Southern Europe. Environmental contraints including climate change. *Ann For Sci* 53:347–358.

Bugalho, M.N., M.C. Caldeira, J.S. Pereira, J. Aronson, and J.G. Pausas. 2011. Mediterranean cork oak savannas require human use to sustain biodiversity and ecosystem services. *Front Ecol Environ* 9:278–286.

Burcham, L.T. 1981. *California Range Land: A Historic-Ecological Study of the Range Resources of California*. U.C. Davis, CA: University of California.

Cabral, M.T., M.C. Ferreira, T. Moreira, E.C. Carvalho, and A.C. Diniz. 1992. Diagnóstico das causas da anormal mortalidade dos sobreiros a Sul do Tejo. *Scientia gerundensis* 18:205–214.

Camilo-Alves, Constança de Sampaio e Paiva, M.I. Esteves da Clara, and N.M. Cabral de Almeida Ribeiro. 2013. Decline of Mediterranean oak trees and its association with *Phytophthora cinnamomi*: A review. *Eur J For Res* 132 (3):411–432.

Campos, P., D. Martín, and G. Montero. 2003. Economías de la reforestación del alcornoque y de la regeneración natural del alcornocal. In *La Gestión Forestal de las Dehesas. ICMC. Junta de Extremadura*, eds., F. Pulido, P. Campos, and G. Montero. Mérida, Spain: Junta de Extremadura.

Cañellas, I. 1992. Producción de bellotas en alcornocales. In *Simposio Mediterráneo sobre Regeneración del Monte Alcornocal y Seminario de Política sobre el Alcornocal en el área Mediterránea*. Mérida-Portugal-Sevilla: Instituto de Promoción del Corcho (Junta de Extremadura), 1994.

Cañellas, I. and G. Montero. 2002. The influence of pruning on the yield of cork oak dehesa wood-land in Extremadura (Spain). *Ann For Sci* 59:753–760.

Cañellas, I., M. Pardos, and G. Montero. 2003. El efecto de la sombra en la regeneración natural del alcornoque (*Quercus suber* L.). *Cuadernos de la Sociedad Española de Ciencias Forrestales* 15:107–112.

Caparros, A., E. Cerda, P. Ovando, and P. Campos. 2010. Carbon Sequestration with Reforestations and Biodiversity-scenic Values. *Environ Res Econ* 45 (1):49–72.

Castro, M. 2009. Silvopastoral systems in Portugal: Current status and future prospects. In *Agroforestry in Europe: Current Status and Future Prospects*, eds. Rigueiro-Rodríguez A et al. Dordrecht, Netherlands: Springer Science + Business Media B.V.

CIWTG. 2005. California Wildlife Habitat Relationships (CWHR) System version 8.1, personal computer program. *California Department of Fish and Game Sacramento.* http://www.dfg.ca.gov/ biogeodata/cwhr/ [Accessed August 5, 2012].

Cortina, J., M. Pérez-Devesa, A. Vilagrosa, M. Abourouh, M. Messaoudène, N. Berrahmouni, L.N. Sousa, M.H. Almeida, and A. Khaldi. 2009. Field Techniques to improve cork oak establishment. In *Cork Oak Woodlands: Ecology, Adaptive Management, and Restoration of an Ancient Mediterranean Ecosystem*, eds. J. Aronson, J.S. Pereira, and J.G. Pausas. Washington, DC: Island Press.

Cortina, J., J.L. Peñuelas, J. Puértolas, R. Savé, and A. Vilagrosa. 2006. *Calidad de planta forestal para la restauración en ambientes mediterraneos. estado actual de conocimientos.* Madrid: Organismo Autónomo Parques Nacionales, Ministerio de Medio Ambiente.

Cortina, J. and V.R. Vallejo. 1999. Restoration of mediterranean ecosystems. In *Perspectives in Ecology*, ed. A. Farina. Leiden: Backhuys Publishers.

Costa, A., H. Pereira, and M. Madeira. 2009. Landscape dynamics in endangered cork oak wood-lands in Southwestern Portugal (1958–2005). *Agroforest Syst* 77:83–96.

Costa, A., H. Pereira, and M. Madeira. 2010. Analysis of spatial patterns of oak decline in cork oak woodlands in Mediterranean conditions. *Ann For Sci* 67 (2):204.

Chirino, E., A. Vilagrosa, E.I. Hernández, A. Matos, and R. Vallejo. 2008. Effects of a deep container on morpho-functional characteristics and root colonization in Quercus suber L. seedlings for reforestation in Mediterranean climate. *For Ecol Manage* 256:779–785.

Díaz, M., P. Campos, and F. Pulido. 1997. The Spanish dehesas: A diversity in land-use and wildlife. In *Farming and Birds in Europe, Academic Press*, eds. D.J. Pain and M.W. Pienkowski. London, UK: Academic Press.

Díaz, M., F. Pulido, and T. Marañón. 2003. Diversidad biológica y sostenibilidad ecológica y económica de los sistemas adehesados. *Ecosistemas* 3:10. Available at: www.aeet.org/ecosistemas/033/ investigacion.htm

Díaz, M., W.D. Tietje, and R.H. Barrett. 2013. Effects of management on biological diversity and endangered species. In , *Mediterranean Oak Woodland Working Landscapes, Landscape Series 16*, eds. P. Campos et al. Dordrecht: Springer Science + Business Media.

Elena-Roselló, M., M. Kelly, A.M. González-Avila, D. Sánchez de Ron, and J. García del Barrio. 2013. Recent oak woodland dynamics: A comparative ecological study at the landscape scale. In *Mediterranean Oak Woodland Working Landscapes. Deheseas of spain and Rachlands of Califormia*, eds. P. Campos, L. Huntsinger, J.L. Oviedo, P.F. Starrs, M. Díaz, R.B. Standiford and G. Montero. Dordrecht, Netherlands: Springer.

Ezquerra, F.J. and L. Gil. 2009. La transformación histórica del paisaje forestal en Extremadura, Tercer Inventario Forestal Nacional: Ministerio de Medio Ambiente, Madrid, 304 pp.

Fuentes, D., A. Valdecantos, and R. Vallejo. 2004. Plantación de Pinus halepensis Mill. y *Quercus ilex* subsp. ballota (Desf) Samp. en condiciones mediterráneas secas utilizandomicrocuencas. *Cuadernos de la Sociedad Española de Ciencias Forestales* 17:157–161.

Garbelotto, M., J.M. Davidson, K. Ivors, P. Maloney, S.T. Koike, and D.M. Rizzo. 2003. Non-native plants are main hosts for sudden oak death in California. *Calif. Agric.* 57(1):18–23.

García del Barrio, J.M., F. Bolaños, M. Ortega, and R. Elena-Roselló. 2004. Dynamics of land use and land cover change in dehesa Landscapes of the "Redpares" network between 1956 and 1998. *Adv Geoecol* 37:47–54.

Gea-Izquierdo, G., P. Cherubini, and I. Cañellas. 2011. Tree-rings reflect the impact of climate change on *Quercus ilex* L. along a temperature gradient in Spain over the last 100 years. *For Ecol Manage* 262:1807–1816.

Gea-Izquierdo, G., L. Fernández de Uña, and I. Cañellas. 2013. Growth projections reveal local vulnerability of Mediterranean oaks with rising temperatures. *For Ecol Manage* 305:282–293.

Gómez-Aparicio, L. 2009. The role of plant interactions in the restoration of degraded ecosystems a meta-anlysis across life-forms and ecosystems. *J Ecol* 97:1202–1214.

Gómez-Gutiérrez, J.M. and M. Pérez-Fernández. 1996. The "dehesas": Silvopastoral systems in semi-arid Mediterranean regions with poor soils, seasonal climate and extensive utilisation. In *The "dehesas": Silvopastoral Systems in Semiarid Mediterranean Regions with Poor Soils, Seasonal Climate and Extensive Utilisation*, ed. M. Etienne. Paris: INRA Editions.

González Alonso, C. 2008. Analysis of the oak decline in spain: La Seca. Bachelor thesis in Forest Management, Swedish University of Agricultural Sciences, Uppsala.

González-Rodríguez, V., R.M. Navarro-Cerrillo, and R. Villar. 2011. Artificial regeneration with Quercus ilex L. and Quercus suber L. by direct seeding and planting in southern Spain. *Ann For Sci* 68:637–646.

Gordon, D.R., J.M. Welker, J.M. Menke, and K.J. Rice. 1989. Competition for soil water between annual plants and blue oak (Quercus douglasii) seedlings. *Oecologia* 79:533–541.

Griffin, J.R. 1971. Oak regeneration in the upper Carmel Valley, California. *Ecology* 52:862–868.

Huntsinger, L. and J.W. Bartolome. 1992. Ecological dynamics of Quercus dominated woodlands in California and southern Spain: A state transition model. *Vegetation* 99–100:299–305.

Jiménez, M.N., E. Fernández-Ondoño, M.A. Ripoll, J. Castro-Rodríguez, L. Huntsinger, and F.B. Navarro. 2013. Stones and organic mulches improve the *Quercus ilex* l. afforestation success under Mediterranean climatic conditions. *Land Degradation & Development*.

Kertis, J.A., R. Gross, D.L. Peterson, R.B. Standiford, and D.D. McCreary. 1993. Growth trends of blue oak (Quercus douglasii) in California. *Can J For Res* 23:1720–1724.

Krelle, Bill. and D. McCreary. 1992. Propagating California native oaks in bareroot nurseries. In: Proceedings, Intermountain forest nursery association. *USDA Forest Service RM. Forest and Range Experiment Station. Gen Tech. Rep. RM-211*, 117–119.

Lawson, D.M. 1993. The effects of fire on stand structure of mixed Quercus agrifolia and Q. engelmannii woodlands. unpublished MS Thesis, San Diego State University.

Leiva, M.J., J.M. Mancilla-Leyton, and Á. Martín-Vicente. 2013. Methods to improve the recruitment of holm-oak seedlings in grazed Mediterranean savanna-like ecosystems (dehesas). *Annals of Forest Science* 70:11–20.

Linares, AM. and S. Zapata. 2003. Una visión panorámica de ocho siglos. In *La Gestión Forestal de las Dehesas*, eds. F. Pulido, P. Campos and G. Montero. Mérida: Instituto del Corcho, la Madera y el Carbón.

Maestre, F.T. and J. Cortina. 2004. Are Pinus halepensis plantations useful as a restoration tool in semiarid Mediterranean areas? *For Ecol Manage* 198:303–317.

Mansourian, S. and D. Vallauri. 2012. *Lessons Learnt from WWF's Worldwide Field Initiatives Aiming at Restoring Forest Landscapes*. Marseille: WWF France.

Manuel, C. and L. Gil. 1996. *La transformación histórica del paisaje forestal en España*. Ministerio de Medio Ambiente. Madrid.

MAPA. 2004. *Anuario de Estadística Agroalimentaria*. Madrid: MAPA.

Marañón, T., F.I. Pugnaire, and R.M. Callaway. 2009. Mediterranean-climate oak savannas: The interplay between abiotic environment and species interactions. *Web Ecol* 9:30–43.

MARM. 2008. Diagnóstico de las Dehesa Ibéricas Mediterráneas. Available online http://www.magrama.gob.es/es/biodiversidad/temas/montes-y-oliticaforestal/anexo_3_4_coruche_2010_tcm7-23749.pdf Tomo I.

Martín, A. and R. Fernández-Alés. 2006. Long term persistence of dehesas. Evidences from history. *Agroforest Syst* 67:19–28.

McClaran, M.P. 1986. Age structure of Quercus douglasii in relation to livestock grazing and fire. Ph.D. Dissertation University of California, Berkeley, 119.

McClaran, M.P. and J.W. Bartolome. 1989. Fire-related recruitment in stagnant Quercus douglasii populations. *Can J For Res* 23:580–585.

McCreary, D.D. 1990. Acorn sowing date affects field performance of blue and valley oaks. *Tree Planters Notes* 41(2):6–9.

McCreary, D.D. 1995. Auguring and fertilization stimulate growth of blue oak seedlings planed from acorns, but not from containers. *Western J Appl Forest* 10(4):133–137.

McCreary, D.D. 2001. Regenerating rangeland oaks in California. University of California Agriculture and Natural Resources Publication 21601:62 p.

McCreary, D.D. and Z. Koukoura. 1990. The effects of collection date and prestorage treatment of the germination of blue oak acorns. *New Forests* 3:303–310.

McCreary, D.D. and J. Tecklin. 1994. Lifting and storing bareroot blue oak seedlings. *New Forests* 8:89–103.

Mensing, S. 1991. The effect of land use changes on blue oak regeneration and recruitment. In *Proceedings, Symposium on Oak Woodlands and Hardwood Rangeland Management*, ed. Standiford R., USDA Forest Service Pacific Southwest Forest and Range Experiment Station Gen. Tech. Rep. PSW-126.

Mensing, S. 1992. The impact of European settlement on blue oak (Quercus douglasii) regeneration and recruitment in the Tehachapi Mountains, California. *Madroño* 39:36–46.

Merouani, H., M. Branco, M.H. Almeida, and J.S. Pereira. 2001. Effects of acorn storage duration and parental tree on emergence and physiological status of Cork oak (Quercus suber L.) seedlings. *Ann For Sci* 58:543–554.

Montero, G. and I. Cañellas. 1999. Manual de Reforestación y cultivo de alcornoque (Quercus suber L.): Monografía INIA.

Montero, G. and I. Cañellas. 2001. Silviculture and sustainable management of cork oak forest in Spain. In *IUFRO meeting on Silviculture of cedar (Cedrus atlantica) and cork oak (Quercus suber)*. Rabat, Morocco. Vienna: IUFRO.

Montero, G., A. San Miguel, and I. Cañellas. 2000. *Systems of Mediterranean Silviculture "La dehesa."* Madrid: Grafistaff. S.L., 48pp.

Moreno, G., J.J. Obrador, E. García, E. Cubera, M.J. Montero, F. Pulido, and C. Dupraz. 2007. Driving competitive and facilitative interactions in oak dehesas through management practices. *Agroforest Syst* 70:25–40.

Moreno, G. and F. Pulido. 2009. The functioning, management and persistence of Dehesas. In *Agroforestry in Europe: Current Status and Future Prospects*, eds. A. Rigueiro-Rodriguez, J. McAdam and M.R. Mosquera-Losada. Berlin: Springer Science +Bussiness Media B.V.

Muick, P.C. and J.W. Bartolome. 1987. Factors associated with oak regeneration in California. In *Proceedings of the Symposium on Multiple-Use Management of California's Hardwood Resources*, ed. P. TR and P. NH. USDA Forest Service Pacific Southwest Forest and Range Experiment Station Gen. Tech. Rep. PSW-100.

Navarro Cerrillo, R.M., P. Fernandez Rebollo, A. Trapero, P. Caetano, M.A Romero, M.E. Sánchez, A. Fernández Cancio, I. Sánchez, and G. López Pantoja. 2004. *Los procesos de decaimiento de encinas y alcornoques*. Direccion General del Medio ambiente, Consejería de Medio Ambiente, Junta de Andalucia.

Navarro, F.B., M.N. Jiménez, M.A. Ripoll, E. Fernández-Ondoño, E. Gallego, and E. De Simón. 2006. Direct sowing of holm oak acorns: Effects of acorn size and soil treatment. *Ann For Sci* 63:961–967.

Nixon, K.C. 2002. The oak (Quercus) biodiversity of California and adjacent regions. In *Proceedings, Fifth Oak Symposium: Oaks in California's Changing Environment*, eds. R.B. Standiford, D.D. McCreary and T.C.K. Purcell. San Diego, CA: USDA Forest Service Pacific SW Forest and Range Experiment Station Gen. Tech. Rep. PSW-184.

Olson, D.M. and E. Dinerstein. 2002. The global 200: Priority ecoregions for global conservation. *Ann Missouri Bot Garden* 89:199–224.

Ovando, P., P. Campos, and G. Montero. 2007. Forestaciones con encina y alcornoque en el área de la dehesa en el marco del Reglamento (CE) 2080/92 (1993–2000). *Revista Española de Estudios Agrosociales y Pesqueros* 214:173–186.

Pardos, M., M.D. Jimeneza, I. Aranda, J. Puertolas, and J.A. Pardos. 2005. Water relations of cork oak (*Quercus suber* L.) seedlings in response to shading and moderate drought. *Ann For Sci* 62:377–384.

Passof, P.C. 1987. Developing and educational program to adress the management of California's hardwood rangelands, ed. T.R. Plumb and N.H. Pillsbury. Berkeley, CA: U.S. Department of Agriculture, Forest Service, Pacific Southwest Research Station, General Techncal Report, PSW-100.

Pausas, J.G., T. Marañón, M. Caldeira, and J. Pons. 2009. Natural regeneration. In *Cork Oak Woodlands on the Edge. Ecology, Adaptive Management, and Restoration*, ed. J. Aronson, Pereira, J.S., and Pausas, J.G. Washington DC: Society for Ecological Restoration International. Island Press.

Pavlik, B.M., P.C. Muick, S. Johnson, and M. Popper. 1991. *Oaks of California*. Los Olivos, CA: Cachuma Press, Inc.

Peñuelas, J.L. and L. Ocaña. 1996. *Cultivo De Plantas Forestales En Contenedor*. Madrid: Ministerio de Agricultura, Pesca y Alimentación, Mundi-Prensa.

Pérez-Devesa, M., J. Cortina, A. Vilagrosa, and R. Vallejo. 2008. Shrubland management to promote Quercus suber L. establishment. *For Ecol Manage* 255:374–382.

Pérez-Ramos, I.M., L. Gómez-Aparicio, R. Villar, L.V. García, and T. Marañón. 2010. Seedling growth and morphology of three oak species along field resource gradients and seed-mass variation: A seedling-age-dependent response. *J Veg Sci* 21:419–437.

Pinto-Correia, T. and J. Mascarenhas. 1999. Contributing to the extensification/intensification debate: New trends in the Portuguese Montado. *Landsc Urban Plann* 46:125–131.

Pinto-Correia, T., N. Ribeiro, and P. Sá-Sousa. 2011. Introducing the montado, the cork and holm oak agroforestry system of Southern Portugal. *Agroforest Syst* 82:99–104.

Pinto-Correia, T. and W. Vos. 2004. Multifunctionality in Mediterranean landscapes: Past and future. In *The New Dimensions of the European Landscape*, ed. R. Jongman. Dordrecht: Wagenigen UR Frontis Series vol. 4. Springer.

Plieninger, T. 2006. Habitat loss, fragmentation, and alteration—Quantifying regional landscape transformation in Spanish holm oak savannas (dehesas) by use of aerial photography and GIS. *Landscape Ecol* 21:91–105.

Plieninger, T., P. Fernando, and H. Schaich. 2004. Effects of land-use and landscape structure on holm oak recruitment and regeneration at farm level in Quercus ilex L. dehesas. *J Arid Environ* 57:345–364.

Plieninger, T. and F. Pulido. 2009. Livestock grazing in dehesas: A self-destruct mechanism? In *Cork Oak Woodlands on the Edge. Ecology, Adaptive Management, and Restoration*, eds. J. Aronson, J. S. Pereira and J. G. Pausas. Washington DC: Society for Ecological Restoration International. Island Press.

Plieninger, T., F. Pulido, and W. Konold. 2003. Effects of land use history on size structure of holm oak stands in Spanish dehesas: Implications for conservation and restoration. *Environ Cons* 30:61–70.

Plieninger, T., V. Rolo, and G. Moreno. 2010. Large-scale patterns of Quercus ilex, Quercus suber, and Quercus pyrenaica regeneration in central-western Spain. *Ecosystems* 13:644–660.

Plumb, T.R. and P.M. McDonald. 1981. Oak management in California. General Technical Report. PSW-54. Pacific Sourchwest Forest and Rangelan Research Station.12.

Puig de Fábregas, J. and T. Mendizabal. 1998. Perspectives on desertification: Western Mediterranean. *J Arid Environ* 39:209–224.

Pulido, F., M. Díaz. 2005. Recruitment of a Mediterranean oak: A whole-cycle approach. *Ecoscience* 12:99–112.

Pulido, F., M. Díaz, and S. Hidalgo. 2001. Size structure and regeneration of Spanish holm oak Quercus ilex forests and dehesas: Effects of agroforestry use on their long-term sustainability. *Forest Ecol Manage* 146:1–13.

Pulido, F., E. García, J.J. Obrador, and G. Moreno. 2010. Multiple pathways for regeneration in anthropogenic savannas: Incorporating abiotic and biotic drivers into management schemes. *J Appl Ecol* 47:1272–1281.

Ramírez, J.A. and M. Díaz. 2008. The role of temporal shrub encroachment for the maintenance of Spanish holm oak Quercus ilex dehesas. *Forest Ecol Manage* 255:1976–1983.

San Miguel, A. 1994. *La dehesa española. Origen, tipología, características y gestión*. Madrid: Fundación Conde del Valle de Salazar.

Smit, C, J. den Ouden, and M. Díaz. 2008. Facilitation of holm oak recruitment by shrubs in Mediterranean open woodlands. *J Veg Sci* 19:193–200.

Standiford, R.B. and R.E. Howitt. 1992. Solving empirical bioeconomic models: A rangeland management application. *Am J Agric Econ* 74:421–433.

Swiecki, T.J. and E. Bernhardt. 1998. Understanding blue oak regeneration. *Fremontia* 26:19–26.

Tecklin, J., J.M. Connor, and D.D. McCreary. 1997. Rehabilitation of a blue oak restoration project. In: *Proceedings, Symposium on Oak Woodlands: Ecology, Management and Urban Interface Issues*. Gen. Tech. Rep. PSW- 160: USDA Forest Service Pacific SW Research Station.

Torres-Quevedo, M., C. Viejo, and R. Vallejo. 2012. *Anuario de estadística forestal 2009*. Madrid: Ministerio de Agricultura, Alimentación y Medio Ambiente.

Torres, E. 1995. Estudio de los principales problemas selvícolas de los alcornocales del macizo de Aljibe (Cádiz y Málaga). Tesis Doctoral. ESTI Montes.

Tuley, G. 1983. Shelters improve the growth of young trees in the forest. *Quart J Forest 77*:77–87.

Tyler, C., B. Kuhn, and F.W. Davis. 2006. Demography and recruitment limitations of three oak species in California. *Q Rev Biol* 81:127–152.

Vallejo, R., E.B. Allen, J. Aronson, J.G. Pausas, J. Cortina, and J.R. Gutierrez. 2012. Restoration of Mediterranean-Type Woodlands and Shrublands. In *Restoration Ecology: The New Frontier*, 2nd Edition, ed. J. Van Andel and J. Aronson. Oxforx, UK: Wiley-Blackwell.

Vallejo, V.R., J. Aronson, J.G. Pausas, J.S. Pereira, and C. Fontaine. 2009. The way forward. In *Cork Oak Woodlands on the Edge. Ecology, Adaptive Management, and Restoration*, ed. J. Aronson, J.S. Pereira and J.G. Pausas. Washington DC: Society for Ecological Restoration International. Island Press.

Vallejo, V.R., S. Bautista, J.C. Delgado, A. Aradottir, and E. Rojas. 2003. Strategies for land restoration. In *Methodologies and Indicators for the Evaluation of Restoration Projects. Proceedings of the first REACTION Workshop*. Alicante, Spain.

Welker, J.M. and J.M. Menke. 1987. *Quercus douglasii* seedling water relations in mesic and grazing induced xeric environments. In *Proceedings, International Conference on Measurements of Soil and Plant Water Status. vol. 2—Plants*. July 6–10, 1987. Logan, UT, pp. 229–234.

Welker, J.M. and J.W. Menke. 1990. The influence of simulated browsing on tissue water relations, growth and survival of *Quercus douglasii* (Hook and Arn.) seedling under slow and rapid rates of soil drought. *Funct Ecol* 4:807–817.

Whisenant, S.G., T.L. Thurow, and S.J. Maranz. 1995. Initiating autogenic restoration on shallow semiarid sites. *Restor Ecol* 3 (1):61–67.

White, K.L. 1966. Structure and composition of foothill woodland in central-coastal California. *Ecology* 47:229–237.

Restoration of Midwestern Oak Woodlands and Savannas

Daniel C. Dey and John M. Kabrick

CONTENTS

20.1 Introduction

There are various definitions for savanna and woodland in the ecological literature. Characteristic elements of each community are broadly defined and often overlap according to the authorities (Curtis 1959; Nuzzo 1986; Nelson 2010). Some confusion is inevitable when categorizing what is in reality a continuum of states from prairie to forest in which there can be much variation. Additional variation arises within each of these community types producing unique associations where composition and structure are further modified by site factors such as soil texture, depth, drainage, and parent materials, which control water and nutrient availability. Nonetheless, given sufficient distance between two points along the continuum, distinct communities are recognizable.

Ground flora composition and species dominance, overstory tree density, and number of canopy layers are used to distinguish between savannas and woodlands. Community structure and composition are inter-related because overhead tree canopies and shrub layers modify the microenvironment, in particular available light, at the ground, which influences floral diversity and species dominance. Historically, fire is considered to have been the driving disturbance that determined the organization of plants, community structure, resource availability, and thus, defined the natural community. Today, restoration and management of savannas and woodlands is focused on directly managing the structure to influence composition and competitive dynamics that favor desired species or plant functional groups, conserve native biodiversity and mitigate problems from invasive species.

20.1.1 Savannas

Savannas are often characterized as being dominated by grass cover with widely spaced trees (Figure 20.1) (McPherson 1997; Anderson et al. 1999; Nelson 2010). However, there is variation in understory composition and forbs can be prominent in some Midwestern savannas (Anderson et al. 1999; Leach and Givnish 1999). In the past, savannas commonly bordered large prairies on gently rolling hills. Their juxtaposition to prairies, low tree density and high fire frequency, favored ground flora that was a mixture of heliophytic grasses, sedges, and forbs common to prairies, especially at the lower range of overstory crown cover (i.e., 5%–10%) for savannas (Curtis 1959; Nuzzo 1986; Haney and Apfelbaum 1990; Nelson 2010). However, increasing tree density and cover creates a more diverse range of light environments compared to prairies. This leads to increasing species richness in savanna ground flora as species that prosper under the partial shade of the oaks cooccur with prairie species growing in the more open areas. In fact, Packard (1988) has identified species that are endemic to tallgrass savannas, and he argues that savannas are natural communities in their own right and not just transitional, ecotonal communities between prairies and forests (Packard 1993). Tree diversity in Midwestern savannas is relatively low, and is dominated by oak species. In contrast, ground flora diversity is high with, for example, 300–500 species being present over a hundred hectares or so in Missouri and Wisconsin savannas/woodlands, respectively (Leach and Givnish 1999; Nelson 2010).

In the continuum from prairie to forest, the number of vegetation canopy layers increases. In savannas, shrubs, and tree reproduction may be present in isolated patches associated

FIGURE 20.1

Ground flora composition and woody structure help to define (left to right) savanna, open woodland, and closed woodland ecosystems. Savanna is from Ha Ha Tonka State Park, Missouri, where frequent fire has been used for over 30 years. Open woodland is from Bennett Spring State Park, Missouri, where annual burning has been done for 40 years. Closed woodland is from Chilton Creek, Missouri, that has been burned periodically (four times) in 10 years. (Photos by Dan Dey [savanna and closed woodland] and Paul Nelson [open woodland].)

with refuges from fire created by topographic breaks or hydrologic features, but typically, savannas have two main canopy layers, that is, ground flora and overstory trees (Nelson 2004). Woody vegetation is largely oak sprouts and shrubs typical of prairies. Prairie grasses are often dominant in full sunlight and are diminished under increasing tree shade. The threshold of tree crown cover that suppresses dominance by prairie grasses becomes a distinguishing characteristic between savanna from woodland. However, some ecologists set the upper limit of tree cover in savannas at 30% (<7 m²/ha basal area) and classify systems with 30%–50% crown cover (7–11 m²/ha basal area) as open woodlands (e.g., MTNF 2005; Nelson 2010). Much of the nuances in differences between savanna and woodland depend on the composition of the ground flora and the associated range of overstory density and light environments that promote the desired composition. Bader (2001) and Packard and Mutel (1997) have developed lists of shrubs, grasses, sedges, and forbs that are characteristic of oak savannas and woodlands. Other conservation organizations and state agencies have produced similar lists for their states and regions. These lists can be used to assess the potential for restoration, define desired future conditions, and monitor treatment effectiveness.

20.1.2 Woodlands

Woodlands have greater tree cover than savannas, and the dominance of grasses decreases with increasing overstory density, yet plant diversity is still high in the ground layer (Figure 20.1). In woodlands, a midstory canopy is sparse or lacking resulting in a fairly open understory and sufficient light at ground-level to support a diverse plant assemblage. As crown cover exceeds 50%, forbs and woody species begin to increase in dominance, as do species that are more tolerant of shaded conditions. With increasing tree crown cover (up to 80%–100%, Nelson 2010), the ground flora of woodlands begins to resemble that of a forest. Shrubs and tree seedlings/saplings may increase in abundance along the transition from savannas to woodlands and forests. Development of a shade-tolerant midstory canopy can result in three strata of vegetation.

20.1.3 Important Drivers in Savannas and Woodlands

Savannas and woodlands were largely fire-mediated ecosystems that were locally modified by grazing and browsing, and by site factors including soils, hydrology, geology, landform, and climate (Anderson et al. 1999). Fire was arguably the single most influential driver and the character of savannas and woodlands at any one time was defined by the nature of the fire regime: intensity, season, extent, severity, and frequency. The fire regime was in turn influenced by human culture and land use because historic fire in eastern North America was largely an anthropogenic phenomenon, as it still is today (Gleason 1913; Pyne 1982; Williams 1989; Abrams 1992; Delcourt and Delcourt 1997, 1998). Guyette et al. (2002) established a strong linkage between fire history and human population density, culture, and land use effects on fuels over the past 300 years in the oak/pine dominated Ozark Highlands of Missouri. In the oak-dominated landscapes of the Wichita Mountains, Oklahoma, Stambaugh et al. (2014) observed that fire regimes were temporally and spatially dynamic, progressing through stages corresponding to cultural and land use changes. They concluded that humans were the progenitors of frequent fire regimes that favored open oak ecosystems. The effects of human population and culture on fire regimes over the past 300 years have been documented at various areas in the eastern U.S. (McEwan et al. 2007; Aldrich et al. 2010; Stambaugh et al. 2011; Brose et al. 2013b).

Woodlands and savannas occurred where fire was more frequent. In recent history (circa 1650–1850), fire occurred most frequently in the southern and Midwestern regions of the United States before the onset of landscape transformation by European settlement and agriculture (see Guyette et al. 2012). Variations in topography, human population density and land use, soils, and hydrology modified the fire regime and resulted in an intricate mosaic of natural communities across the landscape (Grimm 1984; Nowacki and Abrams 2008; Hanberry et al. 2012, 2014a,c,d).

In eastern North America, the climate and soils are suitable for development of forests practically everywhere. Fire was the main disturbance that sustained the extensive eastern tallgrass prairies and promoted the development of the Prairie Peninsula region (Transeau 1935) by preventing the encroachment and dominance of trees. Plants indicative of prairies, savannas, and open woodlands are adapted to frequent and even annual burning, and fire promotes their flowering, reproduction, and growth by ensuring adequate light, increasing nutrient availability, removing excessive litter, and retarding woody competitors (Packard and Mutel 1997). Historically, savannas occurred along the edges of prairies, representing a compositional and structural transition from prairie to woodland and forest. Savannas occurred where fires burned often enough to sustain only a very low density of overstory trees, which is necessary for grasses to dominate in the understory (Grimm 1984; Anderson et al. 1999; Batek et al. 1999; Nelson 2010; Mayer and Khalyani 2011; Starver et al. 2011). On more protected sites (north to east aspects and coves) in topographically rough areas, woodlands and forests burned less frequently.

Grazing and browsing animals such as bison (*Bison bison*), elk (*Cervus canadensis*), and white-tailed deer (*Odocoileus virginianus*) were part of savannas and woodlands in the Midwest. Where local population densities were large, they affected grass and shrub densities at landscape levels (Rooney 2001). Large ungulates modified fire regimes through their use of these habitats, removing fuels by consumption, thus, lessening the occurrence and intensity of the next fire. Freshly burned areas attracted large ungulates because of the abundance of nutritious, highly palatable and available forage and browse. This spatially and temporally dynamic interaction between grazers/browsers and fire at a landscape scale created a shifting mosaic and increased heterogeneity of habitats that supported relatively high biodiversity in flora and fauna. The fire–grazer interaction has been termed "pyric herbivory" (fire driven grazing) by Fuhlendorf et al. (2008).

20.1.4 Historic and Modern Extent of Savannas and Woodlands

Savannas and woodlands are major world terrestrial biomes and were once prominent in North America before European settlement (Beerling and Osborne 2006). Nuzzo (1986) estimated that about 12 million hectares of oak savannas occurred throughout the Midwestern United States. Oak woodlands were just as prominent if not more, though there are few estimates of their former extent. Hanberry et al. (2014b) estimated that 65% of the historic forests (circa 1812–1840) in the Missouri Ozark Highlands (about 5.7 million ha) were woodlands (55%–75% stocking, Gingrich 1967; 175–250 tph; for trees ≥12.7 cm dbh). The distribution of oak forests today may be an indication of the former dominance of oak woodlands because historic oak woodlands succeeded to forests in the absence of fire that resulted from modern fire suppression efforts (Cottam 1949; Heikens and Robertson 1994; Nelson 2010; Hanberry et al. 2014a, d). Now, oak forest types represent 51% of all forestland (78.5 million ha) in the eastern United States (Smith et al. 2009).

In contrast to widespread occurrence of oak forests today, Nuzzo (1986) estimated that <0.02% of the original savannas remain, and untold woodlands have succeeded to forests. Noss et al. (1995) have identified oak savannas in the Midwest as critically endangered ecosystems, and restoration of oak woodlands and savannas are increasingly becoming major management goals of public agencies and conservation organizations (e.g., MTNF 2005; OSFNF 2005; Upper Mississippi River and Great Lakes Region Joint Venture 2007; Eastern Tallgrass Prairie and Big Rivers Landscape Scale Conservation Cooperative 2013).

In areas of intermediate rainfall (i.e., 1000–2500 mm per year), transitional states between savanna and forest are ecologically unstable and are rare in nature (Mayer and Khalyani 2011; Starver et al. 2011). The greatest loss of oak savannas occurred when they were converted to agriculture production. Woodlands that escaped deforestation and conversion to pasture or crop fields were openly grazed, burned, and timbered. Following efforts to suppress wildfires and control open range grazing, remnant savannas and woodlands rapidly succeeded to forests. Because of their adaptations to fire, drought, and browsing, oak seedlings with large, well-developed root systems had accumulated in savanna and woodland understories so that when a sufficient fire-free period occurred, these sprouts grew rapidly to form closed tree canopies in about 20 years (Cottam 1949; Curtis 1959; Nelson 2010). If they were able to grow large enough and develop thick bark before another fire, the trees were able to avoid being top killed by fire and recruit into the overstory, thus contributing to the development of closed forest structure.

Much of the eastern forest has been made more homogeneous in composition and structure from the extensive logging that took place over a relatively short period circa 1850–1920 (Williams 1989). Oak species initially benefitted from changes in land use during European settlement, but more recently Fei et al. (2011) quantified a decline in oak density and importance in a spatially explicit analysis of the eastern United States. Also, Fei and Steiner (2007) reported a concurrent rise in red maple (*Acer rubrum*) dominance throughout its range in the East. Likewise, Hanberry (2013) documented a decline of 7%–9% in the proportion (trees ≥12.7 cm dbh) each of northern red oak (*Quercus rubra*), black oak (*Q. velutina*), and white oak (*Q. alba*) in forests of the eastern United States since 1968, which was accompanied by a 7% increase in proportion for each red maple and sugar maple (*Acer saccharum*).

Schulte et al. (2007) reported a loss of diversity in canopy species and large-sized trees, and simplified landscape compositions and structures in the Great Lakes States Region, which included losses of oak in areas that were in the northern tier of the Lake States on mesic and fine-textured soils. Hanberry et al. (2012, 2014a, c, d) reported similar changes in Missouri, where average tree size has decreased, tree density has increased, oak dominance has declined, and landscape diversity of savannas, woodlands, and forests has been diminished by widespread development of closed forests since the mid-1800s. As a consequence of widespread changes in land use, the age structure of oak ecosystems has been simplified and compressed in the Central Hardwood Region (Shifley and Thompson 2011) and the northern United States (Shifley et al. 2012) such that 60%–70% of the forests are between 40 and 100 years old.

20.1.5 Importance of Savannas and Woodlands

In the eastern United States, concerns for the sustainability of the oak resource arise because oaks are a dominant cover type and have such high ecological and economic value (McShea and Healy 2002; Logan 2006; Johnson et al. 2009). Oak savannas and woodlands are aesthetically pleasing and they support a high level of native floral and faunal diversity (Curtis 1959; Bader 2001; Nelson 2010). However, substantial declines in the amount of oak woodlands and savannas, and the decreasing dominance of oak forests may have significant negative impacts on wildlife populations (Rodewald 2003; McShea et al. 2007; Fox et al. 2010).

Oak savannas often have higher plant diversity at a variety of scales (α, β, and γ diversity) than prairies or forests (Leach and Givnish 1999; Peterson and Reich 2008). The irregular distribution of overstory trees in savannas at variable but low density creates high spatial heterogeneity in environmental resources over the range from fully open to canopy closure conditions at spatially local scales, and this promotes increased compositional and structural diversity in vegetation. Increasing tree canopy cover in woodlands and forests reduces this heterogeneity and thus plant diversity. At the other extreme of the gradient in tree canopy cover and density, C_4 grasses suppress forb abundance and diversity (Collins et al. 1998; Leach and Givnish 1999). But in savannas, increasing overstory shade reduces C_4 grasses locally, giving forbs the opportunity to flourish and add to overall diversity (Peterson et al. 2007). The dense shade of closed-canopy forests decreases grass and forb diversity, abundance, and reproductive capacity (Taft et al. 1995; Bowles and McBride 1998; Peterson and Reich 2008). Diversity in plant composition and structure is also influenced by variation in soil characteristics that affect plant productivity, and disturbances such as fire, grazing, and browsing. The interaction of overstory density, soil condition, and disturbance that promotes spatial heterogeneity in environment and selective pressure against species capable of dominance (e.g., C_4 grasses) will most likely result in higher levels of diversity. Savannas provide habitat for rare and endangered plant species by offering a multitude of microenvironments at local spatial scales, and they may support viable populations of more than 25% of all plant species within a region (Leach and Givnish 1999). High diversity in plant species richness and vertical structure (high environmental heterogeneity) in savanna flora is often associated with increasing species diversity in invertebrates, small mammals, birds, and herptofauna (Huston 1994; Haddad et al. 2001). In addition, savannas are often good habitat for rare species such as the Karner blue butterfly (*Lycaeides melissa*) (Leach and Givnish 1999).

Wildlife species often prefer key structural and compositional features of oak woodlands and savannas (Anderson et al. 1999; Davis et al. 2000; Nelson 2010). Starbuck (2013) reported that big brown bat (*Eptesicus fuscus*), eastern red bat (*Lasiurus borealis*), evening bat (*Nycticeius humeralis*), and tricolored bat (*Perimyotis subflavus*) preferred savanna and woodland habitats over closed canopy forests in the Missouri Ozarks. Thompson et al. (2012) observed that restored savannas and woodlands in the Ozark Highlands provided habitat for a diverse mix of grassland and canopy nesting bird species that are of high conservation concern. Blue-winged warbler (*Vermivora cyanoptera*), eastern towhee (*Pipilo erythrophthalmus*), eastern wood-pewee (*Contopus virens*), field sparrow (*Spizella pusilla*), prairie warbler (*Dendroica discolor*), and summer tanager (*Piranga rubra*) were more abundant in savannas and woodlands than in closed canopy forests. Reidy et al. (2014) found that large-scale savanna and woodland restoration in the Missouri Ozarks provided additional habitat for woodland generalists and early successional species, some of which are of conservation concern. In the managed restorations, most of the focal bird species they studied responded positively to a history of fire over the past 20 years. In this largely forested landscape, fire increased the diversity of habitats available to songbirds, with corresponding increases in bird species richness, diversity, and density. Others have demonstrated the importance of having savannas and woodlands on the landscape for the conservation of rare and declining bird species that rely on disturbance and early successional habitats (Davis et al. 2000; Brawn et al. 2001; Brawn 2006; Grundel and Pavlovic 2007; Au et al. 2008). Even bird species that are known to prefer mature, closed-canopied interior forests benefit from early successional habitat in the nearby landscape because juvenile birds forage for food and use the habitat as a refuge from predators (King and Schlossberg 2014).

A lack of early successional habitat characteristic of regenerating forests, woodlands, and savannas is a major wildlife conservation concern (Hunter et al. 2001; Greenberg et al. 2011; King and Schlossberg 2014). For example, the greatest numbers of bird species of conservation concern that are suffering declining populations rely on grasslands, early successional habitats that promote grass and herbaceous vegetation, and savanna and open woodlands where grasses and forbs are abundant in structurally open environments. Several factors act in concert to limit creation of early successional habitat or restoration of savannas and woodlands. Fire suppression has effectively eliminated fire from eastern woodlands and savannas, thus, promoting their densification and transition to forest structure (Hanberry et al. 2014a, d), with subsequent loss of a rich abundance of grasses, legumes, and forbs in the understory, thus altering the fuel dynamics of historic fire regimes. Forest succession in the absence of fire results in system changes (i.e., fuel loading, fuel type, fuel structure, flammability, etc.) that make it harder to restore savanna and woodland character by prescribed burning, and it may necessitate additional silvicultural interventions to initiate the restoration process (Nowacki and Abrams 2008; Kreye et al. 2013). A decline in forest harvesting on public lands, pressure from interest groups who oppose even-aged regeneration methods, and harvesting by high grading, diameter-limit cutting or uneven-aged silviculture on private lands has led to a sharp decline in early successional habitat in the East. Opposition to savanna and woodland restoration has risen to national political interest recently in the Missouri Ozarks due to concern for damage to timber from prescribed burning and to reductions in timber supply by managing for low-density savannas and woodlands (Vaughn 2014). The potential adverse impacts of prescribed fire smoke emissions on human health complicate or may constrain the use of fire in restoration (Fowler 2003; Ryan et al. 2013). All these sociopolitical factors collectively have contributed to the imbalance in forest age structure and a loss of landscape heterogeneity by the reduction and delay in savanna and woodland restoration.

Despite these challenges, the intent of many public agencies and organizations is to promote the conservation of native flora and provide quality habitat for fauna that desire structurally open environments with an abundant, diverse herbaceous ground flora through restoration of fire-dependent savannas and woodlands. If the current trend toward aging forests and lack of open-environments such as savannas and woodlands is not corrected, then there will be serious ramifications for biodiversity due to declining acorn production in old growth oak forests and lack of quality habitat for species that rely on early successional and open-environments at broad scales in the next century.

Loss of acorn production will decrease the ability of oak to regenerate naturally, promote succession toward other species that may have less habitat value for wildlife, and negatively impact the myriad of wildlife species that rely on acorns. Mast production is foundational to food pyramids and ecosystem function because so many species use acorns (McShea and Healy 2002) and they are interrelated through predator–prey relationships (e.g., Clotfelter et al. 2007). Significant increases in forest regeneration by even-aged methods are needed to bring forest age-structure into a more sustainable balance (see Section 20.2.3) and to increase early successional habitat for species of conservation concern. Modifications of traditional even-aged methods in forest management, for example, shelterwood method, are needed to restore savannas and woodlands. Large-scale restoration of savannas and woodlands would also contribute to landscape diversity, provide quality habitat, and promote the conservation of native flora and fauna. Increases in early successional habitat through large-scale restorations of woodlands and savannas are being implemented especially in the prairie-forest transition zone from Texas north to Minnesota, and eastward toward Wisconsin, Missouri, and Arkansas, where hundreds of thousands of hectares are planned for restoration (e.g., MTNF 2005; OSFNF 2005).

Loss of landscape diversity (e.g., Hanberry 2012, 2014a, c, d) poses major conservation, forest health, productivity, and resilience issues and challenges. For example, Schulte et al. (2007) stated that current landscape composition and structure is more simplified in the Great Lakes Region and is associated with lower species richness, functional diversity, structural complexity, and level of ecosystem goods and services compared to preEuropean settlement according to analysis of public land surveys. Loss of mixed conifer-hardwood forests, woodlands, and savannas due to changes in land use including conversion to agriculture, extensive logging, urban development, and fire suppression have contributed to the simplification of the landscape. A more homogeneous landscape is subject to widespread and catastrophic losses and degradation following invasive species introductions, endemic insect and disease outbreaks, and increasing environmental stresses from changing climates. In the past, abiotic stresses and biotic attacks on natural communities were limited in extent because of the diversity in composition and structure of vegetation across the landscape. A landscape that is low in natural community diversity and has a shortage of early successional stages is limited in the biodiversity it can support (Hunter and Schmiegelow 2011; Shifley and Thompson 2011; Shifley et al. 2014) and it has low resilience in a future of inevitable and increasing threats to ecosystem function, health, and productivity.

The presence of oak in savannas, woodlands, and forests has important implications to ecosystem function and productivity. For example, oak foliage in forest canopies and oak litter provide important inputs to terrestrial and aquatic ecosystems and both enhance productivity by supporting a greater diversity and abundance of organisms involved in energy and nutrient cycles than those supported by forested landscapes without oaks (Hansen 2000; Rubbo and Kiesecker 2004; Tallamy and Shropshire 2009; Stoler and Relyea 2011). Certain oak species are considered pyrophytic because their litter facilitates burning in oak and oak/pine ecosystems. Kane et al. (2008) noted that turkey oak (*Quercus laevis*) and post oak (*Q. stellata*) burned with the greatest intensity, sustainability, and consumability of eight southeastern oak species. Hiers et al. (2014) recognized the facilitative role pyrophytic oaks such as blackjack oak (*Q. marilandica*) and post oak play by providing flammable fuel capable of carrying prescribed fire needed for the restoration and management of fire-dependent ecosystems. The litter from these species has fast leaf drying rates and leaf curling habits, which is important to the disturbance processes in fire-dependent ecosystems. In open woodlands and savannas, higher levels of fine fuels from increased diversity and production in ground flora (i.e., grasses and forbs) may add to the flammability of ground fuels necessary to propagate surface fires, especially when pyrophytic oak leaves are suspended in the grass/forb matrix (Loudermilk et al. 2012; Hiers et al. 2014). Also, restoration of historic fuel conditions may act to extend the annual "burning window," that is, the time when conditions are good for conducting prescribed fires.

Managing for disturbance adapted ecosystems and increasing biodiversity at all scales are considered key management strategies to address anticipated impacts due to climate changes (Janowiak et al. 2011, 2014; Brandt et al. 2014). Restoration of savannas and woodlands would contribute to both these mitigation strategies for the range of future climate scenarios predicted. Tree species common to Midwestern savannas and woodlands such as post oak, blackjack oak, and bur oak (*Quercus macrocarpa* Michx.) are expected to be favored by predicted changes in temperature and precipitation, modifications in their seasonal patterns, and frequency of extreme weather events (Brandt et al. 2014). Communities such as these, with high species richness, are considered more resilient to climate change, better able to recover from disturbance such as drought, less vulnerable to environmental stress and biotic threats, and less susceptible to high-severity wildfires (Brandt et al. 2014).

20.2 Restoration and Management of Savannas and Woodlands

20.2.1 Site Assessment and Selection

Selecting sites for savanna and woodland restoration is an important first step that, once made, has a large bearing on the cost of restoration and the realization of conservation and biodiversity benefits. A list of indicator plants for savannas or woodlands can be used to assess the presence of remnant populations in surveys of the candidate areas and the potential for a positive floral response following the reintroduction of fire (e.g., Bader 2001; Farrington 2010). A more systematic floristic inventory can be made using the Floristic Quality Assessment method (Swink and Wilhelm 1994; Taft et al. 1997) to evaluate the quality of the current flora, integrity of the natural community, and potential for restoration. The method is also useful for monitoring change, assessing treatment effectiveness, and providing input for adaptive management.

Stand structure and characteristics of individual trees may indicate previous savanna or woodland conditions. The presence of large "wolf" trees with low and wide spreading crowns and large lateral branches indicate that previously the trees had grown in the open in a savanna or open woodland. Sometimes these trees are also surrounded by dense thickets of smaller diameter saplings or pole-sized trees that invaded the savanna or woodland following fire suppression. Oaks often dominate the initial recruitment of trees in savannas and woodlands during an extended fire-free period following a long-term history of frequent fire (Stambaugh et al. 2014). Encroachment of mesophytic tree species such as red maple requires longer periods of fire suppression and initiates in closed-canopy oak forests, setting the stage for replacement of the oak (Fei and Steiner 2007; Nowacki and Abrams 2008; Johnson et al. 2009).

Variation in topography affected historic fire regimes and hence landscape setting can be indicative of the distribution of remnant savannas and woodlands. Stambaugh and Guyette (2008) found that topographic *roughness* was inversely correlated to fire frequency, that is, plains burned more frequently than did highly dissected, topographically rough areas. Hence, prairies were most common on plains, savannas occurred at the edges of prairies on gently rolling terrain, and closed woodlands and forests were located in heavily dissected areas, on steeper hillslopes (Heikens and Robertson 1994; Robertson et al. 1997; Batek et al. 1999). Mesic forests were most common in *fire shadows* on eastern sides of streams and lakes where fire burned least frequently (Grimm 1984; Batek et al. 1999). A study of the witness tree data from General Land Office (GLO) survey notes that were made in the early 1800s in the Midwest can provide insight into the past locations of prairie, savanna, woodland, and forest, and establish historic baselines to assess change. Others have used GLO surveys of Midwestern states to reconstruct historic vegetation conditions, which can help to assess the potential for restoration (e.g., Rodgers and Anderson 1979; Radeloff et al. 1998; Batek et al. 1999; Schulte et al. 2002; Kilburn et al. 2009). Historic photos or journals can supplement information from GLO survey data, provide anecdotal confirmation of natural community types, and give insights on fire history and land use practices.

Consideration of soils can be helpful to evaluate the likely occurrence of savannas and woodlands. Soil conditions that promote seasonal drought are more prone to burning and consequently have the potential for higher fire frequency given a nearby ignition and fire spread. Savannas and woodlands are more likely to have occurred on droughty sites. Any condition that limits the volume of soil available for roots and available water content such as shallow depth to bedrock, claypans, or fragipans can lead to summer drought conditions. Soil texture affects available water content. Sandy soils are often well-drained and have low capacity to store water. Conversely, clayey soils can store water but much of it may be unavailable to plants as it is bound by the clay. Post oak flatwoods are savanna/woodland systems that occur on soils underlain by a claypan on poorly drained level areas (Anderson 1983; Nelson 2010). Sites that experience seasonal drought and are of low productivity limit the development of tree/shrub crown cover, stand density, and multiple canopy layers. Less-dense woody cover permits the higher light levels in the understory that are needed to support savanna and woodland ground flora. This increases the likelihood that species of high conservatism and floristic index may be present on the site, if it has not been damaged by grazing, erosion, or other factors that would remove the seedbank and remnant populations of plants.

Knowledge of local fire history can provide insight on the occurrence of savannas and woodlands. Site-specific fire histories are increasingly being developed throughout eastern North America (Dey and Guyette 2000; Guyette and Spetich 2003; Guyette et al. 2003; Stambaugh et al. 2006a, 2011; Hoss et al. 2008; Hart 2012; Brose et al. 2013b). A landscape model predicting historic fire frequency for the continental United States has been produced by Guyette et al. (2012) and is a useful guide for areas that lack local fire history data. Annual and biennial fire regimes are more closely associated with prairies and savannas (Anderson et al. 1999; Anderson 2006; Nelson 2010). Less-frequent fire return intervals favor woodland (>5 years) and forest (>10 years) development. Variation in fire occurrence at a given site is equally important as average fire frequency. To support a woodland structure, infrequent but extended fire-free periods of 10–30 years are needed to permit recruitment of tree saplings into the overstory (Arthur et al. 2012; Dey 2014). Fire regimes that lack variability in fire-free periods trend toward producing savanna or prairie communities. Knowledge of land use history is useful for considering how degraded sites may be. Severely degraded sites include those that were cultivated annually or overgrazed for decades, allowed to erode severely, or developed dramatically altered structure and composition resulting from long-term fire suppression. These sites are typically of low floristic quality having lost native seedbank and remnant vegetation, and are dominated by weedy, invasive, exotic or native-generalist species. Historical photos or local journals provide anecdotal insights on historical conditions and land uses.

20.2.2 Defining Desired Future Conditions

In restoration as in any type of management, it is essential to define the desired future condition. This guides development of reasonable alternative prescriptions and selection of the preferred method for moving the current vegetation toward the desired state. It is important that the desired future condition be specified in as quantitative and measurable of terms as possible, and that indicators of success are ecologically and socially meaningful. Intermediate thresholds for indicators at key stages are helpful to aid in monitoring treatment effectiveness and progress toward the desired state, and to identify when management needs to be modified to correct successional trajectories leading to undesirable conditions (Dey and Schweitzer 2014).

Restoration of savannas and woodlands inherently implies that a former, historic condition be recovered to some extent. Analysis of GLO survey data can help guide tree species composition and stand structure, and used to model landscape patterns of natural communities as they vary along environmental gradients. Ecological knowledge of species adaptations to disturbance, physiological requirements, and competitive dynamics can be used to hypothesize a desired future state. Understanding of historic disturbance regimes helps in the setting of reasonable targets. These considerations are normally useful for establishing desired natural community types and defining broad ranges in composition and structural characteristics of high-quality natural areas, but more quantitative measures of key ecosystem structure or function require study of modern examples and analogs. There are usually isolated modern day examples, nondegraded reference sites considered to be high-quality savannas and woodlands. They can be studied to help quantify composition and structural metrics. A number of reference sites are preferable to just one because they help capture the range of natural variation that can be expected in restoration. Caution must be exercised when using reference sites to guide restoration. They may be isolated, atypical examples, or small relic areas that have been modified by invasive species, overbrowsing by white-tailed deer, or have altered disturbance regimes that are deviant from historic scenarios. Thus, they may not be good examples to emulate or represent desirable future outcomes.

Methods of analyzing historic vegetation surveys such as the GLO data to model vegetation composition and structure are becoming more sophisticated and better able to account for surveyor bias in selection of witness trees (Hanberry et al. 2012, 2014a,c,d). These models are spatially explicit, and based on many standard physical and ecological variables related to the distribution and character of natural communities. They are available in spatially geo-referenced inventories that can be integrated in a GIS and used in decision making in site selection and characterization of natural communities. For example, Hanberry (2012, 2014a,c,d) developed spatially explicit models of the probability of occurrence by tree species and estimated tree density, average diameter, basal area, and stocking for the state of Missouri by ecological subsections based on GLO survey data. Estimates of these metrics can be used to quantify the historic composition and structure. Modern forest stocking charts such as that developed by Gingrich (1967) for upland *Quercus-Carya* forests in the Central Hardwood Region (Figure 20.2) or by Larsen et al. (2010) for floodplain *Populus deltoides-Acer saccharinum-Platanus occidentalis* forests in the Midwest can be used to define desired tree structure and to set structural thresholds that would prompt management action in restoration.

20.2.3 Managing Stand Density

Restoration of oak savannas and woodlands entails reducing the density of trees and reintroducing fire. Higher light levels than exist in most forest understories (~3%–5% of full sunlight) are needed to stimulate germination, promote growth, and encourage seed production of the herbaceous species commonly found in savannas and woodlands. For example, for prairie grasses such as big bluestem (*Andropogon gerardii*) and little bluestem (*Schizachyrium scoparium*), to proliferate, tree canopy cover needs to be <50% (Anderson et al. 1999; Nelson 2010; Mayer and Khalyani 2011; Starver et al. 2011). Reducing overstory density to about 6.9 m² per ha, 30%–40% stocking, or 50% crown cover produces about 50% of full sunlight at the ground level in oak forests (Figure 20.3) (Parker and Dey 2008; Blizzard et al. 2013).

Low-intensity dormant season fires, which are commonly used in restoration, have limited capacity to reduce stand density because they can only reliably top kill hardwood stems

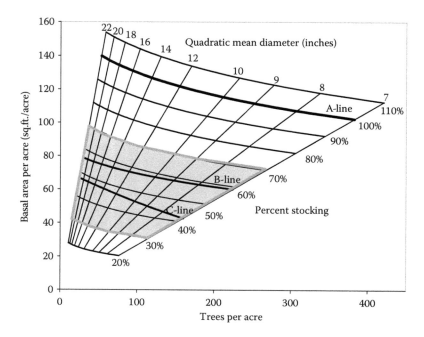

FIGURE 20.2
Stocking chart for upland *Quercus-Carya* in the Central Hardwoods Region (Gingrich 1967) used to manage savanna and woodlands. Stocking in woodlands is maintained between about 30% and 70% stocking. Closed woodlands have higher stocking (70%) compared to open woodlands (30%–40%) and savannas (<30%). When regenerating closed woodlands, stocking is reduced to below B-level stocking.

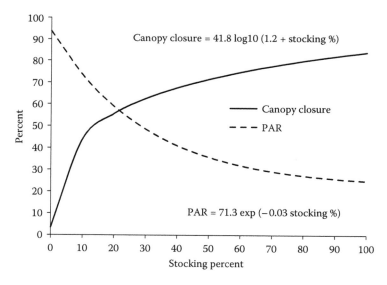

FIGURE 20.3
Relationships between stocking, canopy closure, and photosynthetically active radiation (PAR) in stands that have been thinned from below in the Missouri Ozarks. (Adapted from Blizzard, E.M. et al. 2013. *General Technical Report SRS 75*. Asheville, NC: U.S. Department of Agriculture, Forest Service, Southern Research Station, pp. 73–79.)

less than about 12 cm dbh (Figure 20.4) (Arthur et al. 2012). Thus, they can only increase understory light in a closed-canopy forest up to about 15% of full sunlight by reduction of the midstory canopy/sapling layer (Lorimer et al. 1994; Ostrom and Loewenstein 2006; Motsinger et al. 2010). Moderate- to high-intensity fires are capable of killing larger trees as well as increasing the proportion of smaller stems that do not sprout postfire (Brose and Van Lear 1998; Brose et al. 2013a). Killing of larger, older trees often leads to complete mortality because they are less likely to sprout after death of the stem than are saplings and pole-sized trees (Dey et al. 1996). This can be accomplished at a local scale by placement of downed tree tops that are cured near the boles of live trees, or by broadcast burning when fuel moisture and relative humidity are low, and fuels are sufficient to generate high fire temperatures for the duration needed to kill cambial tissue. Managing the reduction of larger trees with fire can be problematic because burning under weather and fuel conditions to achieve high-intensity fires is risky from a fire escape and safety standpoint. Fire is more indiscriminant in what trees are killed than are alternative methods for reducing stand density. Generally, other practices such as timber harvesting or chemical/mechanical thinning must be used with fire to complete the restoration process in a timely, safe and affordable manner.

Timber harvesting to reduce the density of large trees is an efficient alternative method of large tree removal, and affords a high degree of control of the spatial arrangement and composition of remaining trees. It also provides needed income to pay for other costs of restoration. Mechanical cutting of large diameter trees before burning is an effective way of controlling overstory density. If larger diameter trees produce sprouts after mechanical cutting, their sprouts are then vulnerable to being top killed by a subsequent fire due to their small diameter and thin bark. If large trees are unmerchantable or timber harvesting is not a management option, another effective method is the stem injection of herbicides. When applied correctly, herbicides can kill trees completely preventing them from sprouting. Dead trees can be left standing to fall apart over time, if public safety is not an issue.

FIGURE 20.4
Annual and periodic low-intensity spring prescribed burns in mature *Quercus-Carya* forests are effective in reducing the midstory canopy of woody species, but there has been little effect on the overstory canopy. At the Chilton Creek woodland restoration area, annual burning (left to right) has been conducted for 10 years and has substantially reduced the mid- and understory vegetation. Two years after the last fire in the periodic burning (4 times in 10 years) treatment, ground cover is higher than in the annual or the no burn treatment. Unburned stands have greater vertical woody structure and less diverse understory. (Photos by Dan Dey.)

In the process of decaying, snags can provide critical habitat for a wide variety of wildlife. Methods of mechanically girdling trees are also available for killing larger trees, though they are not as effective as herbicides.

20.2.4 Managing Ground Flora

Reducing tree density in combination with fire will produce the fastest improvements in ground flora diversity and dominance in most cases where the sites were not previously degraded by livestock overgrazing and soil erosion. Prescribed burning alone generally increases herbaceous species coverage and richness when trying to restore oak savannas from mature forests conditions, but improvements are most dramatic where the overstory is opened as a consequence of burning or thinning (Hutchinson et al. 2005; Hutchinson 2006; Waldrop et al. 2008; Kinkead et al. 2013). Even without fire, positive responses in herbaceous richness and coverage have been achieved by mechanical thinning or timber harvesting (Hutchinson 2006; Zenner et al. 2006; Waldrop et al. 2008; Kinkead et al. 2013). However, these gains in diversity are ephemeral as an abundance of woody sprouts grow rapidly to form canopy closure and shade out the ground flora.

Continued use of fire is crucial to control the growth of woody sprouts that can develop quickly following burning or harvesting. Maintaining higher levels of overstory canopy cover (e.g., >70%) or density (e.g., >13.8 m^2 per ha) can suppress the growth of woody sprouts in the understory (Dey and Hartman 2005; Kinkead et al. 2013), but would also delay the restoration of savanna ground flora, which requires more sunlight. Law et al. (1994) published a chart useful for managing stand density to produce desired crown cover in upland oak savannas depending on overstory tree size and density. Blizzard et al. (2013) developed models that are useful for estimating available light in the understory from overstory crown cover, density, basal area or stocking for upland oak-hickory forests in the Missouri Ozarks. These can be used to manage stand density to provide sufficient light to promote desired ground flora compositions.

Managing overstory crown cover between 5% and 50% influences the development of the ground flora. Tree crown cover below 30% is needed to promote grass domination and proliferation of heliophytic forbs commonly found in prairies (Nelson 2010). Greater tree cover begins to favor the abundance and diversity of more shade tolerant forbs characteristic of woodlands as the grasses decrease in dominance. Maintaining tree cover above 50% inhibits domination of grasses in the ground layer (Mayer and Khalyani 2011; Starver et al. 2011).

In addition to using fire to restore and maintain the woody structure of savanna ecosystems, the restoration of fire as a disturbance that shapes the ground flora community is important. Fire promotes germination and growth of herbaceous plants by (1) removing litter that acts as a physical barrier to germination and seedling establishment, (2) preparing a receptive seedbed for colonization by wind and animal dispersed seed, (3) breaking chemical and thermal seed dormancy by producing heat and smoke, (4) releasing nutrients tied up in the litter, and (5) increasing light and temperature at ground-level. Frequent fire, that is, every 1–3 years for example (Anderson et al. 1999; Nelson 2010), can be managed to favor dominance of grasses, legumes or forbs by varying the season, intensity and frequency of prescribed burning. Simple recommendations cannot be given here because vegetation response to fire is complicated by its modification by differences in climate, soils, hydrology, grazing and browsing, and overstory interactions. However, Peterson and Reich (2008) observed that annual burning promoted C$_4$ grasses in open-environments, forb richness was greatest with biennial fires, and shrubs, trees and vines increased in dominance with longer fire-free periods on a sand plain in east-central Minnesota. Spring fires when grasses are still dormant, favor grass domination in general while summer burns favor forb diversity (Nelson 2010). Dormant season fires have the least impact on the herbaceous plant community (Hutchinson 2006). Consistent application of prescribed fire tends to create homogeneity in the vegetation community, therefore it is recommended to vary the frequency, intensity and season of burning when maintaining savanna ecosystems (Nelson 2010). Restoration of severely degraded sites may require artificial regeneration of the ground flora by seeding and planting. Packard and Mutel (1997) published guidelines for actively managing the recovery of ground flora through artificial regeneration.

20.2.5 Invasive Species, Fire and Restoration

Fire, like many disturbances, increases resource (light, moisture, nutrients) availability and can create receptive environments for nonnative invasive species (NNIS) establishment by removing litter and exposing mineral soil for the colonization and expansion of NNIS. Seeds or propagules onsite or nearby are a prerequisite to invasion. It is incumbent therefore to have a good inventory of the proposed restoration site and nearby areas to know if NNIS may be a problem that needs to be factored into restoration plans. Since repeated burning is standard practice in restoration and maintenance of oak woodlands and savannas, vigilance through monitoring is essential for detecting NNIS problems early on. The list of current NNIS that threaten the integrity of woodlands and savannas is large and varies by region (Grace et al. 2001). Some common problem species in oak savannas and woodlands include smooth brome grass (*Bromus inermis*), Canada thistle (*Cirsium arvense*), musk thistle (*Carduus nutans*), sericia lespedeza (*Lespedeza cuneata*), autumn olive (*Elaeagnus umbellata*), multiflora rose (*Rosa multiflora*), teasel (*Dipsacus* spp.), crown vetch (*Coronilla varia*), white sweetclover (*Melilotus alba*), yellow sweetclover (*Melilotus officinalis*), spotted knapweed (*Centaurea biebersteinii*), European buckthorn (*Rhamnus cathartica*), and Japanese honeysuckle (*Lonicera japonica*). The spread and dominance of NNIS over native species may increase with decreasing tree canopy cover in woodlands and savannas as many of them are adapted to high light environments and fire.

It is impossible to generalize about fire and NNIS because there is so much variation in site conditions, fire behavior, fire regime, NNIS species traits, initial floristic composition and structure, competitive dynamics between NNIS and native species, and a myriad of other factors that affect plant survival, growth, spread and rise to dominance. Zouhar et al. (2008) provided a thorough overview of fire effects on NNIS in natural communities and DiTomaso et al. (2006) discussed the use of fire to control NNIS. Each restoration project area must be evaluated on a case-by-case basis to integrate site conditions, planned disturbances, and reproductive modes and phenology of native and NNIS species in deciding what the potential problems may be and how they can be mitigated by modifying the timing, type and combination of management practices. There are some basic things to consider when evaluating the potential for fire to create or exacerbate a NNIS problem.

First consider the modes of regeneration of NNIS. Plants may reproduce by sexual (seed) or asexual (vegetative) mechanisms. Most species use both modes though one may be more prevalent especially under certain disturbance regimes. An array of sexual reproductive traits of NNIS that enable them to exploit postfire environments include

- Mature rapidly in an annual or biennial cycle
  - Increasing the probability of completing seed production and dispersal in between fires

- Able to self-pollinate
  - Can reproduce in low density or sparse populations
- Prolific production of wind and bird dispersed seeds
  - Maximizes numbers of seeds disseminated for opportunistic colonization of ephemerally favorable sites
- Seeds remain viable in seed bank for years or decades
  - Seed can accumulate to high densities far exceeding annual crops and are in place for release by an appropriate disturbance
- Chemical or thermal induced seed dormancy
  - Increases the probability of surviving burning
  - Fire stimulates germination and synchronizes it with creation of favorable seedbed conditions
- Rapid early growth of plants adapted to high light environments
  - Promotes early dominance and acquisition of resources for development toward maturity and completion of life cycle

NNIS can persist in a regime of fire by either surviving the fire intact or by reproducing asexually. The cambium of perennial woody species (e.g., tree of heaven (*Ailanthus altissima*)) is protected from the heat of fire by bark. The cambium's ability to avoid fire damage increases exponentially with increasing bark thickness (Hare 1965). The location of meristematic tissues in the form of vegetative and reproductive buds influences whether they will survive a fire or not. Buds located in soil are in most cases insulated from the heat of fire as most surface fires do not increase temperatures to lethal levels in the top 5 cm of soil (Iverson and Hutchinson 2002; Iverson et al. 2004). Buds at the extremities of branches in tall trees and shrubs are distanced from the direct heat of many surface fires, especially dormant season low intensity burns. Species with root-centric growth, who preferentially store carbohydrates in their roots, may build large energy reserves between fires. This stored energy is used to fuel rapid shoot growth from adventitious buds following a fire that top kills the parent stem.

Fire effects on plants depend, in part, on the specific fire regime, plant life cycle and reproductive strategies, plant phenology at the time of burn, and population demographics. Fire type (ground, surface, or crown), severity and frequency affect the persistence of plants, and determine if populations will be reduced, or spread to dominate after burning. Fire size and inclusion of unburned areas within the restoration area influence NNIS colonization and spread. Larger burns and more completely burned areas limit NNIS invasion from outside seed sources and from previously established individuals and colonies. Season of burning in relation to phenology influences damage and mortality. Fire severity is greater when plants are actively growing at the time of burning. Burning plants before they set seed, or when seed is vulnerable to direct fire injury reduces NNIS populations more effectively. Plants with below ground reproductive structures such as rhizomes, caudices, bulbs, corms or root crown buds are well-protected from fire. Many plants with these traits sprout prolifically after fire and prosper in an environment of readily available resources. Plants that store viable seed for years or decades in organic or soil layers are well-suited to persist as a species after fire. In the winter and spring, moist or frozen organic layers protect seed stored within or beneath that horizon.

The timing and severity of fires is key to controlling NNIS, and determines the fate of native species as well. The easiest NNIS to control are annuals that produce seed after the fire season, whose seed is readily exposed directly to fire's flames and does not persist in the seedbank. Late spring to early summer fires are most likely to control NNIS annuals that set seed later in the summer. Biennial and perennial NNIS are more difficult to control. More severe fires are needed to kill reproductive structures in organic or mineral soil layers. Few NNIS are controlled with a single fire. It takes consecutive, repeated fires to stop seed production by killing existing individuals and to eliminate plants that arise from the seed bank or from vegetative structures, which often are stimulated by the initial fire. Scheduling fires several years apart only allows NNIS to add seed to the seedbank, or build energy reserves in belowground structures.

Burning followed by herbicide application is an effective alternative method (DiTomaso et al. 2006). The fire kills current vegetation, stimulates germination, converts large plants into small concentrated sprout clumps through top kill and sprouting, and removes debris that facilitates herbicide application. Herbicides are effective at killing plants that sprout prolifically from large underground bud banks and stored energy reserves. The succulent growth of seedling sprouts and germinants readily absorbs herbicides, increasing the efficacy of the herbicide. Fire effect on native species must also be considered in planning the prescribed fire regime to ensure they are not adversely impacted and that their response to fire is vigorous. Dominance of native species after fire can help to suppress NNIS establishment or recovery.

20.2.6 The Role of Grazing in Restoration

In the past, bison, elk and deer were free to roam across the landscape, and fires burned without suppression. Our challenge today, is to develop modern analogs of these critical interactions for smaller landscapes. There are few examples in research or management where wild or domestic grazers/browsers have been integrated with fire and other vegetation management to restore savannas and woodlands at the scale of landscapes. However, Collins et al. (1998) found that bison grazing or mowing in late June increased total species richness by promoting C_3 grasses, forbs and woody species in prairie ecosystems that are dominated by C_4 grasses under a regime of annual to frequent fire. Also, Hartnett et al. (1996) demonstrated that bison grazing in combination with various fire frequency burn treatments (i.e., annual to 4-year cycles) significantly increased plant species diversity by preferentially grazing C_4 grasses and creating greater spatial heterogeneity. Similar results may be expected in savanna and woodland ecosystems but this hypothesis needs to be tested.

Restoration of savannas and woodlands is a relatively new management goal, having increased in application over the past 30 years in the Midwest. Initial efforts focused on simply reintroducing fire to reclaim floristic quality. Fire was prescribed fairly often, for example, every 3 years, based on local fire history knowledge. Before long, it became obvious that burning alone was not having its desired effect on regulating stand density or woody sprouts. Prescribed fires were typically low intensity and conducted in the late winter to spring seasons. Only recently have we begun to understand and manage stand density and ground flora interactions, and realize the positive benefits of variability in the fire regime (season, intensity, frequency, severity, size) in creating heterogeneity in habitats and increasing biodiverstiy at the community and landscape level. Introduction of native grazers and browsers is still in the future in research and management of savannas and woodlands in the Midwest. There is resistance by some managers to the idea of introducing grazing in restoration projects because of the long history of ecosystem degradation caused by overgrazing by domestic cattle, which led to soil erosion and compaction, loss of native floral diversity, and proliferation of exotic species. But managed grazing as part of a suite of restoration practices may have a positive role to play in managing vegetation and achieving natural community, biodiversity, and conservation goals (Bronny 1989; Harrington and Kathol 2008). Certainly, there are differences in the grazing habits and behavior of domestic cattle compared to bison and elk that affect structure, composition and diversity of flora (Hartnett et al. 1996), and these must be studied in the future. Integrating grazing may be difficult to nigh impossible to implement on small parcels, but on larger landscape restoration projects, it could contribute to restoring ecosystem function, and promoting heterogeneity in habitats that favor biodiversity for a wide array of native flora and fauna.

20.2.7 Sustaining Savannas and Woodlands

Once the structure, composition, and function have been restored to savannas and woodlands, it eventually becomes necessary to plan for replacing some of the overstory trees to sustain desired stocking. This need arises because some of the trees will succumb to competition-induced mortality, and others will die in old age of physiological stress from injuries suffered through management and other agents causing physical damage and decay. Post oak is one of the longer lived oak species and individuals may live to be over 400 years old, but even they need to be replaced to sustain savanna or woodland overstories.

Harvesting mature trees to regenerate and recruit the next generation of overstory can provide financial benefits. If managed properly, mature oak trees are capable of producing forest products, and the periodic harvest and sale of these can be used to offset some of the management costs of restoration, though potential forest product values are limited in savannas due to low tree density. It is not something that is commonly thought of in restoration since the emphasis has been on the diversity and quality of the ground flora. But, prescribed fire can be a major inciting agent of tree value loss in savanna and woodland restoration, as fire-wounding provides entry to wood decaying fungi. The key is to achieve ecological benefits through prescribed burning without causing unnecessary damage to trees that are capable of producing valuable products.

A comprehensive management system that includes a plan for regenerating trees is recommended for all restoration prescriptions to ensure the sustainability of the ecosystem.

Presently, there are no well-defined silvicultural systems that include a planned series of treatments for regenerating and tending savannas and woodlands. Nonetheless, there are many relevant silvicultural methods potentially applicable for restoration. Most of the regeneration methods used in forest management can be applied to savannas and woodlands. For example, trees can be regenerated with the clear-cut or shelterwood methods (each with reserves) and tended with thinning and prescribed burning in even-aged systems, or regenerated with the uneven-aged group selection method and tended with thinning by mechanical or chemical methods. Application of prescribed fire with the group selection method is complicated because the matrix area may need burning but fire must be excluded from the individual group openings to permit recruitment, and they are scattered throughout the area. With time the overall restoration area becomes an integrated mosaic of different aged group openings that need protecting from fire.

However, these regeneration and tending methods may be applied differently in savannas and woodlands than in forests. For example, retaining residual stocking with reserve trees may be more preferable for regenerating savannas and woodlands than forests. This residual overstory provides habitat for wildlife and provides partial shade to reduce the growth of woody advance regeneration that is released by harvesting. It also influences the composition of ground flora based largely on the shade tolerance of the underlying grasses, sedges, forbs and legumes. Applying a method to develop two-aged stands comprised of a partial overstory (>20%–30% stocking of dominant/codominant trees) and a regenerating subcanopy may reduce intense shading of ground flora by woody vegetation developing in the regeneration layer that would occur at lower overstory stocking.

During the recruitment phase in savannas and woodlands, when tree seedling sprouts grow into larger size classes advancing into the overstory, prescribed fire should be excluded until a portion of the recruiting cohort is sufficiently large to escape being top killed by fire's reintroduction. Here it is important to recognize that in mature woodlands there will only be about 74–99 canopy dominant or codominant trees per hectare and <27 trees per hectare (>50 cm dbh) in a savanna. Thus, managing trees in savannas or woodlands is analogous to the silvicultural practice of crop-tree management in which a small number of trees is selected at an early age as the "crop" trees destined to become the future dominant trees at maturity, which can be harvested for profit when it is time to replace the overstory trees. Savanna and woodland crop trees are carefully cultured from a young age while the vast majority of trees in the stand are given no special attention or protection from fire. Thus, the noncrop trees are subject to removal arbitrarily by burning or deliberately by mechanical thinning, in fact they may need to be removed to maintain desired stocking for ground flora development.

Recruitment of seedling sprouts and grubs into the overstory requires a sufficiently long fire-free period for trees to grow large enough in size to gain resistance to being top killed by the next fire. In general, this may require a 20–30-year fire-free period depending on tree growth rates and source of reproduction (Johnson et al. 2009; Wakeling et al. 2011; Arthur et al. 2012). Oak stump sprouts are the fastest growing source of oak reproduction. One of the fastest growing oak species in the uplands is scarlet oak, and its stump sprouts can achieve diameters averaging 7.6 cm in 10 years when growing in the open (Dey et al. 2008). White oak, bur oak and post oak are slower growing species. Increasing overstory density reduces the growth of oak sprouts and lengthens the time needed for them to achieve a diameter, that is, bark thickness to become resistant to top kill by fire. These longer fire-free periods are not atypical in historical fire history records for the period before European settlement (Guyette et al. 2002; Guyette and Spetich 2003; Stambaugh et al. 2006a,b).

If producing marketable timber is also an objective, the fire-free interval may need to be 30 years or longer to allow a critical number of trees to become large enough to not be severely damaged by prescribed fire. If individual crop trees are given protection, for example, fuels are reduced around trees to minimize scarring or fires are ignited in a way to reduce severity in the immediate vicinity, then the greater area being renewed may be burned for other ecological reasons. Harvested trees can be directly felled away from trees to be retained to maintain overstory structure, thus placing tree crowns that may be left as slash on the site at a distance from retention trees. Felled-trees can also be whole-tree skidded to concentrate slash at the landing where it can be isolated and burned. Slash near retention trees can be cut to lie flat on the ground to minimize fire intensity near the bole of the retention tree, and to promote more rapid decay of the slash.

Fire damage to mature oaks was studied by Marschall et al. (2014) who found that there was a 10% loss in value 14 years after red oak trees (≥28 cm dbh at time of first scar) were fire scarred. They suggested that pole-sized trees (13–28 cm dbh) were at high risk of value loss if fire-scarred due to the length of time decay would have to develop before the trees reached maturity. After the recruitment phase, when sufficient trees >28 cm dbh are in the overstory, care must be practiced when reintroducing prescribed burning to prevent mortality of those trees or to minimize damage to them.

In woodlands, reducing the stand stocking to approximately 20%–30% (Figure 20.2) is important for ensuring that the advance reproduction can recruit rapidly into the overstory, to minimize the time that fire must be withheld from the immediate area. Overstory stocking in savannas may be low enough to promote overstory recruitment if prescribed burning ceases for the requisite length of time for trees to develop resistance to fire induced top kill. Maintaining greater residual stocking levels will substantially reduce the growth rate of the recruiting trees (Dey and Hartman 2005; Dey et al. 2008), increasing the duration of the fire-free interval needed for allowing sufficient numbers of trees to grow larger than the threshold diameters identified above.

In large landscape burns there is less control over local fire behavior, in fact a goal may be to have variability in fire intensity and severity across the restoration area. It is also impractical to protect individual trees over thousands of acres at a time and fire is more indiscriminate in which trees are killed or damaged. In these situations, even-aged regeneration methods with area regulation are better suited for managing savannas and woodlands when it comes time to replace overstory trees. With area regulation, specific stands or land units are selected for regeneration and tending to replace the overstory. In selected areas, prescribed fire can be excluded from stands or land units with fire lines, roads, or natural fire breaks to protect the seedlings and allow for recruitment into the overstory. After a sufficient number of trees have recruited and are no longer in danger of being top killed or severely damaged, fire can be reintroduced along with other tending methods. Because fuels may accumulate for up to 12–15 years in the Central Hardwood Region in the absence of fire before a balance between fuel input and decomposition is achieved (Stambaugh et al. 2006b), care must be exercised when reintroducing fire and the initial burn should be low intensity to begin reducing fuel loading. This is especially important if activities such as tree thinning have contributed additional fuels.

Single-tree selection has not been a successful method for regenerating oak forests, except perhaps in the most xeric regions where shade tolerant competitors are few (Johnson et al. 2009). Also, it may be exceptionally difficult to manage woodlands using single-tree selection because this method requires the continuous establishment and recruitment of seedlings and small trees that are vulnerable to top kill and damage that can lead to substantial value loss by periodic fire. In general, the fire-free interval will need to increase as target stocking increases, since higher stocking will slow the growth of the recruiting trees; thus, lengthening the time it takes for desirable numbers to recruit into size classes less vulnerable to fire damage or top kill.

Longer rotations may be desired in savannas and woodlands than in forests. Rotations of 100 years are commonly used in hardwood forest management for optimizing the sustained production of timber, somewhat shorter (70–80 years) for red oak species to mitigate for oak decline losses. However, a longer rotation can be used for managing long-lived species where timber production is not a primary objective. For example, in the Ozark-St. Francis National Forest (OSFNF 2005), rotations for oak woodlands were extended to 140–160 years and for oak savannas to 180–200 years. Extending the rotation means that savannas and woodlands can be maintained in a mature state and tended with prescribed fire for a longer proportion of the rotation, especially if the overstory is dominated by white oaks. White oak, bur oak and post oak are longer-lived than red oak species, not as susceptible to oak decline, and are better able at compartmentalizing decay arising from fire injury (Fan et al. 2008; Dey and Schweitzer 2015). Lengthening the rotation decreases the size of the area that needs to be regenerated in each harvest period.

20.3 Conclusion

Open-structured oak savannas and woodlands were once prominent natural communities in eastern North America. They existed because of a long-history of frequent fire. Their distribution changed over time with changing climates, and human populations and cultures. With the advent of fire suppression, these communities succeeded to closed forests. Today, they are rare throughout the East. Restoration of oak savannas and woodlands has become a focus of land managers for a myriad of reasons including conservation of native biodiversity, quality habitat for game species and those of conservation concern, diversification of habitats at the landscape scale, for aesthetic and recreational purposes, and to restore ecosystem function. Restoration and maintenance of these systems requires active management. Reintroducing fire is fundamental to restoration, but other silvicultural practices are needed to efficiently manage vegetation composition and structure, and achieve desired future conditions. Research is needed to

- Identify desired future conditions.
- Assess innovative treatment combinations.
- Quantify threshold resource requirements for ground flora species.
- Understand relationships between woody structure and resource availability to ground flora.
- Predict vegetation response to variability in fire regime.
- Develop silviculture methods to prevent or mitigate invasive species.
- Model soil-topography influences on disturbance regimes and community composition and structure.
- Evaluate the interactive effects of fire-grazers/browsers-vegetation.

Management efforts to restore oak savannas and woodlands often precede research, provide early tests of innovative treatment combinations, and help to identify key questions. Monitoring to inform adaptive management is an important source of knowledge and part of the learning process. Restoring oak savannas and woodlands will help to expand the distribution of rare natural communities, conserve native biodiversity, create a more diverse landscape, provide habitat for wildlife species of concern, and should increase our options for responding to uncertain futures due to increasing human population, climate change, and invasive species.

References

Abrams, M.D. 1992. Fire and the development of oak forests. *Bioscience* 42:346–353.

Aldrich, S.R., C.W. Lafon, H.D. Grissino-Mayer, G.G. DeWeese and J.A. Hoss. 2010. Three centuries of fire in montane pine–oak stands on a temperate forest landscape. *Applied Vegetation Science* 13:36–46.

Anderson, R.C. 1983. The eastern prairie-forest transition—An overview. In: R. Brewer (ed.), *Proceedings of the 8th North American Prairie Conference*. Kalamazoo, MI: Western Michigan University, pp. 86–92.

Anderson, R.C. 2006. Evolution and origin of the central grassland of North America: Climate, fire, and mammalian grazers. *Journal of the Torrey Botanical Society* 133(4):626–647.

Anderson, R.C., J.L. Fralish and J.M. Baskin. 1999. *Savannas, Barrens, and Rock Outcrop Plant Communities of North America*. Cambridge, UK: Cambridge University Press, 470 p.

Arthur, M.A., H.D. Alexander, D.C. Dey, C.J. Schweitzer and D.L. Loftis. 2012. Refining the oak–fire hypothesis for management of oak-dominated forests of the Eastern United States. *Journal of Forestry* 110:257–266.

Au, L., D.E. Andersen and M. Davis. 2008. Patterns in bird community structure related to restoration of Minnesota dry oak savannas and across a prairie to oak woodland ecological gradient. *Natural Areas Journal* 28:330–341.

Bader, B.J. 2001. Developing a species list for oak savanna/oak woodland restoration at the University of Wisconsin-Madison Arboretum. *Ecological Restoration* 19(4):242–250.

Batek, M.J., A.J. Rebertus, W.A. Schroeder, T.L. Haithcoat, T.L. E. Compas and R.P. Guyette. 1999. Reconstruction of early nineteenth-century vegetation and fire regimes in the Missouri Ozarks. *Journal of Biogeography* 26:397–412.

Beerling, D.J. and C.P. Osborne. 2006. The origin of the savanna biome. *Global Change Biology* 12:2023–2031.

Blizzard, E.M., J.M. Kabrick, D.C. Dey, D.R. Larsen, S.G. Pallardy and G.P. Gwaze. 2013. Light, canopy closure, and overstory retention in upland Ozark forests. *General Technical Report SRS 75*. Asheville, NC: U.S. Department of Agriculture, Forest Service, Southern Research Station, pp. 73–79.

Bowles, M.L. and J.L. McBride. 1998. Vegetation composition, structure, and chronological change in a decadent Midwestern North America savanna remnant. *Natural Areas Journal* 18:14–27.

Brandt, L., He, H., Iverson, L., Thompson, F.R., Butler, P., Handler, S., Janowiak, M. et al. 2014. Central Hardwoods ecosystem vulnerability assessment and synthesis: A report from the Central Hardwoods Climate Change Response Framework project. Gen. Tech. Rep. NRS-124. Newtown Square, PA: U.S. Department of Agriculture, Forest Service, Northern Research Station, 254 p.

Brawn, J.D. 2006. Effects of restoring oak savannas on bird communities and populations. *Conservation Biology* 20:460–469.

Brawn, J.D., S.K. Robinson and F.R. Thompson, III. 2001. The role of disturbance in the ecology and conservation of birds. *Annual Review of Ecology and Systematics* 32:251–276.

Bronny, C. 1989. One-two punch: Grazing history and the recovery potential of oak savannas. *Restoration Management & Notes* 7(2):73–76.

Brose, P.H., D.C. Dey, R.P. Guyette, J.M. Marschall and M.C. Stambaugh. 2013b. The influences of drought and humans on the fire regimes of northern Pennsylvania, USA. *Canadian Journal of Forest Research* 43:757–767.

Brose, P.H., D.C. Dey, R.J. Phillips and T.A. Waldrop. 2013a. A meta-analysis of the fire-oak literature: Does prescribed burning promote oak reproduction in eastern North America? *Forest Science* 59(3):322–334.

Brose, P.H. and D.H. Van Lear. 1998. Responses of hardwood advance regeneration to seasonal prescribed fires in oak dominated shelterwood stands. *Canadian Journal of Forest Research* 28:331–339.

Clotfelter, E.D., A.B. Pedersen, J.A. Cranford, N. Ram, E.A. Snajdr, V. Nolan, Jr. and E.D. Ketterson. 2007. Acorn mast drives long-term dynamics of rodent and songbird populations. *Oecologia* 154:493–503.

Collins, S.L., A.K. Knapp, J.M. Briggs, J.M. Blair and E.M. Steinauer. 1998. Modulation of diversity by grazing and mowing in native tallgrass prairie. *Science* 280:745–747.

Cottam, G. 1949. The phytosociology of an oak woods in southwestern Wisconsin. *Ecology* 30:271–287.

Curtis, J.T. 1959. *The Vegetation of Wisconsin: An Ordination of Plant Communities*. Madison, WI: University of Wisconsin Press, 657 p.

Davis, M.A., D.W. Peterson, P.B. Reich, M. Crozier, T. Query, E. Mitchell, J. Huntington and P. Bazakas. 2000. Restoring savanna using fire: Impact on the breeding bird community. *Restoration Ecology* 8:30–40.

Delcourt, H.R. and P.A. Delcourt. 1997. Pre-Columbian Native American use of fire on southern Appalachian landscapes. *Conservation Biology* 11:1010–1014.

Delcourt, P.A. and H.R. Delcourt. 1998. The influence of prehistoric human-set fires on oak-chestnut forests in the southern Appalachians. *Castanea* 63(3):337–345.

Dey, D.C. 2014. Sustaining oak forests in eastern North America: Regeneration and recruitment, the pillars of sustainability. *Forest Science* 60(5):926–942.

Dey, D.C. and R.P. Guyette. 2000. Anthropogenic fire history and red oak forests in south-central Ontario. *The Forestry Chronicle* 76(2):339–347.

Dey, D.C. and G. Hartman. 2005. Returning fire to Ozark Highland forest ecosystems: Effects on advance regeneration. *Forest Ecology and Management* 217:37–53.

Dey, D.C., R.G. Jensen and M.J. Wallendorf. 2008. Single-tree selection reduces survival and growth of oak stump sprouts in the Missouri Ozark Highlands. *General Technical Report NRS-P-24.* Newtown Square, PA: USDA Forest Service Northern Research Station, pp. 26–37.

Dey, D.C., P.S. Johnson and H.E. Garrett. 1996. Modeling the regeneration of oak stands in the Missouri Ozark Highlands. *Canadian Journal of Forest Research* 26:573–583.

Dey, D.C. and C.J. Schweitzer. 2014. Restoration for the future: Endpoints, targets and indicators of progress and success. *Journal of Sustainable Forestry* 33(sup1): S43–S65.

Dey, D.C. and C.J. Schweitzer. 2015. Timing fire to minimize damage in managing oak ecosystems. *Proceedings 17th Biennial Southern Silvicultural Research Conference.* Asheville, NC: USDA Forest Service Southern Research Station, e–Gen. Tech. Rep. SRS–203. 11 p. http://www.srs.fs.usda.gov/pubs/47516#sthash.37uPFFy0.dpuf.

DiTomaso, J.M., M.L. Brooks, E.B. Allen, R. Minnich, P.M. Rice and G.B. Kyser. 2006. Control of invasive weeds with prescribed burning. *Weed Technology* 20(2):535–548.

Eastern Tallgrass Prairie and Big Rivers Landscape Scale Conservation Cooperative. 2013. Operations and strategic plan: 2013–2020. Online: http://www.tallgrassprairielcc.org/wp-content/uploads/2013/03/ETPBR-LCC-Strategic-Plan-8-21-13.pdf (accessed September 18, 2013). 120 p.

Fan, Z., J.M. Kabrick, M.A. Spetich, S.R. Shifley and R.G. Jensen. 2008. Oak mortality associated with crown dieback and oak borer attack in the Ozark Highlands. *Forest Ecology Management* 255:2297–2305.

Farrington, S. 2010. *Common Indicator Plants of Missouri Upland Woodlands.* Jefferson City, MO: Missouri Department of Conservation, 16 p.

Fei, S. and K.C. Steiner. 2007. Evidence for increasing red maple abundance in the eastern United States. *Forest Science* 53(4):473–477.

Fei, S., N. Kong, K.C. Steiner, W.K. Moser and E.B. Steiner. 2011. Change in oak abundance in the eastern United States from 1980 to 2008. *Forest Ecology and Management* 262:1370–1377.

Fowler, C.T. 2003. Human health impacts of forest fires in the southern United States: A literature review. *Journal of Ecological Anthropology* 7(1):39–63.

Fox, V.L., C.P. Buehler, C.M. Byers and S.E. Drake. 2010. Forest composition, leaf litter, and songbird communities in oak- vs. maple-dominated forests in the eastern United States. *Forest Ecology and Management* 259:2426–2432.

Fuhlendorf, S.D., D.M. Engle, J. Kerby and R. Hamilton. 2008. Pyric herbivory: Rewilding landscapes through the recoupling of fire and grazing. *Conservation Biology* 23(3):588–598.

Gingrich, S.F. 1967. Measuring and evaluating stocking and stand density in upland hardwood forests in the Central States. *Forest Science* 13:38–53.

Gleason, H.A. 1913. The relation of forest distribution and prairie fires in the middle west. *Torreya* 13(8):173–181.

Grace, J.B., M.D. Smith, S.L. Grace, S.L. Collins and T.J. Stohlgren. 2001. Interactions between fire and invasive plants in temperate grasslands of North America. In: K.E.M. Galley and T.P. Wilson (eds.), *Proceedings of the Invasive Species Workshop: The Role of Fire in the Control and Spread of Invasive Species. Fire Conference 2000: The First National Congress on Fire Ecology, Prevention, and Management.* Tallahassee, FL: Miscellaneous Publication No. 11, Tall Timbers Research Station, pp. 40–65.

Greenberg, C.H., B.S. Collins and F.R. Thompson, III. 2011. *Sustaining Young Forest Communities: Ecology and Management of Early Successional Habitats in the Central Hardwood Region*. New York, NY: Springer, 312 p.

Grimm, E.C. 1984. Fire and other factors controlling the Big Woods vegetation of Minnesota in the mid-nineteenth century. *Ecological Monographs* 54(3):291–311.

Grundel, R. and N.B. Pavlovic. 2007. Response of bird species densities to habitat structure and fire history along a Midwestern open-forest gradient. *Condor* 109:734–749.

Guyette, R.P., D.C. Dey and M.C. Stambaugh. 2003. Fire and human history of a barren-forest mosaic in southern Indiana. *American Midland Naturalist* 149:21–34.

Guyette, R.P., R.M. Muzika and D.C. Dey. 2002. Dynamics of an anthropogenic fire regime. *Ecosystems* 5:472–486.

Guyette, R.P. and M.A. Spetich. 2003. Fire history of oak-pine forests in the lower Boston Mountains, Arkansas, USA. *Forest Ecology and Management* 180:463–474.

Guyette, R.P., M.C. Stambaugh, D.C. Dey and R.M. Muzika. 2012. Predicting fire frequency with chemistry and climate. *Ecosystems* 15:322–335.

Haddad, N.M., D. Tilman, J. Haarstad, M. Ritchie and J.M.H. Knops. 2001. Contrasting effects of plant richness and composition on insect communities: A field experiment. *The American Naturalist* 158(1):17–35.

Hanberry, B.B. 2013. Changing eastern broadleaf, southern mixed, and northern mixed forest ecosystems of the eastern United States. *Forest Ecology and Management* 306:171–178.

Hanberry, B.B., D.C. Dey and H.S. He. 2012. Regime shifts and weakened environmental gradients in open oak and pine ecosystems. *PLoS ONE* 7 (7):e41337.

Hanberry, B.B., D.C. Dey and H.S. He. 2014a. The history of widespread decrease in oak dominance exemplified in a grassland-forest landscape. *Science of the Total Environment* 476–477:591–600.

Hanberry, B.B., D.T. Jones-Farrand and J.M. Kabrick. 2014b. Historical open forest ecosystems in the Missouri Ozarks: Reconstruction and restoration targets. *Ecological Restoration* 32(4):407–416.

Hanberry, B.B., J.M. Kabrick and H.S. He. 2014c. Changing tree composition by life history strategy in a grassland-forest landscape. *Ecosphere* 5(3), art34.

Hanberry, B.B., J.M. Kabrick and H.S. He. 2014d. Densification and state transition across the Missouri Ozarks landcape. *Ecosystems* 17:66–81.

Haney, A. and S.I. Apfelbaum. 1990. Structure and dynamics of midwest oak savannas. In: J.M. Sweeney (ed.), *Management of Dynamic Ecosystems*. West Lafayette, IN: The Wildlife Society, North Central Section, pp. 19–30.

Hansen, R.A. 2000. Effects of habitat complexity and composition on a diverse litter microarthropod assemblage. *Ecology* 81(4):1120–1132.

Hare, R.C. 1965. The contribution of bark to fire resistance of southern trees. *Journal of Forestry* 63:248–251.

Harrington, J.A. and E. Kathol. 2008. Responses of shrub midstory and herbaceous layers to managed grazing and fire in a North American savanna (oak woodland) and prairie landscape. *Restoration Ecology* 17(2):234–244.

Hart, J.L. 2012. History of fire in eastern oak forests and implications for restoration. *Proceedings 4th Fire in Eastern Oak Forests Conference*. General Technical Report NRS-P-102. Newtown Square, PA: USDA Forest Service Northern Research Station, pp. 32–51.

Hartnett, D.C., K.R. Hickman and L.E. Fischer Walter. 1996. Effects of bison grazing, fire, and topography on floristic diversity in tallgrass prairie. *Journal of Range Management* 49:413–420.

Heikens, A.L. and P.A. Robertson. 1994. Barrens of the Midwest: A review of the literature. *Castanea* 59(3):184–194.

Hiers, J.K., J.W. Walters, R.J. Mitchell, J.M. Varner, L.M. Conner, L. Blanc and J. Stowe. 2014. Ecological value of retaining pyrophytic oaks in longleaf pine ecosystems. *The Journal of Wildlife Management* 78(3):383–393.

Hoss, J.A., C.W. Lafon, H.D. Grissno-Mayer, S.R. Aldrich and G.G. DeWeese. 2008. Fire history of a temperate forest with an endemic fire-dependent herb. *Physical Geography* 29(5):424–441.

Hunter, M.L., Jr. and F.K.A. Schmiegelow. 2011. *Wildlife, Forests and Forestry*. New York: Prentice Hall.

Hunter, W.C., D.A. Buehler, R.A. Canterbury, J.L. Confer and P.B. Hamel. 2001. Conservation of disturbance-dependent birds in eastern North America. *Wildlife Society Bulletin* 29(2):440–455.

Huston, M.A. 1994. *Biological Diversity: The Coexistence of Species on Changing Landscapes*. Cambridge, UK: Cambridge University Press, 681 p.

Hutchinson, T.F. 2006. Fire and the herbaceous layer in eastern oak forests. In *General Technical Report NRS-P-1*. Newtown Square, PA: U.S. Department of Agriculture, Forest Service, Northern Research Station, pp. 136–149.

Hutchinson, T.F., R.E.J. Boerner, S. Sutherland, E.K. Sutherland, M. Ortt and L.R. Iverson. 2005. Prescribed fire effects on the herbaceous layer of mixed-oak forests. *Canadian Journal of Forest Research* 35:877–890.

Iverson, L.R. and T.F. Hutchinson. 2002. Soil temperature and moisture fluctuations during and after prescribed fire in mixed-oak forests, USA. *Natural Areas Journal* 22:296–304.

Iverson, L.R., D.A. Yaussy, J. Rebbeck, T.F. Hutchinson, R.P. Long and A.M. Prasad. 2004. A comparison of thermocouples and temperature paints to monitor spatial and temporal characteristics of landscape-scale prescribed fires. *International Journal of Wildland Fire* 13:311–322.

Janowiak, M.K., L.R. Iverson, D.J. Mladenoff, E. Peters, et al. 2014. Forest ecosystem vulnerability assessment and synthesis for northern Wisconsin and western Upper Michigan: A report from the Northwoods Climate Change Response Framework project. General Technical Report NRS-136. Newtown Square, PA: U.S. Department of Agriculture, Forest Service, Northern Research Station, 247 p.

Janowiak, M.K., C.W. Swanston, L.M. Nagel, C.R. Webster, B.J. Palik, M.J. Twery, J.B. Bradford, L.R. Parker, A.T. Hille and S.M. Johnson. 2011. Silvicultural decision making in an uncertain climate future: A workshop-based exploration of considerations, strategies, and approaches. General Technical Report NRS-81. Newtown Square, PA: U.S. Department of Agriculture, Forest Service, Northern Research Station, 14 p.

Johnson, P.S., S.R. Shifley and R. Rogers. 2009. *The Ecology and Silviculture of Oaks* (2nd ed.). New York: CABI Publishing, 580 p.

Kane, J.M., J.M. Varner and J.K. Hiers. 2008. The burning characteristics of southeastern oaks: Discriminating fire facilitators from fire impeders. *Forest Ecology and Management* 256:2039–2045.

Kilburn, P., B. Tutterow and R.B. Brugam. 2009. The tree species composition and history of barrens identified by government land surveyors in southwestern Illinois. *The Journal of the Torrey Botanical Society* 136:272–283.

King, D.I. and S. Schlossberg. 2014. Synthesis of the conservation value of the early-successional stage in forests of eastern North America. *Forest Ecology and Management* 324:186–195.

Kinkead, C.O., J.M. Kabrick, M.C. Stambaugh and K.W. Grabner. 2013. Changes to oak woodland stand structure and ground flora composition caused by thinning and burning. *General Technical Report NRS-P-117*. Newtown Square, PA: U.S. Department of Agriculture, Forest Service, Northern Research Station, pp. 373–383.

Kreye, J.K., J.M. Varner, J.K. Hiers and J. Mola. 2013. Toward a mechanism for eastern North American forest mesophication: Differential litter drying across 17 species. *Ecological Applications* 23(8):1976–1986.

Larsen, D.R., D.C. Dey and T. Faust. 2010. A stocking diagram for Midwestern eastern cottonwood-silver maple-American sycamore bottomland forests. *Northern Journal Applied Forestry* 27(4):132–139.

Law, J.R., P.S. Johnson and G. Houf. 1994. A crown cover chart for oak savannas. *Technical Brief TP-NC-2*. St. Paul, MN: USDA Forest Service North Central Forest Experiment Station, 4 p.

Leach, M.K. and T.J. Givnish. 1999. Gradients in the composition, structure, and diversity of remnant oak savannas in southern Wisconsin. *Ecological Monographs* 69(3):353–374.

Logan, W.B. 2006. *Oak: The Frame of Civilization*. New York, NY: W.W. Norton & Co, 336 p.

Lorimer, C.G., J.W. Chapman and W.D. Lambert. 1994. Tall understorey vegetation as a factor in the poor development of oak seedlings beneath mature stands. *Journal of Ecology* 82:227–237.

Loudermilk, E.L., J.J. O'Brien, R.J. Mitchell, W.P. Cropper, Jr., J.K. Hiers, S. Grunwald and J. Grego. 2012. Linking complex fuel behavior at fine scales. *International Journal of Wildland Fire* 21:882–893.

Marschall, J.M., R.P. Guyette, M.C. Stambaugh, and A.P. Stevenson. 2014. Fire damage effects on red oak timber product value. *Forest Ecology and Management* 320:182–189.

Mayer, A.L. and A.H. Khalyani. 2011. Grass trumps trees with fire. *Science* 334:188–189.

McEwan, R.W., T.F. Hutchinson, R.P. Long, D.R. Ford and B.C. McCarthy. 2007. Temporal and spatial patterns in fire occurrence during the establishment of mixed-oak forests in Eastern North America. *Journal of Vegetation Science* 18:655–664.

McPherson, G.R. 1997. *Ecology and Management of North American Savannas*. Tucson, AZ: The University of Arizona Press, 208 p.

McShea, W.J. and W.M. Healy. 2002. *Oak Forest Ecosystems Ecology and Management for Wildlife*. Baltimore, MD: The Johns Hopkins University Press, 432 p.

McShea, W.J., W.M. Healy, P. Devers, T. Fearer, F.H. Koch, D. Stauffer and J. Waldon. 2007. Forestry matters: Decline of oaks will impact wildlife in hardwood forests. *The Journal of Wildlife Management* 71(5):1717–1728.

Motsinger, J.R., J.M. Kabrick, D.C. Dey, D.E. Henderson and E.K. Zenner. 2010. Effects of midstory and understory removal on the establishment and development of natural and artificial pin oak advance reproduction in bottomland forests. *New Forests* 39:195–213.

MTNF. 2005. *2005 Mark Twain National Forest Land and Resource Management Plan*. Rolla, MO: USDA Forest Service, Mark Twain National Forest.

Nelson, P.W. 2004. Classification and characterization of savannas and woodlands in Missouri. *Proceedings of SRM 2002: Savanna/Woodland Symposium*. Jefferson City, MO: Missouri Department of Conservation, pp. 9–25.

Nelson, P.W. 2010. *The Terrestrial Natural Communities of Missouri*. Jefferson City, MO: Missouri Natural Areas Committee, 550 p.

Noss, R.F., E.T. LaRoe and J.M. Scott. 1995. *Endangered Ecosystems of the United States: A Preliminary Assessment of Loss and Degradation*. Vol. 28. Washington, DC, USA: US Department of the Interior, National Biological Service.

Nowacki, G.J. and M.D. Abrams. 2008. The demise of fire and the mesophication of forests in the eastern United States. *BioScience* 58(2):123–138.

Nuzzo, V.A. 1986. Extent and status of Midwest oak savanna: Presettlement and 1985. *Natural Areas Journal* 6:6–36.

OSFNF. 2005. *Revised Land and Resource Management Plan Ozark-St. Francis National Forests*. Manage. Bull. R8-MB-125A. Asheville, NC: USDA Forest Service Southern Region, 296 p.

Ostrom B.J. and E.F. Loewenstein. 2006. Light transmittance following midstory removal in a riparian hardwood forest. *Gen. Tech. Rep. SRS-92*. Asheville, NC: USDA Forest Service Southern Research Station, pp. 265–268.

Packard, S. 1988. Just a few oddball species: Restoration and the rediscovery of the tallgrass savanna. *Restoration and Management Notes* 6(1):13–22.

Packard, S. 1993. Restoring oak ecosystems. *Restoration and Management Notes* 11(1):5–16.

Packard, S. and C.F. Mutel. 1997. *The Tallgrass Restoration Handbook*. Washington, DC: Island Press, 463 p.

Parker, W.C. and D.C. Dey. 2008. Influence of overstory density on ecophysiology of red oak (*Quercus rubra*) and sugar maple (*Acer saccharum*) seedlings in central Ontario shelterwoods. *Tree Physiology* 28:797–804.

Peterson, D.W. and P.B. Reich. 2008. Fire frequency and tree canopy structure influence plant species diversity in a forest-grassland ecotone. *Plant Ecology* 194:5–16.

Peterson, D.W., P.B. Reich and K.J. Wrage. 2007. Plant functional group responses to fire frequency and tree canopy cover gradients in oak savannas and woodlands. *Journal of Vegetation Science* 18:3–12.

Pyne, S.J. 1982. *Fire in America*. Princeton, NJ: Princeton University Press, 654 p.

Radeloff, V., D.J. Mladenoff, K.L. Maines and M.S. Boyce. 1998. Analyzing forest landscape restoration potential: Pre-settlement and current distribution of oak in the northwest pine barrens. *Transactions of the Wisconsin Academy of Sciences, Arts and Letters* 86:189–205.

Reidy, J.L., F.R. Thompson, III, and S.W. Kendrick. 2014. Breeding bird response to habitat and land-scape factors across a gradient of savanna, woodland, and forest in the Missouri Ozarks. *Forest Ecology and Management* 313:34–46.

Robertson, K.R., R.C. Anderson and M.W. Schwartz. 1997. The tallgrass prairie mosaic. In: M.W. Schwartz (ed.), *Conservation in Highly Fragmented Landscapes*. New York, NY: Chapman & Hall, pp. 55–87.

Rodewald, A.D. 2003. Decline of oak forests and implications for forest wildlife conservation. *Natural Areas Journal* 23(4):368–371.

Rodgers, C.S. and R.C. Anderson. 1979. Presettlement vegetation of two prairie peninsula counties. *Botanical Gazette* 140(2):232–240.

Rooney, T.P. 2001. Deer impacts on forest ecosystems: A North American perspective. *Forestry* 74(3):201–208.

Rubbo, M.J. and J.M. Kiesecker. 2004. Leaf litter composition and community structure: Translating regional species changes into local dynamics. *Ecology* 85(9):2519–2525.

Ryan, K.C., E.E. Knapp and J.M. Varner. 2013. Prescribed fire in North American forests and wood-lands: History, current practice, and challenges. *Frontiers in Ecology* 11 (Online Issue 1):e15–e24.

Schulte, L.A., D.J. Mladenoff, T.R. Crow, L.C. Merrick and D.T. Cleland. 2007. Homogenization of northern U.S. Great Lakes forests due to land use. *Landscape Ecology* 22:1089–1103.

Schulte, L.A., D.J. Mladenoff and E.V. Nordheim. 2002. Quantitative classification of a historic north-ern Wisconsin (U.S.A.) landscape: Mapping forests at regional scales. *Canadian Journal of Forest Research* 32(9):1616–1638.

Shifley, S.R., F.X. Aguilar, N. Song, S.I. Stewart, et al. 2012. *Forests of the Northern United States*. General Technical Report NRS-90. USDA Forest Service Northern Research Station, 202 p.

Shifley, S.R., W.K. Moser, D.J. Nowak, P.D. Miles, B.J. Butler, F.X. Aguilar, R.D. DeSantis and E.J. Greenfield. 2014. Five anthropogenic factors that will radically alter forest conditions and management needs in the northern United States. *Forest Science* 60(5):914–925.

Shifley, S.R. and F.R. Thompson, III. 2011. Chapter 6 spatial and temporal patterns in the amount of young forests and implications for biodiversity. In: C.H. Greenberg, B.S. Collins and F.R. Thompson III (eds.), *Sustaining Young Forests Communities Ecology and Management of Early Successional Habitats in the Central Hardwood Region, USA*. New York, NY: Springer, pp. 73–95.

Smith, W.B., P.D. Miles, C.H. Perry and S.A. Pugh. 2009. *Forest Resources of the United States, 2007*. General Technical Report WO-78. Washington, DC: USDA Forest Service, 336 p.

Stambaugh, M.C. and R.P. Guyette. 2008. Predicting spatio-temporal variability in fire return inter-vals using a topographic roughness index. *Forest Ecology and Management* 254:463–473.

Stambaugh, M.C., R.P. Guyette, K.W. Grabner and J. Kolaks. 2006b. Understanding Ozark forest litter variability through a synthesis of accumulation rates and fire events. *General Technical Report RMRS-P-41*. Fort Collins, CO: U.S. Department of Agriculture, Forest Service, Rocky Mountain Research Station, pp. 321–332.

Stambaugh, M.C., R.P. Guyette, E.R. McMurry, E.R. and D.C. Dey. 2006a. Fire history at the eastern Great Plains margin, Missouri River Loess Hills. *Great Plains Research* 16:149–159.

Stambaugh, M.C., J.M. Marschall and R.P. Guyette. 2014. Linking fire history to successional changes of xeric oak woodlands. *Forest Ecology and Management* 320:83–95.

Stambaugh, M.C., J. Sparks, R.P. Guyette and G. Willson. 2011. Fire history of a relict oak woodland in northeast Texas. *Rangeland Ecology and Management* 64:419–423.

Starbuck, C. 2013. *Bat Occupancy of Forests and Managed Savanna and Woodland in the Missouri Ozark Region*. M. Sc. thesis. Columbia, MO: University of Missouri, 82 p.

Starver, A.C., S. Archibald and S.A. Levin. 2011. The global extent and determinants of savanna and forest as alternative biome states. *Science* 334:230–232.

Stoler, A.B. and R.A. Relyea. 2011. Living in the litter: The influence of tree leaf litter on wetland com-munities. *Oikos* 120:862–872.

Swink, F. and G.Wilhelm. 1994. *Plants of the Chicago Region* (4th ed.). Indianapolis, IN: Indiana Academy of Science, 921 p.

Taft, J.B., M.W. Schwartz and L.R. Philippe. 1995. Vegetation ecology of flatwoods on the Illinoian till plain. *Journal of Vegetation Science* 6:647–666.

Taft, J.B., G.S. Wilhelm, D.M. Ladd and L.A. Masters. 1997. Floristic quality assessments for vegetation in Illinois a method for assessing vegetation integrity. http://wwx.inhs.illinois.edu/files/5413/4021/3268/Wilhelm_Illinois_FQA.pdf (accessed March 28, 2014).

Tallamy, D.W. and K.J. Shropshire. 2009. Ranking Lepidoptera use of native versus introduced plants. *Conservation Biology* 23(4):941–947.

Thompson, F.R., III, J.L. Reidy, S.W. Kendrick and J.A. Fitzgerald. 2012. Songbirds in managed and non-managed savannas and woodlands in the central hardwoods region. *General Technical Report NRS-P-102*. Newtown Square, PA: USDA Forest Service Northern Research Station, pp. 159–169.

Transeau, E.N. 1935. The prairie peninsula. *Ecology* 16(3):423–437.

Upper Mississippi River and Great Lakes Joint Venture. 2007. Landbird habitat conservation strategy. Online: http://www.uppermissgreatlakesjv.org/docs/UMRGLR_JV_LandbirdHCS.pdf (accessed September 18, 2013). 128 p.

Vaughn, D.H. 2014. *Fire from the Sky: Climate Change, Oak Decline Enter Debate over Prescribed Burns to Restore Shortleaf Pine*. April 8, 2014. West Plains, MO: West Plains Daily Quill, p. 1, 12.

Wakeling, J.L., A.C. Staver and W.J. Bond. 2011. Simply the best: The transition of savanna saplings to trees. *Oikos* 120:1448–1451.

Waldrop, T.A., D.A. Yaussy, R.J. Phillips, T.F. Hutchinson, L. Brudnak and R.E.J. Boerner. 2008. Fuel reduction treatments affect stand structure of hardwood forests in western North Carolina and southern Ohio, USA. *Forest Ecology and Management* 255:3117–3129.

Williams, M. 1989. *Americans & Their Forests: A Historical Geography*. Cambridge, UK: Cambridge Univ. Press, 599 p.

Zenner, E.K., J.M. Kabrick, R.G. Jensen, J.E. Peck and J.K. Grabner. 2006. Responses of ground flora to a gradient of harvest intensity in the Missouri Ozarks. *Forest Ecology and Management* 222:326–334.

Zouhar, K., J.K. Smith and S. Sutherland. 2008. Effects of fire on nonnative invasive plants and invisibility of wildland ecosystems. *General Technical Report RMRS-42*. Ft. Collins, CO: USDA Forest Service Rocky Mountain Research Station, pp. 7–31.

21

Restoration of Oak Forests (Quercus humboldtii) *in the Colombian Andes: A Case Study of Landscape-Scale Ecological Restoration Initiatives in the Guacha River Watershed*

Andres Avella, Selene Torres, Luis Mario Cárdenas, and Alejandro A. Royo

CONTENTS

21.1 Introduction

Oak forests of Colombia are characterized by the dominance of two species of Fagaceae, *Quercus humboldtii* and *Colombobalanus excelsa*. Both species comprise the core of various Neotropical montane forest ecosystems. However, past and present use of these ecosystems regionally has led to high levels of deforestation for agriculture, forest fragmentation, and loss of biodiversity and ecosystem services. Indeed, some authors estimate the existing cover of oak forest in Colombia represents only 10%–40% of the original extent (Gentry 1993; Rangel-Ch 2000; Etter et al. 2006). Aronson and Andel (2006) asserted ecological restoration should be prioritized in landscapes exemplified by these remnant Andean oak forests where the human dependence and influence over native ecosystems is strong, and plant and animal diversity remains relatively high.

In Colombia, the Nature Foundation (Fundación Natura) is a nonprofit agency that initiated a regional conservation, restoration, and rehabilitation plan in the northern sector of the eastern range of the Andes, known as the Guantiva-La Rusia-Iguaque (GRI) conservation corridor. Current landscape habitat structure in this region has resulted from the dependence of local rural communities on natural resources. Given the deeply intertwined relationship between the socio-economic and ecological systems (i.e., socio-ecological system) of the region, the conservation and restoration plan seeks to utilize the existing land use mosaic as an opportunity for conserving biodiversity and improving the quality of life in local communities.

In this chapter we summarize the importance to conservation of biodiversity and the ecological characteristics of oak forests in the Colombian Andes, the various benefits they provide to rural communities, as well as the urgent need of develop restoration initiatives in order to improve their current conservation status. We present a description of the GRI conservation corridor area and briefly summarize the criteria that guided the conservation and sustainable use action plan. We summarize a case study of an ecological restoration project carried out in oak (*Q. humboldtii*) forests within the GRI conservation corridor. Finally, we highlight the current need to develop more comprehensive proposals for addressing issues of conservation and ecological restoration within the Colombian Andes region.

21.2 Overview of Oak Forests in Colombia

According to Kapelle (2006), Neotropical *Quercus* species range from central Mexico (23°30′N) to the Andean region of northern Colombia. Thus, existence of the genus in Colombia is biogeographically unique in South America. Palynological records indicate *Quercus* arrived in northern South America approximately 250,000–470,000 years ago (Van der Hammen and Gonzalez 1963; Hooghiemstra 2006). The genus entered South America through the mountainous region in the Darien Gap between Panama and the Choco Department of Colombia. From there *Quercus* expanded southward throughout the central and western branches of the Colombian Andes reaching 1°N. Later, the genus crossed the narrow sectors of the Magdalena River basin and expanded northward through the eastern branch of the Colombian Andes reaching its northernmost distribution in Colombia at 8°N (Van der Hammen et al. 2008).

Currently, *Q. humboldtii* is distributed across all three branches of the Colombian Andes between 750 and 3450 m elevation above sea level (Cuatrecasas 1958; Lozano and Torres 1974; Pulido et al. 2006; Cantillo and Rangel 2011), as well as in some isolated massifs in the Caribbean region of Colombia (Rangel et al. 2009; Rangel and Avella 2011). This same region of the Colombian Andes also contains forests dominated by another Fagaceae species commonly called Black Oak (*Colombobalanus excelsa*)—this species fills an ecological niche similar to *Q. humboldtii* in Andean forests. The distribution of *C. excelsa* forests is estimated at approximately 1100 km^2 with relict populations recorded in the departments of Huila, Valle del Cauca, Antioquia, and Santander (Cárdenas and Salinas 2006). The specific loss of oak ecosystems is still an unknown, as accurate assessments of their historic and actual coverage are still lacking. Nevertheless, it is clear these unique community types are highly threatened ecosystems because approximately 40% of the territory has been transformed by population growth and land use change (Etter et al. 2006). In particular, deforestation has altered at least 60% of the natural forest cover throughout the Andean region (Andrade 1993; Gentry 1993; Rangel 2000).

Oak forests of the Colombian Andes contain high levels of biodiversity and serve as refugia and critical habitat for several plant and animal species. These forests have high conservation value at local, regional, and national scales as they contain numerous endemic, rare, and threatened plant species. Plant species of concern found in these forests include

Aniba perutilis, Cedrela montana, Juglans neotropica, Matudaea colombiana, Podocarpus oleifolius, Sterigmapetalum tachirense, various *Magnolia* spp., *Weinmannia* spp., *Nectandra* spp., *Ocotea* spp., *Persea* spp., and a number of palms (e.g., *Geonoma orbignyana, Ceroxylum quinduense, C. vogelianum, C. parvifrons*) (Avella and Cárdenas 2010). Oak forests support at least 29 globally threatened bird species, including endemics such as Gorgeted Wood-Quail (*Odontophorus strophium*), Chestnut-capped Piha (*Lipaugus weberi*), and the critically endangered Colorful Puffleg (*Eriocnemis mirabilis*). Two rare parrot species, the Rusty-faced Parrot (*Hapalopsittaca amazonia*) and the critically endangered Fuertes's Parrot (*Hapalopsittaca fuertesi*), appear restricted to montane oak forests of Colombia where they depend heavily on *Q. humboldtii* acorns as food Fuertes's Parrot was thought to be extinct until rediscovered in the early part of this century (Stattersfield and Capper 2000; Renjifo 2002). Five other threatened parrot species are associated with Andean oak forests including the Flame-winged Parakeet (*Pyrrhura calliptera*), the Rufous-fronted Parakeet (*Bolborhynchus ferrugineifrons*), the Golden-plumed Parakeet (*Leptosittaca branickii*), the Yellow-eared Parrot (*Ognorhynchus icterotis*), and the Spot-winged Parrotlet (*Touit stictopterus*) (see Hilty and Brown 1986; Stotz et al. 1996). Additionally, Andean oak forests function as the primary wintering habitat for Golden-winged (*Vermivora chrysoptera*) and Cerulean warblers (*Setophaga cerulea*), two Neotropical migrants of very high conservation concern (e.g., Saenz 2010).

In Colombia, oak forests also provide important cultural, social, and economic benefits to communities. Timber harvesting for national use is estimated at 20 million cubic meters per year with approximately 80% of that being used to satisfy domestic needs including construction, agricultural uses, and fuel (Díaz 2010). The World Health Organization (WHO) estimated 15% of the Colombian population (approximately 2.5 million people) is dependent on firewood and charcoal as their primary source of heating and cooking fuel (World Health Organization 2006). In total, an estimated 8.43 million cubic meters of wood are utilized annually for fuel wood with the average peasant family consuming approximately 6.2 tons of fuel wood yearly, much of it coming from oak forests (Aristizabal 2010). Additionally, some communities have artisans and crafters that utilize wood from oak and associated species to create products that are sold in order to supplement their livelihoods. The importance of these ancillary activities to rural families cannot be disregarded considering the relatively low productivity of the land for agriculture and livestock, and the poor condition and low access of transportation networks to get perishable goods to market. Finally, Andean oak forests provide various important ecosystem services nationwide, including sustained and continued hydrologic flow, mitigation of soil erosion, buffering against natural disasters, and various timber and nontimber forest products (e.g., Botero et al. 2009).

21.3 Study Area

On the western slopes of the eastern branch of the Colombian Andes, between the departments (i.e., states) of Boyacá and Santander, is an area that includes rural Andean agricultural landscapes, arid regions, páramos (i.e., tundras), and oak-dominated forests. According to Solano et al. (2005), this region contains the most extensive oak forests of the country. Beginning in 2003, the Natura Foundation supported development of the Guantiva-La Rusia-Iguaque (GRI) conservation corridor: a 1,073,000 hectare region ranging in elevation from 350–4100 m and encompassing 67 municipalities in the departments of Boyacá, Santander, and Cudinamarca. Spatially, the GRI spans the continuum between montane forests and páramo and is exemplified by a matrix of both natural and anthropogenically converted habitats dominated by the Guantiva, La Russia, and Iguaque massifs.

From a policy standpoint, the management plan for the corridor (*Estrategia de Desarrollo Sostenible, Corredor de Conservación Guantiva-La Rusia-Iguaque*; Solano et al. 2005), developed under the leadership of the Natura Foundation, defines clear targets and strategies focused on conservation of Andean oak ecosystems. The importance of these guidelines were strengthened by the passing of the national ban on *Quercus humboltii* harvesting enacted by the Ministry of the Environment (Resolution 096-2006) and further bolstered by Resolution 095-2008, which designated the Integrated Management District of Guantiva-La Rusia a protected area. Resolution 0161-2010 specifically mandated that 7828 ha of the region be designated for restoration (Corporación Autónoma de Santander (CAS) 2011).

Nevertheless, these decrees have not significantly changed conditions on the ground, in part, due to the inconsistent application and interpretation by organizations and personnel. In stark contrast, the corridor management plan has succeeded both in emphasizing the socio-ecological importance of the region and securing resources towards its conservation. For example, since 2004, the Fund for Environmental Action and Childhood (*Fondo para la Acción Ambiental y la Niñez*) has financed projects aimed at fomenting sustainable agriculture and grazing, strengthening the capacity of local organizations, and supporting the establishment of private nature reserves.

Avella and Rangel (2014) used seven criteria to design a sustainable forest management plan that concurrently achieved conservation goals while improving the quality of life dependent on the land. The criteria were based on the forest ecosystem management approaches of Lammerts and Bloom (1997) and OIMT (2005), and the ecosystem approach advocated by SCDB (2004) and Herrera and Chaverri (2006). These criteria, their definitions, and regional significance are presented in Table 21.1.

With funding from the MacArthur Foundation, we developed a landscape level ecological restoration project in the central portion of the GRI conservation corridor from 2006 until 2011. The area of focus encompassed the greater Guacha river watershed region of the Encino, Belén, Duitama, and Santa Rosa de Viterbo Municipalities of the Santander and Boyacá departments (Figure 21.1). The project area is 27,545 ha and encompassed an elevation gradient from 1650–4240 m. Within the greater watershed are the sub-watersheds of the Guacha River (14,877 ha), Minas River (10,025 ha), and La Lejía Brook (2625 ha). The landscape is primarily mountainous with abrupt homoclinal ridges, or cuestas.

21.4 Methods

We followed approaches advocated by Reynolds and Hessburg (2005); Aronson and Andel (2006); Rodrígues et al. (2011) and the ITTO (2002) to develop an ecological restoration plan that consisted of four phases: diagnosis, planning, implementation, and monitoring. In the diagnostic phase, we conducted vegetation composition and structure surveys to determine the current state of the system, identify key functions provided to rural communities, and learn of the ecological and social barriers to natural regeneration of forests. In the planning phase, we assessed the conservation status of the landscape, defined specific management units, and established baseline reference areas. This information allowed us to plan restoration

TABLE 21.1

Forest Ecosystem Management Criteria for the Guantiva-La Rusia-Iguaque (GRI) Conservation Corridor Designed to Improve Sustainable Forest Management, Conserve Natural Forest, and Improve Quality of Life of the Local Communities Dependent on these Forests

Criteria	Definition and Significance
Legal, political, and institutional framework supporting sustainable forest management	This criterion comprises the institutional variables needed to guarantee sustainable structures for determining the extent of the national political commitment to sustainable forest management. In this case, the framework encompasses the Colombian Political Constitution of 1991, The National Biodiversity Policy, the Decree-Law on Forest Harvesting, the National Forest Policy, the National forestry Development Plan, the National Environmental Research Policy (particularly the MAVDT Resolution #096 of 2006 that prohibits harvesting oaks).
Conservation of oak forest biodiversity	This criterion is related to the conservation and management of biological diversity, including ecosystems, species, and genetic diversity. It establishes the assessment of biodiversity in areas suitable for productive forests.
Maintenance of ecological integrity	The healthy functioning of forest ecosystems can be affected by human activities such as illegal extraction, fires, livestock grazing, mining and illegal hunting, and exotic invasives; as well as by natural phenomena like fires, insect attacks, diseases, strong winds, floods and droughts. The main activities found to affect the health of oak forests were: wood extraction for mining activities, timber, and charcoal production. Currently, the factors causing alterations and degradation are the expansion of agriculture through incremental and almost imperceptible actions, and the demand for wood and firewood for domestic use.
Soil and water conservation	This criterion relates to the maintenance of soil and water quality in both forests and productive areas. Soil and water quality are also variables that influence the ecological integrity of forests. Furthermore, the criterion addresses water flow in lowland rivers and mitigation of the effects of floods and sedimentation.
Sustainability of timber and nontimber forest products	This criterion is related to the management of the forest for the extraction of timber and nontimber products. Production will be sustainable in the long term only if it is economically feasible, ecologically rational, and accepted by society. Productive forests can also offer important environmental services such as carbon storage and preservation of fauna and flora. These functions can only be maintained through the application of rational forest management practices that ensure the sustainability of all ecosystem services.
Incorporation of cultural views, uses, and oak management practices	Local knowledge of the management and preservation of oak forests exists within the region in what has been termed the "oak culture" (Solano et al., 2005). Considering the slow development of scientific and technical knowledge about Andean oak forest management, this grassroots cultural knowledge could prove highly informative to the design of management alternatives. The incorporation of this criterion should generate future processes in which local communities and institutions would recognize and value these cultural strengths as a means to foment a sustainable relationship with the environment.
Social and economic proposals to achieve oak forest sustainability and improve quality of life	This criterion addresses the economic and social aspects of oak forests. Properly managed, natural renewable resources that produces products and services required to meet the needs of surrounding communities. Similarly, they contribute to human well-being and quality of life by offering ecotourism, recreational, and environment-related job opportunities.

strategies that served, maintained or enhanced the interests of the local community, as well as the attributes of the reference areas. The third phase consisted of implementing ecological restoration strategies following formalized conservation-production agreements signed between the Nature Foundation and 80 landowners. These agreements were enacted with the aim of directly involving local landowners as central players in the management of the region (Solano 2012). The selection criteria for inclusion into the project included, among others, the presence

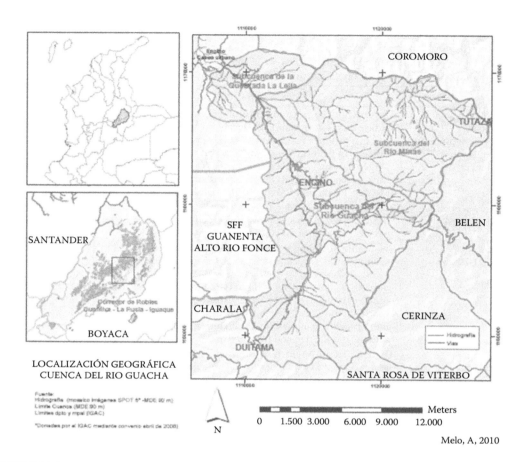

FIGURE 21.1
Map of the study region. (Adapted from Melo, A. 2010. Sistema de monitoreo a escala de paisaje de la conservación y uso del recurso forestal de roble en el corredor a través de sistemas de información geográfica cuenca del río guacha (Encino, Santander y Belén, Boyacá). (GIS analyses of the Rio Guacha watershed to monitor conservation áreas and forest use within the conservation corridor). Bogotá: Fundación Natura.)

of representative forest cover, position within priority areas of habitat connectivity, land ownership status, and the cooperation of the landowner as evidenced by willing participation in other local organizations (Melo and Arango 2012). The final phase established a monitoring system to assess long-term changes in composition, structure and function of the forests.

21.5 Implementation and Results

The results include the first two years of development of this case study and they are presented as described above: diagnosis (Phase I) and planning (Phase II), which were developed during the first year, and implementation (Phase III) and monitoring (Phase IV), which were developed during the second year. Additionally, we report survey data characterizing the views of the local community and participating landowners in regards to the restoration process.

21.5.1 Diagnosis

The Guacha watershed is comprised of several types of ecosystems including sub-Andean forest, Andean forest, and páramo. Forests cover 22% of the total watershed area (6318 ha) and páramos cover 39% of the watershed (10,745 ha). In total, approximately 68% of the watershed (18,704 ha) is categorized as natural vegetation associations whereas 30% (8399 ha) is categorized as transformed landscapes (e.g., pastures, plantations, degraded bare lands). Interestingly, successional processes associated with land abandonment following anthropogenic use have yielded areas of early successional landscapes, including shrub land and young secondary forest. Overall, the landscape is considered fragmented as only 42% of the land area is covered by forests and patch connectivity is low (Melo 2010).

In general, there are two types of forest fragments. The first are large fragments (≥50 ha) with low levels of human disturbance. These are typically found on steeper slopes (≥60% slope) at higher elevations. The second type represents small forest relicts (<5 ha) that are subject to frequent human perturbation from fuel-wood cutting and grazing. These are typically found at lower elevations where slopes are not very steep and the soils have high agricultural productivity (Avella and Garcia 2011). We delineated three elevation zones in the study area: (i) High (2800–3400 m), (ii) Middle (2200–2800 m), and (iii) Low (1800–2200 m). Avella (2010) speculated that the original ecosystems on sites between 1650 and 3400 m elevation were primarily oak-dominated forests whereas páramos were found at the higher elevations. Currently, as a result of anthropogenic disturbance, the landscape is a mosaic of production systems (i.e., agriculture, livestock), extant forest patches ranging from fairly intact to degraded, early successional shrub lands, and páramo. Work by Devia and Arenas (2000), and Avella and Cárdenas (2010), suggested the current landscape structure is the direct result of the two predominant socio-economic development processes: (i) the overexploitation of original forest cover for high-value timber and sleepers for railway construction and (ii) agricultural expansion. Primary barriers to regeneration of tree species in this forest type include grazing and agriculture, availability of propagules, invasive species, a paucity of pollinators and dispersers, and a lack of a seed bank. These barriers have prevented natural regeneration in small forest fragments and other degraded forest areas. While many of these barriers are less pronounced or nonexistent in larger forest fragments, the lack of landscape-level connectivity may limit gene flow between fragments and endanger long-term sustainability.

21.5.2 Planning

We established three distinct management units based on five criteria: (i) number of vegetation strata, (ii) percentage of canopy cover, (iii) presence of native forest species, (iv) degree of human impact, and (v) fragment size.

- Management Unit 1: Intact Forests (IF): These represent relict forest patches greater than 50 ha, with 4–5 vegetation strata and tree species composition is dominated by late-successional species like including *Quercus humboldtii* and others belonging to the genera *Billia, Ocotea, Nectandra, Podocarpus, Compsoneura, Virola* and *Aniba*. Human activity in these management units is low and has not altered the regeneration dynamics of the system. At the landscape level, the main barrier to long-term conservation of these forests is low connectivity with other similar patches. They represent high quality habitats that ensure continued provision of ecosystem services.

- Management Unit 2: Degraded Forests (DF): These are forest relicts of less than 10 ha with less continuous forest cover and fewer vegetation strata than the IF. In particular, the shrub and herb layer is highly degraded. Forests in these management units lack several late-successional species. Barriers to regeneration are widespread and include grazing, low propagule availability, low landscape connectivity, and invasion by exotic plant species such as *Pteridium aquilinum* and *Pennisetum clandestinum*. Resource extraction in these forests has been unsustainable resulting in altered structure, function, and processes that limit short- to medium-term recovery.

- Management Unit 3: Converted Forest (CF): These represent ecosystems where the level of degradation prevents a quick return to the original forest cover. These management units were typically deforested for agricultural production. These areas are characterized by pastures or openings dominated by a dense cover of shade-intolerant pioneer herbs that inhibit tree establishment. Barriers to natural regeneration include agriculture, grazing, a lack of dispersed and dormant seeds, exotic invasive species, a lack of favorable microsites for seedling establishment, inadequate habitat for fauna resulting in a general lack of pollinators and seed dispersal agents, and a lack of patch connectivity.

In highly modified landscapes, the desired ecological reference state may be defined by plant compositional and structural trajectories established by management rather than by a static ecosystem state (Rodrígues et al. 2011). In our study region, intact forest remnants (IF) were defined as the reference system to guide our ecological restoration process for degraded forests (DF). To guide restoration goals on CF, we considered attributes including tree density and size, native species richness, and restoration of some measure of productivity and ecosystem services necessary to meet long-term wood consumption requirements of the local communities.

The overarching goal of our watershed restoration strategy was to maintain and increase forest cover while simultaneously meeting the demands of local communities including hydrological regulation, soil protection, a sustained source of timber, fuel wood, and other nontimber forest products. Given that the main cause of forest loss in the region has been the expansion of livestock grazing, the restoration strategy considers the establishment of agroforestry systems that maintain current livestock productivity while averting their expansion. Another process that has contributed to the degradation of forests is the use of wood as an energy source for cooking and heating. For these reasons, the restoration strategy focused on reducing dependence on natural forests as a source of wood supply by establishing multipurpose plantation forests to ensure energy independence of the family unit in the medium- and long-term.

21.5.3 Implementation of the Strategy

The management objective for IF is to ensure ecosystem protection for continuity of ecological processes and flow of ecosystem services. To date, we have secured protection of 3433 ha of forest cover through signing formal conservation-production agreements or easements. Of the total, 125 ha were protected in the high elevation zone, 863 ha in the middle elevation zone and 2444 ha in the lower elevation zone. Individual ownerships at higher elevations were typically 30–50 ha, but parcels typically had low (<30%) forest cover (Avella et al. 2013). Individual ownerships at mid-elevations were smaller (7–20 ha) with approximately 30% of each ownership containing forests in varying states of conservation. The area protected in the lower elevation zone is largely comprised of the 1200 ha Cachalú Biological Reserve, which is 71% forested and has no pastures or agricultural lands.

To enhance seedling regeneration, we eliminated grazing pressure from all IF patches by constructing fences or herding cattle to other areas. Management of intact forests focused exclusively on enhancing growth, survival, and reproduction of ecologically important seedlings, saplings, and mature trees. Following the silvicultural guidelines proposed by Avella and Garcia (2011), we conducted pruning on individuals of *Q. humboltii, W. tomentosa, P. oleifolius* and *Callophylum brasilense* in order to enhance growth, improve tree form and nurture seed production.

The main restoration objective in DF is to initiate successional trajectories that enhance ecological integrity relative to the structural and floristic composition of the IF reference. It was necessary to halt timber extraction and grazing pressure from these forests remnants before initiating silvicultural treatments. Conservation agreements that limited wood extraction and excluded grazing animals via fencing were signed on 34 ha of DF, 18 ha in the high elevation zone and 16 ha in the middle zone.

The DF stands often lacked adequate advance regeneration in the form of saplings or pole-sized trees and, indeed, the structure and species composition of these forest understories was severely impacted by grazing. Therefore, silvicultural treatments were designed to enhance natural regeneration, nurture growth and development of existing and mostly intermediate-shade tolerant regeneration, promote the establishment of a dense and diverse understory, and maintain a minimum of 70% forest cover. As a complementary treatment we pruned the crowns of ecologically important species including *Q. humboldtii, M. guianensis, W. tomentosa, Scheflera fontiana, P. oleifolius, F. andicola, C. brasilense*, and *Dacryodes* spp. in both the overstory and midstory. Treatment goals were to reduce tree density by 10%–20% and thereby increase light availability into the understory, improve tree form and nurture seed production and growth. We established three community-run tree nurseries that produced over 128,000 seedlings of 41 species (35 native and 6 exotic). Seedlings were outplanted into our restoration (DF) and rehabilitation (CF) sites, and to augment natural regeneration where it was inadequate.

In contrast to the ecological conservation and restoration approaches used in IF and DF fragments, rehabilitation approaches in CF areas do not emphasize restoration of biotic integrity in terms of species composition and structure due to the high level of degradation and the low likelihood of restoring components of the original system (Society for Ecological Restoration International (SER) 2004). Such is the case for CF areas in the Guacha watershed, where local communities subsist on agriculture and livestock production. Local communities require fuel wood to meet their energy needs, and lumber for various domestic and agricultural purposes (e.g., fence posts, construction, tool handles). The objectives for the CF management unit were to restore some measure of productivity and ecosystem services to meet long-term wood consumption requirements of local communities and diminish pressure on existing forests. To achieve these objectives, we established multipurpose forest plantations and developed silvopastoral systems appropriate to these areas (Zapata and Diaz 2012).

Plantations were established on 60 individual farms with each landowner setting aside between 0.25 and 0.5 ha of their properties for the creation of these multipurpose forests. A total of 20 ha of plantations were established. Species composition of each plantation was determined through consultations between the technical team and individual landowners that identified specific needs as well as potential medium- and long-term resource potential. In some cases, multipurpose plantations were established along margins of existing forest patches to increase forest cover and augment the system with species that confer benefits to the production system. Plantings included 24,000 seedlings of 20 native species and five exotic species.

Improved silvopastural practices were implemented across 130 ha in the region with the goal of improving agricultural productivity and promoting sound agricultural practices. These approaches enhance ecosystem services such as biodiversity conservation, carbon sequestration, soil conservation, and protection of water resources. The project sought to develop a model of sustainable grazing through the establishment of hedgerow trees, improved grassland, forage banks, and fruit crops.

21.5.4 Monitoring Ecological Restoration Strategies and Evaluating the Process by Local Communities

A monitoring system was designed and implemented to analyze natural dynamics and responses of forests to ecological restoration treatments. We established 100 permanent monitoring plots: 39 in IF, 38 in DF, and 23 in multipurpose plantations on CF lands. Plots in IF and DF stands were 1000 m^2, and plots in multipurpose plantations on CF lands were 25 m^2. In IF stands, all trees ≥10 cm dbh were identified and tagged throughout the 1000 m^2 plot. For each individual, we measured height, diameter, and phenology (flowering and fruiting), as well as assessed their survival and health status. Additionally, we measured the number of vegetation strata, tree cover by strata, richness and abundance of natural regeneration, litter cover and a qualitative assessment of anthropogenic stressors that might affect restoration goals (e.g., evidence of cattle grazing, timber/fuel wood extraction).

Because the implemented management plan and its permanent monitoring plots remain in their initial stages, we cannot assess whether our restoration activities are enhancing ecological integrity of the DF and CV compared to the IF. Nevertheless, our monitoring provides critical baseline information that will ultimately permit us to evaluate success in enhancing the ecological and socio-economic values of an entire region.

It is clear, however, that implementation of the management plan has yielded important social benefits as evidenced by the active and continued participation of various landowner participants in a variety of activities initiated by the Natura Foundation including: (i) a shift towards silvopastoral systems from conventional grazing, (ii) the conservation and management of the remnant oak forests including the restoration of connectivity in former pastures, and (iii) the initiation in 2010 of coordinated forest planning and management designed to enhance adaptation to climate change within the oak forest corridor.

Collaboration between the Natura Foundation and private landowners over seven years has yielded important socioecological impacts at both the ownership and the landscape level. At the level of the ownership, each landowner designated an average of 1.2 ha to silvopastoral systems. 0.2 ha for fuel wood forests, and protected remnant forests and water sources. At the landscape level, the impact is evident in the conservation of 3433 ha of extant forest, the establishment of 20 ha of fuel wood forests, and the implementation of 130 ha of silvopastoral systems (Table 21.2). In sum, these actions have helped conserve and enhance functional aspects of the ecosystems that support much of the local economy and thereby promote food security, protect soils, improve the beef industry, with the potential for a nontraditional forest products industry. Our continued monitoring will elucidate whether or not management to restore species composition, richness, and structure to levels that are characteristic of IF can be done in a manner that is compatible with the socio-economic needs of inhabitants of the GRI region (see below).

TABLE 21.2

Total and Average Area Contained within Each Restoration and Rehabilitation Management Unit

Management Unit	Restoration Strategy	Area (ha)		Number of Species Conserved	Average Remnant Area (ha)	Total Area of Extant Forest Remnants (ha)
Intact Forests (IF) and Degraded Forests (DF):	Conservation of extant forest remnants	High Elevation Zone	124.5	21 in IF 13 in DF	21	3,433
		Mid Elevation Zone	863	41 in IF 31 in DF		
		Low Elevation Zone	2,444	92 in IF		

Management Unit	Rehabilitation Strategy	Area (ha)	Number of Species Planted	Total Area Protected (ha)
Intact Forests and Degraded Forests	Protection of water resources	40	2	40

Management Unit	Rehabilitation Strategy	Extent		Average Extent	Number of Species Planted	Total Area Multipurpose forests (ha)
Converted Forest	Multipurpose forests	20 ha		0.2 ha	25	20
	Silvopastoral systems	Living Fences	24 km	0.3 km	20	130
		Forage banks	16 ha	0.2 ha	11	
		Improved grassland	80 ha	1 ha	1	

Peñaloza (2012) surveyed the residents of the watershed and landowners who signed cooperative agreements with the Nature Foundation and identified multiple benefits of the restoration and rehabilitation actions including: (i) increased quantity and quality of livestock forage, (ii) gradual recovery of soil conditions by the increasing pasture rotation length, (iii) reduced wind and solar impact on meadows from increased tree planting or establishment, and (iv) availability of shade for livestock due to the establishment of live fences. Technical expertise and assistance were identified as the most useful components of the process.

Surveys also identified several aspects that limited the potential efficacy of the project. These included: (i) the fact that the distribution of participating properties across the landscape could not be consolidated via conservation corridors, (ii) a lack of coordination between private organizations and government authorities, and (iii) the ingrained belief in the community that forests are considered as private property whose use is determined exclusively by the landowners rather than a resource that provides services for the common good.

Peñaloza (2012) provided specific recommendations to improve ecological restoration practices in the region and to encourage greater acceptance and adoption of these practices. These include: (i) adding an agricultural component that focuses on enhancing food security, (ii) explicit consideration and integration of local knowledge regarding the traditional use of natural resources, (iii) increased outreach by the Nature Foundation or other NGOs to enhance local understanding of ecosystem services, management and decision-making processes, and (iv) inclusion of youth and educational organizations in the process from design to implementation and monitoring.

Complementing the initiatives described above, it is worth noting that the national government declared two protected areas within the corridor: *Sanctuario de Flora y Fauna Iguaque* and the *Sanctuario de Flora y Fauna Guanentá Alto Río Fonce*. Additional restoration initiatives, both actual and planned, are being developed following management plans decreed by governmental resolutions 044 and 045 of January 26th, 2007 (RUNAP 2014).

21.6 Final Considerations

Results obtained during the restoration strategy process can be useful to inform new restoration initiatives carried out in Andean oak forests. First, our approach explicitly recognized stresses on the remnant forest patches caused by humans (e.g., grazing and fuel wood collection) and integrated these factors into the strategy. Thus, restoration activities should consider the needs of the local communities, evaluate and assess how the activities have led to the current state of the forest, and develop management strategies that include them.

Within this context, any restoration actions should grow from collaborative agreements between local communities and management/conservation organizations that consider the joint goals of conservation along with sustainable production on the landscape. In this manner, ecosystems with high conservation value and zones with steep slopes and high soil erosion risk are prioritized for protection through conservation easements so they can continue to regulate hydrology, protect soils, and safeguard plant and animal species diversity. For residents of the region, the perceived benefits gained by the maintenance of hydrological regulation allow the establishment of these initial agreements and provided additional incentives that increased the likelihood of success of any additional restoration and rehabilitation activities including plantation systems and silvicultural manipulations.

The enhanced pasture and grazing conditions created by living fences, the prospect of a constant and sustainable fuel wood supply from the multipurpose plantations, and the increased forage quality and quantity are perceived as incentives towards conservation and are an essential part of the restoration strategy. In summary, our case study demonstrates that grounding restoration strategies on the fundamental condition that needs of local communities should be met and improved using their traditional production activities, ultimately facilitates achieving the overall conservation and restoration objectives and increases the flow of ecosystem services.

Keeping in mind the significant human, social, institutional, and financial investments made to achieve our objectives, the challenge now is to assure its continued viability. Our hope is that local communities will not revert to their prior practices that degraded the forests, but rather take ownership of the new conservation model and maintain it. Given that this work takes place in impoverished, rural communities which will require external funds to continue and grow this process, it remains critical to actively seek external funding and continued technical assistance from the Natura Foundation in the short-term in order to improve the socio-economic conditions of the local families and thereby increase the possibility that they can maintain and expand restoration activities in the long-term.

In order to achieve this goal, we propose to: (i) openly discuss and define the desired future socioecological scenarios for their region, (ii) create and/or strengthen local institutions that can design and carry out their own management programs (Torres and Dueñas 2014), (iii) establish a market for a variety of sustainably produced products in order to provide economic benefits to local communities and thus continue to promote conservation and restoration, and (iv) integrate these initiatives into the process of land-use planning and management in a manner that explicitly recognizes the relationship among socio-economic needs and ecosystem services, the ecological structure of the area, and ecosystem risk assessment (including climate change) and management (Andrade et al. 2013).

The preceding initiatives must occur concomitantly with improvements in project implementation by environmental agencies such that they interact in a more effective manner with local landowners. Such projects, informed by local suggestions and needs should conform to national conservation and restoration mandates such as the National Restoration Plan and the decree for environmental compensation (MADS 2012) in order to ensure a sustained leveraging of resources, strengthening of habitat management, and establish further restoration initiatives such as the nationally mandated habitat mitigation that should occur following infrastructure development (Ministerio de Ambiente y Desarrollo Sostenible (MADS) 2012).

Andean oak systems are typically located in areas of human habitation with agriculture and livestock production very similar to the ones in this case study. As remnant oak systems become increasingly fragmented by human-use, our study highlights how restoration and conservation strategies should be developed in cooperation with local communities and with explicit consideration of their needs. Ultimately, this approach will ensure that the road to oak forest conservation in the Andes is achieved with the inhabitants of the region and not at their expense.

References

Andrade, G. I. 1993. Biodiversidad y conservación en Colombia *(Biodiversity and conservation in Colombia)*. In: S. Cárdenas y H. Correa (eds). Nuestra Diversidad Biológica. *Fundación Alejandro Escobar, Colección María Restrepo Ángel:* pp. 23–42. Bogotá: CEREC.

Andrade, G., A. Avella, and W. Gómez. 2013. Construcción de un paisaje resiliente en la cuenca del Río Guacha (Encino, Santander): Conceptos, diagnóstico y lineamientos para su implementación (Establishment of a resilient landscape within the Rio Guacha watershed (Encino, Santander): concepts, diagnosis, and guidelines for implementation). Colombia. Fundación Natura. Unpublished report on file at Fundación Natura, Bogotá.

Aristizabal, J. 2010. Estufas mejoradas y bancos de leña: una alternativa de autoabastecimiento energético a nivel de finca para comunidades dependientes de los bosques de roble de la cordillera oriental. *(Improved stoves and fuelwood reserves: a self-sufficient alternative energy for farms in communities dependent on oak forests in the Eastern range).* Revista Colombia Forestal 13 (2): 5–30.

Aronson, J. and J. Andel. (eds). 2006. *Restoration Ecology: The New Frontier.* United Kingdom: Blackwell Publishing.

Avella, M. 2010. *Aproximaciones al ecosistema de referencia para los procesos de restauración ecológica y manejo ecosistémico de los bosques de roble de la cuenca del río guacha Ecosistema de referencia para la cuenca del rio Guacha. (Reference ecosystem benchmarks for ecological restoration and sustainable management in oak forests of the Rio Guacha watershed).* Bogotá, Colombia: Unpublished report on file at Fundación Natura.

Avella, A. and L. M. Cárdenas. 2010. Conservación y uso sostenible de los bosques de roble en el corredor de conservación Guantiva-La Rusia—Iguaque, departamentos de Santander y Boyacá, Colombia. *(Conservation and sustainable use of oak forests in the Guantiva-LaRusia-Iguaque Conservation Corridor).* Revista Colombia Forestal 13 (2): 5–30.

Avella, A. and N. Garcia. 2011. *Planes de manejo forestal sostenible para la Cuenca del Río Guacha. (Sustainable forestry practices for the Rio Guacha watershed).* Bogotá, Colombia: Fundación Natura.

Avella, A. and O. Rangel. 2014. Oak forests types of Quercus humboldtii and their sustainable use and conservation in the Guantiva-La Rusia-Iguaque corridor (Santander-Boyacá, Colombia). *Revista Colombia Forestal* 17 (1): 100–116.

Avella, M. A., N. Rodríguez, and O. Rangel-Ch. 2013. Lineamientos para la conservación y uso sostenible de los bosques de roble (*Quercus humboldtii*) del sector central del Corredor de Conservación Guantiva-La Rusia-Iguaque (Departamentos de Santander y Boyacá). *(Guidelines for conservation and sustainable use of oak forests (Quercus humboldtii) in the central región of the Guantiva-LaRusia-Iguaque Conservation Corridor)*. In: Rodríguez, N. (ed). *Desarrollo y ambiente: Contribuciones teóricas y metodológicas IDEAS* 24: pp. 289–333. Universidad Nacional de Colombia. Instituto de Estudios Ambientales (IDEA). Bogotá, Colombia: Programa de Maestría en Medio Ambiente y Desarrollo PMAD.

Botero, J., N. Aguirre, J. Paiba, J. Palacio, D. Barrios, A. López, R. Espinosa, N. Franco, and C. Parra. 2009. Robles y café en el sur de Huila *(Oak and coffee the south of Huila)*. En: Parrado-Roselli, A. and L. M. Cárdenas. (eds). *Libro de Resúmenes II Simposio Internacional de Bosques de Robles y Ecosistemas asociados*: pp. 10. Bogotá: Fundación Natura Colombia—Universidad Distrital Francisco José de Caldas.

Cantillo, E. and J. Rangel. 2011. La Estructura y el Patrón de la Riqueza de la Vegetación del Parque Nacional Natural Los Nevados *(Vegetation structure and species richness within the Los Nevados National Park)*. En: Rangel-Ch, J. O. (ed). *Colombia Diversidad Biótica XI. Patrones de la Estructura y de la Riqueza de la Vegetación en Colombia*. Bogotá DC: Universidad Nacional de Colombia-Instituto de Ciencias Naturales, 69–125.

Cárdenas, D. and N. Salinas (eds). 2006. *Libro rojo de plantas de Colombia. Volumen 4. (Red book of plants of Colombia. Volume 4). Especies maderables amenazadas: Primera parte. Serie de libros rojos de especies amenazadas en Colombia*: pp. 232. Bogotá, Colombia: Instituto Amazónico de Investigaciones Científicas SINCHI—Ministerio de Ambiente, Vivienda y Desarrollo Territorial.

Corporación Autónoma de Santander (CAS). 2011. Acuerdo 182 de 2011, por el cual se modifica parcialmente el acuerdo No. 0161-10, que declara y alinda el Distrito Regional de Manejo Integral de los Recursos Naturales Renovables DMI en el territorio que comprende los páramos de Guantiva y la Rusia, bosques de roble y zonas aledañas, localizado en los municipios de Charalá, Coromoro, Encino, Gambita, Mogotes Onzaga, San Joaquín y Suaita, Departamento de Santander.

Cuatrecasas, J. 1958. Aspectos de la vegetación natural de Colombia *(Aspects of the natural vegetation of Colombia). Revista de la Academia Colombiana de Ciencias Exactas (Bogotá)*. 10 (40): 221–268.

Devia, C. and H. Arenas. 2000. Evaluación del estatus ecosistémico y de manejo de los bosques de fagáceas (*Quercus humboldtii y Trigonobalanus excelsa*) en el norte de la Cordillera Oriental (Cundinamarca, Santander y Boyacá). *(Evaluation of the ecosystem status and forest management in Fagaceae forests in the north of the Eastern range)*. En: Cárdenas, F. (ed). *Desarrollo sostenible en los Andes de Colombia (Provincias de Norte, Gutierrez y Valderrama) Boyacá, Colombia*. Bogotá: IDEAD-Universidad Javeriana con el apoyo de la Unión Europea.

Díaz, S. 2010. Uso de especies forestales asociadas a bosques de roble (*Quercus humboldtii* Bompl.) con fines energéticos, en tres veredas del municipio de Encino-Santander. *(On the use of species associated to oak forests (Q. humboldtii) for energy goals, in three villages of the Encino-Santander municipality). Revista Colombia Forestal* 13 (2): 237–243.

Etter, A., C. MacAlpine, D. Pullar, and H. Possingham. 2006. Modelling the conversion of Colombian lowland ecosystems since 1940: Drivers, patterns and rates. *Journal of Environmental Management* 79: 74–87.

Gentry, A. 1993. Vistazo general a los ecosistemas nublados andinos y la flora de Carpanta *(Overview of andean cloud forests and the flora of Carpanta)*. Páginas. In: Andrade, G. I. (ed). *Carpanta: Selva nublada y páramo. Fundación Natura Colombia*: pp. 67–80. Santafé de Bogotá: Edit. Presencia.

Herrera, B. and A. Chaverri. 2006. Criteria and indicators for sustainable management of Central American Montane Oak Forests. In: Kapelle, M. (ed). *Ecology and Conservation of Neotropical Montane Oak Forests*. Ecological Studies 185: pp. 421–432. Berlin, Heidelberg: Springer-Verlag.

Hilty, S. L. and W. L. Brown. 1986. *A Guide to the Birds of Colombia*. Princeton, New Jersey: Princeton University Press.

Hooghiemstra, H. 2006. Immigration of oak into northern South America: A Paleo-Ecological Document. In: M. Kappelle (Ed.), *Ecology and Conservation of Neotropical Montane Oak Forests*. Ecological Studies 185: pp. 17–28. Berlin, Heidelberg: Springer-Verlag.

Kapelle, M. 2006. Neotropical montane oak forest: Overview and outlook. In: Kapelle, M. (ed). *Ecology and Conservation of Neotropical Montane Oak Forests*. Ecological Studies 185: pp. 449–463. Berlin, Heidelberg: Springer-Verlag.

Lammerts van Bueren, E. M. and E. Blom. 1997. *Hierarchical Framework for the Formulation of Sustainable Forest Management Standards*: pp. 108. Netherlands: Veeman Drukkers.

Lozano, G. and J. H. Torres. 1974. Aspectos generales sobre la distribución, sistemática fitosociológica y clasificación ecológica de los bosques de robles (Quercus) en Colombia. *(Overview on the distribution, plant associations, and ecological classification of Colombian oak forests)*. *Ecología Tropical* 1: 45–79.

Melo, A. 2010. *Sistema de monitoreo a escala de paisaje de la conservación y uso del recurso forestal de roble en el corredor a través de sistemas de información geográfica cuenca del río guacha (Encino, Santander y Belén, Boyacá)*. *(GIS analyses of the Rio Guacha watershed to monitor conservation areas and forest use within the conservation corridor)*. Bogotá: Fundación Natura.

Melo, A. and H. Arango. 2012. Análisis de la intervención: Criterios de elegibilidad de finca y escenarios de finca (Intervention analysis: Eligibility criteria and farm scenarios). En: C. Giraldo, F. Díaz, R.L. Gómez (eds.). Ganadería sostenible en el trópico de altura en el corredor de conservación de robles (pp. 21–28). Fundación Natura, Fundación CIPAV. Cali, Colombia.

Ministerio de Ambiente y Desarrollo Sostenible (MADS). 2012. Manual para la asignación de compensaciones por pérdida de biodiversidad. *(Manual for determine compensation for biodiversity losses)*, Bogotá.

Organización Internacional de las Maderas Tropicales (ITTO). 2002. *Directrices de la OIMT para la restauración, ordenación y rehabilitación de bosques tropicales secundarios y degradados*: (International Tropical Timber Organization's guidelines for restoration, management and rehabilitation of secondary and degraded tropical forests). Serie OIMT de políticas forestales N° 13. 88 pp. Yokohama, Japón.

Organización Internacional de las Maderas Tropicales (ITTO). 2005. *Criterios e indicadores revisados de la OIMT para la ordenación sostenible de los bosques tropicales con inclusión de un formato de informes*. *(International Tropical Timber Organization's revised criteria and indicators guidelines for the sustainable management of tropical forests)*. Serie OIMT de políticas forestales N° 15. 40 pp. 40 Yokohama, Japan.

Peñaloza, L. 2012. *Percepciones y aportes de la comunidad de la cuenca del río Guacha para establecer una estrategia de restauración ecológica en sus territorios*. *(Community perceptions and support for the establishment of a ecological restoration strategy within the Guacha River watershed)*. Bogotá, Colombia: Fundación Natura.

Pulido, M. T., J. Cavelier, and S. P Cortés. 2006. Structure and Composition of Colombian Montane Oak Forest. En: Kapelle, M. (ed). *Ecology and conservation of Neotropical montane oak forest*. Ecological Studies 185. Berlin, Heidelberg: Springer-Verlag.

Rangel-Ch, J. O. 2000. La Megadiversidad Biológica de Colombia: ¿Realidad o Ilusión? *(Biological megadiversity in Colombia: Fact or Fiction?)*. En: Aguirre, J. (ed). *Memorias del Primer Congreso Colombiano de Botánica (Versión en CD-Rom)*: Bogotá.

Rangel, O. and A. Avella. 2011. Oak forest *Quercus humboldtii* in the Caribbean region and distribution patterns related with environmental factors in Colombia. *Plant Biosystems* 145: 186–198.

Rangel, O., A. Avella, and H. Garay. 2009. Caracterización florística y estructural de los relictos boscosos del sur del departamento del Cesar. *(Floristic and structural characterization of relict forests in the south of the Cesar department)*. En: Rangel-Ch, J. O. (ed). *Colombia Diversidad Biótica VIII. Media y baja montaña de la Serranía del Perijá*. Bogotá: pp. 365–392. Instituto de Ciencias Naturales. Universidad Nacional de Colombia-CORPOCESAR.

Renjifo, L. M. 2002. *Hapalopsittaca fuerteri*. In: Renfijo, L. M., A. M. Franco-Amaya, J. D. Amaya-Espinel, G. H. Kattan, and B. Lopez-Lanus. (eds). *Libro rojo de Aves de Colombia. (Red book of the birds of Colombia)*. Serie Libros Rojos de Especies Amenazadas de Colombia. Bogotá, Colombia: Instituto Alexander von Humboldt y Ministerio del Medio Ambiente.

Reynolds, M. and P. F. Hessburg. 2005. Decision support for integrated landscape evaluation and restoration planning. *Forest Ecology and Management* 207: 263–278.

Rodrígues, R., S. Gandolfi, A. Nave, and J. Aronson. 2011. Large-scale ecological restoration of high-diversity tropical forest in SE Brazil. *Forest Ecology and Management* 261: 1605–1613.

RUNAP, 2014. *Resolución 044 del 26 de enero de 2007. por medio de la cual se adopta el plan de manejo del santuario de fauna y flora Iguaque*. Disponible en Registro Único Nacional de Áreas Protegidas RUNAP. http://runap.parquesnacionales.gov.co, Noviembre 19 de 2014.

Saenz, F. A. 2010. Aproximación a la fauna asociada a los bosques de roble del Corredor Guantiva—La Rusia—Iguaque (Boyacá—Santander, Colombia) *(Characterization of the faunal assemblages found within the oak forests of the Guantiva-LaRusia-Iguaque Conservation Corridor)*. *Revista Colombia Forestal* 13 (2): 299–334.

Secretaría del Convenio de Diversidad Biológica (SCDB). 2004. *Enfoque por Ecosistemas, (Directrices del CDB) (Focus on Ecosystems: Directives from the Biological Diversity Agreement)*: pp. 50. Holanda: Secretaria del Convenio de Diversidad Biológica.

Society for Ecological Restoration International (SER). 2004. *The SER International Primer on Ecological Restoration*. Science and Policy Working Group. http://www.ser.org.

Solano, C., C. Roa, and Z. Calle. 2005. *Estrategia de Desarrollo Sostenible del Corredor de Conservación Guantiva—La Rusia—Iguaque privados (Volunteer Conservation agreement at private lands)*: pp. 87. Bogotá: Fundación Natura.

Solano, C. 2012. Acuerdos de conservación—producción: herramienta de negociación recíproca para la conservación voluntaria en predios privados. En: C. Giraldo, F. Díaz, R. L. Gómez (eds.). Ganadería sostenible en el trópico de altura en el corredor de conservación de robles (pp. 73–84). Fundación Natura, Fundación CIPAV. Calí, Colombia.

Stattersfield, A. J. and D. C. Capper (eds). 2000. *Threatened birds of the world: the official source for birds on the IUCN red list*. Cambridge, UK: Birdlife International.

Stotz, D. F., J. W. Fitzpatrick, T. A. Parker, and D. K. Moskovits. 1996. *Neotropical Birds: Ecology and Conservation*. Chicago: University of Chicago Press.

Torres, S. and J. Dueñas. 2014. *Lineamientos para la gestión de paisajes socio-ecológicamente resilientes frente al cambio climático en la Cuenca del Río Guacha. (Guidelines for the arrangement of socio-ecologically resilient landscapes in the face of climate change within the Rio Guacha Watershed)*. Tesis Maestría en Gerencia Ambiental. Universidad de los Andes. Facultad de Administración. Bogotá, Colombia: Unpublished report on file at Fundación Natura.

Van der Hammen, T. and E. Gonzalez. 1963. Historia del clima y vegetación del Pleistoceno Superior y del Holoceno de la Sabana de Bogotá *(Climate and vegetation of the Bogotá Savannah during the Late Pleistocene and Holocene)*. *Boletín Geológico* 40: 189–266.

Van der Hammen, T., R. Jaramillo, and M. T. Murillo. 2008. Oak forests of the Andean Forest zone of the Eastern Cordillera of the Colombian Andean. In: Van der Hammen, T. (ed). *Estudios de Ecosistemas Tropandinos—La cordillera Oriental colombiana, transecto Sumapaz, Ecoandes* 7: pp. 594–614. Berlín-Stuttgaart: J. Cramer.

World Health Organization. 2006. *Fuel for Life: Household Energy and Health*: pp. 42. Geneva, Switzerland: WHO Press.

Zapata, P. and F. Díaz. 2012. Reconversión productiva en la Cuenca del Rio Guacha. In: Giraldo, C., F. Díaz, and R. L. Gómez. (eds). *Ganadería sostenible en trópico de altura en el corredor de conservación de robles . (Reconversion of productive systems in the Rio Guacha Watershed)*: pp. 200. Cali, Colombia: *Fundación Natura, Fundación Cipav*.

22

Restoring Longleaf Pine Forest Ecosystems in the Southern United States

Dale G. Brockway, Kenneth W. Outcalt, Donald J. Tomczak, and Everett E. Johnson

CONTENTS

22.1 Southern Forest Environment

Longleaf pine (*Pinus palustris*) ecosystems are native to nine states of the southern region of the United States. Longleaf pine can grow on a variety of site types including wet flatwoods and savannas along the Atlantic and Gulf coastal plain, higher droughty sand deposits from the fall line sandhills to the central ridge of Florida (Stout and Marion 1993), and the montane slopes and ridges of Alabama and northwest Georgia up to 600 m elevation (Boyer 1990b). This region has a humid subtropical climate (Bailey 1995). Maximum July temperatures average 29°C to >35°C while minimums during January range from 0°C to 13°C. The mean annual precipitation is 1040–1750 mm and is well distributed through the year. Growing season is comparatively long, ranging from 300+ days in Florida to 220 days at the northern limit of longleaf pine's range. During the late summer and fall, hurricanes can develop over the Atlantic Ocean, move westward, and impact coastal plain forests. Such tropical storms are a principal large-scale disturbance agent for longleaf pine forests near the seacoast.

Longleaf pine grows on soils derived from marine sediments ranging from deep, coarse, and excessively drained sands to finer textured clays (Boyer 1990b). Entisols and Spodosols, two of the major orders occupied by longleaf pine, are generally sandy, acidic, low in organic matter, and relatively infertile. Quartzipsamments, the most prevalent Entisols on xeric sandhills, are deep sands with weak horizon development. Spodosols, principally Aquods, are found on lower coastal plain flatwoods. These are wet sandy soils with a shallow water table that is at or near the ground surface during the rainy season. Longleaf pine is also found on more fertile clay soils (Ultisols) like the red hills region of southern Georgia. Typic Paleudults and Plinthic Paleudults are the Ultisols most frequently supporting longleaf pine.

22.2 Longleaf Pine Ecology

22.2.1 Longleaf Pine Ecosystems

Longleaf pine forests were once among the most extensive ecosystems in North America (Landers et al. 1995). Prior to European settlement, these forests occupied 37 million ha in the southeastern United States; 23 million of longleaf dominated forests and 14 million of longleaf mixed with other pines and hardwoods (Frost 1993). Travelers in this region during the late eighteenth and early nineteenth centuries reported vast areas of longleaf pine that in some places covered >90% of the landscape (Bartram 1791; Williams 1837). The native range of longleaf pine (Figure 22.1) encompasses an area along the Gulf and Atlantic Coastal Plains from Texas to Virginia, extending well into central Florida and the Piedmont and mountains of northern Alabama and northwest Georgia (Boyer 1990b; Stout and Marion 1993). Longleaf forests occupied 59% of the total land area within the 412 counties of this range (Frost 2006).

An open, park-like stand structure (Figure 22.2) is a distinguishing characteristic of longleaf pine ecosystems (Schwarz 1907; Wahlenberg 1946). Naturally occurring longleaf pine forests contain numerous embedded special habitats such as stream bottoms, wetlands and seeps (Platt and Rathbun 1993; Brockway and Outcalt 1998; Hilton 1999). In the western Gulf Coastal Plain, bluestem grasses (*Schizachyrium scoparium* and *Andropogon* spp.) dominate longleaf pine understories. From Florida north and eastward, longleaf pine is typically associated with wiregrass (*Aristida stricta* and *Aristida beyrichiana*), also known as pineland three-awn. Fallen pine needles and understory grasses facilitate the ignition and spread of fire, which limits woody shrubs and hardwood trees (Landers 1991). While such woody plants may be more numerous on mesic sites, their stature is typically limited by frequent burning. At various locations within the native range, *Quercus*, *Ilex* and

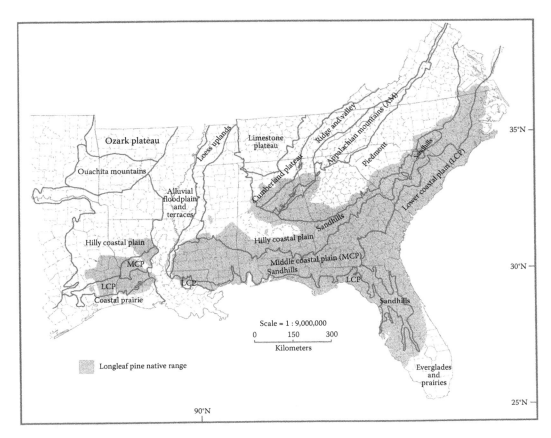

FIGURE 22.1
Native range of longleaf pine and physiographic provinces of the southeastern United States. (Adapted from Little, E.L. *Atlas of United States Trees. Volume 1. Conifers and Important Hardwoods*, U.S. Department of Agriculture, Forest Service, Miscellaneous Publication 1146, Washington, DC, 320 pp., 1971, Miller and Robinson 1995.)

Serenoa may be common tree and shrub associates. Longleaf pine ecosystems support a great variety of herbaceous plant species. The high diversity of understory plants per unit area makes these ecosystems among the most species-rich plant communities outside the tropics (Peet and Allard 1993).

Longleaf pine is closely associated with frequent surface fires (Garren 1943; Brockway and Lewis 1997; Outcalt 2000). The longleaf pine and bunchgrasses function together as keystone species that facilitate but are resistant to fire (Platt et al. 1988b; Noss 1989). Their longevity and nutrient and water retention ability reinforce their site dominance and minimize change in the plant community following disturbance (Landers et al. 1995). The long and highly flammable needles of longleaf pine together with the living and dead leaves of bunchgrasses constitute a fine-fuel matrix that facilitates the rapid spread of fire (Abrahamson and Hartnett 1990; Landers 1991).

Prior to landscape fragmentation, natural fires occurred every 2–8 years throughout most of the region (Christensen 1981; Abrahamson and Hartnett 1990). Longleaf pine dominated this large expanse primarily because it tolerates frequent fire better than seedlings of thinner-barked competitors. Longleaf pine seedlings are susceptible to fire-caused mortality during the first year following germination, but they become increasingly resistant

FIGURE 22.2
Longleaf pine bunchgrass ecosystem on xeric sandhills.

to fire in subsequent years. A unique adaptation of longleaf pine to a fire-prone environment is a seedling "grass stage," during which root growth is favored and the seedling top remains a tuft of needles surrounding and protecting a large terminal bud (Brockway et al. 2006). The lack of a stem limits exposure to damage from surface fires. When sufficient root reserves have accumulated, grass stage longleaf pine seedlings "bolt" by rapidly growing 1–2 m in a short time period, putting their terminal bud beyond the lethal reach of most surface fires. Larger longleaf pine trees have relatively thick bark that protects cambial tissue from the lethal heating of surface fires (Wahlenberg 1946). Fires assist in the natural pruning of longleaf pine, creating a clear bole between the crown and any accumulated surface fuels. Surface fires are thereby prevented from easily moving into the canopy. Longleaf pine also tends to regenerate more successfully in forest openings than directly beneath mature trees (Brockway and Outcalt 1998); thus, keeping ladder fuels away from the crowns of adult trees.

Longleaf pine evolved in an environment influenced by frequent disturbance, principally fire (Palik and Pederson 1996; Engstrom et al. 2001). Damaging tropical storms, such as hurricanes and associated tornadoes, may fell trees over an extensive area and open gaps in the canopy of longleaf pine forests (Croker 1987). Lightning is another important agent in overstory tree mortality, typically impacting individual trees but also sometimes striking small groups of trees (Komarek 1968; Taylor 1974; Palik and Pederson 1996), which creates or enlarges canopy gaps and was the ignition source for the frequent fires (Outcalt 2008). Insect infestations are uncommon; however, annosus root rot (*Heterobasidion annosum*), pitch canker (*Fusarium moniliforme* var. *subglutinans*), and cone rust (*Cronartium strobilinum*) are among the pathogens that may infect longleaf pine (Boyer 1990b). Epidemics of brown-spot disease (*Mycosphaerella dearnessii*) occasionally occur in young longleaf pines; this pathogen is usually fatal unless a surface fire consumes the infected needles to cleanse the stand of inoculum (Boyer 1990b).

Longleaf pine is a shade-intolerant tree species and regenerations naturally only in canopy gaps (Wahlenberg 1946). Seedlings developing in gaps created at differing time intervals result in a network of forest patches at various stages of development dispersed across the landscape (Pickett and White 1985). Such gap-phase regeneration dynamics produce a forest structure commonly observed in longleaf pine ecosystems of even-aged patches distributed within an uneven-aged mosaic (Palik et al. 1997).

22.2.2 Ecological Significance

The complex natural pattern and disturbance-mediated processes of longleaf pine forests cause extraordinarily high levels of biological diversity in these ecosystems, with as many as 140 species of vascular plants in a 1000 m² area. Counts of more than 40 species per m² have been recorded in many longleaf pine communities (Peet and Allard 1993). A large number of these plant species are restricted to, or found principally in, longleaf pine habitats. Not surprisingly, many animal species also depend on longleaf pine ecosystems for much of their habitat, including two increasingly rare animals that are important primary excavators. Tree cavities created by red-cockaded woodpeckers (*Picoides borealis*) and ground burrows dug by gopher tortoises (*Gopherus polyphemus*) provide homes for a variety of secondary users such as insects, snakes, birds and mammals (Jackson and Milstrey 1989; Engstrom 2001).

The longleaf pine forests and savannas of the southeastern coastal plain are among the most critically endangered ecosystems in the United States, now occupying less than 3% of their original extent (Ware et al. 1993; Noss et al. 1995). Extreme habitat reduction is the primary cause for increasing rarity of 191 taxa of vascular plants and several terrestrial vertebrate species that are endemic to or exist largely in longleaf pine communities (Hardin and White 1989; Walker 1993). Habitat loss principally has resulted from conversion of longleaf pine forests to other land uses (i.e., agriculture, industrial pine plantations, and urban development), landscape fragmentation and interruption of natural fire regimes (Landers et al. 1995; Wear and Greis 2002). Long-term exclusion of fire typically results in depressed species diversity, a substantial hardwood understory and midstory and a thick layer of forest litter (Brockway and Lewis 1997; Kush and Meldahl 2000) (Figure 22.3). Such extraordinary buildup of forest fuel poses a serious wildfire hazard and, rather than naturally occurring surface fires, crown fires with potentially catastrophic effects on the rare plants and animals are likely. Safe and effective reintroduction of fire into long-unburned forests remains the critical conservation challenge (Wear and Greis 2002).

Longleaf pine bunchgrass ecosystems are also vital to the maintenance of numerous biotic communities embedded within the southern forest landscape matrix (Landers et al. 1990). Many of these adjacent communities require periodic fire to maintain their ecological structure and health (Kirkman et al. 1998). Wildfires typically begin in pyrogenic longleaf pine forests and spread into adjoining habitats such as seepage slopes, canebrakes, treeless savannas, and sand pine scrub. Without periodic fire, these communities change in ways that make them less suitable habitats for other fire-adapted plants and animals.

22.3 History of Longleaf Pine Ecosystems

Longleaf pine, moving northward and eastward from its ice age refuge in southern Texas or northern Mexico (Schmidtling and Hipkins 1998), established in the lower coastal plain ~8000 years ago (Watts et al. 1992) and during the ensuing 4000 years spread throughout

FIGURE 22.3
Former longleaf pine site invaded and occupied by oak.

the southeast (Delcourt and Delcourt 1987). Interestingly, this time period coincides with increasing population levels of Native Americans throughout the region; their use of fire is thought to be related to the spread of longleaf pine forests (Schwartz 1994; Pyne 1997; Landers and Boyer 1999). Native Americans frequently used fire to manipulate their environment (Robbins and Myers 1992; Anderson 1996; Carroll et al. 2002; Stanturf et al. 2002). Recognizing the benefits of fire on the landscape, early European settlers adopted the practice of periodically burning nearby forests and woodlands to improve forage quality for cattle grazing and discouraged the encroachment of shrubby undergrowth.

European settlement had little impact on longleaf pine forests initially, with harvesting limited to areas near towns and villages for building log structures (Croker 1987). By the 1700s, water-powered sawmills became common, but log transportation was inefficient and confined to rivers (Frost 1993). After 1830, removal of the longleaf pine resource accelerated significantly with the arrival of steam railroads and skidders. By 1880, most of the longleaf pine forests along streams and railroads had been harvested (Frost 1993). During the next 40 years, the great forests of longleaf and other southern yellow pines were harvested, with temporary railroad spur lines laid down every quarter mile (Croker 1987). Timber extraction peaked in 1907, when 39 million m³ were removed (Wahlenberg 1946). By 1930, nearly all old-growth longleaf pine was harvested and lumber companies migrated west.

Although well adapted to frequent disturbance from surface fires, longleaf pine was not well suited to disturbances brought by European settlement. As a result of cumulative impacts over three centuries of changing land use, longleaf pine forests declined dramatically. By 1900, logging, naval stores extraction, and agriculture had reduced the area dominated by longleaf pine by more than half (Frost 1993). Second-growth longleaf pine stands became reestablished on only one-third of the sites previously occupied (Wahlenberg 1946). Harvest of these second growth stands, often followed by conversion to other southern pines or urban development, continued through 1985 (Kelly and Bechtold 1990) and beyond, until longleaf pine was reduced to 1.2 million ha; less than 5% of its original area (Outcalt and Sheffield 1996).

22.4 Social and Political Context

Longleaf pine ecosystems have been very important in the southern United States, providing an environmental setting and raw materials for social development in this region. Wild game, forage grasses, wood, and naval stores (i.e., chemicals derived from pine resin) were principal products of these forests (Franklin 1997). During the early twentieth century, affluent landowners, recognizing the value of longleaf pine forests as habitat for bobwhite quail (*Colinus virginianus*) and white-tailed deer (*Odocoileus virginianus*), acquired large tracts to serve as private hunting reserves. Many large areas of longleaf pine exist today because of the opportunities for hunting and timber harvest provided by these lands. Nevertheless, economic exploitation has played a major role in the decline of these forests. Recent developments show that these negative trends have been reversed with longleaf dominated area increasing from a low of 1.2 million ha in 1995 (Outcalt and Sheffield 1996) to 1.74 million ha in 2010 (Oswalt et al. 2012). Conversion of longleaf pine to other tree species has slowed on private lands, federal and state agencies are regenerating longleaf pine on their lands following harvest, and they are rehabilitating degraded longleaf pine forests with fire and other appropriate techniques (Hilliard 1998; McMahon et al. 1998). Longleaf pine reforestation and afforestation have increased on private lands because of incentives provided by the federal government, like the Conservation Reserve Program—Longleaf Pine Initiative (United States Department of Agriculture (USDA) 2014) with 54,412 ha planted in 2013 (Longleaf Partnership Council 2014).

In the mid-1990s, the southern forestry community gained an improved understanding of longleaf pine ecosystems and an appreciation of the natural heritage that could be lost. Since no single entity dominated land ownership in longleaf pine ecosystems, a common sense of urgency among numerous groups fostered formation of partnerships. The Nature Conservancy, Tall Timbers Research Station, Joseph W. Jones Ecological Research Center, USDA Forest Service, USDI Fish and Wildlife Service, U.S. Department of Defense, Cooperative Extension Service, state agencies, private landowners, universities, and forest industry worked together to promote longleaf pine ecosystem restoration. In 1995, the Longleaf Alliance was formed to serve as a regional clearinghouse for a broad range of information on the regeneration, restoration, and management of longleaf pine ecosystems. The Alliance is housed at Auburn University Solon Dixon Forestry Education Center in Alabama (http://www.longleafalliance.org/). It facilitates communication among these groups and provides training sessions for public land managers and private landowners concerning successful longleaf pine regeneration, management, and restoration.

22.5 Restoration Perspectives

Longleaf pine still occurs over most of its natural range, albeit in isolated fragments; thus, restoration is feasible (Landers et al. 1995). Restoration to historical authenticity may not be desirable or even possible, but natural authenticity is a reasonable goal (Brockway et al. 2005), meaning that compositional, structural, and functional components are present within an appropriate physical environment. Thus, ecological processes can be sustained in restored longleaf pine ecosystems, providing for native species perpetuation and evolution, ecosystem resiliency to disturbance, and adaptation to long-term environmental change, goods and services for human societies, and safe harbors for rare and endangered species (Clewell 2000).

Specific restoration goals need to be stated (Society for Ecological Restoration International Science (SERI) & Policy Working Group 2004), which can range from the complete restoration of a naturally functioning longleaf pine ecosystem to establishing a stand of longleaf pine. A survey of your potential restoration area is needed that should include a prism or plot cruise of any existing trees by species in the overstory and midstory layers, a list of understory species and notes on disturbances. If the goal is complete restoration, the understory survey needs to be more intensive; including species density and frequency. If you have more than one site, you may want to start with the least degraded site because time and expense of understory restoration increases with level of degradation. Walker and Silletti (2006) developed a graphical means for determining site degradation based on fire frequency and agricultural history, which were also shown by Brudvig et al. (2014) to be the factors that influenced understory degradation most. Augmenting existing longleaf pine fragments and creating new connecting habitat patches should also be considered when prioritizing sites for treatment, as this will reduce habitat fragmentation, population isolation, and species rarity. Private landowners with multiple sites also need to consider requirements of government cost share programs, because only pasture and agricultural areas qualify for some programs.

Because natural longleaf pine forest ecosystems are so variable, the range of conditions that fall within natural variability are correspondingly broad. Overall, full restoration would mean an overstory dominated by longleaf pine, occurring as uneven-aged stands or even-aged patches across an uneven-aged landscape mosaic. On some site types and locations in the native range, a lesser component of other tree species may be present, such as slash pine, loblolly pine, or oaks, which may occur singly or in clusters. In the historically mixed zone, longleaf, loblolly, and shortleaf pine (*Pinus echinata*) should be overstory codominants (Frost 2006). The midstory should generally be absent or mostly composed of ascending longleaf pines. Native grasses and forbs should dominate the understory, with lesser cover of shrubs and vines. Long-term ecosystem recovery and sustainability will be fostered by properly functioning ecological processes such as periodic surface fires, natural regeneration that leads to normal stand replacement dynamics, nutrient cycling that maintains primary productivity and suitable habitat that facilitates life cycle completion by numerous native organisms.

Once a site has been selected for ecological restoration a monitoring plan will be required to determine progress toward stated goals. A good source of information is the Society for Ecological Restoration International Primer (Society for Ecological Restoration International Science (SERI) & Policy Working Group 2004), which provides a definition of ecological restoration, outlines attributes of restored ecosystems, explains construction of reference ecosystems descriptions and lists methods of evaluation and monitoring. At a minimum a qualitative description of overstory, midstory, and understory communities is required. If a suitable reference site or historical description is not available, a general description can be developed by classifying the site using the ecological system developed by Peet (2006). This description combined with the general attributes on structure and composition given in the previous paragraph and the attributes of restored ecosystems from the Primer can be used to evaluate restoration progress.

22.6 Restoration Methods

22.6.1 Restoration Framework

Despite a wealth of knowledge and experience concerning longleaf pine restoration, much uncertainty still exists, fostering a healthy debate about the best approaches. Desirable changes in longleaf pine communities can be achieved by using a variety of methods, machines and products, either singly or in combination. Prescribed fire may be used to reduce midstory, understory, and occasionally overstory layers and encourage fire-tolerant plants. Because frequent fire is crucial for ecosystem restoration, other treatments should be used to facilitate the eventual application of prescribed fire (Menges and Gordon 2010; Outcalt and Brockway 2010). Physical or mechanical treatments include complete overstory harvest, selective thinning of overstory and midstory trees, and shredding or mowing midstory and understory plant layers. Chemical treatments, principally herbicide application, can be used to selectively induce mortality of undesirable plant groups. In highly degraded ecosystems, biological approaches such as reintroducing extirpated species will likely be required for full restoration.

22.6.2 Selecting Techniques

Historical events and changing land use provide an array of candidate sites in various conditions for restoration of longleaf pine ecosystems. While about 1.74 million ha currently have an overstory of longleaf pine (Oswalt et al. 2012), only 0.5–0.8 million ha of these have native intact understories (Noss 1989). Other candidate areas with little overstory longleaf pine have understories that range from having most of the native species to highly altered understories with no native species (Outcalt 2000). This variety of existing vegetation exists across the range of sites that longleaf pine can occupy, from dry sandhills to wet savannas. Suitable restoration techniques depend on the site type and degree of ecosystem degradation (Table 22.1). The types of longleaf pine ecosystems discussed are based on the classification of Peet and Allard (1993), with sandhills corresponding to their xeric and subxeric series, flatwoods and wet lowlands to their seasonally wet series, and uplands to their mesic series. However, we include their Piedmont/upland subxeric woodland community in the uplands rather than sandhills. An alternate classification method with restoration prescriptions, based on commonly occurring forest conditions from old fields to longleaf stands with recent fire, is found in Johnson and Gjerstad (2006). They also provide extensive information on planting longleaf seedlings, that is, handling, planting depth, season, and cost.

TABLE 22.1

Prescriptions for Restoring Longleaf Pine Ecosystems in Varying Degrees of Degradation

Degree of Degradation	Stand Condition	Landscape Position		
		Xeric and Subxeric Sandhills	Flatwoods and Wet Lowlands	Montane and Mesic Uplands
Moderately	Longleaf pine overstory, woody understory	Initial Restoration Prescription Reduce fuel loads, introduce summer burns	Reduce fuel loads, introduce summer burns	Reduce fuel loads, remove other pines in overstory, introduce summer burns
Very	Other trees now in overstory, native plant understory	Chop and burn broadleaves; remove other pine; plant longleaf pine; no or minimal site preparation	Reduce fuel loads, remove other pines, chop; reduce slash, no bedding; plant longleaf pine	Reduce fuel loads, Remove other pines in overstory, plant longleaf pine
Severely	Former longleaf pine site, other trees now in overstory, nonnative plant understory	Remove other trees; chop and burn; plant longleaf pine; establish *Aristida stricta* (a native grass to facilitate reintroduction of fire) by direct seeding or, if a longleaf pine overstory is present, plant grass seedlings	Remove other trees; chop and burn; plant longleaf pine; plant or direct seed native grasses	Remove other trees; chop and burn; plant longleaf pine plant or direct seed native grasses

22.6.3 Restoration Prescriptions

22.6.3.1 Xeric and Subxeric Sandhills Dominated by Longleaf Pine with Native Understory

In many existing xeric and subxeric sandhills longleaf pine forests, the absence of frequent fire allowed turkey oak (*Quercus laevis*), bluejack oak (*Quercus incana*), sand live oak (*Quercus geminata*), and sand post oak (*Quercus stellata* var. *margaretta*) to develop into a scrub oak midstory. Repeated applications of fire during the growing season are effective at restoring these sites, by gradually reducing the density of the midstory scrub oaks (Glitzenstein et al. 1995). Fires stimulate grasses and forbs to produce flowers and seeds (Christensen 1981; Platt et al. 1988a; Clewell 1989; Outcalt 1994), which aid in colonization of newly exposed microsites.

Reintroducing growing-season fires into xeric longleaf pine forests that have not been burned for a prolonged period may result in increased mortality among older trees, during the 1–3 year interval following the initial burn. This is caused by fine root mortality from smoldering combustion of the accumulated forest litter around the base of larger longleaf pines (O'Brien et al. 2010). To decrease this mortality, a series of dormant-season fires should be applied to gradually reduce the accumulated litter before switching to growing-season burning. During burns duff moisture levels must be high enough to prevent ignition of litter at the base of larger longleaf pines, to avoid excessive fine root damage.

Usually three or four growing-season fires are sufficient to control scrub oak on these sites, but supplemental treatments can accelerate the restoration process. Mechanical methods like chainsaw felling, girdling or chipping on site can reduce midstory hardwoods (Provencher et al. 2001); following these treatments with prescribed burning will stimulate grasses and forbs and reduce growth of hardwood sprouts. If woody material from the midstory is not chipped or removed from the site, it should be allowed to decay before introducing the first prescribed fire. Mechanical methods are expensive and are most appropriate for critical areas in need of rapid restoration, like red-cockaded wood-pecker colony sites or along the urban–wildland interface where it is difficult to schedule the series of prescribed fires required for restoration.

Hexazinone herbicide can also be useful in accelerating the restoration process com-pared with using fire alone (Brockway et al. 1998). Application rates of 1–2 kg a.i./ha liquid formulation in a 2×2 m grid pattern will produce 80%–90% oak mortality without long-term damage to herbaceous understory species (Brockway and Outcalt 2000). Because hexazinone does impact woody species, desirable nontarget species, such as gopher apple (*Licania michauxii*), may be reduced for a time. During dry periods, liquid hexazinone may photo-degrade before sufficient rainfall transports it into the soil for oak roots to absorb it (Berish 1996); hence, application should be just prior to rainfall. Granular hexazinone is less subject to this problem, but it potentially causes a greater reduction in the cover of grasses and forbs when it is uniformly applied across the entire site (Brockway et al. 1998). Broadcast application must also be conducted with care to avoid distribution overlap that could inadvertently double the applied rate.

22.6.3.2 Xeric and Subxeric Sandhills Dominated by Other Trees with Native Understory

Substantial areas exist where scrub oaks became dominant following the harvest of long-leaf pine. Although somewhat suppressed in the absence of frequent fire, the understory plant community still contains many of the native species. Other areas were converted to slash pine plantations following the removal of longleaf pine. Although understory spe-cies, especially the important grasses, are susceptible to severe mortality from soil dis-turbance on dry sandhills sites (Grelen 1962; Outcalt and Lewis 1990), some of these slash pine plantations have intact understory communities due to less intense site preparation or fortuitous rainfall and higher soil moisture levels following soil disturbance. There are also extensive areas in western Florida where Choctawhatchee sand pine (*Pinus clausa* var. *immuginata*) invaded former longleaf pine lands following harvest. Unlike slash pine, sand pine is much more adapted to dry sites, forming a nearly continuous canopy that severely reduces understory density. However, plant diversity in these stands is generally unaffected, with native species surviving, but much reduced in number (Provencher et al. 2001). Restoration under these conditions requires invigorating the herbaceous understory, if present, removing off-site slash pine or sand pine, reducing the scrub oak tree layer, and establishing longleaf pine seedlings.

Areas dominated by scrub oak can be treated with a small (3–5 t) single-drum roller–chopper with no offset. Heavier choppers with offset rollers should be avoided, because they can cause excessive soil disturbance that will harm understory plants. The objective of this treatment is to knock down the oaks and compress them into a layer that will carry a prescribed burn, preferably in the growing season (Walker and Silletti 2006). By con-trast, slash pine plantations often have enough needle litter to support a prescribed burn. Burning these plantations will invigorate the grasses allowing them to accumulate root reserves and thereby increase their ability to recover from the impacts associated with removal of the slash pine and establishment of longleaf pine seedlings. A second fire fol-lowing harvest will remove logging slash, help control oak sprouts and increase the cover of herbaceous species. If slash pine plantations contain numerous scrub oaks, hexazinone can be applied as outlined earlier. Application can be made prior to harvest, then logging will knock down many of the standing dead stems, which then serve as additional fuel for prescribed burning. If herbicide is applied after the logging, most dead oak stems should be allowed to fall before burning, as this will remove debris and facilitate planting long-leaf pine seedlings. Sand pine often grows so densely that it must be removed to release surviving understory species. Sites can then be burned to remove logging slash, reduce abundant sand pine seedlings, and consume sand pine seed.

Options for establishing longleaf pine seedlings include manual or machine planting of either bareroot or container seedlings (Barnett et al. 1990; Barnett and McGilvray 1997). Site preparation, other than that discussed above, should be avoided to protect the understory plant community. It is much less expensive to plant additional longleaf pine seedlings to compensate for lower survival, than it is to reestablish key understory species lost to exces-sive soil disturbance. If grass competition is vigorous (≥60% cover) and bareroot seedlings are being used, a planting machine with a small scalper blade can be used to increase seedling survival (Outcalt 1995). Although this removes a strip of vegetation ~1 m wide, native grasses and forbs will recolonize these strips within 3–5 years, as long as invasive woody plants are discouraged by periodic growing-season fire. Planting containerized longleaf pine seedlings results in acceptable survival rates with no required site prepara-tion other than burning, although hexazinone application may increase survival on areas with vigorous scrub oak competition.

22.6.3.3 Xeric and Subxeric Sandhills without a Native Understory

Highly altered sites that once supported native longleaf pine ecosystems may have no longleaf pine trees and a much altered understory, or longleaf pines may be present but the native understory is not. Most of these sites were once used for agriculture or intensively managed plantations of other pines. Restoration of the understory is a formidable and therefore expensive task. In most cases, the first step is removal of trees other than longleaf pine from the overstory. Since there are few understory plants to protect, many options are available for site preparation. Chopping with a double-drum offset roller–chopper effec-tively controls all competition and produces a clean site for restoration (Burns and Hebb 1972). This treatment can be combined with burning, if there are significant quantities of woody residue. Much of the nutrient capital on these sites is in the litter layer and upper soil horizon, therefore, soil and litter movement should be minimized and root raking and shearing, if used, must be carefully applied. Longleaf pine bareroot or containerized seedlings can then be planted on the site after the soil has settled.

Restoration of the understory plants is best done simultaneously with replanting longleaf pine seedlings to take advantage of the reduced competition and ease of onsite operability. The most critical part in this process is reestablishing grasses, because of their important role as fuel to support recurrent burns. To date, most work on reestablishment of wiregrass has focused on the eastern portion of the range (Means 1997; Seamon 1998; Mulligan et al. 2002). A planting density of 0.5–1 seedling m^{-2} is recommended for restoration of wire-grass with plugs (Outcalt et al. 1999). To successfully establish wiregrass under existing plantations of longleaf pine, repeated burning, mechanical felling, herbicide application, or some combination must remove most hardwood midstory. A heavy-duty woods-har-row is then used to disk strips between trees with plugs planted using 1×1 m spacing. Avoid spring planting, unless there is an existing tree canopy and select a time when moisture and heat stress are not likely (Trusty and Ober 2011). Application of fertilizer during the second or third growing season will stimulate wiregrass growth (Outcalt et al. 1999), but should be applied only around wiregrass plants to avoid stimulating growth of competing vegetation. In pastures occupied by bahia grass (*Paspalum notatum*), cultivation will break up the old sod and use of herbicide will improve both the survival and growth of wiregrass (Uridel 1994).

Direct seeding to reestablish wiregrass between rows of trees in newly planted or existing plantations is less expensive than planting seedlings or plugs (Hattenbach et al. 1998). Small quantities of seed can be collected by hand or with a hand held seed stripper. For larger quantities, a tractor-mounted flail-vac is effective. Seeds can be stored in woven bags or sown immediately, by hand or with a small bale chopper. Rolling seed into the soil can improve wiregrass establishment and survival (Hattenbach et al. 1998). Other grass species are part of the native understory in sandhills longleaf pine forests and should be included in seed mixes. Pineywoods dropseed (*Sporobolus junceus*), for example, is common on many sites and, like wiregrass, will produce seed following fire. Its seed can be collected and mixed with wiregrass when sowing restoration sites.

An extensive program for direct seeding of understory species has been conducted at Fort Stewart, where seed has been collected and sown on site-prepared areas since 1997. Seed is collected using a tractor-mounted flail-vac from areas burned during the growing season yielding from 750 to 1100 kg yr^{-1}. At a mean sowing rate of 13.2 kg ha^{-1}, enough seed is collected to sow 57–83 ha yr^{-1}. Seed is spread using a platform-mounted bale chopper on the back of a farm tractor. Their goal is to restore 8100 ha of former agricultural fields to functioning longleaf pine ecosystems (Hilliard 1998).

Many understory species on sandhills sites survive extreme disturbance as propagules in the soil, or reinvade sites after the disturbance ends (Hattenbach et al. 1998). In one comparison, understories of remnant xeric longleaf pine stands and 30- to 40-year-old plantations on old-field sites were similar (Smith et al. 2002). Although the remnant longleaf pine stands had higher species diversity, nearly 90% of the understory species in the plantations were native to natural longleaf pine communities. Similar comparisons for the sandhills of South Carolina showed that species abundance was the same in plantations and reference stands, except for wiregrass and dwarf huckleberry (*Gaylussacia dumosa*), which were significantly reduced in plantations (Walker 1998). Thus, restoration may not require that every plant species be reintroduced. In addition, to certain common species that do not easily reinvade or survive, reintroduction of some rare species will likely be required (Glitzenstein et al. 1998, 2001; Walker 1998).

22.6.3.4 *Flatwoods and Wet Lowlands Dominated by Longleaf Pine with Native Understory*

Some stands have been degraded by years of fire suppression. Rehabilitation using prescribed burning to reduce woody understory and midstory species and allow grasses and forbs to increase can be effective. Growing-season burns are as useful and often more effective than dormant-season burns for readjusting understory composition. One or two dormant-season fires will gradually reduce litter buildup and is advisable before the first growing-season burn. Initial burns should be conducted when the Keetch–Byram Drought Index (KBDI) (Keetch and Byram 1968) is less than 250 (Miller and Bossuot 2000). Flatwoods understories dominated by saw palmetto, gallberry (*Ilex glabra*), waxmyrtle (*Myrica cerifera*) and sweetgum (*Liquidambar styraciflua*) are quite resistant to fire. Only repeated fires at short return intervals over a long period significantly reduce these woody species (Waldrop et al. 1987; Outcalt and Wade 2004). Thus, burning every 2 years for a period of 10–20 years may be required to readjust the understory composition on wet sites.

Lightweight choppers or heavy-duty mowers may be used to reduce saw-palmetto coverage and dominance (Huffman and Dye 1994). Both methods cause limited soil disturbance and thus do not reduce native grass species. Findings from research at Myakka

River State Park in Florida indicate that the chopping treatment is more effective for reducing saw-palmetto cover. Prescribed burning 3–6 months before or after these mechanical treatments, and follow-up burns every 2 years are needed to prevent recovery of shrubs (Schwilk et al. 2009).

22.6.3.5 Flatwoods and Wet Lowlands Dominated by Other Trees with Native Understory

Longleaf pine overstory on wet sites may have been replaced by other pines, leaving a native understory. Such sites include naturally regenerated stands that were invaded by slash pine and loblolly pine after the removal of native longleaf pine and site-prepared plantations that were planted with other southern pines. Restoration requires replacement of at least some of the loblolly or slash pine overstory with longleaf pine via establishment of longleaf pine seedlings. One system is prescribed burning to reduce woody competition and stimulate growth of herbaceous understory species 1 or 2 years prior to harvest. A site-preparation fire following logging may be needed to remove debris and discourage hardwood trees and shrubs. Between the harvest and site preparation burn, chopping may be used to control woody competitors and reduce density of loblolly and slash pine seedlings. A single-drum chopper should be used to avoid excessive soil disturbance. Herbicides can also be effective for reducing competition, promoting longleaf seedling growth, and stimulating herbaceous understory (Freeman and Jose 2009). Some managers prescribe bedding on these wet sites before planting to increase survival rates of bareroot or container longleaf pine seedlings. Bedding will improve seedling survival during wetter years by about 15%. However, this survival gain comes at a cost, not only of the operation but also from damage to the native groundcover. Bedding may also alter site moisture relations and nutrient distribution for more than 30 years (Schultz 1976). Alternatively, planting additional longleaf seedlings would be more economical and ecologically advantageous, as a hedge against lower survival. Expected seeding survival may also be increased by postponement of planting when soils are saturated.

An alternative method is to create gaps in the overstory and plant them with longleaf seedlings (Kirkman and Mitchell 2002). The site should be prescribed burned, the understory vegetation allowed to recover, and spots selected that have no midstory and are dominated by grasses and forbs. If there are no spots with little woody competition, multiple prescribed burns may be needed before cutting gaps to avoid creating shrub openings that are difficult to burn. If this is not done it will be necessary to use herbicide to control woody species or herbaceous cover will be suppressed (Harrington 2011). Prescribed burning will be needed every 2 or 3 years to control regeneration of other pines (Knapp et al. 2011).

22.6.3.6 Upland and Montane Sites Dominated by Longleaf Pine with Native Understory

Few upland and mountain sites remain in longleaf pine because these were preferred areas for agricultural, urban, and residential development. However, there are upland areas mostly in Alabama, Mississippi, Louisiana and Texas, and montane sites in Alabama and Georgia that have developed unnaturally dense hardwood midstories. Because these are among the most biologically productive longleaf pine sites, they change the most rapidly, quickly developing midstory layers in the absence of frequent fire. In addition to a very dense midstory and a shrub-dominated understory, these sites also accumulate significant quantities of potentially hazardous fuel. Frequent growing-season fires are needed on upland sites with better soils to adequately control competition from woody plants. Like flatwoods sites, frequent growing-season fires over many years are required to reduce the hardwood rootstocks (Boyer 1990a). As noted for other longleaf pine ecosystem types, a series of dormant-season fires may be necessary to gradually reduce fuel levels before growing-season burning begins.

A variant of this ecosystem type, where longleaf pine is present but, other southern pines occur as co-dominants, is common. Prescribed burning can be used as outlined above, but restoration will be enhanced when combined with selective harvesting to reduce the presence of other southern pines and hardwoods in the overstory (Outcalt and Brockway 2010). The objective is not total elimination of other tree species, but rather a proportional readjustment of overstory composition, recognizing that these other species are part of the natural longleaf pine community. Understory burning should begin prior to selective harvest to initiate control of competition from woody plants so they do not proliferate and form a shrub thicket in openings created by harvestings. Herbicide application and mechanical treatment of nonmerchantable woody species used in conjunction with prescribed burning will accelerate reduction of midstory hardwoods and understory shrubs while increasing understory herb cover and diversity (Outcalt and Brockway 2010).

22.6.3.7 Upland and Montane Sites Dominated by Other Species

Only limited research information or management experience is available to guide restoration on upland sites dominated by other overstory species. The few sites that show no evidence of severe soil disturbance contain scattered natural longleaf pine trees in a mixture dominated by loblolly pine, shortleaf pine, and hardwoods. Some native understory likely still exists in the soil seed bank or as suppressed individuals (Varner et al. 2000). Therefore, restoration would consist of burning to reduce fuel and initiate control of woody shrubs and hardwoods. Repeated and prolonged treatment with prescribed fire should eventually reduce the abundance and cover of woody plants in the understory. Selective harvest can be used to release any native longleaf pine and reduce the hardwood component. Other pines may need to be retained onsite to furnish sufficient needlefall for prescribed burning and to avoid release of woody competition. Once prescribed burning and other mechanical or chemical methods have reduced the woody midstory and understory layers, some of these other pines could be removed and replaced with longleaf pine seedlings. This is probably best done by creating canopy gaps in areas where the understory has become dominated by grasses and forbs.

These sites can also be clear-cut and regenerated with longleaf pine seedlings. Herbicide treatment with imazapyr/glyphosate or hexazinone will enhance longleaf seedling growth and increase the herbaceous fuel levels (Addington et al. 2012). After this treatment, these sites should be successful restored with only frequent prescribed burns.

Special techniques are required for establishing longleaf seedlings on wet lowlands and upland sites once used for agriculture or intensive forestry (Johnson and Gjerstad 2006). Few of the many native understory grasses and forbs were able to survive intensive soil disturbance; however, there is a large soil seedbank of herbaceous weeds that must be controlled. In pastures, grasses need to be killed before seedlings are planted and both old fields and pastures usually require scalping and often subsoiling. Follow-up herbicides may also be needed to control competition. There is little fine fuel and burning is usually not possible. A technique for establishing native ground cover on old fields is multiple-pass harrowing to reduce weeds followed by planting wiregrass plugs. High survival rates have been attained with this method, but long-term growth rates are still uncertain (Mulligan and Kirkman 2002).

22.7 Costs and Benefits Associated with Restoration

22.7.1 Estimating Restoration Costs

Reestablishing longleaf pine as the dominant tree species on a site is often the first and, in many ways the easiest step in the restoration process. Reforestation costs vary according to ambient conditions and the type and amount of site preparation needed to achieve successful tree seedling establishment. On previously harvested or old-field sites, costs typically range from $370 to 740 ha^{-1}, depending on site conditions and whether bareroot seedlings or containerized seedlings are selected. This range reflects the current costs for site preparation, seedlings and planting. To control competing vegetation, increase survival and stimulate early growth, an additional $85–100 ha^{-1} might be expended for herbicide application. Since these costs occur early in the investment, they cannot be discounted over time when calculating net present value (NPV), internal rate of return (IRR) or other economic indices. Despite these expenditures, the average IRR for such an investment is estimated at 10.1%, with a range from 8% to 12% (Busby et al. 1996). The Longleaf Alliance offers a free online interactive course allowing you to see how different management scenarios, cost share options, expenses and product prices change IRR and NPV (http://lleconomics.sref.info/).

Restoring groundcover plants can be very expensive, with costs sharply rising when quick success is desired. In relatively undisturbed forests, many plants native to the site may be reestablished by the reintroduction of fire, particularly growing-season burning, through stimulating residual seed banks and inducing flowering and seed production in plants. The cost of fire reintroduction varies with existing site conditions, especially the number of fuel reduction burns needed. Where seed banks are depleted from severe soil disturbance, restoring the plant community is more difficult. Reseeding can cost $7000 ha^{-1} while replanting selected understory plant species can be accomplished successfully (Aschenbach et al. 2010), but will cost over $20,000 ha^{-1} (Walker and Silletti 2006). Seed collection, cultivation, distribution, planting techniques, and other steps in the process are being refined and are generally focused on pyrophytic graminoids (e.g., wiregrass), species consumed by wildlife (e.g., legumes), and species of special concern (e.g., American chaffseed, *Schwalbea americana*).

22.7.2 Benefits of Restored Longleaf Pine Ecosystems

The material and intangible benefits of restoring longleaf pine ecosystems are substantial. The economic value of longleaf pine forests is considerable and commercial products can be extracted from a properly functioning forest without significantly disrupting ecological processes. Longleaf pine is the most versatile of all the southern pines and provides a wide variety of high valued products. Longleaf pine forests typically produce up to five times more tree stems of sufficient quality to be used as utility poles than stands of slash pine or loblolly pine (Boyer and White 1990). Stumpage values for such poles exceed prices for sawtimber by ~40% in local wood markets. When the high value of pine straw (i.e., fallen needles used as landscaping material which may be harvested from stands as early as age 10) is added, the economic value of longleaf pine forests becomes increasingly obvious. Surveys consistently indicate the value of hunter access to private lands as a tradable commodity throughout the natural range of longleaf pine. Where longleaf pine forests are maintained in open park-like condition, the higher quality of this habitat for quail, turkey and deer brings premium economic returns in the form of hunting leases and related services to private landowners.

Acknowledgments

The authors express their appreciation to Becky Estes for searching the literature to identify numerous relevant publications. We are also grateful to John Stanturf, Palle Madsen, Dave Haywood, Steve Jack, John Kush, Dave Borland, and Ric Jeffers for comments helpful in improving this manuscript. This chapter was updated by K.W. Outcalt.

References

Abrahamson, W.G. and Hartnett, D.C., Pine flatwoods and dry prairies, in *Ecosystems of Florida*, Myers, R.L. and Ewel, J.J., Eds., University of Central Florida Press, Orlando, FL, p. 103, 1990.

Addington, R.N., Greene, T.A., Elmore, M.L et al. Influence of herbicide site preparation on longleaf pine ecosystem development and fire Management, *South. J. Appl. For.*, 36, 173–180, 2012.

Anderson, K.M., Tending the wilderness. *Restor. Manage. Notes* 14, 154, 1996.

Aschenbach, T.A., Foster, B.L., and Imm, D.W., The initial phase of a longleaf pine-wiregrass savanna restoration: Species establishment and community responses, *Restor. Ecol.*, 18, 762, 2010.

Bailey, R.G., *Description of the Ecoregions of the United States*. USDA Forest Service, Miscellaneous Publication 1391, Washington, DC, 108 pp., 1995.

Barnett, J.P., Lauer, D.K., and Brissette, J.C., Regenerating longleaf pine with artificial methods, in *Management of Longleaf Pine*, Farrar, R.M., Ed., U.S. Department of Agriculture, Forest Service, Southern Forest Experiment Station, New Orleans, LA, General Technical Report SO-75, p. 72, 1990.

Barnett, J.P. and McGilvray, J.M., *Practical Guidelines for Producing Longleaf Pine Seedlings in Containers.* U.S. Department of Agriculture, Forest Service, Southern Research Station, Asheville, NC, General Technical Report SRS-14, 28 pp., 1997.

Bartram, W., *Travel Through North and South Carolina, Georgia and East and West Florida*, Dover Publishers, New York, NY, 414 pp., 1791.

Berish, S.J., *Efficacy of Three Formulations of the Forest Herbicide Hexazinone as an Aid to Reforestation of Longleaf Pine (Pinus palustris) Sandhills at Eglin Air Force Base, Florida*, M.S. Thesis, University of Florida, Gainesville, FL, 51 pp., 1996.

Boyer, W.D., *Growing-season Burns for Control of Hardwoods in Longleaf Pine Stands*, U.S. Department of Agriculture, Forest Service, Southern Forest Experiment Station, Research Paper SO-256, New Orleans, LA, 7 pp., 1990a.

Boyer, W.D., *Pinus palustris*, Mill. Longleaf pine, in *Silvics of North America, Vol. 1, Conifers*, Burns, R.M. and Honkala, B.H., Tech. Coordinators, U.S. Department of Agriculture, Forest Service, Washington, DC, p. 405, 1990b.

Boyer, W.D. and White, J.B., Natural regeneration of longleaf pine, in *Management of Longleaf Pine*, Farrar, R.M. Ed., U.S. Department of Agriculture, Forest Service, Southern Forest Experiment Station, New Orleans, LA, General Technical Report SO-75, p. 94, 1990.

Brockway, D.G. and Lewis, C.E., Long-term effects of dormant-season prescribed fire on plant community diversity, structure and productivity in a longleaf pine wiregrass ecosystem, *For. Ecol. Manage.*, 96, 167, 1997.

Brockway, D.G. and Outcalt, K.W., Gap-phase regeneration in longleaf pine wiregrass ecosystems, *For. Ecol. Manage.*, 106, 125, 1998.

Brockway, D.G. and Outcalt, K.W., Restoring longleaf pine wiregrass ecosystems: Hexazinone application enhances effects of prescribed fire, *For. Ecol. Manage.*, 137, 121, 2000.

Brockway, D.G., Outcalt, K.W., and Boyer, W.D., Longleaf pine regeneration ecology and methods, in *The Longleaf Pine Ecosystem: Ecology, Silviculture, and Restoration,* Jose, S., Jokela, E.J., and Miller, D.L., Eds., Springer, New York, NY, p. 95, 2006.

Brockway, D.G., Outcalt, K.W., Tomczak, D.J et al. *Restoration of Longleaf Pine Ecosystems,* U.S. Department of Agriculture, Forest Service, Southern Research Station, Asheville, NC, General Technical Report SRS-83, 34 pp., 2005.

Brockway, D.G., Outcalt, K.W., and Wilkins, R.N., Restoring longleaf pine wiregrass ecosystems: Plant cover, diversity and biomass following low-rate hexazinone application on Florida sandhills, *For. Ecol. Manage.*, 103, 159, 1998.

Brudvig, L.A., Orrock, J.L., Damschen, E.I et al. Land-use history and contemporary management inform an ecological reference model for longleaf pine woodland understory plant communities, *PLoS ONE*, 9, e86604, doi:10.1371/journal.pone.0086604, 2014.

Burns, R.M. and Hebb, E.A., *Site Preparation and Reforestation of Droughty, Acid Sands*, U.S. Department of Agriculture, Forest Service, Washington, DC, Agriculture Handbook No. 426, 61 pp., 1972.

Busby, R.L., Thomas, C.E., and Lohrey, R.E., The best kept secret in southern forestry: Longleaf pine plantation investments, in *Longleaf Pine: A Regional Perspective of Challenges and Opportunities,* Kush, J.S. Comp., Longleaf Alliance Report No. 1, Solon Dixon Forestry Education Center, Andalusia, AL, p. 26, 1996.

Carroll, W.D., Kapeluck, P.R., Harper, R.A et al. Historical overview of the southern forest landscape and associated resources, in *The Southern Forest Resource Assessment,* Wear, D.N., and Greis, J.G., Eds, U.S. Department of Agriculture, Forest Service, Southern Research Station, Asheville, NC, General Technical Report SRS-53, p. 583, 2002.

Christensen, N.L., Fire regimes in southeastern ecosystems, in *Fire Regimes and Ecosystem Properties,* Mooney, H.A., Bonnicksen, T.M., Christensen, N.L., Lotan, J.E., and Reiners, W.A., Eds., U.S. Department of Agriculture, Forest Service, Washington, DC, General Technical Report WO-26, p. 112, 1981.

Clewell, A.F., Natural history of wiregrass (*Aristida stricta* Michx., Gramineae), *Nat. Areas J.*, 9, 223, 1989.

Clewell, A.F., Restoring for natural authenticity, *Ecol. Restor.*, 18, 216, 2000.

Croker, T.C., *Longleaf Pine: A History of Man and a Forest*, U.S. Department of Agriculture, Forest Service, Southern Region, Atlanta, GA, Forestry Report R8-FR7, 37 pp., 1987.

Delcourt, P.A. and Delcourt, H.R., *Long-Term Forest Dynamics of the Temperate Zone,* Springer-Verlag Publishers, New York, NY, 439 pp., 1987.

Engstrom, R.T., Red-cockaded woodpeckers: Prospects for recovery, in *The Fire Forest: Longleaf Pine-Wiregrass Ecosystems,* Wilson, J.R., Ed., Georgia Wildlife, vol. 8, p. 12, 2001.

Engstrom, R.T., Kirkman, L.K., and Mitchell, R.J., The natural history of the fire forest, in *The Fire Forest: Longleaf Pine-Wiregrass Ecosystems,* Wilson, J.R., Ed., Georgia Wildlife, vol. 8, pp. 5, 14, 2001.

Franklin, R.M., *Stewardship of Longleaf Pine Forests: A Guide for Landowners,* Longleaf Alliance Report No. 2, Longleaf Alliance, Solon Dixon Forestry Education Center, Andalusia, AL, 44 pp., 1997.

Freeman, J.E. and Jose, S., The role of herbicide in savanna restoration: Effects of shrub reduction treatments on the understory of a longleaf pine flatwoods, *For. Ecol. Manage.*, 257, 978, 2009.

Frost, C.C., Four centuries of changing landscape patterns in the longleaf pine ecosystem, in *Proceedings of the 18th Tall Timbers Fire Ecology Conference,* Herman, S.M. Ed., Tall Timbers Research Station, Tallahassee, FL, p. 17, 1993.

Frost, C.C., History and future of the longleaf pine ecosystem, in *The Longleaf Pine Ecosystem: Ecology, Silviculture, and Restoration,* Jose, S., Jokela, E.J., and Miller, D.L., Eds., Springer, New York, NY, p. 9, 2006.

Garren, K.H., Effects of fire on vegetation of the southeastern United States. *Bot. Rev.*, 9, 617, 1943.

Glitzenstein, J.S., Platt, W.J., and Streng, D.R., Effects of fire regime and habitat on tree dynamics in north Florida longleaf pine savannas, *Ecol. Monogr.*, 65, 441, 1995.

Glitzenstein, J.S., Streng, D.R., and Wade, D.D., A promising start for a new population of *Parnassia caroliniana* Michx., in *Ecological Restoration and Regional Conservation Strategies*, Kush, J.S., Comp., Longleaf Alliance Report No. 3, Solon Dixon Forestry Education Center, Andalusia, AL, p. 44, 1998.

Glitzenstein, J.S., Streng, D.R., Wade, D.D et al. Starting new populations of longleaf pine ground-layer plants in the Outer Coastal Plain of South Carolina, USA., *Nat. Areas J.*, 21, 89, 2001.

Grelen, H.E., Plant succession on cleared sandhills in northern Florida, *Am. Midl. Nat.*, 67, 36, 1962.

Hardin, E.D. and White, D.L., Rare vascular plant taxa associated with wiregrass (*Aristida stricta*) in the southeastern United States, *Am. Midl. Nat.*, 9, 234, 1989.

Harrington, T.B., Overstory and understory relationships in longleaf pine plantations 14 years after thinning and woody control, *Can. J. For. Res.*, 41, 2301, doi:10.1139/X11-140, 2011.

Hattenbach, M.J., Gordon, D.R., Seamon, G.S et al. Development of direct-seeding techniques to restore native groundcover in a sandhill ecosystem, in *Ecological Restoration and Regional Conservation Strategies*, Kush, J.S., Comp., Longleaf Alliance Report No. 3, Solon Dixon Forestry Education Center, Andalusia, AL, p. 64, 1998.

Hilliard, T., Longleaf-wiregrass restoration at Fort Stewart, Georgia: The military's role in restoration, in *Ecological Restoration and Regional Conservation Strategies*, Kush, J.S. Comp., Longleaf Alliance Report No. 3, Solon Dixon Forestry Education Center, Andalusia, AL, p. 32, 1998.

Hilton, J., Biological diversity in the longleaf pine ecosystem, *Alabama's Treasured For*, 18, 28, 1999.

Huffman, J.M. and Dye, R., Summary of wiregrass ecosystem restoration projects at Myakka River State Park, in *Proceedings of the Wiregrass Ecosystem Restoration Workshop*, Clewell, A.F., Cleckley, W., Eds., State of Florida, Northwest Florida Water Management District, Tallahassee, FL, p. 21, 1994.

Jackson, D.R. and Milstrey, E.R., The fauna of gopher tortoise burrows, in *Proceedings of the Gopher Tortoise Relocation Symposium*, Diemer, J.E., Ed., State of Florida, Game and Freshwater Fish Commission, Tallahassee, FL, p. 86, 1989.

Johnson, R. and Gjerstad, D., Restoring the ovestory of longleaf pine ecosystems, in *The Longleaf Pine Ecosystem: Ecology, Silviculture, and Restoration*, Jose, S., Jokela, E.J., and Miller, D.L., Eds., Springer, New York, NY, p. 271, 2006.

Keetch, J.J. and Byram, G.M., *A Drought Index for Forest Fire Control*, U.S. Department of Agriculture, Forest Service, Southeastern Forest Experiment Station, Research Paper SE-38, Asheville, NC, 35 pp., 1968.

Kelly, J.F. and Bechtold, W.A., The longleaf pine resource, in *Management of Longleaf Pine*, Farrar, R.M., Ed., U.S. Department of Agriculture, Forest Service, Southern Forest Experiment Station, General Technical Report SO-75, New Orleans, LA, p. 11, 1990.

Kirkman, L.K., Drew, M.B., West, L.T et al. Ecotone characterization between upland longleaf pine/wiregrass stands and seasonally-ponded isolated wetlands, *Wetlands*, 18, 346, 1998.

Kirkman, L.K. and Mitchell, R., A forest gap approach to restoring longleaf pine-wiregrass ecosystems (Georgia and Florida), *Ecol. Rest.*, 20, 50, 2002.

Knapp, B.O., Wang, G.G., Hua, H et al. Restoring longleaf pine (*Pinus palustris* Mill.) in loblolly pine (*Pinus taeda* L.) stands: Effects of restoration treatments on loblolly pine regeneration, *For. Ecol. Manage.*, 262, 1157, 2011.

Komarek, E.V., Lightning and lightning fires as ecological forces, in *Proceedings of the 9th Tall Timbers Fire Ecology Conference*, Tall Timbers Research Station, Tallahassee, FL, p. 169, 1968.

Kush, J.S. and Meldahl, R.S., Composition of a virgin stand of longleaf pine in south Alabama, *Castanea*, 65, 56, 2000.

Landers, J.L., Disturbance influences on pine traits in the southeastern United States in *Proceedings of the 17th Tall Timbers Fire Ecology Conference*, Tall Timbers Research Station, Tallahassee, FL, p. 61, 1991.

Landers, J.L. and Boyer, W.D., *An Old-growth Definition for Upland Longleaf and South Florida Slash Pine Forests, Woodlands and Savannas*, U.S. Department of Agriculture, Forest Service, Southern Research Station, General Technical Report SRS-29, Asheville, NC, 15 pp., 1999.

Landers, J.L., Byrd, N.A., and Komarek, R., A holistic approach to managing longleaf pine communities, in *Management of Longleaf Pine*, Farrar, R.M., Ed., U.S. Department of Agriculture, Forest Service, Southern Forest Experiment Station, General Technical Report SO-75, New Orleans, LA, p. 135, 1990.

Landers, J.L., Van Lear, D.H., and Boyer, W.D., The longleaf pine forests of the Southeast: Requiem or renaissance? *J. For.*, 93, 39, 1995.

Little, E.L., *Atlas of United States Trees. Volume 1. Conifers and Important Hardwoods*, U.S. Department of Agriculture, Forest Service, Miscellaneous Publication 1146, Washington, DC, 320 pp., 1971.

Longleaf Partnership Council, *America's Longleaf Restoration Initiative, 2013 Range-wide Accomplishment Report*, (http://www.longleafalliance.org/publications/2013RangewideAccomplishmentReport_2_12_FINAL.pdf, accessed March 2014).

McMahon, C.K., Tomczak, D.J., and Jeffers, R.M., Longleaf pine ecosystem restoration: The role of the USDA Forest Service, in *Ecological Restoration and Regional Conservation Strategies*, Kush, J.S., Comp., Longleaf Alliance Report No. 3, Solon Dixon Forestry Education Center, Andalusia, AL, p. 20, 1998.

Means, D.B., Wiregrass restoration: Probable shading effects in a slash pine plantation, *Rest. Manage. Notes*, 15, 52, 1997.

Menges, E.S. and Gordon, D.R., Should mechanical treatments and herbicides be used as fire surrogates to manage Florida's uplands? A review, *Fla. Sci.*, 73, 147, 2010.

Miller, S.R. and Bossuot, W.R., Flatwoods restoration on the St. Johns River Water Management District, Florida: A prescription to cut and burn, in *Proceedings of the 21st Tall Timbers Fire Ecology Conference*, Tall Timbers Research Station, Tallahassee, FL, p. 212, 2000.

Miller, J.H. and Robinson, K.S., A regional perspective of the physiographic provinces of the Southeastern United States, in *Proceedings of the Eighth Biennial Southern Silvicultural Research Conference, Edwards*, M.B., Ed., U.S. Department of Agriculture, Forest Service, Southern Research Station, Asheville, NC, General Technical Report SRS-1, p. 581, 1995.

Mulligan, M.K. and Kirkman, L.K., Competition effects on wiregrass (*Aristida beyrichiana*) growth and survival, *Plant Ecol.*, 167, 39, 2002.

Mulligan, M.K., Kirkman, L.K., and Mitchell, R.J., *Aristida beyrichiana* (wiregrass) establishment and recruitment: Implications for restoration, *Rest. Ecol.*, 10, 68, 2002.

Noss, R.F., Longleaf pine and wiregrass: Keystone components of an endangered ecosystem, *Nat. Areas J.*, 9, 211, 1989.

Noss, R.F., LaRoe, E.T., and Scott, J.M., *Endangered Ecosystems of the United States: A Preliminary Assessment of Loss and Degradation*. U.S. Department of Interior, National Biological Service, Washington, DC, Biological Report 28, 59 pp., 1995.

O'Brien, J.J., Hiers, J.K., Mitchell, R.J et al. Acute physiological stress and mortality following fire in a long-unburned longleaf pine ecosystem, *Fire Ecol.*, 6, 12, doi: 10.4996/fireecology.0602001, 2010.

Oswalt, C.M., Cooper, J.A., Brockway, D.G et al. *History and Current Condition of Longleaf Pine in the Southern United States*, U.S. Department of Agriculture, Forest Service, Southern Research Station, Asheville, NC, General Technical Report SRS-166, 51 pp., 2012.

Outcalt, K.W., Seed production of wiregrass in central Florida following growing season prescribed burns. *Int. J. Wildland Fire*, 4, 123, 1994.

Outcalt, K.W., *Maintaining the Native Plant Community during Longleaf Pine Establishment*. Forestry Research Institute, Rotorua, New Zealand, Bulletin No. 192, p. 283, 1995.

Outcalt, K.W., Occurrence of fire in longleaf pine stands in the southeastern United States, in *Proceedings of the 21st Tall Timbers Fire Ecology Conference*, Moser W.K., and Moser, C.F., Eds., Tall Timbers Research Station, Tallahassee, FL, p. 178, 2000.

Outcalt, K.W., Lightning, fire and longleaf pine: Using natural disturbance to guide management. *For. Ecol. Manage.*, 255, 3351, 10.1016/j.foreco.2008.02.016, 2008.

Outcalt, K.W. and Brockway, D.G., Structure and composition changes following restoration treatments of longleaf pine forests on the Gulf Coastal Plain of Alabama, *For. Ecol. Manage.*, 259, 1615, 2010.

Outcalt, K.W. and Lewis, C.E., Response of wiregrass (*Aristida stricta*) to mechanical site preparation, in *Wiregrass Biology and Management*, Duever, L.C. and Noss, R.F., Eds., KBN Engineering and Applied Sciences, Gainesville, FL, p. 60, 1990.

Outcalt, K.W. and Sheffield, R.M., The longleaf pine forest: Trends and current conditions, U.S. Department of Agriculture, Forest Service, Southern Research Station, Asheville, NC, Resource Bulletin SRS-9, 23 pp., 1996.

Outcalt, K.W. and Wade, D.D., Response of a longleaf pine (*Pinus palustris*) flatwoods community to long-term dormant season prescribed burning, in *Proceedings of 89th Annual Ecological Society of America Meeting*, Ecological Society of America, Washington, DC, p. 384, 2004.

Outcalt, K.W., Williams, M.E., and Onokpise, O., Restoring *Aristida stricta* to *Pinus palustris* ecosystems on the Atlantic Coastal Plain, USA. *Rest. Ecol.*, 7, 262, 1999.

Palik, B.J., Mitchell, R.J., Houseal, G et al. Effects of canopy structure on resource availability and seedling responses in a longleaf pine ecosystem, *Can. J. For. Res.*, 27, 1458, 1997.

Palik, B.J. and Pedersen, N., Overstory mortality and canopy disturbances in longleaf pine ecosystems, *Can. J. For. Res.*, 26, 2035, 1996.

Peet, R.K., Ecological classification of longleaf pine woodlands, in *The Longleaf Pine Ecosystem: Ecology, Silviculture, and Restoration*, Jose, S., Jokela, E.J., and Miller, D.L., Eds., Springer, New York, NY, p. 51, 2006.

Peet, R.K. and Allard, D.J., Longleaf pine-dominated vegetation of the southern Atlantic and eastern Gulf Coast region, USA, in *Proceedings of the 18th Tall Timbers Fire Ecology Conference*, Herman, S.M., Ed., Tall Timbers Research Station, Tallahassee, FL, p. 45, 1993.

Pickett, S.T.A. and White, P.S., *The Ecology of Natural Disturbance and Patch Dynamics*, Academic Press, Orlando, FL, p. 1985.

Platt, W.J., Evans, G.W., and Davis, M.M., Effects of fire season on flowering of forbs and shrubs in longleaf pine forests, *Oecologia* 76, 353, 1988a.

Platt, W.J., Evans, G.W., and Rathbun, S.L., The population dynamics of a long-lived conifer (*Pinus palustris*). *Am. Nat.*, 131, 491, 1988b.

Platt, W.J. and Rathbun, S.L., Dynamics of an old-growth longleaf pine population, in *Proceedings of the 18th Tall Timbers Fire Ecology Conference*, Herman, S.M., Ed., Tall Timbers Research Station, Tallahassee, FL, p. 275, 1993.

Provencher, L., Litt, A.R., Galley, K.E.M et al. *Restoration of Fire Suppressed Longleaf Pine Sandhills at Eglin Air Force Base, Niceville, Florida*, The Nature Conservancy, Science Division, Gainesville, FL, 294 pp., 2001.

Pyne, S.J., *Fire in America: A Cultural History of Wildland and Rural Fire*, Princeton University Press, Princeton, NJ, 654 pp., 1997.

Robbins, L.E. and Myers, R.L., *Seasonal Effects of Prescribed Burning in Florida: A Review*. Tall Timbers Research Station, Miscellaneous Publication No. 8, Tallahassee, FL, 96 pp., 1992.

Schmidtling, R. and Hipkins, V., Genetic diversity in longleaf pine (*Pinus palustris*): influence of historical and prehistorical events. *Can. J. For. Res.*, 28, 1135, 1998.

Schultz, R.P., *Environmental Change after Site Preparation and Slash Pine Planting On a Flatwoods Site*. U.S. Department of Agriculture, Forest Service, Southeastern Forest Experiment Station, Asheville, NC, Research Paper SE-156, 20 pp., 1976.

Schwartz, M.W., Natural distribution and abundance of forest species and communities in northern Florida, *Ecology*, 75, 687, 1994.

Schwarz, G.F., *The Longleaf Pine Virgin Forest: A Silvical Study*, John Wiley and Sons, New York, NY, 135 pp., 1907.

Schwilk, D.W., Keeley, J.E., Knapp, E.E et al. The national Fire and Fire Surrogate study: Effects of fuel reduction methods on forest vegetation structure and fuels. *Ecol. Appl.*, 19, 285, 2009.

Seamon, G., A longleaf pine sandhill restoration in northwest Florida, *Rest. Manage. Notes*, 16, 46, 1998.

Smith, G.P., Shelburne, V.B., and Walker, J.L. Structure and composition of vegetation of longleaf pine plantations compared to natural stands occurring along an environmental gradient at the Savannah River Site, in *Proceedings of The 11th Biennial Southern Silvicultural Research Conference*, Outcalt, K.W., Ed., U.S. Department of Agriculture, Forest Service, Southern Research Station, Asheville, NC, General Technical Report SRS-48, p. 481, 2002.

Society for Ecological Restoration International Science (SERI) & Policy Working Group, *The SER International Primer on Ecological Restoration*. www.ser.org & Tucson: Society for Ecological Restoration International, 8 pp., 2004.

Stanturf, J.A., Wade, D.D., Waldrop, T.A et al. Fire in southern landscapes, in *The Southern Forest Resource Assessment*, Wear, D.N. and Greis, J.G., Eds., U.S. Department of Agriculture, Forest Service, Southern Research Station, Asheville, NC, General Technical Report SRS-53, p. 607, 2002.

Stout, I.J. and Marion, W.R., Pine flatwoods and xeric pine forests of the southern lower coastal plain, in *Biodiversity of the Southeastern United States: Lowland Terrestrial Communities*, Martin, W.H., Boyce, S.G., and Echternacht, A.C., Eds., Wiley, New York, NY, p. 373, 1993.

Taylor, A.R., Ecological aspects of lightning in forests, in *Proceedings of the 13th Tall Timbers Fire Ecology Conference*, Komarek, E.V. Sr., Ed., Tall Timbers Research Station, Tallahassee, FL, p. 455, 1974.

Trusty, J.L. and Ober, H.K., Determinants of successful groundcover restoration in forests of the southeastern United States, *J. Nat. Con.*, 19, 34, 2011.

United States Department of Agriculture (USDA), *Farm Service Agency*, (http://www.fsa.usda.gov/FSA/, accessed 2014).

Uridel, K.W., *Restoration of Native Herbs in Abandoned Paspalum notatum (Bahia Grass) Pastures*, M.S. Thesis, University of Florida, Gainesville, 61 pp., 1994.

Varner, J.M., Kush, J.S., and Meldahl, R.S., Ecological restoration of an old-growth longleaf pine stand utilizing prescribed fire, in *Proceedings of the 21st Tall Timbers Fire Ecology Conference*, Moser, W.K., Moser, C.F., Eds., Tall Timbers Research Station, Tallahassee, FL, p. 216, 2000.

Wahlenberg, W.G., *Longleaf Pine: Its Use, Ecology, Regeneration, Protection, Growth and Management*, C.L. Pack Forestry Foundation and U.S. Department of Agriculture, Forest Service, Washington, DC, 429 pp., 1946.

Waldrop, T.A., Van Lear, D.H., Lloyd, F.T et al. *Long-Term Studies of Prescribed Burning in Loblolly Pine Forests of the Southeastern Coastal Plain*, U.S. Department of Agriculture, Forest Service, Southeastern Forest Experiment Station, Asheville, NC, General Technical Report SE-45, 23 pp., 1987.

Walker, J.L., Rare vascular plant taxa associated with the longleaf pine ecosystem, in *Proceedings of the 18th Tall Timbers Fire Ecology Conference*. Hermann, S.M., Ed., Tall Timbers Research Station, Tallahassee, FL, p. 105, 1993.

Walker, J.L., Ground layer vegetation in longleaf pine landscapes: An overview for restoration and management, in *Ecological Restoration and Regional Conservation Strategies*, Kush, J.S., Comp., Longleaf Alliance Report No. 3, Solon Dixon Forestry Education Center, Andalusia, AL, p. 2, 1998.

Walker, J.L. and Silletti, A.M., Restoring the ground layer of longleaf pine ecosystems, in *The Longleaf Pine Ecosystem: Ecology, Silviculture, and Restoration*, Jose, S., Jokela, E.J., and Miller, D.L., Eds., Springer, New York, NY, p. 297, 2006.

Ware, S., Frost, C.C., and Doerr, P.D., Southern mixed hardwood forest: The former longleaf pine forest, in *Biodiversity of the Southeastern United States: Lowland Terrestrial Communities*, Martin, W.H., Boyce, S.G., and Echternacht, A.C., Eds., John Wiley and Sons, New York, NY, p. 447, 1993.

Watts, W.A., Hansen, B.C.S., and Grimm, E.C., Camel Lake: A 40,000-yr record of vegetational and forest history from north Florida, *Ecology*, 73, 1056, 1992.

Wear, D.N. and Greis, J.G., Summary report, in *The Southern Forest Resources Assessment*, Wear, D.N. and Greis, J.G., Eds., U.S. Department of Agriculture, Forest Service, Southern Research Station, Asheville, NC, General Technical Report SRS-53, 114 pp., 2002.

Williams, J.L., *The Territory of Florida, 1962 Edition*, University of Florida Press, Gainesville, FL, 304 pp., 1837.

23

Restoring Dry and Moist Forests of the
Inland Northwestern United States

Theresa B. Jain and Russell T. Graham

CONTENTS

23.1 Introduction

The complex topography of the Inland Northwestern United States (58.4 million ha) interacts with soils and a highly variable climate to provide a mosaic of dry and moist mixed conifer forest settings. Approximately 20% of the area is covered by dry forests dominated by *Pinus ponderosa, Pseudotsuga menziesii* and contains a diversity of lower vegetation ranging from a grass savannah on the less-productive sites to shrub and forb dominated vegetation on the more-productive sites. An estimated 18% of the area is covered by moist mixed conifer forests with some places growing up to 10 different conifer species, in addition to a diversity of surface vegetation often dominated by shrub and forb vegetation (e.g., with *Pinus monticola* and *Tsuga heterophylla*).

In the dry forests, historically frequent surface and mixed fires burned over 75% of the area of dry forests; however, successful fire exclusion, harvesting and a cool and moist period allowed dense stands of *Abies grandis, P. menziesii*, and small *P. ponderosa* to develop. Historically, forest canopies and their nutrients were located well above the soil surface; fine roots and microbial activity were located deep in mineral soils, thus protecting them from wildfire. In contrast, the *Abies* and *Pseudotsuga* dominated forests of today contain nutrient-rich crowns that extend to the forest floor. Nutrients and microbial activity are located near the soil surface, increasing their susceptibility to loss from wildfire.

In the moist forests, fire exclusion, harvesting, and the introduction of *Cronartium ribicola* (a stem disease) from Europe are the primary change agents. In the Northern Rocky Mountain moist forests, early-seral *P. monticola* has nearly been extirpated and mid- to late-seral conifers now dominate. In the moist forest of the eastern Cascades Mountains in Washington and Oregon, an increase in homogeneity of mid-seral forests containing *A. grandis*, *T. heterophylla*, and *P. menziesii* has occurred, encouraged by the harvesting of *Larix occidentalis* and *P. ponderosa*. These changes have elevated the risk to large-scale insect and disease epidemics and uncharacteristic wildfires.

Successful restoration strategies in dry and moist forests should apply concepts learned from the past but we must also be cognizant of the changes that have occurred not only in the tree component but also those occurring in the lower vegetation and soil and across landscapes. The reintroduction of fire alone is not the answer to restore these forests, because today we have ever-changing social desires, changes in soil microbial and chemical properties, potential changes in long-term climate, and both native and exotic diseases and insects that prevent reverting to the past. Rather a multiscale approach applied that integrate these conditions over short- and long-term temporal (decades to centuries) and spatial (site to landscape) scales may provide a template for restoring the moist and dry forests of the Inland Northwestern United States.

23.2 Forests of the Inland Northwest

The Inland Northwestern United States (58.4 million ha) is defined by the Bitterroot, Selkirk, Cabinet, Salmon River, Lemhi, Steens, Purcell, Cascade, and Blue mountain ranges with elevations over 1500 m (Figure 23.1). Within these ranges, the valley bottoms can be low (225 m) and the topography steep. This rough and complex topography results in a variety of forest settings ranging from steep slopes, in narrow V-cut canyons, to gentle rolling slopes, in wide river valleys. During the Pleistocene, alpine glaciers shaped the canyons and valleys; today, a mantle of glacial till covers these glaciated landscapes. Much of the fine silt washed out by the glaciers was redeposited by winds, leaving deep layers of loess over many landscapes. Some 12,000 to 15,000 years ago, Glacial Lake Missoula repeatedly filled and emptied, flooding most of northern Idaho and eastern Washington. The eruption of prehistoric Mt. Mazama 7500 years ago formed Crater Lake in Oregon and deposited a layer of ash up to 62 cm thick across the area. Disturbance events continually modify the granitic and metasedimentary rocks, ash, and loess deposits, giving rise to diverse soils (Quigley et al. 1996).

Moist marine air originating from the Pacific Ocean moderates temperatures within the Inland Northwestern United States, while continental dry and cold air from the north and east brings cold weather in winter and hot weather in summer. During the summer, these air masses interact and bring convective precipitation, lightning, and cool periods. Dry Arctic air in the winter brings damaging frosts and cold temperatures ($\leq 5°C$) that alternate with wet warm periods. This highly variable climate interacts with the heterogeneous and rugged topography to create mosaics of dry and moist mixed conifer forests (Franklin and Dryness 1973; Foiles et al. 1990; Graham 1990; Hann et al. 1997).

Until 1900, forests covered over 47% of the Inland Northwest (Figure 23.1). Dry forests occupied an estimated 11 million ha, dominated by *P. ponderosa*, and moist forests covered

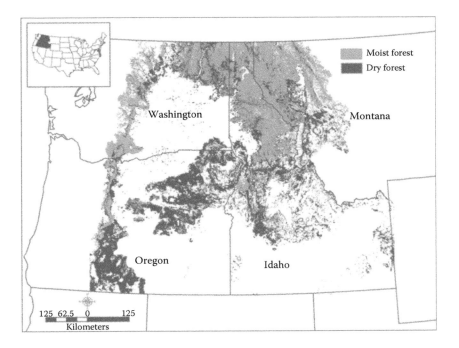

FIGURE 23.1
There are 58,361,400 ha in the Inland Northwestern United States framed by the Columbia River Basin. The topography is rugged, ranging from the Cascade Mountains in the West to the Bitterroot and Salmon River Mountains in Idaho. Elevations in the region range from 225 m to over 3000 m. Dry and moist forests make up 90% of the forests occurring in the Inland Northwestern United States. The moist forests occur primarily in northern Idaho, northwestern Montana, and northeastern Washington and along the eastern slopes of the Cascade Mountains in Washington and Oregon. The dry forests are dispersed throughout the region. (From Hann, W.J., J.L. Jones, M.G. Karl et al. In *An Assessment of Ecosystem Components in the Interior Columbia Basin and Portions of the Klamath and Great Basins: Volume II*, edited by T.M. Quigley and S.J. Arbelbide, 338–1055. PNW-GTR-405. Portland, OR: U.S. Department of Agriculture Forest Service, Pacific Northwest Research Station, 1997.)

an estimated 10.5 million ha (18%). The United States Forest Service and the Bureau of Land Management administer more than 50% of both the dry and moist forests (Quigley et al. 1996). Other federal and state agencies administer approximately 5% of these forests and several industrial and nonindustrial owners manage smaller tracts. Both the moist and dry forests have lost many native structures (large early-seral tree component) and processes (native fire regimes) that were integral in maintaining these systems and the myriad plants, animals, and uses they supported (Quigley et al. 1996). Similar series of events occurred in other locales with similar forest types located in British Columbia, Canada, and throughout the western United States (e.g., Hessburg et al. 1994; Burton and MacDonald 2011; Franklin and Johnson 2012).

23.3 Dry Forests

Dry forests occur across a wide range of elevations in northeastern Washington, northeastern Oregon, central and southern Idaho, and south-central Oregon (Figure 23.1) (Hann et al. 1997). These forests are complex, depending on the weather, physical setting, disturbances, forest succession, and potential vegetation. Depending on the combination of these components, multiple tree, shrub, and forb species can vary within a given site and across landscapes. Soil parent materials include granites, metasedimentaries, glacial tills, and basalts. Tree and plant communities that are considered dry forests occur in places that are water-limited and often are subject to drought and these forests can also occur on shallow soils which also influence the nutrient and water holding capacity. *P. menziesii*, *P. ponderosa*, and dry *A. grandis/Abies concolor* potential vegetation types (PVTs) dominate these settings (Hann et al. 1997). Potential vegetation type is a classification system based on the physical and biological environment characterized by the abundance and presence of vegetation in the absence of disturbance (Daubenmire and Daubenmire 1968; Pfister et al. 1977; Cooper et al. 1991). Potential vegetation types are defined by and named using indicator species (tree and surface vegetation) that grow in similar environmental conditions. When *L. occidentalis* is present it is always an early successional species (dominant after disturbance). *A. grandis/A. concolor* are late-successional species and are more shade-tolerant than *P. ponderosa* and *L. occidentalis*. *P. ponderosa* and *P. menziesii* can play both late- and early-successional roles, depending on the PVT (Daubenmire and Daubenmire 1968). Places that support forests but have limited water availability, *P. ponderosa* plays a late-successional role. In contrast on productive sites with more water availability, *P. ponderosa* plays an early-successional role in *A. grandis/A. concolor* and *P. menziesii* PVTs. *P. menziesii* plays a mid-successional role in *A. grandis/A. concolor* and also in moist forests. Surface vegetation in the dry forests includes shrubs (*Arctostaphylos uva-ursi*, *Ceanothus* spp., *Purshia tridentata*, *Symphoricarpos albus*, *Physocarpus malvaceus*), grasses (*Calamagrostis rubescens*, *Bromus vulgaris*), and sedges (*Carex* spp.) (Foiles et al. 1990; Hermann and Lavender 1990; Oliver and Ryker 1990).

Disturbances and physical setting historically maintained a variety of structural and successional stages (Table 23.1). Fire, insects, diseases, snow, ice, and competition thinned these forests, and surface fires provided opportunities for regeneration (Foiles et al. 1990; Hermann and Lavender 1990; Oliver and Ryker 1990; Jain et al. 2012; Stine et al. 2014). Approximately 18% of the area was in a grass, forb, and shrub stage for long (100s years) periods and 15% contained early-seral *P. ponderosa* with diameters ranging from 5 to 80 cm (Meyer 1938). As these forests aged, mid-seral multistoried forest structures developed. Three percent of the area contained late-seral *P. menziesii* and *A. grandis/A. concolor* with multiple canopies. Large, widely-spaced (~250 trees per ha) *P. ponderosa* often dominated 21% of the dry forests, with the plurality of diameters ranging from 30 to 60 cm (Figure 23.2a) (Daubenmire and Daubenmire 1968; Hann et al. 1997). Late-seral single-storied forests containing *P. menziesii* and *A. grandis/A. concolor* complexes dominated some settings (2%).

23.3.1 Dry Forest Change

The dry forests were adapted to a wide range of site conditions and short-term climate variation. These characteristics created an ecosystem that appeared to be long-lived and relatively resilient to disturbances (fire, insect, and disease) (Harvey et al. 1994). Since 1900, approximately 8% (600,000 ha) of the dry forests have been converted to agriculture, urbanization, and industry (Hann et al. 1997). Fire exclusion, harvesting, and changes in fire regime altered the composition and structure of the remaining dry forests (Everett et al. 1994; Hann et al. 1997; Lewis 2005; Jain et al. 2012). The area burned by surface fires has decreased from an estimated 80% to less than 50% of the area. The mean fire return

TABLE 23.1

Historical (1850–1900) and 1991 Distribution of Forest Structures within the Dry and Moist Forests of the Inland Northwest

Forest Structure	Historical (%)	1991 (%)	Change (%)
Dry Forests			
Grass/forb/shrub	18	1	−17
Early seral intolerant	15	14	−1
Early seral tolerant	3	3	0
Mid-seral intolerant	21	35	+14
Mid-seral tolerant	8	22	+14
Late seral—intolerant single story	21	5	−16
Late seral—tolerant single story	2	3	+1
Late seral—intolerant multistory	9	8	−1
Late seral—tolerant multistory	3	9	+6
Moist Forests			
Northern Rocky Mountain Region (NRM)			
Early seral—single story	29	20	−9
Mid-seral	41	69	+28
Late seral—single story	11	3	−8
Late seral—multistory	19	8	−11
Eastern Cascade Region (Northern Cascade/Southern Cascade)			
Early seral—single story	23/25	32/15	+9/−10
Mid-seral	37/34	48/37	+11/+3
Late seral—single story	9/9	4/29	−5/+20
Late seral—multistory	31/32	16/19	−15/−13

Source: Adapted from Hann, W.J. et al. In *An Assessment of Ecosystem Components in the Interior Columbia Basin and Portions of the Klamath and Great Basins: Volume II*, edited by T.M. Quigley and S.J. Arbelbide, 338–1055. PNW-GTR-405. Portland, OR: U.S. Department of Agriculture Forest Service, Pacific Northwest Research Station, 1997.

interval has also increased from less than 20 years to 40–80 years. Mixed-fires (combination of surface and crown fires) have increased from 5% to an estimated 35% of burned area and the mean fire return interval has increased from 45 to 60 years. A similar increase in crown fires has also occurred (Hann et al. 1997). Mid-seral structures have increased (from an estimated 29%–57% of the area), often containing dense areas of small *P. ponderosa*, *P. menziesii*, or *A. grandis/A. concolor* (Table 23.1; Figure 23.2b). The proportion of the dry forests occupied by late-seral single-storied *P. ponderosa* has declined from 21% to 5% (Figure 23.2a). In addition, small diameter trees have encroached and now occupy all but one percent of the dry forests that formerly were covered by grasses, forbs, and shrubs (Figure 23.2b). The dominant tree species has changed from *P. ponderosa* to *P. menziesii* or *A. grandis/A. concolor,* changing the character and canopy architecture of the forest.

The shift in species composition from *P. ponderosa* to *Abies* and *Pseudotsuga* dominated forests changed litter type and quantity, which changed soil chemistry, microbial processes, and ectomycorrhizal relationships (Rose et al. 1983). For example, decomposed true firs create white rotten wood, which rapidly disperses into the soil and is quickly consumed by decomposers. In contrast, decomposed *P. ponderosa* and *L. occidentalis* create brown rotten wood, which can persist in soil for centuries and has been shown to retain nutrients

(a)

(b)

FIGURE 23.2
A historical (1850–1900) (*Pinus ponderosa*) stand exhibiting a lush understory layer of forbs and grasses (a). These conditions were maintained by frequent (<20 year) nonlethal surface fires. Within the dry forests, successful fire exclusion and harvesting have allowed dense stands of vegetation to develop (b). (USDA Forest Service photographs.)

and hold water (Larsen et al. 1980; Harvey et al. 1987). *L. occidentalis* and *P. ponderosa* tend to be deep-rooted, in contrast to the relatively shallow-rooted *Pseudotsuga* and *Abies*, which have abundant feeder roots and ectomycorrhizae in the shallow soil organic layers (Minore 1979; Harvey et al. 1987). *P. ponderosa* and *L. occidentalis* forests are generally tall and self-pruning, even in moderately dense areas. They have large branches high in the crowns and the base of the crowns is well above surface fuels. In general, this crown architecture protects the nutrients stored in the canopy from surface fires. In contrast, young- to mid-aged (<150 years) *P. menziesii* and *A. grandis/A. concolor* generally do not self-prune. This canopy architecture favors lower crown base heights, higher crown densities, and canopies with higher nutrient (especially potassium) content than it occurs in *L. occidentalis* and *P. ponderosa* dominated forests (Figure 23.2b) (Minore 1979; Harvey et al. 1999).

In the dry forests, biological decomposition is more limited than biological production. When fire return intervals reflected historical fire frequencies, the accumulation of thick organic layers was minimized and nutrient storage and nutrient turnover was dispersed in the mineral soils (Marschner and Marschner 1996; Harvey et al. 1999). In the absence of fire, bark slough, needles, twigs, and small branches accumulated on the forest floor allowing ectomycorrhizae and fine roots of all species to concentrate in the surface mineral soil and thick organic layers (Harvey et al. 1994; Hood 2010; Jain et al. 2012).

Harvesting the *L. occidentalis* and *P. ponderosa* and the ingrowth of *A. grandis/A. concolor* and *P. menziesii* in the dry forests together facilitated the accumulation of both above- and below-ground biomass and their nutrient content close to the soil surface (Harvey et al. 1986; Hood 2010). Even low-intensity surface fires now consume the surface organic layers, killing fine roots, volatilizing nutrients, killing trees, and increasing soil erosion potential (Debano 1991; Hungerford et. al. 1991; Ryan and Amman 1996; Robichaud et al. 2000; Hood 2010). In addition, fir ingrowth creates nutrient-rich ladder fuels that facilitate crown-fire initiation, increasing the likelihood of nutrient loss (Van Wagner 1977; Minore 1979; Harvey et al. 1999). The risk of nutrient loss is great on infertile sites, because dense areas of late-seral species are more demanding of nutrients and water than the historical areas dominated by widely-spaced early-seral species (Minore 1979; Harvey et al. 1999).

23.4 Moist Forests

Moist forests of the Inland Northwestern United States occur in two locations, the eastern Cascade Mountains (east of the Cascade Crest in Washington and Oregon) and the Northern Rocky Mountains (northeastern Washington and Oregon, northern Idaho, and western Montana) (Figure 23.1). They grow at elevations ranging from 460 to 1600 m and occasionally occur at elevations up to 1800 m (Foiles et al. 1990; Graham 1990; Packee 1990; Schmidt and Shearer 1990; Hann et al. 1997) (Figure 23.1). These forests are influenced by a maritime climate with wet winters and dry summers. Most precipitation occurs during November through May, with amounts ranging from 500 to 2300 mm (Foiles et al. 1990; Graham 1990; Packee 1990; Schmidt and Shearer 1990). Precipitation comes as snow and prolonged gentle rains, accompanied by cloudiness, fog, and high humidity. Rain-on-snow events are common from January to March. A distinct warm and sunny drought period occurs in July and August with rainfall in some places averaging less than 25 mm per month.

Soils that maintain these forests include, but are not limited to, Spodosols, Inceptisols, and Alfisols. A defining characteristic of the Northern Rocky Mountains is the layer of fine-textured ash (up to 62 cm thick) that caps the residual soils. The ash soils and loess deposits throughout the moist forests are continually being modified by disturbance events giving rise to soils with differing levels of productivity (Foiles et al. 1990; Graham 1990; Packee 1990; Schmidt and Shearer 1990). The combination of climate, topography, parent material, soils, weathering, and ash depth (unique to the Northern Rocky Mountains) creates the most productive of all forests occurring within the Inland Northwest.

The historical vegetation complexes in the Cascades and Northern Rocky Mountains ranged from early- to late-seral, and occurred within a landscape mosaic possessing all possible combinations of species and seral stages. The PVT's in the Northern Rocky Mountains include *Thuja plicata*, *T. heterophylla* and *A. grandis* with *P. monticola*, *L. occidentalis*, *Pinus contorta*, *P. menziesii* and *P. ponderosa* are always the early- and mid-seral species (Daubenmire and Daubenmire 1968; Hann et al. 1997). The eastern Cascades PVTs include *T. plicata*, *T. heterophylla*, *A. grandis*, *Abies amabilis*, and *Abies procera*. The early- and mid-seral species include *P. contorta*, *P. menziesii* and *P. ponderosa* while *P. monticola* and *L. occidentalis* are less abundant when compared to the Northern Rocky Mountains (Franklin and Dyrness 1973; Lillybridge et al. 1995).

Lush ground-level vegetation is the norm in the moist forests. The vegetation complexes are similar to those occurring on the west-side of the Cascade Mountains and in some Pacific coastal areas. Tall shrubs include *Acer circinatum*, *Achylys triphylla*, *Acer glabrum*, *Alnus sinuata*, *Oplopanax horridus*, *Rosa* spp., *Ribes* spp., *Vaccinium* spp., and *Salix* spp. Forbs include *Actaea rubra*, *Adenocaulon bicolor*, *Asarum caudatum*, *Clintonia uniflora*, *Cornus canadensis*, and *Coptis occidentalis*. Phytogeographic evidence indicates that some plant populations on the west side of the Cascade Mountains also occur as disjunct populations in the moist forests. For example, low-elevation riparian areas in northern Idaho contain disjunct populations of *Alnus rubra*, *Cornus nuttallii*, *Symphoricarpos mollis*, *Selaginella douglasii*, and *Physocarpus capitatus* (Foiles et al. 1990; Graham 1990; Packee 1990; Schmidt and Shearer 1990).

Snow, ice, insects, disease, and fire, when combined, created heterogeneity in patch sizes, forest structures, and compositions. Ice and snow created small gaps and openings, thinning forest densities and altering species composition (Figure 23.3a). Native insects (e.g., *Dendroctonus* spp.) and diseases (e.g., *Armillaria* spp., *Arceuthobium* spp.) infected and killed the very old or stressed individuals, which tended to diversify vegetation communities (Figure 23.3b) (Hessburg et al. 1994; Rippy et al. 2005). A mixed-fire regime best defines the role fire played in creating a mosaic of forest compositions and structures. Nonlethal surface-fires occurred at relatively frequent intervals (15–25 years) in a quarter of the area (Figure 23.4a). Lethal crown-fires burned about a quarter of the area at intervals of 20–150 years but occasionally extended to 300 years (Figure 23.4a and c). The mixed-fire regime occurred across the rest of the moist forests at 20 to150 year intervals. Fires typically started burning in July and were usually out by early September (Hann et al. 1997).

23.4.1 Moist Forest Change

The current distribution of successional-stage, forest structure, species composition, and disturbance regimes differs from the historic (1850–1900) patterns of the moist forest (Hann et al. 1997). In some settings, the mixed-fire regime maintained closed canopy conditions, which allowed for the mid-seral stage to develop into late-seral multistory stages (Hann et al. 1997). The late-seral multistory structure, which typically developed in cool, moist bottoms and basins, has decreased by about half in the last century (Table 23.1). The early-seral single-story stands that once occupied an estimated 25%–30% of the area now occupy only 9%–10% of the area, except in the northern Cascades (Washington) where they increased in abundance. The mid-seral stages have generally increased in abundance in the Northern Rocky Mountains and to a lesser degree in the eastern Cascades.

Species composition has shifted in the Northern Rocky Mountains (Hann et al. 1997; Neuenschwander et al. 1999; Fins et al. 2001); before 1900, *P. monticola* (early- to mid-seral species) dominated, often representing 15%–80% of the trees within stands (Figure 23.5)

FIGURE 23.3
Ice and snow damage create small gaps and openings, decrease forest densities, and alter species composition (a). Historically (1850–1900), in the moist forests, diseases (e.g., *Armillaria* spp., *Arceuthobium* spp.) attacked the very old, unthrifty, or stressed individuals (b). These disturbances stabilized and diversified vegetation communities. Currently, in the changed systems, epidemics of these disturbances often occur. (USDA Forest Service photographs.)

(Fins et al. 2001). This species is resistant to many endemic insects and diseases; it is long-lived (300 years) and can grow across 90% of the moist forest environments. It is a prolific seed producer after age 70 and is the only moist-forest conifer with seed remaining viable for up to three years, which allows it to regenerate abundantly after disturbance (Haig et al. 1941; Graham 1990). It is broadly adapted genetically (an ecological generalist) to the environment (Rehfeldt et al. 1984) and it is moderately shade-tolerant, allowing it to establish and develop within a wide range of canopy openings (Haig et al. 1941; Graham 1990; Jain et al. 2002, 2004). *P. monticola* often reaches 30 m in height or greater within 50 years of establishment (Graham 1990). *L. occidentalis* and *P. ponderosa* also occurred in the early- and mid-seral structures, but declined along with *P. monticola* and were succeeded by *A. grandis*, *P. menziesii*, and *T. heterophylla* (Hann et al. 1997; Atkins et al. 1999).

(a)   (b)   (c)

FIGURE 23.4
Historically (1850–1900), in the moist forests, nonlethal surface fires (a) occurred at relatively frequent intervals (15–25 years) in 25% of the moist forests while lethal crown fires (b and c) occurred over 25% of the forests at 300 years intervals. (USDA Forest Service photographs.)

FIGURE 23.5
Historically (1850–1900), 25%–50% of the moist forests were dominated by *Pinus monticola* and 15%–80% of the trees within stands were *Pinus monticola*. This photo shows 150–180-year-old *Pinus monticola* (circa 1935) growing in northern Idaho. (USDA Forest Service photograph.)

The eastern Cascades had limited amounts of *P. monticola* and *L. occidentalis*; therefore, *P. ponderosa*, *P. contorta*, and *P. menziesii* played more of a role in occupying the early- to mid-seral successional stages.

Native insects and pathogens occurred in these forests, but recent (1991) activity levels far exceed those of the past (Hessburg et al. 1994). Within the *A. grandis/A. concolor*

PVT, fire maintained landscapes that contained a plurality of early-seral *P. ponderosa* and *L. occidentalis*; the insects, *Dendroctonus pseudotsugae*, *Choristoneura occidentalis*, and *Orgyia pseudotsugata* were generally endemic. But they are often epidemic in the current forests dominated by *A. grandis/A. concolor* and *P. menziesii* (Hessburg et al. 1994). Similarly, the diseases *Armillaria* spp. and *Phellinus weirii* were historically endemic, but the current fir-dominated forests make epidemics of these diseases more common (Hessburg et al. 1994; Hann et al. 1997). As in many forest ecosystems in the western United States, effective fire exclusion contributed to these changes. Historically, 25% of the area had surface fires, 50% mixed fires, and 25% stand-replacing crown fires. Today, crown fires burn approximately 60% of the areas in these forests and only 15% are burned by surface fires and 20% are burned by mixed fires (Hann et al. 1997).

Although fire exclusion played a role in altering forests in the Northern Rocky Mountains, introduction of a European stem rust, *C. ribicola*, caused the greatest change (Figure 23.6a). The rust infects all five-needle pines, and subsequently decimated the abundant *P. monticola* (Figure 23.6b). Because the rust killed so many trees, the majority of surviving pines were harvested under the assumption they too would succumb to the rust (Ketcham et al. 1968). *A. grandis* and *P. menziesii* readily filled the niche *P. monticola* once held. In the eastern Cascades, blister rust was less severe since *P. monticola* was not the dominant species, thus fire exclusion and harvesting were more important agents in altering these forests. Harvesting removed the early-seral, shade-intolerant species (e.g., *P. ponderosa*, *L. occidentalis*) that were resistant to fire and other disturbances. Partial canopy removal and minimal soil surface disturbance in these harvests were ideal for *P. menziesii* and *A. grandis* which regenerated aggressively, rather than the shade-intolerant *Pinus* and *L.* species. Fire exclusion also prevented the creation of canopy openings and receptive seedbeds for the regeneration of *Pinus* and *Larix*. Similar to the dry forests, high canopies (>50 m) of *P. monticola*, *L. occidentalis*, *P. ponderosa* and other early- and mid-seral species currently are absent. In their place, the present forest structure and composition (*A. grandis* and *P. menziesii*) favor the compression of nutrients, microbial

(a)   (b)

FIGURE 23.6
White pine blister rust (*Cronartium ribicola*) canker occurring on a young *Pinus monticola* (a). A mid-aged (70–80 years) stand of *Pinus monticola* experiencing extreme mortality from blister rust (b). (Photograph A is USDA Forest Service and photograph B is from USDA. Forest Service, Ogden, Archive, Bugwood.org)

processes, and root activity toward the soil surface (Harvey et al. 1999). When wildfires occur, surface organic layers can be consumed, decreasing the nutrition and microbial processes important for sustaining these forests.

23.5 Restoration Approaches

When forests fail to recover after disturbances; human intervention may be required to alter a forest's trajectory and create opportunities for self-renewal and postdisturbance recovery (Stanturf 2004). Unlike timber management methods, treatments directed at forest restoration are not always similar, thus there is a need to develop techniques that reinstate processes that lead to restored ecosystems. Some scientists hypothesize that returning forest structures and compositions (vegetative characteristics, fire regimes, and species mixes) similar to those that existed prior to European settlement will reset a forest to a more functional condition (Moore and Covington 1999; Caprio and Graber 2000; Keane et al. 2009; Churchill et al. 2013). They use these reference conditions to help in formulating restoration targets and management strategies. However, simply returning a forest to a state that occurred at specific point-in-time and assume a forest will function as it did in the past is imperfect for several reasons. A serendipitous series of events (moisture, seed availability, and disturbance) occurred decades to centuries ago that created the environments for a forest to regenerate and develop (Haig et al. 1941; Oliver and Larson 1990). These events are not static but they fluctuate through time and space (Marshall 1928). The climate and disturbances that created a forest's structure and composition in the past differs from those of today and most likely will differ from those of the future. For example, American Indian burning contributed to historical fire regimes and such fires will most likely not occur now or in the future because of massive changes that have occurred over the last 100 years to the biophysical, social, and economic environments of western forests (Pyne 2010; Jain et al. 2012). For example, hunting, fishing, and gathering were primary historical land uses and now society values roads, recreation, forest products, and wildlife habitat and dislike many aspects associated with forest clearcutting (Reynolds et al. 1992; Bliss 2000; DellaSala 2003; Mercer 2005; De Groot et al. 2013). For these reasons, replicating forest conditions of the past may not be the best approach to use in restoration strategies but rather scientists and managers need to develop a pragmatic view of forest restoration that recognizes how forests and their social and economic context have changed over the last 100 years (Stanturf 2004).

This philosophy holds true when developing restoration strategies for the dry and moist mixed-conifer forests in the northern Rocky Mountains. In the last 100 years, these forests have changed and in some cases it may take centuries to achieve restoration goals. However, by integrating past and present knowledge, scientists have identified some ecological elements that if restored can increase opportunities for self-renewal (e.g., Harvey et al. 1994; Neuenschwander et al. 1999; Hessburg 2000; Fins et al. 2001; Hood 2010; Jain et al. 2014). Increasing landscape heterogeneity alters how disturbances move across a landscape and allows for wildlife corridors that will help maintain genetic diversity. Within stand, heterogeneity provides vegetative regeneration opportunities, snags, variable canopy cover, and down woody debris which are important for many wildlife species. Because early- and mid-seral species tend to be resistant to both endemic and introduced disturbances, their increased presence and abundance are usually central to restore both the dry and moist forests of the northern Rocky Mountains. Also, the abundance of late-seral regeneration in both forests confounds most restoration strategies as does the changes to the forest floor (deep duff) and species shifts that have altered soil properties (from red rot to white rot dominated systems). The challenge for those developing restoration strategies is to develop them within the context of changing societal values, economic uncertainty, and an indeterminate future climate.

23.5.1 Increase Landscape Heterogeneity in Dry and Moist Forests

In moist and dry mixed conifer forests, endemic disturbances (wind, ice, snow, disease, insects, and wildfire) coupled with the native vegetation, complex topography, and soils historically created and sustained various patch sizes and patch mosaics throughout the landscape (Marshall 1928; Hessburg et al. 2001; Moritz et al. 2011; Perry et al. 2011). These disturbances also created fine scale (≤ 1.0 ha) mosaics of canopy gaps, burned surfaces, live trees, snags, decadent large trees, and many other forest structures and vegetative compositions. Conversely, timber management that was practiced for several decades in the Inland Northwest tended to homogenize and regulate forests to achieve orderly and predictable amounts of forest products (Puettmann et al. 2008). This homogeneity created conditions where disturbances were no longer self-regulating, which has led to large and severe disturbances (wildfires and bark beetle epidemics) (Hessburg et al. 2007; Bentz et al. 2009; Littell et al. 2009; Mortz et al. 2011; Stephens et al. 2014). Moreover, in moist mixed-conifer forests, *A. grandis, T. heterophylla,* and *P. menziesii* replaced *P. monticola, L. occidentalis,* and *P. ponderosa* making these forests highly susceptible to epidemics of insects and diseases and catastrophic wildfire. It will undoubtedly take from decades to centuries to restore these early- and mid-seral species to their historic presence and abundance, but by increasing landscape heterogeneity and structural diversity within these forests the extent and severity of disturbances will likely diminish. However, the economic and social constraints prevent restoring an entire landscape creating a situation where scientists and managers need to develop approaches for prioritizing treatment areas (Hessburg et al. 2000; Hann and Bunnell 2001; Reynolds and Hessburg 2005; De Groot et al. 2013; Bollenbacher et al. 2014).

Several scientists suggest using a multiscale planning process to identify restoration opportunities and treatment priorities (e.g., Hann and Burnell 2001; Reynolds and Hessburg 2005; De Groot et al. 2013). At the broadest scale (e.g., subbasin ≈ 750,000 ha), managers can identify endangered species habitat, old growth, stream networks for anadromous fish, and social and economic infrastructure (population trends, cities, towns, and communities) (Hessburg et al. 2000). This analysis can identify areas that already have good road access; areas of concern that need protected, and identify priority areas for watershed restoration. At the mid-scale or river basins forest composition, distributions, and similar elements can be addressed. For example, in the Coeur d'Alene River Basin (≈350,000 ha) of the northern Rocky Mountains, Jain et al. (2002) related historical patterns of *P. monticola* and physical characteristics that identified places for restoring the species. At the fine scale (≈1000 ha), Camp et al. (1997) identified fire refugia as a function of physical landscape attributes in the eastern Cascades of Washington. Managers could use such techniques in both dry and moist forests to plan and prioritize restoration opportunities reflecting the key issues and values that may affect forest management.

Once landscape priorities are identified, landscape heterogeneity can be addressed by the strategic placement of forest treatments to create treatment mosaics that can be dispersed across the large landscape. For example, at Priest River Experimental Forest in northern Idaho, USA, scientists created forest mosaics of different sized vegetative patches distributed across the landscape (Figure 23.7) (Jain et al. 2008, 2012). The objective of the treatments was to increase landscape heterogeneity but also integrate other objectives such as maintaining stream buffers and when appropriate maintain and enhance old-growth forests. Simultaneously, a goal of the treatments was to create fuel patterns as to modify fire growth and behavior if one was to occur and decrease crown fuel homogeneity (Finney 2001). As a result of the treatments, a much more heterogeneous landscape was created that will most likely be more resilient to disturbances if and when they may occur compared to the landscape prior to the treatment.

Characteristics that made dry forests disturbance resilient included low overstory densities, fire-resistant tree species, and continuous intervention of disturbance (fire, insect, and disease) returning at various intervals. The complex interaction between disturbance and weather created a series of regeneration phases over time that was serendipitous and varied over the landscape making them difficult to replicate. These disturbances and regeneration process tended to develop forests containing a fine scale mosaic that included tree groups, single trees, and openings distributed throughout the landscape. Recent guides have attempted to quantify this historical pattern for developing restoration management strategies (Moore and Covington 1999; Larson and Churchill 2012; Churchill et al. 2013). Target forest conditions in these guides primarily focus on creating historical structure

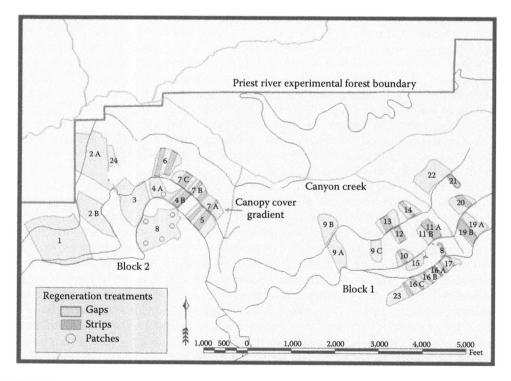

FIGURE 23.7
At Priest River Experimental Forest in northern Idaho, USA, researchers develop, implement, and evaluate silvicultural methods to introduce landscape heterogeneity by applying treatment mosaics (Jain et al. 2012). Within each treatment mosaic we leave stream corridors, canopy cover gradients alongside harvested units. Within harvested units when appropriate *Pinus monticola*, *Larix occidentalis*, or *Pinus ponderosa* pine remain to provide seed and contribute to variation in canopy cover within the strips (dark gray strips).

and composition. This specific pattern has been suggested as a plausible reference condition that made these dry forests resilient to fire and insects.

However, these characteristics alone do not provide for numerous contemporary values important to wildlife where fires were suppressed. This is exemplified by wildlife which may require late-seral vegetation, down wood, snags (of all sizes), and deep organic layers (fungi habitat) that are not integral to such rigid restoration prescriptions (e.g., Reynolds et al. 1992; Long and Smith 2000; Spies et al. 2006). Similarly, reintroducing fire may be preferred, but smoke, burn windows, risk, and liability may limit the use of prescribed or wildfire in restoration strategies. As such, most wildfires will be suppressed and particularly those which are located near homes, towns, and within municipal watersheds. Some wildfires and the environments they create will still occur but most likely not to the same extent, location, or frequency. More importantly, multiple land ownerships occur throughout these areas, probably with different objectives; limiting large landscape restoration activities (millions of hectares).

On the Boise Basin and Black Hills Experimental Forests, treatments are being evaluated that will produce wildfire and insect resilient forests while also providing timber and recreation opportunities (Jain et al. 2008, 2014). For example, on Boise Basin, an irregular silvicultural system is being assessed for its application in maintaining old forest structures. In this system, old *P. ponderosa* (>400 years old) were left and younger trees (approximately 180 years old) were retained based on their crown ratio (>40%) and diameter to height ratio (>80), both indicators of tree vigor. Dense patches of *P. menziesii* established after the last wildfire (Graham and Jain 2005). The objective was not to recreate a historical forest structure but to separate crowns and remove ladder fuels to diminish crown fire potential, and also to leave areas with interlocking crowns to provide habitat for the northern goshawk and its prey (Reynolds et al. 1992). To encourage grass, forb, and shrub regeneration, large openings were created to allow sufficient sunlight and growing space for their establishment. This increase in understory diversity would provide habitat for ground dwelling birds and mammals (Reynolds et al. 1992). Some thickets of *P. menziesii* were dispersed through the forest to provide hiding cover for wildlife, but some of them were located away from the large trees and on steep north facing aspects where they would thrive. Delayed tree mortality has increased snag abundance, which provides wildlife habitat and as they decay and fall, they will produce large woody debris. When opportunities arise, we will use prescribed fire, but we will also use mastication, mowing, cutting, or harvesting to maintain the desired forest floor conditions (Jain et al. 2012, 2014).

23.5.2 Increase within Stand Heterogeneity in Moist Forests

Site-specific restoration strategies should not be isolated to an individual stand but employed within the context of creating diverse and heterogeneous landscapes with a variety of stand compositions and structures. Silvicultural treatments that create within stand heterogeneity requires opening sizes large enough to favor shade-intolerant tree species, shrubs, forbs, and grasses and creating a variety of forest floor and mineral soil surfaces including blackened, mineral, and organic (Graham et al. 2004; Oliver and O'Hara 2004). Such treatments reflect the environments that were created by windstorms, root disease, ice, snow, and low intensity surface fires (Figures 23.3 and 23.4a). Again, these treatments did not necessarily emulate historical conditions but these treatments were initiated as to increase the abundance of early- and mid-seral species, produce wildlife habitat, and diversify the fuel matrix.

A variety of irregular uneven-aged and two-aged silvicultural systems can be developed that have application in both the dry and moist forests (Graham et al. 1983; Graham and Jain 2005; Graham et al. 2007; Churchill et al. 2013; Franklin et al. 2013). Even-aged (e.g., clearcut, seed tree, and shelterwood) silvicultural systems with varying amounts of reserve trees left in the cuttings create two-aged structures, and uneven-aged systems can introduce spatial and vertical heterogeneity. Even though uneven-aged selection systems are most often associated with the rigid reverse-j shaped diameter distributions, this is only one approach applicable for describing uneven-aged structures (Smith et al. 1997; Puettmann et al. 2008). These systems maintain multiaged (diameter distributions) forest structures by planning for and executing frequent (e.g., 10–20 years) entries (treatments). The heterogeneity introduced in these systems requires the timely establishment and tending of regeneration during each entry. Also, horizontal diversity in crowns is preferred to avoid crown fire but it is allowed for a diversity of postfire outcomes if a fire outbreaks. The majority of these systems differ from the top down "command and control" approach designed to produce timber as they allow for and encourage adaptive management (Drever 2006; Puettmann et al. 2008). A key to their success is maintaining the long-term (decades to centuries) vision while implementing short-term silvicultural methods (e.g., regenerating, weeding, cleaning, thinning, and releasing) aimed at producing and sustaining the desired forest structures and compositions.

At Priest River, Experimental Forest within stand heterogeneity was created using several silvicultural techniques (Jain et al. 2008). Because the area occurred on steep slopes, it was harvested using a cable (sky-line) system and cutting units of varying widths were created. Along the harvest edge, a canopy cover gradient was used to enhance the regeneration of both overstory and understory species (Figure 23.7). This variable density edge also increased the abundance of snags, down wood, and other structures that was favored by wildlife. We placed the largest openings (<5 ha) where *L. occidentalis*, *P. monticola*, and *P. ponderosa* were present to promote natural regeneration of these species. In all cases, early- and mid-seral species with good tree vigor were chosen as leave-trees. After harvest, a diversity of forest floor conditions (site preparation) was produced which ensured seed germination substrates that would favor a wide range of tree species. For example, in one stand, we masticated, prescribed burned, and grapple piled the postharvest slash and combined with the variation in canopy cover a plethora of vegetative establishment and development conditions were created. As such, highly heterogeneous forest floor vegetative and fuel complexes will be created.

In the dry forests, on the Black Hills Experimental Forest, we applied two different cleaning and weeding prescriptions to create highly variable ground-level vegetation complexes (Jain et al. 2014). The first approach used a 4 × 4 m tree spacing as a separate stratum ignoring the overstory trees. This method tends to produce ladder fuels and favors regularly spaced trees. The second approach was to clean the advanced regeneration to 4 × 4 m spacing but include overstory trees in the prescription. This technique favored maximum growing space for each crown class, eliminated ladder fuels from beneath the large trees, and produced horizontal separation between the overstory and understory.

23.5.3 Change Species Composition

In the moist forests, disturbances provided opportunities for early- and mid-seral species to regenerate, and reestablishing such species is central to restoring the moist forests (Everett et al. 1994; Hann et al. 1997; Harvey et al. 1999; Neuenschwander et al. 1999; Fins et al. 2001; Franklin et al. 2013). *P. monticola*, *L. occidentals*, and *P. ponderosa* are resilient to many endemic insects and diseases. *P. monticola* is an opportunistic species and can regenerate in a variety of openings. It, along with *L. occidentalis*, which is deciduous, compete well with shade-tolerant species and their crown architectures shed snow readily and diminish wind damage (Graham 1990; Schmidt and Shearer 1990; Jain et al. 2004, 2012). These species along with late-seral *T. plicata* are well-suited to feature in forest restoration strategies especially in face of climate change. In dry mixed-conifer forests, *P. ponderosa* and *L. occidentalis* have thick bark, especially as they age, which allows them to survive both prescribed and wild surface fires. These species, in both moist and dry forests, when combined with minor amounts of late-seral species (*A. grandis/A. concolor*, *P. menziesii*, and *T. heterophylla*) create greater species diversity than what is present today.

Early- and mid-seral conifer species were historically managed with even-aged silvicultural systems and in the moist forests, clearcutting was the dominant method used. For various reasons including cumulative watershed effects, wildlife habitat destruction, and just being unsightly clearcutting is being used minimally and systems that maintain high forest cover are being favored. To facilitate, the maintenance of high forest cover yet increase the presence and plurality of early and mid-seral species; Jain et al. (2004) identified four opening size thresholds that favor *P. monticola* establishment, competitive advantage over other species, in which it is free-to-grow, and conditions when optimum growth can be achieved. At Priest River Experimental, we used these thresholds to design the variety of opening sizes as we developed the treatment mosaic (Jain et al. 2002). In this study, the range of opening sizes were specifically designed to favor free-to-grow (0.5 hectare) to optimum growth (4.5 hectares) for *P. monticola*. However, in many places, seed sources did not exist and *P. monticola*, *L. occidentalis*, and *T. plicata* were regenerated artificially.

In the dry mixed-conifer forests, *P. ponderosa* is the preferred species on the driest sites (*P. ponderosa* and *P. menziesii* PVTs) with *Populus tremuloides* and *Picea* spp. occurring in wet areas or in cold pockets. On more productive mesic sites, *L. occidentalis* is also preferred. These forests can be quite topographically diverse and depending on the location and topographic diversity, inclusions of *T. plicata* PVT can occur in the draws that contained *P. monticola*, *P. contorta*, and *T. plicata* such as those located in northern Idaho and southern British Columbia (Daubenmire 1980) (Figure 23.8). This mosaic of pines, other conifers, and hardwood, all contributed to create disturbance resilient forests. Moreover, the

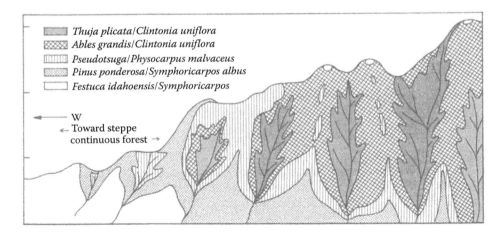

FIGURE 23.8
Topographic diverse south facing aspects have a diversity of vegetation. In draws, *Thuja plicata* dominates while on ridges *Pinus ponderosa* and *Pseudotsuga menziesii* dominate. Ecotones most often contain *Abies grandis*. *Armellaria* species and other root diseases often occur in these areas that create pockets that contain shrubs. This diversity in vegetation contributed to the mixed and variable fire regime common in moist mixed-conifer forests. (Reprinted with permission from Daubenmire, R. *Northwest Science* 54, 1980: 146–52.)

hardwood species are critical habitat for wildlife species (Reynolds et al. 2012). Fortunately, on dry forests, seed sources are plentiful but continuous disturbance is required to keep pine regeneration controlled and to provide opportunities for hardwoods to sprout and dominate. Thus restoration strategies in these forests need to control pine regeneration. On the Black Hills, both mechanical treatments and prescribed fire are being used to remove or kill *P. ponderosa* regeneration (Battaglia et al. 2008; Jain et al. 2012, 2014).

23.5.4 Integrate Disturbance into Restoration Strategies

Different from timber producing treatments, treatments designed to restore forests need to integrate disturbance into their management strategies, including introduced disturbances such as *C. ribicola* (a stem disease, blister rust) and its influence on *P. monticola* (Figure 23.6). Integrating introduced disease offer additional challenges, but a breeding program currently produces rust-resistant *P. monticola* seedlings for reforestation (68% of which exhibit some resistance to *Cronartium*; Fins et al. 2001) and continued breeding will ensure that rust mutations will not compromise the resistant material. Fortunately, *P. monticola* native populations do contain some natural resistance to the rust, thus in addition to breeding programs, mass selection presents an opportunity to utilize the rust-resistance occurring in natural stands (Hoff and McDonald 1980; Graham et al. 1994). In stands where blister rust has killed the majority (over 70%) of the *P. monticola*, approximately 7%–10% of the progeny, often thousands per ha, of the survivors exhibit rust-resistance (Hoff and McDonald 1980). Regenerating these wild populations can supplement the genetic diversity contained in breeding programs and provide prospects for restoring *P. monticola* when planting opportunities do not exist.

For example, on the Deception Creek Experimental Forest, we evaluated the amount of *C. ribicola* resistance occurring in naturally regenerated *P. monticola*. Using *P. monticola* planted in 1936 in the Forest, we collected cones from trees randomly, collected from trees exhibiting no rust, and from trees exhibiting no rust after all other *P. monticolas* were removed from the plantation. The last collection occurred after a full cone cycle to ensure that the cones which are collected reflected the reduced pollen cloud that would have fertilized the flowers. Seeds were extracted from each of these collections and seedlings were grown and subjected to an artificial inoculation by *C. ribicola* using *C. ribicola* infected *Ribes* spp. leaves. The *P. monticola* seedlings were planted in nursery beds and inspected biannually for rust infections during a four-year period. Trees from all collections were successfully exposed to blister rust as all trees exhibited needle spots indicating that *C. ribicola* spores had germinated on the tree needles. After four years, approximately 2% of the trees originating from the last collection were totally free from rust and approximately 1% was free from rust from the other collections. Natural regeneration of *P. monticola* can exceed over 10,000 seedlings ha^{-1}; if only 1% of seedlings survive, 100 trees ha^{-1} would remain and because this species is resilient to many endemic diseases, several species could survive to produce cones (Haig et al. 1941). These preliminary results show promise that through silvicultural treatments, *P. monticola* rust-resistance can be increased and utilized in restoration strategies.

23.5.5 Surface Vegetation and Forest Floor

Although trees are most frequently removed to modify fire behavior, the entire fuel matrix influences fires and, in particular, surface fuels (Rothermel 1983). The most effective strategy for reducing crown fire occurrence and severity is to (1) reduce surface and ladder fuels, (2) increase height to live crown, (3) reduce canopy bulk density, and (4) reduce continuity of the forest canopy, in that order of importance (Graham et al. 2004) (Figure 23.9a). More importantly, in dry mixed-conifer forests where soil surface organic layers were historically reduced by surface fires, fire suppression has allowed thick and deep layers of needle and bark slough to accumulate at the base of live trees. Burning these surface layers can result in smoldering combustion that can last for days. As such, these types of fires can girdle the tree by killing the tree cambium or kill fine roots that could cause delayed tree mortality (Hood 2010; Graham et al. 2012; Jain et al. 2012).

In dry forests, restoring conditions in which surface fires (whether prescribed or wild) can burn, requires approaches that conserve nutrients, microbial activity, and fine roots

(a)

(b)

FIGURE 23.9
There are six layers of canopy fuels (a). They are canopy (A), shrubs and small trees (B), low, nonwoody vegetation (C), woody fuels (D), and ground fuels (E). Fire exclusion has caused deep organic layers to develop around old *Pinus ponderosa* boles. Applying prescribed fire (A) or raking when the lower organic layers are high in moisture (>100%), and soil temperatures are low (−2°C), the depth of these layers can be decreased with minimal damage to the fine roots they contain. (USDA Forest Service photographs.)

that often develop in the uncharacteristically deep surface organic layers (Jain et al. 2012). Gradually, decreasing the depth and volume of these organic layers by repeating treatments over a series of years will force the fine roots to migrate down into the mineral soil. Depending on the dry forest setting, it may take one to multiple combinations of mechanical and carefully executed prescribed burns before fuel loads, species composition, and forest structures allow prescribed fires to burn freely in the dry forests (Hood 2010; Jain et al. 2012). Burning, when moisture is high (>100%) in the lower layers, will preserve them, but it allows the drier upper layers to be consumed (Brown et al. 1985). Moreover, burning when soil temperatures are low (less than 2°C) and when fine root growth is minimal results in minimal root damage (Kramer and Kozlowski 1979). These conditions occur most often in early spring (Figure 23.9b). Nutrient volatilization is minimized by burning surface fuels when the lower layers have high amounts of moisture (>100% moisture by weight) (Hungerford et al. 1991). If fire is not an option for decreasing these layers, raking or removing them using a leaf blower is suggested; however, if fine roots are present in the organic layers, treatments need to be applied when soil temperatures are less than 2°C (Hood 2010; Jain et al. 2012).

23.6 Economic and Social Aspects

Ecological information is available to restore the moist and dry forests of the US Pacific Northwest (Reynolds et al. 1992; Covington and Moore 1994; Hann et al. 1997; Harvey et al. 1999; Nuenschwander et al. 1999; Long and Smith 2000; Finns et al. 2001). Interweaving the social, economic, and political needs of the society present the greater challenge in restoring these forests. Costs can be specified for restoration activities such as harvesting, thinning, planting, weeding/cleaning, prescribed burning, exotic plant control, riparian area treatments, and for planning, monitoring and analysis. Specific costs will vary depending on site characteristics and stand structure. For example, the cost per unit area ranges from $75 (U.S. dollars) per ha for vegetation management to $750 per ha depending on slope, material removed and intensity of surface treatments (Jain et al. 2012). However, current extent and severity of disturbances also have an economic trade-off when considering factors such as fire suppression and resource damage from extreme and large wildfires (Holms et al. 2008).

These cost estimates do not reflect some of the benefits of restoring moist and dry forests (De Groot et al. 2013). For example, an estimated 83% of the recreational benefits within the Interior Columbia Basin come from federally administered Forest Service and Bureau of Land Management lands (Phillips and Williams 1998; Reyna 1998). These recreation benefits include trail use, hunting, fishing, camping, boating, wildlife viewing, winter sports, day use, and motor viewing. When restoration positively influences these activities, the benefit may exceed the costs. Converting forests dominated by late-seral structures to forests dominated by early- and mid-seral structures most likely will benefit these recreational activities. Moreover, restored forest conditions may improve the habitat of legally threatened or endangered wildlife species, but all restoration treatments have to be implemented to avoid negative impacts on other protected species (such as, the bull trout, grizzly bear, and Canadian lynx). Nevertheless, a vocal segment of society prefers the status quo and resists actively nmanaging forests.

Active management including timber harvest and restoring forests near timber-dependent and isolated communities most likely would be a positive benefit. Reyna (1998) and Phillips and Williams (1998) reported that 137 communities specialized in logging and wood products manufacturing in the Interior Columbia River Basin, with 64 being isolated. Timber harvesting and wood products manufacturing generally has been important in these communities since the 1800s when many towns were established. However, timber harvesting is sometimes a controversial issue, but collaborative groups are working with public land managers to identify socially acceptable management techniques that will promote restoration opportunities (Cheng and Sturtevant 2012). Moreover, the infrastructure for manufacturing wood products has decreased in recent years, increasing transportation costs for the harvested material (Haynes 2002; Jain et al. 2012). Perhaps the greatest economic challenge is that the small trees (especially *P. ponderosa*) available for harvesting in most restoration activities have low value (Lippke 2002; McKetta 2002). However, small-diameter (15–25 cm in diameter) *A. grandis* and *P. menziesii* seem to be an exception, and are of value in producing construction materials (McKetta 2002). Although there is value in this material, out of the 5.7 million forested hectares in the Inland Northwest not in wilderness, national parks, or in other reserved status, it is estimated that there are only 800,000 hectares that have material of value that needs to be removed (Jain et al. 2012).

In general, private landowners have more flexibility for conducting restoration activities where timber production is often the primary objective (Blatner et al. 1994). Unlike their federal counterparts, managers of private lands have fewer requirements for analysis and planning prior to conducting activities, which allows them to respond quickly to insects, diseases, wildfires, and storms, as well as changing markets. Moreover, forests containing *P. ponderosa*, *P. menziesii*, *P. monticola*, and *L. occidentalis* tend to have high commercial value and are resilient to native disturbances (Nuenschwander et al. 1999; McKetta 2002). However, just species presence would not necessarily indicate that a forest is restored; the entire suite of forest structures (biological and physical properties of vegetation, soils, microbes, and water) and compositions distributed in a mosaic over the landscape would most likely constitute a restored forest (Nuenschwander et al. 1999).

23.7 Conclusions

A combination of harvesting and the introduction of *C. ribicola* greatly impacted the inland moist forests of the US Pacific Northwest. Fire exclusion played a role in changing these forests, but not to the same extent that it did in the dry forests. Because of current (recreation, scenic, and wildlife habitat) and past (harvesting and road construction) human uses and values, restoring both the dry and moist forests will be challenging, but it is not impossible. By viewing and developing management strategies using a multiscale approach, landscape plans and silvicultural systems can be designed that move these forests on a trajectory toward their desired restored compositions and structures.

Majestic stands of *P. monticola*, *P. ponderosa*, *L. occidentalis*, *T. plicata*, and all possible combinations of these species and their associates once populated the forests of the Inland Northwest. Once these systems reflect the historical composition and structure, endemic levels of other disturbances can aid in sustaining these forests into the future. However, because of human presence, the extent and intensity of endemic disturbances plus exotic introductions into these forests will make restoration activities challenging. If society determines that the dry and moist forests should be restored, it will take time, patience, perseverance, and commitment by both public and private individuals and organizations to accomplish the task.

References

Atkins, D., J. Byler, L. Livingston, P. Rogers, and B. Bennett. *Health of Idaho's Forests: A Summary of Conditions, Issues and Implications*. Missoula, MT: U.S. Department of Agriculture Forest Service, Northern Region, 1999.

Battaglia, M.A., F.W. Frederick, and W.D. Sheppard. Can prescribed fire be used to maintain fuel treatment effectiveness over time in Black Hills ponderosa pine forests? *Forest Ecology and Management* 256, 2008: 2029–2038.

Bentz, B., J. Logan, J. MacMahon et al., *Bark Beetle Outbreaks in Western North America: Causes and Consequences*. Salt Lake, UT: University of Utah Press, 2009.

Blatner, K.A., C.E. Keegan III, J. O'Laughlin, and D.L. Adams. Forest health management policy. *Journal of Sustainable Forestry* 2, 1994: 317–37.

Bliss, J.C. Public perceptions of clearcutting, *Journal of Forestry* 98, 2000: 4–9.

Bollenbacher, B.L., R.T. Graham, and K.M. Reynolds. Regional forest landscape restoration priorities: Integrating historical conditions and an uncertain future in the northern Rocky Mountains. *Journal of Forestry* 112, 2014: 474–83.

Brown, J.K., E.D. Reinhardt, and W.C. Fischer. *Predicting Duff and Woody Fuel Consumed by Prescribed Fire in the Northern Rocky Mountains*. INT-RP-337. Ogden, UT: U.S. Department of Agriculture Forest Service, Intermountain Research Station, 1985.

Burton, P.J. and S.E. Macdonald. The restorative imperative: Assessing objectives, approaches, and challenges to restoring naturalness in forests. *Silva Fennica*, 45 (5), 2011: 843–63.

Camp, A., C. Oliver, P. Hessburg, and R. Everett. Predicting late-successional fire refugia pre-dating European settlement in the Wenatchee Mountains. *Forest Ecology and Management* 95, 1997: 63–77.

Caprio, A.C. and D.M. Graber. Returning fire to the mountains: Can we successfully restore the ecological role of pre-euro American fire regimes to the Sierra Nevada? In *Wilderness Science in a Time of Change Conference: Volume 5- Wilderness Ecosystems, Threats, and Management*, compiled by N.C. David, S.F. McCool, W.T. Borrie, and J. O'Loughlin, 233–41. RMRS-P-15-VOL-5. Fort Collins, CO: U.S. Department of Agriculture Forest Service, Rocky Mountain Research Station, 2000.

Cheng, A.S. and V.E. Sturtevant. A framework for assessing collaborative capacity in community-based public forest management. *Environmental Management* 49, 2012: 675–89.

Churchill, D.J., A.J. Larson, M.C. Dahlgreen, J.F. Franklin, P.F. Hessburg, and J.A. Lutz. Restoring forest resilience: From reference spatial patterns to silvicultural prescriptions and monitoring. *Forest Ecology and Management* 291, 2013: 442–57.

Cooper, S.V., K.E. Neiman, and D.W. Roberts. *Forest Habitat Types of Northern Idaho: A Second Approximation*. INT-GTR-236. Ogden, UT: U.S. Department of Agriculture Forest Service, Intermountain Research Station, 1991.

Covington, W.W. and M.M. Moore. Post settlement changes in natural fire regimes and forest structure. In *Assessing Forest Ecosystem Health in the Inland West*, edited by R.N. Sampson and D.L. Adams, 153–81. New York: Haworth Press, 1994.

Daubenmire, R. Mountain topography and vegetation patterns. *Northwest Science* 54, 1980: 146–52.

Daubenmire, R. and J. Daubenmire. *Forest Vegetation of Eastern Washington and Northern Idaho*. Bull. 60. Pullman, WA: Washington State Agricultural Experiment Station, 1968.

Debano L.F. Effects of fire on soil properties. In *Proceedings: Management and Productivity of Western-Montane Forest Soils*, compiled by A.E. Harvey and F.L. Neuenschwander, 151–6. Ogden, UT: U.S. Department of Agriculture Forest Service, Intermountain Research Station, 1991.

De Groot, R.S., J. Blignaut, S. Van Der Ploeg, J. Aronson, T. Elmqvist, and J. Farley. Benefits of investing in ecosystem restoration. *Conservation Biology* 27, 2013: 1286–1293.

DellaSala, D.A., A. Martin, and R. Spivak et al. A citizen's call for ecological forest restoration: Forest restoration principles and criteria. *Ecological Restoration* 21, 2003: 14–23.

Drever, C.R., G. Peterson, C. Messier, Y. Bergeron, and M. Flannigan, 2006. Can forest management based on natural disturbances maintain ecological resilience? *Canadian Journal of Forest Research* 36, 2006: 2285–99.

Everett, R.L., P. Hessburg, M. Jensen, and B. Bormann. *Eastside Forest Ecosystem Health Assessment: Volume I, Executive Summary*. PNW-GTR-317. Portland, OR: U.S. Department of Agriculture Forest Service, Pacific Northwest Research Station, 1994.

Finney, M.A. Design of regular landscape fuel treatment patterns for modifying fire growth and behavior. *Forest Science* 47, 2001: 219–28.

Fins, L., J. Byler, D. Ferguson et al. *Return of the Giants: Restoring White Pine Ecosystems by Breeding and Aggressive Planting of Blister Rust-Resistant White Pines*. Moscow, ID: University of Idaho, 2001.

Foiles, M.W., R.T. Graham, and D.F. Olson Jr. *Abies grandis* (Dougl. ex D. Don.) Lindl. In *Silvics of North America, Volume 1, Conifers*, technical coordinators R.M. Burns and B.H. Honkala, 52–9. Agric. Handb. 654. Washington, DC: U.S. Department of Agriculture Forest Service, 1990.

Franklin, J.F. and C.T. Dryness. *Natural Vegetation of Oregon and Washington*. Corvallis, OR: Oregon State University Press, 1973.

Franklin, J.F. and K.N. Johnson. 2012. A restoration framework for federal forests in the Pacific Northwest. *Journal of Forestry* 110, 2012: 429–39.

Franklin, J.F., K.N. Johnson, D.J. Churchill, K. Hagmann, D. Johnson, and J. Johnston. 2013. *Restoration of Dry Forests in Eastern Oregon: A Field Guide*. Portland, OR: The Nature Conservancy.

Graham, R., M. Finney, C. McHugh et al., *Fourmile Canyon Fire Findings*. RMRS-GTR-289. Fort Collins, CO: U.S. Department of Agriculture Forest Service, Rocky Mountain Research Station, 2012.

Graham, R.T., *Pinus monticola* (Dougl. Ex. D. Don.). In *Silvics of North America, Volume 1, Conifers*, technical coordinators R.M. Burns and B.H. Honkala, 385–94. Agric. Handb. 654. Washington, DC: U.S. Department of Agriculture Forest Service, 1990.

Graham, R.T., C.A. Wellner, and R. Ward. Mixed conifers, western white pine, and western redcedar. In *Silvicultural Systems for the Major Forest Types of the United States*, compiled by Burns, R.M., 67–9. Agric. Handb. 445. Washington, DC: U.S. Department of Agriculture Forest Service, 1983.

Graham, R.T., J.R. Tonn, and T.B. Jain. Managing western white pine plantations for multiple resource objectives. In *Proceedings—Interior Cedar-Hemlock-White Pine Forests: Ecology and Management*, edited by D.M. Baumgartner, J.E. Lotan, and J.R. Tonn, 269–75. Pullman WA: Washington State University, 1994.

Graham, R.T., S. McCaffrey, and T.B. Jain, tech. Eds. *Science Basis for Changing Forest Structure to Modify Wildfire Behavior and Severity*. RMRS-GTR-120. Fort Collins, CO: U.S. Department of Agriculture Forest Service, Rocky Mountain Research Station, 2004.

Graham, R.T. and T.B. Jain. Application of free selection in mixed forests of the inland northwestern United States. *Forest Ecology and Management* 209, 2005: 131–45.

Graham, R.T., T.B. Jain, and J. Sandquist. Free selection: A silvicultural option. In *Restoring Fire-Adapted Forested Ecosystems. 2005 National Silviculture Workshop*, edited by R. Powers. PSW-GTR-203. Albany, CA: U.S. Department of Agriculture Forest Service, Pacific Southwest Research Station, 2007.

Haig, I.T., K.P. Davis, and R.H. Weidman. *Natural Regeneration in the Western White Pine Type*, Tech. Bull. No. 767. Washington, DC: U.S. Department of Agriculture Forest Service, 1941.

Hann W.J. and D.L. Bunnell. Fire and land management planning and implementation across multiple scales. *International Journal of Wildland Fire*. 10, 2001: 389–403.

Hann, W.J., J.L. Jones, M.G. Karl et al. Landscape dynamics of the basin. In *An Assessment of Ecosystem Components in the Interior Columbia Basin and Portions of the Klamath and Great Basins: Volume II*, edited by T.M. Quigley and S.J. Arbelbide, 338–1055. PNW-GTR-405. Portland, OR: U.S. Department of Agriculture Forest Service, Pacific Northwest Research Station, 1997.

Harvey, A.E., M.F. Jurgensen, M.J. Larsen, and J.A. Schlieter. *Distribution of Active Ectomycorrhizal Short Roots in Forest Soils of the Inland Northwest: Effects of Site and Disturbance.* INT-RP-374. Ogden, UT: U.S. Department of Agriculture Forest Service, Intermountain Research Station, 1986.

Harvey, A.E., M.F. Jurgensen, M.J. Larsen, and R.T. Graham. *Decaying Organic Materials and Soil Quality in the Inland Northwest: A Management Opportunity.* INT-GTR-225. Ogden, UT: U.S. Department of Agriculture Forest Service, Intermountain Research Station, 1987.

Harvey, A.E., P.F. Hessburg, J.W. Byler, G.I. McDonald, J.C. Weatherby, and B.E. Wickman. "Health declines in interior forests: Symptoms and solutions. In *Ecosystem Management in Western Interior Forests*, edited by D.M. Baumgartner and R.L. Everett, 163–70. Pullman, WA: Washington State University, 1994.

Harvey, A.E., R.T. Graham, and G.M. McDonald. Tree species composition change: Forest soil organism interaction potential effects on nutrient cycling and conservation processes in interior forests. In *Pacific Northwest Forest and Rangeland Soil Organism Symposium*, edited by R. Meurisse, W.G. Ypsllantis and C. Seybold, 137–45. PNW-GTR-46.1 Portland, OR: U.S. Department of Agriculture Forest Service, Pacific Northwest Research Station, 1999.

Haynes, R.W. U.S. timber supply and demand in the United States, 1996 to 2050. In *Small-Diameter Timber: Resource Management, Manufacturing, and Markets, Symposium Proceedings*, compiled by D.M. Baumgartner, L.R. Johnson and E.J. Depuit, 33–6. Pullman, WA: Washington State University, 2002.

Hermann, R.K. and D.P. Lavender. *Pseudotsuga menziesii* (Mirb.) Franco. In *Silvics of North America, Volume 1, Conifers*, technical coordinators R.M. Burns and B.H. Honkala, 527–40. Agric. Handb. 654. Washington, DC: U.S. Department of Agriculture Forest Service, 1990.

Hessburg, P.F., J.K. Agee, and J.F. Franklin. Dry forests and wildland fires of the Inland Northwest USA: Contrasting the landscape ecology of the pre-settlement and modern eras. *Forest Ecology and Management* 211, 2001: 117–39.

Hessburg, P.F., R.G. Mitchell, and G.M. Filip. *Historical and Current Roles of Insects and Pathogens in Eastern Oregon and Washington Forested Landscapes.* PNW-GTR-327. Portland, OR: U.S. Department of Agriculture Forest Service, Pacific Northwest Research Station, 1994.

Hessburg, P.F., B.R. Salter, and K.M. James. Re-examining fire severity relations in pre-management era mixed conifer forests: Inferences from landscape patterns of forest structure. *Landscape Ecology* 22, 2007: 5–24.

Hessburg, P.F., R.B. Salter, R.B. Richman, and B.G. Smith. Ecological subregions of the Interior Columbia Basin, US. *Applied Vegetation Science* 3, 2000: 163–80.

Hoff, R.J. and G.I. McDonald. *Improving Rust-Resistant Strains of Inland Western White Pine.* INT-RP-245. Ogden, UT: U.S. Department of Agriculture Forest Service, Intermountain Forest and Range Experiment Station, 1980.

Holmes, T.P., Prestemon J.P., and K.L. Abt. (eds.). Chapter 1: An introduction to the economics of forest disturbance. In *The Economics of Forest Disturbances: Wildfires, Storms, and Invasive Species.* 3–14, Dordrecht: Springer, 2008.

Hood, S.M. *Mitigating Old Tree Mortality in Long-Unburned, Fire-Dependent Forests: A Synthesis.* RMRS-GTR-238. Fort Collins, CO: U.S. Department of Agriculture Forest Service, Rocky Mountain Research Station, 2010.

Hungerford, R.D., M.G. Harrington, W.H. Frandsen, K.C. Ryan, and G.J. Niehoff. Influence of fire on factors that affect site productivity. In *Proceedings—Management and productivity of western-montane forest soils*, compiled by A.E. Harvey and F.L. Neuenschwander, 32–50. Ogden, UT: U.S. Department of Agriculture Forest Service, Intermountain Research Station, 1991.

Jain, T.B., M.A. Battaglia, and R.T. Graham. Northern Rocky Mountain experimental forests: settings for science, management, and education alliances. *Journal of Forestry* 112, 2014: 534–41.

Jain, T.B., M.A. Battaglia, H-S. Han et al. *A Comprehensive Guide to Fuel Management Practices for Dry Mixed Conifer Forests in the Northwestern United States.* RMRS-GTR-292. Fort Collins, CO: U.S. Department of Agriculture, Forest Service, Rocky Mountain Research Station, 2012.

Jain, T.B., R.T. Graham, and P. Morgan. Western white pine development in relation to biophysical characteristics across different spatial scales in the Coeur D'Alene River Basin in northern Idaho, U.S.A. *Canadian Journal of Forest Research* 32, 2002: 1109–25.

Jain, T.B., R.T. Graham, and P. Morgan. Western white pine growth relative to forest openings. *Canadian Journal of Forest Research* 34, 2004: 2187–98.

Jain, T.B., R.T. Graham, J. Sandquist et al. Restoration of Northern Rocky Mountain moist forests: Integrating fuel treatments from the site to the landscape. In *Integrated Restoration Efforts for Harvested Forest Ecosystems*, edited by R. Deal, 147–172. PNW-GTR-733. Portland, OR: U.S. Department of Agriculture Forest Service, Pacific Northwest Research Station, 2008.

Keane, R.E., P.F. Hessburg, P.B. Landres, and F.J. Swanson. The use of historical range of variability (HRV) in the landscape management. *Forest Ecology and Management* 258, 2009: 1025–37.

Ketcham D.A., C.A. Wellner, and S.S. Evans Jr. Western white pine management programs realigned on Northern Rocky Mountain National Forests. *Journal of Forestry* 66, 1968: 329–32.

Kramer, P.J. and T.T. Kozlowski. *Physiology of Woody Plants*. New York: Academic Press, 1979.

Larson, A.J. and D. Churchill. Three spatial patterns in fire-frequent forests of western North America, including mechanisms of pattern formation and implications for designing fuel reduction and restoration treatments. *Forest Ecology and Management* 267, 2012: 74–92.

Larsen, M.J., A.E. Harvey, and M.F. Jurgensen. Residue decay processes and associated environmental functions in northern Rocky Mountain forests. In *Environmental Consequences of Timber Harvesting in Rocky Mountain Coniferous Forests: Symposium Proceedings*, 157–74. INT-GTR-90. Ogden, UT: U.S. Department of Agriculture Forest Service, 1980.

Lewis, J.G. *The Forest Service and the Greatest Good: A Centennial History*. Durham, NC: Forest History Society, 2005.

Lillybridge, T.R., B.L. Kovalchik, C.K. Williams, and B.G. Smith. *Field Guide for Forested Plant Associations of the Wenatchee National Forest*, PNW-GTR-359. Portland, OR: U.S. Department of Agriculture Forest Service, Pacific Northwest Research Station, 1995.

Lippke, B. Technology transfer opportunities for small diameter timber problems. In *Small-Diameter Timber: Resource Management, Manufacturing, and Markets, Symposium Proceedings*, compiled by D.M. Baumgartner, L.R. Johnson, and E.J. Depuit, 53–56. Pullman, WA: Washington State University, 2002.

Littell, J.S., D. McKenzie, D.L. Peterson, and A.L. Westerling. Climate and wildfire area burned in western U.S. ecoprovinces, 1916–2003. *Ecological Applications* 19, 2009: 1003–21.

Long, J.N. and F.W. Smith. Restructuring the forest: Goshawks and restoration of southwestern ponderosa pine. *Journal of Forestry* 98, 2000: 25–30.

Marschner, G.E. and H. Marschner. Nutrient and water uptake by roots of forest trees. *Pflanzenernahr Bodenk* 159, 1996: 11–21.

Marshall, R. The life history of some western white pine stands on the Kaniksu National Forest. *Northwest Science* 2, 1928: 48–53.

McKetta, C. Why grow large trees anyway? A timber grower's perspective. In *Small-Diameter Timber: Resource Management, Manufacturing, and Markets, Symposium Proceedings*, compiled by D.M. Baumgartner, L.R. Johnson, and E.J. Depuit, 133–8. Pullman, WA: Washington State University, 2002.

Mercer, D.E. Chapter 6: Policies for encouraging forest restoration. In *Restoration of Boreal and Temperate Forests*, edited by J.A. Stanturf and P. Madsen, 97–110. Boca Raton, FL: CRC Press, 2005.

Meyer, W.H. *Yield of Even-Aged Stands of Ponderosa Pine*. Tech. Bull. 630. Washington, DC: U.S. Department of Agriculture Forest Service, 1938.

Minore, D. *Comparative Autecological Characteristics of Northwestern Tree Species-A Literature Review*. PNW-GTR-87. Portland, OR: U.S. Department of Agriculture Forest Service, Pacific Northwest Forest and Range Experiment Station, 1979.

Moore, M.M., W.W. Covington, and P.Z. Fule. Reference conditions and ecological restoration: A southwestern ponderosa pine perspective. *Ecological Applications* 9, 1999: 1266–77.

Moritz, M.A., P.F. Hessburg, and N.A. Povak. Native fire regimes and landscape resilience. In *The Landscape Ecology of Fire*, edited by D. McKenzie, C. Miller and D.A. Falk, 51–86. New York: Springer, 2011.

Neuenschwander, L.F., J.W. Byler, A.E. Harvey et al. *White Pine in the American West: A Vanishing Species—Can We Save It?* RMRS-GTR-35. Ogden, UT: U.S. Department of Agriculture Forest Service, Rocky Mountain Research Station, 1999.

Oliver C.D. and K.L. O'Hara. Effects of restoration at the stand level. In *Restoration of Boreal and Temperate Forests*, 31–60. Boca Raton, FL: CRC Press, 2004.

Oliver C.D. and B.C. Larson. *Forest Stand Dynamics*. New York: John Wiley and Sons, 1990.

Oliver, W.W. and R.A. Ryker, *Pinus ponderosa* (Dougl. ex Laws). In *Silvics of North America, Volume 1, Conifers*, technical coordinators R.M. Burns and B.H. Honkala, 413–24. Agric. Handb. 654. Washington, DC: U.S. Department of Agriculture Forest Service, 1990.

Packee, E.C. *Tsuga heterophylla* (Raf.) Sarg. In *Silvics of North America, Volume 1, Conifers*, technical coordinators R.M. Burns and B.H. Honkala, 613–22. Agric. Handb. 654. Washington, DC: U.S. Department of Agriculture Forest Service, 1990.

Perry, D.A., P.F. Hessburg, C.N. Skinner et al. The ecology of mixed severity fire regimes in Washington, Oregon, and Northern California. *Forest Ecology and Management* 262, 2011: 703–17.

Pfister, R.D., B.L. Kovalchik, S.F. Arno, and R.C. Presby. *Forest Habitat Types of Montana*. INT-GTR-34. Ogden, UT: U.S. Department of Agriculture Forest Service, Intermountain Research Station, 1977.

Phillips, R.H. and G.W. Williams. An estimation of effects of the draft EIS alternatives on communities. In *Economic and Social Conditions of Communities: Economic and Social Characteristics of Interior Columbia Basin Communities and an Estimation of Effects on Communities from the Alternatives of the Eastside and Upper Columbia River Basin Draft Environmental Impact Statements*, 83–116. Portland, OR: U.S. Department of Agriculture Forest Service, U.S. Department of Interior Bureau of Land Management, 1998.

Puettmann, K.J., K.D. Coates, and C. Messier. *A Critique of Silviculture: Managing for Complexity*. Washington, DC: Island Press, 2008.

Pyne, S.J. *America's Fires: A Historical Context for Policy and Practice*, Revised edition. Durham, NC: Forest History Society, 2010.

Quigley, T., R.W. Haynes, and R.T. Graham. *Integrated Scientific Assessment for Ecosystem Management in the Interior Columbia Basin*, PNW-GTR-382. Portland, OR: U.S. Department of Agriculture Forest Service, Pacific Northwest Research Station, 1996.

Rehfeldt, G.E., R.J. Hoff, and R.J. Steinhoff. Geographic patterns of genetic variation in *Pinus monticola*. *Botanical Gazette* 145, 1984: 229–39.

Reyna, N.E. Economic and social characteristics of communities in the Interior Columbia Basin. In *Economic and Social Conditions of Communities: Economic and Social Characteristics of Interior Columbia Basin Communities and an Estimation of Effects on Communities from the Alternatives of the Eastside and Upper Columbia River Basin Draft Environmental Impact Statements*, 3–82. Portland, OR: U.S. Department of Agriculture Forest Service, U.S. Department of Interior Bureau of Land Management, 1998.

Reynolds, R.T., D.A. Boyce Jr., and R.T. Graham. Ponderosa pine forest structure and Northern Goshawk reproduction: Response to Beier et al. (2008). *Wildlife Society Bulletin* 36, 2012: 147–52.

Reynolds, K.M. and P.F. Hessburg, 2005. Decision support for integrated landscape evaluation and restoration planning. *Forest Ecology and Management* 207, 2005: 263–78.

Reynolds, R.T., R.T. Graham, and R.M. Hildegard. *Management Recommendations for the Northern Goshawk in the Southwestern United States*. RM-217. Fort Collins, CO: U.S. Department of Agriculture Forest Service, Rocky Mountain Forest and Range Experiment Station and Southwestern Region, 1992.

Rippy, R.C., J.E. Stewart, P.J. Zambino et al. *Root Diseases in Coniferous Forests of the Inland West: Potential Implications of Fuels Treatments.* RMRS-GTR-141. Fort Collins, CO: U.S. Department of Agriculture, Forest Service, Rocky Mountain Research Station, 2005.

Robichaud, P.R., J.L. Byers, and D.G. Neary. *Evaluating the Effectiveness of Postfire Rehabilitation Treatments.* RMRS-GTR-63. Fort Collins, CO: U.S. Department of Agriculture Forest Service, Rocky Mountain Research Station, 2000.

Rose, S.L., D.A. Perry, D. Pilz, and M.M. Schoeneberger. Allelopathic effects of litter on the growth and colonization of mycorrhizal fungi. *Journal of Chemical Ecology* 9, 1983: 1153–62.

Rothermel, R.C. *How to Predict the Spread and Intensity of Forest and Range Fires.* INT-GTR-143. Ogden, UT: U.S. Department of Agriculture, Forest Service, Intermountain Forest and Range Experiment Station, 1983.

Ryan, K.C. and G.D. Amman. Bark beetle activity and delayed tree mortality in the Greater Yellowstone area following the 1988 fires. In *Ecological Implications of Fire in Greater Yellowstone USA*, edited by J.M. Greenlee, 151–8. Fairfield, WA: International Association of Wildland Fire, 1996.

Schmidt, W.C. and R.C. Shearer. *Larix occidentalis* (Nutt.). In *Silvics of North America*, Volume 1, Conifers, technical coordinators R.M. Burns and B.H. Honkala, 160–72. Agric. Handb. 654. Washington, DC: U.S. Department of Agriculture Forest Service, 1990.

Smith, D.M., B.C. Larson, M.J. Kelty, and P.M.S. Ashton. *The Practice of Silviculture: Applied Forest Ecology.* New York: John Wiley & Sons, 1997.

Spies, T.A., M.A. Hemstrom, A. Youngblood, and S. Hummel. Conserving old-growth forest diversity in disturbance-prone landscapes. *Conservation Biology* 20, 2006: 351–62.

Stanturf, J.A. What is forest restoration? In *Restoration of Boreal and Temperate Forests*, 3–11. Boca Raton, FL: CRC Press, 2004.

Stephens, S.L., N. Burrows, A. Buyantuyev et al. Temperate and boreal forest mega-fires: Characteristics and challenges. *Frontiers in Ecology and the Environment* 12, 2014: 115–22.

Stine, P.A., P. Hessburg, T. Spies et al. *The Ecology and Management of Moist Mixed-Conifer Forests in Eastern Oregon and Washington: A Synthesis of the Relevant Biophysical Science and Implications for Future Land Management.* PNW-GTR-897. Portland, OR: U.S. Department of Agriculture Forest Service, Pacific Northwest Research Station, 2014.

Van Wagner, C.E. Conditions for the start and spread of crown fire. *Canadian Journal of Forest Research* 7, 1977: 23–34.

24

Options for Promoting the Recovery and Rehabilitation of Forests Affected by Severe Insect Outbreaks

Philip J. Burton, Miroslav Svoboda, Daniel Kneeshaw, and Kurt W. Gottschalk

CONTENTS

24.1 Introduction

Natural disturbances to forests around the world are on the rise, constituting a bellweather of global change associated with a warming climate (Parry et al. 2007; Weed et al. 2013). The rapid loss of mature forest cover as a result of wind storms, forest fires, and insect outbreaks often begs for a management response, much like that designed to reverse ecological degradation induced by resource exploitation. That management response may be as simple as an effort to recover economic value from the dead and damaged trees (i.e., salvage logging; Prestemon et al. 2013), or as complex as detailed prescriptions to enhance resilience at stand and landscape levels (Puettmann et al. 2009).

Among those disturbances are outbreaks of herbivorous insect species that kill trees over millions of hectares every year. In temperate and boreal forests, the insect pests that undergo population explosions include Lepidopterans (the larvae of which can consume large quantities of tree foliage), Coleopterans (particularly scolytids—bark beetles—the larvae of which burrow under the bark of conifers and compromise the function of vascular tissues), and to a lesser extent, some Hemipterans (adelgids and aphids, which have sucking mouthparts). While bark beetles can kill trees in a single season, it typically takes several years of herbivory by large numbers of defoliators and other insects before trees die. Most insect pests tend to have clear preferences in terms of the species (or genus) and the size or canopy position of trees that serve as their hosts and primary food source. Consequently, they act as selective thinning agents, often with little direct effect on non-host species, seedlings and saplings (at least in the case of bark beetles and most insect species at non-outbreak population levels), the understory plant community, or the forest floor. Nonetheless, it is recognized that insect outbreaks can influence a wide range of ecological processes and values, including water balance and streamflow regulation, nutrient cycling, wildlife habitat, and forest fire behavior (Lovett et al. 2006; Martin et al. 2006; Houle et al. 2009; Mikkelson et al. 2013).

One might question whether management interventions after insect outbreaks qualify as forest restoration, particularly if we adopt the perspective that ecological restoration only applies to situations where ecosystems have been degraded by human activities (e.g., Jackson et al. 1995). On the other hand, more recent interpretations of the restoration paradigm include interventions designed to promote the recovery of particular ecosystem values and processes (Society for Ecological Restoration International (SERI) 2004; Stanturf et al. 2014). Consequently, activities that accelerate reforestation or direct forest recovery along trajectories inspired by previous or analogous undisturbed templates can be considered particular cases of ecological restoration (Burton and MacDonald 2011). As insect outbreaks in and of themselves do not denote a loss of forest land cover or a change of land use, postoutbreak restoration includes a range of options characterized as "self-recovery" or "rehabilitation" by Stanturf (2015).

The purpose of this chapter is not to advocate for or against the active restoration of forests disturbed by insect outbreaks. Elaborating upon considerations introduced by Beatty and Owen (2005), it explores the conditions under which restorative actions are likely to be appropriate and effective, and what kinds of intervention are likely to be constructive under different scenarios. While insect outbreaks can be significant agents of change in forests throughout the Southern Hemisphere (e.g., Ohmart and Edwards 1991), we limit our discussion and examples to the forests of Europe and North America with which we have experience.

24.2 Effects of Insect Population Explosions on Forests

Insect outbreaks by diverse species and of varying severity have molded temperate and boreal forests for millennia, though reliable statistics have been compiled only over the last few decades. Data assembled for 2005 by the Food and Agriculture Organization (FAO) (2014) indicate that 17.3 million ha in Canada was affected by forest insects, 5.6 million ha in the United States, 3.2 million ha in China, 1.7 million ha in Russia, and 1.3 million ha in Romania. In Europe as a whole, bark beetles alone affected 8.2 million ha or 3.0% of the forest area in 2005 (European Forest Institute (EFI) n.d.). These levels of forest disturbance are greater than those caused by fire and logging combined, and is exceeded only by wind and storm damage in Europe.

Some insect outbreaks in historic times have had major influences on forest resources and regional economies. For example, a widespread nun moth (*Lymantria monacha*) outbreak ravaged spruce (*Picea abies*) forests in western Russia and eastern Prussia (modern-day Poland) for a decade in the mid-1800s, followed by bark beetle (*Ips typographus*; Figure 24.1a) attack and widespread salvage operations (Bejer 1988). Bark beetle outbreaks have

FIGURE 24.1
Insect species responsible for widespread tree death and occasional forest-altering population outbreaks:
(a) Adult European spruce bark beetle, *Ips typographus* L. (Photo by Gyorgy Csoka, Hungary Forest Research
Institute, Bugwood.org; (b) Larva of eastern North America's spruce budworm, *Choristoneura fumiferana*
(Clemens) (Photo by Jorge Monerris-Llopis); (c) Larvae and adult of western North America's mountain pine
beetle, *Dendroctonus ponderosae* Hopkins (Photo by Dion Manastyrski); and (d) Adult Asian gypsy moth
(*Lymantria dispar* L. (Photo by John H. Ghent, USDA Forest Service, Bugwood.org), introduced to North America
from Eurasia.)

recurred occasionally in central Europe. For example, between 1868 and 1880, more than
10,000 ha of *Picea abies* forest was cut down in the Bohemian Forest (border region between
the modern-day Czech Republic and Germany) as a result of a severe windstorm followed
by bark beetle outbreak (Bruna et al. 2013). Interestingly, the same region was affected by a
severe bark beetle outbreak 150 years later, when more than 10,000 ha of *Picea abies* forests
were affected in the Bohemian and Bavarian Forest National Parks (Svoboda et al. 2010,
2012; Lausch et al. 2011).

Recurrent spruce budworm (*Choristoneura fumiferana*; Figure 24.1b) outbreaks have defoliated balsam fir (*Abies balsamea*) forests from the Great Lakes region to the Maritime region
of eastern North America every 30–40 years for millennia (Boulanger and Arsenault
2004). An unprecedented outbreak of the mountain pine beetle (*Dendroctonus ponderosae*;
Figure 24.1c) in central British Columbia grew from the mid-1990s to affect a cumulative
area of more than 18 million ha by 2013 (Westfall and Ebata 2014). Invasive populations of
the Eurasian gypsy moth (*Lymantria dispar*; Figure 24.1d), first introduced to North America
in the 1860s, started to explode in the mixed oak forests of New England and Pennsylvania
in the 1970s and 1980s, and has continuously expanded its range since its introduction
(Davidson et al. 2001). The hemlock woolly adelgid (*Adelges tsugae*) is also a foreign invader,

expanding through forests of the northeastern United States since the 1980s (Preisser et al. 2008). Massive outbreaks of spruce beetles (*Dendroctonus rufipennis*) attacked *Picea sitchensis* forests on the Kenai Peninsula of Alaska and *P. glauca* forests in southwestern Yukon in the 1990s (Berg et al. 2006). The mountain pine beetle and the spruce beetle are just two of the many bark beetles that attack a wide range of conifer species across western North America (Raffa et al. 2008; Bentz et al. 2010). Outbreaks of the aspen serpentine leaf miner (*Phyllocnistis populiella*) and the forest tent caterpillar (*Malacosoma disstria*) recently have been occurring across the boreal and hemiboreal forests of North America, unexpectedly resulting in the death of trembling aspen (*Populus tremuloides*) trees where they have already been stressed by drought or prior defoliation (Frey et al. 2004). The Asian longhorned beetle (*Anoplophora glabripennis*) attacks a wide range of hardwood trees, causing widespread damage in both its native China and throughout its expanding range in the United States (Dodds and Orwig 2011). Most recently (since 2002), population expansion by the exotic emerald ash borer (*Agrilus planipennis*) has been a major source of mortality in *Fraxinus* trees in both urban and forested settings in eastern and mid-western North America (Poland and McCullough 2006; Pugh et al. 2011).

All insect outbreaks have a number of features in common. It is usually the larval forms of insect species that are the voracious herbivores, while it is the behavior of adults that governs population spread and dispersal. Insect species vary in the degree to which they are generalists or specialists in their feeding behavior. The hemlock looper (*Lambdina fiscellaria*) feeds on all age classes of foliage of several conifer species, killing trees in one to two years; in contrast, the spruce budworm consumes only new foliage, and even heavily infested trees take several years to die (Cooke et al. 2007). Most insect species exhibit some degree of selectivity in determining the species and size of tree on which they lay eggs, and on which the larvae feed, "spilling over" into less preferred trees (of other species or smaller sizes) only at high population levels (White and Whitham 2000; Nealis and Régnière 2004). For example, the eastern spruce budworm causes greater defoliation and mortality of balsam fir, followed by white spruce, followed by red spruce (*Picea rubens*) or black spruce (*Picea mariana*) (Hennigar et al. 2008). Trees die within 5–6 years if severely defoliated, or over 10–12 years following consecutive years of light or moderate defoliation (Nealis and Régnière 2004). The mountain pine beetle has typically attacked lodgepole pine (*Pinus contorta* var. *latifolia*), but also feeds on other western pine species such as ponderosa pine (*P. ponderosa*) and whitebark pine (*Pinus albicaulis*), preferring stems greater than approximately 23 cm in diameter in order to find phloem thick enough to support the larvae that will soon be developing under the bark (Safranyik and Carroll 2006; Björklund and Lindgren 2009). Recently found in jack pine (*Pinus banksiana*) trees in Alberta, and in the process of spreading eastward, it can now be considered a "native invasive" species in response to warming winters (Fuentealba et al. 2013).

At endemic (nonoutbreak) or low-severity outbreak population levels, insects attack and kill small pockets of trees, serving to thin the forest and release the growth of the survivors, including advance regeneration of seedlings and saplings that might be in the understory (Osawa 1994; Berg et al. 2006). Where host trees are found in mixed species stands, as is often the case, non-host species persist and thrive upon the death of competing trees. In this manner, whether at low insect population levels or in predominantly mixed stands, tree defoliation and mortality caused by forest insects initiate gap-level tree replacement processes (Figure 24.2a) and serves as a releasing disturbance (Kneeshaw and Bergeron 1998; Hawkins et al. 2013). When insect populations are very high and dense, and where stands consist almost entirely of susceptible trees, forest cover loss is much more

FIGURE 24.2
Contrasting scales of tree mortality and canopy disruption caused by forest insect outbreaks: (a) A single dead oak tree resulting from gypsy moth defoliation in a mixed-species hardwood forest in West Virginia, United States (Photo by Kurt Gottschalk); (b) Stand-level mortality caused by European spruce bark beetle in the Sumava Mountains, Czech Republic (Photo by J. Wild); (c) Recent landscape-level mortality (gray trees) of lodgepole pine caused by mountain pine beetle in central British Columbia, Canada. (Photo by Jeff Burrows.)

dramatic (Figure 24.2b). In these situations, insect outbreaks can serve as stand-replacing disturbances, initiating even-aged stand regeneration and development (McCarthy and Weetman 2007; Dhar and Hawkins 2011) over large areas (Figure 24.2c). Depending on the severity of the outbreak, the composition and structure of stands, and the silvics of tree species in the area, an insect outbreak can thus serve as the disturbance trigger for forest regeneration and diverse successional trajectories (Bouchard et al. 2006a). Different outbreaking insect species and different outbreak severities create different stand compositions and structures (e.g., Figure 24.3.1, Oliver et al. 2015).

All insect outbreaks are characterized by a suite of environmental and ecological impacts on forests. Unlike other disturbances such as fires and storm damage, the loss of overstory canopy cover tends to be gradual: defoliators may require several years of attack in order to kill trees, and dead foliage is retained on trees killed by bark beetles for several years after the tree is killed. Then it takes a number of years for standing dead trees to lose twigs and branches, finally collapsing over a series of years or decades. Those standing dead trees serve as important habitat features, especially for birds and other cavity nesters (Lindenmayer et al. 2008). The gradual loss of canopy foliage and branches increases light penetration to the understory, which not only promotes the growth of young trees and shrubs, but also increases average temperatures, temperature extremes (including the risk of frost damage) and wind speeds within the stand. Combined with the frass (insect feces) and litterfall associated directly with defoliators or indirectly with tree death, those microclimate changes also contribute to accelerated nutrient cycling and release in the forest floor (Lovett et al. 2006; Cobb 2010). The loss of living trees—and their interception of precipitation and their transpiration of water from the soil—also means that more water reaches and persists in the soil, often resulting in greater overland flow, greater stream-flow, and reduced capacity of the site to support the operation of heavy machinery (Swank et al. 1988; Redding et al. 2008). Forest fires that follow the death and collapse of insect-killed forests may exhibit more extreme fire behavior than in green timber, and greater intensity where jack-strawed boles (criss-crossed and elevated; e.g., of pine trees following mountain pine beetle attack) have been drying out for years (Page et al. 2013). In parts of Canada, balsam fir fuel types that have been altered by spruce budworm exhibit the most intense and rapidly spreading fire behavior under dry conditions (Wotton et al. 2009), with the rate of spread for fires in beetle-affected lodgepole pine forests being somewhat elevated over the usual rate of spread in green lodgepole pine (Perrakis et al. 2014). Yet, as a whole, there is little evidence that insect damage results in a greater probability of a forested area being burned than would be expected from variations in fire weather conditions (Romme et al. 2006; Simard et al. 2011; Harvey et al. 2013). Finally, it is worth noting that the carbon dynamics of insect outbreaks, like all disturbances, means that stocks of carbon switch from being incremented (by active photosynthesis and growth) to being important sources of CO_2 released into the atmosphere as litter and wood starts to decompose (Hicke et al. 2012).

24.3 Representative Responses to Severe Insect Outbreaks

24.3.1 Spruce Beetle Outbreaks in Central Europe

Severe bark beetle (*Ips typographus* L. Coleoptera: Scolytidae) outbreaks in Central Europe over recent decades have achieved a high public profile and posed many challenges to the managers of national parks of Germany and the Czech Republic (Jonasova and Prach 2004; Jonasova and Matejkova 2007; Muller et al. 2008; Hais et al. 2009; Jonasova et al. 2010; Svoboda et al. 2010). *Picea abies* (Norway spruce) trees dominating the upper forest stratum over more than 10,000 ha have died in the core zone of two national parks between 1996 and 2010 (Hais et al. 2009; Lausch et al. 2011). Death of the main canopy raised major concerns, especially by foresters and local communities, regarding future development of the stands. There was widespread fear bark beetle populations would continue to grow and overwhelm nearby commercial forests, or that the forest ecosystems would not recover, or that recovery would take such a long time that many ecosystems services would be threatened. As a result, salvage logging was conducted in many areas, especially in the Czech Republic and even in the core zones of the Šumava (Bohemian Forest) National Park (Jonasova and Prach 2004; Jonasova and Matejkova 2007). After a decade of intensive research, a majority of studies have shown that, even with stand-replacing disturbance, that is, the death of the majority of canopy and subcanopy trees, Norway spruce forest regenerates well (Jonasova and Matejkova 2007; Jonasova et al. 2010; Svoboda et al. 2010). A decade and a half following the bark beetle outbreak, a new tree cohort has reached densities that will most probably secure the predisturbance tree layer density. Most interestingly, Norway spruce was again the most dominant species, when no other early successional tree species gained dominance. In retrospect, it is now evident, that the bark beetle disturbance actually induced a positive effect, prompting forest renewal through spruce regeneration.

Regarding ecosystem services, no threats were found to underground water quality with respect to high levels of nitrogen, for example. Moreover, the bark beetle outbreak had a positive effect on the species diversity of many organism groups, especially insects (Muller et al. 2008). Areas that had been treated with salvage logging showed lower regeneration densities of Norway spruce as a result of the operation of heavy machinery and had to be restocked using planting (Jonasova and Matejkova 2007). The logged areas also showed overall lower structural diversity, so this treatment is considered inappropriate in core zones of national parks in the region. In conclusion, results from the case studies in the Bohemian Forest region have shown that bark beetle outbreaks are an important natural process, and need to be incorporated into forest management. In commercial forests, where wood production is the major concern, salvage logging operations are still commonly practiced, and there is great concern that bark beetles will spread from infested areas. However, if all ecosystem services—especially biodiversity—are considered, commercial forest management would benefit from accepting or emulating the natural disturbance regime to some degree.

24.3.2 Spruce Budworm Outbreaks in Eastern North America

A widespread outbreak of the spruce budworm (*Choristoneura fumiferana* (Clemens) Lepidoptera: Tortricidae) occurred in the 1970s and 1980s, defoliating 52 million ha of forest in eastern North America (Volney and Fleming 2007). In 2005, a new outbreak began with the defoliated area reaching almost 4.3 million ha in 2014. Since 2011, tree mortality has begun in the most severely defoliated sectors, and salvage logging is again a forest strategy that is being used as a response. During the last outbreak, the primary concern was to harvest the most wood fiber possible while forest structure and potential effects on biodiversity were not considered (Jette and Chabot 2013). In the last 25 years since the previous outbreak ended, guidelines have been developed for salvage logging following fire, (Nappi et al. 2011) but none have been developed for postbudworm management. As mortality in the current outbreak is beginning, there is new concern that salvage logging be conducted so as to minimize effects on biodiversity and soil fertility, as well as on the recruitment of the future forest.

Because fire and spruce budworm outbreaks have spatiotemporal differences as well as differences in species effects, guidelines developed for harvest following fire cannot be directly transferred to postbudworm forests. Spruce budworm outbreaks are species-specific, affecting balsam fir to a greater extent than any other species, but also damaging and killing spruce species (Hennigar et al. 2008). During the last outbreak, most salvage operations were clearcuts in which both dead trees and live trees of secondary hosts and non-hosts were also harvested (Angers et al. 2011). Harvesting during outbreaks occurred in three phases: preventive harvesting of the most vulnerable stands before the stands were defoliated, but when the outbreak was active elsewhere; pre-salvage harvesting during the first five years of defoliation but before mortality had begun; and salvage logging that began when more than 10% of the volume of a stand had been killed (Jette and Chabot 2013). Naturally, many of the stands disturbed by budworm underwent partial mortality, and had a patchy structure. In fact, in all forested regions containing fir, except the most eastern fir forests, less than 10% of forests could be considered reinitiated due to an almost complete mortality of the overstory.

In contrast, the ubiquitous use of clearcut harvesting in all salvaged stands created an even-aged structure that favored a shift to greater dominance by balsam fir (Bouchard et al. 2009). This compositional shift is due to the greater proportion of balsam fir in the advance regeneration and the harvest of overstory black spruce. In areas with pure fir forests and stand reinitiating mortality, clearcutting did not lead to compositional changes, but in mixed species forests, spruce budworm outbreaks were natural mechanisms for the recruitment of non-host species (Bouchard et al. 2006b). In this manner, spruce budworm outbreaks are potentially important in maintaining or increasing secondary host abundance, and hence in conferring improved ecosystem resilience to future outbreaks. Salvage logging, on the other hand, led to a shift in composition due to the preference of the forest products industry for spruce over fir, and thus a systematic harvesting of spruce species that naturally would have experienced little mortality during outbreaks. In recipes for pulp, there is a limit to the proportion of balsam fir relative to black spruce that is accepted in order to ensure paper quality. Mixed spruce-fir stands that naturally underwent partial mortality following outbreaks were thus preferred stands for clearcut salvage harvesting. Further composition changes could also be expected, as stands harvested by clearcutting with protection of advance regeneration have been found to support a greater abundance of noncommercial species compared to budworm-disturbed stands (Belle-Isle and Kneeshaw 2007).

The postbudworm salvage harvests also had a large impact on forest structure (Belle-Isle and Kneeshaw 2007), decreasing the quantity of dead wood by almost two-thirds and homogenising the diversity of decomposition classes compared to unsalvaged stands 20 years after the outbreak (Norvez et al. 2013). Although the abundance of saproxylic beetles was similar between salvaged and unsalvaged stands, reduced deadwood in salvaged stands led to differences in beetle communities (Norvez et al. 2013). Changes in insectivorous forest birds feeding on caterpillars have been observed throughout outbreaks (Venier et al. 2009) and there are concerns that reductions in dead wood following salvage logging could reduce the abundance of species such as woodpeckers that feed on saproxylic insects found in dead wood (Morris 1963). Bouchard et al. (2009) suggest that large-scale conifer mortality would lead to major alterations in habitat that could lead to the migration of wildlife species, perhaps favoring genetic diversity through increasing interactions between metapopulations.

Given the millions of hectares affected over a number of years, it is logistically impossible for all disturbed stands to be harvested. Globally, this may ensure that not all standing dead wood for biodiversity is removed. However, it also results in a prioritization of stands to be salvaged, which, as described above, led to compositional and structural shifts due to the targeting of partially disturbed spruce-fir stands with clearcut operations during the last outbreak.

24.3.3 Mountain Pine Beetle Outbreaks in Western North America

The mountain pine beetle (*Dendroctonus ponderosae* Hopkins Coleoptera: Scolytidae) is native to the pine forests of western North America. It is a bark beetle that reproduces through pheromone-mediated mass attacks of living pine trees, primarily *P. contorta* var. *latifolia*, but also *P. ponderosa*, *P. monticola* and the subalpine *Pinus albicaulis*. Because the larvae that emerge from eggs laid under the bark require phloem 5–6 mm thick in which to feed, and overwinter, they typically attack trees >23 cm in diameter at breast height (Björklund and Lindgren 2009). At normal or endemic population levels, hundreds or thousands of these insects will "mass attack" and overcome the defenses of individual large trees, principally those that are weakened by drought or other stress (Safranyik and Carroll 2006). The attacking beetles carry spores of symbiotic fungi that establish and grow in the sapwood of the pine tree, disrupting water flow, and resulting in tree death within the year. After overwintering under the bark, larvae emerge the next spring as adults that typically disperse to nearby host trees, repeating the cycle and contributing to the death of small patches of trees. However, mountain pine beetle populations occasionally erupt when constraints (such as limited food supply and overwintering mortality) are overcome, resulting in widespread tree mortality, especially in monotypic even-aged stands of mature and old lodgepole pine (Raffa et al. 2008; Bentz et al. 2010). Since the late 1990s, however, mountain pine beetle outbreaks from Colorado to British Columbia and Alberta have grown from gap- or stand-level disturbances to becoming landscape-wide agents of forest disturbance, affecting some 9 million ha in the western United States (Fettig et al. 2014) and 18 million ha of forest in British Columbia alone by 2013 (Westfall and Ebata 2014). Under such superoutbreak conditions, mountain pine beetles often spilled over to attack and killed trees as small as 8–12 cm in diameter.

The management response to bark beetle outbreaks in the United States and Canada has generally been similar to that described in the European example above, namely sanitation cutting in efforts to control growing insect populations, followed by salvage logging when outbreaks outpace control efforts. Management actions are constrained by the fact that visible signs of successful beetle attack—foliage turning yellow and then red—are only visible from the air after beetles have already killed the tree and spread into nearby green trees. Labor-intensive ground surveys referred to as "beetle probes" identify green-attack trees, often found around the visibly dying trees, and can focus sanitation cutting to harvest or burn trees that had been attacked (Maclauchlan and Brooks 1998). These operations typically generate some spatial and structural variability in large even-aged stands of lodgepole pine that regenerated following forest fires. But once beetles are found throughout a stand and are spreading across the landscape, salvage logging (or even "pre-emptive salvage" in advance of an unstoppable wave of insects) is typically conducted by clearcutting. The case for widespread salvage logging as a single and uniform response to mountain pine beetle outbreaks has been promoted by the widespread perception that "the forest is dead," further accentuated by the visual impact of bright red hillsides of dead pine trees, and by maps portraying the advance of outbreaks over time as a uniform and complete agent of mortality (e.g., http://www.for.gov.bc.ca/hre/bcmpb/graphics.htm). Despite these impressions, most lodgepole pine in central British Columbia, for example, is not found in pure stands, but is mixed with other conifers and broadleaf tree species that are not attacked by the pine beetle (Taylor and Carroll 2004).

In order to salvage solid timber before its value deteriorated, allowable cut levels were elevated on public lands throughout interior B.C., with forest companies responsible to regenerate (typically by planting) those areas that had been logged. Where beetle-attacked stands were too far from processing facilities or too poor in quality to warrant commercial harvesting, the Government of British Columbia offered incentives to rehabilitate stands by knocking them down, burning the trees in windrows, and planting (Lindenmayer et al. 2008). As with the spruce budworm outbreak in eastern Canada, several assessments pointed out that a large quantity of green trees was being harvested or destroyed along with the insect-killed trees (Coates et al. 2006; Vyse et al. 2009; Forest Practices Board (FPB) 2014). Some efforts at careful logging and partial cutting were undertaken to protect this "secondary structure"—green trees in the overstory and found as advance regeneration (seedlings and saplings in the understory; Nishio 2010). The Chief Forester of B.C. also issued guidelines for improved retention of green trees, dead trees, and well developed understories in the areas where timber harvesting was most accelerated, but these guidelines have generally been implemented on a stand by stand basis (rather than through stratification within stands) due to operational practicalities.

Efforts at controlling mountain pine beetle outbreaks and renewing the forest using prescribed fire have been tried on a limited basis, primarily in protected areas such as provincial and national parks. In other areas, particularly in and around urban municipalities and First Nations communities, timber harvesting and stand manipulation focused on fuel reduction to offset the perceived risk of greater fire danger associated with the buildup of dead flammable material from the dying trees. After more than a decade of this unprecedented insect outbreak, the central interior of B.C. is left with a legacy of logging companies and sawmills that geared up to harvest levels that cannot be sustained, and a landscape with reduced mature forest cover, and simplified forest and stand structure (Burton 2010). It could be argued that the impacts of salvage logging require ecological restoration much more than did the original disruption caused by the mountain pine beetle. To truly be restored (in the ecological sense of a return to preexisting composition and structure), beetle-attacked stands should be burned once again (Burton 2006).

24.3.4 Gypsy Moth in the Eastern United States

Gypsy moth (*Lymantria dispar* L. Lepidoptera: Lymantriidae) is an imported pest to North America, having been brought over from Europe for silk production in 1869 (Liebhold et al. 1989). It is a polyphagous species with more than 300 potential host species, but has feeding preferences and only grows to outbreak level populations in stands with moderate to high amounts of highly preferred or susceptible hosts (Liebhold et al. 1995, 1997). Outbreaks generally last from 1 to 3 years and occur with either approximately 4–5 or 8–10 year periodicity, depending upon forest type and site conditions (Johnson et al. 2006). Oak-dominated forests are the most susceptible forests. In the mid-1900s, gypsy moth had occupied about 23% of its potential host type range in the eastern United States (Morin et al. 2005) but the success of the gypsy moth "Slow-The-Spread" (STS) program is estimated to have reduced the rate of spread by 50–80% (Tobin and Blackburn 2007). The STS Program is a regional integrated pest management strategy that aims to minimize the rate of gypsy moth range expansion into uninfested areas. The premise is to deploy extensive grids of pheromone-baited traps along the expanding population front to identify newly establishing population spots, which are then eradicated using mating disruption or microbial pesticides to prevent them from growing, coalescing, and contributing to the progression of the population front.

A gypsy moth defoliation results in a wide variety of ecological effects, but tree mortality is often the greatest concern for forest management (Mason et al. 1989). Tree mortality varies widely due to variation in defoliation intensity and duration, tree species, and site and environmental conditions (Gansner et al. 1987; Davidson et al. 1999, 2001; Morin et al. 2004). Defoliation of hundreds of thousands to millions of hectares of oak-dominated forests in outbreak years has created thousands of forest stands with average mortality of 25%–35% of the basal area while oak mortality can range from 10% to 90% of the basal area (Davidson et al. 1999, 2001). These large areas create forest management opportunities for both preventative and restoration treatments.

Early researchers postulated some silvicultural treatments for dealing with gypsy moth, but it was not until the 1980s when silvicultural guidelines were developed and landowners started implementing them (Gottschalk 1982, 1993; Gansner et al. 1987). The ability to predict with some degree of accuracy which trees had higher probabilities of dying following defoliation led to the development of thinning prescriptions for forest stands prior to defoliation (Gansner et al. 1987; Gottschalk and MacFarlane 1993; Gottschalk et al. 1998). Several public and large private landowners implemented silvicultural treatments prior to gypsy moth in a preventative mode (Fosbroke and Hicks 1987; Hall 1995). Subsequent research has shown that thinning can partially offset growth loss due to defoliation (Fajvan et al. 2008; Fajvan and Gottschalk 2012). Silvicultural guidelines also included salvage thinning and salvage harvest (regeneration) treatments for dealing with forest stands postdefoliation (Fosbroke and Hicks 1987; Gottschalk 1993). Several large public and private landowners in the mid-Atlantic region implemented large-scale salvage operations during the 1980s and 1990s. Smaller private landowners also conducted salvage operations but usually did so without a silvicultural plan. Such large volumes of salvage were put up for sale that in some areas that the total volume exceeded the annual capacity of local sawmills by a factor of ten or more (Fosbroke and Hicks 1987). Approximately 25% of the defoliated area of an outbreak results in stand-replacing disturbances while the other 75% of the area receives varying degrees of thinning. Regeneration was generally via natural regeneration which resulted in succession to less susceptible hardwood species such as red maple (*Acer rubrum*) and tulip poplar (*Liriodendron tuplipifera*) on high quality sites, and black birch (*Betula lenta*) and blackgum (*Nyssa sylvatica*) on lower quality sites (Allen and Bowersox 1989; Fajvan and Wood 1996). One advantage of these mixed hardwood stands is their capacity for natural regeneration which reduces the need for restoration treatments. However, the Pennsylvania Bureau of Forestry did plant seedlings, especially conifers such as white pine (*Pinus strobus*), in some areas of heavy mortality, to reduce susceptibility to future defoliation. Restoration of oak rather than other mixed hardwoods would require additional preparatory treatments such as prescribed burning, herbicide spraying, or (in areas with no remaining seed sources) artificial regeneration (Gottschalk 1993; Brose et al. 2008).

24.4 Prioritizing Restorative Management Responses

24.4.1 General Considerations

We proceed from the premise that the appropriate management response to disruptions such as insect outbreaks is a joint function of the degree of ecological degradation and of land use objectives (Burton 2014). Ecological degradation has been defined as any change or disturbance to the environment perceived to be deleterious or undesirable (Australian Capital Territory (ACT) 2011), while in the context of disturbance ecology, an ecosystem can be considered degraded if it has been disrupted beyond the range of natural variability from which it has been able to recover in the past (Burton 2005). In some cases, disturbances such as wildfire or insect outbreaks may shift ecosystems into alternative stable states that may not be desirable from the perspective of land management goals (Scheffer et al. 2001; Jasinski and Payette 2005). Among land use objectives, the key consideration is the relative weight land owners and managers place on natural ecosystem processes versus specific goods and services such as timber, berries and mushrooms, habitat for target wildlife, watershed protection, and so on. From a biodiversity perspective, we might evaluate the degree to which an insect outbreak has pushed stand structures and compositions beyond their range of historic variability, or beyond the habitat requirements of vulnerable species. Particular consideration might be given to the rate of wood deterioration in insect-killed trees, the potential for increased risk of fire and treefall, and the relative rarity of naturally disturbed early seral ecosystems in the landscape (Lindenmayer et al. 2008). Restoration treatments may be undertaken to accelerate recovery of individual forest values or functions, or may have general objectives of promoting biodiversity, predisturbance conditions, or overall naturalness (Burton and MacDonald 2011).

In the case of insect outbreaks, perhaps the key challenge is how to judge the degree of ecological degradation: Is the current state of the forest truly beyond its natural range of variability (Burton 2005, 2014), can the forest recover in a reasonable period of time without management, how will forest goods and services be affected? An understanding of the historic range of variability in disturbance regimes, their residual structures, and the relationship of forest composition and structure to climate can provide guidance to restoration prescriptions at a coarse level (Millar 2014; Beatty and Owen 2005). For managed forests in particular, we might also ask whether the current state of the forest, and its likely trajectory for recovery, are beyond the range of socioeconomic acceptability.

Outbreaks of indigenous forest insects can be considered natural disturbances, with forests and insect species typically having undergone many cycles of attack, recovery, and genetic adjustment over the course of evolutionary history. However, there are occasions when human actions have precipitated outbreaks, making the distinction between "natural" and "anthropogenic" disturbances less clear. For example, the Eurasian gypsy moth and the emerald ash borer were introduced to North America by human commerce, so the forests they affect quickly assume a state for which there is no natural precedent. At least from a causal perspective, those forests can be considered altered as a direct consequence of human actions, and would normally be considered appropriate candidates for restoration. But many indirect human actions precipitate or accentuate insect outbreaks too: spruce logs stored at roadside in Norway apparently promoted population build-ups of spruce beetle, which then attacked green trees too (Niciforuk 2011); forest fragmentation has resulted in amplified edge effects that have compromised the parasitoids that normally limit the growth and duration of forest tent caterpillars outbreaks in central Canada (Roland and Taylor 1997). It also has been suggested that forest management has increased the proportion of fir in the landscape in eastern North America and subsequently the frequency and severity of spruce budworm outbreaks (Blais 1983), although this interpretation has been challenged by others (Boulanger and Arsenault 2004). The recent mountain pine beetle outbreak grew in large expanses of even-aged pine forest that arose as a consequence of forest fires initiated during the era of European settlement in central British Columbia (1880–1920). Those forests were then protected from fires and subject to warming winters resulting from a greenhouse effect amplified by worldwide consumption of fossil fuels (Burton 2006, 2010). These cases may or may not rank neatly along a continuum of human causation, so other dimensions of degradation and the potential for unaided resilience must be considered as well.

The greatest indication of inherent resilience and the capacity for a forest stand to recover without assistance is the abundance of living trees left in the wake of a disturbance, and the prospects for natural regeneration. Many trees that are lightly or moderately defoliated, even over multiple years, can eventually recover (Campbell and Valentine 1972; Gross 1991; MacLean and MacKinnon 1997). Likewise, the defense mechanisms of conifers can sometimes limit attack by bark beetles to "strip attacks" that let a tree survive (Holsten et al. 1999; Bleiker et al. 2014). With insect herbivores so specialized in their choice of host trees, there are typically many surviving trees in mixed species and multilayered stands. In addition to the potential for non-host trees to undergo growth release and (depending on tree architecture) expand their crowns once freed from neighboring competition (Campbell and Garlo 1982; Muzika and Liebhold 1999), their presence also provides resistance to the growth of insect populations and can accelerate outbreak collapse by hosting a greater variety of insect parasites and predators than are found in pure stands (Jactel et al. 2005).

FIGURE 24.3
Contrasting potential for unassisted, rapid forest recovery after canopy mortality: (a) Dense advance regeneration of balsam fir in a stand defoliated by spruce budworm (photo by Dan Kneeshaw); (b) Limited advance regeneration and dense competition from hayscented fern (*Dennstaedtia punctilobula*) resulting from interactions of deer browsing and defoliation by gypsy moth (Photo by Kurt Gottschalk); (c) No advance regeneration and a thick carpet of feathermoss (*Pleurozium schreberi*) in an even-aged stand of lodgepole pine under attack by mountain pine beetle. (Photo by Phil Burton).

Released growth of advance regeneration—seedlings and saplings (typically of shade tolerant species) already established in the understory (Figure 24.3a)—is the means by which many forests recover from insect-induced mortality of overstory trees (e.g., DeRose and Long 2010; Kayes and Tinker 2012; Spence and MacLean 2012; Hawkins et al. 2013). It has also been shown that outbreaks may be the mechanism by which less tolerant species and complex stand structures can be maintained in the landscape, with peaks of recruitment following outbreaks (Bouchard et al. 2006b; Reinikainen et al. 2012). On the other hand, some understory trees can be just as susceptible as mature trees, as is the case for uneven-aged stands of Douglas-fir (*Pseudotsuga menziesii*) attacked by western spruce budworm (*Choristoneura occidentalis*), the larvae of which balloon down from the canopy and feed on suppressed trees (Hadley and Veblen 1993). Absence of advance regeneration, coupled with dense nontree cover (Figure 24.3b) or a thick forest floor (Figure 24.3c), means that postoutbreak succession may instead be dominated by shrubs, ferns, or grass rather than trees for a considerable period of time (Royo and Carson 2006; Griffin and Turner 2012).

24.4.2 The Intervention Continuum

We here explore options for responding to forest insect outbreaks, considering both resource management objectives and factors such as outbreak origins and severity, and stand and landscape composition. We posit that any decision to undertake active management must be explicitly weighed against the option of doing nothing—of letting ecosystem recovery proceed unaided (Foster and Orwig 2006; Stanturf 2015)—for which a solid understanding of forest stand dynamics is required. Identifying the appropriate criteria for, and balance of, active restoration and passive recovery is an ongoing issue in the field of ecological restoration (Prach and Hobbs 2008; Walker et al. 2014). We point out that forests are often heterogeneous and patchy with respect to overstory and understory composition, and urge that postoutbreak management options—including prescriptions to intervene or leave for natural recovery—be applied at much finer scales than standard forest harvesting and silvicultural practices.

There is typically no need or incentive for active forest rehabilitation after an insect outbreak if overstory mortality is low, or if the understory is already well stocked with vigorous seedlings and saplings or is soon expected to be so (Table 24.1). What constitutes

TABLE 24.1

Principal Considerations in Responding to Forest Insect Outbreaks through Passive Management or Active Intervention

Natural Recovery Only	Assisted Recovery/Rehabilitation
• Logging forbidden, natural processes required to dominate (as in strict nature reserves)	• Active management (e.g., through logging, thinning, site preparation, sowing or planting, brushing, burning) permissible
• No personal or community safety concerns[a] associated with treefall or fuel accumulation	• Removal of dead trees desirable[a] (for revenue, safety, or fuel reduction)
• Satisfactory[b] levels of overstory survival and/or understory stocking[b] by desired[b] tree species	• High[b] levels of tree mortality with low[b] stocking densities and poor[b] prospects for natural regeneration by desired[b] tree species

[a] Criteria and thresholds depend on cultural norms, regulations, and socioeconomic conditions.
[b] Criteria and thresholds depend on forest type, site characteristics, species growth attributes, and management objectives.

"sufficient" overstory recovery and growth potential, "adequate" stocking and "desired" tree species inevitably depends on species attributes and management objectives. For example, some tree species can recover more readily than others from heavy defoliation (Gottschalk et al. 1998), broadleaf species have a greater ability than conifers to expand their crowns in response to the death of neighboring trees (Muzika and Liebhold 1999), and land owners will typically have economic, habitat, or aesthetic preferences for some tree species over others. If active management is permitted, there may be good reasons to remove the perceived danger associated with dead trees falling (e.g., along roads, walking trails, or power lines; Manning and Deans 2010) and accumulations of elevated coarse woody debris that dry out rapidly and would burn with great intensity (Table 24.1). Redirecting the speed, density, and composition of forest recovery may also be reasons to rehabilitate the disturbed forest with thinning and cleaning operations, or through artificial regeneration. Those silvicultural operations often require prior treatments that may include tree felling and yarding, or site preparation treatments such as broadcast burning. When timber production is the primary land use objective, salvage logging is typically carried out to capture value and renew the stand as quickly as possible. Even where timber production is not a priority, revenues from timber sales can fund or subsidize any subsequent restoration treatments that accelerate or redirect the recovery of other forest values. These treatments are most appropriate when applied where they can be more effective than passive recovery, often in limited pockets of insect-affected stands. Processes of natural ecosystem recovery typically are more desirable, less intrusive, and less costly than active intervention (Walker et al. 2014).

Once forest stand rehabilitation is identified as a management option (e.g., according to the considerations in Table 24.1), this intervention can be conducted with varying degrees of intensity and selectivity (see Table 24.2). Greater intervention may be justified where the forest ecosystem is more severely degraded (as per the right column in Table 24.2), and/or where affected forest stands are no longer in a strong position to provide the forest values (timber or otherwise, as per the left column in Table 24.2) desired or mandated by the landowner. The degree of forest simplification or potential degradation during an insect outbreak is generally a function of insect population numbers and the proportion of the stand that is susceptible to that insect species.

TABLE 24.2

Example of a Scorecard to Identify the Urgency and Intensity of Appropriate Forest Rehabilitation Actions after Insect Outbreaks

Land Use Emphasis, Values and Priorities		Indicators of Disturbance Severity and Ecological Degradation	
Timber emphasis	−1 = low/none 0 = medium +1 = high/exclusive	Outbreak origins and context	−1 = indigenous sp. in wild forest 0 = indigenous sp. in managed stands +1 = exotic sp.
Priority to other values dependent on live trees (e.g., mushrooms, wildlife, watershed protection)	−1 = low 0 = medium +1 = high	Insect population density or outbreak severity	−1 = low 0 = medium +1 = high
Harvesting can be accommodated by redirecting a predetermined level of sustainable cut[a]	−1 = not within 5 yrs[a] 0 = within 3–5 yrs +1 = within 1–2 yrs	Proportion of stand preferred or susceptible (or the converse of the proportion expected to survive)	−1 = low 0 = medium +1 = high
Emphasis on biodiversity and natural processes at the stand level	+1 = low 0 = medium −1 = high	Stocking by acceptable advance regeneration	+1 = low 0 = medium −1 = high
Rarity and value of early seral, naturally disturbed stands in the landscape	+1 = low 0 = medium −1 = high	Prospects[b] for natural regeneration by seeds or sprouts	+1 = low 0 = medium −1 = high
Rate of wood (or other resource value) deterioration	− 1 = low (>10 yrs) 0 = medium (6–10 yrs) +1 = high (2–5 yrs)	Environmental impact[c] of timber harvesting or silvicultural practices	+1 = low 0 = medium −1 = high

Note: Sum the scores in all 12 categories: higher values (>~ +5) denote a stronger case for intensive silvicultural intervention (which may include salvage logging, site preparation, planting, etc.), while lower values (<~ −4) denote a greater potential for natural recovery and (for scores ~−3 to ~+5) greater importance to be placed on the retention of biological legacies.

[a] That is, where sustained yield forestry with annual harvesting is practiced, elevated harvest levels (redirected to restoration or salvage treatments) could return to normal levels within 1–2 yrs, 3–5 yrs, or would require a longer period of time; replace "yrs" with "entries" where the cutting cycle is longer, as in small holdings.

[b] Includes consideration of forest floor thickness, spp. reproduction potential, competing shrub or herbaceous cover.

[c] Associated with any road construction, ground access, soil compaction or degradation, etc.

We might also consider conditions under which a particular outbreak is truly unprecedented—namely when due to an invasive insect species that is exotic to the continent or to the biome—which, from the perspective of ecosystem integrity, may make restoration more urgent but also more difficult if not combined with exotic species control. Along with the prospects for natural recovery through growth release by surviving green trees and new regeneration, we must also consider whether intervention activities (e.g., if new roads need to be constructed, or if heavy machinery would be operated on sensitive soils) might cause more damage than aid (Table 24.2).

Standard industrial silvicultural practices are less appropriate where there is a management mandate to conserve biodiversity or for natural processes to dominate, such as in national parks and designated wilderness areas. In such cases, which may also include intensively managed landscapes where naturally disturbed forests and standing dead trees are rare, it is important to leave a greater amount of residual structure, coarse woody debris, and a variety of microhabitats (Lindenmayer et al. 2008). Conversely, where a stand is managed solely for timber, and especially where the timber (or other values associated with live trees) is in danger of deteriorating, it is understandable for forest restoration to proceed by standard salvage logging and renewal methods appropriate to the silvics and ecology of the forest type and its socioeconomic context (Table 24.2). On the other hand, it is also possible to initiate forest renewal without first removing dead trees in the overstory, by means of tree planting directly under the standing dead trees (Manning and Deans 2010; Jonas et al. 2012). A cool surface fire (underburning) can also be effective in controlling insect infestations of understory saplings (e.g., by western spruce budworm), or to stimulate seedling sprouts or prepare receptive seedbeds for natural regeneration (Gordon 1989). Partial cutting or "selective logging" can generate a wide range of stand structures, with an array of horizontal and vertical patterning that is more or less similar to that generated by intermediate severities of natural disturbance (Kneeshaw et al. 2011). A key objective in management decisions after insect outbreaks should be to reduce susceptibility to future insect attack (Xi et al. 2013), so care must be taken to promote rather than to compromise the inherent resilience of temperate and boreal forests. In some cases, proactive restoration planning can be a component or contingent phase of an integrated pest management plan.

24.5 Conclusions

The value, resilience, and adaptive capacity of many forests are not compromised by insect outbreaks. Indeed, insect outbreaks are part of the natural disturbance regime of most forest types, stimulating important compositional and structural diversity in the landscape. On the other hand, there is no question that severe outbreaks can drastically alter forest structure, composition, and function in the short run (Fajvan et al. 2008; Kneeshaw et al. 2011). Where those changes compromise the delivery of forest values (such as timber or wildlife habitat), with poor prospects for their rapid recovery, management interventions can serve to accelerate and direct that process of recovery. Forest rehabilitation methods may first depend on salvage logging and/or site preparation, and the sale of timber (for milling, pulping, or as fuelwood) may serve to underwrite the costs of forest renewal. Where intervention is warranted, and even where timber is a primary objective, the retention of some residual structure and a variety of biological legacies should be a priority, as should be the promotion of species diversity. Forest rehabilitation programs provide an opportunity to direct forest composition and structure to be more resilient to future disturbances, stresses, and uncertainties.

Acknowledgments

We thank John Stanturf, Chad Oliver, and an anonymous reviewer for their constructive input to this chapter. Numerous colleagues and students have contributed to the analysis and understanding of the case studies reported here, for which we are grateful.

References

Allen, D. and T. Bowersox. 1989. Regeneration in oak stands following gypsy moth defoliation. In: *Proceedings, 7th Central Hardwood Forest Conference*, 1989 March 5–8, Carbondale, IL. Gen. Tech. Rep. NC-132: pp. 67–73. St. Paul, MN: US Department of Agriculture, Forest Service, North Central Forest Experiment Station.

Angers, V.-A., H. Varady-Szabo, A. Malenfant, and M. Bosquet. 2011. *Mesure des écarts des attributs de bois mort entre la forêt naturelle et la forêt aménagée en Gaspésie*: 51 pp. Québec, QC: Consortium en foresterie Gaspésie Les îles, Gaspé.

Australian Capital Territory (ACT). 2011. *ACT State of the Environment Report 2011*: 598 pp. Canberra, Australia: Office of the Commissioner for Sustainability and the Environment. http://www.environmentcommissioner.act.gov.au/publications/annual_reports (accessed December 2, 2014).

Beatty, S. W. and B. S. Owen. 2005. Incorporating disturbance into forest restoration. In Stanturf, J. A. and Madsen, P. (Eds.) *Restoration of Boreal and Temperate Forests*, pp. 61–76, Boca Raton: CRC Press.

Bejer, B. 1988. The nun moth in European spruce forests. In: Berryman, A. A. (ed). *Dynamics of Forest Insect Populations: Patterns, Causes, Implications*: pp. 211–231. New York, NY: Plenum Press.

Belle-Isle, J. and D. Kneeshaw. 2007. A stand and landscape comparison of the effects of a spruce budworm (*Choristoneura fumiferana* (Clem.)) outbreak to the combined effects of harvesting and thinning on forest structure. *Forest Ecology and Management* 246: 163–174.

Bentz, B. J., J. Régnière, C. J. Fettig, E. M. Hansen, J. L. Hayes, J. A. Hicke et al. 2010. Climate change and bark beetles of the western United States and Canada: Direct and indirect effects. *BioScience* 60: 602–613.

Berg, E., J. D. Henry, C. L. Fastie, A. D. De Volder, and S. M. Matsuoka. 2006. Spruce beetle outbreaks on the Kenai Peninsula, Alaska, and Kluane National Park and Reserve, Yukon Territory: Relationship to summer temperatures and regional differences in disturbance regimes. *Forest Ecology and Management* 227: 219–232.

Björklund, N. and B. S. Lindgren. 2009. Diameter of lodgepole pine and mortality caused by the mountain pine beetle: Factors that influence their relationship and applicability for susceptibility rating. *Canadian Journal of Forest Research* 39: 908–916.

Blais, J. R. 1983. Trends in the frequency, extent, and severity of spruce budworm outbreaks in eastern Canada. *Canadian Journal of Forest Research* 13: 539–547.

Bleiker, K. P., M. R. O'Brien, G. D. Smith, and A. L. Carroll. 2014. Characterisation of attacks made by the mountain pine beetle (Coleoptera: Curculionidae) during its endemic population phase. *The Canadian Entomologist* 146: 271–284.

Bouchard, M., D. Kneeshaw, and Y. Bergeron. 2006a. Forest dynamics after successive spruce budworm outbreaks in mixedwood forests. *Ecology* 87: 2319–2329.

Bouchard, M., D. D. Kneeshaw, and Y. Bergeron. 2006b. Tree recruitment pulses and long-term species coexistence in mixed forests of western Québec. *Ecoscience* 13: 82–88.

Bouchard, M., D. D. Kneeshaw, and Y. Bergeron. 2009. Ecosystem management based on large-scale, episodic disturbances: A case study from sub-boreal forests. *Forest Ecology and Management* 256: 1734–1742.

Boulanger, Y. and D. Arsenault. 2004. Spruce budworm outbreaks in eastern Quebec over the last 450 years. *Canadian Journal of Forest Research* 34: 1035–1043.

Brose, P. H., K. W. Gottschalk, S. B. Horsley, P. D. Knopp, J. N. Kochendorfer, B. J. McGuiness et al. 2008. *Prescribing Regeneration Treatments for Mixed-Oak forests in the Mid-Atlantic Region*. Gen. Tech. Rep. NRS-33: 100 pp. Newtown Square, PA: USDA Forest Service, Northern Research Station.

Bruna, J., J. Wild, M. Svoboda, M. Heurich, and J. Mullerova. 2013. Impacts and underlying factors of landscape-scale, historical disturbance of mountain forest identified using archival documents. *Forest Ecology and Management* 305: 294–306.

Burton, P. J. 2005. Ecosystem management and conservation biology. In: Watts, S. B. and Tolland, L. (eds). *Forestry Handbook for British Columbia*, fifth edition: pp. 307–322. Vancouver, BC: Faculty of Forestry, University of British Columbia.

Burton, P. J. 2006. Restoration of forests attacked by mountain pine beetle: Misnomer, misdirected, or must-do? *Journal of Ecosystems and Management* 7 (2): 1–10. http://jem.forrex.org/index.php/jem/article/view/537/444.

Burton, P. J. 2010. Striving for sustainability and resilience in the face of unprecedented change: The case of the mountain pine beetle outbreak in British Columbia. *Sustainability* 2: 2403–2423.

Burton, P. J. 2014. Considerations for monitoring and evaluating forest restoration. *Journal of Sustainable Forestry* 33: S149–S160.

Burton, P. J. and S. E. MacDonald. 2011. The restorative imperative: Assessing objectives, approaches and challenges to restoring naturalness in forests. *Silva Fennica* 45 (5): 843–863.

Campbell, R. W. and A. S. Garlo. 1982. Gypsy moth in New Jersey pine-oak. *Journal of Forestry* 80 (2): 89–90.

Campbell, R. W. and H. T. Valentine. 1972. *Tree Condition and Mortality Following Defoliation by the Gypsy Moth*. Res. Pap. NE-236: 331 pp. Upper Darby, PA: U.S. Department of Agriculture Forest Service.

Coates, K. D., C. DeLong, P. J. Burton, and D. L. Sachs. 2006. *Abundance of Secondary Structure in Lodgepole Pine Stands Affected by Mountain Pine Beetle: Report for the Chief Forester*: 18 pp. Smithers, BC: Bulkley Valley Research Centre. http://bvcentre.ca/files/SORTIE-ND_reports/Report_for_Chief_Forester_Secondary_Structure_in_Pine_Stands_Final.pdf, accessed December 2, 2014.

Cobb, R. C. 2010. Species shift drives decomposition rates following invasion by hemlock woolly adelgid. *Oikos* 119: 1291–1298.

Cooke, B. J., V. G. Nealis, and J. Régnière. 2007. Insect defoliators as periodic disturbances in northern forest ecosystems. In: Johnson, E. A. and Miyanishi K. (ed.) *Plant Disturbance Ecology: The Process and the Response*: pp. 487–525. San Diego, CA: Academic Press.

Davidson, C. B., K. W. Gottschalk, and J. E. Johnson. 1999. Tree mortality following defoliation by the European gypsy moth (*Lymantria dispar* L.) in the United States: A review. *Forest Science* 45 (1): 74–84.

Davidson, C. B., K. W. Gottschalk, and J. E. Johnson. 2001. *European Gypsy Moth (*Lymantria dispar *L.) Outbreaks: A Review of the Literature*. Gen. Tech. Rep. NE-278: 15 pp. Newtown Square, PA: USDA Forest Service, Northeastern Research Station.

DeRose, R. J. and J. N. Long. 2010. Regeneration response and seedling bank dynamics on a *Dendroctonus rufipennis*-killed *Picea engelmannii* landscape. *Journal of Vegetation Science* 21 (2): 377–387.

Dhar, A. and C. D. B. Hawkins. 2011. Regeneration and growth following mountain pine beetle attack: A synthesis of knowledge. *Journal of Ecosystems and Management* 12 (2): 1–16. http://jem.forrex.org/index.php/jem/article/view/22/84.

Dodds, K. J. and D. A. Orwig. 2011. An invasive urban forest pest invades natural environments—Asian longhorned beetle in northeastern U.S. hardwood forests. *Canadian Journal of Forest Research* 41 (9): 1729–1742.

European Forest Institute (EFI). n.d. *Database on Forest Disturbances in Europe*. Joensuu, Finland: European Forest Institute. http://www.efi.int/databases/dfde/, accessed March 26, 2014.

Fajvan, M. A. and K. W. Gottschalk. 2012. The effects of silvicultural thinning and *Lymantria dispar* L. defoliation on wood volume growth of *Quercus* spp. *American Journal of Plant Sciences* 3: 276–282.

Fajvan, M. A., J. Rentch, and K. Gottschalk. 2008. The effects of thinning and gypsy moth defoliation on wood volume growth in oaks. *Trees* 22: 257–268.

Fajvan, M. A. and J. M. Wood. 1996. Stand structure and development after gypsy moth defoliation in the Appalachian Plateau. *Forest Ecology and Management* 89: 79–88.

Fettig, C. J., K. E. Gibson, A. S. Munson, and J. F. Negrón. 2014. Cultural practices for prevention and mitigation of mountain pine beetle infestations. *Forest Science* 60 (3): 450–463.

Food and Agriculture Organization (FAO). 2014. *Disturbance Affecting Forest Health and Vitality (1000 ha) by FRA Categories, Year, Country*. Rome, Italy: Food and Agriculture Organization of the United Nations. Rome, Italy. http://countrystat.org/home.aspx?c=FOR&tr=5, accessed March 26, 2014.

Forest Practices Board (FPB). 2014. *Timber Harvesting in Beetle-Affected Areas: Is it Meeting Government's Expectations?* Special Report 44: 37 pp. Victoria, BC: British Columbia Forest Practices Board. http://www.fpb.gov.bc.ca/SR44_Timber_Harvesting_in_Beetle-Affected_Areas.pdf, accessed April 27, 2014.

Fosbroke, S. and R. R. Hicks, Jr. (eds). 1987. *Proceedings of Coping with the Gypsy Moth in the New Frontier*: 153 pp. Morgantown, WV: West Virginia University Books.

Foster, D. R. and D. A. Orwig. 2006. Preemptive and salvage harvesting of New England forests: When doing nothing is a viable alternative. *Conservation Biology* 20: 959–970.

Frey, B. R., V. J. Lieffers, E. H. Hogg, and S. M. Landhäusser. 2004. Predicting landscape patterns of aspen dieback: Mechanisms and knowledge gaps. *Canadian Journal of Forest Research* 34: 1379–1390.

Fuentealba, A., R. Alfaro, and É Bauce. 2013. Theoretical framework for assessment of risks posed to Canadian forests by invasive insect species. *Forest Ecology and Management* 302: 97–106.

Gansner, D. A., O. W. Herrick, G. N. Mason, and K. W. Gottschalk. 1987. Coping with the gypsy moth on new frontiers of infestation. *Southern Journal of Applied Forestry* 11 (4): 201–209.

Gordon, C. 1989. Silviculture and forest health on the Pike and San Isabel National Forest. In: *Proceedings of the National Silviculture Workshop: Silvicultural Challenges and Opportunities in the 1990s*, July 10–13, 1989, Petersburg, Alaska: pp. 96–100. Washington, DC: USDA Forest Service.

Gottschalk, K. W. 1982. Silvicultural alternatives for coping with the gypsy moth. In: Finley J. C., R. S. Cochran, and M. J. Baughman (eds). *Proceedings, Coping with the Gypsy Moth*, February 17–18, 1982, Pennsylvania State University: pp. 147–156. University Park, PA: Pennsylvania State University.

Gottschalk, K. W. 1993. *Silvicultural Guidelines for Forest Stands Threatened by the Gypsy Moth*. Gen. Tech. Rep. NE-171: 49 pp. Radnor, PA: USDA Forest Service, Northeastern Forest Experiment Station.

Gottschalk, K. W., J. J. Colbert, and D. L. Feicht. 1998. Tree mortality risk of oak due to gypsy moth. *European Journal of Forest Pathology* 28: 121–132.

Gottschalk, K. W. and W. R. MacFarlane. 1993. *Photographic Guide to Crown Condition of Oaks: Use for Gypsy Moth Silviculture*. Gen. Tech. Rep. NE-168: 8 pp. Radnor, P. A.: USDA Forest Service, Northeastern Forest Experiment Station.

Griffin, J. M. and M. G. Turner. 2012. Changes to the N cycle following bark beetle outbreaks in two contrasting conifer forest types. *Oecologia* 170 (2): 551–565.

Gross, H. L. 1991. Dieback and growth loss of sugar maple associated with defoliation by the forest tent caterpillar. *The Forestry Chronicle* 67 (1): 33–42.

Hadley, K. S. and T. T. Veblen. 1993. Stand response to western spruce budworm and Douglas-fir bark beetle outbreaks, Colorado Front Range. *Canadian Journal of Forest Research* 23 (3): 479–491.

Hais, M., M. Jonasova, J. Langhammer, and T. Kucera. 2009. Comparison of two types of forest disturbance using multitemporal Landsat TM/ETM plus imagery and field vegetation data. *Remote Sensing of Environment* 113: 835–845.

Hall, D. 1995. Silvicultural guidelines for oak stands threatened by gypsy moth in Wisconsin. In: Hilburn, D. J., K. J. R. Johnson, and A. D. Mudge (eds). *Proceedings of the 1994 Annual Gypsy Moth Review*, Portland, OR, October 30–November 2, 1994: pp. 118–123. Salem, OR: Oregon Department of Agriculture.

Hawkins, C. D., A. Dhar, and N. A. Balliet. 2013. Radial growth of residual overstory trees and understory saplings after mountain pine beetle attack in central British Columbia. *Forest Ecology and Management* 310: 348–356.

Harvey, B. J., D. C. Donato, W. H. Romme, and M. G. Turner. 2013. Influence of recent bark beetle outbreak on fire severity and postfire tree regeneration in montane Douglas-fir forests. *Ecology* 94: 2475–2486.

Hennigar, C. R., D. A. MacLean, D. T. Quiring, and J. A. Kershaw. 2008. Differences in spruce budworm defoliation among balsam fir and white, red, and black spruce. *Forest Science* 54: 158–166.

Hicke, J. A., C. D. Allen, A. R. Desai, M. C. Dietze, R. J. Hall, D. M. Kashian et al. 2012. Effects of biotic disturbances on forest carbon cycling in the United States and Canada. *Global Change Biology* 18: 7–34.

Holsten, E., R. Their, A. Munson, and K. Gibson. 1999. *The Spruce Beetle*. Forest Insect and Disease Leaflet 127: 11 pp. Anchorage, AK: USDA Forest Service.

Houle, D., L. Duchesne, and R. Boutin. 2009. Effects of a spruce budworm outbreak on element export below the rooting zone: A case study for a balsam fir forest. *Annals of Forest Science* 66 (7): 1–9.

Jackson, L. L., N. Lopoukhine, and D. Hillyard. 1995. Ecological restoration: A definition and comments. *Restoration Ecology* 3 (2): 71–75.

Jactel, H., E. Brockerhoff, and P. Duelli. 2005. A test of the biodiversity-stability theory: Meta-analysis of tree species diversity effects on insect pest infestations, and re-examination of responsible factors. *Ecological Studies* 176: 235–262.

Jasinski, J. P. and S. Payette. 2005. The creation of alternative stable states in the southern boreal forest, Quebec, Canada. *Ecological Monographs* 75 (4): 561–583.

Jette, J. P. and M. Chabot 2013. *Modulation des activités forestières pour faire face à une épidémie de la tordeuse des bourgeons de l'épinette dans un contexte d'aménagement écosystémique*: 72 pp. Québec, QC: Gouvernement du Québec, Ministère des ressources naturelles, Direction de l'aménagement et de l'environnement forestiers.

Johnson, D. M., A. M. Liebhold, and O. N. Bjørnstad. 2006. Geographical variation in the periodicity of gypsy moth outbreaks. *Ecography* 29: 367–374.

Jonas, S. Z., W. Xi, J. D. Waldron, and R. N. Coulson. 2012. Impacts of hemlock decline and ecological considerations for hemlock stand restoration following hemlock woolly adelgid outbreaks. *Tree and Forestry Science and Biotechnology* 6 (SI1): 22–26.

Jonasova, M. and I. Matejkova. 2007. Natural regeneration and vegetation changes in wet spruce forests after natural and artificial disturbances. *Canadian Journal of Forest Research* 37: 1907–1914.

Jonasova, M. and K. Prach. 2004. Central-European mountain spruce (*Picea abies* (L.) Karst.) forests: Regeneration of tree species after a bark beetle outbreak. *Ecological Engineering* 23: 15–27.

Jonasova, M., E. Vavrova, and P. Cudlin. 2010. Western Carpathian mountain spruce forest after a windthrow: Natural regeneration in cleared and uncleared areas. *Forest Ecology and Management* 259: 1127–1134.

Kayes, L. J. and D. B. Tinker. 2012. Forest structure and regeneration following a mountain pine beetle epidemic in southeastern Wyoming. *Forest Ecology and Management* 263: 57–66.

Kneeshaw, D. D. and Y. Bergeron. 1998. Canopy gap characteristics and tree replacement in the southeastern boreal forest. *Ecology* 79: 783–794.

Kneeshaw, D. D., B. D. Harvey, G. P. Reyes, M.-N. Caron, and S. Barlow. 2011. Spruce budworm, windthrow and partial cutting: Do different partial disturbances produce different forest structures? *Forest Ecology and Management* 262: 482–490.

Lausch, A., L. Fahse, and M. Heurich. 2011. Factors affecting the spatio-temporal dispersion of *Ips typographus* (L.) in Bavarian Forest National Park: A long-term quantitative landscape-level analysis. *Forest Ecology and Management* 261: 233–245.

Liebhold, A. M., K. W. Gottschalk, D. A. Mason, and R. R. Bush. 1997. Forest susceptibility to the gypsy moth. *Journal of Forestry* 95: 20–24.

Liebhold, A. M., K. W. Gottschalk, R. M. Muzika, M. E. Montgomery, R. Young, K. O'Day et al. 1995. *Suitability of North American Tree Species to the Gypsy Moth: A Summary of Field and Laboratory Tests*. Gen. Tech. Rep. GTR-NE-211: 45 pp. Radnor, PA: USDA Forest Service.

Liebhold A., V. Mastro, and P. W. Schaefer. 1989. Learning from the legacy of Leopold Trouvelot. *Bulletin of the Entomological Society of America* 35: 20–22.

Lindenmayer, D. B., P. J. Burton, and J. F. Franklin. 2008. *Salvage Logging and its Ecological Consequences*: 227 pp. Washington, DC: Island Press.

Lovett G. M., C. D. Canham, M. A. Arthur, K. C. Weathers, and R. D. Fitzhugh. 2006. Forest ecosystem responses to exotic pests and pathogens in eastern North America. *BioScience* 56: 395–405.

Maclauchlan, L. E. and J. E. Brooks. 1998. *Strategies and Tactics for Managing the Mountain Pine Beetle, Dendroctonus ponderosae*: 55 pp. Kamloops, BC: British Columbia Forest Service, Kamloops

Forest Region. http://www.for.gov.bc.ca/hfd/library/documents/bib87292.pdf, accessed November 25, 2014.

MacLean, D. A. and W. E. MacKinnon. 1997. Effects of stand and site characteristics on susceptibility and vulnerability of balsam fir and spruce to spruce budworm in New Brunswick. *Canadian Journal of Forest Research* 27: 1859–1871.

Manning, E. T. and A. Deans. 2010. Evaluation of tree condition and tree safety assessment procedures in beetle-killed and fire-damaged lodgepole pine stands in central interior British Columbia. *Journal of Ecosystems and Management* 10 (3): 90–103. http://jem.forrex.org/index.php/jem/article/view/37/10.

Martin, K., A. Norris, and M. Drever. 2006. Effects of bark beetle outbreaks on avian biodiversity in the British Columbia interior: Implications for critical habitat management. *Journal of Ecosystems and Management* 7 (3): 10–24. http://jem.forrex.org/index.php/jem/article/view/354/269.

Mason, G. N., K. W. Gottschalk, and J. S. Hadfield. 1989. Effects of timber management practices on insects and diseases. In: Burns, R. M. (Tech. Coord). *The Scientific Basis for Silvicultural and Management Decisions in the National Forest System*. Gen. Tech. Rep. GTR-WO-55: pp. 152–171. Washington, DC: USDA Forest Service.

McCarthy, J. W. and G. Weetman. 2007. Stand structure and development of an insect-mediated boreal forest landscape. *Forest Ecology and Management* 241: 101–114.

Mikkelson, K. M., L. A. Bearup, R. M. Maxwell, J. D. Stednick, J. E. McCray, and J. O. Sharp. 2013. Bark beetle infestation impacts on nutrient cycling, water quality and interdependent hydrological effects. *Biogeochemistry* 115 (1–3): 1–21.

Millar, C. I. 2014. Historic variability: Informing restoration strategies, not prescribing targets. *Journal of Sustainable Forestry* 33: S28–S42.

Morin, R. S. Jr., A. M. Liebhold, and K. W. Gottschalk. 2004. Area-wide analysis of hardwood defoliator effects on tree conditions in the Allegheny Plateau. *Northern Journal of Applied Forestry* 21 (1): 31–39.

Morin, R. S., A. M. Liebhold, E. R. Luzader, A. J. Lister, K. W. Gottschalk, and D. B. Twardus. 2005. Mapping host-species abundance of three major exotic forest pests. Res. Pap. NE-726: 11 pp. Newtown Square, PA: USDA Forest Service, Northeastern Research Station.

Morris, R. F. 1963. The dynamics of epidemic spruce budworm populations. *Memoirs of the Entomological Society of Canada* 95 (S31): 1–12.

Muller, J., H. Bussler, M. Gossner, T. Rettelbach, and P. Duelli. 2008. The European spruce bark beetle *Ips typographus* in a national park: From pest to keystone species. *Biodiversity and Conservation* 17: 2979–3001.

Muzika, R. M. and A. M. Liebhold. 1999. Changes in radial increment of host and nonhost tree species with gypsy moth defoliation. *Canadian Journal of Forest Research* 29 (9): 1365–1373.

Nappi, A., S. Dery, F. Bujold, M. Chabot, M.-C. Dumont, J. Duval et al. 2011. *La récolte dans les forêts brulées—Enjeux et orientations pour un aménagement écosystémique*: 51 pp. Québec, QC: Ministère des ressources naturelles du Québec.

Nealis, V. G. and J. Régnière. 2004. Insect–host relationships influencing disturbance by the spruce budworm in a boreal mixedwood forest. *Canadian Journal of Forest Research* 34: 1870–1882.

Niciforuk, A. 2011. *Empire of the Beetle: How Human Folly and a Tiny Bug Are Killing North America's Great Forests*: 230 pp. Vancouver, BC: Greystone Books.

Nishio, G. 2010. *Harvesting Mountain Pine Beetle-Killed Pine While Protecting the Secondary Structure: A Comparison of Partial Harvesting and Clearcutting Methods*: 12 pp. Vancouver, BC: FP Innovations, FERIC Division.

Norvez, O., C. Hébert, and L. Bélanger. 2013. Impact of salvage logging on stand structure and beetle diversity in boreal balsam fir forest, 20 years after a spruce budworm outbreak. *Forest Ecology and Management* 302: 122–132.

Ohmart, C. P. and P. B. Edwards. 1991. Insect herbivory on eucalyptus. *Annual Review of Entomology* 36 (1): 637–657.

Oliver, C. D., K. L. O'Hara, and P. J. Baker. 2015. Effects of restoration at the stand level. In: Stanturf, J. A. (Ed.) *Restoration of Boreal and Temperate Forests*, 2nd. ed., pp. 37–68, Boca Raton: CRC Press.

Osawa, A. 1994. Seedling responses to forest canopy disturbance following a spruce budworm outbreak in Maine. *Canadian Journal of Forest Research* 24: 850–859.

Page, W. G., M. E. Alexander, and M. J. Jenkins. 2013. Wildfire's resistance to control in mountain pine beetle-attacked lodgepole pine forests. *The Forestry Chronicle* 89: 783–794.

Parry, M. L., O. F. Canziani, J. P. Palutikof, P. J. van der Linden, and C. E. Hanson (eds). 2007. *Contribution of Working Group II to the Fourth Assessment Report of the Intergovernmental Panel on Climate Change.* Cambridge, UK: Cambridge University Press.

Perrakis, D. D. B, R. A. Lanoville, S. W. Taylor, and D. Hicks. 2014. Modeling wildfire spread in mountain pine beetle-affected forest stands, British Columbia, Canada. *Fire Ecology* 10 (2): 10–35. http://fireecologyjournal.org/docs/Journal/pdf/Volume10/Issue02/010.pdf.

Poland, T. M. and D. G. McCullough. 2006. Emerald ash borer: Invasion of the urban forest and the threat to North America's ash resource. *Journal of Forestry* 104 (3): 118–123.

Prach, K. and R. J. Hobbs. 2008. Spontaneous succession versus technical reclamation in the restoration of disturbed sites. *Restoration Ecology* 16: 363–366.

Preisser, E. L., A. G. Lodge, D. A. Orwig, and J. S. Elkinton. 2008. Range expansion and population dynamics of co-occurring invasive herbivores. *Biological Invasions* 10: 201–213.

Prestemon, J. P., K. L. Abt, K. M. Potter, and F. H. Koch. 2013. An economic assessment of mountain pine beetle timber salvage in the West. *Western Journal of Applied Forestry* 28 (4): 143–153.

Puettmann K. J., K. D. Coates, and C. Messier. 2009. *A Critique of Silviculture: Managing for Complexity*: 189 pp. Washington, DC: Island Press.

Pugh, S. A., A. M. Liebhold, and R. S. Morin. 2011. Changes in ash tree demography associated with emerald ash borer invasion, indicated by regional forest inventory data from the Great Lakes. *Canadian Journal of Forest Research* 41: 2165–2175.

Raffa, K. F., B. H. Aukema, B. J. Bentz, A. L. Carroll, J. A. Hicke, M. G. Turner et al. 2008. Cross-scale drivers of natural disturbances prone to anthropogenic amplification: The dynamics of bark beetle eruptions. *BioScience* 58: 501–517.

Redding, D., R. Winkler, P. Teti, D. Spittlehouse, S. Boon, J. Rex et al. 2008. Mountain pine beetle and watershed hydrology. *Journal of Ecosystems and Management* 9 (3): 33–50. http://jem.forrex.org/index.php/jem/article/view/402/317.

Reinikainen, M., A. W. D'Amato, and S. Fraver. 2012. Repeated insect outbreaks promote multicohort aspen mixedwood forests in northern Minnesota, USA. *Forest Ecology and Management* 266: 148–159.

Roland, J. and P. D. Taylor. 1997. Insect parasitoid species respond to forest structure at different spatial scales. *Nature* 386 (6626): 710–713.

Romme, W. H., J. Clement, J. Hicke, D. Kulakowski, L. H. MacDonald, T. L. Schoennagel et al. 2006. *Recent Forest Insect Outbreaks and Fire Risk in Colorado forests: A Brief Synthesis of Relevant Research*: 24 pp. Fort Collins, CO: Colorado Forest Restoration Institute. http://www.colorado.edu/geography/courses/geog_1001_lab/Webpage_material/Left%20Navigator_subjects/Labs/lab_material/Reading_for_Lab_K.pdf, accessed November 25, 2014.

Royo, A. A. and W. P. Carson. 2006. On the formation of dense understory layers in forests worldwide: Consequences and implications for forest dynamics, biodiversity, and succession. *Canadian Journal of Forest Research* 36: 1345–1362.

Safranyik, L. and A. L. Carroll. 2006. The biology and epidemiology of the mountain pine beetle in lodgepole pine forests. In: Safranyik, L. and B. Wilson (eds). *The Mountain Pine Beetle: A Synthesis of Biology, Management, and Impacts on Lodgepole Pine*: pp. 3–66. Victoria, BC: Canadian Forest Service, Pacific Forestry Centre.

Scheffer, M., S. Carpenter, J. A. Foley, C. Folke, and B. Walker, B. 2001. Catastrophic shifts in ecosystems. *Nature* 413 (6856): 591–596.

Simard, M., W. H. Romme, J. M. Griffin, and M. G. Turner. 2011. Do mountain pine beetle outbreaks change the probability of active crown fire in lodgepole pine forests? *Ecological Monographs* 81 (1): 3–24.

Society for Ecological Restoration International (SERI). 2004. *The SER International Primer on Ecological Restoration, Version 2.* Washington, DC: Society for Ecological Restoration International,

Science & Policy Working Group. http://www.ser.org/docs/default-document-library/english.pdf, viewed November 25, 2014.

Spence, C. E. and D. A. MacLean. 2012. Regeneration and stand development following a spruce budworm outbreak, spruce budworm inspired harvest, and salvage harvest. *Canadian Journal of Forest Research* 42 (10): 1759–1770.

Stanturf, J. A. 2015. What is forest restoration? In: *Restoration of Boreal and Temperate Forests*, 2nd. ed., pp. 1–16, Boca Raton: CRC Press.

Stanturf, J. A., B. J. Palik, M. I. Williams, R. K. Dumroese, and P. Madsen. 2014. Forest restoration paradigms. *Journal of Sustainable Forestry* 33: S161–S194.

Svoboda, M., S. Fraver, P. Janda, R. Bace, and J. Zenahlikova. 2010. Natural development and regeneration of a Central European montane spruce forest. *Forest Ecology and Management* 260: 707–714.

Svoboda, M., P. Janda, T. A. Nagel, S. Fraver, J. Rejzek, and R. Bače. 2012. Disturbance history of an old-growth sub-alpine *Picea abies* stand in the Bohemian Forest, Czech Republic. *Journal of Vegetation Science* 23 (1): 86–97.

Swank, W. T., L. W. Swift, and J. E. Douglass. 1988. Streamflow changes associated with forest cutting, species conversions, and natural disturbances. In: Swank, W. T. and D. A. Crossley (eds). *Forest Hydrology and Ecology at Coweeta*: pp. 297–312. New York, NY: Springer.

Taylor, S. W. and A. L. Carroll. 2004. Disturbance, forest age, and mountain pine beetle outbreak dynamics in BC: A historical perspective. In: Shore, T. L., J. E. Brooks, and J. E. Stone (eds). *Proceedings of the Mountain Pine Beetle Symposium: Challenges and Solutions*, Kelowna, British Columbia, Canada, 30–31 October 2003: pp. 41–51. Info. Rep. BC-X-399. Victoria, BC: Canadian Forest Service, Pacific Forestry Centre.

Tobin P. C. and L. M. Blackburn (eds). 2007. *Slow the Spread: A National Program to Manage the Gypsy Moth*. Gen. Tech. Rep. NRS-6: 109 pp. Newtown Square, PA: USDA Forest Service.

Venier, L. A., J. L. Pearce, D. R. Fillman, and D. K. McNicol. 2009. Effects of spruce budworm (*Choristoneura fumiferana* (Clem.)) outbreaks on boreal mixed-wood bird communities. *Avian Conservation and Ecology* 4 (1): 3. http://www.aceeco.org/vol4/iss1/art3/.

Volney, W. and R. A. Fleming. 2007. Spruce budworm (*Choristoneura* spp.) biotype reactions to forest and climate characteristics. *Global Change Biology* 13 (8): 1630–1643.

Vyse, A., C. Ferguson, D. J. Huggard, J. Roach, and B. Zimonick. 2009. Regeneration beneath lodgepole pine dominated stands attacked or threatened by the mountain pine beetle in the south central Interior, British Columbia. *Forest Ecology and Management* 258: S36–S43.

Walker, L. R., N. Hölzel, R. Marrs, R. del Moral, and K. Prach. 2014. Optimization of intervention levels in ecological succession. *Applied Vegetation Science* 17 (2): 187–192.

Weed, A. S., M. P. Ayres, and J. A. Hicke. 2013. Consequences of climate change for biotic disturbances in North American forests. *Ecological Monographs* 83 (4): 441–470.

Westfall, J. and T. Ebata. 2014. *2013 Summary of Forest Health Conditions in British Columbia*: 80 pp. Pest Manage. Rep. No. 15. Victoria, BC: BC Ministry of Forests, Lands and Natural Resource Operations, Forest Practices Branch. http://www.for.gov.bc.ca/hfp/health/overview/overview.htm, viewed November 25, 2014.

White, J. A. and T. G. Whitham. 2000. Associational susceptibility of cottonwood to a box elder herbivore. *Ecology* 81: 1795–1803.

Wotton, B. M., M. E. Alexander, and S. W. Taylor. 2009. *Updates and Revisions to the 1992 Canadian Forest Fire Behavior Prediction System*. Info. Rep. GLC-X-10: 42 pp. Sault Ste. Marie, ON: Natural Resources Canada, Canadian Forest Service. http://cfs.nrcan.gc.ca/pubwarehouse/pdfs/31414.pdf, viewed November 25, 2014.

Xi, W., J. D. Waldron, D. M. Cairns, C. W. Lafon, S. G. Birt, M. D. Tchakerian et al. 2013. Restoration of southern pine forests after the southern pine beetle. In: Stanturf, J., P. Madsen, and D. Lamb (eds). *A Goal-Oriented Approach to Forest Landscape Restoration*: pp. 321–354. Dordrecht, Germany: Springer.

25

Restoring Boreal and Temperate Forests: A Perspective

John A. Stanturf

CONTENTS

25.1 Introduction

Throughout the temperate and boreal zones, human intervention has influenced landscapes and forests for thousands of years (Ellis et al., 2013; Foley et al., 2005; Kareiva et al., 2007). The degree of human disturbance has only been constrained by the technology and resources available to different cultures and by time since initial habitation. For millennia, humans have influenced forests by regulating populations of ungulate browsers or keeping domestic livestock, clearing for agriculture, cutting trees for fuel, building material and fiber, introducing new species, using or suppressing fire (Kretch III, 1999; Sanderson et al., 2002). Today's forests are the result of all these disturbances, along with climatic change and species migration into postglacial landscapes (e.g., Bradshaw, 2015). The ability of humans to affect forest ecosystems increased dramatically after the Industrial Revolution. Engineering works, including mining, dams, and roads, are both more widespread and more intensive (Sanderson et al., 2002). Management has been extended to native forests over larger areas; at the same time, the switch from biomass to fossil fuels changed traditional forest management. Industrial, residential, and automotive emissions of combustion products affect forests directly and through their effects on climate.

The present is certainly challenging but global climate change presents even greater challenges (Kolström et al., 2011; Liu et al., 2010; Turner, 2010; Weed et al., 2013) and likely there will be an even greater need for restoration in the future (Steffen et al., 2007; Zalasiewicz et al., 2010). Climate change may degrade forests through increased variability and extreme events (Allen, 2009; 2010; Meehl et al., 2000, 2005; Reichstein et al., 2013) and altered mean conditions (Aitken et al., 2008; Liu et al., 2010, 2013; Weltzin et al., 2003; Williams and Jackson, 2007). The needs for restoration under climate change are fourfold: restoring already degraded lands in a manner that increases their adaptation to future climatic conditions, restoring forests catastrophically disturbed by climatic extreme events, reducing vulnerability of forests undergoing regeneration to altered climate ("normal silvicultural" adaptation to climate change (Bolte et al., 2009a,b), and the options of carbon sequestration and bioenergy for mitigating climate change.

25.2 Why Restore?

25.2.1 Vague Goals

Forest restoration is under way in order to counteract the negative effects of human activities (Lamb, 2010; Stanturf et al., 2012a,b, 2014a). Restoration is important for many interest groups in all countries, and in most countries occurs on both publicly and privately owned land. Nevertheless, the motivations for attempting forest restoration are diverse and often vague (Clewell and Aronson, 2006), reflecting the complexity of ecosystems, their current state and past land-use, and the human context of culture, economics, and governance (Stanturf et al., 2014b). Restoration goals vary, but in most countries, restoration is undertaken within the policy framework of increasing sustainability by enlarging the area of specific ecosystems (Thorpe and Stanley, 2011), enhancing biodiversity (Seabrook et al., 2011), or repairing ecosystem functions (Stanturf et al., 2014a,b). The most common specific objectives for restored forests include timber, wildlife habitat for game species, or aesthetics. Increasingly, other objectives are considered, including carbon sequestration, biological diversity, nongame mammals and birds, endangered animals and plants, protection of water quality and aquatic resources, and recreation (Ciccarese et al., 2012).

Afforestation is an important component of forest restoration, particularly in Europe and the US. Forest land area has expanded in Europe in the twentieth century through afforestation and natural recolonization (Balandier and Prévosto, 2015; Jögiste et al., 2015; Madsen et al., 2015; Renou-Wilson and Byrne, 2015; Weber and Liebel, 2015), driven initially by the need of the war industry and the fear of a wood shortage and recently as a way to address problems of agricultural overproduction and rural unemployment. Increasingly, afforestation is undertaken for ecological and amenity reasons (Gardiner and Oliver, 2005). Perhaps the most important lesson learned from the European experience is that not everyone welcomes afforestation, even when programs are voluntary. In predominantly agricultural areas, planting trees may be seen as an assault on rural culture and traditional landscapes. Certainly, there has been a backlash in many countries to the former afforestation practice of planting a few conifer species (Balandier and Prévosto, 2015; Harmer et al., 2015;

Madsen et al., 2015; Lee et al., 2015). In Japan (Nagaike, 2015), the economics of timber production have changed in favor of importing wood, rather than pay the high labor costs for managing native conifer plantations or coppiced stands of mostly oak species.

In the Republic of Korea, restoration goals have evolved in the last half of the twentieth century following wartime turmoil and a shift from biomass to fossil fuels. Early restoration goals focused on afforestation and reforestation, usually with fast-growing nonnative species; this shifted toward timber production and the establishment of multifunctional forests with current goals emphasizing native species and sustainability (Lee et al., 2015). In Denmark, the primary driver has been changing agricultural policy (Madsen et al., 2015); in the Baltic States, more sweeping changes in governance from a centrally planned economy under the former communist regime to a more open, market-oriented economy has led to the need for alternatives to farming for large areas of land (Jögiste et al., 2015).

25.2.2 Naturalness Paradigm

A pervasive construct in the restoration literature is to restore to more "natural" conditions in terms of native species, complex stand structure, and natural disturbance regimes. The defining characteristics of naturalness are lack of major human interference for all or most of the lifespan of the oldest trees, native species composition, complex vegetative structures, and historical fidelity in disturbance regimes and their proportion in the landscape (Frelich and Reich, 2003; Hunter, 1993).

25.2.2.1 Native Species

A recurring theme in the restoration literature is returning forests to more natural conditions (Burton and Macdonald, 2011; Hallett et al., 2013). Native species are emphasized, although there can be compelling reasons for using nonnative species, at least as the initial intervention (Lee et al., 2015; Madsen et al., 2015; Stanturf et al., 2014a). Even native species may be discriminated against if they are not site-adapted, which has motivated conversion of Norway spruce in Europe (Hansen and Spiecker, 2015) or other pines to longleaf pine in the southern US (Brockway et al., 2015). With the shift in emphasis from timber to ecological services in afforestation programs, even site-adapted native species may be deemphasized for other species, especially to develop mixed-species stands (Balandier and Prévosto, 2015; Harmer et al., 2015; Madsen et al., 2015).

Current definitions of native species are vague (Smith and Winslow, 2001) but two aspects are common: historical presence and locality, in the sense of native range (Davis et al., 2011). Climate change is likely to cause nativity to undergo redefinition as climatic extremes cause extirpation of local populations, thereby validating assisted migration (Shackelford et al., 2013; Williams and Dumroese, 2013).

25.2.2.2 Natural Disturbance Regimes

Another aspect of the naturalness paradigm that crosses continental as well as national boundaries is restoration of natural disturbance regimes (Attiwill, 1994; Stanturf et al., 2014a; Turner, 2010). Besides the obvious link of disturbance to the need for restoration (Stanturf, 2015; Stanturf et al., 2014a), the dynamic nature of forest ecosystems (Oliver et al., 2015) shows that even without anthropogenic disturbances, it is very difficult to say what would be a natural forest in a given place and time. In addition to the effects of climate change on geological time scales (Bradshaw, 2015; Millar, 2014), examples of changes in forest composition and structure due to human intervention in disturbance regimes abound. In particular, reversing the negative effects of fire suppression and altered hydroperiod are important topics in restoration (Agee, 2002; Hughes et al., 2012; Nagy and Lockaby, 2012; Stanturf et al., 2014a).

Restoring fire regimes may include altering stand structure and species composition (Beatty and Owen, 2005; Brockway et al., 2015; Jain and Graham, 2015; Van Lear and Wurtz, 2005). Recreating historic forest types, structures, and fire regimes dominate thinking about forest restoration in the western US and boreal Europe (Kaufmann et al., 2005; Kuuluvainen et al., 2015; Phillips et al., 2012). Although logging, grazing, and fire suppression have heavily impacted these forests, ample remnants of relatively undisturbed conditions remain. Alternatively, such "natural" forests can be discerned from the historical record or through stand reconstruction techniques such as dendrochronology.

Pests outbreaks, particularly insect borers, are degrading agents that engender a restoration response (e.g., Burton et al., 2015; O'Hara and Waring, 2005; Waring and O'Hara, 2005; Xi et al., 2012). Climate change will affect native insects and may cause novel outbreaks, particularly if temperature increases alter insect phenology (Marini et al., 2012; Weed et al., 2013).

25.2.2.3 Complex Structure

Restoring structural heterogeneity in stands and across the landscape often is a component of a restoration program (Stanturf et al., 2014a). Defining a "natural" stand structure can be arbitrary (Oliver et al., 2015). The old steady-state paradigm of succession to a climax led many to equate complex structure with old forests (O'Hara, 1998; O'Hara and Waring, 2005). To some, this meant that uneven-aged management was preferred over even-aged management because it led to more "natural" forests. Given the luxury of large areas of contiguous forest under a single or a few owners, it is possible to restore to a diversity of stand structures on the landscape in roughly the same proportions as occurred historically with little human influence; that is to say, under mostly natural disturbance regimes. These conditions appear to exist in lightly populated regions such as boreal North America, Fennoscandia, and Siberia (Gower et al., 2015; Kuuluvainen et al., 2015; Van Lear and Wurtz, 2005). In more populated regions, however, nature-based silviculture that emphasizes restoring complex stand structures should stress the positive aspects of structural diversity in terms of stand stability (Hahn et al., 2005; Hansen and Spiecker, 2015).

25.2.2.4 Limits of the Naturalness Paradigm

The notion of what are natural forests is a social construct (Emborg et al., 2012), defined in terms of social values (Cole and Yung, 2010), which presents difficulties for using the concept to drive restoration programs. Where remnants of the putative natural forest remain, or can be reconstructed from the historical record, the question of what is natural is still hotly debated (e.g., Bradshaw, 2015; Kuuluvainen et al., 2015). Examples certainly exist of forests where extensive human alteration is absent or historically recent (e.g., Matuszkiewicz et al., 2013). Nevertheless, even these forests have been altered (Bradshaw, 2015; Ellis et al., 2013). It would be easy to conclude that there is no scientific basis for restoring natural forests because humans and climate change have so drastically changed the whole biosphere. Indeed, some restorationists regard the choice of endpoint as inherently political, not scientific (Simula and Mansur, 2011). Nevertheless, such choices are necessary in a restoration program and the rationale for the choice must be conveyed to the public (Daniels et al., 2012). The task for restorationists is to interpret the scattered scientifically based knowledge of forest history, stand development, and natural processes, and combine it with practical experience to design objectives that improve sustainability.

Generally, restoration connotes some transition from a degraded state to a more "natural" condition (Stanturf, 2015). In the narrowest sense, restoration requires a return to an ideal natural or historic ecosystem with the same species diversity, composition, and structure as occurred before human intervention (SERI, 2004). Because this ideal state is probably impossible to attain (Hobbs, 2013; Hobbs et al., 2011), a more feasible approach is to regard forest restoration broadly, applying it to situations where forested land-use and land cover are restored (reconstruction or reclamation), as well as instances when an existing forest is rehabilitated (no change in land cover). Broadly defining restoration to include the diverse approaches described in this volume allows for diverse goals and greater participation by more landowners than narrowly defined historic conditions (Stanturf, 2015). This pragmatic approach will also better meet future restoration needs (Choi, 2007).

25.2.3 Defined Expectations

Well-defined expectations, or measures of success, are critical to restoration practice and their lack often leads to failure (Dey and Schweitzer, 2014; Kapos et al., 2008). Expectations are the prediction of the postrestoration state and the mechanism for change from the degraded state or baseline (Toth and Anderson, 1998). Expectations express the causal mechanism for change and the trajectory of moving from the baseline to the endpoint (Stanturf, 2015). Progress can be measured by indicators of various ecosystem components, although some components change faster than others (Stanturf et al., 2014a) and intermediate conditions must be considered in evaluating success (Oliver et al., 2015).

A frequent assumption is that restoring desired stand structure and, if necessary, species composition will restore ecological functions. This is certainly true for afforestation; a tree plantation functions more like a forest than a cotton field. Nevertheless, even in gross terms, a plantation does not offer the same degree of functioning as a multispecies forest with complex structure but how much they differ depends on many factors (Brockerhoff et al., 2008). Restoration of functions such as biogeochemical processes is even more difficult to evaluate than altered structure or composition, because these processes operate at multiple temporal and spatial scales. By focusing on emergent properties of ecosystem processes, such as nutrient budgets, the task is more manageable. Because pools and processes that depend primarily on biological agents may vary from those dependent upon physical agents, responses will be variable and evaluation of restoration success will depend upon the choice of key indicators (Dey and Schweitzer, 2014; Vose et al., 2005).

25.3 Social Context

The social context for restoration varies between countries and the legacy of the past forest management and land-use practices determine the major restoration needs and influence what is feasible. Restoration is not a well-developed concept in all countries and motivations are as varied as the climate. In general, the motivation for restoration is to produce or enhance ecological services, which are usually public goods or externalities. Encouraging restoration involves activity on public land and controlling or encouraging activity on private land through regulations and incentives. Because no markets exist for most of the ecological outcomes from restoration, policymakers have no easy criteria for deciding optimum levels of restoration (Pullar and Lamb, 2012; Wilson et al., 2011). There is a further spatial dimension to restoration that complicates allocation decisions. For example, within a degraded watershed, restoring some areas will produce a higher level of ecological benefit than others, given the same level of restoration effort (Allan et al., 2008; Mercer, 2005). The best combination of policies will use market solutions, government intervention, and combinations (Wilson et al., 2012). When government intervenes, policymakers have a range of carrots (incentives) and sticks (regulation) at their disposal. Different approaches will be more feasible in some countries than others but, in general, the approach that seems the easiest, concentrating restoration on an expanded public land base, is the least feasible (Lamb, 2010; Lamb et al., 2012; Mercer, 2005). Targeted voluntary and mandatory programs, combined with community-based participatory approaches, have the greatest likelihood of successfully providing long-term benefits. Such approaches will be complicated, costly, and will challenge our understanding of social dynamics (Emborg et al., 2012; Mercer, 2005).

One such targeted voluntary approach is payments for ecosystem services; the most developed program is for increased carbon sequestration (Ciccarese et al., 2012). Forests are both sources and sinks for atmospheric carbon, and forestry and land-use change activities can mitigate carbon emissions. Restoration of watershed services, and the willingness of down-stream water users such as irrigators to pay, is another potential source of funding (Mueller et al., 2013). These alternatives to taxation-derived public funding, added to financial returns from market goods such as timber, edible, or medicinal products, could expand the appeal of forest restoration to private landowners (Harper et al., 2012; Pejchar and Press, 2006; Townsend et al., 2012).

25.4 Key Issues in Practice

The key issues in forest restoration practice are obtaining appropriate plant materials, optimizing the biophysical environment, taking the necessary steps beyond the initial phase, and increasing diversity at the landscape-scale (Stanturf et al., 2014a).

25.4.1 Appropriate Material

Appropriate genetic material for restoration must be decided within a national context. In some countries, restoration may include nonnative species, but in all countries appropriate provenances will be critical for short-term success (seedling survival and stocking density) and for long-term productivity, resistance to pests, and resilience following disturbance (Kjær et al., 2005; Stanturf et al., 2014a). Forest restoration usually involves significant, if not total, change of species composition. The restoration process provides a unique opportunity to obtain genetically high-quality material to support multifunctional goals. This is most clearly seen when trees are planted, but genetics, in terms of plant adaptations to site conditions and disturbance regimes, is an underlying factor in all forest restoration.

Plant materials for restoration may require different characteristics than commercially produced material for reforestation (Schröder and Prasse, 2013). Desirable attributes for restoration may include precocious flowering or the ability to re-sprout after fire, rather than good form or wood quality (Stanturf et al., 2014a). Nevertheless, the same propagation methods used in nursery practice can produce quality material for restoration plantings. The need for low-cost methods increasingly is being met by use of natural regeneration or direct seedling, rather than planting bare root or container seedlings (e.g., Fischer et al., 2015). Active methods provide greater control of species composition and stocking levels than passive methods that rely on natural dispersal and recolonization (Stanturf et al., 2014a). For active methods to be successful, however, vigorous, site-adapted material must be properly planted or sown within the outplanting window when soil moisture and temperature are favorable for establishment.

25.4.2 Biophysical Environment

Most restoration efforts include a regeneration phase and choices made at this stage have far-reaching effects on restoration success. The microsite environment of the newly established propagules may result from intervention such as site preparation to enhance the rooting zone of planted seedlings (Löf et al., 2015), or may receive no manipulation such as passive restoration from natural recolonization (Balandier and Prévosto, 2015) or natural regeneration (Fischer et al., 2015). Nevertheless, the principles of how life cycle stage, site, and silvicultural operations interact are universally applicable (Stanturf et al., 2014a). Understanding requirements of new plant materials and available microsite conditions will help restorationists specify needed intervention treatments. Restoration will have to proceed with very imperfect knowledge and researchers must attempt continuously to create a better foundation of documented knowledge to support restoration decisions.

Rehabilitation of degraded stands has been the traditional focus of restoration; even when entire landscapes are being restored, prescriptions focus on stand-sized patches. The shift from viewing forest ecosystems as closed, steady-state systems with predictable development patterns to the present view of open systems that operate opportunistically, with multiple developmental pathways following disturbance has been critical to identifying appropriate operations for restoration (Oliver et al., 2015). To be successful, restoration efforts must adopt this dynamic view of forests. To be effective, restorationists will have to educate the nontechnical restoration enthusiasts as well as the general public in this new paradigm (Stanturf et al., 2014b). Restoring structural and age diversity to forests that have been homogenized by intensive management or degraded by extreme events should increase biodiversity and render these forests better adapted to altered climate (Millar et al., 2007). Large woody debris provides important habitat and is usually depleted in managed forests; restoring deadwood structures is another important goal in rehabilitation of degraded stands (Jonsson et al., 2005). Diversity of structures across the landscape is also important (Oliver et al., 2012); early successional structures are under-represented in some landscapes (Greenberg et al., 2011).

25.4.3 Multiple Interventions

A weakness of many restoration efforts is that although success ultimately depends on a series of interventions, only the short-term operations are addressed in detail. Because restoration requires ongoing management, the silviculturist plays an important role in bringing to bear time-tested techniques to shape the development trajectory of a stand toward the desired condition, and to maintain the restored ecosystem (Oliver et al., 2015; Stanturf et al., 2014a).

To be successful, restoration requires sequential intervention and ongoing management. Appropriate silvicultural operations can be designed for any forest restoration objective. Nevertheless, forest restoration requires creativity and flexibility in applying silvicultural tools. In the initial restoration phase, decisions must be made on whether to add or remove species, by active or passive means. The desired species density and spatial arrangement to meet restoration objectives determines cultural methods, along with the amount of over-story present (i.e., no, partial, or full) and how much of the area is to be treated (Stanturf et al., 2014a). For example, partial removals of overstory or mid-story trees may be needed to develop or release advance regeneration and increase species diversity. Transforming simple to complex structures requires multiple entries into stands over an extended time period (Nyland, 2003; Pommerening, 2006).

Forests that are blown down by severe winds, or killed by insect and disease outbreaks, frequently are salvage logged, because of their economic value. These extreme events provide opportunities to change species or structure over large areas that otherwise would not be feasible because of cost or public opposition. For example, damage from windstorms, faltering timber economics, and changes in public perception have contributed to the present desire across Europe to change species composition of existing stands. The predominance of conifer monocultures in Europe reflects great variation in historical development in these countries; nevertheless, the common theme is that high productivity and value of products from conifers such as Norway spruce, and the ease of their establishment, caused them to be planted even on sites where they were not adapted (e.g., Balandier and Prévosto, 2015; Hansen and Spiecker, 2015).

Retention of some live trees during harvest and even creation of deadwood structures by damaging residual trees may be desirable to further restoration objectives of increasing diversity (e.g., Lilja, 2005). Creating structural diversity in intact stands using variable density thinning has gained increasing use (e.g., Ribe et al., 2013; Vanha-Majamaa and Jalonen, 2001) but may not meet some restoration objectives (see Kuuluvainen et al., 2015). Methods of reducing overstory trees, such as creating artificial gaps, can be combined with underplanting to increase species diversity (Dey et al., 2012; Paquette et al., 2006).

25.4.4 Scale and Diversity

Landscape-level restoration has gained increased emphasis, in part due to the desire to incorporate ecosystem processes (Stanturf et al., 2014a), landscape-level concerns such as biodiversity and more natural disturbance regimes (Lindenmayer et al., 2008), and the daunting amount of degraded forest land (Minnemayer et al., 2011). Many of the treatments designed for restoring individual stands can be scaled-up to the landscape; planting designs such as those for creating complex mixtures can be spread over the entire landscape or different patches planted in simple or complex designs (Stanturf et al., 2014a). Methods for transforming stands to more complex structure (Pommerening, 2006), variable retention harvests (Gustafsson et al., 2012), and underplanting (Dey et al., 2012; Paquette et al., 2006) can be applied in various configurations with proper regard to site adaptation. The goal is to create a diversity of vegetation types and ages across the landscape (Lamb et al., 2012). Creating asynchrony of stand ages and developmental stages is critical to long-term sustainability, especially in large areas that have been damaged in a natural disaster (Millar et al., 2007). Temporal diversity can be created by staggering planting or thinning, density manipulation, or interplanting (Stanturf et al., 2014a).

25.5 Adaptation to Future Environments

Restoration focusing on the historic conditions of the past is increasingly difficult, if indeed, it is even possible given the rapidly changing present and anticipated future (Aitken et al., 2008; Diffenbaugh and Field, 2013; Hiers et al., 2012; Park et al., 2014; Turner, 2010). The daunting challenge is how to restore under great uncertainty and rapid change. Several incremental approaches are available (Kates et al., 2012).

- Pursue endpoints based on the best available understanding of contemporary conditions, that is, reference points or stands (Fulé 2008; Stephens et al., 2010). This approach would focus on using native species of local sources and strive to reduce current stressors.
- Do so in a way that facilitates adaptation to new conditions. For example, managing stem density at low levels of stocking (but within the natural range of variability), is thought to be adaptive to drought that may occur infrequently now and become more frequent in a hotter, drier future climate (Kohler et al., 2010; Linder, 2000).
- Restore diversity of composition, structure, and function to simplified, production-oriented stands. This diversity provides the flexibility to intervene and shift development in better adapted directions (Yemshanov et al., 2013).

These incremental approaches exemplify "no regrets" adaptation to climate change (Heltberg et al., 2009), in that the results of taking these actions are beneficial regardless of whether global change alters the future environment of the current stands. More anticipatory or even transformative approaches (Kates et al., 2012) include accepting or intentionally using nonnative species that are functional equivalents to natives (Davis et al., 2011), employing assisted migration of not only populations but also to extend ranges or translocate species outside their native range (Stanturf et al., 2014a; Williams and Dumroese, 2013). The objective in all cases is to reduce vulnerability to current and future stressors (Pielke et al., 2013). Truly transformative approaches have been advocated, such as intervention ecology to replace ecological restoration (Hobbs et al., 2011), using biotechnology to create transgenic species to replace extinct keystone species or genotypes better adapted to future climate (Jacobs et al., 2013; Strauss and Bradshaw, 2004), or synthetic biology to create designer organisms with heretofore unknown capabilities (Rautner, 2001). Whether these novel (emergent or no-analog) ecosystems arise from human intervention or through natural processes in response to altered climate, dynamic and pragmatic approaches to management will be required.

References

Agee JK. The fallacy of passive management managing for firesafe forest reserves. *Conservation in Practice* 2002, 3:18–26.

Aitken SN, Yeaman S, Holliday JA, Wang T, Curtis-McLane S. Adaptation, migration or extirpation: Climate change outcomes for tree populations. *Evolutionary Applications* 2008, 1:95–111.

Allan C, Curtis A, Stankey G, Shindler B. Adaptive management and watersheds: A social science perspective. *Journal of the American Water Resources Association* 2008, 44:166–174.

Allen C. Climate-induced forest dieback: An escalating global phenomenon. *Unasylva* 2009, 231:60.

Allen CD et al. A global overview of drought and heat-induced tree mortality reveals emerging climate change risks for forests. *Forest Ecology and Management* 2010, 259:660–684.

Attiwill PM. The disturbance of forest ecosystems: The ecological basis for conservative management. *Forest Ecology and Management* 1994, 63:247–300.

Balandier, P, Prévosto, B. Forest restoration in the French Massif Central Mountains. In: *Restoration of Boreal and Temperate Forests*, 2nd edition. Stanturf, JA, ed. 2015: Boca Raton: CRC Press. Chapter 16, 337–354.

Beatty S, Owen B. Incorporating disturbance into forest restoration. In: *Restoration of Boreal and Temperate Forests*. Stanturf J, Madsen P, eds. 2005: Boca Raton, FL: CRC Press. 61–76.

Bolte A et al. Adaptive forest management in central Europe: Climate change impacts, strategies and integrative concept. *Scandinavian Journal of Forest Research* 2009a, 24:473–482.

Bolte A, Ammer C, Löf M, Nabuurs G-J, Schall P, Spathelf P. Adaptive forest management: A prerequisite for sustainable forestry in the face of climate change. In: *Sustainable Forest Management in a Changing World*. Spathelf P, ed. 2009b: Netherlands: Springer. 115–139.

Bradshaw, RHW. What is a natural forest? In: *Restoration of Boreal and Temperate Forests*, 2nd edition. Stanturf JA, ed. 2015: Boca Raton: CRC Press. Chapter 2, 17–68.

Brockerhoff EG, Jactel H, Parrotta JA, Quine CP, Sayer J. Plantation forests and biodiversity: Oxymoron or opportunity? *Biodiversity and Conservation* 2008, 17:925–951.

Brockway DG, Outcalt KW, Tomczak DJ, Johnson EE. Restoring longleaf pine forest ecosystems in the southern United States. In: *Restoration of Boreal and Temperate Forests*, 2nd edition. Stanturf JA, ed. 2015: Boca Raton: CRC Press. Chapter 22, 445–466.

Burton PJ, Macdonald SE. The restorative imperative: Challenges, objectives and approaches to restoring naturalness in forests. *Silva Fennica* 2011, 45:843–863.

Burton PJ, Svoboda M, Kneeshaw D, Gottschalk KW. Options for promoting the recovery and rehabilitation of forests affected by severe insect outbreaks. In: *Restoration of Boreal and Temperate Forests*, 2nd edition. Stanturf JA, ed. 2015: Boca Raton: CRC Press. Chapter 24, 495–517.

Choi YD. Restoration ecology to the future: A call for new paradigm. *Restoration Ecology* 2007, 15:351–353.

Ciccarese L, Mattsson A, Pettenella D. Ecosystem services from forest restoration: Thinking ahead. *New Forests* 2012, 43:543–560.

Clewell AF, Aronson J. Motivations for the restoration of ecosystems. *Conservation Biology* 2006, 20:420–428.

Cole DN, Yung L. *Beyond Naturalness: Rethinking Park and Wilderness Stewardship in an Era of Rapid Change*. 2010: Washington, DC: Island Press.

Daniels SE, Walker GB, Emborg J. The unifying negotiation framework: A model of policy discourse. *Conflict Resolution Quarterly* 2012, 30:3–31.

Davis MA et al. Don't judge species on their origins. *Nature* 2011, 474:153–154.

Dey DC, Gardiner ES, Schweitzer CJ, Kabrick JM, Jacobs DF. Underplanting to sustain future stocking of oak (*Quercus*) in temperate deciduous forests. *New Forests* 2012, 43:955–978.

Dey DC, Schweitzer CJ. Restoration for the future: Endpoints, targets, and indicators of progress and success. *Journal of Sustainable Forestry* 2014, 33:S43–S65.

Diffenbaugh NS, Field CB. Changes in ecologically critical terrestrial climate conditions. *Science* 2013, 341:486–492.

Ellis EC, Kaplan JO, Fuller DQ, Vavrus S, Goldewijk KK, Verburg PH. Used planet: A global history. *Proceedings of the National Academy of Sciences* 2013, 110:7978–7985.

Emborg J, Walker G, Daniels S. Forest landscape restoration decision-making and conflict management: Applying discourse-based approaches. In: *Forest Landscape Restoration*. Stanturf J, Lamb D, Madsen P, eds. 2012: Dordrecht: Springer. 131–153.

Fischer H, Huth F, Hagemann U, Wagner S. Developing restoration strategies for temperate forests using natural regeneration processes. In: *Restoration of Boreal and Temperate Forests*, 2nd edition. Stanturf JA, ed. 2015: Boca Raton: CRC Press. Chapter 6, 103–164.

Foley JA et al. Global consequences of land use. *Science* 2005, 309:570–574.

Frelich LE, Reich PB. Perspectives on development of definitions and values related to old-growth forests. *Environmental Reviews* 2003, 11:S9–S22.

Fulé PZ. Does it make sense to restore wildland fire in changing climate? *Restoration Ecology* 2008, 16:526–531.

Gardiner ES, Oliver JM. Restoration of bottomland hardwood forests in the Lower Mississippi Alluvial Valley, USA. In: *Restoration of Boreal and Temperate Forests*. Stanturf J, Madsen P, eds. 2005: Boca Raton, FL: CRC Press. 235–251.

Gower TL, Burton PJ, Fenger M. Integrating forest restoration into mainstream land management in British Columbia, Canada. In: *Restoration of Boreal and Temperate Forests*, 2nd edition. Stanturf JA, ed. 2015: Boca Raton: CRC Press. Chapter 13, 271–297.

Greenberg C, Collins B, Thompson F. *Sustaining Young Forest Communities*. 2011: Berlin: Springer.

Gustafsson L et al. Retention forestry to maintain multifunctional forests: A world perspective. *BioScience* 2012, 62:633–645.

Hahn K, Emborg J, Larsen J, Madsen P. Forest rehabilitation in Denmark using nature-based forestry. In: *Restoration of Boreal and Temperate Forests*. Stanturf J, Madsen P, eds. 2005: Boca Raton, FL: CRC Press. 299–317.

Hallett LM et al. Do we practice what we preach? Goal setting for ecological restoration. *Restoration Ecology* 2013, 21:312–319.

Hansen J, Spiecker H. Conversion of Norway spruce (*Picea abies* [L.] Karst.) forests in Europe. In: *Restoration of Boreal and Temperate Forests*, 2nd edition. Stanturf JA, ed. 2015: Boca Raton: CRC Press. Chapter 17, 355–364.

Harmer R, Watts K, Ray D. A hundred years of woodland restoration in Great Britain: Changes in the drivers that influenced the increase in woodland cover. In: *Restoration of Boreal and Temperate Forests*, 2nd edition. Stanturf JA, ed. 2015: Boca Raton: CRC Press. Chapter 14, 299–320.

Harper R, Smettem K, Townsend P, Bartle J, McGrath J. Broad-scale restoration of landscape function with timber, carbon and water investment. In: *Forest Landscape Restoration*, Stanturf JS, Lamb D, Madsen P, eds. 2012: Berlin: Springer. 275–292.

Heltberg R, Siegel PB, Jorgensen SL. Addressing human vulnerability to climate change: Toward a "no-regrets" approach. *Global Environmental Change* 2009, 19:89–99.

Hiers JK et al. The dynamic reference concept: Measuring restoration success in a rapidly changing no-analogue future. *Ecological Restoration* 2012, 30:27–36.

Hobbs RJ. Grieving for the past and hoping for the future: Balancing polarizing perspectives in conservation and restoration. *Restoration Ecology* 2013, 21:145–148.

Hobbs RJ, Hallett LM, Ehrlich PR, Mooney HA. Intervention ecology: Applying ecological science in the twenty-first century. *BioScience* 2011, 61:442–450.

Hughes FM, del Tánago MG, Mountford JO. Restoring floodplain forests in Europe. In: *A Goal-Oriented Approach to Forest Landscape Restoration*. Stanturf J, Madsen P, Lamb D, eds. 2012: Dordrecht: Springer. 393–422.

Hunter ML. Natural fire regimes as spatial models for managing boreal forests. *Biological Conservation* 1993, 65:115–120.

Jacobs DF, Dalgleish HJ, Nelson CD. A conceptual framework for restoration of threatened plants: The effective model of American chestnut (*Castanea dentata*) reintroduction. *New Phytologist* 2013, 197:378–393.

Jain TB, Graham RT. Restoring dry and moist forests of the Inland Northwestern United States. In: *Restoration of Boreal and Temperate Forests*, 2nd edition. Stanturf JA, ed. 2015: Boca Raton: CRC Press. Chapter 23, 467–493.

Jõgiste K, Metslaid M, Uri V. Afforestation and land use dynamics in the Baltic States. In: *Restoration of Boreal and Temperate Forests*, 2nd edition. Stanturf JA, ed. 2015: Boca Raton: CRC Press. Chapter 8, 187–199.

Jonsson BG, Kruys N, Ranius T. Ecology of species living on dead wood—lessons for dead wood management. *Silva Fennica* 2005, 39:289–309.

Kapos V et al. Calibrating conservation: New tools for measuring success. *Conservation Letters* 2008, 1:155–164.

Kareiva P, Watts S, McDonald R, Boucher T. Domesticated nature: Shaping landscapes and ecosystems for human welfare. *Science* 2007, 316:1866–1869.

Kates RW, Travis WR, Wilbanks TJ. Transformational adaptation when incremental adaptations to climate change are insufficient. *Proceedings of the National Academy of Sciences* 2012, 109:7156–7161.

Kaufmann MR, Fulé PZ, Romme WH, Ryan KC. Restoration of ponderosa pine forests in the interior western US after logging, grazing, and fire suppression. In: *Restoration of Boreal and Temperate Forests*. Stanturf J, Madsen P, eds. 2005: Boca Raton, FL: CRC Press.

Kjær ED, Hansen CP, Roulund H, Graudal L. Procurement of plant material of good genetic quality. In: *Restoration of Boreal and Temperate Forests*. Stanturf J, Madsen P, eds. 2005: Boca Raton, FL: CRC Press. 139–152.

Kohler M, Sohn J, Nägele G, Bauhus J. Can drought tolerance of Norway spruce (*Picea abies* (L.) Karst.) be increased through thinning? *European Journal of Forest Research* 2010, 129:1109–1118.

Kolström M et al. Reviewing the science and implementation of climate change adaptation measures in European forestry. *Forests* 2011, 2:961–982.

Kretch III S. *Myth and History: The Ecological Indian.* 1999: New York: WW Norton.

Kuuluvainen T, Bergeron Y, Coates KD. Restoration and ecosystem-based management in the circumboreal forest: Background, challenges and opportunities. In: *Restoration of Boreal and Temperate Forests*, 2nd edition. Stanturf JA, ed. 2015: Boca Raton: CRC Press. Chapter 12, 251–270.

Lamb D. *Regreening the Bare Hills: Tropical Forest Restoration in the Asia-Pacific Region.* 2010: Berlin: Springer.

Lamb D, Stanturf J, Madsen P. What is forest landscape restoration? In: *Forest Landscape Restoration*. Stanturf J, Lamb D, Madsen P, eds. 2012: Dordrecht: Springer. 3–23.

Lee DK, Park PS, Park YD. Forest restoration and rehabilitation in the Republic of Korea. In: *Restoration of Boreal and Temperate Forests*, 2nd edition. Stanturf JA, ed. 2015: Boca Raton: CRC Press. Chapter 10, 217–231.

Lilja S. dCM, Kuuluvainen T., Vanha-Majamaa I., Puttonen P. Restoring natural characteristics in managed Norway spruce (*Picea abies*) stands with partial cutting, dead wood creation and fire: Immediate treatment effects. *Scandinavian Journal of Forest and Research.* 2005, 6:68–78.

Lindenmayer D et al. A checklist for ecological management of landscapes for conservation. *Ecology Letters* 2008, 11:78–91.

Linder M. Developing adaptive forest management strategies to cope with climate change. *Tree Physiology* 2000, 20:299–307.

Liu Y, Goodrick SL, Stanturf JA. Future US wildfire potential trends projected using a dynamically downscaled climate change scenario. *Forest Ecology and Management* 2013, 294:120–135.

Liu Y, Stanturf J, Goodrick S. Trends in global wildfire potential in a changing climate. *Forest Ecology and Management* 2010, 259:685–697.

Löf M, Ersson BT, Hjältén J, Nordfjell T, Oliet JA, Willoughby I. Site preparation techniques for forest restoration. In: *Restoration of Boreal and Temperate Forests*, 2nd edition. Stanturf JA, ed. 2015: Boca Raton: CRC Press. Chapter 5, 85–102.

Madsen P, Jensen FS, Fodgaard S. Afforestation in Denmark. In: *Restoration of Boreal and Temperate Forests*, 2nd edition. Stanturf JA, ed. 2015: Boca Raton: CRC Press. Chapter 9, 201–216.

Marini L, Ayres MP, Battisti A, Faccoli M. Climate affects severity and altitudinal distribution of outbreaks in an eruptive bark beetle. *Climatic Change* 2012, 115:327–341.

Matuszkiewicz JM, Kowalska A, Kozłowska A, Roo-Zielińska E, Solon J. Differences in plant-species composition, richness and community structure in ancient and post-agricultural pine forests in central Poland. *Forest Ecology and Management* 2013, 310:567–576.

Meehl GA et al. How much more global warming and sea level rise? *Science* 2005, 307:1769–1772.

Meehl GA, Zwiers F, Evans J, Knutson T, Mearns L, Whetton P. Trends in extreme weather and climate events: Issues related to modeling extremes in projections of future climate change. *Bulletin of the American Meteorological Society* 2000, 81:427–436.

Mercer DE. Policies for encouraging forest restoration. In: *Restoration of Boreal and Temperate Forests*. Stanturf J, Madsen P, eds. 2005: Boca Raton, FL: CRC Press. 97–109.

Millar CI. Historic Variability: Informing restoration strategies, not prescribing targets. *Journal of Sustainable Forestry* 2014, 33:S28–S42.

Millar CI, Stephenson NL, Stephens SL. Climate change and forests of the future: Managing in the face of uncertainty. *Ecological Applications* 2007, 17:2145–2151.

Minnemayer S, Laestadius L, Sizer N. *A World of Opportunity*. 2011: Washington, DC: World Resource Institute.

Mueller JM, Swaffar W, Nielsen EA, Springer AE, Lopez SM. Estimating the value of watershed services following forest restoration. *Water Resources Research* 2013, 49:1773–1781.

Nagaike T. Restoration of conifer plantations in Japan: Perspectives for stand and landscape management and for enabling social participation. In: *Restoration of Boreal and Temperate Forests*, 2nd edition. Stanturf JA, ed. 2015: Boca Raton: CRC Press. Chapter 18, 365–376.

Nagy RC, Lockaby BG. Hydrologic connectivity of landscapes and implications for forest restoration. In: *Forest Landscape Restoration*. Stanturf J, Lamb D, Madsen P, eds. 2012: Berlin: Springer. 69–91.

Nyland RD. Even-to uneven-aged: The challenges of conversion. *Forest Ecology and Management* 2003, 172:291–300.

O'Hara KL. Silviculture for structural diversity: A new look at multiaged systems. *Journal of Forestry* 1998, 96:4–10.

O'Hara K, Waring K. Forest restoration practices in the Pacific Northwest and California. In: *Restoration of Boreal and Temperate Forests*. Stanturf J, Madsen P, eds. 2005: Boca Raton, FL: CRC Press. 569.

Oliver CD et al. Landscape management. In: *Forest Landscape Restoration*. Stanturf J, Lamb D, Madsen P, eds. 2012: Dordrecht: Springer. 39–65.

Oliver CD, O'Hara KL, Baker PJ. Effects of restoration at the stand level. In: *Restoration of Boreal and Temperate Forests*, 2nd edition. Stanturf JA, ed. 2015: Boca Raton: CRC Press. Chapter 3, 37–68.

Park A, Puettmann K, Wilson E, Messier C, Kames S, Dhar A. Can boreal and temperate forest management be adapted to the uncertainties of 21st century climate change? *Critical Reviews in Plant Sciences* 2014, 33:251–285.

Paquette A, Bouchard A, Cogliastro A. Survival and growth of under-planted trees: A meta-analysis across four biomes. *Ecological Applications* 2006, 16:1575–1589.

Pejchar L, Press DM. Achieving conservation objectives through production forestry: The case of *Acacia koa* on Hawaii Island. *Environmental Science and Policy* 2006, 9:439–447.

Phillips RJ, Waldrop TA, Brose PH, Wang GG. Restoring fire-adapted forests in eastern North America for biodiversity conservation and hazardous fuels reduction. In: *A Goal-Oriented Approach to Forest Landscape Restoration*. Stanturf JA, Madsen P, Lamb D, eds. 2012: Berlin: Springer. 187–219.

Pielke RA et al. Dealing with complexity and extreme events using a bottom-up, resource-based vulnerability perspective. In: *Extreme Events and Natural Hazards: The Complexity Perspective*. Sharma AS, Bunde A, Dimri VP, Baker DN, eds. 2013: Washington, DC: American Geophysical Union. 345–359.

Pommerening A. Transformation to continuous cover forestry in a changing environment. *Forest Ecology and Management* 2006, 224:227–228.

Pullar D, Lamb D. A tool for comparing alternative forest landscape restoration scenarios. In: *A Goal-Oriented Approach to Forest Landscape Restoration*. Stanturf J, Madsen P, Lamb D, eds. 2012: Dordrecht: Springer. 3–20.

Rautner M. Designer trees. *Biotechnology and Development Monitor* 2001, 44:2–7.

Reichstein M et al. Climate extremes and the carbon cycle. *Nature* 2013, 500:287–295.

Renou-Wilson F, Byrne KA. Irish peatland forests: Lessons from the past and pathways to a sustainable future. In: *Restoration of Boreal and Temperate Forests*, 2nd edition. Stanturf JA, ed. 2015: Boca Raton: CRC Press. Chapter 15, 321–335.

Ribe RG, Ford RM, Williams KJ. Clearfell controversies and alternative timber harvest designs: How acceptability perceptions vary between Tasmania and the US Pacific Northwest. *Journal of Environmental Management* 2013, 114:46–62.

Sanderson EW, Jaiteh M, Levy MA, Redford KH, Wannebo AV, Woolmer G. The human footprint and the last of the wild: The human footprint is a global map of human influence on the land surface, which suggests that human beings are stewards of nature, whether we like it or not. *BioScience* 2002, 52:891–904.

Schröder R, Prasse R. Cultivation and hybridization alter the germination behavior of native plants used in revegetation and restoration. *Restoration Ecology* 2013, 21:793–800.

Seabrook L, Mcalpine CA, Bowen ME. Restore, repair or reinvent: Options for sustainable landscapes in a changing climate. *Landscape and Urban Planning* 2011, 100:407–410.

SERI. The SER International Primer on ecological restoration 2004. http://www.ser.org/resources/resources-detail-view/ser-international-primer-on-ecological-restoration (Type of Medium).

Shackelford N, Hobbs RJ, Heller NE, Hallett LM, Seastedt TR. Finding a middle-ground: The native/non-native debate. *Biological Conservation* 2013, 158:55–62.

Simula M, Mansur E. A global challenge needing local response. *Unasylva* 2011, 62:238.

Smith SE, Winslow SR. Comparing perceptions of native status. *Native Plants Journal* 2001, 2:5–11.

Stanturf JA. What is forest restoration? In: *Restoration of Boreal and Temperate Forests*, 2nd edition. Stanturf JA, ed. 2015: Boca Raton: CRC Press. Chapter 1, 1–16.

Stanturf J, Lamb D, Madsen P. *Forest Landscape Restoration* 2012a: Berlin: Springer.

Stanturf J, Madsen P, Lamb D. *A Goal-Oriented Approach to Forest Landscape Restoration* 2012b: Berlin: Springer.

Stanturf J, Palik B, Dumroese RK. Contemporary forest restoration: A review emphasizing function. *Forest Ecology and Management* 2014a, 331:292–323.

Stanturf JA, Palik BJ, Williams MI, Dumroese RK, Madsen P. Forest restoration paradigms. *Journal of Sustainable Forestry* 2014b, 33:S161–S194.

Steffen W, Crutzen PJ, McNeill JR. The Anthropocene: Are humans now overwhelming the great forces of nature. *Ambio: A Journal of the Human Environment* 2007, 36:614–621.

Stephens SL, Millar CI, Collins BM. Operational approaches to managing forests of the future in Mediterranean regions within a context of changing climates. *Environmental Research Letters* 2010, 5:024003.

Strauss SH, Bradshaw HD. *The Bioengineered Forest: Challenges for Science and Society*. 2004: Washington, DC: Resources for the Future. 245 pp.

Thorpe AS, Stanley AG. Determining appropriate goals for restoration of imperilled communities and species. *Journal of Applied Ecology* 2011, 48:275–279.

Toth LA, Anderson DH. Developing expectations for ecosystem restoration. In: *Transactions of the North American Wildlife and Natural Resources Conference*, Orlando, FL, 1998: Gardners, PA: Wildlife Management Institute. 122–134.

Townsend P et al. Multiple environmental services as an opportunity for watershed restoration. *Forest Policy and Economics* 2012, 17:45–58.

Turner MG. Disturbance and landscape dynamics in a changing world. *Ecology* 2010, 91:2833–2849.

Van Lear DH, Wurtz TL. Cultural practices for restoring and maintaining ecosystem function. In: *Restoration of Boreal and Temperate Forests*. Stanturf J, Madsen P, eds. 2005: Boca Raton, FL: CRC Press.

Vanha-Majamaa I, Jalonen J. Green tree retention in Fennoscandian forestry. *Scandinavian Journal of Forest Research* 2001, 16:79–90.

Vose JM, Geron C, Walker J, Rauland-Rasmussen K. Restoration effects on N cycling pools and processes. In: *Restoration of Boreal and Temperate Forests*. Stanturf J, Madsen P, eds. 2005: Boca Raton, FL: CRC Press. 77–94.

Waring KM, O'Hara KL. Silvicultural strategies in forest ecosystems affected by introduced pests. *Forest Ecology and Management* 2005, 209:27–41.

Weber N, Liebal S. Plantations: Forests: Wilderness: The diversity of forest landscapes in Europe as a consequence of social change, technological progress and disturbance. In: *Restoration of Boreal and Temperate Forests*, 2nd edition. Stanturf JA, ed. 2015: Boca Raton: CRC Press. Chapter 7, 165–185.

Weed AS, Ayres MP, Hicke JA. Consequences of climate change for biotic disturbances in North American forests. *Ecological Monographs* 2013, 83:441–470.

Weltzin JF et al. Assessing the response of terrestrial ecosystems to potential changes in precipitation. *BioScience* 2003, 53:941–952.

Williams JW, Jackson ST. Novel climates, no-analog communities, and ecological surprises. *Frontiers in Ecology and the Environment* 2007, 5:475–482.

Williams MI, Dumroese RK. Preparing for climate change: Forestry and assisted migration. *Journal of Forestry* 2013, 114:287–297.

Wilson K, Lulow M, Burger J, McBride M. The Economics of Restoration. In: *Forest Landscape Restoration*. Stanturf J, Lamb D, Madsen P, eds. 2012: Netherlands: Springer. 215–231.

Wilson KA et al. Optimal restoration: Accounting for space, time and uncertainty. *Journal of Applied Ecology* 2011, 48:715–725.

Xi W et al. Restoration of southern pine forests after the southern pine beetle. In: *A Goal-Oriented Approach to Forest Landscape Restoration*. Stanturf J, Madsen P, Lamb D, eds. 2012: Dordrecht: Springer. 321–354.

Yemshanov D, Koch FH, Ducey M, Koehler K. Mapping ecological risks with a portfolio-based technique: Incorporating uncertainty and decision-making preferences. *Diversity and Distributions* 2013, 19:567–579.

Zalasiewicz J, Williams M, Steffen W, Crutzen P. The New World of the Anthropocene 1. *Environmental Science and Technology* 2010, 44:2228–2231.

Index

Milton Keynes UK
Ingram Content Group UK Ltd.
UKHW052027071024
449327UK00027B/2454